PHARMACOLOGY OF PURINE AND PYRIMIDINE RECEPTORS

PHARMACOLOGY OF PURINE AND PYRIMIDINE RECEPTORS

Edited by

Kenneth A. Jacobson

Laboratory of Bioorganic Chemistry & Molecular Recognition Section
National Institute of Diabetes & Digestive & Kidney Diseases
National Institutes of Health
Bethesda, MD, USA

Joel Linden

Division of Inflammation Biology, La Jolla Institute for Allergy and Immunology
La Jolla, CA, USA

Serial Editor

S.J. Enna

Department of Molecular and Integrative Physiology
Department of Pharmacology, Toxicology and Therapeutics
University of Kansas Medical Center, Kansas City, Kansas, USA

ADVANCES IN

PHARMACOLOGY

VOLUME 61

ELSEVIER

AMSTERDAM • BOSTON • HEIDELBERG • LONDON
NEW YORK • OXFORD • PARIS • SAN DIEGO
SAN FRANCISCO • SINGAPORE • SYDNEY • TOKYO
Academic Press is an imprint of Elsevier

Academic Press is an imprint of Elsevier
525 B Street, Suite 1900, San Diego, CA 92101-4495, USA
225 Wyman Street, Waltham, MA 02451, USA
32 Jamestown Road, London NW1 7BY, UK
Radarweg 29, PO Box 211, 1000 AE Amsterdam, The Netherlands

First edition 2011

British Library Cataloguing in Publication Data
A catalogue record for this book is available from the British Library

Library of Congress Cataloging-in-Publication Data
A catalog record for this book is available from the Library of Congress

ISBN: 978-0-12-385526-8

For information on all Academic Press publications visit our website at elsevierdirect.com

Printed and bound in USA
11 12 13 14 10 9 8 7 6 5 4 3 2 1

Contents

P2 Receptor Signaling in Neurons and Glial Cells of the Central Nervous System 441

Laszlo Köles, Anna Leichsenring, Patrizia Rubini, and Peter Illes

Role of Purinergic Receptors in CNS Function and Neuroprotection 495

Hidetoshi Tozaki-Saitoh, Makoto Tsuda, and Kazuhide Inoue

Contributors

Numbers in parentheses indicate the pages on which the authors' contributions begin.

Italo Biaggioni (115) Departments of Medicine and Pharmacology, Vanderbilt University, Division of Clinical Pharmacology, Nashville, Tennessee, USA

Pier Andrea Borea (41) Department of Clinical and Experimental Medicine, Pharmacology Section, University of Ferrara, Ferrara, Italy

G. Burnstock (333) Autonomic Neuroscience Centre, University College Medical School, London, United Kingdom

Silvia Deaglio (301) Department of Genetics, Biology, and Biochemistry, University of Turin & Human Genetics Foundation, Turin, Italy

Tobias Eckle (145) Mucosal Inflammation Program, Department of Anesthesiology, University of Colorado Denver, Aurora, Colorado, USA

Holger K. Eltzschig (145) Mucosal Inflammation Program, Department of Anesthesiology, University of Colorado Denver, Aurora, Colorado, USA

David Erlinge (417) Department of Cardiology, Lund University, Skane University Hospital, Lund, Sweden

Igor Feoktistov (115) Departments of Medicine and Pharmacology, Vanderbilt University, Division of Cardiovascular Medicine, Nashville, Tennessee, USA

Bertil B. Fredholm (77) Department of Physiology and Pharmacology, Karolinska Institutet, Stockholm, Sweden

Anikó Göblyös (187) Division of Medicinal Chemistry, Leiden/Amsterdam Center for Drug Research, Leiden University, Leiden, The Netherlands

Zhan-Guo Gao (187) Molecular Recognition Section, Laboratory of Bioorganic Chemistry, National Institute of Diabetes and Digestive and Kidney Diseases, National Institutes of Health, Bethesda, Maryland, USA

Stefania Gessi (41) Department of Clinical and Experimental Medicine, Pharmacology Section, University of Ferrara, Ferrara, Italy

T. Kendall Harden (373) Department of Pharmacology, University of North Carolina School of Medicine, Chapel Hill, North Carolina, USA

Adriaan P. IJzerman (1, 187) Division of Medicinal Chemistry, Leiden/Amsterdam Centre for Drug Research, Leiden University, Leiden, The Netherlands

Peter Illes (441) Rudolph-Boehm-Institute of Pharmacology and Toxicology, University of Leipzig, Leipzig, Germany

Kazuhide Inoue (495) Department of Molecular and System Pharmacology, Graduate School of Pharmaceutical Sciences, Kyushu University, Higashi, Fukuoka, Japan

Veli-Pekka Jaakola (1) Oulu Biocenter and Department of Biochemistry, University of Oulu, Oulu, Finland

Kenneth A. Jacobson (187) Molecular Recognition Section, Laboratory of Bioorganic Chemistry, National Institute of Diabetes and Digestive and Kidney Diseases, National Institutes of Health, Bethesda, Maryland, USA

Stina Johansson (77) Department of Physiology and Pharmacology, Karolinska Institutet, Stockholm, Sweden

Laszlo Köles (441) Rudolph-Boehm-Institute of Pharmacology and Toxicology, University of Leipzig, Leipzig, Germany; Department of Pharmacology and Pharmacotherapy, Semmelweis University, Budapest, Hungary

C. Kennedy (333) Strathclyde Institute of Pharmacy and Biomedical Sciences, University of Strathclyde, Glasgow, United Kingdom

Michael Koeppen (145) Mucosal Inflammation Program, Department of Anesthesiology, University of Colorado Denver, Aurora, Colorado, USA

Silvia M. Kreda (221) Cystic Fibrosis/Pulmonary Research & Treatment Center, School of Medicine, University of North Carolina, Chapel Hill, North Carolina, USA

Filip Kukulski (263) Centre de Recherche en Rhumatologie et Immunologie, Centre Hospitalier Universitaire de Québec (pavillon CHUL); Département de Microbiologie-Infectiologie et d'Immunologie, Faculté de Médecine, Université Laval, Québec, Québec, Canada

Sébastien A. Lévesque (263) Centre de Recherche en Rhumatologie et Immunologie, Centre Hospitalier Universitaire de Québec (pavillon CHUL); Département de Microbiologie-Infectiologie et d'Immunologie, Faculté de Médecine, Université Laval, Québec, Québec, Canada

J. Robert Lane (1) Division of Medicinal Chemistry, Leiden/Amsterdam Centre for Drug Research, Leiden University, Leiden, The Netherlands; Current address: Drug Discovery Biology, Monash Institute of Pharmaceutical Sciences, Melbourne, Victoria, Australia

Eduardo R. Lazarowski (221) Cystic Fibrosis/Pulmonary Research & Treatment Center, School of Medicine, University of North Carolina, Chapel Hill, North Carolina, USA

Anna Leichsenring (441) Rudolph-Boehm-Institute of Pharmacology and Toxicology, University of Leipzig, Leipzig, Germany

Joel Linden (95) Division of Inflammation Biology, La Jolla Institute of Allergy and Immunology, La Jolla, California, USA

Stefania Merighi (41) Department of Clinical and Experimental Medicine, Pharmacology Section, University of Ferrara, Ferrara, Italy

Simon C. Robson (301) Department of Medicine, Transplant Institute, Beth Israel Deaconess Medical Center, Harvard Medical School, Boston, Massachusetts, USA

Patrizia Rubini (441) Rudolph-Boehm-Institute of Pharmacology and Toxicology, University of Leipzig, Leipzig, Germany

Jean Sévigny (263) Centre de Recherche en Rhumatologie et Immunologie, Centre Hospitalier Universitaire de Québec (pavillon CHUL): Département de Microbiologie-Infectiologie et d'Immunologie, Faculté de Médecine, Université Laval, Québec, Québec, Canada

Lucia Seminario-Vidal (221) Cystic Fibrosis/Pulmonary Research & Treatment Center, School of Medicine, University of North Carolina, Chapel Hill, North Carolina, USA

Juliana I. Sesma (221) Cystic Fibrosis/Pulmonary Research & Treatment Center, School of Medicine, University of North Carolina, Chapel Hill, North Carolina, USA

Hidetoshi Tozaki-Saitoh (495) Department of Molecular and System Pharmacology, Graduate School of Pharmaceutical Sciences, Kyushu University, Higashi, Fukuoka, Japan

Makoto Tsuda (495) Department of Molecular and System Pharmacology, Graduate School of Pharmaceutical Sciences, Kyushu University, Higashi, Fukuoka, Japan

Katia Varani (41) Department of Clinical and Experimental Medicine, Pharmacology Section, University of Ferrara, Ferrara, Italy

Ivar von Kügelgen (373) Department of Pharmacology and Toxicology, University of Bonn, Bonn, Germany

Ying-Qing Wang (77) Department of Physiology and Pharmacology, Karolinska Institutet, Stockholm, Sweden; Present address: College of Pharmacy, Jilin University, Changchun City, China

Preface

The aim of this volume is to highlight recent discoveries on the signaling roles of extracellular nucleosides and nucleotides and to summarize new pharmaceutical research and development efforts based on these concepts. Several chapters are devoted to a review tissue responses mediated by P1 (adenosine) and P2 (P2X and P2Y nucleotide) receptors. New insights are provided on the release and extracellular metabolism of purines and pyrimidines and the induction of receptors and ecto-enzymes in response to stressful stimuli. Subjects covered go beyond purines alone in recognition of the important roles of pyrimidines for at least four subtypes of the P2Y receptors. Given the universality of purinergic signaling in organ systems, it is impossible to provide a comprehensive account of all tissue responses. Not included in this work is coverage of kinases, intracellular purine metabolism, and purinergic antiviral agents.

Since their initial cloning 20 years ago, the list of adenosine and nucleotide receptors has expanded to 19. Currently, there are 12 known G protein-coupled receptors that respond to nucleosides, including four subtypes of adenosine receptors and eight subtypes of P2Y receptors. There are seven subtypes of nucleotide-gated ion channels, that is, P2X receptors. Many new chemical entities that act through purine-related mechanisms are now being utilized as pharmacological probes. As there have been several reviews on these receptors from the perspective of medicinal chemistry, including exhaustive lists of new compounds, the present work focuses on the pharmacology of these sites. Those wishing to obtain additional information on these receptor systems are referred to other contemporary reviews on this topic (Fredholm et al., 2011; Gunosewoyo & Kassiou, 2010; Jacobson & Boeynaems, 2010).

In the present work, recent advances on the mechanisms of nucleotide release from cells are reviewed by Lazarowski et al., and the newly appreciated importance of nucleotide metabolism in the extracellular space is discussed by Kukulski et al., and Robson and Deaglio. The recent advance in the structural biology of adenosine receptors and the determination of the crystallographic structure of the A_{2A} site is covered by Lane et al. In the absence of precise structural information on the P2Y receptors, von Kügelgen and Harden discuss this class from the perspective of molecular pharmacology and physiology. Chapters by Gessi et al. (adenosine), Burnstock and Kennedy (P2X), and Erlinge (P2Y) describe new therapeutic concepts related to each family of receptors, including newly identified therapeutic targets in neurons and glia. The actions of purines and pyrimidines in the nervous system are summarized in chapters by Köles et al. and Hidetoshi et al. Purinergic receptor allosteric agonists are reviewed by Jacobson et al., with the importance of purines in regulating immunity and inflammation detailed in two chapters (Linden and Feoktistov and Biagionni). The latest information on the regulation of metabolism and temperature by central and peripheral purinergic mechanisms is considered by Fredholm et al., with the role of A_{2B} receptors in tissue responses to hypoxia and ischemia reviewed by Koeppen et al.

This volume is a testament to the fact that knowledge of the major roles played by purine and pyrimidine signaling in normal and pathophysiology continues to grow. In view of the therapeutic promise of many new agents within this class, and the rapidly expanding literature in this field, we believe basic and clinical scientists will find these chapters of particular value in synthesizing and interpreting these advances.

References

Fredholm, B. B., IJzerman, A. P., Jacobson, K. A., Linden, J., & Müller, C. E. (2011). International Union of Basic and Clinical Pharmacology. LXXXI. Nomenclature and classification of adenosine receptors—an update. *Pharmacological Reviews*, *63*(1), 1–34.

Gunosewoyo, H., & Kassiou, M. (2010). P2X purinergic receptor ligands: Recently patented compounds. *Expert Opinion on Therapeutic Patents*, *20*, 625–646.

Jacobson, K. A., & Boeynaems, J.-M. (2010). P2Y nucleotide receptors: Promise of therapeutic applications. *Drug Discovery Today*, *15*, 570–578.

Kenneth A. Jacobson, Ph.D.
Bethesda, MD

Joel Linden, Ph.D.
San Diego, CA

J. Robert Lane*,[1], Veli-Pekka Jaakola[†], and Adriaan P. IJzerman*

*Division of Medicinal Chemistry, Leiden/Amsterdam Centre for Drug Research, Leiden University, Leiden, The Netherlands

[†]Oulu Biocenter and Department of Biochemistry, University of Oulu, Oulu, Finland

[1]Current address: Drug Discovery Biology, Monash Institute of Pharmaceutical Sciences, Melbourne, Victoria, Australia.

The Structure of the Adenosine Receptors: Implications for Drug Discovery

Abstract

Extracellular adenosine mediates most of its physiological effects via an interaction with four G protein-coupled receptors (GPCRs), the adenosine receptors (ARs). These ARs are important pharmacological targets in the treatment of a wide variety of diseases from central nervous system disorders to ischemic injury. As for other GPCRs, drug development for the ARs has been hampered by the lack of structural data for this class of membrane proteins. However, in the past 3 years, this situation has changed with the elucidation of structures for the turkey β_1-adrenoceptor, the human β_2-adrenoceptor, squid rhodopsin, the activated form of bovine (rhod)opsin, the human adenosine A_{2A} receptor, and most recently the CXCR4 chemokine receptor. In this review, the structural features of the human adenosine A_{2A} receptor will be discussed with a particular focus on the ligand binding site. Further, the implications of this structural information for AR ligand selectivity, drug screening, homology modeling, and virtual ligand screening will be discussed.

I. Introduction

Extracellular adenosine has an important physiological role both as a signal of metabolic stress and as a modulator of neurotransmitter release (Fredholm et al., 2001, 2005). Its effects are predominantly mediated via its interaction with adenosine receptors (ARs), members of the G protein-coupled receptor (GPCR)

Advances in Pharmacology, Volume 61

1054-3589/11 $35.00
10.1016/B978-0-12-385526-8.00001-1

superfamily (Fredholm et al., 2001). Four subtypes of ARs have been identified in humans, A_1 AR, A_{2A} AR, A_{2B} AR, and A_3 AR, and each AR subtype possesses distinct pharmacological properties, tissue/cellular distribution, and secondary effector coupling (Fredholm et al., 2001). The A_1 AR and A_3 AR couple to G_i family G proteins and therefore act to decrease intracellular cAMP levels, whereas the A_{2A} AR and A_{2B} AR couple to G_s family G proteins and increase intracellular cAMP levels. The A_1 AR and A_{2A} AR subtypes play an important role in the central nervous system (CNS) (Fredholm et al., 2005). Contrastingly, the A_{2B} AR and A_3 AR subtypes are located mainly peripherally and play roles in inflammation and immune responses, although in peripheral tissues the A_{2A} AR is also involved in inflammation (Gessi et al., 2008; Hasko et al., 2009; Ohta & Sitkovsky, 2001; Sitkovsky & Ohta, 2005). As a consequence, these ARs represent attractive therapeutic targets for the treatment of CNS disorders such as Parkinson's disease, inflammatory diseases, asthma, kidney failure, ischemic injuries, and cancer (Fishman et al., 2009; Jacobson & Gao, 2006; Schapira et al., 2006; Stone et al., 2009; Wilson et al., 2009). In addition, caffeine (1,3,7-trimethylxanthine), which as the active ingredient of coffee and tea is the most widely consumed stimulant in the world, acts as a competitive antagonist of ARs (Fredholm et al., 2001). It is much more potent on the A_1 AR, A_{2A} AR, and A_{2B} AR than on the fourth, the A_3 AR (Fredholm et al., 2001). Interestingly, epidemiological studies have shown a correlation between a coffee intake and a slower onset of Parkinson's disease (Hernan et al., 2002). This effect has been linked to the blockade of the A_{2A} AR subtype by caffeine, as this controls locomotor behavior and neurotransmitter release in basal ganglia together with dopamine D_2 receptors, and metabotropic glutamate receptors (Ferre et al., 2008a, 2008b). *In vitro* experiments suggest that the A_{2A} AR may form an oligomeric complex with these two receptors (Canals et al., 2003; Fuxe et al., 2005). Not surprisingly, many drug discovery teams in industry and academia have developed antagonists for the A_{2A} AR as potential therapeutic agents to control the motor effects in Parkinson's disease. For instance, it has recently been observed that the synthetic A_{2A} AR antagonist ZM241385, the ligand cocrystallized with the A_{2A} AR, enhanced L-DOPA-derived dopamine release, again evidence for its potential use in the treatment of Parkinson's symptoms (Golembiowska & Dziubina, 2004; Golembiowska et al., 2009). Preladenant (SCH-420814) is an AR antagonist with a high affinity and selectivity for the A_{2A} AR. Phases I and II clinical trials indicated that Preladenant met its major endpoints by reducing OFF time and increasing ON time in L-DOPA-treated patients with Parkinson's disease, without worsening dyskinesias (Salamone, 2010). Further, the U.S. Food and Drug Administration recently approved the A_{2A} AR agonist Regadenoson (CTV-3146, Lexiscan™) as a coronary vasodilator for use in myocardial perfusion imaging (Thomas et al., 2009).

Historically, GPCR drug discovery has relied on known natural ligands or screening assay hits as starting points for optimization of affinity, subtype selectivity, and pharmacokinetic properties (Klabunde & Hessler, 2002).

The most useful scaffolds for design of AR ligands have been provided by adenosine and xanthine chemotypes (Cristalli et al., 2009; Jacobson & Gao, 2006; Kalla et al., 2009). Thus, adenosine derivatives with various substitutions in position 2 or N6 of the adenine ring and 3′, 4′, or 5′ position of the ribose ring have been developed as selective agonists for all four AR subtypes; only a few other chemotypes have been found with agonist activity (Beukers et al., 2004a). Since the early discovery of caffeine and theophylline as nonselective AR antagonists, derivatization of the xanthine scaffold yielded a number of high-affinity subtype selective antagonists (Jacobson & Gao, 2006). Several other chemotypes for AR antagonists have been discovered over the past decade using a combination of experimental screening and ligand-based methods (Colotta et al., 2009; Mantri et al., 2008; van Veldhoven et al., 2008). These ligand-based approaches, however, require preexisting knowledge of ligand structure–activity relationships (SARs) and are largely limited to relatively close analogs of known ligands. The reason behind these limitations was of course the relative paucity of three-dimensional (3D) structural data for the GPCR super family. Indeed, until recently, our atomic-level understanding of GPCRs has been based on rhodopsin in its inactive state (Palczewski et al., 2000). Although groundbreaking, it became apparent over the ensuing 8 years that rhodopsin is a highly specialized member of the GPCR family that might not be the ideal representative for drawing generalized conclusions about the other family members. We will discuss this later in the chapter with regard to homology modeling of the ARs.

In the past 3 years, the field of GPCR structural biology has experienced a renaissance, with five new members of the superfamily yielding to crystallization efforts (Cherezov et al., 2007; Jaakola et al., 2008; Rasmussen et al., 2007; Rosenbaum et al., 2007; Warne et al., 2008, Wu et al. 2010; Fig. 1). Crucially, all of these novel structures feature a cocrystallized diffusible ligand, as opposed to the covalently bound retinal in rhodopsin (Fig. 1E). Intriguingly, the CXCR4 structure suggests that this receptor exists as a homodimeric complex, adding another layer of complexity to the GPCR structural picture (Wu et al., 2010; Fig. 1D). Further, significant inroads have been made in resolving the activated state of bovine rhodopsin (Park et al., 2008; Scheerer et al., 2008). Of particular note for this chapter is the determination of a 2.6-Å resolution structure of the A_{2A} AR bound to the high-affinity selective nonxanthine antagonist ZM241385. Therefore, this chapter summarizes our current understanding of AR structure in the light of this exciting development.

II. Primary Sequence and Covalent Modifications of the Adenosine A_{2A} Receptors

The overall sequence similarity between human ARs is relatively high (Fig. 2). The adenosine A_{2A} AR has higher sequence identity with the adenosine A_{2B} AR (46%) than with either the adenosine A_1 AR (37%) or A_3 AR

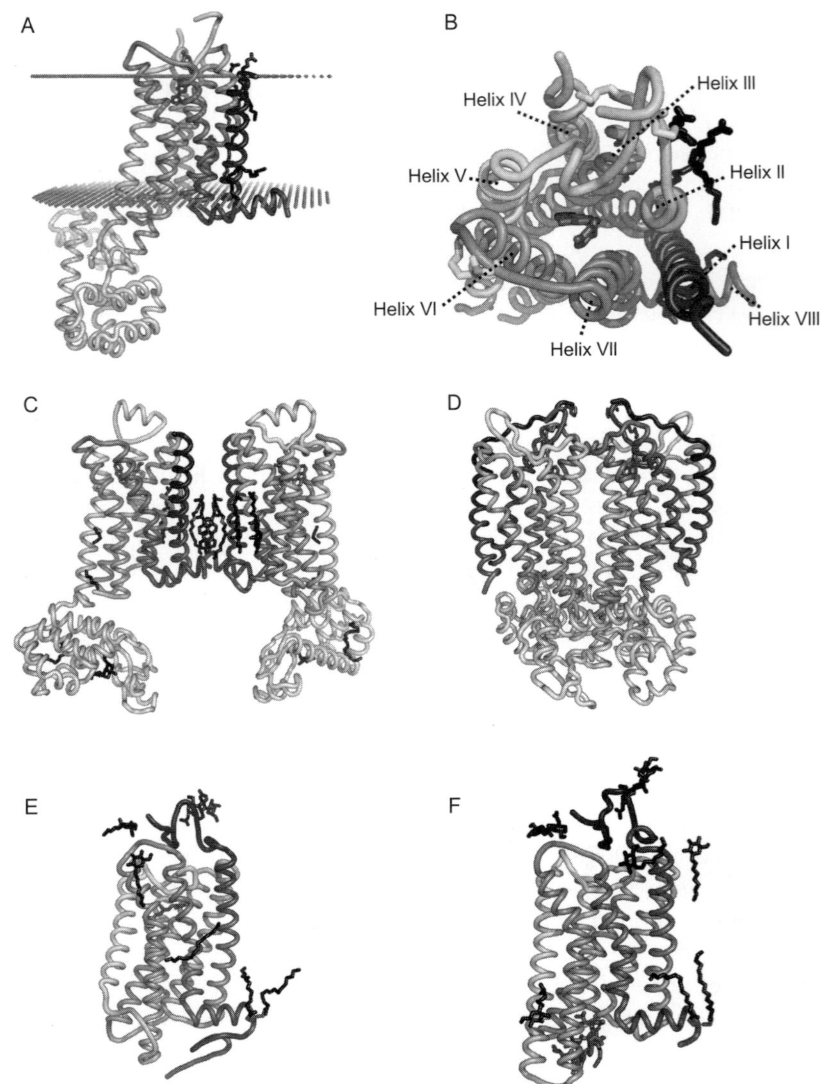

FIGURE 1 A summary of each class of GPCR crystal structure determined to date. (A) A_{2A} AR T4 lysozyme receptor construct cocrystalized with the antagonist ZM241385. (B) Overhead view of the A_{2A} AR crystal structure showing ZM241385 predominantly making interactions with helices 5, 6, and 7. (C) The β_2-adrenoceptor cocrystalized with the inverse agonist carazolol. (D) The CXCR4 chemokine receptor crystalized as dimers (E) rhodopsin and (F) opsin.

(31%). All subtypes have potential N-linked glycosylation sites. However, the adenosine A_{2A} AR lacks the potential palmitoylation sites at the end of helix 8 that are present in the adenosine A_1 AR, A_{2B} AR, and A_3 AR receptor subtypes. Glycosylation may be important in targeting receptors to the

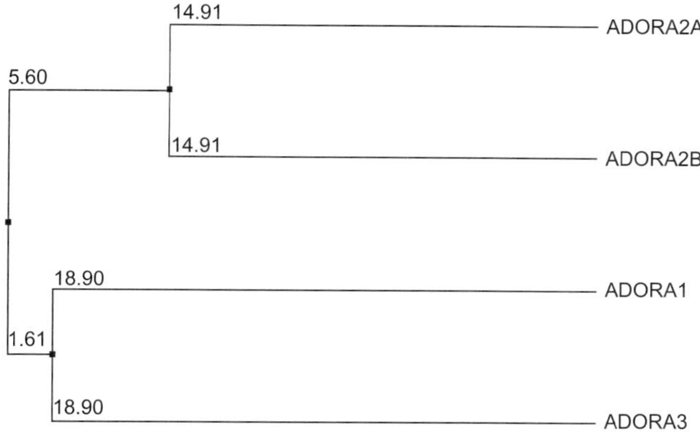

FIGURE 2 Phylogenetic tree of the human adenosine receptors.

plasma membrane. Removal of palmitoylation sites by mutagenesis had no effect on G protein coupling or receptor downregulation after activation but did affect receptor degradation after synthesis (Gao et al., 1999a, 1999b). In contrast to the adenosine A_1 AR receptor, phosphorylation of the receptor adenosine A_3 AR and adenosine A_{2A} AR receptors by GPCR kinases (GRKs) causes rapid desensitization of the receptor (Klaasse et al., 2008; Palmer & Stiles, 1999, 2000; Palmer et al., 1995, 1996). There are several predicted phosphorylation sites in both the cytoplasmic and carboxy terminal domains of all four AR subtypes (Piirainen et al., 2010).

III. The A_{2A}AR Crystal Structure

A. Receptor Engineering Is Required for GPCR Structure Determination

Except for rhodopsin, wild-type GPCRs tend to be too thermolabile to allow prolonged solubilization, purification, and X-ray studies (Kobilka & Schertler, 2008). Therefore, before embarking on a discussion about the structure of the A_{2A} AR, it is important to highlight that all 3D structures of GPCRs with diffusible ligands to date have used engineered receptors. Tate and coworkers elegantly addressed this issue by mutating the β_1-adrenergic receptor sequentially to improve its thermostability and successfully determined the crystal structure (Warne et al., 2008, 2009). This mutagenesis approach was pursued for the A_{2A} AR by Magnani et al. (2008). They managed to create relatively stable receptor variants with up to five amino acid substitutions. Interestingly, they determined a different pattern of mutants to stabilize either agonist or antagonist binding. In a different approach to crystallize the A_{2A} AR, Jaakola and coworkers adopted a route very similar to the one developed by Kobilka and coworkers for the β_2-adrenergic receptor by using a T4 lysozyme (T4L) fusion protein (Jaakola et al., 2008; Rosenbaum et al., 2007). As for the β_2-adrenergic receptor, this approach proved to be successful. Most of the flexible and potentially disordered third cytosolic/intracellular loop of the A_{2A} AR (Leu209-Ala221) was replaced with the very stable and easily crystallized lysozyme protein from T4 bacteriophage, increasing the available surface area potential for crystal contacts (Fig. 1A). The receptor was further stabilized by deletion of the larger part of the very flexible C-terminus (Ala317-Ser412), and a histidine purification epitope (6X-His tag) was added. This C-terminal truncation removed the predicted disordered regions and improved the likelihood of crystal formation. During purification this construct (A_{2A}-T4L-ΔC), the presence of a high concentration of sodium chloride, cholesterol, and a receptor-saturating concentration of the antagonist theophylline was essential. The latter was replaced by the high-affinity antagonist ZM241385 in the final purification step. Using differential scanning fluorometry-based and analytical-size exclusion chromatography-based stability screening prior to crystallization, the addition of all of these ligands or modulators was identified as vital (Alexandrov et al., 2008; Kawate & Gouaux, 2006; Vedadi et al., 2006). In addition, the engineered receptor construct was modified at the amino terminus by addition of a signal sequence (hemagglutinin) and a detection tag (FLAG-M2). The hemagglutinin signal sequence and FLAG-M2 tag as well as the first two residues of the receptor are not visible in the electron density maps. Likewise, the residues between Gln311$^{8.62}$ and Ala316$^{8.67}$ as well as the purification tag are also not visible in the electron density map. A glycan on Asn15$^{4.75}$ was enzymatically removed during purification. Such a highly engineered receptor construct necessitated a thorough pharmacological characterization with respect to signaling and ligand binding properties. Signaling, measured as modulation of cAMP production, was completely abrogated, most probably due to the insertion of T4L in the third intracellular loop of the receptor, a region crucial for G protein coupling (Rosenbaum et al., 2007). Compared to the wild-type receptor, the construct displayed virtually identical affinity for the antagonist ZM241385 in radioligand binding studies, whereas agonist affinity was somewhat higher. Thus, the incorporation of T4L may have shifted the receptor conformational equilibrium toward a high-affinity agonist state (Jaakola et al., 2008). Indeed, a similar agonist-induced shift was reported for the crystallized β_2-adrenoceptor-T4L(ΔC) fusion receptor (Rosenbaum et al., 2007; Fig. 1C). High sodium chloride concentrations, essential for the generation of crystals, did not affect antagonist affinity, whereas agonist affinity was reduced to a similar value for wild type and engineered receptor, in line with earlier observations (Gao & IJzerman, 2000). This suggests that the antagonist binding site in the crystal structure was not affected by the substantial modifications to the receptor protein. It is noteworthy that the T4L method was also successfully applied in the recent determination of the CXCR4 chemokine receptor structure, illustrating the utility of this approach (Wu et al., 2010).

B. Overall Architecture of the A_{2A} AR as Determined by the Crystal Structure: Similar but Crucially Different

1. The Similarities

Diffraction data from 13 of the best crystals were combined to yield a 2.6-Å data set from which a model was constructed. The final refined model included residues Ile3–Gln310 of the human A_{2A} AR, residues 2–161 of the T4L, five lipid hydrocarbon chains modeled as stearic acid, eight sulfate ions, and the antagonist ZM241385 bound in the ligand binding cavity. The overall structure of the A_{2A} AR structure is relatively similar to that of the previously determined GPCRs. This is particularly true within the seven transmembrane domains as the root mean square deviation (RMSD) of the transmembrane helices is relatively small (<3 Å) (Fig. 1; Jaakola et al., 2008). Indeed, by selecting the most conserved residues, the transmembrane alignment can be improved further, close to an RMSD value of 1.3 Å (Cα of 97 residues) (Hanson & Stevens, 2009). This suggests a similar mechanism of activation within the class A, rhodopsin-like, GPCR subfamily (Hanson & Stevens, 2009; Schwartz et al., 2006). The residues constituting the transmembrane α helices are $Gly5^{1.31}$–$Trp32^{1.58}$ (helix I); $Thr41^{2.39}$–$Ser67^{2.65}$ (helix II); $His75^{3.23}$–$Arg107^{3.55}$ (helix III); $Thr119^{4.40}$–$Leu140^{4.61}$ (helix IV); $Asn175^{5.36}$–$Ala204^{5.65}$ (helix V); $Arg222^{6.24}$–$Phe258^{6.60}$ (helix VI); and $Leu269^{7.34}$–$Arg291^{7.56}$ (helix VII) (Jaakola et al., 2008). Helix VIII, a small helix that does not cross the cell membrane, is located at the membrane cytoplasm interface and comprises $Arg296^{8.47}$–$Leu308^{8.59}$. As discussed earlier, the A_{2A} AR does not contain the canonical palmitoylation site(s) found in the majority of GPCRs; instead, helix VIII is stabilized by interactions with helix I. The residues defining intracellular and extracellular loops (ICLs and ECLs) are $Leu33^{1.59}$–$Val40^{2.38}$ (ICL1); $Ile108^{3.56}$–$Gly118^{4.39}$ (ICL2); $Leu208^{5.69}$–$Ala221^{6.23}$ (ICL3); $Thr68^{2.66}$–$Cys74^{3.22}$ (ECL1); $Leu141^{4.62}$–$Met174^{5.35}$ (ECL2); and $Cys259^{6.61}$–$Trp268^{7.33}$ (ECL3). In the structure, ICL3 has been replaced by 160 residues from T4L. In addition, the N-linked glycan associated with $Asn154^{4.75}$ has been removed enzymatically to improve crystallization.

2. The Key Differences

Importantly, this crystallographic model of A_{2A}-T4L-ΔC bound to ZM241385 reveals three features distinct from the previously reported GPCR structures (Jaakola & IJzerman, 2010; Jaakola et al., 2008). First, and most surprisingly, the cocrystallized antagonist ZM241385 binds in an extended conformation perpendicular to the plane of the membrane and colinear with transmembrane helix VII, while interacting with both ECL2 and ECL3 (Fig. 3A). This cannot be reconciled with earlier molecular modeling studies based on rhodopsin homology models in which ZM241385 and other antagonists were docked into a binding site emulating that of the β_2-adrenoceptor and rhodopsin (Martinelli & Tuccinardi, 2008; Yuzlenko & Kiec-Kononowicz, 2009). Second, the organization of the extracellular loops is markedly different from those of the β_1-adrenoceptor, the β_2-adrenoceptor, or the bovine and squid rhodopsins. Finally, the antagonist binding cavity is redefined by subtle changes in the orientation and position of the transmembrane helices relative to those of rhodopsin and the β-adrenoceptors so that it is located closer to helices VI and VII and allows only limited interactions with helices III and V (Fig. 1B; Hanson & Stevens, 2009; Jaakola et al., 2008). Such divergence in the ligand binding site of class A GPCRs is exemplified by the novel CXCR4 chemokine receptor structure in complex with both a small molecule and a peptide antagonist. Again in this structure large shifts are observed in the extracellular ends of the transmembrane helices (up to 9 Å for helix 1 as compared to the β_2-adrenoceptor structure) (Wu et al., 2010). Such differences illustrate the challenge facing the GPCR homology modeling community. We shall focus on some of these key areas of structural divergence observed in the A_{2A} AR structure in the following sections.

C. Extracellular Domain Architecture

Compared to the transmembrane helical domains, there is relatively little conservation of primary structure for the extracellular domains across the family A GPCRs (Hanson & Stevens, 2009). Accordingly, the organization of the extracellular domains differs significantly between solved GPCR structures. The extracellular loops of the A_{2A} AR form a domain structure that is structurally distinct from that of the rhodopsin or β-adrenoceptor structures (Figs. 1 and 3A) (Jaakola et al., 2008; Palczewski et al., 2000; Rosenbaum et al., 2007; Warne et al., 2008). The extracellular domain of rhodopsin has a long amino terminal part and second extracellular loop (Palczewski et al., 2000). The amino terminus and loops are stacked as antiparallel β-sheet structures and effectively cover extracellular space, acting as a lid over the 7TM region. The turkey β_1-adrenoceptor and human β_2-adrenoceptor both have extracellular domain structures that contain a short α-helical segment (Rosenbaum et al., 2007; Warne et al., 2008). An intraloop cysteine bridge between amino acids $184^{4.76}$ and $190^{5.29}$ may serve to keep the binding cavity open for diffusible ligands to enter. In comparison, the extracellular domain in the A_{2A} AR is largely constituted by ECL2 (Fig. 3A). This loop is essentially random coil, and the middle part of ECL2 is disordered and is not defined in the electron density maps (residues between $Gln148^{4.69}$ and $Ser156^{4.77}$). The entire domain is constrained by four cysteine bridges and

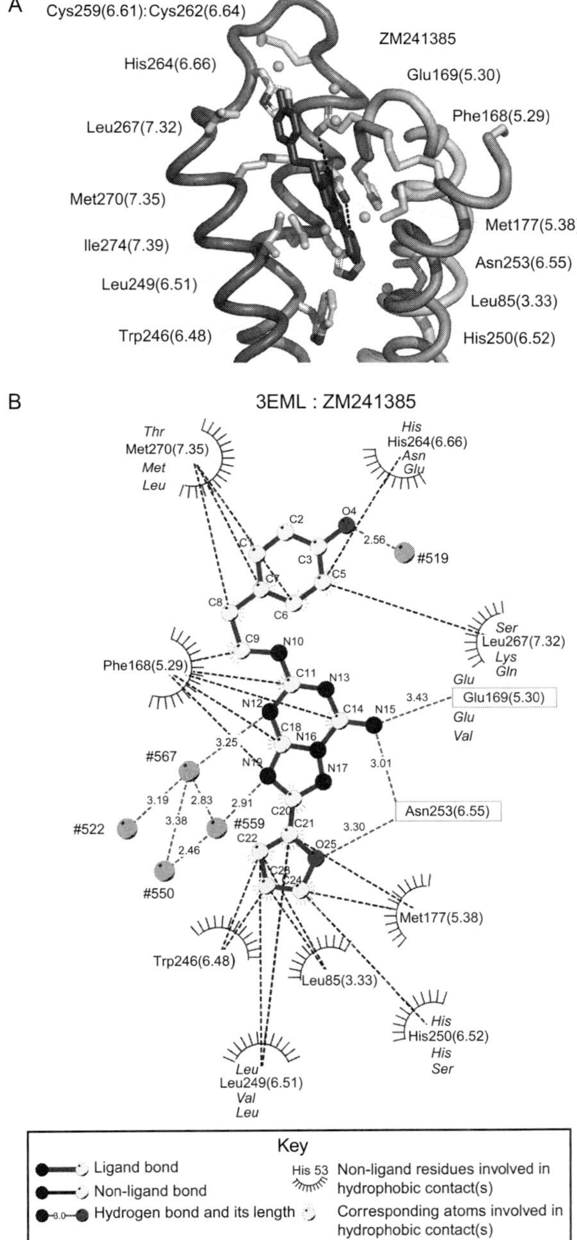

FIGURE 3 (A) Three-dimensional representation of the binding site of ZM241385 in the A_{2A} AR showing the important role of ECL2 in binding the antagonist. (B) Two-dimensional representation of the binding site of ZM241385 illustrating the nature of the key ligand–receptor interactions and ligand–water interactions. For residues not conserved across the human adenosine receptor subtypes, the residues for each receptor are shown in the following order from top to bottom: A_1 AR, A_{2A} AR, A_{2B} AR, A_3 AR.

van der Waals interactions between residues from different loops. Three cysteine bridges anchor the extracellular loops and domain together and constrain the essential ligand binding residues Phe168$^{5.29}$ and Glu169$^{5.30}$ from ECL2 in a defined position. Two of these cysteine bridges between 71$^{2.69}$:159$^{5.20}$ and 74$^{3.22}$:146$^{4.67}$ are not present in the rhodopsin or β-adrenergic receptor structures. The third cysteine bridge is between Cys77$^{3.25}$ and Cys166$^{5.27}$ of ECL3 and is conserved across the family A GPCRs (Rosenbaum et al., 2009). In addition, a fourth intraloop disulfide bond is formed in ECL3 between Cys259$^{6.61}$ and Cys262$^{6.64}$ with the sequence Cys-Pro-Asp-Cys (CPDC), which creates a kink in the loop that constrains the position of ECL3 and orients His264$^{6.66}$ at the top of the ligand binding site. It will be interesting to compare this organization of the A_{2A} AR extracellular domain to other AR subtype structures and see how this domain is structurally evolved.

The human A_1 AR and A_{2B} AR have the corresponding primary sequence for at least two of the above disulfide bridges and may also form a coil stabilized by these disulfides. However, based on primary sequence, the adenosine A_3 AR subtype has only the family A conserved 3.25:5.27 cysteine bridge. Interestingly, the A_3 AR has a distinct pharmacology compared to the other AR subtype, particularly, in terms of its low affinity for xanthine-based antagonists (Fredholm et al., 2001). Given the key role of extracellular loop residues in ligand binding at the A_{2A} AR, it is clear that an understanding of the structure of this region across the receptor family may hold the key to understanding ligand selectivity. It should be noted that for both the A_{2A} AR and the β-adrenoceptors, potential glycosylation sites in the ECL2 were mutated or enzymatically removed in the recombinant crystallized proteins. The precise function of this glycan moiety is unknown. In a recent paper from Kobilka and coworkers, nuclear magnetic resonance (NMR) spectroscopy was used to provide clues as regard to the mechanism and regulation of ligand entry to the β_2-adrenoceptor. The authors demonstrated a conformational coupling between the extracellular domain and the orthosteric binding site, impacting ligand entry (Bokoch et al., 2010). Further, recent solid-state NMR studies provide evidence for conformational changes that disrupt a hydrogen-bond network between ECL2 and the extracellular ends of TM4, TM5, and TM6 in metarhodopsin II before the dissociation of retinal and the formation of opsin (Hornak et al., 2010). Such biophysical experiments highlight the importance of this region for both ligand binding and receptor activation. Similar experiments on the ARs will be revealing in terms of the mechanisms behind both receptor activation and ligand selectivity. In conclusion, rhodopsin's retinal binding pocket is closed, thus effectively protecting retinal from solvent molecules, whereas the A_{2A} AR and β-adrenoceptor have found distinct extracellular domain folding solutions to provide open access to diffusible ligands allowing entrance into the binding cavity.

D. Cytoplasmic Architecture and the C-Terminus

The intracellular domain of the A_{2A} AR mediates G protein binding and signaling (Fredholm et al., 2001; Zezula & Freissmuth, 2008). In the A_{2A} AR structure, this region is perturbed by insertion of T4L and truncation of the carboxy terminal domain, both of which were necessary for highly ordered crystal formation (Jaakola et al., 2008). The same is true for β_2-adrenoceptor (with T4L and antibody complex) and β_1-adrenoceptor (truncations and receptor mutations) (Rosenbaum et al., 2007; Warne et al., 2008). Therefore, any discussion concerning the structure of this region must be made with caution. However, there are several features of interest that should be highlighted. An interaction between a highly conserved sequence motif at the cytoplasmic end of helix III (Asp$^{3.49}$-Arg$^{3.50}$-Tyr$^{3.51}$:Asp/Glu-Arg-Tyr:-D/ERY) and helix VI (Glu$^{6.30}$) has been proposed to constitute an "ionic lock" that may play a role in restraining the fully inactive conformation of rhodopsin and other class A receptors (Altenbach et al., 2008; Scheerer et al., 2008, 2009; Sheikh et al., 1996). However, with the exception of the rhodopsins, none of the GPCR structures published to date have the ionic-lock interaction, including the A_{2A} AR (Jaakola et al., 2008; Rosenbaum et al., 2007; Warne et al., 2008). Instead, as in the β_1- and β_2-adrenoceptors, the D/ERY motif in the A_{2A} AR participates in interactions that restrain the conformation of ICL2. Whether this is a true indication of the inactive structure of the receptor or a structure that has been influenced by the insertion of the T4-lysozyme remains a point of discussion (Hanson & Stevens, 2009; Kobilka & Schertler, 2008). In the A_{2A} AR, Asp101$^{3.49}$ forms a hydrogen bond with Tyr112$^{3.60}$ in ICL2 and Thr41$^{2.39}$ at the base of helix II (Jaakola et al., 2008). A similar hydrogen-bonding interaction was reported in the turkey β_1-adrenoceptor structure, and in both receptor structures, a short helical section is present in the ICL2 (Warne et al., 2008). This is not present in any of the β_2-adrenoceptor structures where Asp130$^{3.49}$ forms a hydrogen bond with Ser143$^{3.62}$, although there is a tyrosine at the 3.60 position (Rasmussen et al., 2007; Rosenbaum et al., 2007). It has been proposed that ICL2 serves as a control switch facilitating G protein activation through a select set of interactions (Burstein et al., 1998). To date, the influence upon basal receptor activity of residues Asp101$^{3.49}$, Tyr112$^{3.60}$, or Thr41$^{2.39}$ in the A_{2A} AR has not been investigated using a mutagenesis approach. However, in the highly homologous A_{2B} AR mutation of Thr42$^{2.39}$, the equivalent residue of Thr41 in the A_{2A} AR caused a 12-fold increase in receptor basal activity (Beukers et al., 2004b). Interestingly, for the β_1-adrenoceptor and A_{2A} AR, both of which have low basal activity, the hydrogen-bonding interaction between this short helix in ICL2 and Asp$^{3.49}$ is present. This set of interactions may have direct implications in G protein activation (Rosenbaum et al., 2009). As a contrast, β_2-adrenoceptor exhibits high basal activity and lacks helical structure within its ICL2, which results in altered interactions with the DRY motif. In agreement with this high basal activity, a recent molecular dynamics study has suggested that the β_2-adrenoceptor can equilibrate between conformations in which the ionic lock is present and absent independently of the presence of agonist ligand (Dror et al., 2009). At several scientific meetings, Heptares have presented an A_{2A} AR crystal structure determined using an engineered A_{2A} AR construct with thermostabilizing mutations (termed a StaR construct) in which the ionic lock remains intact, distinct from that of the published A_{2A}AR-T4L structure. It will be fascinating for the GPCR field to compare these two structures.

The adenosine A_{2A} AR has a relatively long cytoplasmic C-terminus. As described earlier, this region was truncated in the crystallization construct with the residues from Ala317 to Ser412 removed. The remaining part of the C-terminus formed helix VIII as described earlier. The portion of C-terminus removed is a long, 96 amino acid sequence that may lack 3D structure in the absence of accessory/regulatory proteins. It has been also demonstrated in many studies that the carboxy terminus of the A_{2A} AR interacts with α-actinin (type 2), dopamine receptors (types 2 and 3), glutamate mGlu5 receptors, and many other regulatory proteins such as ARNO (ARF-nucleotide binding site opener), TRAX (translin-associated factor X), and calmodulin (Navarro et al., 2009; Zezula & Freissmuth, 2008).

E. The Ligand Binding Pocket

1. The Crystal Structure Revealed an Unexpected Antagonist Binding Pocket

The most interesting region for a molecular pharmacologist and medicinal chemist is the binding pocket of a receptor. Given the interest in the A_{2A} AR as a therapeutic target for a number of diseases, it is perhaps not surprising that, prior to the determination of the crystal structure, considerable mutagenesis (Table I) and computer modeling studies had focused on the elucidation of the ligand binding site (Fredholm et al., 2001; Ivanov et al., 2009; Kim et al., 2003). With the determination of the crystal structure, this gives us an opportunity to revisit these biochemical studies with the crystal structure as a template. The bound antagonist, ZM241385, is a prototypical AR antagonist. ZM241385 binds to the engineered adenosine A_{2A} AR in an extended conformation perpendicular to the plane of the membrane bilayer (Fig. 3A). This orientation is very different from the retinal:rhodopsin and the β-blocker: β-adrenoceptor structures in which ligands bind parallel to the plasma membrane. Therefore, the ligand binding site of the A_{2A} AR is of particular interest given this unexpected orientation. ZM241385 consists of a bicyclic triazolotriazine core (located roughly in the middle of the binding cavity), a furan ring (located in the lower part of the binding cavity), and a 4-hydroxyphenylethyl side chain (located in the upper part of the binding cavity). The bicyclic core unit makes hydrophobic interaction with Ile274$^{7.39}$;

TABLE I Mutational Analysis of the Adenosine Receptors Focused on Residues Implicated in Ligand Binding or Function (Adapted from Fredholm et al., 2001; Piirainen et al., 2010)

	Mutation analysis for A_{2A}		*Mutation analysis on A_1, A_{2B}, or A_3*	
Residue	*Mutation*	*Results*	*Mutation*	*Results*
$E13^{1.39}$	Q	Slight reduction ag. but not ant. affinity (IJzerman et al., 1996)	A_1:E16A/G	Slight reduction ag. but not ant. (Klaasse et al., 2005)
$V84^{3.32}$	L	Decrease in ant. activity, no effect ag. (Jiang et al., 1997)	A_1:V87A	No effect (Rivkees et al., 1999)
$T88^{3.36}$	A/S/R	Decrease in ag. activity, no effect ant. (Jiang et al., 1997)	A_1:T91A	Substantial decrease in ag. Decrease in ant. N0840 (Rivkees et al., 1999)
$Q89^{3.37}$	A	Marginal decrease in ag. and ant. activity (Jiang et al., 1997)	A_1:Q92A	Substantial decrease in ag. Decrease in ant. N0840 (Rivkees et al., 1999)
	D	Increase in ag. but not ant. affinity (Jiang et al., 1997)		
$S90^{3.38}$	A	Marginal changes (Jiang et al., 1997)	A_1:S93A	No change in ant. or ag. binding (Klaasse et al., 2005)
$S91^{3.39}$	A	Marginal changes (Jiang et al., 1997)	A_1:S94A	No detectable ag. or ant. binding (Klaasse et al., 2005)
$E151^{4.72}$ (ECL2)	A/D/Q	Loss of ag. and ant. binding (Kim et al., 1996)		
$E161^{5.22}$ (ECL2)	A	Increase in ant. affinity, no effect ag. affinity (Kim et al., 1996)		
$F168^{5.29}$ (ECL2)	A	No ant. binding, substantial decrease in ag. potency (Jaakola et al., 2010)		
	Y/W	Marginal changes in ant. binding and ag. potency (Jaakola et al., 2010)		

(continued)

TABLE I *(continued)*

Residue	Mutation analysis for A_{2A}		Mutation analysis on A_1, A_{2B}, or A_3	
	Mutation	*Results*	*Mutation*	*Results*
E169 (ECL2)	A	Loss of ag. and ant. binding (Kim et al., 1996)		
	Q	Gain in N^6-substituted Ag affinity (Kim et al., 1996)		
D170	K	No effect (Kim et al., 1996)		
$P173^{5.34}$	R	No effect (Kim et al., 1996)		
$M177^{5.38}$	A	Modest reduction in ant. and ag. affinity (Jaakola et al., 2010)		
$F180^{5.41}$	A	No effect (Kim et al., 1995)		
$N181^{5.42}$	S	Reduction in affinity for N^6 or C-2 but not C-5 substituted ag. (Kim et al., 1995)		
$F182^{5.43}$	A	No detectable binding (Kim et al., 1995)		
	Y/W	Modest reduction of ag. but not ant. binding (Kim et al., 1995)		
$L249^{6.51}$	A	No detectable ant. binding Reduction in ag. potency (Jaakola et al., 2010)		
$H250^{6.52}$	A	No detectable binding (Jiang et al., 1997; Kim et al., 1995)	A_1:H251L (bovine)	Fourfold decrease in ant. affinity (Olah et al., 1992)
	F,Y, N	Modest decrease ag. affinity (Jiang et al., 1997; Kim et al., 1995)		
$N253^{6.55}$	A/D	Loss of ag. and ant. binding	A_3:N250A	No binding (Gao et al., 2002)

		(Kim et al., 1995)		
$C254^{6.56}$	A	No effect (Kim et al., 1995)	A_1:C255A	No effect (Scholl and Wells, 2000)
$F257^{6.59}$	A	Loss of ag. and ant. binding (Kim et al., 1995)		
$C262^{6.64}$	G	No effect (Kim et al., 1996)		
$I274^{7.39}$	A	Loss of ag and ant. binding (Kim et al., 1995)	A_1:I274C	Decrease in ant. binding affinity, implicated in ag. and ant. binding using MTSET-protection experiments (Dawson and Wells, 2001)
$S277^{7.42}$	A	No effect ant. decrease in ag. affinity and potency	A_1:T277A	Decrease in ag. binding affinity
				No change in ant. affinity (Dalpiaz et al., 1998; Townsend-Nicholson and Schofield, 1994)
	N,T,E	Marginal changes (Jiang et al., 1996; Kim et al., 1995)	A_1:T277S	Modest decrease in ag. affinity
				No change in ant. affinity (Townsend-Nicholson and Schofield, 1994)
$H278^{7.43}$	A	No binding	A_1:H278	Loss of ag. and ant. binding (Olah et al., 1992)
	Y	No change in ant.	A_3:H272E	Substantial decrease of ag. and ant. binding (Gao et al., 2002)
		Modest reduction of ag.		
	D,E	Marginal changes (Gao et al., 2000; Kim et al., 1995)		
$S281^{7.46}$	A	Loss of ag. and ant. binding		
	T	Increased activity for ag.		
	N	Marginal changes (Gao et al., 2000; Kim et al., 1995)		

Ant., antagonist; ag., agonist.

accordingly, mutation of Ile274$^{7.39}$ to alanine results in negligible antagonist binding and an order of magnitude reduction in agonist potency (Kim et al., 1995). The exocyclic nitrogen of the triazolotriazine core makes two polar interactions, one with the conserved Asn253$^{6.55}$ and the other with the Glu169$^{5.30}$ from ECL2. Mutation of Glu169$^{5.30}$ to alanine reduced the affinity for both antagonists and agonists and causes a reduction in agonist efficacy by three orders of magnitude (Kim et al., 1996). However, mutating this position to glutamine did not have a substantial impact on antagonist binding affinity, which suggested hydrogen bonding as the predominant means of interacting with this exocyclic nitrogen. The mutation of Asn253$^{6.55}$ to alanine was shown to cause a complete loss of both agonist and antagonist binding (Kim et al., 1995). The interaction of A_{2A} AR antagonists such as ZM241385 with the residues Glu169$^{5.30}$, His250$^{6.52}$, Asn253$^{6.55}$, and Ile274$^{7.39}$ (Table I) had been recognized as important residues for ligand binding prior to the crystal structure by mutagenesis studies and computer modeling studies (Ivanov et al., 2009; Kim et al., 1995, 1996, 2003). However, the crystal structure indicated a number of interactions that had not been identified before, involving eight uncharacterized residues (Fig. 3B). This is not surprising when considering the unexpected location of the ZM241385 binding pocket compared to earlier computational receptor homology models. Given the highly engineered receptor–lysozyme fusion construct used for these crystallization studies, it is important to assess the importance of these residues in ligand binding. Jaakola et al. performed a systematic mutagenesis study of the binding cavity residues to validate their role(s) in antagonist and agonist binding (Jaakola et al., 2010). The crystal structure allows the measured effect of such mutations upon ligand binding affinity to be related to the relative contribution of these residues toward ligand binding calculated from the crystal structure. The importance of Phe168$^{5.29}$ to ligand binding had not been fully recognized prior to the determination of the crystal structure of A_{2A} AR, though this amino acid is conserved between all known sequences of AR subtypes/species, and homology modeling studies provided some hints for its involvement in ligand binding (Ivanov et al., 2009; Kim et al., 2003). Interestingly, based on normalized occluded surface (NOS) calculations in the crystal structure, Phe168$^{5.29}$ has the highest contact area with ZM241385 and contributes an aromatic π-stacking interaction with the central triazolotriazine unit of ZM241385 (Fig. 3). The calculated contribution of Phe168$^{5.29}$ to binding is about 25% of total binding energy for ZM241385 and the most significant interaction (Jaakola et al., 2010). Further, radioligand binding and functional experiments using receptors with mutations at Phe168$^{5.29}$ showed the importance of aromatic stacking to ligand binding. The Phe168$^{5.29}$Trp mutation retained wild-type agonist- and antagonist-binding properties and signaling function even though tryptophan has a much bulkier side chain. In contrast, mutation of this phenylalanine to alanine resulted in a complete inability to measurably bind the radiolabeled antagonist [^{3}H]ZM241385. These results demonstrate the essential role of the Phe168$^{5.29}$ side chain in ECL2 in ligand binding. Further, as described earlier, both Glu169 and Phe168 are located in ECL2 underlining the importance of this region for high-affinity binding.

Another significant interaction between a receptor residue and ZM241385 is with Leu249$^{6.51}$, which is calculated to be 70% of the Phe168$^{5.29}$ NOS area (Jaakola et al., 2010). Leu249$^{6.51}$ is located almost opposite to Phe168$^{5.29}$ with respect to ZM241385 and makes hydrophobic interactions with the central triazolotriazine unit of the ligand. Although a role for this residue in ligand binding was suggested by Kim and coworkers, the specific nature of the interaction was not described (Kim et al., 2003). Substitution of Leu249$^{6.51}$ to the smaller and less hydrophobic alanine residue abolished radioligand binding suggesting strong structural requirements at this position in the triazolotriazine binding cavity. The bicyclic core makes also several hydrogen-bonding interactions (directly or indirectly) with crystallographic water molecules (#559, #567, #550, and #522) in the binding cavity (Jaakola et al., 2008).

When considering the interaction of the 4-hydroxyphenylethyl side chain with the upper part of the binding cavity of the A_{2A} AR, it is important to note that this part of the structure has higher temperature factors than other parts of the ligand and protein structure, indicating larger structural flexibility/disorderedness. The phenolic hydroxyl group extending from the ethylamine chain of ZM241385 forms a hydrogen bond with an ordered water molecule. The phenyl ring forms hydrophobic interactions with Leu267$^{7.32}$ and Met270$^{7.35}$ that would suggest hydrophobicity rather than aromaticity as means of interaction with the phenolic substituent. Indeed, a ZM241385 derivative, with a cycloalkyl substituent (LUF5477) instead of phenylmethylene, also has high affinity for the A_{2A} AR receptor. Further, in a recent study by Mantri and coworkers on new antagonists for the A_{2A} AR, it was demonstrated that tremendous substituent flexibility exists in this area of the pharmacophore (Mantri et al., 2008). This observation correlates well with the directionality of the phenylethylamine substituent in ZM241385 because it extends toward the more solvent-exposed extracellular region (ECL2 and ECL3) rather than toward the transmembrane domain of the receptor, as was previously proposed (Martinelli & Tuccinardi, 2008; Yuzlenko & Kiec-Kononowicz, 2009).

The other substituent in ZM241385 is the furan ring, a feature that occurs in many A_{2A} AR antagonists. This moiety is located deep in the ligand binding cavity and directed toward helices V and VI, where it hydrogen bonds to Asn253$^{6.55}$ and forms a water-mediated interaction with His250$^{6.52}$. Hydrophobic interactions of the furan ring system include those with His250$^{6.52}$ and Leu249$^{6.51}$. Mutation of His250$^{6.52}$ to alanine completely abolishes both agonist and antagonist binding, mutation to phenylalanine or tyrosine residues modestly affects agonist binding but not antagonist binding and replacement with an asparagine slightly increases ligand affinity (Kim et al., 1995). A third previously uncharacterized residue in the lower part of the ZM241385 binding cavity is Met177$^{5.38}$, which is conserved throughout the AR family. Based on the crystal structure, Met177$^{5.38}$ interacts with the furan ring of ZM241385 and is calculated to have 27% of the Phe168$^{5.29}$ binding surface. As predicted by this more modest contact area, the alanine mutation only moderately reduced [^{3}H] ZM241385 binding affinity.

The furan ring is approximately 3 Å away from Trp246$^{6.48}$. This amino acid is part of the CW$^{6.48}$XP$^{6.50aa}$ motif, which is conserved throughout the family A GPCRs (Holst et al., 2010; Shi et al., 2002). A rotamer switch of this tryptophan and the nearby aromatic residues is predicted to be coupled during activation. This assumption is largely based on the position of retinal in rhodopsin, where it is in close proximity to the tryptophan residue keeping rhodopsin in an inactive form (Palczewski et al., 2000). The furan ring interactions may directly and indirectly restrict the movement of the "toggle switch" Trp246$^{6.48}$. However, compared to the 32 Å^2 in the rhodopsin: retinal (inactive) structure, ZM241385 has a more modest 14 Å^2 contact area with this residue. Further in the β_2-adrenoceptor:carazolol structure, this interaction is even less prominent which suggesting that for these structures this interaction is less important. It should be noted that the apo-β_2-adrenoceptor has a high basal activity.

2. *The Agonist Binding Pocket*

It should be noted that the structure of the A_{2A} AR cocrystallized with ZM241385 represents an inactive receptor conformation. Indeed, to date, all of the published structures of GPCRs with diffusible ligands have been cocrystallized with antagonists or inverse agonists and represent inactive structures. The fundamental question of the mechanism for ligand-activated GPCRs remains: how does binding of an agonist, and the resulting changes in interactions at the ligand binding pocket, lead to conformational changes that are propagated from the extracellular portion of the molecule to the cytoplasmic surface involved in G protein binding. The recent structures of opsin could provide clues to the transmembrane helix rearrangements that can be expected as a result of agonist binding. Opsin is the retinal-free photoreceptor protein generated after photoactivation and Schiff base hydrolysis of rhodopsin. In the crystal structure of opsin at low pH, there are several subtle changes in the conformations of binding-pocket residues, relative to rhodopsin (Scheerer et al., 2008). Most importantly, the side chain of Trp265$^{6.48}$ (the toggle switch) moves into space previously occupied by the ionone ring of retinal. More dramatic structural changes are observed at the cytoplasmic surface of the molecule in which the cytoplasmic end of TM6 is shifted more than 6 Å outward from the center of the bundle relative to its position in the inactive state, and at the same time moves closer to TM5. The new position of the cytoplasmic end of TM6 is accompanied by changes in several key interactions most notably the breaking of the ionic lock described earlier.

In the light of these structural changes, any discussion regarding the binding of agonists to ARs is highly speculative. However, using results from mutagenesis experiments with the ZM241385:A_{2A} AR structure as a template, we can at least speculate on the role of various residues within the binding pocket for both agonist and antagonist binding and the orientation of an agonist within the binding site. It is important therefore to first compare the structures of various A_{2A} AR agonists with that of the cocrystallized antagonist ZM241385 (Fig. 4A). It is noteworthy that both the structures of the nonselective agonist NECA and the selective A_{2A} AR agonist CGS21680 include a heterocyclic purine core with an exocyclic nitrogen analogous to that of the bicyclic triazolotriazine core unit of ZM241385. In addition, both the selective agonist CGS21680 and the selective antagonist ZM241385 have a phenylethylamine moiety. Such structural similarities between agonist and antagonist imply that the heterocyclic core of such adenosine-derived agonists occupies at least an overlapping region within the receptor binding site. The crystal structure identifies the key role of several residues for antagonist binding, namely Asn253$^{6.55}$, Phe168$^{5.29}$, Glu169$^{5.30}$, Leu249$^{6.51}$, His250$^{6.52}$, His278$^{7.43}$, Asn181$^{5.48}$, Ile274$^{7.42}$, and Leu267$^{7.32}$. All of the above residues abrogated or decreased both antagonist and agonist binding when mutated to alanine suggesting that both agonist and antagonists do indeed occupy a very similar binding site within the A_{2A} AR (Table I). More specifically, a Phe168$^{5.29}$ to alanine mutation resulted in a receptor which could bind the A_{2A} AR agonist CGS21680 and activate receptor-mediated G protein signaling with a 65-fold lower potency than the wild-type receptor. Mutation of this residue to an aromatic Trp retained wild-type agonist affinity implying that the aromatic stacking interactions between Phe168$^{5.29}$ and the heterocyclic core of "classical" AR ligands are essential. Leu249 makes a hydrophobic interaction with the heterocyclic core of ZM241385 and mutation of this residue to alanine also decreased the potency of ZM241385 by 10-fold, again underlining the conserved position of the heterocyclic core of agonists and antagonists within the ligand binding site. An attempt to dock the agonists NECA or CGS21680 into an A_{2A} AR model based on the antagonist bound crystal structure showed the exocyclic nitrogen of these agonists interacting with both Asn253$^{6.55}$ and Glu169$^{5.30}$ (Fig. 4). Accordingly, mutation of both Asn253 and Glu169$^{5.30}$ to alanine abrogated binding of the agonist CGS21680 and the antagonist CGS15943. There are also a number of residues reported in mutagenesis studies that influence ligand binding but are not in direct contact or in close proximity to the bound antagonist in the A_{2A} AR crystal structure. Based on molecular modeling, it was hypothesized that Thr88$^{3.36}$, Phe180$^{5.41}$, Asn181$^{5.48}$, Phe182$^{5.43}$,

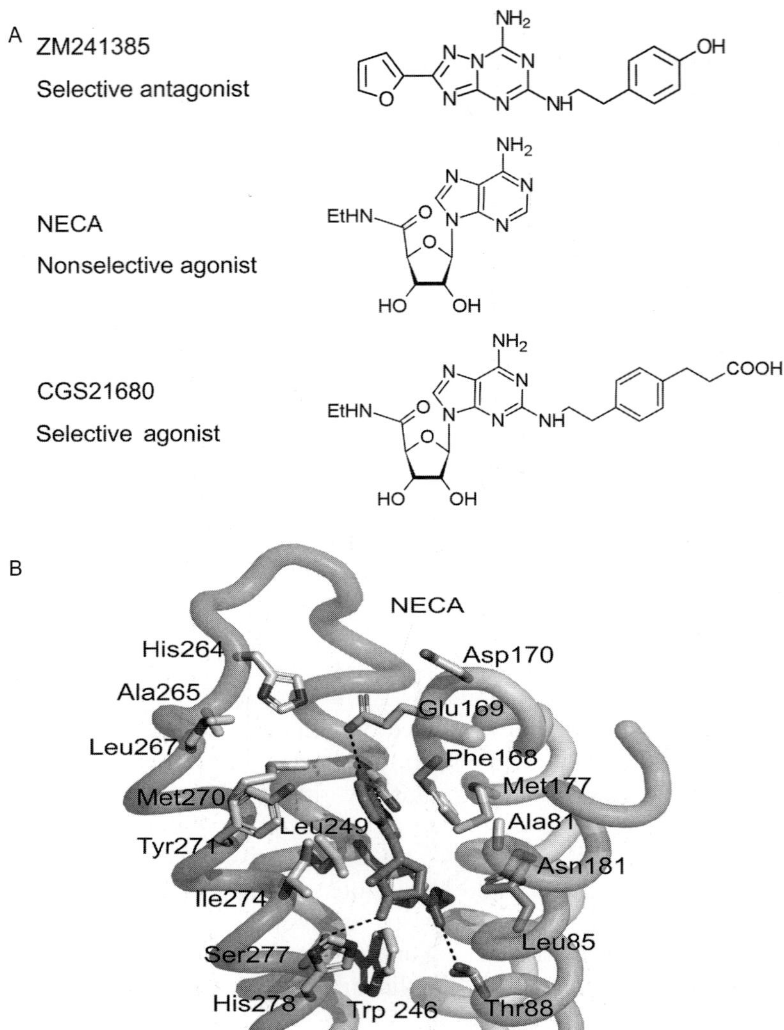

FIGURE 4 (A) Two-dimensional structures of the selective A_{2A} AR antagonist ZM241385, the nonselective agonist NECA and the selective A_{2A} AR agonist CGS21680. (B) Crystallographic structure-based molecular model of the human A_{2A} AR containing "docked" agonist NECA bound in the antagonist-binding cavity. Only parts of TM3, TM5, TM6, TM7, and a selected set of side chains in these TMs are shown.

Ser277$^{7.42}$, His278$^{7.43}$, and Ser281$^{7.46}$ residues might interact with determinants specific for AR agonists. However, this is highly speculative: a relatively small-molecule agonist such as NECA may fit perfectly in several orientations (poses) in the relatively large binding cavity of ZM241385. However, mutagenesis data can again provide useful clues. Based on the crystal structure,

Met177$^{5.38}$ which is conserved throughout the AR family interacts with the furan ring of ZM241385. The mutation of this residue to alanine only moderately reduced [^{3}H]ZM241385 binding affinity and had no significant effect on the affinity of the agonist NECA. These mutagenesis data along with the docking results suggest that the agonist ribose moiety is not in the same location as the furan ring of ZM241385. It is tempting to speculate that the very hydrophilic ribose moiety, so important for receptor activation, would be located where the crystallographic water network in the lower part of the binding pocket resides, and at least two docking studies support this localization (Ivanov et al., 2009; Kim et al., 2003). In both studies, the ribose moiety is in an orientation in which significant interactions are made with residues Thr88$^{3.36}$, Ser277$^{7.42}$, and His278$^{7.43}$, close to the water network mentioned earlier. All three residues have previously been mutated and have been shown to be critically involved in agonist binding (Jiang et al., 1996, 1997; Kim et al., 2003). Indeed, mutation of Thr88$^{3.36}$ or Ser277$^{7.42}$ to alanine results in a substantial decrease in agonist but not antagonist binding and potency. Further, in one such study, a "neoceptor" was generated by mutation of Thr88$^{3.36}$ to aspartate which responded to a positively charged amino sugar agonist derivative, again confirming the important interaction between this residue and the ribose moiety of agonist ligands (Jacobson et al., 2005; Kim et al., 2003).

Interestingly, residues in TM3 have been shown to be essential for agonist binding and function in the β_2- and β_1-adrenoceptors (Strader et al., 1988). For these receptors, it has been proposed that the upper region of TM5, which contains several catechol-binding serines, moves closer to TM3 (Liapakis et al., 2000; Rosenbaum et al., 2009; Strader et al., 1989). Simultaneous engagement of the agonist by TM5–catechol hydrogen bonding and TM3/TM7–amine polar contacts (also essential for agonist binding) would facilitate changes in the packing of nearby aromatic amino acids that shield Trp268$^{6.48}$. It has been speculated that for the adenosine A_{2A} AR agonists with the ribose functional group would promote the engagement of TM3 residues resulting in small changes in the relative transmembrane helix dispositions that could activate the rotamer toggle switch (Rosenbaum et al., 2009). However, without an agonist-occupied structure, it remains unclear to what extent structural rearrangements in the binding cavity occur upon binding of an agonist and, consequently, what the exact atomic interactions would be.

F. The Role of Water in Ligand Binding and Receptor Activation

So far we have concentrated on the role of residues toward ligand binding. However, the antagonist ZM241385 also makes a number of important contacts with water molecules within the binding site (Jaakola &

IJzerman, 2010; Jaakola et al., 2008). Moreover, a number of crystallographic water molecules are located within the cytosolic half of the A_{2A} AR (Fig. 5). Interestingly, water molecules in the cytosolic half of the A_{2A} AR occupy similar positions in the high-resolution structures of rhodopsin and the β-adrenoceptors (Angel et al., 2009; Nygaard et al., 2009). Several of these conserved water clusters make interactions with highly conserved residues or motifs within the family A GPCRs such as the NPXXY motif and the WxP6.50F/Y motif which includes the toggle switch tryptophan. It has been postulated that these water clusters may play a crucial role in receptor activation (Angel et al., 2009; Nygaard et al., 2009). These water molecules could not be defined in the low-resolution structures of opsin, and the opsin:G protein peptide complex that would show how these water molecules are associated with signal transduction. In the adenosine A_{2A} AR structure, the primary ZM241385 binding cavity includes seven ordered water molecules (Figs. 3 and 5). The exact functional role of the waters for

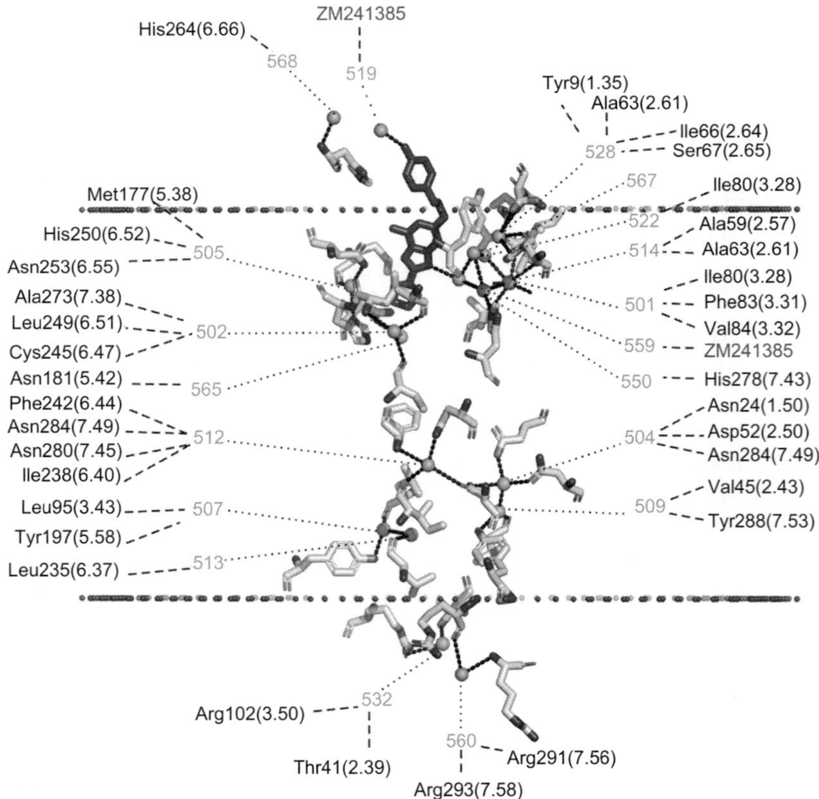

FIGURE 5 Structural details of key regions of the human adenosine A_{2A} AR bound to the antagonist ZM241385. Water molecules are shown as dots and interacting side chains as sticks.

antagonist versus agonist binding is still unclear; however, they might have an important role in ligand specificity. For instance, an *in silico* ligand screening study (discussed later) was significantly improved when three crystallographic waters were left in the template structure (Katritch et al., 2010a). Further, subsequent to the publication of the A_{2A} AR crystal structure, one molecular dynamic simulation coupled with a statistical thermodynamic analysis of water molecules within the receptor binding site was used to explain an otherwise unintuitive SAR across a series of triazolylpurine derivatives (Higgs et al., 2010). In this chapter, as the 2-substituted aliphatic groups were extended to methyl and propyl, a decrease in affinity was observed followed by an increase when the substituent was extended to an *n*-butyl or *n*-pentyl. This pattern was predicted using solvent thermodynamics in which the smaller substituents occupy space containing stable waters, whereas the longer substituents compensate for this by also displacing unstable waters as the ligands extend out of the binding site.

IV. Binding Selectivity Across the ARs

The AR family represents a clear example of a family of closely related GPCRs activated by a single endogenous ligand (adenosine). As discussed earlier, all four AR subtypes have been implicated in the treatment of a wide range of disease states from neurodegenerative, cardiac, and inflammatory disorders to cancer (Fishman et al., 2009; Mustafa et al., 2009; Sebastiao & Ribeiro, 2009; Wilson et al., 2009). However, the functional importance of ARs in a wide range of physiological functions across numerous tissues imposes a clear need for subtype selectivity for both AR agonists and antagonists as candidate drugs (Jacobson & Gao, 2006). The A_{2A} AR crystal structure gave us an insight into the interaction between an A_{2A} AR selective antagonist and the receptor at atomic resolution (Jaakola et al., 2008). Despite this major advance, the mechanism behind subtype selectivity for ligands is not immediately apparent. As discussed earlier, the interaction between the heterocyclic core of ZM241385 and so-called "core" residues makes the major contribution to ligand affinity. This interaction is defined by the aromatic stacking interaction between Phe168$^{5.29}$ and the heterocyclic core of the ligand, hydrophobic interactions with Leu249$^{6.51}$, Ile274$^{7.39}$, and M177$^{5.38}$, and hydrogen-bonding interactions with both Asn253$^{6.55}$ and Glu169$^{5.30}$ (Ivanov et al., 2009; Jaakola et al., 2010). It should be noted that all of the above core pocket amino acid side chains are fully conserved in all four subtypes with the exception of Leu249$^{6.51}$ which is replaced by a similar valine in the A_{2B} AR (Fig. 3B). From the perspective of the ligand, previous SAR studies for AR binding chemotypes such as xanthine and adenine derivatives suggest that the core chemical scaffolds (namely the conserved heterocyclic core) do not themselves provide significant selectivity.

This is exemplified by a virtual ligand screening (VLS) study in which all of the ligands selected had a comparatively low affinity for the A_3 AR but only 1 of 23 showed a greater than 10-fold selectivity for the A_{2A} AR over the A_1 AR (Katritch et al., 2010a). This suggests that ligand subtype selectivity does not come hand in hand with high-affinity binding to one of the receptor subtypes.

If we move up the binding pocket into the extracellular domain, then we observe less conservation of residues. The 4-hydroxyphenyl ring of ZM241385 makes largely hydrophobic interactions with Ile267$^{7.32}$, Met270$^{7.35}$, and His264$^{6.66}$ in the "upper" region of the binding cavity and a polar interaction with a crystallographic water molecule (Jaakola et al., 2008). It should be noted however that this 4-hydroxyphenyl moiety has high crystallographic *B*-factors pointing to its high conformational flexibility even in the receptor-bound state. These observations suggest that interactions in the "upper" region of the binding pocket are less important for ligand binding affinity, but rather contribute to A_{2A} AR ligand selectivity (Jaakola et al., 2010).

Certain suppositions can be made such as the modest selectivity of ZM241385 between A_{2A} AR and A_{2B} AR has risen from very small amino acid variations in the binding cavity; potentially in the lower part of cavity in which there is a subtle variation of Leu$^{6.51}$Val and in the upper part of the cavity in which there is a variation of Leu$^{7.32}$Lys—this residue is not conserved across the AR subtypes. Clearer differences are seen between A_{2A} AR and A_3 AR, which is the most divergent from other AR subtypes. In this case, Glu169$^{5.30}$ in the A_{2A} AR, a residue that provides a hydrogen-bonding interaction with the exocyclic nitrogen of ZM241385, is replaced by the hydrophobic residue valine in the A_3 AR. Similarly, His250$^{6.52}$ which in the A_{2A} AR makes a hydrophobic interaction with the furan ring of ZM241385 (a substituent conserved in many A_{2A} selective antagonists) is replaced by the smaller polar serine. Finally, other key A_3 AR selectivity residues may include Ser$^{5.42}$ which is a conserved Asn across the AR subtypes apart from the A_3 and Gln$^{7.32}$ within the extracellular binding domain which differs for all AR subtypes. These relatively large differences are also mirrored by the pharmacology of the A_3 AR, particularly, with respect to the low affinity of this receptor for xanthines. In comparison, the differences between the A_1 AR and the A_{2A} AR are more subtle. Met270$^{7.35}$ that makes a hydrophobic interaction with the hydroxyphenylethylamine group of ZM241385 in the A_{2A} AR structure is replaced by a Thr in the A_1 AR. However, this single change in an interaction does not fully explain the A_{2A} selectivity of ZM241385 over A_1 AR. Table II shows a number of ZM241385 derivatives. In all cases, a high degree of selectivity for the A_1 AR, A_{2B} AR, and A_{2A} AR receptors over the A_3 AR is observed, in agreement with the lower conservation of binding site residues observed at the A_3 AR (De Zwart et al., 1999). However, a phenylethylamine substituent shows no selectivity across the A_1 AR, A_{2A} AR, and A_{2B} AR receptors. Consequently, the hydrophobic interaction

TABLE II The Affinity of Various ZM241385 Derivatives for the Adenosine Receptors (Adapted from de Zwart et al., 1999)

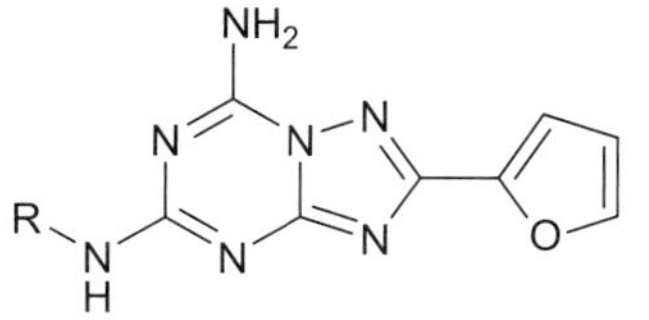

	K_i (nM)			
R	A_{2A} AR	A_1 AR	A_{2B} AR	A_3 AR
4-OH-Ph-Et-	1.8	257	17	3090
Ph-Et	5.4	7.1	9.9	1450
Ph-Me	13	13	7.6	207
Ph	3.0	23	33	376

of Met270$^{7.35}$ is not the only determinant of selectivity. Selectivity for the A_{2A} AR is only gained with the addition of the hydroxyl group at the *para* position on the phenyl ring. In the crystallographic model of the A_{2A} AR, this hydroxyl group is shown to interact predominantly with a conserved water molecule. Consequently, the full spectrum of subtype selectivity determinants for the A_{2A} AR is not apparent from the crystal structure.

A very recent study by Katritch and coworkers rose to the challenge of trying to predict AR ligand selectivity in a homology modeling and molecular docking approach (Katritch et al., 2010b). Using the crystal structure of adenosine A_{2A} AR as a template and using ligand-guided receptor optimization in combination with subtype-selective ligands optimized models were developed that showed the ability to discriminate between subtype-selective ligands. Interestingly, using this approach, the authors also were able to highlight potential determinants of ligand selectivity for each of the receptor subtypes. Notably for the A_3 AR, the valine at position 5.30 allows bulky substitutions to protrude toward the extracellular domain of the binding pocket. Further, the small serine residues at positions 6.52 and 5.42 (His and Asn, respectively, in the A_{2A} AR) create an additional subpocket that could be exploited to create subtype-selective ligands. For the A_1 AR, they highlighted the role of residue 7.35 (Met in A_{2A}R, Thr in A_1 AR) as described earlier. Interestingly, the need for a slightly shifted conformation of Glu$^{5.30}$ was emphasized to accommodate the small aliphatic substitutions made at most A_1 selective ligands. The authors postulated that this may be mediated by the EL3 loop which is one residue shorter in the A_1 AR compared to the A_{2A} AR (Katritch et al., 2010b). This suggests that the structure of the extracellular loops plays a key role in determining ligand selectivity, and underlines the importance of correctly predicting their conformation. Such predictions remain a challenge to the homology modeling community as discussed in the next section (Michino et al., 2009). A similar role is suggested for $Asn^{6.66}$ in the EL3 of the A_{2B} AR which again is thought to affect the interaction of $Glu^{5.30}$ with the exocyclic amine of ligands such as ZM241385 making it less important. Interestingly, most ligands with A_{2B} AR versus A_{2A} AR selectivity lack this amino group. Last, the importance of $Lys^{7.32}$ (a leucine in the A_{2A} AR) was highlighted. This basic residue is proposed to interact with the acidic group of A_{2B} AR ligands such as sulfonyl moieties (Borrmann et al., 2009; Stefanachi et al., 2008).

This computational study has highlighted residues that may be key players in AR ligand selectivity. As such, there is a clear need for the role of these residues to be validated in systematic mutagenesis studies. It should also be noted that the success of this type of study depends on the quality of the ligand set used to optimize the receptor models. As such, a data set dominated by a particular chemotype such as the pyrazolo-triazolo-pyrimidine scaffold used for the A_3 receptor in the above study may bias the results (Katritch et al., 2010b). In this case, it may not be possible to account for the entire spectrum of mechanisms behind ligand selectivity. Similarly, the above study used an inactive receptor model. Given the structural changes in terms of the movement of both transmembrane helices and the extracellular loop regions associated with agonist binding, it is likely that such an approach would be unable to predict subtype-selective agonists.

V. Receptor Structure and Receptor Homology Modeling

Despite the innate tractability of GPCRs as drug targets, the 3D modeling of GPCRs and the use of these models as starting points for drug discovery programs were hampered by the lack of structural data. Indeed, until 2007, rhodopsin was the only GPCR with a solved structure. More recently, this situation has improved with the publication of the β-adrenoceptors, the adenosine A_{2A} AR, and the chemokine CXCR4 receptor crystal structures (Jaakola et al., 2008; Rosenbaum et al., 2007; Warne et al., 2008; Wu et al., 2010). However, these publications have also illustrated that the format of the ligand binding cavity may vary considerably between receptors. As discussed earlier, this is particularly true for the adenosine A_{2A} AR structure. This divergence was somewhat unexpected: the dogma was that ligands use the same binding site orientation based on the hypothesis that variation in both ligand structure and binding site residues would provide sufficient selectivity for any GPCR. If so, this approach would validate the use of the rhodopsin template for homology modeling of other GPCRs, in which the typical retinal binding site serves as the fingerprint for any ligand binding site. Martinelli and Tuccinardi have summarized all published homology models for ARs; all of these use the retinal binding site for the docking of AR agonists and antagonists (Martinelli and Tuccinardi, 2008). Given the somewhat unexpected binding orientation of the cocrystallized antagonist ZM241385 in the A_{2A} AR structure, the determination of this structure also presented an opportunity to assess the ability of GPCR homology modeling to predict both protein architecture and ligand binding orientation. Prior to the release of the A_{2A} AR crystal structure into the public domain, 29 research groups submitted over 200 receptor models in an assessment called GPCRDock 2008 (Michino et al., 2009). The quality of the models was evaluated with respect to both protein architecture and the ligand "poses." The quality of the predictions for the receptor structure alone seems relatively good: 4.2 ± 0.9 Å for the receptor Cα RMSD and 2.8 ± 0.5 Å for the TM helices Cα RMSD. This is perhaps not surprising given that overall deviation between the solved receptor structures to date is less than 3 Å. It is also noteworthy that accurate prediction of the TM region does not necessarily lead to accurate prediction of the ligand binding mode. This indicates that the methods for modeling the receptor and docking of the ligand can be generally considered as distinct steps in the generation of models for the receptor–ligand complex. However, the ligand binding site and the organization of the extracellular loops were not predicted well. The majority of models had RMSD values greater than 10 Å for the ligand binding site; in only one, the typical perpendicular orientation of ZM241385 was suggested. Inaccuracies in homology models can arise from errors in side-chain packing, main chain shifts in aligned regions, errors in unaligned loop regions, misalignments, and incorrect templates (Marti-Renom et al., 2000; Moult et al., 2007). These errors are also applicable to GPCR modeling. In particular, the interactions of the ligand with residues located in structurally divergent regions from the template(s) were consistently not modeled accurately in even the best predictions in this study. For example, the hydrogen-bonding interaction between Glu169$^{5.30}$ in ECL2 and the exocyclic N15 atom of the ligand was not captured despite considerable evidence from mutagenesis experiments underlining its importance (Kim et al., 1996; Michino et al., 2009). Another important interaction from this region of the receptor is the aromatic stacking between F168$^{5.29}$ in ECL2 and the bicyclic ring of the ligand. This was only modeled correctly in one quarter of the submitted models. Interestingly, F168$^{5.29}$ is structurally homologous to F193$^{5.32}$, which interacts with the carbazole heterocycle of the ligand carazolol in the β_2-adrenoceptor structure; hence, correct modeling of this interaction may have been guided by homology (Katritch et al., 2010c; Michino et al., 2009). Another region of structural divergence is an extended bulge at the extracellular end of TMV. This means that, even in the best predictions, the Met177$^{5.38}$ is not orientated toward the ligand and as a result the ligand was situated too far down in the binding pocket. As described earlier, the helical shifts in TM helices I, II, and III alter the location of the binding pocket and redefine the pocket size and shape. The inaccuracy in the orientation of the ligand binding pose—for example, the parallel orientation with the phenolic substituent positioned close to TM helices II and III—may in part be due to the inaccurate modeling of these helical shifts. The helical shifts were most accurately modeled by an effective use of multiple template structures of rhodopsin and β-adrenoceptors or an all-atom refinement approach implemented by the ROSeTTA program (Barth et al., 2007). Finally, in the AR crystal structure, water molecules provide key interactions with the ligand; yet none of the submitted predictions included these water molecules. From a different perspective, Mobarec et al. concluded that based on sequence identity the currently available structures (rhodopsin, the β-adrenoceptor, and the A_{2A} ARs) may only provide reasonable models for a rather restricted subset of GPCRs (Mobarec et al., 2009). Indeed, taking the premise that a suitable template must have more than 30% sequence identity with the intended target, their results showed that the β_1-adrenoceptor receptor, the β_2-adrenoceptor, and the A_{2A} AR are suitable templates for homology modeling of the 7TMs of 18%, 16%, and 12% of non-orphan non-olfactory human class A GPCRs, respectively (Mobarec et al., 2009). Interestingly, they also calculated that the rhodopsin structures can significantly contribute to the modeling of only 2% of non-olfactory GPCRs. Such a finding correlates well with the results of GPCRDock 2008. It is clear then that further structures are needed to produce sensible models for the majority of the class A GPCRs.

Subsequent to the publication of both the A_{2A} AR crystal structure and GPCRDock 2008, Ivanov et al. (2009) reexamined the homology modeling of the ARs. Indeed, it is interesting to note that the most accurate solution presented to GPCRDock 2008 was that of Costanzi et al., a group with considerable experience in the modeling of adenosine GPCRs (Michino et al., 2009). The authors emphasized that domain knowledge such as receptor mutation data yields useful constraints for the positioning of ZM241385 into a receptor homology model. In this case, they highlighted the hydrogen-bonding interaction of the exocyclic nitrogen of ZM241385 with Asn253$^{6.55}$ as a key interaction that allowed a good prediction of ligand binding orientation (Kim et al., 1995). However, such mutational data may also be difficult to interpret. For example, several residues shown by mutagenesis to be critical for antagonist binding such as Phe182$^{5.43}$ are not involved in direct contact with ZM241385 in the recent A_{2A} AR crystal structure (Kim et al., 1995). Thus, the involvement might conceivably be through intervening amino acid residues or a conserved bound water molecule.

Ivanov et al. (2009) also demonstrated that the inclusion of ordered water molecules as present in the crystal structure into the receptor model proved equally important to generate acceptable ligand poses. The importance of these binding site water molecules are also highlighted in the discussion regarding *in silico* screening below. However, the accurate modeling of structurally divergent regions such as extracellular loops and the positioning of water molecules within the binding site present significant challenges to the modeling community. With the publication of the CXCR4 receptor, our knowledge of GPCR receptor structure is extended further across the phylogenetic tree. However, this structure has also underlined the structural divergences between family A GPCRs and the challenge that modelers face to predict them. Further, the conformational differences observed between the structures with a cyclic peptide or small-molecule bound illustrate the structural plasticity of GPCR ligand binding sites (Wu et al., 2010).

VI. *In Silico* Screening

Until recently, ligand-based approaches to drug discovery were the only ones feasible for GPCRs. They have been and continue to be very successful, but, by definition, rely on a repertoire of already available chemical structures, generally derived from the structure of the endogenous ligand. Work from several research teams demonstrated the success of pharmacophore modeling for both the adenosine A_1 AR and A_{2A} AR in particular (Chang et al., 2004; Mantri et al., 2008; Moro et al., 2006; van Galen et al., 1991). However, using a pharmacophore approach means that there will be bias toward the existing chemotypes and hurdles such as receptor subtype selectivity can be difficult to overcome. With the recently published GPCR crystal structures, a more structure-based approach to GPCR drug discovery can be taken. This has been exemplified by the work of Kobilka and coworkers who identified 25 potential "hits" against the β_2-adrenoceptor-T4L:carazolol crystal structure, by computationally screening a 1 million compound library (Kolb et al., 2009). In the subsequent radioligand binding assay, six compounds displayed apparent dissociation constants (K_i values) of less than 4 μM; one compound had a particularly high (9 nM) affinity. The six validated hits fall into two classes: four compounds adopted a carazolol-like docking geometry and two compounds appeared to interact with the key binding site residue Asp113$^{3.22}$ in a mode unexpected for β-adrenergic receptor ligands. Importantly, in a functional assay all tested compounds behaved as low-efficacy inverse agonists. Recently, a similar screen has been performed by two separate groups using the adenosine A_{2A}-T4L: ZM241385 complex as a template structure, one study was undertaken by a joint collaboration between the groups of Stevens, IJzerman, and Abagyan and another performed by the groups of Shoichet and Jacobson (Carlsson et al., 2010; Katritch et al., 2010a). In the study by Katrich and coworkers, the A_{2A} AR crystal structure (PDB code: 3EML) and ligand-refined models of the A_{2A} AR were evaluated for their ability to select 23 known A_{2A} AR specific antagonists in a pool of 2000 random decoy compounds in a docking and VLS benchmark test (Katritch et al., 2010a). The predicted binding poses for the majority of the known A_{2A} AR antagonists in the diverse benchmark set displayed key similarities with the binding mode of ZM241385 in the crystal structure. This common binding motif involved stacking between aromatic moieties of the ligands and the conserved Phe168$^{5.29}$ side chain of the receptor, as well as polar interactions with the conserved Asn253$^{6.55}$ side chain. Further, most compounds had an aromatic group extending deeper into the binding pocket and flexible extensions toward the extracellular opening of the pocket. The model based simply on the crystal structure showed a good overall performance comparable to that of the β_2-adrenergic receptor. However, the model with three structured water molecules in the binding pocket achieved a significantly higher initial enrichment. The three waters selected for the 3EML:W3 model (wa, wa14, and wa5 in the 3EML PDB entry) have the lowest *B* factor values and form an extended hydrogen-bonding network with the binding pocket residues, suggesting their highly structured nature. Rather than improving the binding scores of the 23 known ligands these selected structured water molecules occupy highly polar sub-pockets in the A_{2A} AR and apparently prevent adverse binding of some decoy compounds into these subpockets. Further modest improvement of the model was achieved by ligand-guided optimization of side chains in the binding site of the A_{2A} AR structure. Into this optimized receptor model, a huge set of over 4,000,000 commercially available compounds was computationally docked, and prioritized on the basis of interaction energy values. Fifty-six compounds of various chemical classes that scored well were purchased and tested in radioligand binding assays. Twenty-three of these compounds showed activity in the micromolar to nanomolar range, representing a very high hit rate of 40%. These studies were extended to the ability of the selected compounds to bind to other ARs. Although all tested compounds were moderately selective for the A_{2A} over the A_3 AR, no such selectivity was observed over the A_1 AR. In functional assays, all tested compounds were antagonists. Carlsson et al. also used the A_{2A} structure as the basis of a model to perform a virtual screen for A_{2A} ligands (Carlsson et al., 2010). However, in this case, a different approach was used to optimize the model. The best scoring conformation was calculated as the sum of the receptor–ligand electrostatic and van der Waals interaction energy and corrected for ligand desolvation. Crucially, in this calculation, partial charges were used for all receptor atoms except the side-chain amide of Asn253, for which the dipole moment was increased to favor hydrogen bonding to this residue. This approach was also used by Ivanov et al. as discussed in the homology modeling section above (Ivanov et al., 2009). A 1.4-million compound database was screened, and 20 high ranking compounds were tested. Of these, a similar proportion of 35% showed substantial activity (with affinities between 9 μM and 200 nM) as compared to the previous study of Katritch et al., although it should be noted that the selection criteria differed between the two studies with that of Carlsson et al. being slightly less stringent. Interestingly, although the affinities of these compounds for the A_{2A} AR were in general lower than those of the compounds selected in our study, they displayed a degree of selectivity over both the A_3 AR and A_1 AR (up to 50-fold) that we did not observe. Although comparisons are difficult due to the use of different ligand databases, it is tempting to speculate on the reason for both the slight decrease in affinity and gain of specificity. As discussed in the section covering ligand selectivity, it appears that receptor selectivity is gained via the interaction of the ligand with the upper (extracellular) part of the binding site of the A_{2A} AR. Conversely, the core of the binding site and, in particular, the interaction of the heterocyclic core of ZM241385 with Phe168 and Asn253 provide ligand binding affinity. By increasing the dipole moment of Asn253, the interaction of screened ligands with other "core" residues becomes less important. This may allow the selection of ligands which sit higher up in the binding site gaining selectivity but perhaps losing some affinity. Accordingly, the compounds selected by Carlsson et al. do show a lower affinity for the receptor as compared to the selected compounds of Katritch et al. In both cases, functional assays revealed only antagonists for the A_{2A} AR were discovered. Further, the VLS using the β_2-adrenoceptor structure also exclusively yielded antagonists (Kolb et al., 2009). This may be a combination of two factors. First of all, it may be that in the compound collections antagonists rather than agonists are overrepresented. Second, all the receptor structures to date were cocrystallized with an antagonist or an inverse agonist. Consequently, these structures represent inactive structures and it may be that agonists cannot be reliably docked into them (Hanson and Stevens, 2009; Rosenbaum et al., 2009). In agreement with this, Carlsson et al. included in their 1.4 million compound screening database both known A_{2A} AR antagonists and two known agonists, adenosine and NECA. While all of the known antagonists appeared in the top 500 ranked compounds, the two agonists were ranked as 951057 and 919993, respectively. This bias of the A_{2A} AR:ZM241385 structure to select antagonists over agonists underlines its inactive nature.

It is also interesting to note the extremely high hit rate for all three of the VLS experiments performed with the β_2-adrenoceptor, and A_{2A} AR receptor structures (24%, 40%, and 35%, respectively). As illustrated by Carlsson et al., this is in far excess of the hit rate that this group experienced for enzyme targets (~5%). Further, the affinity of selected hit compounds for GPCRs was 100-fold higher than those of the best selected hits for enzyme targets. Carlsson et al. point out that this reflects the amenability of GPCR binding sites for specific recognition of small molecules. Largely buried from bulk solvent, these sites can almost completely enclose a "drug-like" molecule and can do so with a mixture of nonpolar and polar interactions. Given this amenability and the therapeutic relevance of GPCRs for the treatment of a wide range of diseases, there has been a considerable medicinal chemistry effort directed at these targets. Consequently, ligand libraries, such as ZINC, may have become populated with molecules bearing "GPCR-like" chemotypes. This also reflects a bias toward naturally occurring molecules in our screening libraries (Hert et al., 2009) Indeed, Kolb et al. estimated that there were 3–12 times as many small molecules that were similar to GPCR ligands in the ZINC lead-like set compared to other common drug targets such as kinases, proteases, and ligand-gated ion channels (Kolb et al., 2009). Accordingly, the hit rates for the GPCR targets to date have been 10- to 100-fold higher than that for the latter targets. Consequently, both the "small-molecule-ligand-friendly" nature of GPCR orthosteric sites and the GPCR-like chemotype bias in available compound libraries will make GPCR targets a fruitful avenue for structure-based discovery. However, in view of the caveats associated with homology modeling discussed earlier, this also highlights the urgent need for new GPCR crystal structures as templates for such VLS programs.

VII. Conclusion

With publication of the crystal structures of a handful of GPCRs with diffusible ligands, it is an exciting time for GPCR research. Given the unexpected binding orientation of the antagonist ZM241385 revealed by the A_{2A} AR structure, the major impact of this structure was to illustrate that ligand selectivity across GPCRs is engendered not simply by differences in binding site residues or ligand structure but also by a variation in the topography of the ligand binding site. It also highlighted the structural diversity of GPCRs particularly within divergent regions such as the extracellular loops. These observations have profound consequences for GPCR homology modeling highlighted by the GPCRDock 2008 study. However, the description of the antagonist binding site will have the most profound impact on the AR field itself. This has been exemplified by the demonstration that such structures are highly amenable to VLS studies and that such studies can reveal novel chemotypes that represent starting points for the development of novel therapeutic agents. However, such approaches still lag behind ligand-based SAR approaches in terms of predicting receptor subtype affinity. Indeed, the A_{2A} AR structure has not provided a clear answer to this problem. The current set of GPCR structures, including that of the A_{2A} AR, is however biased toward the inactive state(s) of receptors as they are bound to high-affinity antagonist or inverse agonist small molecules. The propensity for the inactive state is also highlighted in the *in silico* screening results followed by experimental validation, as all selected compounds were either partial inverse agonist or antagonist (i.e., receptor blockers). This might suggest that activated receptors have structures that significantly differ from the inactive state. Therefore, an agonist bound receptor structure would have an extremely high impact across the GPCR field. Such results also highlight the limitations of X-ray crystallography. This approach provides a static structure of a protein complex at near atomic resolution, a snapshot of the life of a GPCR. These static structures need to be complemented and validated by other biophysical techniques such as electron paramagnetic resonance, NMR, and other spectroscopic approaches to appreciate the dynamics of GPCRs. Such studies have already begun on both rhodopsin and the β-adrenoceptors. There is clearly then a need for such studies to be carried over to the A_{2A} AR structure. Given the divergence of the A_{2A} AR structure compared to other receptor structures, such studies may also answer important questions as regard the conservation of activation mechanisms across the GPCR receptor family. Crystal structures of the four AR subtypes with subtype-selective antagonists bound in combination with such biophysical approaches may be needed to gain a true understanding as regards receptor subtype selectivity. Further, the A_1 AR receptor and the A_3 AR receptor have been shown to be allosterically modulated by small-molecule ligands. Such ligands represent attractive targets for the development of drugs not least because they interact with the receptor at a site distinct from the highly conserved orthosteric binding site and therefore have the potential to achieve greater subtype selectivity. The development of such ligands would be greatly aided by a receptor structure in which such a ligand is cocrystallized. All approaches used to date to obtain GPCR crystal structures have involved engineered, that is non-wild type, receptors in which for instance mutations in or near the ligand binding pocket or perturbations of the intracellular domain of the GPCR were introduced to achieve thermostability of the receptor. Thus research to perform structural studies with wild-type receptors should be encouraged. Finally, to understand the mechanism of interaction of (adenosine) receptors with G proteins and other signaling proteins, crystal structures with the receptor in complex with these proteins are required.

Conflict of Interest: The authors have no conflicts of interest to declare.

Abbreviations

AR	adenosine receptor
CGS15943	9-Chloro-2-(2-furanyl)-[1,2,4]triazolo[1,5-c]quinazolin-5-amine
CGS21680	2-*p*-(2-carboxyethyl)phenethylamino-5′-*N*-ethylcarboxamidoadenosine hydrochloride
ECL	extracellular loop
GPCR	G protein-coupled receptor
ICL	intracellular loop
NECA	5′-(*N*-ethylcarboxamido)adenosine
TM	transmembrane domain
ZM241385	4-{2-[7-amino-2-(2-furyl)-1,2,4-triazolo[1,5-*a*]1,3,5]triazin-5-yl-amino]ethyl}phenol

References

Alexandrov, A. I., Mileni, M., Chien, E. Y., Hanson, M. A., & Stevens, R. C. (2008). Microscale fluorescent thermal stability assay for membrane proteins. *Structure*, *16*(3), 351–359.

Altenbach, C., Kusnetzow, A. K., Ernst, O. P., Hofmann, K. P., & Hubbell, W. L. (2008). High-resolution distance mapping in rhodopsin reveals the pattern of helix movement due to activation. *Proceedings of the National Academy of Sciences of the United States of America*, *105*(21), 7439–7444.

Angel, T. E., Chance, M. R., & Palczewski, K. (2009). Conserved waters mediate structural and functional activation of family A (rhodopsin-like) G protein-coupled receptors. *Proceedings of the National Academy of Sciences of the United States of America*, *106*(21), 8555–8560.

Barth, P., Schonbrun, J., & Baker, D. (2007). Toward high-resolution prediction and design of transmembrane helical protein structures. *Proceedings of the National Academy of Sciences of the United States of America*, *104*(40), 15682–15687.

Beukers, M. W., Chang, L. C. W., von Frijtag Drabbe Kunzel, J. K., Mulder-Krieger, T., Spanjersberg, R. F., Brussee, J., et al. (2004). New, non-adenosine, high-potency agonists for the human adenosine A_{2B} receptor with an improved selectivity profile compared to the reference agonist N-ethylcarboxamidoadenosine. *Journal of Medicinal Chemistry*, *47*(15), 3707–3709.

Beukers, M. W., van Oppenraaij, J., van der Hoorn, P. P. W., Blad, C. C., Dulk, H., et al. (2004). Random mutagenesis of the human adenosine A_{2B} receptor followed by growth selection in yeast. Identification of constitutively active and gain of function mutations. *Molecular Pharmacogy*, *65*(3), 702–710.

Bokoch, M. P., Zou, Y., Rasmussen, S. G. F., Liu, C. W., Nygaard, R., Rosenbaum, D. M., et al. (2010). Ligand-specific regulation of the extracellular surface of a G-protein-coupled receptor. *Nature*, *463*(7277), 108–112.

Borrmann, T., Hinz, S., Bertarelli, D. C., Li, W., Florin, N. C., Scheiff, A. B., & Müller, C. E. (2009). 1-Alkyl-8-(piperazine-1-sulfonyl)phenylxanthines: Development and characterization of adenosine A_{2B} receptor antagonists and a new radioligand with subnanomolar affinity and subtype specificity. *Journal of Medicinal Chemistry*, *52*(13), 3994–4006.

Burstein, E. S., Spalding, T. A., & Brann, M. R. (1998). The second intracellular loop of the M5 muscarinic receptor is the switch which enables G-protein coupling. *The Journal of Biological Chemistry*, *273*(38), 24322–24327.

Canals, M., Marcellino, D., Fanelli, F., Ciruela, F., de Benedetti, P., Goldberg, S. R., et al. (2003). Adenosine A_{2A}-dopamine D_2 receptor-receptor heteromerization: Qualitative and quantitative assessment by fluorescence and bioluminescence energy transfer. *The Journal of Biological Chemistry*, *278*(47), 46741–46749.

Carlsson, J., Yoo, L., Gao, Z. G., Irwin, J. J., Shoichet, B. K., & Jacobson, K. A. (2010). Structure-based discovery of A_{2A} adenosine receptor ligands. *Journal of Medicinal Chemistry*, *53*(9), 3748–3755.

Chang, L. C., Spanjersberg, R. F., von Frijtag Drabbe Kunzel, J. K., Mulder-Krieger, T., van den Hout, G., Beukers, M. W., et al. (2004). 2,4,6-Trisubstituted pyrimidines as a new class of selective adenosine A1 receptor antagonists. *Journal of Medicinal Chemistry*, *47*(26), 6529–6540.

Cherezov, V., Rosenbaum, D. M., Hanson, M. A., Rasmussen, S. G. F., Thian, F. S., Kobilka, T. S., et al. (2007). High-resolution crystal structure of an engineered human beta2-adrenergic G protein-coupled receptor. *Science*, *318*(5854), 1258–1265.

Colotta, V., Lenzi, O., Catarzi, D., Varano, F., Filacchioni, G., Martini, C., et al. (2009). Pyrido [2,3-e]-1,2,4-triazolo[4,3-a]pyrazin-1-one as a new scaffold to develop potent and selective human A3 adenosine receptor antagonists. Synthesis, pharmacological evaluation, and ligand-receptor modeling studies. *Journal of Medicinal Chemistry*, *52*(8), 2407–2419.

Cristalli, G., Müller, C. E., & Volpini, R. (2009). Recent developments in adenosine A_{2A} receptor ligands. *Handbook of Experimental Pharmacology*, *193*, 59–98.

Dalpiaz, A., Townsend-Nicholson, A., Beukers, M. W., Schofield, P. R., & IJzerman, A. P. (1998). Thermodynamics of full agonist, partial agonist, and antagonist binding to wild-type and mutant adenosine A_1 receptors. *Biochemical Pharmacology*, *56*(11), 1437–1445.

Dawson, E. S., & Wells, J. N. (2001). Determination of amino acid residues that are accessible from the ligand binding crevice in the seventh transmembrane-spanning region of the human A_1 adenosine receptor. *Molecular Pharmacology*, *59*(5), 1187–1195.

De Zwart, M., Vollinga, R. C., Beukers, M. W., Sleegers, D. F., Von Frijtag Drabbe Kunzel, J., de Groote, M., et al. (1999). Potent antagonists for the human adenosine A_{2B} receptor. Derivatives of the triazolotriazine adenosine receptor antagonist ZM241385 with high affinity. *Drug Development Research*, *48*, 95–103.

Dror, R. O., Arlow, D. H., Borhani, D. W., Jensen, M. O., Piana, S., & Shaw, D. E. (2009). Identification of two distinct inactive conformations of the β_2-adrenoceptor reconciles structural and biochemical observations. *Proceedings of the National Academy of Sciences of the United States of America*, *106*(12), 4689–4694.

Ferre, S., Ciruela, F., Borycz, J., Solinas, M., Quarta, D., Antoniou, K., et al. (2008). Adenosine A_1-A_{2A} receptor heteromers: New targets for caffeine in the brain. *Frontiers in Bioscience*, *13*, 2391–2399.

Ferre, S., Quiroz, C., Woods, A. S., Cunha, R., Popoli, P., Ciruela, F., et al. (2008). An update on adenosine A_{2A}-dopamine D_2 receptor interactions: Implications for the function of G protein-coupled receptors. *Current Pharmaceutical Design*, *14*(15), 1468–1474.

Fishman, P., Bar-Yehuda, S., Synowitz, M., Powell, J. D., Klotz, K. N., Gessi, S., et al. (2009). Adenosine receptors and cancer. *Handbook of Experimental Pharmacology*, *193*, 399–441.

Fredholm, B. B., Chen, J. F., Masino, S. A., & Vaugeois, J. M. (2005). Actions of adenosine at its receptors in the CNS: Insights from knockouts and drugs. *Annual Review of Pharmacology and Toxicology*, *45*, 385–412.

Fredholm, B. B., IJzerman, A. P., Jacobson, K. A., Klotz, K. N., & Linden, J. (2001). International Union of Pharmacology. XXV. Nomenclature and classification of adenosine receptors. *Pharmacological Reviews*, *53*(4), 527–552.

Fuxe, K., Ferre, S., Canals, M., Torvinen, M., Terasmaa, A., Marcellino, D., et al. (2005). Adenosine A_{2A} and dopamine D_2 heteromeric receptor complexes and their function. *Journal of Molecular Neuroscience*, *26*(2–3), 209–220.

Gao, Z. G., & IJzerman, A. P. (2000). Allosteric modulation of A(2A) adenosine receptors by amiloride analogues and sodium ions. *Biochemical Pharmacology*, *60*(5), 669–676.

Gao, Z.-G., Chen, A., Barak, D., Kim, S.-K., Müller, C. E., & Jacobson, K. A. (2002). Identification by site-directed mutagenesis of residues involved in ligand recognition and activation of the human A_3 adenosine receptor. *The Journal of Biological Chemistry*, *277*(21), 19056–19063.

Gao, Z. G., Jiang, Q., Jacobson, K. A., & IJzerman, A. P. (2000). Site-directed mutagenesis studies of human A_{2A} adenosine receptors: Involvement of glu(13) and his(278) in ligand binding and sodium modulation. *Biochemical Pharmacology*, *60*(5), 661–668.

Gao, Z., Ni, Y., Szabo, G., & Linden, J. (1999). Palmitoylation of the recombinant human A_1 adenosine receptor: Enhanced proteolysis of palmitoylation-deficient mutant receptors. *The Biochemical Journal*, *342*(Pt. 2), 387–395.

Gao, Z., Robeva, A. S., & Linden, J. (1999). Purification of A_1 adenosine receptor-G-protein complexes: Effects of receptor down-regulation and phosphorylation on coupling. *The Biochemical Journal*, *338*(Pt. 3), 729–736.

Gessi, S., Merighi, S., Varani, K., Leung, E., MacLennan, S., & Borea, P. A. (2008). The A_3 adenosine receptor: An enigmatic player in cell biology. *Pharmacology & Therapeutics, 117*(1), 123–140.

Golembiowska, K., & Dziubina, A. (2004). Striatal adenosine A_{2A} receptor blockade increases extracellular dopamine release following L-DOPA administration in intact and dopamine-denervated rats. *Neuropharmacology, 47*(3), 414–426.

Golembiowska, K., Dziubina, A., Kowalska, M., & Kaminska, K. (2009). Effect of adenosine A_{2A} receptor antagonists on L-DOPA-induced hydroxyl radical formation in rat striatum. *Neurotoxicity Research, 15*(2), 155–166.

Hanson, M. A., & Stevens, R. C. (2009). Discovery of new GPCR biology: One receptor structure at a time. *Structure, 17*(1), 8–14.

Hasko, G., Csoka, B., Nemeth, Z. H., Vizi, E. S., & Pacher, P. (2009). A_{2B} adenosine receptors in immunity and inflammation. *Trends in Immunology, 30*(6), 263–270.

Hernan, M. A., Takkouche, B., Caamano-Isorna, F., & Gestal-Otero, J. J. (2002). A meta-analysis of coffee drinking, cigarette smoking, and the risk of Parkinson's disease. *Annals of Neurology, 52*(3), 276–284.

Hert, J., Irwin, J. J., Laggner, C., Keiser, M. J., & Shoichet, B. K. (2009). Quantifying biogenic bias in screening libraries. *Nature Chemical Biology, 5*(7), 479–483.

Higgs, C., Beuming, T., & Sherman, W. (2010). Hydration site thermodynamics explain SARs for triazolylpurines analogues binding to the A_{2A} receptor. *ACS Medicinal Chemistry Letters, 1*(4), 160–164.

Holst, B., Nygaard, R., Valentin-Hansen, L., Bach, A., Engelstoft, M. S., Petersen, P. S., Frimurer, T. M., & Schwartz, T. W. (2010). A conserved aromatic lock for the tryptophan rotameric switch in TM-VI of seven-transmembrane receptors. *The Journal of Biological Chemistry, 285*(6), 3973–3985.

Hornak, V., Ahuja, S., Eilers, M., Goncalves, J. A., Sheves, M., Reeves, P. J., et al. (2010). Light activation of rhodopsin: Insights from molecular dynamics simulations guided by solid-state NMR distance restraints. *Journal of Molecular Biology, 396*(3), 510–527.

IJzerman, A. P., Von Frijtag Drabbe Kunzel, J. K., Kim, J., Jiang, Q., & Jacobson, K. A. (1996). Site-directed mutagenesis of the human adenosine A_{2A} receptor. Critical involvement of Glu13 in agonist recognition. *European Journal of Pharmacology, 310*(2–3), 269–272.

Ivanov, A. A., Barak, D., & Jacobson, K. A. (2009). Evaluation of homology modeling of G-protein-coupled receptors in light of the A_{2A} adenosine receptor crystallographic structure. *Journal of Medicinal Chemistry, 52*(10), 3284–3292.

Jaakola, V. P., Griffith, M. T., Hanson, M. A., Cherezov, V., Chien, E. Y., Lane, J. R., et al. (2008). The 2.6 angstrom crystal structure of a human A2A adenosine receptor bound to an antagonist. *Science, 322*(5905), 1211–1217.

Jaakola, V. P., & IJzerman, A. P. (2010). The crystallographic structure of the human adenosine A_{2A} receptor in a high-affinity antagonist-bound state: Implications for GPCR drug screening and design. *Current Opinion in Structural Biology, 20*(4), 401–414.

Jaakola, V. P., Lane, J. R., Lin, J. Y., Katritch, V., IJzerman, A. P., & Stevens, R. C. (2010). Ligand binding and subtype selectivity of the human A_{2A} adenosine receptor: Identification and characterization of essential amino acid residues. *The Journal of Biological Chemistry, 285*(17), 13032–13044.

Jacobson, K. A., & Gao, Z. G. (2006). Adenosine receptors as therapeutic targets. *Nature Reviews. Drug Discovery, 5*(3), 247–264.

Jacobson, K. A., Ohno, M., Duong, H. T., Kim, S. K., Tchilibon, S., Cesnek, M., et al. (2005). A neoceptor approach to unraveling microscopic interactions between the human A_{2A} adenosine receptor and its agonists. *Chemistry & Biology, 12*(2), 237–247.

Jiang, Q., Lee, B. X., Glashofer, M., van Rhee, A. M., & Jacobson, K. A. (1997). Mutagenesis reveals structure activity parallels between human A_{2A} adenosine receptors and biogenic amine G protein-coupled receptors. *Journal of Medicinal Chemistry, 40*(16), 2588–2595.

Jiang, Q., Van Rhee, A. M., Kim, J., Yehle, S., Wess, J., & Jacobson, K. A. (1996). Hydrophilic side chains in the third and seventh transmembrane helical domains of human A_{2A} adenosine receptors are required for ligand recognition. *Molecular Pharmaceutics*, *50*(3), 512–521.

Kalla, R. V., Zablocki, J., Tabrizi, M. A., & Baraldi, P. G. (2009). Recent developments in A_{2B} adenosine receptor ligands. *Handbook of Experimental Pharmacology*, (193), 99–122.

Katritch, V., Jaakola, V.-P., Lane, J. R., Lin, J., IJzerman, A. P., Yeager, M., et al. (2010). Structure-based discovery of novel chemotypes for adenosine A_{2A} receptor antagonists. *Journal of Medicinal Chemistry*, *53*(4), 1799–1809.

Katritch, V., Kufareva, I., & Abagyan, R. (2010). Structure based prediction of subtype-selectivity for adenosine receptor antagonists. *Neuropharmacology*, *60*(1), 108–115.

Katritch, V., Rueda, M., Lam, P. C.-H., Yeager, M., & Abagyan, R. (2010). GPCR 3D homology models for ligand screening: Lessons learned from blind predictions of adenosine A_{2A} receptor complex. *Proteins*, *78*(1), 197–211.

Kawate, T., & Gouaux, E. (2006). Fluorescence-detection size-exclusion chromatography for precrystallization screening of integral membrane proteins. *Structure*, *14*(4), 673–681.

Kim, S.-K., Gao, Z.-G., Van Rompaey, P., Gross, A. S., Chen, A., Van Calenbergh, S., et al. (2003). Modeling the adenosine receptors: Comparison of the binding domains of A_{2A} agonists and antagonists. *Journal of Medicinal Chemistry*, *46*(23), 4847–4859.

Kim, J., Jiang, Q., Glashofer, M., Yehle, S., Wess, J., & Jacobson, K. A. (1996). Glutamate residues in the second extracellular loop of the human A_{2A} adenosine receptor are required for ligand recognition. *Molecular Pharmaceutics*, *49*(4), 683–691.

Kim, J., Wess, J., van Rhee, A. M., Schoeneberg, T., & Jacobson, K. A. (1995). Site-directed mutagenesis identifies residues involved in ligand recognition in the human A_{2A} adenosine receptor. *The Journal of Biological Chemistry*, *270*(23), 13987–13997.

Klaasse, E. C., IJzerman, A. P., de Grip, W. J., & Beukers, M. W. (2008). Internalization and desensitization of adenosine receptors. *Purinergic Signaling*, *4*(1), 21–37.

Klaasse, E. C., van den Hout, G., Roerink, S. F., de Grip, W. J., IJzerman, A. P., & Beukers, M. W. (2005). Allosteric modulators affect the internalization of human adenosine A_1 receptors. *European Journal of Pharmacology*, *522*(1–3), 1–8.

Klabunde, T., & Hessler, G. (2002). Drug design strategies for targeting G-protein-coupled receptors. *Chembiochem*, *3*(10), 928–944.

Kobilka, B., & Schertler, G. F. X. (2008). New G-protein-coupled receptor crystal structures: Insights and limitations. *Trends in Pharmacological Sciences*, *29*(2), 79–83.

Kolb, P., Rosenbaum, D. M., Irwin, J. J., Fung, J. J., Kobilka, B. K., & Shoichet, B. K. (2009). Structure-based discovery of β2-adrenoceptor ligands. *Proceedings of the National Academy of Sciences of the United States of America*, *106*(16), 6843–6848.

Liapakis, G., Ballesteros, J. A., Papachristou, S., Chan, W. C., Chen, X., & Javitch, J. A. (2000). The forgotten serine. A critical role for Ser-$203^{5.42}$ in ligand binding to and activation of the β2-adrenoceptor. *The Journal of Biological Chemistry*, *275*(48), 37779–37788.

Magnani, F., Shibata, Y., Serrano-Vega, M. J., & Tate, C. G. (2008). Co-evolving stability and conformational homogeneity of the human adenosine A_{2A} receptor. *Proceedings of the National Academy of Sciences of the United States of America*, *105*(31), 10744–10749.

Mantri, M., de Graaf, O., van Veldhoven, J., Goblyos, A., von Frijtag Drabbe Kunzel, J. K., Mulder-Krieger, T., et al. (2008). 2-Amino-6-furan-2-yl-4-substituted nicotinonitriles as A_{2A} adenosine receptor antagonists. *Journal of Medicinal Chemistry*, *51*(15), 4449–4455.

Martinelli, A., & Tuccinardi, T. (2008). Molecular modeling of adenosine receptors: New results and trends. *Medicinal Research Reviews*, *28*(2), 247–277.

Marti-Renom, M. A., Stuart, A. C., Fiser, A., Sanchez, R., Melo, F., & Sali, A. (2000). Comparative protein structure modeling of genes and genomes. *Annual Review of Biophysics and Biomolecular Structure*, *29*, 291–325.

Michino, M., Abola, E., Brooks, C. L., Dixon, J. S., Moult, J., & Stevens, R. C. (2009). Community-wide assessment of GPCR structure modelling and ligand docking: GPCRDock 2008. *Nature Reviews. Drug Discovery*, *8*(6), 455–463.

Mobarec, J. C., Sanchez, R., & Filizola, M. (2009). Modern homology modeling of G-protein coupled receptors: Which structural template to use? *Journal of Medicinal Chemistry*, *52*(16), 5207–5216.

Moro, S., Gao, Z. G., Jacobson, K. A., & Spalluto, G. (2006). Progress in the pursuit of therapeutic adenosine receptor antagonists. *Medicinal Research Reviews*, *26*(2), 131–159.

Moult, J., Fidelis, K., Kryshtafovych, A., Rost, B., Hubbard, T., & Tramontano, A. (2007). Critical assessment of methods of protein structure prediction-Round VII. *Proteins*, *69*(Suppl. 8), 3–9.

Mustafa, S. J., Morrison, R. R., Teng, B., & Pelleg, A. (2009). Adenosine receptors and the heart: Role in regulation of coronary blood flow and cardiac electrophysiology. *Handbook of Experimental Pharmacology*, *193*, 161–188.

Navarro, G., Aymerich, M. S., Marcellino, D., Cortes, A., Casado, V., Mallol, J., et al. (2009). Interactions between calmodulin, adenosine A_{2A}, and dopamine D_2 receptors. *The Journal of Biological Chemistry*, *284*(41), 28058–28068.

Nygaard, R., Frimurer, T. M., Holst, B., Rosenkilde, M. M., & Schwartz, T. W. (2009). Ligand binding and micro-switches in 7TM receptor structures. *Trends in Pharmacological Sciences*, *30*(5), 249–259.

Ohta, A., & Sitkovsky, M. (2001). Role of G-protein-coupled adenosine receptors in downregulation of inflammation and protection from tissue damage. *Nature*, *414* (6866), 916–920.

Olah, M. E., Ren, H., Ostrowski, J., Jacobson, K. A., & Stiles, G. L. (1992). Cloning, expression, and characterization of the unique bovine A_1 adenosine receptor. Studies on the ligand binding site by site-directed mutagenesis. *The Journal of Biological Chemistry*, *267*(15), 10764–10770.

Palczewski, K., Kumasaka, T., Hori, T., Behnke, C. A., Motoshima, H., & Fox, B. A. (2000). Crystal structure of rhodopsin: A G protein-coupled receptor. *Science*, *289*(5480), 739–745.

Palmer, T. M., Benovic, J. L., & Stiles, G. L. (1995). Agonist-dependent phosphorylation and desensitization of the rat A_3 adenosine receptor. Evidence for a G-protein-coupled receptor kinase-mediated mechanism. *The Journal of Biological Chemistry*, *270*(49), 29607–29613.

Palmer, T. M., Benovic, J. L., & Stiles, G. L. (1996). Molecular basis for subtype-specific desensitization of inhibitory adenosine receptors. Analysis of a chimeric A_1-A_3 adenosine receptor. *The Journal of Biological Chemistry*, *271*(25), 15272–15278.

Palmer, T. M., & Stiles, G. L. (1999). Stimulation of A_{2A} adenosine receptor phosphorylation by protein kinase C activation: Evidence for regulation by multiple protein kinase C isoforms. *Biochemistry*, *38*(45), 14833–14842.

Palmer, T. M., & Stiles, G. L. (2000). Identification of threonine residues controlling the agonist-dependent phosphorylation and desensitization of the rat A_3 adenosine receptor. *Molecular Pharmacology*, *57*(3), 539–545.

Park, J. H., Scheerer, P., Hofmann, K. P., Choe, H. W., & Ernst, O. P. (2008). Crystal structure of the ligand-free G-protein-coupled receptor opsin. *Nature*, *454*(7201), 183–187.

Piirainen, H., Ashok, Y., Nanekar, R. T., & Jaakola, V. P. (2010). Structural features of adenosine receptors: From crystal to function. *Biochimica et Biophysica Acta* (In press).

Rasmussen, S. G. F., Choi, H.-J., Rosenbaum, D. M., Kobilka, T. S., Thian, F. S., Edwards, P. C., et al. (2007). Crystal structure of the human β_2-adrenergic G-protein-coupled receptor. *Nature*, *450*(7168), 383–387.

Rivkees, S. A., Barbhaiya, H., & IJzerman, A. P. (1999). Identification of the adenine binding site of the human A_1 adenosine receptor. *The Journal of Biological Chemistry*, *274*(6), 3617–3621.

Rosenbaum, D. M., Cherezov, V., Hanson, M. A., Rasmussen, S. G. F., Thian, F. S., Kobilka, T. S., et al. (2007). GPCR engineering yields high-resolution structural insights into β_2-adrenergic receptor function. *Science*, *318*(5854), 1266–1273.

Rosenbaum, D. M., Rasmussen, S. G. F., & Kobilka, B. K. (2009). The structure and function of G-protein-coupled receptors. *Nature*, *459*(7245), 356–363.

Salamone, J. D. (2010). Preladenant, a novel adenosine A_{2A} receptor antagonist for the potential treatment of parkinsonism and other disorders. *IDrugs*, *13*(10), 723–731.

Schapira, A. H. V., Bezard, E., Brotchie, J., Calon, F., Collingridge, G. L., Ferger, B., et al. (2006). Novel pharmacological targets for the treatment of Parkinson's disease. *Nature Reviews. Drug Discovery*, *5*(10), 845–854.

Scheerer, P., Heck, M., Goede, A., Park, J. H., Choe, H. W., Ernst, O. P., et al. (2009). Structural and kinetic modeling of an activating helix switch in the rhodopsin-transducin interface. *Proceedings of the National Academy of Sciences of the United States of America*, *106*(26), 10660–10665.

Scheerer, P., Park, J. H., Hildebrand, P. W., Kim, Y. J., Krauss, N., Choe, H. W., et al. (2008). Crystal structure of opsin in its G-protein-interacting conformation. *Nature*, *455*(7212), 497–502.

Scholl, D. J., & Wells, J. N. (2000). Serine and alanine mutagenesis of the nine native cysteine residues of the human A_1 adenosine receptor. *Biochemical Pharmacology*, *60*(11), 1647–1654.

Schwartz, T. W., Frimurer, T. M., Holst, B., Rosenkilde, M. M., & Elling, C. E. (2006). Molecular mechanism of 7TM receptor activation-a global toggle switch model. *Annual Review of Pharmacology and Toxicology*, *46*, 481–519.

Sebastiao, A. M., & Ribeiro, J. A. (2009). Adenosine receptors and the central nervous system. *Handbook of Experimental Pharmacology*, *193*, 471–534.

Sheikh, S. P., Zvyaga, T. A., Lichtarge, O., Sakmar, T. P., & Bourne, H. R. (1996). Rhodopsin activation blocked by metal-ion-binding sites linking transmembrane helices C and F. *Nature*, *383*(6598), 347–350.

Shi, L., Liapakis, G., Xu, R., Guarnieri, F., Ballesteros, J. A., & Javitch, J. A. (2002). β2 adrenergic receptor activation. Modulation of the proline kink in transmembrane 6 by a rotamer toggle switch. *The Journal of Biological Chemistry*, *277*(43), 40989–40996.

Sitkovsky, M. V., & Ohta, A. (2005). The 'danger' sensors that STOP the immune response: The A_2 adenosine receptors? *Trends in Immunology*, *26*(6), 299–304.

Stefanachi, A., Nicolotti, O., Leonetti, F., Cellamare, S., Campagna, F., Loza, M. I., et al. (2008). 1,3-Dialkyl-8-(hetero)aryl-9-OH-9-deazaxanthines as potent A2B adenosine receptor antagonists: Design, synthesis, structure-affinity and structure-selectivity relationships. *Bioorganic & Medicinal Chemistry*, *16*(22), 9780–9789.

Stone, T. W., Ceruti, S., & Abbracchio, M. P. (2009). Adenosine receptors and neurological disease: Neuroprotection and neurodegeneration. *Handbook of Experimental Pharmacology*, *193*, 535–587.

Strader, C. D., Candelore, M. R., Hill, W. S., Sigal, I. S., & Dixon, R. A. (1989). Identification of two serine residues involved in agonist activation of the beta-adrenergic receptor. *The Journal of Biological Chemistry*, *264*(23), 13572–13578.

Strader, C. D., Sigal, I. S., Candelore, M. R., Rands, E., Hill, W. S., & Dixon, R. A. (1988). Conserved aspartic acid residues 79 and 113 of the beta-adrenergic receptor have different roles in receptor function. *The Journal of Biological Chemistry*, *263*(21), 10267–10271.

Thomas, G. S., Thompson, R. C., Miyamoto, M. I., Ip, T. K., Rice, D. L., Milikien, D., et al. (2009). The RegEx trial: a randomized, double-blind, placebo- and active-controlled pilot study combining regadenoson, a selective A(2A) adenosine agonist, with low-level exercise, in patients undergoing myocardial perfusion imaging. *Journal of Nuclear Cardiology*, *16*(1), 63–72.

Townsend-Nicholson, A., & Schofield, P. R. (1994). A threonine residue in the seventh transmembrane domain of the human A_1 adenosine receptor mediates specific agonist binding. *The Journal of Biological Chemistry*, *269*(4), 2373–2376.

van Galen, P. J., Nissen, P., van Wijngaarden, I., IJzerman, A. P., & Soudijn, W. (1991). 1H-imidazo[4,5-c]quinolin-4-amines: Novel non-xanthine adenosine antagonists. *Journal of Medicinal Chemistry*, *34*(3), 1202–1206.

van Veldhoven, J. P., Chang, L. C., von Frijtag Drabbe Kunzel, J. K., Mulder-Krieger, T., Struensee-Link, R., Beukers, M. W., et al. (2008). A new generation of adenosine receptor antagonists: From di- to trisubstituted aminopyrimidines. *Bioorganic & Medicinal Chemistry*, *16*(6), 2741–2752.

Vedadi, M., Niesen, F. H., Allali-Hassani, A., Fedorov, O. Y., Finerty, P. J., Wasney, G. A., et al. (2006). Chemical screening methods to identify ligands that promote protein stability, protein crystallization, and structure determination. *Proceedings of the National Academy of Sciences of the United States of America*, *103*(43), 15835–15840.

Warne, T., Serrano-Vega, M. J., Baker, J. G., Moukhametzianov, R., Edwards, P. C., Henderson, R., et al. (2008). Structure of a β_1-adrenergic G-protein-coupled receptor. *Nature*, *454*(7203), 486–491.

Warne, T., Serrano-Vega, M. J., Tate, C. G., & Schertler, G. F. (2009). Development and crystallization of a minimal thermostabilised G protein-coupled receptor. *Protein Expression and Purification*, *65*(2), 204–213.

Wilson, C. N., Nadeem, A., Spina, D., Brown, R., Page, C. P., & Mustafa, S. J. (2009). Adenosine receptors and asthma. *Handbook of Experimnetal Pharmacology*, (193), 329–362.

Wu, B., Chien, E. Y., Mol, C. D., Fenalti, G., Liu, W., Katritch, V., et al. (2010). Structures of the CXCR4 chemokine GPCR with small-molecule and cyclic peptide antagonists. *Science*, *330*, 1066–1071.

Yuzlenko, O., & Kiec-Kononowicz, K. (2009). Molecular modeling of A_1 and A_{2A} adenosine receptors: Comparison of rhodopsin- and β_2-adrenergic-based homology models through the docking studies. *Journal of Computational Chemistry*, *30*(1), 14–32.

Zezula, J., & Freissmuth, M. (2008). The A_{2A} adenosine receptor: A GPCR with unique features? *British Journal of Pharmacology*, *153*(S1), S184–S190.

Stefania Gessi, Stefania Merighi, Katia Varani, and Pier Andrea Borea

Department of Clinical and Experimental Medicine, Pharmacology Section, University of Ferrara, Ferrara, Italy

Adenosine Receptors in Health and Disease

Abstract

The adenosine receptors A_1, A_{2A}, A_{2B}, and A_3 are important and ubiquitous mediators of cellular signaling, which play vital roles in protecting tissues and organs from damage. In particular, adenosine triggers tissue protection and repair by different receptor-mediated mechanisms, including an increase of oxygen supply/demand ratio, preconditioning, anti-inflammatory effects, and stimulation of angiogenesis. Considerable advances have been recently achieved in the pharmacological and molecular characterization of adenosine receptors, which have been proposed as targets for drug design and discovery. At the present time, it can be speculated that adenosine A_1, A_{2A}, A_{2B}, and A_3 receptor-selective ligands may show utility in the treatment of pain, ischemic conditions, glaucoma, asthma, arthritis, cancer, and other disorders in which inflammation is a feature. This chapter documents the present state of knowledge of adenosine receptors' role in health and disease.

I. Introduction

Adenosine is an endogenous nucleoside modulator released from almost all cells and is generated in the extracellular space by breakdown of ATP through a series of ectoenzymes, including the ENTDase (CD39) and 5′-nucleotidase (CD73; Linden, 2001). The latter dephosphorylates extracellular AMP into adenosine and constitutes a limiting step in its formation.

Advances in Pharmacology, Volume 61

1054-3589/11 $35.00
10.1016/B978-0-12-385526-8.00002-3

Extracellular adenosine concentration is kept in equilibrium by reuptake mechanisms operated through the action of specific transporters. This nucleoside mediates its effects through activation of a family of four G-protein-coupled receptors (GPCRs) named A_1, A_{2A}, A_{2B}, and A_3. These receptors differ in their affinity for adenosine, in the type of G proteins that they recruit and finally in the downstream signaling pathways that are activated in the target cells (Fredholm et al., 2001). It is estimated that the levels of adenosine in the interstitial fluid are in the range of 20–200 nM (Fredholm, 2010). Adenosine concentrations increase under metabolically unfavorable conditions. Tissue hypoxia, for example, leads to an enhanced breakdown of ATP and an increased generation of adenosine. In addition to this route, the release of adenosine might be potentiated by hypoxia-dependent inhibition of the salvage enzyme adenosine kinase which rephosphorylates the nucleoside to AMP (Ciruela et al., 2010). As adenosine is unstable and its half-life is limited by deamination or cellular reuptake, hypoxia-induced increase typically affects only local adenosine receptor (AR) signaling. Many pathophysiological conditions are believed to be associated with changes of adenosine levels such as asthma, neurodegenerative disorders, chronic inflammatory diseases, and cancer. The primary undertaking of adenosine is to reduce tissue injury and promote repair by different receptor-mediated mechanisms, including an increase of oxygen supply/demand ratio, preconditioning, anti-inflammatory effects, and stimulation of angiogenesis (Linden, 2005). Adenosine effects are widespread and pleiotropic. The cellular response to this autacoid strictly depends on the expression of the different AR subtypes, which can be coexpressed by the same cell and serve as active modulators in signal transduction. ARs have been actively studied as potential therapeutic targets in several disorders such as Parkinson's disease, schizophrenia, analgesia, ischemia, and cancer.

Until now, adenosine has mainly been used for terminating paroxysmal supraventricular tachycardia and Wolff–Parkinson–White syndrome. In addition, adenosine has been indicated as a diagnostic agent, that is, a coronary vasodilator, to assess coronary artery function in conjunction with radionuclide myocardial perfusion imaging (Haskó et al., 2008). Selective agonists and antagonists are now available for all four AR subtypes, enabling the examination of ARs function in health and disease. In this chapter, an overview on the role played by each of the ARs in mediating important physiopathological processes will be discussed; further, a list of the molecules under active development will be given as well. Some of these compounds are in preclinical investigation, whereas others have already entered clinical trials for various indications. Although it is only the beginning of a more intense, expensive, and challenging work, it is likely that purine scientists are getting closer to their goal: the use of AR ligands as drugs with the ability to save lives and improve human health.

II. A_1 Adenosine Receptors

The A_1AR has been cloned from several animal species including humans and is characterized by a close similarity at least for mammals (Ralevic and Burnstock, 1998). As for signal transduction, the A_1AR is coupled to members of the Gi/Go family of G proteins inducing inhibition of adenylyl cyclase (AC) activity (Van Calker et al., 1979). In addition, it might activate phospholipase-C (PLC)-β, which is known to increase inositol 1,4,5-triphosphate and intracellular Ca^{2+}. The A_1AR is coupled to pertussis toxine-sensitive potassium channels as well as K_{ATP} channels, essentially in cardiac tissue and neurons. Moreover, it can inhibit Q-, P-, and N-type Ca^{2+} channels and modulate extracellular signal-regulated protein kinases (ERKs; Fredholm et al., 2001). Recently, a role of β-arrestin1/ERK MAP kinase pathway in regulating A_1AR desensitization and recovery has been reported (Jajoo et al., 2010).

The A_1AR is widely distributed in the central nervous system (CNS), with high levels in brain, cortex, cerebellum, hippocampus, and dorsal horn of spinal cord. It modulates the activity of the nervous system at the cellular level and is present in both pre- and postsynaptic terminals. At the presynaptical level, it mediates inhibition of neurotransmitter release, while at the postsynaptical level, it induces neuronal hyperpolarization. Therefore, activation of A_1ARs is responsible for sedative, anticonvulsant, anxiolytic, and locomotor depressant effects induced by adenosine. The endogenous levels of adenosine are sufficient to tonically activate inhibitory A_1ARs, and caffeine, perhaps the most commonly used drug in the world, mediates its excitatory effects through the antagonism of this inhibition. Importantly, adenosine has an important role also in analgesia (Eisenach et al., 2002). It is known that spinal or systemic administration of adenosine and its analogs produces antinociception in a variety of animal models by A_1AR activation (Boison, 2007; Gong et al., 2010; Nascimento et al., 2010; Sowa et al., 2010). Antinociceptive effects of adenosine may be related to the inhibition of intrinsic neurons by an increase in K^+ conductance and presynaptic inhibition of sensory nerve terminals, decreasing the release of substance P and glutamate. Further, attenuation by NMDA-induced production of nitric oxide (NO) also may be involved. Adenosine has been shown to mediate opioid analgesia (Gan and Habib, 2007). Recently, it has been reported that allopurinol, a potent inhibitor of the enzyme xanthine oxidase, used primarily in the treatment of hyperuricemia and gout, induces antinociception related to adenosine accumulation. This effect is completely prevented after A_1AR blockade (Schmidt et al., 2009). Compounds that are able to enhance the activity of the A_1ARs by the endogenous ligand within specific tissues may have potential therapeutic advantages over nonendogenous agonists. Such an opportunity for intervention is provided by the concept of allosteric modulation of GPCRs. Therefore, the use of allosteric enhancers to increase the responsiveness of the A_1 receptors to endogenous adenosine at sites of its production is an appealing alternative to activation by exogenous agonists (Romagnoli et al., 2010). This approach minimizes side effects because allosteric enhancers act only on the agonist–A_1AR–G protein ternary complex, limiting their action to sites and times of adenosine accumulation. T-62 (1-(2-amino-4,5,6,7-tetrahydrobenzo[*b*]thiophen-3-yl)(4-chlorophenyl) methanone), discovered by Baraldi et al., is actually under development for the treatment of chronic pain (Baraldi et al., 2000). In particular, King Pharmaceuticals has initiated a phase II clinical trial program evaluating the efficacy and safety of T-62, an oral tablet investigational drug for the treatment of neuropathic pain. The trial is a multicenter, randomized, double-blind, placebo-controlled study assessing the analgesic efficacy and safety of T-62 in subjects with postherpetic neuralgia and its associated pain. The study is now terminated and has evaluated two doses of T-62 and placebo utilizing a parallel design. Some patients experienced asymptomatic, transient elevations in liver transaminases. Results are not yet available (ClinicalTrials.gov Identifier: NCT00809679).

A_1ARs are responsible for many effects induced by adenosine in the CNS as well as in peripheral tissues (Baraldi et al., 2008; Russo et al., 2006). Adenosine is a signaling nucleoside that has been implicated in the regulation of asthma and chronic obstructive pulmonary disease (COPD; Russo et al., 2006). Levels of adenosine are increased in the lungs of asthmatics, in which elevations correlate with the degree of inflammatory insult (Driver et al., 1993). The expression of A_1ARs is increased in the epithelium and airway smooth muscle of airways of human asthmatics (Brown et al., 2008a). The early evidence that the A_1AR is involved in asthma derived from studies on allergic rabbit models, where the adenosine-induced acute bronchoconstrictor response was reduced by pretreatment with A_1AR antagonists. In human airway tissue and human bronchial smooth muscle cells, activation of A_1ARs produces effects that cause airway hyperresponsiveness. On human airway epithelial cells, activation of A_1ARs causes an increase in expression of the MUC 2 gene responsible for mucus hypersecretion. Moreover, activation of A_1ARs on a number of different human cells produces proinflammatory effects (Ponnoth et al., 2010). Taken together, these effects of A_1ARs in humans suggest that the A_1AR is an important target in human asthma (Baraldi et al., 2008; Ethier & Madison, 2006; Wilson, 2008). Indeed, investigational bronchodilators for respiratory disorders such as asthma include the nonselective AR antagonists theophylline and doxofylline (Press et al., 2007). Paradoxically, findings in adenosine deaminase (ADA)-deficient mice suggest the occurrence of anti-inflammatory actions of adenosine in the lung, through chronic A_1AR activation in macrophages (Sun et al., 2005). Accordingly, it has been recently reported that in a murine model of lipopolysaccharide (LPS)-induced lung injury, A_1AR activation inhibits transendothelial and transepithelial polymorphonuclear cells (PMN) migration, most likely by reducing the release of chemotactic cytokines into the alveolar airspace. In addition, A_1ARs on endothelial cells are involved in decreasing microvascular permeability and leukocyte transmigration (Ngamsri et al., 2010). These results suggest a protective and anti-inflammatory role for A_1ARs (Gazoni et al., 2010).

At the cardiovascular level, A_1ARs mediate negative chronotropic, dromotropic, and inotropic effects. A_1 subtypes located on sinoatrial and atrioventricular nodes cause bradycardia and heart block, respectively, while the negative inotropic effects include a decrease in atrial contractility and action potential duration. Recently, it has been shown that selective deletion of the A_1AR abolishes heart-rate slowing effects of intravascular adenosine *in vivo* (Koeppen et al., 2009). Stimulation of A_1ARs in the heart exerts cardioprotective effects by inhibiting norepinephrine release from sympathetic nerve endings (Schutte et al., 2006). Adenosine also protects tissues through its effect in ischemic preconditioning (IPC), a brief period of ischemia and reperfusion, that can protect myocardium against infarction from a subsequent prolonged ischemic insult. Activation of A_1ARs, protein kinase C (PKC), and mitochondrial K_{ATP} channels is responsible for this response (Shneyvays et al., 2005; Solenkova et al., 2006). IPC has been most widely investigated in the heart but also occurs in other tissues (Grenz et al., 2007; Yldiz et al., 2007). A_1 receptor agonists, for example, Tecadenoson (*N*6-[3 (*R*)-tetrahydrofuranyl]adenosine), are in development for arrhythmias and atrial fibrillation. Clinical studies with intravenous Tecadenoson suggest that it may slow the speed of AV nodal conduction by selectively stimulating the A_1AR and may avoid blood pressure lowering by not stimulating the adenosine A_2AR (Kiesman et al., 2009).

In the kidney, A_1ARs mediate vasoconstriction, decrease glomerular filtration rate, inhibit renin secretion, and inhibit neurotransmitter release. A_1AR antagonists represent a novel class of agents for potential use in the treatment of hypertension and edema (Vallon et al., 2006). A_1AR antagonists produced diuresis and natriuresis of greater magnitude than thiazide diuretics but without significant potassium wasting or reductions of renal blood flow and glomerular filtration rate (Zhou & Kost, 2006). Further, evidence obtained from genetically altered mice indicates that transcellular NaCl transport induces the generation of adenosine that, in conjunction with angiotensin II, elicits afferent arteriolar constriction through A_1 receptor activation (Schnermann and Briggs, 2008; Sun et al., 2001). Clinical trials in a limited number of subjects demonstrated that A_1AR antagonists produced natriuretic and hypotensive effects in essential hypertensive patients and attenuated the furosemide-induced decline of renal hemodynamic function in heart failure patients. Selective adenosine A_1AR antagonists targeting renal microcirculation are novel pharmacologic agents that are currently under development for the treatment of acute heart failure as well as for chronic heart failure. Rolofylline (1,3-dipropyl-8-(2-nor-1-adamantyl)xanthine,

KW-3902) is an A_1AR antagonist that facilitates diuresis and preserves renal function in patients with acute decompensated heart failure and renal dysfunction. Pilot data also suggest beneficial effects on symptoms and short-term outcomes (Slawsky and Givertz, 2009). Despite several studies showing improvement of renal function and/or increased diuresis with adenosine A_1AR antagonists, particularly in chronic heart failure, these findings were not confirmed by large phase III trials PROTECT 1 and 2 (Placebo-controlled Randomized study of the selective A_1AR antagonist Rolofylline for patients hospitalized with acute heart failure and volume Overload to assess Treatment Effect on Congestion and renal function) in acute heart failure patients. The pooled/meta-analysis of two studies demonstrated that treatment with Rolofylline was associated with poor outcomes due to worsening renal function in patients with acute decompensated heart failure (Weatherley et al., 2010). However, lessons can be learned from these and other studies, and there might still be a potential role for the clinical use of adenosine A_1 antagonists (Hocher, 2010).

Otherwise, it is relevant that several studies have demonstrated that A_1AR activation is protective *in vivo* by inhibiting necrosis, inflammation, and apoptosis. A_1ARs have been implicated as potent anti-inflammatory mediators in various injury models of kidney, heart, liver, lung, and brain (Kim et al., 2009; Ngamsri et al., 2010). In particular, A_1AR activation protects against hepatic injury by upregulation and phosphorylation of heat shock protein 27, a member of family of chaperone proteins that serves to defend against cell damage (Chen et al., 2009).

A critical role for adenosine in bone homeostasis via interaction with adenosine A_1ARs has been recently reported. In particular, due to the stimulatory effect played by A_1ARs on osteoclast function and formation, antagonists of this receptor may be important to prevent the bone loss associated with inflammatory diseases and menopause (Kara et al., 2010a, 2010b).

Activation of A_1ARs inhibits lipolysis and lowers plasma-free fatty acid concentrations by inhibiting adenylyl cyclase (AC) and downstream cyclic AMP (cAMP) formation. It is unfortunate that the majority of full A_1 agonists also have significant cardiovascular effects. For this reason, selective but partial A_1AR agonists have been developed (Dhalla et al., 2007a, 2007b). CVT-3619 (2-{6-[((1*R*,2*R*)-2-hydroxycyclopentyl) amino]purin-9-yl}(4*S*,5*S*,2*R*,3*R*)-5-[(2-fluorophenylthio)-methyl]oxolane-3,4-diol) is a partial A_1AR agonist that has antilipolytic effects at concentrations that are not accompanied by significant cardiovascular effects (Dhalla et al., 2007a, 2007b; Fatholahi et al., 2006). Desensitization of ARs to chronic exposure is also minimal with partial agonists (Kiesman et al., 2009; Shearer et al., 2009).

A variety of studies investigating the role of A_1ARs in tumor development have been performed with contrasting anti- and protumoral effects (Gessi et al., 2010b). In particular, A_1AR activation has been found to inhibit proliferation of different types of tumor cells including human LoVo metastatic, TM4 Sertoli-like, MOLT-4 leukemia, T47D, HS578T, MCF-7 breast, and glioblastoma cancer cells. Further, it has been reported that in rat astrocytoma cells, extracellular adenosine appears to activate caspase-9 followed by the effector caspase-3, at least via two independent pathways linked to A_1AR-mediated AC inhibition and adenosine uptake into cells. More recently, it has been shown that extracellular adenosine induces apoptosis of CW2 human colonic cancer cells by activating caspase-3, -8, and -9, through A_1ARs. For mice inoculated with CW2 cells, intraperitoneal injection with adenosine reduced tumor growth by inducing apoptosis mediated via A_1ARs. Consistent with a protumoral effect, it has been reported that A_1AR activation increases the chemotaxis of tumor melanoma cells. Further, A_1ARs increase both cell growth and cell proliferation in MDA-MB-468 human breast carcinoma cells. Cell cycle analysis indicated that depletion of A_1ARs by small interfering RNA (siRNA) impairs G1 checkpoint, leading to marked accumulation of cells in G2/M phase. Further, A_1AR stimulation increases cyclin-dependent kinase (CDK)4 and cyclin E protein expression, while decreasing the CDK inhibitor p27 in HeLa cervical cancer cells. Further, in a proof-of-principle study, the adenosine A_1AR antagonist 7-chloro-4-hydroxy-2-phenyl-1,8-naphthyridine, an inhibitor of blood vascular and lymphatic development in Xenopus, was shown to act also as a potent antagonist of vascular endothelial growth factor (VEGF)-induced adult neovascularization in mice (Kälin et al., 2009).

A list of A_1 receptor ligands in clinical studies for novel therapeutic treatments is reported in Table I.

III. A_{2A} Adenosine Receptors

Of the four ARs, A_{2A}ARs have taken center stage as the primary anti-inflammatory effectors of extracellular adenosine (Haskó and Pacher, 2008). The gene for the A_{2A}AR has been cloned from several species including dog, rat, human, guinea-pig, and mouse and has demonstrated a high degree of homology among human, mouse, and rat (Baraldi et al., 2008). The A_{2A}AR stimulates AC activity through coupling with Gs proteins leading to the activation of cAMP-dependent protein kinase A (PKA). This in turn phosphorylates and activates various receptors, ion channels, phosphodiesterases, and phosphoproteins like CREB and DARPP-32. Activation of PKC has been also reported by A_{2A}AR activation. In brain striatum, the A_{2A} subtype stimulates Golf, another member of the Gs subfamily of G proteins. In addition, the A_{2A}AR can interact with different types of Ca^{2+} channels to either increase intracellular Ca^{2+} or decrease Ca^{2+} influx and is involved like the other adenosine subtypes in the modulation of ERKs activity. Due to a long carboxy terminal domain, the A_{2A}AR has a greater molecular weight (45 kDa) in comparison to the other subtypes (36–37 kDa).

TABLE I A_1 Receptor Ligands in Clinical Studies for Novel Therapeutic Treatments

Pathology		*Drug name*	*Mechanism of action*	*Status of development*
Cardiovascular diseases	Arrhythmia, atrial fibrillation	Capadenoson	Agonist	Phase II
		Selodenoson	Agonist	Phase II
		Tecadenoson	Agonist	Phase III
	Heart failure, congestive	Derenofylline	Antagonist	Phase II
	Chronic heart failure	Tonapofylline	Antagonist	Phase III
Respiratory disorders	Asthma	Doxofylline	Antagonist	Launched—1987
		Theophylline	Antagonist	Launched—1939
	COPD	Theophylline	Antagonist	Phase II
Renal disorders	Renal failure	Rolofylline	Antagonist	Phase III
		Derenofylline	Antagonist	Phase II
Endocrine disorders	Diabetes	CVT-3619	Partial agonist	Phase I
Eye disorders	Glaucoma	PJ-875	Agonist	Phase II
Metabolic diseases	Lipoprotein disorders	CVT-3619	Partial agonist	Phase I
Pain	Neuropathic pain	T-62	Allosteric enhancer	Phase II

The $A_{2A}AR$ C-terminus has been defined as a crowded place where different accessory proteins may interact such as D_2-dopamine receptors, α-actinin, ADP-ribosylation factor nucleotide site opener (ARNO), ubiquitin-specific protease (USP4), and translin-associated protein X (TRAX). The lack or the presence of such different partners may explain conflicting results deriving from $A_{2A}AR$ activation, for example, neuroprotection versus neurotoxicity (Sun et al., 2006a).

$A_{2A}ARs$ are found ubiquitously in the body, and their expression is highest in the immune system and the striatopallidal system in the brain (Fredholm et al., 2001).

$A_{2A}AR$ localization in basal ganglia is restricted to GABAergic neurons of the indirect pathway, projecting from the caudate putamen to the globus pallidus, which also selectively expresses the D_2 dopamine receptor and the peptide enkephalin (Jenner et al., 2009). $A_{2A}ARs$ are observed primarily at asymmetric synapses, suggesting that adenosine may be important in modulating excitatory input to striatal neurons. Several studies have suggested the possible involvement of $A_{2A}ARs$ in the pathogenesis of neuronal disorders, including Huntington's and Parkinson's diseases.

Changes in $A_{2A}AR$ expression and signaling have been reported in various experimental models of Huntington's disease. It has been reported that an aberrant amplification of $A_{2A}AR$-stimulated AC response in striatal-derived cells engineered to express mutant Huntington. A subsequent study demonstrated an aberrant increase of $A_{2A}AR$ density in peripheral blood cells of Huntington's disease patients in comparison with age-matched healthy subjects. This opened the possibility that the aberrant $A_{2A}AR$ phenotype may represent a novel potential biomarker of Huntington's disease, useful for monitoring disease progression and assessing the efficacy of novel neuroprotective approaches. Analysis of striatal $A_{2A}AR$ binding and AC activity in one of the best-characterized animal models of Huntington's disease, R6/2 mice, of different developmental ages in comparison with age-matched wild-type animals showed a transient increase in $A_{2A}AR$ density and A_{2A} receptor-dependent cAMP production at early presymptomatic ages (Tarditi et al., 2006).

$A_{2A}ARs$ present in the CNS have been implicated in the modulation of motor functions. Accordingly, A_{2A} antagonists currently constitute an attractive nondopaminergic option for use in the treatment of Parkinson's disease (Simola et al., 2008). Istradefylline (KW-6002) is an A_{2A} antagonist that is now preregistered by Kyowa Hakko Kirin in North America for Parkinson's disease (LeWitt et al., 2008). It has been extensively demonstrated that A_{2A} antagonists can reverse motor deficits or enhance dopaminergic treatments in animal models of Parkinson's disease. Istradefylline in combination therapy with levodopa or dopamine agonists has been shown to improve the symptoms of the disease in a parkinsonian monkey model, without increasing the incidence or severity of dopaminergic-related side effects or inducing or worsening dyskinesia. Moreover, A_{2A} antagonists have been shown to attenuate neurotoxicity induced by kainate and quinolinate (Baraldi et al., 2008). Recently, it has been highlighted that there is the presence of an $A_{2A}AR$ alteration in postmortem putamen of Parkinson's disease patients when compared with healthy controls, confirming that $A_{2A}AR$ plays a key role in this neurological pathology. Further, a selective increase of A_{2A} density in the peripheral circulating cells of patients affected by Parkinson's disease was observed. These data indicate that A_{2A} alteration is a property common to both peripheral circulating cells and putamen in Parkinson's disease, confirming that lymphocytes or neutrophils could represent a mirror of the CNS (Varani et al., 2010b).

Adenosine has important protective effects on the cardiovascular system. Regadenoson is a short-acting, selective adenosine A_{2A} agonist which was approved and launched by Astellas Pharma in the USA in 2008 as an adjunctive pharmacological stress agent for myocardial perfusion imaging studies. A phase III study started in November 2009 and expected to be completed by September 2011 will compare the safety and efficacy of adenosine versus *trans*-4-[3-[6-amino-9-[(2*R*,3*R*,4*S*,5*S*)-5-(*N*-ethylcarbamoyl)-

3,4-dihydroxytetrahydrofuran-2-yl]-9H-purin-2-yl]-2-propynyl]cyclohexane-carboxylic acid methyl ester (Apadenoson) when used in single photon emission computed tomography myocardial perfusion imaging in patients with coronary artery diseases (ClinicalTrials.gov Identifier: NCT00990327; Bayes, 2007; Kern et al., 2006).

An upregulation of A_{2A}ARs was found in peripheral circulating cells of end-stage chronic heart failure patients with respect to sex- and age-matched healthy subjects. Upon heart transplantation, peripheral A_{2A}AR density gradually normalizes in parallel with the normalization of hemodynamic parameters. Hence, the evaluation of the expression and function of these receptors in peripheral blood cells may be useful for monitoring hemodynamic changes and the efficacy of pharmacological and nonpharmacological treatments in chronic heart failure patients (Varani et al., 2003).

Activation of the A_{2A}AR subtype on coronary smooth muscle cells, endothelial cells, monocytes/macrophages, and foam cells results in vasodilation, neoangiogenesis, inhibition of proinflammatory cytokines production, and reduction of plaque formation (Belardinelli et al., 1998; Bingham et al., 2010; Blackburn et al., 2009; Gessi et al., 2000). Substantial evidence suggests that A_{2A}ARs are able to mediate the majority of anti-inflammatory effects of endogenous adenosine (Blackburn et al., 2009; Ohta and Sitkovsky, 2009). In particular, A_{2A} activation to suppress cytokine and chemokine expression by immune cells is likely the dominant mechanism involved. In neutrophils, adenosine, acting at A_{2A}ARs, regulates the production of tumor necrosis factor-α (TNF-α), macrophage inflammatory protein (MIP)-1α, MIP-1β, MIP-2α, and MIP-3α (McColl et al., 2006). Studies using A_{2A}-knockout (KO) models have shown that A_{2A}AR activation inhibits interleukin (IL)-2 secretion by naive CD4+ T cells thereby reducing their proliferation, confirming the immunosuppressive effects of A_{2A}AR stimulation (Naganuma et al., 2006; Sevigny et al., 2007). One of the mechanisms used by T regulatory cells to induce immunosuppression is the expression of CD39 in order to generate adenosine (Borsellino et al., 2007; Deaglio et al., 2007). A_{2A}Rs are generally viewed as negative regulators of immune cells, including activated T cells. However, recently, it has been reported that adenosine A_{2A}AR activation protects CD4+ T lymphocytes against activation-induced cell death. Because activation-induced cell death can be viewed as a process that terminates an immune response, the fact that A_{2A}AR activation prevents it indicates that the role of A_{2A}Rs in regulating immune responses is more complex than previously thought and that A_{2A}R activation can actually prolong immune processes. Clearly, further studies will be necessary to dissect the precise role of the antiapoptotic effect of A_{2A}R activation in regulating T cell-mediated immune responses (Himer et al., 2010). It has been also demonstrated that A_{2A}ARs play an important role in the promotion of wound healing and angiogenesis (Ahmad et al., 2009; Ernens et al., 2010). Moreover, A_{2A} and A_3ARs are responsible for the anti-inflammatory actions of methotrexate (MTX) in the treatment of inflammatory arthritis (Chan and Cronstein, 2010; Montesinos et al., 2006). In rheumatoid arthritis patients, adenosine has been reported to suppress the elevated levels of proinflammatory cytokines, including TNF-α and IL-1β. In a recent study, an upregulation of A_{2A} and A_3 receptors was found in lymphocytes and neutrophils obtained from early rheumatoid arthritis patients and MTX-treated patients. This alteration was associated with high levels of TNF-α and nuclear factor kappa B (NF-κB) activation. Interestingly, treatment with anti-TNF-α drugs normalized A_{2A} and A_3AR expression and functionality (Varani et al., 2009). These data consolidate the involvement of A_{2A} and A_3ARs in rheumatoid arthritis and support the importance of these receptors in human diseases characterized by a marked inflammatory component. Adenosine has been reported to reduce inflammation in several *in vivo* models, suggesting a potential value of this purine nucleoside as a therapeutic mediator of inflammatory joint disease able to limit articular cartilage degeneration. In synoviocytes obtained from osteoarthritis patients, the activation of A_{2A} receptors inhibited p38 mitogen-activated protein kinase (MAPK) and NF-κB pathways as well as the production of TNF-α and IL-8 (Varani et al., 2010c). These results indicate that A_{2A} receptors may represent a potential target in therapeutic modulation of joint inflammation.

Activation of the A_{2A}ARs during reperfusion of various tissues has been found to markedly reduce ischemia–reperfusion injury. In particular, in a model of ischemia–reperfusion liver injury, A_{2A} stimulation with the selective agonist Apadenoson is associated with decreased inflammation and profoundly protects mouse liver from injury when administered at the time of reperfusion (Gazoni et al., 2010). Adenosine, acting at A_{2A}ARs, plays an important role in the pathogenesis of hepatic fibrosis in response to hepatotoxins. In particular, it has been demonstrated that A_{2A}ARs are expressed on human hepatic stellate cell lines and A_{2A} occupancy promotes collagen production by these cells. Further, A_{2A}AR KO mice are protected from developing hepatic fibrosis in two different hepatic fibrosis models (Chan et al., 2006).

It is well reported that hypoxia-induced accumulation of adenosine may represent one of the most fundamental and immediate tissue-protecting mechanisms, with A_{2A}ARs triggering off signals in activated immune cells. In these regulatory mechanisms, oxygen deprivation and extracellular adenosine accumulation serve as "reporters," while A_{2A}ARs serve as "sensors" of excessive tissue damage (Sitkovsky et al., 2004). The hypoxia–adenosinergic tissue-protecting mechanism is triggered by inflammatory damage to blood vessels, interruption in oxygen supply, low oxygen tension (i.e., hypoxia), and also by the hypoxia-driven accumulation of extracellular adenosine acting via immunosuppressive, cAMP-elevating A_{2A}ARs (Sitkovsky, 2009). Another area where A_{2A}AR signaling has received attention as a potential therapeutic target is the gastrointestinal tract. Studies have highlighted the protective effects of A_{2A}AR activation in various animal models of colitis, and these protective effects can be ascribed to two major mechanisms: decrease of inflammatory-cell infiltration and increased activity of regulatory T cells (Haskó et al., 2008; Naganuma et al., 2006). A_{2A} stimulation was found to attenuate gastric mucosal inflammation induced by indomethacin. This effect was obtained by blocking secondary injury due to stomach inflammation, through a reduction of myeloperoxidase and proinflammatory cytokines (Koizumi et al., 2009).

Adenosine levels are increased in the lungs of individuals with asthma or COPD and ARs are known to be expressed on most if not all inflammatory and stromal cell types involved in the pathogenesis of these diseases (Polosa and Blackburn, 2009). In addition, pharmacological treatment of allergic rats with an A_{2A} agonist resulted in diminished pulmonary inflammation. A recent study in ADA-deficient model demonstrated that genetic removal of A_{2A} leads to enhanced pulmonary inflammation, mucus production, and alveolar airway destruction (Mohsenin et al., 2007). Further, A_{2A}ARs induced on iNKT and natural killer (NK) cells reduced pulmonary inflammation in mice with sickle cell disease, improving baseline pulmonary function and preventing hypoxia–reoxygenation-induced exacerbation of pulmonary injury (Wallace & Linden, 2010). These data further confirm the involvement of A_{2A}ARs in the anti-inflammatory networks in the lung. A study performed in peripheral lung parenchyma demonstrated that affinity and/or density of ARs are altered in patients with COPD compared with smokers having normal lung function. Moreover, there was a significant correlation between the density and affinity of ARs and the forced expiratory volume in 1 s/forced vital capacity (FEV_1/FVC) ratio, an established index of airflow obstruction. In particular, A_{2A}ARs, as well as A_3ARs, were found to be upregulated in COPD patients (Varani et al., 2006). This alteration may represent a compensatory response mechanism and may contribute to the anti-inflammatory effects mediated by the stimulation of these receptors. Given the central role of inflammation in asthma and COPD, substantial preclinical research activity targeted at understanding the function of A_{2A}ARs in models of airway inflammation has been performed.

A list of A_{2A}ARs ligands in clinical studies for novel therapeutic treatments is reported in Table II.

IV. A_{2B} Adenosine Receptors

A_{2B}ARs were cloned from rat hypothalamus, human hippocampus, and mouse mast cells. Following initial studies indicating selective induction of A_{2B}ARs by hypoxia, analysis of the cloned human A_{2B} promoter identified a functional hypoxia-responsive region, including a functional binding site for

TABLE II A_{2A} Receptor Ligands in Clinical Studies for Novel Therapeutic Treatments

Pathology		*Drug name*	*Mechanism of action*	*Status of development*
Neurological disorders	Huntington's disease	[^{123}I]MNI-420	Antagonist	Phase I
	Parkinson's disease	Istradefylline	Antagonist	Preregistered
		Preladenant	Antagonist	Phase III
		ST-1535	Antagonist	Phase I
		SYN-115	Antagonist	Phase II
		[^{123}I]MNI-420	Antagonist	Phase I
Cardiovascular diseases	Coronary artery disease diagnosis	Regadenoson	Agonist	Launched—2008
		Apadenoson	Agonist	Phase III
		Binodenoson	Agonist	Preregistered
	Hypertension	YT-146	Agonist	Phase II
	Sickle cell disease	Regadenoson	Agonist	Phase I
Respiratory disorders	COPD	Apadenoson	Agonist	Phase I
	Asthma	Apadenoson	Agonist	Phase I
Dermatological disorders	Ulcer, diabetic	Sonedenoson	Agonist	Phase II

hypoxia-inducible factor (HIF) within the A_{2B} promoter (Kong et al., 2006; Yang et al., 2010b). These results demonstrated transcriptional coordination of A_{2B}ARs by HIF-1α and amplified adenosine signaling during hypoxia, suggesting an important link between hypoxia and metabolic conditions related with inflammation and angiogenesis (Gessi et al., 2010a).

A_{2B}ARs have long been known to couple to AC activation through Gs proteins. However, other intracellular signaling pathways have been demonstrated to be associated to A_{2B}ARs including Ca^{2+} mobilization through Gq proteins and MAPK activation (Cohen et al., 2010; Linden, 2001). A_{2B}AR-induced stimulation of PLC results in mobilization of intracellular calcium in human mast cells (HMC)-1 and promotion of IL-8 production (Feoktistov & Biaggioni, 1995).

Initially, tissue distribution of A_{2B}ARs was reported in peripheral organs like bowel, bladder, lung, and vas deferens. As for the brain, mRNA and protein were detected in hippocampal neurons and glial cells but not in microglial cells (Linden, 2001). Stimulation of A_{2B}ARs mediates the release of IL-6 from astrocytes. Due to the neuroprotective effect of IL-6 against hypoxia and glutamate neurotoxicity, activation of A_{2B} subtype provides a damage-control mechanism during CNS injury (Haskó et al., 2005).

Functional studies have identified A_{2B}ARs in airway smooth muscle, fibroblasts, glial cells, gastrointestinal tract, vasculature, and platelets. Vascular A_{2B}ARs may be associated with vasodilatation in both smooth muscle and endothelium. This subtype is important in modulation of vasodilatation in certain vessels such as mesenteric, pulmonary, and coronary arteries but not in others, where the A_{2A} effect predominates (Haskó et al., 2005). A_{2A}/A_{2B}ARs are functionally expressed in juxtamedullary afferent arterioles, where A_{2B}ARs exert the powerful vasodilating action of adenosine, which counteracts A_1ARs-mediated vasoconstriction (Feng and Navar, 2010). Further, adenosine stimulates Cl^- secretion through the cystic fibrosis transmembrane conductance regulator in mIMCD-K2 cells, a murine model system for the renal inner medullary collecting duct, by activating apical A_{2B}ARs and signaling through cAMP/PKA. This suggests that the A_{2B}AR pathway may provide one mechanism for enhancing urine NaCl excretion in the setting of high dietary NaCl intake (Rajagopal & Pao, 2010).

Activation of A_{2B}ARs may prevent cardiac remodeling after myocardial infarction (Wakeno et al., 2006). A protective effect from infarction has also been attributed to A_{2B}ARs in ischemic postconditioning, through a pathway involving PKC and phosphatidylinositol-3-kinase (PI3K; Kuno et al., 2007, 2008; Methner et al., 2010; Philipp et al., 2006). Further, A_{2B}/A_3ARs mediate the cardioprotective effects induced by IPC through PKCepsilon, aldehyde dehydrogenase type-2 (ALDH2) activation, and renin inhibition (Koda et al., 2010). Finally, a new role for the A_{2B}AR has been discovered in the regulation of platelet function. In particular, A_{2B}ARs have been found upregulated under stress *in vivo* in platelets where they modulate ADP receptor expression and inhibit agonist-induced aggregation (Yang et al., 2010a).

According to mRNA analysis revealing high amounts of A_{2B}AR message in the cecum and large intestine, it has been reported that A_{2B}ARs in intestinal epithelial cells trigger an increase in cAMP levels that is responsible for Cl^- secretion. This pathway results in movement of isotonic fluid into the lumen, a process that naturally serves to hydrate the mucosal surface but, in extreme, produces secretory diarrhea (Strohmeier et al., 1995). Recently, it has been reported that adenosine increases HCO_3^- secretion in an intact epithelium *in vivo* through the activation of A_{2B} receptors expressed in the brush border membrane of duodenal villi (Ham et al., 2010). Further, A_{2B}AR stimulation of intestinal epithelial cells increases intracellular cAMP levels, which in turn leads to IL-6 transcription via activation of the ATF, CREB and C/EBPb (NF-IL6) transcription factor systems. The physiological relevance of this response is that it provides an amplification mechanism for intestinal inflammation because neutrophils transmigrating through the epithelial cell layer release adenosine, which in turn induces the production of the neutrophil-activating IL-6. This amplification loop is further enhanced by a rapid increase in the surface expression of A_{2B}ARs after stimulation of the cells with adenosine, which is made possible by prompt recruitment of preformed A_{2B}ARs from intracellular stores (Haskó et al., 2009). Accordingly, it has been found that A_{2B}AR gene deletion in mice attenuates murine colitis (Kolachala et al., 2008). However, recent studies combining pharmacological and genetic approaches demonstrated that adenosine signaling via the A_{2B}ARs dampens mucosal inflammation and tissue injury during intestinal ischemia or experimental colitis (Eltzschig et al., 2009). It has been reported that A_{2B}ARs play a central regulatory role on IL-10 modulation in the acute inflammatory phase of dextran sodium sulfate colitis, thereby implicating A_{2B}ARs as an endogenously protective protein expressed on intestinal epithelial cells (Frick et al., 2009). Epithelial adenosine A_{2B}AR mRNA and protein have been found upregulated in colitis, via TNF-α through a posttranscriptional mechanism involving microRNA (Kolachala et al., 2010).

A_{2B}ARs have been reported to mediate degranulation and activation of canine and human mast cells, thus suggesting a possible role in allergic and inflammatory disorders (Haskó et al., 2009). Adenosine constricts airways of asthmatic patients through the release of histamine and leukotrienes from sensitized mast cells (Polosa et al., 2002). The receptor involved seems to be the A_{2B} in humans, or the A_3 in rats. Recently, A_{2B}ARs have been reported to mediate several proinflammatory effects of adenosine in inflammatory cells of the lung. In addition to mast cells, functional A_{2B}ARs have been found in bronchial smooth muscle cells and lung fibroblasts. In these cells, adenosine, through stimulation of A_{2B}AR subtype, increases the release of various inflammatory cytokines, supporting the evidence that A_{2B}ARs play a key role in the inflammatory response associated with asthma. Further, it has been reported that dendritic cells differentiated with adenosine have impaired allostimulatory activity and express high levels of angiogenic, proinflammatory, immune suppressor, and tolerogenic factors, including VEGF, IL-8, IL-6, IL-10, COX-2, TGF-β, and IDO, through A_{2B} receptors activation (Ben Addi et al., 2008; Novitskiy et al., 2008). Using ADA KO animals, it has been shown that DCs with a proangiogenic phenotype are highly abundant under conditions associated with elevated levels of extracellular adenosine *in vivo*. The first evidence for the involvement of A_{2B}ARs in asthma derived from studies concerning the selectivity of enprofylline, a methylxanthine structurally related to theophylline (Feoktistov et al., 1998). Further support for the role of A_{2B}ARs in asthma comes from studies demonstrating their presence on different type of cells important for the cytokine release in asthmatic disease such as smooth muscle cells, lung fibroblasts, endothelial cells, bronchial epithelium, and mast cells. Expression of A_{2B}ARs was also found in mast cells and macrophages of patients affected by COPD (Varani et al., 2006). Activation of A_{2B} in the HMC1 mast cell line stimulated IL-8 release (Feoktistov & Biaggioni, 1995). Further, A_{2B} antagonists potently inhibited activation, and degranulation of HMC cells induced by adenosine (Haskó et al., 2009). Recently, it has been reported that ADA-deficient mice treated with the selective A_{2B} antagonist 3-ethyl-1-propyl-8-[1-

[3-(trifluoromethyl)benzyl]-1H-pyrazol-4-yl]xanthine (CVT-6883) showed reduced elevations in proinflammatory cytokines and chemokines as well as mediators of fibrosis and airway destruction (Sun et al., 2006b). Interestingly, other authors investigated the role of A_{2B}ARs in inflammation *in vivo* (Yang et al., 2006). This study was carried out on A_{2B} KO mice in which exon 1 of the A_{2B} was replaced by a reporter gene, allowing examination of endogenous A_{2B}ARs expression in various tissues and cell types *in vivo*. Results show that there is abundant reporter expression in the vasculature and in macrophages. This new animal model emphasizes a role for the A_{2B}ARs in attenuating inflammation through regulation of proinflammatory cytokines production and in inhibiting leukocyte adhesion to the vasculature. In contrast with the function of A_{2B}ARs in vasodilation, the A_{2B} KO mice have normal blood pressure (Yang et al., 2006). The apparent contradiction about the pro- or anti-inflammatory effects exerted by A_{2B} receptors may be related to a difference between acute and chronic inflammation: an A_{2B} agonist would protect against acute endotoxin-mediated lung toxicity, whereas chronic accumulation of adenosine will induce lung lesions.

Studies with a specific A_{2B}AR agonist 2-[6-amino-3,5-dicyano-4-[4-(cyclopropylmethoxy)phenyl]pyridin-2-ylsulfanyl]acetamide (BAY 60-6583) demonstrated attenuation of lung inflammation and pulmonary edema in wild-type but not in A_{2B}AR KO mice. These studies suggest the A_{2B}AR is a potential therapeutic target for the treatment of endotoxin-induced forms of acute lung injury (Schingnitz et al., 2010).

A_{2B}ARs play a role in cancer development by modulation of both anti- and protumoral effects. In particular, A_{2B} receptor stimulation inhibits ERK1/2 phosphorylation in breast cancer cells while it increases angiogenesis, proliferation, IL-8, VEGF, and basic fibroblast growth factor (bFGF) in endothelial, foam, and tumor cells (Gessi et al., 2010a). Recently, it has been reported that hypoxia-induced apoptosis of T cells was mediated by A_{2A}- and A_{2B}ARs and that the blockade of A_2AR signaling can increase the antiapoptotic function of T cells and may become a new strategy to improve antitumor potential (Sun et al., 2010).

A list of A_{2B} receptor ligands in clinical studies for novel therapeutic treatments is reported in Table III.

V. A_3 Adenosine Receptors

The A_3AR is the only adenosine subtype cloned before its pharmacologic identification. It was originally isolated as an orphan receptor from rat testis, having 40% sequence homology with canine A_1 and A_{2A} subtypes. Homologs of the rat striatal A_3AR have been cloned from sheep and human. Interspecies differences in A_3AR structure are large, showing the rat A_3AR has only 74% sequence homology with sheep and human (Baraldi and Borea, 2000; Baraldi et al., 2008).

TABLE III A_{2B} Receptor Ligands in Clinical Studies for Novel Therapeutic Treatments

Pathology		*Drug name*	*Mechanism of action*	*Status of development*
Respiratory disorders	Asthma	CVT-6883	Antagonist	Phase I
		GS-6201	Antagonist	Phase I
Gastrointestinal disorders	Diarrhea	CVT-6883	Antagonist	Phase I
Cardiovascular disorders		CVT-6883	Antagonist	Phase I

A_3AR activation inhibits AC activity by coupling with G_i proteins. In the rat mast cell line RBL-2H3 and rat brain, A_3AR stimulation activates PLC through Gq proteins. Further, depending on the cell type studied, A_3ARs may also signal through the activation of MAPK signaling pathway, critically important in the regulation of cell proliferation and differentiation (Raman et al., 2007). The A_3AR is widely distributed and its mRNA is expressed in testis, lung, kidneys, placenta, heart, brain, spleen, liver, uterus, bladder, jejunum, proximal colon, and eye of rat, sheep, and humans (Jacobson & Gao, 2006).

A dual role of A_3ARs has been reported in the brain. In particular, it seems that chronic preischemic administration of the agonist 1-deoxy-1-[6-[[(3-iodophenyl)methyl]amino]-9H-purin-9-yl]-*N*-methyl-*β*-D-ribofuranuronamide (CF-101, IB-MECA) induces a significant neuronal protection and reduction of the subsequent mortality, while acute administration of the drug results in a pronounced worsening of neuronal damage and postischemic mortality. Mice with deletions of the A_3AR reveal a number of CNS functions where A_3ARs play a role, including nociception, locomotion, behavioral depression, and neuroprotection. Consistent with previous reports of the neuroprotective actions of A_3AR agonists, A_3AR KO mice show an increase in neurodegeneration in response to repeated episodes of hypoxia suggesting the possible use of A_3 agonists in the treatment of ischemic and degenerative conditions of the CNS (Fedorova et al., 2003).

Several lines of evidence suggest that activation of A_3ARs is crucial for cardioprotection during and following ischemia–reperfusion and it is likely that a consistent part of the cardioprotective effects exerted by adenosine, once largely attributed to the A_1AR, may be now in part ascribed to A_3AR activation (Ge et al., 2006). The molecular mechanism of A_3AR cardioprotection has been attributed to regulation of K_{ATP} channels. The cardioprotective effects of A_3AR overexpression were also detected in mice expressing low levels of A_3ARs without detectable adverse effects, while higher levels of A_3 expression led to the development of a dilated cardiomyopathy (Black et al., 2002). Similar data were observed in the case of A_1AR overexpression (Funakoshi et al., 2006). As stated for A_{2B}AR overexpression, a signaling cascade initiated by A_{2B}/A_3 subtypes, which triggers PKC epsilon-mediated ALDH2 activation in cardiac mast cells, contributes to IPC-induced cardioprotection by preventing mast cell renin release and the dysfunctional consequences of local renin angiotensin system (RAS) activation. Thus, unlike classic IPC in which cardiac myocytes are the main target, cardiac mast cells are the critical site at which the cardioprotective anti-RAS effects of IPC develop (Koda et al., 2010). A role of NO in A_3AR-mediated cardioprotection has been also reported. In particular, the involvement of inducible NO synthase as a downstream effector of the PI3K signaling cascade after activation of A_3ARs at reperfusion has been demonstrated (Hussain et al., 2009; Karjian et al., 2006, 2008). Further, A_3AR stimulation restores vascular reactivity after hemorrhagic shock through a ryanodine receptor-mediated and calcium-activated potassium channel-dependent pathway (Zhou et al., 2010).

In addition to reducing injury in myocardial and vascular tissues, other beneficial antiinflammatory actions have been attributed to the A_3 subtype. For example, A_3ARs are expressed in human neutrophils where they are involved together with A_{2A}AR in the reduction of superoxide anion generation and have been implicated in the suppression of TNF-α release induced by endotoxin from human monocytes (Gessi et al., 2002). In neutrophils, A_3ARs also play a role in chemotaxis together with P2Y receptors (Chen et al., 2006; Linden, 2006). Moreover, A_3 activation seems to inhibit degranulation and superoxide anion production in human eosinophils. Transcript levels for the A_3 subtype are elevated in the lungs of asthma and COPD patients, where expression is localized to eosinophilic infiltrates. Similar results were observed in the lungs of ADA KO mice that exhibited adenosine-mediated lung disease. Treatment of ADA KO mice with 3-propyl-6-ethyl-5-[(ethylthio)carbonyl]-2 phenyl-4-propyl-3-pyridine carboxylate (MRS 1523), a selective A_3 antagonist, prevented airway eosinophilia and mucus production. These results are in contrast to experiments performed with human eosinophils *ex vivo*, where chemotaxis was reduced by A_3AR activation, suggesting that significant differences exist between the impact of A_3 signaling on eosinophil migration *ex vivo* and in the whole animal. More recently, the involvement of the A_3AR in a bleomycin model of pulmonary inflammation and fibrosis has been addressed. Results demonstrated that A_3AR KO mice exhibit enhanced pulmonary inflammation that included an increase in eosinophils. Accordingly, there was a selective upregulation of eosinophil-related chemokines and cytokines in the lungs of A_3AR KO mice exposed to bleomycin, thus suggesting that the A_3AR has anti-inflammatory functions in the bleomycin model (Morschl et al., 2008). The role of the A_3 receptor in the human lung and indeed in asthma remains to be clarified. What is clear, however, is that the expression of the A_3AR in asthmatic airways is predominantly located in eosinophils (Brown et al, 2008b; Gessi et al., 2008; Wilson, 2008).

A very interesting area of application of A_3 ligands concerns cancer therapies. The possibility that the A_3AR plays a role in the development of cancer has aroused considerable interest in recent years

(Merighi et al., 2003). The A_3AR subtype has been implicated in the regulation of the cell cycle, and both pro- and antiapoptotic effects have been reported depending on the level of receptor activation (Gessi et al., 2007; Jacobson, 1998; Kim et al., 2010; Merighi et al., 2005; Taliani et al., 2010; Varani et al., 2010a). The A_3AR activation has been demonstrated to be involved in inhibition of tumor growth both *in vitro* and *in vivo* leading to the development of A_3 agonists in clinical trials for cancer. The molecular mechanisms involved in the anticancer effects induced by A_3 agonists include regulation of the WNT pathway (Fishman et al., 2004). However, it has been reported that adenosine upregulates HIF-1α protein expression and VEGF protein accumulation by activating the A_3AR subtype in tumor cells (Merighi et al., 2006, 2007a, 2007b). It is interesting to note that etoposide and doxorubicin affect VEGF and HIF-1 expression in human melanoma cancer cells. In particular, blockade of A_3ARs potentiates inhibition of VEGF secretion induced by etoposide and doxorubicin in melanoma cells. This finding suggests the possibility of using adenosine AR antagonists to improve the ability of chemotherapeutic drugs to block angiogenesis (Merighi et al., 2009).

Overexpression of the A_3AR subtype has been demonstrated in colon cancer tissues obtained from patients undergoing surgery in comparison to normal mucosa. Overexpression in tissues was also reflected at the level of peripheral blood cells rendering this AR subtype a possible marker for cancer detection (Gessi et al., 2004). In a further study, it has been shown that A_3AR mRNA expression was upregulated in hepatocellular carcinoma (HCC) tissues in comparison to adjacent normal tissues (Bar-Yehuda et al., 2008). Remarkably, upregulation of the A_3AR was also noted in peripheral blood mononuclear cells (PBMCs) derived from the HCC patients compared to healthy subjects. These results further show that A_3ARs on PBMCs reflect receptor status on the remote tumor tissue (Gessi et al., 2004). The A_3AR reduces the ability of prostate cancer cells to migrate *in vitro* and metastasize *in vivo*. In particular, it has been reported that activation of the A_3AR in prostate cancer cells reduced PKA-mediated stimulation of ERK1/2, leading to lower NADPH oxidase activity and cancer cell invasiveness (Jajoo et al., 2009). In a different study, the biological functions of adenosine on matrix metalloproteinase-9 (MMP-9) regulation in U87MG human glioblastoma cells were investigated. In this case, it was revealed that A_3AR stimulation induced an increase of MMP-9 levels in U87MG cells by phosphorylation of ERK1/2, c-Jun N-terminal kinase/stress-activated protein kinase (pJNK/SAPK), PKB/Akt, and finally activator protein 1 (AP-1). This effect was responsible for an increase of glioblastoma cells invasion (Gessi et al., 2010c).

In the field of cancer and ARs, the results of *in vivo* studies have been reported only for the A_3AR (Press et al., 2007). Below, the principal findings obtained in animal studies will be summarized. In all experimental models, the A_3AR agonists were administered orally due to their stability and bioavailability profile. The studies included syngeneic, xenograft, orthotopic, and metastatic experimental animal models utilizing IB-MECA and 2-chloro-N^6-(3-iodobenzyl)adenosine-5′-*N*-methylcarboxamide (CF-102, Cl-IB-MECA) as the therapeutic agents in melanoma, colon, prostate, and HCCs. Oral administration of 10–100 μg kg^{-1} IB-MECA and Cl-IB-MECA once or twice daily inhibited the growth of primary B16-F10 murine melanoma tumors in syngeneic models. Moreover, in an artificial metastatic model, IB-MECA inhibited the development of B16-F10 murine melanoma lung metastases. The specificity of the response was demonstrated by the administration of an A_3AR antagonist that reversed the effect of the agonist. Further, IB-MECA or Cl-IB-MECA in combination with the chemotherapeutic agent cyclophosphamide induced an additive antitumor effect on the development of B16-F10 melanoma lung metastatic foci. Oral administration of 10–100 μg kg^{-1} IB-MECA once or twice daily inhibited the growth of primary CT-26 colon tumors. Further, in xenograft models, IB-MECA inhibited the development of HCT-116 human colon carcinoma in nude mice. In these studies, the combined treatment of IB-MECA and 5-fluorouracil resulted in an enhanced antitumor effect. IB-MECA was also efficacious in inhibiting liver metastases of CT-26 colon carcinoma cells inoculated in the spleen. IB-MECA inhibited the development of PC3 human prostate carcinoma in nude mice. Additionally, IB-MECA increased the cytotoxic index of Taxol in PC3 prostate carcinoma-bearing mice. Finally, Cl-IB-MECA treatment dose dependently inhibited hepatocellular tumor growth (Bar-Yehuda et al., 2008; Fishman et al., 2009; Gessi et al., 2010b).

In addition, a role for A_3AR in inflammation has been reported in the literature. In arthritis, A_3 activation shows beneficial effects by suppression of TNF-α production (Fishman et al., 2006). A_3ARs suppress TNF-α release induced by the endotoxin/CD14 receptor signal transduction pathway from human monocytes and murine J774.1 macrophages. Moreover, in a macrophage model, the A_3AR was the prominent subtype implicated in the inhibition of LPS-induced TNF-α production (Sajjadi et al., 1996). This effect was associated with changes in stimulation of the AP-1 transcription factor, whereas it was independent on MAPK, NF-κB, PKA, PKC, and PLC. This was not confirmed in BV2 microglial cells where A_3-mediated inhibition of LPS-induced TNF-α expression was associated with the inhibition of LPS-induced activation of PI3K/Akt and NF-κB pathway (Lee et al., 2006a). The inhibitory effect induced by A_3AR on TNF-α production was also assessed in A_3KO mice where the A_3 agonist was unable to reduce TNF-α levels in contrast with its effect in wild-type animals (Salvatore et al., 2000). Recently, it has been reported that in mouse RAW 264.7 cells, the A_3 subtype inhibits LPS-stimulated TNF-α release by reducing calcium-dependent activation of NF-κB and ERK 1/2 (Martin et al., 2006). In contrast, in peritoneal macrophages, isolated from A_3KO mice, the ability of IB-MECA to inhibit TNF-α release was not altered in comparison to wild-type mice (Kreckler et al., 2006). In this study, the inhibitory effect was exerted through the activation of A_{2A} and A_{2B} agonists as recently demonstrated also in human monocytes (Haskó et al., 2007). The discrepancy observed among these papers might not depend on species differences, being in both cases mouse cells, but by other factors including the source of the cells, and/or the inflammatory stimulus used. However, in spite of these contrasting results, one of the best potential therapeutic applications of the regulatory role of A_3 activation on TNF-α release has been found in the treatment of arthritis. A_3AR agonists exert significant antirheumatic effects in different autoimmune arthritis models by suppression of TNF-α production. The molecular mechanism involved in the inhibitory effect of IB-MECA on adjuvant-induced arthritis included receptor downregulation and deregulation of the PI3K–NF-κB signaling pathway (Fishman et al., 2006; Madi et al., 2007). Previous studies also demonstrated that A_3AR activation inhibited MIP-1α, that is, a C-C chemokine with potent inflammatory effects, in a model of collagen-induced arthritis, providing the first proof of concept of the adenosine agonists utility in the treatment of arthritis (Gessi et al., 2008).

In agreement with an anti-inflammatory role for the A_3AR, it has been recently demonstrated that A_3AR activation decreases mortality and renal and hepatic injury in murine septic peritonitis (Lee et al., 2006b). Higher levels of endogenous TNF-α were observed in A_3KO mice after sepsis induction, in comparison to wild-type animals, and IB-MECA significantly reduced mortality in mice lacking the A_1 or A_{2A} but not the A_3AR, demonstrating specificity of the A_3 agonist in activating A_3 subtype and mediating protection against sepsis-induced mortality (Lee et al., 2006b). Recently, in a mouse model of cecal ligation and puncture-induced sepsis, A_3AR blockade reduced acute lung injury and polymorphonuclear leukocytes accumulation in lung tissue (Inoue et al., 2010). A similar mortality reduction associated with a decrease of IL-12 and interferon-γ production induced by A_3AR activation was observed in endotoxemic mice. In addition, it has been reported that there is a reduced inflammation and increased survival following A_3AR activation in two murine models of colitis. Further, a protective role for A_3ARs in lung injury following *in vivo* reperfusion has been observed (Matot et al., 2006). This effect has been attributed to the stimulation of A_1AR, A_2AR, and A_3ARs leading to increased lung compliance and oxygenation, decreased pulmonary artery pressure, decreased neutrophil infiltration, decreased edema, and reduced TNF-α production (Gazoni et al., 2010).

Recently, it has been shown that adenosine in hypoxic foam cells stimulates HIF-1α accumulation by activating all ARs. HIF-1α modulation involved ERK 1/2, p38 MAPK, and Akt phosphorylation in the case of A_1AR, A_{2A}AR, and A_{2B}AR, while only ERK 1/2 activation in the case of A_3ARs. Further, adenosine, through the activation of A_3ARs and A_{2B}ARs, stimulates VEGF secretion in a HIF-1α-dependent way. Finally, adenosine stimulates foam cell formation, and this effect is strongly reduced by A_3AR and A_{2B}AR blockers and by HIF-1α silencing. This study provides the first evidence that A_3AR, A_{2B}AR, or mixed

$A_3/A_{2B}AR$ antagonists may be useful to block important steps in the atherosclerotic plaque development (Gessi et al., 2010a).

ARs have been implicated in many ocular and systemic ischemic diseases (e.g., retinal ischemia). The A_3 KO mouse showed lower intracellular pressure suggesting a role for A_3 antagonists in the therapy of glaucoma (Yang et al., 2005). Accordingly, nucleoside-derived antagonists to A_3ARs lower mouse intraocular pressure and act across species (Wang et al., 2010). Further, retinal ganglion cells express A_3ARs. Agonists for the A_3AR prevented the Ca^{2+} rise and cell death which accompanied activation of the P2X7 and NMDA receptors suggesting a neuroprotective potential of the A_3AR on retinal ganglion cells. These findings have been confirmed in *in vivo* experiments (Hu et al., 2010; Zhang et al., 2006a, 2006b, 2010). Importantly, studies from a phase II clinical trial reveal that CF-101, given orally, was well tolerated and induced a statistically significant improvement in patients with moderate to severe dry eye syndrome. These data and the anti-inflammatory characteristic of CF-101 support further studies of the drug as a potential treatment for the signs and symptoms of dry eye syndrome (Avni et al., 2010).

A list of A_3 receptor ligands in clinical studies for novel therapeutic treatments is reported in Table IV.

VI. Conclusion

The investigation of ARs and their ligands is rapidly growing with an increasing impact on the drug discovery process. There is now extensive evidence for the involvement of ARs in the physiological regulation of several homeostatic processes and in the etiology of many diseases. A considerable body of research over the past 30 years in the AR field has resulted in the identification of clinical candidates for AR agonism, partial agonism, and antagonism. It is exciting to underline that some molecules coming from the purine world have been developed as drugs. In particular, Regadenoson is already commercially available while Binodenoson and Istradefylline are now preregistered. Further, other molecules are in advanced clinical phase such as Tecadenoson, Rolofylline, Tonapofylline, and CF-101 that could easily, barring setbacks, become new drugs.

Although the ARs are increasingly being recognized for their growing number of biological roles through the body and many AR ligands have proven useful in elucidating peripheral and central pathologies, many issues remain unresolved. Therefore, research activity in this field continues to grow exponentially, resulting in a flow of new information (Fig. 1).

Based on the important scientific and clinical advances overviewed in this chapter, purine scientists do seem to be getting closer to their goal: the incorporation of adenosine ligands into drugs with the ability to save lives and improve human health.

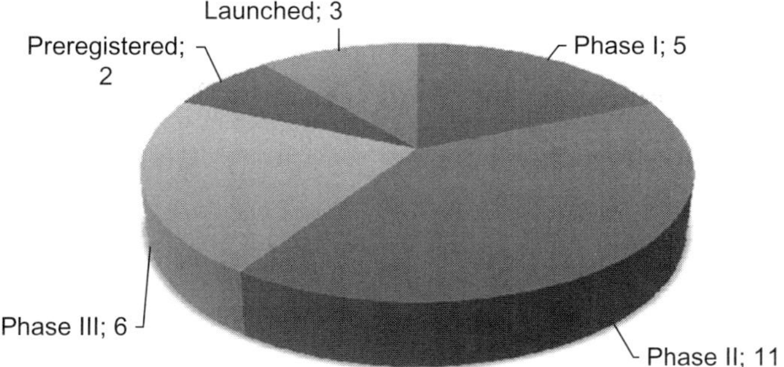

FIGURE 1 Pie chart showing number of ARs ligands in clinical development.

TABLE IV A_3 Receptor Ligands in Clinical Studies for Novel Therapeutic Treatments

Pathology		*Drug name*	*Mechanism of action*	*Status of development*
Musculoskeletal and connective tissue disorders	Osteoarthritis	CF-101	Agonist	Phase II
	Rheumatoid arthritis	CF-101	Agonist	Phase I
Cancer	Hepatocellular carcinoma	CF-102	Agonist	Phase I/II
Eye disorders	Dry eyes	CF-101	Agonist	Phase II
	Glaucoma	CF-101	Agonist	Phase II
	Uveitis	CF-101	Agonist	Phase I
Gastrointestinal disorders	Liver disorders	CF-102	Agonist	Phase I
	Hepatitis C	CF-102	Agonist	Phase I/II
Dermatological disorders	Psoriasis	CF-101	Agonist	Phase II/III

Conflict of Interest: The authors have no conflict of interest to declare.

Database: Part of the information reported in this chapter derives from the following database: http://Integrity.prous.com.

Abbreviations

[^{123}I]MNI-420	7-[2-[4-[2-fluoro-4-[^{123}I]iodophenyl]piperazin-1-yl]ethyl]-2-(2-furyl)-7H-pyrazolo[4,3-e][1,2,4]triazolo[1,5-c]pyrimidin-5-amine
AC	adenylyl cyclase
ADA	adenosine deaminase
ALDH2	aldehyde dehydrogenase type-2
AP-1	activator protein 1

Apadenoson	*trans*-4-[3-[6-amino-9-[(2*R*,3*R*,4*S*,5*S*)-5-(*N*-ethylcarbamoyl)-3,4-dihydroxytetrahydrofuran-2-yl]-9H-purin-2-yl]-2-propynyl]cyclohexanecarboxylic acid methyl ester
ARs	adenosine receptors
bFGF	basic fibroblast growth factor
Binodenoson	2-[2(cyclohexylmethyl)hydrazino]adenosine
Capadenoson	2-amino-6-[2-(4-chlorophenyl)thiazol-4-ylmethylsulfanyl]-4-[4-(2-hydroxyethoxy)phenyl]pyridine-3,5-dicarbonitrile
CDK	cyclin-dependent kinase
CF-101	*N*6-(3-iodobenzyl)adenosine-5′-(*N*-methyluronamide)
CF-102	(2*S*,3*S*,4*R*,5*R*)-5-[2-chloro-6-(3-iodobenzylamino)-9H-purin-9-yl]-3,4-dihydroxy-*N*-methyltetrahydrofuran-2-carboxamide
CNS	central nervous system
COPD	chronic obstructive pulmonary disease
CVT-3619	CVT-3619 (2-{6-[((1*R*,2*R*)-2-hydroxycyclopentyl) amino] purin-9-yl}(4*S*,5*S*,2*R*,3*R*)-5-[(2-fluorophenylthio)-methyl] oxolane-3,4-diol)
CVT-6883	3-ethyl-1-propyl-8-[1-[3-(trifluoromethyl)benzyl]-1H-pyrazol-4-yl]xanthine
ERKs	extracellular signal-regulated protein kinases
GPCRs	G-protein-coupled receptors
HCC	hepatocellular carcinoma
HIF	hypoxia-inducible factor
HMC	human mast cells
IL	interleukin
IPC	ischemic preconditioning
Istradefylline	8-[2(E)-(3,4-dimethoxyphenyl)vinyl]-1,3-diethyl-7-methylxanthine
KO	knockout
LPS	lipopolysaccharide
MAPK	mitogen-activated protein kinase
MIP	macrophage inflammatory protein
MMP-9	matrix metalloproteinase-9
MRS 1523	3-propyl-6-ethyl-5-[(ethylthio)carbonyl]-2 phenyl-4-propyl-3-pyridine carboxylate
MTX	methotrexate
NF-κB	nuclear factor kappa B
NK	natural killer
NO	nitric oxide
PBMCs	peripheral blood mononuclear cells
PI3K	phosphatidylinositol-3-kinase
PKA	protein kinase A

PKC	protein kinase C
PLC	phospholipase-C
Preladenant	2-(2-furyl)-7-[2-[4-[4-(2-methoxyethoxy)phenyl]piperazin-1-yl]ethyl]-7H-pyrazolo[4,3-e][1,2,4]triazolo[1,5-c] pyrimidin-5-amine
RAS	renin angiotensin system
Regadenoson	2-[4-(*N*-methylcarbamoyl)-1H-pyrazol-4-yl]adenosine
Rolofylline	1,3-dipropyl-8-(2-nor-1-adamantyl)xanthine
Selodenoson	*N*6-cyclopentyl-N5′-ethyladenosine-5′-uronamide
siRNA	small interfering RNA
Sonedenoson	2-[2-(4 chlorophenyl)ethoxy]adenosine
ST-1535	2-butyl-9-methyl-8-(2H-1,2,3-triazol-2-yl)adenine
T62	1-(2-amino-4,5,6,7-tetrahydrobenzo[*b*]thiophen-3-yl)-1-(4-chlorophenyl)methanone
Tecadenoson	*N*6-[3(*R*)-tetrahydrofuranyl]adenosine
TNF-α	tumor necrosis factor-α
Tonapofylline	3-[4-(1,3-dipropylxanthin-8-yl)bicyclo[2.2.2]oct-1-yl] propionic acid
VEGF	vascular endothelial growth factor
YT-146	2-(1-octynyl)adenosine

References

Ahmad, A., Ahmad, S., Glover, L., Miller, S. M., Shannon, J. M., Guo, X., et al. (2009). Adenosine A_{2A} receptor is a unique angiogenic target of HIF-2alpha in pulmonary endothelial cells. *Proceedings of the National Academy of Sciences of the United States of America*, *106*, 10684–10689.

Avni, I., Garzozi, H. J., Barequet, I. S., Segev, F., Varssano, D., Sartani, G., et al. (2010). Treatment of dry eye syndrome with orally administered CF101: Data from a phase 2 clinical trial. *Ophthalmology*, *117*, 1287–1293.

Baraldi, P. G., & Borea, P. A. (2000). New potent and selective human adenosine $A_{(3)}$ receptor antagonists. *Trends in Pharmacological Sciences*, *21*, 456–459.

Baraldi, P. G., Tabrizi, M. A., Gessi, S., & Borea, P. A. (2008). Adenosine receptor antagonists: Translating medicinal chemistry and pharmacology into clinical utility. *Chemical Reviews*, *108*, 238–263.

Baraldi, P. G., Zaid, A. N., Lampronti, I., Fruttarolo, F., Pavani, M. G., Tabrizi, M. A., et al. (2000). Synthesis and biological effects of a new series of 2-amino-3-benzoylthiophenes as allosteric enhancers of A_1-adenosine receptor. *Bioorganic & Medicinal Chemistry Letters*, *10*, 1953–1957.

Bar-Yehuda, S., Stemmer, S. M., Madi, L., Castel, D., Ochaion, A., Cohen, S., et al. (2008). The A_3 adenosine receptor agonist CF102 induces apoptosis of hepatocellular carcinoma via de-regulation of the Wnt and NF-kappaB signal transduction pathways. *International Journal of Oncology*, *33*, 287–295.

Bayes, M. (2007). Gateways to clinical trials. *Methods and Findings in Experimental and Clinical Pharmacology*, *29*, 153–173.

Belardinelli, L., Shryock, J. C., Snowdy, S., Zhang, Y., Monopoli, A., Lozza, G., et al. (1998). The A_{2A} adenosine receptor mediates coronary vasodilation. *The Journal of Pharmacology and Experimental Therapeutics*, *284*, 1066–1073.

Ben Addi, A., Lefort, A., Hua, X., Libert, F., Communi, D., Ledent, C., et al. (2008). Modulation of murine dendritic cell function by adenine nucleotides and adenosine: Involvement of the A_{2B} receptor. *European Journal of Immunology*, *38*, 1610–1620.

Bingham, T. C., Fisher, E. A., Parathath, S., Reiss, A. B., Chan, E. S., & Cronstein, B. N. (2010). A2A adenosine receptor stimulation decreases foam cell formation by enhancing ABCA1-dependent cholesterol efflux. *Journal of Leukocyte Biology*, *87*, 683–690.

Black, R. G., Guo, Y., Ge, Z. D., Murphree, S. S., Prabhu, S. D., Jones, W. K., et al. (2002). Gene dosage-dependent effects of cardiac-specific overexpression of the A_3 adenosine receptor. *Circulation Research*, *91*, 165–172.

Blackburn, M. R., Vance, C. O., Morschl, E., & Wilson, C. N. (2009). Adenosine receptors and inflammation. *Handbook of Experimental Pharmacology*, *193*, 215–269.

Boison, D. (2007). Adenosine as a neuromodulator in neurological diseases. *Current Opinion in Pharmacology*, *8*, 1–6.

Borsellino, G., Kleinewietfeld, M., Di Mitri, D., Sternjak, A., Diamantini, A., Giometto, R., et al. (2007). Expression of ectonucleotidase CD39 by Foxp3+ Treg cells: Hydrolysis of extracellular ATP and immune suppression. *Blood*, *110*, 1225–1232.

Brown, R. A., Clarke, G. W., Ledbetter, C. L., Hurle, M. J., Denyer, J. C., Simcock, D. E., et al. (2008a). Elevated expression of adenosine A_1 receptor in bronchial biopsies from asthmatic subjects. *European Respiratory Journal*, *31*, 1–9.

Brown, R. A., Spina, D., & Page, C. P. (2008b). Adenosine receptors and asthma. *British Journal of Pharmacology*, *153*, S446–S456.

Chan, E. S., & Cronstein, B. N. (2010). Methotrexate—How does it really work? *Nature Reviews. Rheumatology*, *6*, 175–178.

Chan, E. S., Montesinos, M. C., Fernandez, P., Desai, A., Delano, D. L., Yee, H., et al. (2006). Adenosine A_{2A} receptors play a role in the pathogenesis of hepatic cirrhosis. *British Journal of Pharmacology*, *148*, 1144–1155.

Chen, Y., Corriden, R., Inoue, Y., Yip, L., Hashiguchi, N., Zinkernagel, A., et al. (2006). ATP release guides neutrophil chemotaxis via P2Y2 and A_3 receptors. *Science*, *314*, 1792–1795.

Chen, S. W., Park, S. W., Kim, M., Brown, K. M., D'Agati, V. D., & Lee, H. T. (2009). Human heat shock protein 27 overexpressing mice are protected against hepatic ischemia and reperfusion injury. *Transplantation*, *87*, 1478–1487.

Ciruela, F., Albergaria, C., Soriano, A., Cuffí, L., Carbonell, L., Sánchez, S., et al. (2010). Adenosine receptors interacting proteins (ARIPs): Behind the biology of adenosine signaling. *Biochimica et Biophysica Acta*, *1798*, 9–20.

Cohen, M. V., Yang, X., & Downey, J. M. (2010). A_{2B} adenosine receptors can change their spots. *British Journal of Pharmacology*, *159*, 1595–1597.

Deaglio, S., Dwyer, K. M., Gao, W., Friedman, D., Usheva, A., Erat, A., et al. (2007). Adenosine generation catalyzed by CD39 and CD73 expressed on regulatory T cells mediates immune suppression. *The Journal of Experimental Medicine*, *204*, 1257–1265.

Dhalla, A. K., Melissa, S., Smith, M., Wong, M., Shryock, J. C., & Beraldinelli, L. (2007a). Antilipolytic activity of a novel partial A_1 adenosine receptor agonist devoid of cardiovascular effects: Comparison with nicotinic acid. *The Journal of Pharmacology and Experimental Therapeutics*, *321*, 327–333.

Dhalla, A. K., Wong, M. Y., Voshol, P. J., Belardinelli, L., & Reaven, G. M. (2007b). A_1-adenosine receptor partial agonist lowers plasma FFA and improves insulin resistance induced by high-fat diet in rodents. *American Journal of Physiology. Endocrinology and Metabolism*, *292*, E1358–E1363.

Driver, A. G., Kukoly, C. A., Ali, S., & Mustafa, S. J. (1993). Adenosine in bronchoalveolar lavage fluid in asthma. *The American Review of Respiratory Disease*, *148*, 91–97.

Eisenach, J. C., Hood, D. D., & Curry, R. (2002). Preliminary efficacy assessment of intrathecal injection of an American formulation of adenosine in humans. *Anesthesiology*, *96*, 29–34.

Eltzschig, H. K., Rivera-Nieves, J., & Colgan, S. P. (2009). Targeting the A_{2B} adenosine receptor during gastrointestinal ischemia and inflammation. *Expert Opinion on Therapeutic Targets*, *13*, 1267–1277.

Ernens, I., Léonard, F., Vausort, M., Rolland-Turner, M., Devaux, Y., & Wagner, D. R. (2010). Adenosine up-regulates vascular endothelial growth factor in human macrophages. *Biochemical and Biophysical Research Communications*, *392*, 351–356.

Ethier, M. F., & Madison, J. M. (2006). Adenosine A_1 receptors mediate mobilization of calcium in human bronchial smooth muscle cells. *American Journal of Respiratory Cell and Molecular Biology*, *35*, 496–502.

Fatholahi, M., Xiang, Y., Wu, Y., Li, Y., Wu, L., Dhalla, A. K., et al. (2006). A novel partial agonist of the A_1-adenosine receptor and evidence of receptor homogeneity in adipocytes. *The Journal of Pharmacology and Experimental Therapeutics*, *317*, 676–684.

Fedorova, I. M., Jacobson, M. A., Basile, Q. A., & Jacobson, K. A. (2003). Behavioral characterization of mice lacking the A_3 adenosine receptor: Sensitivity to hypoxic neurodegeneration. *Cellular and Molecular Neurobiology*, *23*, 431–447.

Feng, M. G., & Navar, L. G. (2010). Afferent arteriolar vasodilator effect of adenosine predominantly involves adenosine A_{2B} receptor activation. *American Journal of Physiology. Renal Physiology*, *299*, F310–F315.

Feoktistov, I., & Biaggioni, I. (1995). Adenosine A_{2B} receptors evoke interleukin-8 secretion in human mast cells. An enprofylline-sensitive mechanism with implications for asthma. *Journal of Clinical Investigation*, *96*, 1979–1986.

Feoktistov, I., Polosa, R., Volgate, S. T., & Biaggioni, I. (1998). Adenosine A_{2B} receptors: A novel therapeutic target in asthma? *Trends in Pharmacological Sciences*, *1998*(19), 148–153.

Fishman, P., Bar-Yehuda, S., Madi, L., Rath-Wolfson, L., Ochaion, A., Cohen, S., et al. (2006). The PI3K-NF-kappaB signal transduction pathway is involved in mediating the anti-inflammatory effect of IB-MECA in adjuvant-induced arthritis. *Arthritis Research & Therapy*, *8*, R33.

Fishman, P., Bar-Yehuda, S., Ohana, G., Barer, F., Ochaion, A., Erlanger, A., et al. (2004). An agonist to the A_3 adenosine receptor inhibits colon carcinoma growth in mice via modulation of GSK-3 beta and NF-kappa B. *Oncogene*, *23*, 2465–2471.

Fishman, P., Bar-Yehuda, S., Synowitz, M., Powell, J. D., Klotz, K. N., Gessi, S., et al. (2009). Adenosine receptors and cancer. *Handbook of Experimental Pharmacology*, *193*, 399–441.

Fredholm, B. B. (2010). Adenosine receptors as drug targets. *Experimental Cell Research*, *316*, 1284–1288.

Fredholm, B. B., IJzerman, A. P., Jacobson, K. A., Kloz, K. N., & Linden, J. (2001). International Union of Pharmacology. XXV. Nomenclature and classification of adenosine receptors. *Pharmacological Reviews*, *53*, 527–552.

Frick, J. S., MacManus, C. F., Scully, M., Glover, L. E., Eltzschig, H. K., & Colgan, S. P. (2009). Contribution of adenosine A_{2B} receptors to inflammatory parameters of experimental colitis. *Journal of Immunology*, *182*, 4957–4964.

Funakoshi, H., Chan, T. O., Good, J. C., Libonati, J. R., Piuhola, J., Chen, X., et al. (2006). Regulated overexpression of the A_1-adenosine receptor in mice results in adverse but reversible changes in cardiac morphology and function. *Circulation*, *114*, 2240–2250.

Gan, T. J., & Habib, A. S. (2007). Adenosine as a non-opioid analgesic in the perioperative setting. *Anesthesia and Analgesia*, *105*, 487–494.

Gazoni, L. M., Walters, D. M., Unger, E. B., Linden, J., Kron, I. L., & Laubach, V. E. (2010). Activation of A_1, A_{2A}, or A_3 adenosine receptors attenuates lung ischemia-reperfusion injury. *The Journal of Thoracic and Cardiovascular Surgery*, *140*, 440–446.

Ge, Z. D., Peart, J. N., Kreckler, L. M., Wan, T. C., Jacobson, M. A., Gross, G. J., et al. (2006). Cl-IB-MECA [2-chloro-N6-(3-iodobenzyl)adenosine-5′-N-methylcarboxamide] reduces ischemia/reperfusion injury in mice by activating the A_3 adenosine receptor. *Journal of Pharmacology and Experimental Therapeutics*, *319*, 1200–1210.

Gessi, S., Cattabriga, E., Avitabile, A., Gafa', R., Lanza, G., Cavazzini, L., et al. (2004). Elevated expression of A_3 adenosine receptors in human colorectal cancer is reflected in peripheral blood cells. *Clinical Cancer Research*, *10*, 5895–5901.

Gessi, S., Fogli, E., Sacchetto, V., Merighi, S., Varani, K., Preti, D., et al. (2010a). Adenosine modulates HIF-1alpha, VEGF, IL-8, and foam cell formation in a human model of hypoxic foam cells. *Arteriosclerosis, Thrombosis, and Vascular Biology*, *30*, 90–97.

Gessi, S., Merighi, S., Sacchetto, V., Simioni, C., & Borea, P. A. (2010b). Adenosine receptors and cancer. *Biochimica et Biophysica Acta*, in press.

Gessi, S., Merighi, S., Varani, K., Cattabriga, E., Benini, A., Mirandola, P., et al. (2007). Adenosine receptors in colon carcinoma tissues and colon tumoral cell lines: Focus on the A_3 adenosine subtype. *Journal of Cellular Physiology*, *211*, 826–836.

Gessi, S., Merighi, S., Varani, K., Leung, E., Mac Lennan, S., & Borea, P. A. (2008). The A_3 adenosine receptor: An enigmatic player in cell biology. *Pharmacology & Therapeutics*, *117*, 123–140.

Gessi, S., Sacchetto, V., Fogli, E., Merighi, S., Varani, K., & Baraldi, P. G. (2010c). Modulation of metalloproteinase-9 in U87MG glioblastoma cells by A_3 adenosine receptors. *Biochemical Pharmacology*, *79*, 1483–1495.

Gessi, S., Varani, K., Merighi, S., Cattabriga, E., Iannotta, V., Leung, E., et al. (2002). A_3 adenosine receptors in human neutrophils and promyelocytic HL60 cells: A pharmacological and biochemical study. *Molecular Pharmacology*, *61*, 415–424.

Gessi, S., Varani, K., Merighi, S., Ongini, E., & Borea, P. A. (2000). A_{2A} adenosine receptors in human peripheral blood cells. *British Journal of Pharmacology*, *129*, 2–11.

Gong, Q. J., Li, Y. Y., Xin, W. J., Wei, X. H., Cui, Y., Wang, J., et al. (2010). Differential effects of adenosine A_1 receptor on pain-related behavior in normal and nerve-injured rats. *Brain Research*, *1361*, 23–30.

Grenz, A., Zhang, H., Eckle, T., Mittelbronn, M., Wehrmann, M., Kohle, C., et al. (2007). Protective role of ecto-5′-nucleotidase (CD73) in renal ischemia. *Journal of the American Society of Nephrology*, *18*, 833–845.

Ham, M., Mizumori, M., Watanabe, C., Wang, J., Inoue, T., Nakano, T., et al. (2010). Endogenous luminal surface adenosine signaling regulates duodenal bicarbonate secretion in rats. *The Journal of Pharmacology and Experimental Therapeutics*, *335*, 607–613.

Haskó, G., Csóka, B., Németh, Z. H., Vizi, E. S., & Pacher, P. (2009). A_{2B} adenosine receptors in immunity and inflammation. *Trends in Immunology*, *30*, 263–270.

Haskó, G., Linden, J., Cronstein, B., & Pacher, P. (2008). Adenosine receptors: Therapeutic aspects for inflammatory and immune diseases. *Nature Reviews. Drug Discovery*, *7*, 759–770.

Haskó, G., & Pacher, P. (2008). A_{2A} receptors in inflammation and injury: Lessons learned from transgenic animals. *Journal of Leukocyte Biology*, *83*, 447–455.

Haskó, G., Pacher, P., Deitch, E. A., & Vizi, E. S. (2007). Shaping of monocyte and macrophage function by adenosine receptors. *Pharmacology & Therapeutics*, *113*, 264–275.

Haskó, G., Pacher, P., Vizi, E. S., & Illes, P. (2005). Adenosine receptor signaling in the brain immune system. *Trends in Pharmacological Sciences*, *26*, 511–516.

Himer, L., Csóka, B., Selmeczy, Z., Koscsó, B., Pócza, T., Pacher, P., et al. (2010). Adenosine A_{2A} receptor activation protects CD4+ T lymphocytes against activation-induced cell death. *The FASEB Journal*, *24*, 2631–2640.

Hocher, B. (2010). Adenosine A_1 receptor antagonists in clinical research and development. *Kidney International*, *78*, 438–445.

Hu, H., Lu, W., Zhang, M., Zhang, X., Argall, A. J., Patel, S., et al. (2010). Stimulation of the P2X7 receptor kills rat retinal ganglion cells in vivo. *Experimental Eye Research*, *91*, 425–432.

Hussain, A., Karjian, P., & Maddock, H. (2009). The role of nitric oxide in A_3 adenosine receptor-mediated cardioprotection. *Autonomic & Autacoid Pharmacology*, *29*, 97–104.

Inoue, Y., Tanaka, H., Sumi, Y., Woehrle, T., Chen, Y., Hirsh, M. I., et al. (2010). A3 adenosine receptor inhibition improves the efficacy of hypertonic saline resuscitation. *Shock*, *35*, 178–183.

Jacobson, K. A. (1998). Adenosine A_3 receptors: Novel ligands and paradoxical effects. *Trends in Pharmacological Sciences*, *19*, 184–191.

Jacobson, K. A., & Gao, Z. G. (2006). Adenosine receptors as therapeutic targets. *Nature Reviews*, *5*, 247–264.

Jajoo, S., Mukherjea, D., Kumar, S., Sheth, S., Kaur, T., Rybak, L. P., et al. (2010). Role of beta-arrestin1/ERK MAP kinase pathway in regulating adenosine A_1 receptor desensitization and recovery. *American Journal of Physiology. Cell Physiology*, *298*, C56–C65.

Jajoo, S., Mukherjea, D., Watabe, K., & Ramkumar, V. (2009). Adenosine A_3 receptor suppresses prostate cancer metastasis by inhibiting NADPH oxidase activity. *Neoplasia*, *11*, 1132–1145.

Jenner, P., Mori, A., Hauser, R., Morelli, M., Fredholm, B. B., & Chen, J. F. (2009). Adenosine, adenosine A_{2A} antagonists, and Parkinson's disease. *Parkinsonism & Related Disorders*, *15*, 406–413.

Kälin, R. E., Bänziger-Tobler, N. E., Detmar, M., & Brändli, A. W. (2009). An in vivo chemical library screen in Xenopus tadpoles reveals novel pathways involved in angiogenesis and lymphangiogenesis. *Blood*, *114*, 1110–1122.

Kara, F. M., Chitu, V., Sloane, J., Axelrod, M., Fredholm, B. B., Stanley, E. R., et al. (2010a). Adenosine A_1 receptors (A_1Rs) play a critical role in osteoclast formation and function. *The FASEB Journal*, *24*, 2325–2333.

Kara, F. M., Doty, S. B., Boskey, A., Goldring, S., Zaidi, M., Fredholm, B. B., et al. (2010b). Adenosine A(1) receptors regulate bone resorption in mice: Adenosine A(1) receptor blockade or deletion increases bone density and prevents ovariectomy-induced bone loss in adenosine A(1) receptor-knockout mice. *Arthritis and Rheumatism*, *62*, 534–541.

Karjian, P., Al-Rajaibi, H., Hussain, A., & Maddock, H. L. (2006). Activation of A_3 adenosine receptors protects the ischemic reperfused myocardium via recruitment of PI3K-AKT-iNOS cell survival pathway. *Circulation*, *114*, II_1204.

Karjian, P., Al-Rajaibi, H., Hussain, A., & Maddock, H. L. (2008). A_3 adenosine receptor activation via 2-Cl-IBMECA protects the myocardium via recruitment of PI3K-AKT-iNOS intracellular signalling pathway during reperfusion. *Journal of Molecular and Cellular Cardiology*, *44*, 738.

Kern, M. J., Hodgson, J. M., Dib, N., Mittleman, R. S., & Crane, P. D. (2006). Effects of apadenoson, a selective adenosine A_{2A} receptor agonist for myocardial perfusion imaging, on coronary blood flow velocity in conscious patients. *Circulation*, *114*, 2780.

Kiesman, W. F., Elzein, E., & Zablocki, J. (2009). A_1 adenosine receptor antagonists, agonists, and allosteric enhancers. In C. N. Wilson & S. J. Mustafà (Eds.), *Adenosine receptors in health and disease* (pp. 25–58). *Handbook of experimental pharmacology* (Vol. 193, pp. 25–58). Berlin, Heidelberg: Springer-Verlag.

Kim, M., Chen, S. W., Park, S. W., Kim, M., D'Agati, V. D., Yang, J., et al. (2009). Kidney-specific reconstitution of the A_1 adenosine receptor in A_1 adenosine receptor knockout mice reduces renal ischemia-reperfusion injury. *Kidney International*, *75*, 809–823.

Kim, H., Kang, J. W., Lee, S., Choi, W. J., Jeong, L. S., Yang, Y., et al. (2010). A_3 adenosine receptor antagonist, truncated Thio-Cl-IB-MECA, induces apoptosis in T24 human bladder cancer cells. *Anticancer Research*, *30*, 2823–2830.

Koda, K., Salazar-Rodriguez, M., Corti, F., Chan, N. Y., Estephan, R., Silver, R. B., et al. (2010). Aldehyde dehydrogenase activation prevents reperfusion arrhythmias by inhibiting local renin release from cardiac mast cells. *Circulation*, *122*, 771–781.

Koeppen, M., Eckle, T., & Eltzschig, H. K. (2009). Selective deletion of the A_1 adenosine receptor abolishes heart-rate slowing effects of intravascular adenosine in vivo. *PLoS One*, *4*, e6784.

Koizumi, S., Odashima, M., Otaka, M., Jin, M., Linden, J., Watanabe, S., et al. (2009). Attenuation of gastric mucosal inflammation induced by indomethacin through activation of the A_{2A} adenosine receptor in rats. *Journal of Gastroenterology*, *44*, 419–425.

Kolachala, V. L., Vijay-Kumar, M., Dalmasso, G., Yang, D., Linden, J., Wang, L., et al. (2008). A_{2B} adenosine receptor gene deletion attenuates murine colitis. *Gastroenterology*, *135*, 861–870.

Kolachala, V. L., Wang, L., Obertone, T. S., Prasad, M., Yan, Y., Dalmasso, G., et al. (2010). Adenosine A_{2B} receptor expression is post-transcriptionally regulated by microRNA. *The Journal of Biological Chemistry*, *285*, 18184–18190.

Kong, T., Westerman, K. A., Faigle, M., Eltzschig, H. K., & Colgan, S. P. (2006). HIF-dependent induction of adenosine A_{2B} receptor in hypoxia. *The FASEB Journal*, *20*, 2242–2250.

Kreckler, L. M., Wan, T. C., Ge, Z. D., & Auchampach, J. A. (2006). Adenosine inhibits tumor necrosis factor-α release from mouse peritoneal macrophages via A_{2A} and A_{2B} but not the A_3 adenosine receptor. *The Journal of Pharmacology and Experimental Therapeutics*, *317*, 172–180.

Kuno, A., Critz, S. D., Cui, L., Solodushko, V., Yang, X. M., Krahn, T., et al. (2007). Protein kinase C protects preconditioned rabbit hearts by increasing sensitivity of adenosine A_{2B}-dependent signaling during early reperfusion. *Journal of Molecular and Cellular Cardiology*, *43*, 262–271.

Kuno, A., Solenkova, N. V., Solodushko, V., Dost, T., Liu, Y., Yang, X. M., et al. (2008). Infarct limitation by a protein kinase G activator at reperfusion in rabbit hearts is dependent on sensitizing the heart to A_{2B} agonists by protein kinase C. *American Journal of Physiology. Heart and Circulatory Physiology*, *295*, H1288–H1295.

Lee, J. Y., Jhun, B. S., Oh, Y. T., Lee, J. H., Choe, W., Baik, H. H., et al. (2006a). Activation of adenosine A_3 receptor suppresses lipopolysaccaride-induced TNF-α production through inhibition of PI-3-kinase/Akt and NF-κB activation in murine BV2 microglial cells. *Neuroscience Letters*, *396*, 1–6.

Lee, H. T., Kim, M., Joo, J. D., Gallos, G., Chen, J. F., & Emala, C. W. (2006b). A_3 adenosine receptor activation decreases mortality and renal and hepatic injury in murine septic peritonitis. *American Journal of Physiology: Regulatory, Integrative and Comparative Physiology*, *291*, R959–R969.

LeWitt, P. A., Guttman, M., Tetrud, J. W., Tuite, P. J., Mori, A., Chaikin, P., et al. (2008). Adenosine A_{2A} receptor antagonist istradefylline (KW-6002) reduces "off" time in Parkinson's disease: A double-blind, randomized, multicenter clinical trial (6002-US-005). *Annals of Neurology*, *63*, 295–302.

Linden, J. (2001). Molecular approach to adenosine receptors: Receptor-mediated mechanisms of tissue protection. *Annual Review of Pharmacology and Toxicology*, *41*, 775–787.

Linden, J. (2005). Adenosine in tissue protection and tissue regeneration. *Molecular Pharmacology*, *67*, 1385–1387.

Linden, J. (2006). Purinergic chemotaxis. *Science*, *314*, 1689–1690.

Madi, L., Cohen, S., Ochaion, A., Bar-Yehuda, S., Barer, F., & Fishman, P. (2007). Overexpression of A_3 adenosine receptor in peripheral blood mononuclear cells in rheumatoid arthritis: Involvement of nuclear factor-kappaB in mediating receptor level. *The Journal of Rheumatology*, *34*, 20–26.

Martin, L., Pingle, S. C., Hallam, D. M., Rybak, L. P., & Ramkumar, V. (2006). Activation of the adenosine A_3 receptor in RAW 264.7 cells inhibits lipopolysaccharide-stimulated tumor necrosis factor-α release by reducing calcium-dependent activation of nuclear factor-kB and extracellular signal-regulated kinase 1/2. *The Journal of Pharmacology and Experimental Therapeutics*, *316*, 71–78.

Matot, I., Weiniger, C. F., Zeira, E., Galun, E., Joshi, B. V., & Jacobson, K. A. (2006). A_3 adenosine receptors and mitogen-activated protein kinases in lung injury following in vivo reperfusion. *Critical Care*, *10*, R65.

McColl, S. R., St-Onge, M., Dussault, A. A., Laflamme, C., Bouchard, L., Boulanger, J., et al. (2006). Immunomodulatory impact of the A_{2A} adenosine receptor on the profile of chemokines produced by neutrophils. *The FASEB Journal*, *20*, 187–189.

Merighi, S., Benini, A., Mirandola, P., Gessi, S., Varani, K., Leung, E., et al. (2005). A_3 adenosine receptor activation inhibits cell proliferation via phosphatidylinositol 3-kinase/Akt-dependent inhibition of the extracellular signal-regulated kinase 1/2 phosphorylation in A375 human melanoma cells. *The Journal of Biological Chemistry*, *280*, 19516–19526.

Merighi, S., Benini, A., Mirandola, P., Gessi, S., Varani, K., Leung, E., et al. (2006). Adenosine modulates vascular endothelial growth factor expression via hypoxia-inducible factor-1 in human glioblastoma cells. *Biochemical Pharmacology*, *72*, 19–31.

Merighi, S., Benini, A., Mirandola, P., Gessi, S., Varani, K., Leung, E., et al. (2007a). Hypoxia inhibits paclitaxel-induced apoptosis through adenosine-mediated phosphorylation of bad in glioblastoma cells. *Molecular Pharmacology*, *72*, 162–172.

Merighi, S., Benini, A., Mirandola, P., Gessi, S., Varani, K., Simioni, C., et al. (2007b). Caffeine inhibits adenosine-induced accumulation of hypoxia-inducible factor-1alpha, vascular endothelial growth factor, and interleukin-8 expression in hypoxic human colon cancer cells. *Molecular Pharmacology*, *72*, 395–406.

Merighi, S., Mirandola, P., Varani, K., Gessi, S., Leung, E., Baraldi, P. G., et al. (2003). A glance at adenosine receptors: Novel target for antitumor therapy. *Pharmacology & Therapeutics*, *100*, 31–48.

Merighi, S., Simioni, C., Gessi, S., Varani, K., Mirandola, P., Tabrizi, M. A., et al. (2009). A_{2B} and A_3 adenosine receptors modulate vascular endothelial growth factor and interleukin-8 expression in human melanoma cells treated with etoposide and doxorubicin. *Neoplasia*, *11*, 1064–1073.

Methner, C., Schmidt, K., Cohen, M. V., Downey, J. M., & Krieg, T. (2010). Both A_{2A} and A_{2B} adenosine receptors at reperfusion are necessary to reduce infarct size in mouse hearts. *American Journal of Physiology. Heart and Circulatory Physiology*, *299*, H1262–H1264.

Mohsenin, A., Mi, T., Xia, Y., Kellems, R. E., Chen, J. F., & Blackburn, M. R. (2007). Genetic removal of the A_{2A} adenosine receptor enhances pulmonary inflammation, mucin production, and angiogenesis in adenosine deaminase-deficient mice. *American Journal of Physiology. Lung Cellular and Molecular Physiology*, *293*, 753–761.

Montesinos, M. C., Desai, A., & Cronstein, B. N. (2006). Suppression of inflammation by low-dose methotrexate is mediated by adenosine A_{2A} receptor but not A_3 receptor activation in thioglycollate-induced peritonitis. *Arthritis Research & Therapy*, *8*, R53.

Morschl, E., Molina, J. G., Volmer, J. B., Mohsenin, A., Pero, R. S., Hong, J. S., et al. (2008). A_3 adenosine receptor signaling influences pulmonary inflammation and fibrosis. *American Journal of Respiratory Cell and Molecular Biology*, *39*, 697–705.

Naganuma, M., Wiznerowicz, E. B., Lappas, C. M., Linden, J., Worthington, M. T., & Ernst, P. B. (2006). Cutting edge: Critical role for A_{2A} adenosine receptors in the T cell-mediated regulation of colitis. *Journal of Immunology*, *177*, 2765–2769.

Nascimento, F. P., Figueredo, S. M., Marcon, R., Martins, D. F., Macedo, S. J., Jr., Lima, D. A., et al. (2010). Inosine reduces pain-related behavior in mice: Involvement of adenosine A_1 and A_{2A} receptor subtypes and protein kinase C pathways. *The Journal of Pharmacology and Experimental Therapeutics*, *334*, 590–598.

Ngamsri, K. C., Wagner, R., Vollmer, I., Stark, S., & Reutershan, J. (2010). Adenosine receptor A_1 regulates polymorphonuclear cell trafficking and microvascular permeability in lipopolysaccharide-induced lung injury. *Journal of Immunology*, *185*, 4374–4384.

Novitskiy, S. V., Ryzhov, S., Zaynagetdinov, R., Goldstein, A. E., Huang, Y., Tikhomirov, O. Y., et al. (2008). Adenosine receptors in regulation of dendritic cell differentiation and function. *Blood*, *112*, 1822–1831.

Ohta, A., & Sitkovsky, M. (2009). The adenosinergic immunomodulatory drugs. *Current Opinion in Pharmacology*, *9*, 501–506.

Philipp, S., Yang, X. M., Cui, L., Davis, A. M., Downey, J. M., & Cohen, M. V. (2006). Postconditioning protects rabbit hearts through a protein kinase C-adenosine A_{2B} receptor cascade. *Cardiovascular Research*, *70*, 308–314.

Polosa, R., & Blackburn, M. R. (2009). Adenosine receptors as targets for therapeutic intervention in asthma and chronic obstructive pulmonary disease. *Trends in Pharmacological Sciences*, *30*, 528–535.

Polosa, R., Rorke, S., & Volgate, S. T. (2002). Evolving concepts on the value of adenosine hyperresponsiveness in asthma and chronic obstructive pulmonary disease. *Thorax*, *57*, 649–654.

Ponnoth, D. S., Nadeem, A., Tilley, S., & Mustafa, S. J. (2010). Involvement of A_1 adenosine receptors in altered vascular responses and inflammation in an allergic mouse model of asthma. *American Journal of Physiology. Heart and Circulatory Physiology*, *299*, H81–H87.

Press, N. J., Gessi, S., Borea, P. A., & Polosa, R. (2007). Therapeutic potential of adenosine receptor antagonists and agonists. *Expert Opinion on Therapeutic Patents*, *17*, 979–991.

Rajagopal, M., & Pao, A. C. (2010). Adenosine activates A_{2B} receptors and enhances chloride secretion in kidney inner medullary collecting duct cells. *Hypertension*, *55*, 1123–1128.

Ralevic, V., & Burnstock, G. (1998). Receptors for purines and pyrimidines. *Pharmacological Reviews*, *50*, 413–492.

Raman, M., Chen, W., & Cobb, M. H. (2007). Differential regulation and properties of MAPKs. *Oncogene*, *26*, 3100–3112.

Romagnoli, R., Baraldi, P. G., Tabrizi, M. A., Gessi, S., Borea, P. A., & Merighi, S. (2010). Allosteric enhancers of A_1 adenosine receptors: State of the art and new horizons for drug development. *Current Medicinal Chemistry*, *17*, 3488–3502.

Russo, C., Arcidiacono, G., & Polosa, R. (2006). Adenosine receptors: Promising targets for the development of novel therapeutics and diagnostics for asthma. *Fundamental & Clinical Pharmacology*, *20*, 9–19.

Sajjadi, F. G., Takabayashi, K., Foster, A. C., Domingo, R. C., & Firestein, G. S. (1996). Inhibition of TNF-α expression by adenosine. Role of A_3 receptors. *Journal of Immunology*, *156*, 3435–3442.

Salvatore, C. A., Tilley, S. L., Latour, A. M., Fletcher, D. S., Koller, B. H., & Jacobson, M. A. (2000). Disruption of the A_3 adenosine receptor gene in mice and its effect on stimulated inflammatory cells. *The Journal of Biological Chemistry*, *275*, 4429–4434.

Schingnitz, U., Hartmann, K., Macmanus, C. F., Eckle, T., Zug, S., Colgan, S. P., et al. (2010). Signaling through the A_{2B} adenosine receptor dampens endotoxin-induced acute lung injury. *Journal of Immunology*, *184*, 5271–5279.

Schmidt, A. P., Böhmer, A. E., Antunes, C., Schallenberger, C., Porciúncula, L. O., Elisabetsky, E., et al. (2009). Anti-nociceptive properties of the xanthine oxidase inhibitor allopurinol in mice: Role of A_1 adenosine receptors. *British Journal of Pharmacology*, *156*, 163–172.

Schnermann, J., & Briggs, J. P. (2008). Tubuloglomerular feedback: Mechanistic insights from gene-manipulated mice. *Kidney International*, *74*, 418–426.

Schutte, F., Burgdorf, C., Richardt, G., & Kurz, T. (2006). Adenosine A_1 receptor-mediated inhibition of myocardial norepinephrine release involves neither phospholipase C nor protein kinase C but does involve adenylyl cyclase. *Canadian Journal of Physiology and Pharmacology*, *84*, 573–577.

Sevigny, C. P., Li, L., Awad, A. S., Huang, L., McDuffie, M., Linden, J., et al. (2007). Activation of adenosine A_{2A} receptors attenuates allograft rejection and alloantigen recognition. *Journal of Immunology*, *178*, 4240–4249.

Shearer, J., Severson, D. L., Su, L., Belardinelli, L., & Dhalla, A. K. (2009). Partial A_1 adenosine receptor agonist regulates cardiac substrate utilization in insulin-resistant rats in vivo. *The Journal of Pharmacology and Experimental Therapeutics*, *328*, 306–311.

Shneyvays, V., Leshem, D., Zinman, T., Mamedova, L. K., Jacobson, K. A., & Shainberg, A. (2005). Role of adenosine A_1 and A_3 receptors in regulation of cardiomyocyte homeostasis after mitochondrial respiratory chain injury. *American Journal of Physiology. Heart and Circulatory Physiology*, *288*, H2792–H2801.

Simola, N., Morelli, M., & Pinna, A. (2008). Adenosine A_{2A} receptor antagonists and Parkinson's disease: State of the art and future directions. *Current Pharmaceutical Design*, *14*, 1475–1789.

Sitkovsky, M. V. (2009). T regulatory cells: Hypoxia-adenosinergic suppression and re-direction of the immune response. *Trends in Immunology*, *30*, 102–108.

Sitkovsky, M. V., Lukashev, D., Apasov, S., Kojima, H., Koshiba, M., Caldwell, C., et al. (2004). Physiological control of immune response and inflammatory tissue damage by hypoxia-inducible factors and adenosine A_{2A} receptors. *Annual Review of Immunology*, *22*, 657–682.

Slawsky, M. T., & Givertz, M. M. (2009). Rolofylline: A selective adenosine A_1 receptor antagonist for the treatment of heart failure. *Expert Opinion on Pharmacotherapy*, *10*, 311–322.

Solenkova, N. V., Solodushko, V., Cohen, M. V., & Downey, J. M. (2006). Endogenous adenosine protects preconditioned heart during early minutes of reperfusion by activating Akt. *American Journal of Physiology. Heart and Circulatory Physiology*, *290*, H441–H449.

Sowa, N. A., Street, S. E., Vihko, P., & Zylka, M. J. (2010). Prostatic acid phosphatase reduces thermal sensitivity and chronic pain sensitization by depleting phosphatidylinositol 4,5-bisphosphate. *The Journal of Neuroscience*, *30*, 10282–10293.

Strohmeier, G. R., Reppert, S. M., Lencer, W. I., & Madara, J. L. (1995). The A_{2B} adenosine receptor mediates cAMP responses to adenosine receptor agonists in human intestinal epithelia. *The Journal of Biological Chemistry*, *270*, 2387–2394.

Sun, C. N., Cheng, H. C., Chou, J. L., Lee, S. Y., Lin, Y. W., Lai, H. L., et al. (2006a). Rescue of p53 blockage by the A_{2A} adenosine receptor via a novel interacting protein, translin-associated protein X. *Molecular Pharmacology*, *70*, 454–466.

Sun, D., Samuelson, L. C., Yang, T., Huang, Y., Paliege, A., Saunders, T., et al. (2001). Mediation of tubuloglomerular feedback by adenosine: Evidence from mice lacking adenosine 1 receptors. *Proceedings of the National Academy of Sciences of the United States of America*, *98*, 9983–9988.

Sun, C. X., Young, H. W., Molina, J. G., Volmer, J. B., Schnermann, J., & Blackburn, M. R. (2005). A protective role for the A1 adenosine receptor in adenosine-dependent pulmonary injury. *Journal of Clinical Investigation*, *115*, 35–43.

Sun, J., Zhang, Y., Yang, M., Zhang, Y., Xie, Q., Li, Z., et al. (2010). Hypoxia induces T-cell apoptosis by inhibiting chemokine C receptor 7 expression: The role of adenosine receptor A_2. *Cellular & Molecular Immunology*, *7*, 77–82.

Sun, C. X., Zhong, H., Mohsenin, A., Morschi, E., Chung, J. L., Molina, J. G., et al. (2006b). Role of A_{2B} adenosine receptor signaling in adenosine-dependent pulmonary inflammation and injury. *Journal of Clinical Investigation*, *116*, 2173–2182.

Taliani, S., La Motta, C., Mugnaini, L., Simorini, F., Salerno, S., Marini, A. M., et al. (2010). Novel N2-substituted pyrazolo[3,4-d]pyrimidine adenosine A3 receptor antagonists: Inhibition of A3-mediated human glioblastoma cell proliferation. *Journal of Medicinal Chemistry*, *53*, 3954–3963.

Tarditi, A., Camurri, A., Varani, K., Borea, P. A., Woodman, B., Bates, G., et al. (2006). Early and transient alteration of adenosine A_{2A} receptor signalling in a mouse model of Huntington disease. *Neurobiology of Disease*, *23*, 44–53.

Vallon, V., Mühlbauer, B., & Osswald, H. (2006). Adenosine and kidney function. *Physiological Reviews*, *86*, 901–940.

Van Calker, D. M., Muller, M., & Hamprecht, B. (1979). Adenosine regulates via two different types of receptors, the accumulation of cyclic AMP in cultured brain cells. *Journal of Neurochemistry*, *33*, 999–1005.

Varani, K., Caramori, G., Vincenzi, F., Adcock, I., Casolari, P., Leung, E., et al. (2006). Alteration of adenosine receptors in patients with chronic obstructive pulmonary disease. *American Journal of Respiratory and Critical Care Medicine*, *173*, 398–406.

Varani, K., Laghi-Pasini, F., Camurri, A., Capecchi, P. L., Maccherini, M., Diciolla, F., et al. (2003). Changes of peripheral A_{2A} adenosine receptors in chronic heart failure and cardiac transplantation. *The FASEB Journal*, *17*, 280–282.

Varani, K., Maniero, S., Vincenzi, F., Targa, M., Stefanelli, A., Maniscalco, P., et al. (2010a). A3 receptors are overexpressed in pleura from mesothelioma patients and reduce cell growth via Akt/NF-kB pathway. *American Journal of Respiratory and Critical Care Medicine*, *183*, 522–530.

Varani, K., Massara, A., Vincenzi, F., Tosi, A., Padovan, M., Trotta, F., et al. (2009). Normalization of A_{2A} and A_3 adenosine receptor up-regulation in rheumatoid arthritis patients by treatment with anti-tumor necrosis factor alpha but not methotrexate. *Arthritis and Rheumatism*, *60*, 2880–2891.

Varani, K., Vincenzi, F., Tosi, A., Gessi, S., Casetta, I., Granieri, G., et al. (2010b). A_{2A} adenosine receptor overexpression and functionality, as well as TNF-alpha levels, correlate with motor symptoms in Parkinson's disease. *The FASEB Journal*, *24*, 587–598.

Varani, K., Vincenzi, F., Tosi, A., Targa, M., Masieri, F. F., Ongaro, A., et al. (2010c). Expression and functional role of adenosine receptors in regulating inflammatory responses in human synoviocytes. *British Journal of Pharmacology*, *160*, 101–115.

Wakeno, M., Minamino, T., Seguchi, O., Okazaki, H., Tsukamoto, O., Okada, K., et al. (2006). Long-term stimulation of adenosine A_{2B} receptors begun after myocardial infarction prevents cardiac remodeling in rats. *Circulation*, *114*, 1923–1932.

Wallace, K. L., & Linden, J. (2010). Adenosine A_{2A} receptors induced on iNKT and NK cells reduce pulmonary inflammation and injury in mice with sickle cell disease. *Blood*, *116*, 5010–5020.

Wang, Z., Do, C. W., Avila, M. Y., Peterson-Yantorno, K., Stone, R. A., Gao, Z. G., et al. (2010). Nucleoside-derived antagonists to A_3 adenosine receptors lower mouse intraocular pressure and act across species. *Experimental Eye Research*, *90*, 146–154.

Weatherley, B. D., Cotter, G., Dittrich, H. C., DeLucca, P., Mansoor, G. A., Bloomfield, D. M., et al. (2010). PROTECT Steering Committee, Investigators, and Coordinators. Design and rationale of the PROTECT study: A placebo-controlled randomized study of the selective A_1 adenosine receptor antagonist rolofylline for patients hospitalized with acute decompensated heart failure and volume overload to assess treatment effect on congestion and renal function. *Journal of Cardiac Failure*, *16*, 25–35.

Wilson, C. N. (2008). Adenosine receptors and asthma in humans. *British Journal of Pharmacology*, *155*, 475–486.

Yang, H., Avila, M. Y., Peterson-Yantorno, K., Coca-Prados, M., Stone, R. A., Jacobson, K. A., & Civan, M. M. (2005). The cross-species A_3 adenosine-receptor antagonist MRS 1292 inhibits adenosine-triggered human nonpigmented ciliary epithelial cell fluid release and reduces mouse intraocular pressure. *Current Eye Research*, *30*, 747–754.

Yang, D., Chen, H., Koupenova, M., Carroll, S. H., Eliades, A., Freedman, J. E., et al. (2010a). A new role for the A_{2B} adenosine receptor in regulating platelet function. *Journal of Thrombosis and Haemostasis*, *8*, 817–827.

Yang, M., Ma, C., Liu, S., Shao, Q., Gao, W., Song, B., et al. (2010b). HIF-dependent induction of adenosine receptor A_{2B} skews human dendritic cells to a Th2-stimulating phenotype under hypoxia. *Immunology and Cell Biology*, *88*, 165–171.

Yang, D., Zhang, Y., Nguyen, H. G., Koupenova, M., Chauhan, A. K., Makitalo, M., et al. (2006). The A_{2B} adenosine receptor protects against inflammation and excessive vascular adhesion. *Journal of Clinical Investigation*, *116*, 1913–1923.

Yldiz, G., Demiryurek, A. T., Gumusel, B., & Lippton, H. (2007). Ischemic preconditioning modulates ischemia-reperfusion injury in the rat lung: Role of adenosine receptors. *European Journal of Pharmacology*, *556*, 144–150.

Zhang, M., Budak, M. T., Lu, W., Khurana, T. S., Zhang, X., Laties, A. M., et al. (2006a). Identification of the A_3 adenosine receptor in rat retinal ganglion cells. *Molecular Vision*, *12*, 937–948.

Zhang, M., Hu, H., Zhang, X., Lu, W., Lim, J., Eysteinsson, T., et al. (2010). The A_3 adenosine receptor attenuates the calcium rise triggered by NMDA receptors in retinal ganglion cells. *Neurochemistry International*, *56*, 35–41.

Zhang, X., Zhang, M., Laties, A. M., & Mitchell, C. H. (2006b). Balance of purines may determine life or death of retinal ganglion cells as A_3 adenosine receptors prevent loss following P2X7 receptor stimulation. *Journal of Neurochemistry*, *98*, 566–575.

Zhou, R., Chen, F., Li, Q., Hu, D. Y., & Liu, L. M. (2010). Stimulation of the adenosine A_3 receptor reverses vascular hyporeactivity after hemorrhagic shock in rats. *Acta Pharmacologica Sinica*, *31*, 413–420.

Zhou, X., & Kost, C. K. (2006). Adenosine A_1 receptor antagonist blunts urinary potassium excretion, but not renal hemodynamic effects, induced by carbonic anhydrase inhibitor in rats. *The Journal of Pharmacology and Experimental Therapeutics*, *316*, 530–538.

Bertil B. Fredholm, Stina Johansson, and Ying-Qing Wang[1]

Department of Physiology and Pharmacology, Karolinska Institutet, Stockholm, Sweden
[1]Present address: College of Pharmacy, Jilin University, Changchun city, China

Adenosine and the Regulation of Metabolism and Body Temperature

Abstract

Adenosine levels are increased under conditions of energy deprivation, both because intracellular energy stores are reduced and because ATP is released. The adenosine thus formed can serve to influence energy homeostasis in a number of different ways, besides alterations in blood supply and cellular work (including contraction, maintenance of membrane potential, and biosynthesis), which will be covered in other chapters. Here, effects on energy homeostasis will be briefly reviewed. Adenosine acting at the A_1 receptor is a powerful and nonredundant inhibitor of lipolysis. It increases glucose uptake in fat and muscle, but its effects on insulin secretion may be even more important than the actions at insulin target tissues. Glucagon is also influenced. In addition to these peripheral actions, adenosine acts in the brain to regulate sleep–wakefulness, food intake, and body temperature. These effects are both direct at the relevant neurons and indirect by influences on regulatory transmitters and hormones.

I. Introduction. The Concept of Adenosine as an Emergency Signal

Since the early days of adenosine research, the role of this endogenous purine nucleoside as a signal of relative energy deficiency has been highlighted (for early references, see Arch & Newsholme, 1978). Thus, adenosine

Advances in Pharmacology, Volume 61

1054-3589/11 $35.00
10.1016/B978-0-12-385526-8.00003-5

formation intracellularly increases when there is a discrepancy between the rate of ATP synthesis and the rate of ATP utilization. ATP levels tend to decrease, and adenosine levels tend to increase manyfold more, when cellular work increases or when the supply of, for example, oxygen and glucose decreases. Intracellularly formed adenosine would then, by means of equilibrative transporters, exit the affected cell(s) and act on receptors in the local environment generally producing changes that tend to reduce the energetic imbalance. Oxygen supply would, for example, be increased by increases in the rate of respiration (Barraco & Janusz, 1989; Monteiro & Ribeiro, 1987) and in the local blood supply (Berne, 1963); excessive work could be shut down, for example, by reducing the release of excitatory neurotransmitters (Dunwiddie et al., 1981; Hedqvist & Fredholm, 1976).

Recently, research focus may have shifted from intracellular adenosine formation to the generation of extracellular adenosine from released adenine nucleotides. Even though some of the processes underlying release of ATP must be considered physiological rather than pathophysiological, there clearly are instances where release of ATP can be regarded as a distress signal. The most obvious is of course necrotic cell death and membrane rupture when a substantial part of the cellular adenine nucleotides is released into the medium causing very substantial increases in adenosine levels. It is important to remember that many preparations used to examine physiological functions in fact harbor some cells that are severely damaged. The brain slice is one obvious example, but also the smooth muscle preparations used in classical pharmacology and most cell cultures possess a small number of dead or damaged cells. Recently, it has been shown that *acidosis* can provide a potent stimulus for ATP release, both in the brain (Gourine et al., 2010) and in other organs such as the skeletal muscle vasculature (Tu et al., 2010). Cellular *swelling* is a powerful stimulus on, for example, epithelia (Musante et al., 1999) and astrocytes (Darby et al., 2003), but probably most cells can respond with ATP release to hypotonic swelling. The precise mechanisms involved are unclear, and possibly variable between cells.

There is also evidence that not only adenosine but also ATP can be released from tissues subjected to hypoxia (Erlichman et al., 2010; Vial et al., 1987). Part of this is no doubt due to ATP release from erythrocytes due to a combination of hypoxia and deformation (Faris & Spence, 2008). Hypoxia-induced release of ATP is also seen in oxygen-sensing organs (Buttigieg & Nurse, 2004; Gourine et al., 2005). ATP can be released from immune competent cells (Eltzschig et al., 2006), and given that hypoxia and reperfusion often leads to infiltration of immune cells, this provides another link.

The above results and considerations, coupled with the fact that adenosine receptors are widely expressed, suggest that adenosine could be one of the factors that adjust metabolism to the needs of the organism. This chapter highlights some results relevant to this idea. I will briefly examine the consequences of the well-known role of adenosine in regulating adipose tissue and also some of the effects on glucose metabolism. I will finally discuss the possibility that adenosine could be a factor in the central regulation of metabolism, focusing especially on its role in temperature regulation.

II. Adenosine and the Regulation of Lipid Metabolism

This section deals almost exclusively with adenosine and adipose tissue. The very important effects in the liver are not discussed here. The topic of adenosine in adipose tissue and obesity has been dealt with in a recent authoritative review (Dhalla et al., 2009), and the reader is referred there for more extensive coverage.

The insulin-like effect of adenosine on adipose tissue was discovered by Dole (1962). The isolated fat cell preparation devised by Martin Rodbell was probably the first isolated cell type where important functions of adenosine could be conclusively demonstrated (Fain et al., 1972). Further, work on fat cells was of critical importance in defining the receptor subtype we now denote A_1 (Londos et al., 1980). Adenosine A_1 receptors are abundant in adipose tissue, and they are known to act by a pertussis toxin-sensitive mechanism to reduce activity of adenylyl cyclase and thereby cAMP, which in turn reduces activity of the hormone-sensitive lipase/adipose triglyceride lipase(s) (Dhalla et al., 2009).

The critically important role of A_1 receptors was recently shown using the A_1 receptor knockout mouse (Johansson et al., 2007b, 2008). As shown in Fig. 1, the nonselective adenosine receptor agonist 2-chloro-adenosine caused a dose-dependent inhibition of cAMP formation in fat cells prepared from wild-type mice, but was essentially ineffective in fat cells prepared from A_1R knockout mice (Johansson et al., 2008). Interestingly, the dose–response curve in fat cells prepared from hemizygous mice was shifted significantly to the right, with no effect on the maximal response. This is compatible with the presence of a "receptor reserve" in fat cells, and for this, there is independent good support (Liang et al., 2002). The antilipolytic effect of adenosine derivatives was also eliminated in the A_1R knockout mouse adipocytes (Johansson et al., 2007b).

The antilipolytic action of adenosine is not only of pharmacological interest, but may play a physiological role. It was shown early (Fain & Wieser, 1975; Hjemdahl & Fredholm, 1976), and has been confirmed repeatedly, that removal of adenosine from the medium by adenosine deaminase or antagonism of receptors by low doses of methylxanthines causes increased lipolysis (e.g., Fig. 2). Even though this may partly be an *in vitro* artifact (adenosine does not accumulate so readily *in vivo*, and generation of adenosine from cell lysis is a typical *in vitro* phenomenon), this does suggest that adenosine is an endogenous regulator of lipid metabolism. As seen in

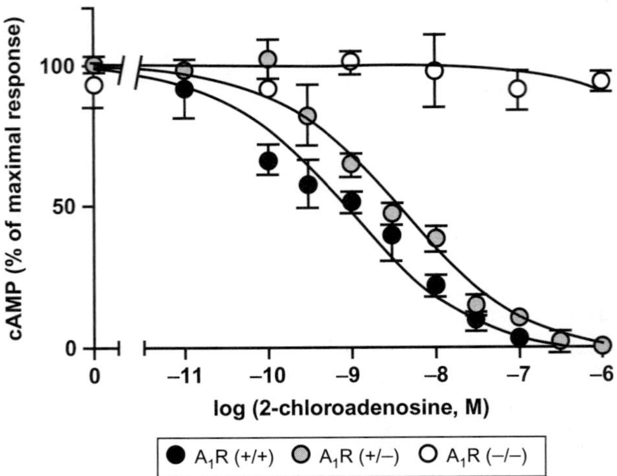

FIGURE 1 Adenosine A_1 receptor-mediated inhibition of NA-stimulated cAMP accumulation in mouse fat cells. The effect of increasing concentrations of a stable adenosine analogue on the cAMP accumulation in noradrenaline-stimulated fat cells prepared from wild-type mice, A_1R knockout mice, and mice with a hemizygous gene deletion. From Johansson et al. (2008) with permission.

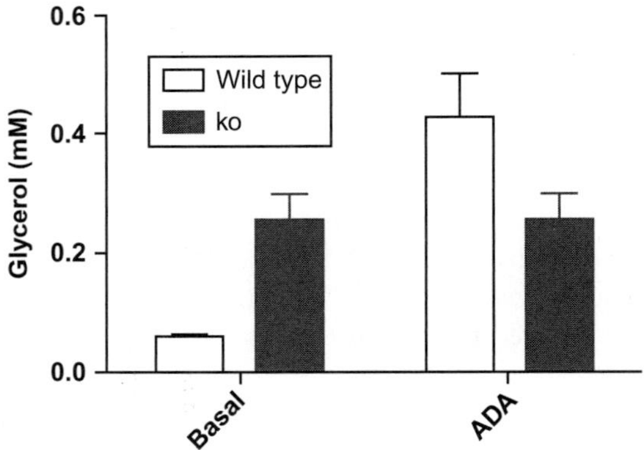

FIGURE 2 Lipolysis induced by removal of adenosine is dependent on A_1R. Lipolysis, measured by release of glycerol, can be markedly stimulated in fat cells by the removal of adenosine from the medium by adenosine deaminase. Fat cells prepared from A_1R (−/−) mice do not show this effect of adenosine deaminase, but instead exhibit a markedly enhanced basal lipolysis. Redrawn using data from Johansson et al. (2007b) with permission.

Fig. 2, the lipolytic effect of adenosine removal was completely lost in the fat cells from A_1R (−/−) mice. Interestingly, adenosine removal can also enhance the lipolysis induced by a submaximal dose of noradrenaline, and this enhancement was also absent in the fat cells prepared from mice lacking A_1R. Together, the results with the knockout mice also suggest that the A_1R

is the only adenosine receptor that is important in regulating lipolysis. This is also the conclusion drawn from an extensive body of pharmacological studies over many decades (Dhalla et al., 2009). mRNA for all adenosine receptors can be found in adipose tissue (Johansson et al., 2007b), but apparently the other receptors are irrelevant for lipolysis.

Given the major physiological role of A_1 receptors in regulating lipolysis, one might expect major adaptations following targeted deletion of the receptor, including upregulation of alternative inhibitory pathways. However, this has not been observed. First, as in several other places, we cannot detect any compensatory change in the expression of the other adenosine receptors (Johansson et al., 2007b). There was no change in the antilipolytic potency of nicotinic acid or prostaglandin E_2 that acts on the two major alternative antilipolytic G protein-mediated pathways (Johansson et al., 2007b). There was also no significant change in the amount of G_i proteins. This is interesting because G_i downregulation has been shown to be a mechanism of desensitization and cross desentization (Green et al., 1992). Thus, the data suggest that endogenous levels of adenosine are not sufficient to cause desentization. Similarly, the antilipolytic effect of insulin was apparently unaffected by the removal of A_1 receptors, even though insulin and adenosine act together (Johansson et al., 2008).

Adenosine acting at A_1 receptors influences not only lipolysis but also lipogenesis. However, the ability of adenosine to stimulate incorporation of glucose into fat was only seen in the presence of insulin, and this interaction was dose dependent demonstrating true synergy (Johansson et al., 2008). The effect of adenosine was completely dependent on A_1 receptors. There was no clear effect of the A_1R knockout alone on lipogenesis, and removal of adenosine by adenosine deaminase did not significantly alter the dose–response curve to insulin (Johansson et al., 2008). Therefore, adenosine effects on lipogenesis may be a high dose phenomenon and not of major physiological significance. One might expect that significantly enhanced lipolysis and smaller lipogenesis should result in a reduced body (and fat) weight. However, body weight is not significantly altered in A_1 (−/−) mice compared to their wild-type littermates over the first 4 months of life, and after that, it is actually slightly higher (Johansson et al., 2008). Fat weight was not significantly influenced in these studies.

It is well known that increasing circulating levels of fatty acids can reduce glucose tolerance and this has been suggested to be a factor in the development of diabetes. Thus, treatment with nicotinic acid was attempted, but tolerance development, side effects, and rebound phenomena limited the acceptance of such therapy. Recent developments may open the nicotinic acid receptor as a target (Vosper, 2009), but there is also evidence that A_1 receptors may be targeted (Dhalla et al., 2003, 2009). The ability of adenosine analogues to lower FFA and triglyceride levels *in vivo* is due to actions at A_1 receptors (Johansson et al., 2008). Thus, several agonists at A_1 receptors, including partial agonists have been tried with good results

(Dhalla et al., 2009). The reason that the partial agonists work is that the fat cell has many A_1 "spare" receptors.

Less is known about the role of A_1 receptors in brown adipose tissue. It is possibly relevant that tolerance to cold is increased by an A_1 antagonist (Lee et al., 1990). We also have evidence that prolonged cold exposure leads to a decrease in the expression of A_1R (Fredholm, unpublished). There is also some evidence that obesity may be associated with overactive A_1 receptors in brown and white adipose tissue (LaNoue & Martin, 1994). Given the recently increased interest in brown adipose tissue in man, this issue deserves additional experimental attention. A lowering of the ability of brown fat cells to generate heat would be an undesirable side effect of adenosine A_1 agonists aiming to improve the metabolic status of risk patients.

The importance of A_1 receptors in brown fat cells is less well known. It is well known that caffeine can increase thermogenesis in man by some 10% (Astrup & Toubro, 1993; Dulloo et al., 1989). However, these acute effects may be related to the well-known sympathoadrenal activation by acute caffeine, and indeed, rapid tolerance appears to occur for the thermogenic effect of caffeine (Bracco et al., 1995). Consistent with this are the significant but minimal effects of long-term caffeine intake on body weight (Lopez-Garcia et al., 2006).

III. Adenosine and Glucose Homeostasis

Glucose homeostasis is partly linked to lipid metabolism. Obesity-induced glucose tolerance was reduced in animals overexpressing A_1 receptors in adipose tissue (Dong et al., 2001). Similarly, treatment with an A_1 agonist improved glucose tolerance in animals on a diabetogenic diet (Dhalla et al., 2007). This suggests that activation of A_1 receptors could significantly enhance insulin sensitivity. Indeed, as mentioned earlier, this is certainly true in adipose tissue. However, this may not be physiologically very important as the glucose uptake in adipose tissue (Johansson et al., 2008) and skeletal muscle (Johansson et al., 2007b) is unaltered in A_1 (−/−) mice compared to control animals. Glucose tolerance was also unaffected. Surprisingly, it was reported that an antagonist at A_1 receptors could improve glucose tolerance (Xu et al., 1998). However, this antagonist (BW-1433) is also an antagonist at A_{2B} receptors, and this may be the relevant receptor. Indeed, BW-1433 reduces insulin sensitivity in adipose tissue, where A_1 receptors are the relevant receptors, but increases it in muscle (Crist et al., 1998), where A_{2B} receptors may be more important. Reducing ATP conversion to adenosine results in reduced hepatic insulin sensitivity (Enjyoji et al., 2008), but it is not known to what extent adenosine and its receptors are responsible.

The interest in the topic of adenosine receptors as potential regulators of glucose metabolism has been boosted by the findings of several large epidemiological studies that caffeine reduces the risk for type II diabetes (see Beaudoin & Graham, 2011). The effect is dose dependent, and risk reduction is marked at higher doses (risk ratio between 0.6 and 0.7 with more than four cups/day). However, acute administration appears to impair glucose homeostasis in man. An insulin sensitivity index based on AUC's for glucose and insulin was consistently reduced by 14–25% over several studies (see Beaudoin & Graham, 2011). Although less complete, the evidence suggests that acute administration of caffeinated coffee has a similar detrimental effect as caffeine.

Long-term coffee consumption, however, does consistently improve glucose tolerance, but surprisingly, this effect is largely shared by caffeine-free coffee. It is also seen in drinkers of tea. An experimental study in rats confirms an effect of decaffeinated coffee (Shearer et al., 2007). One hypothesis is that the effect should be attributed to other components of coffee than caffeine, for example, antioxidants (see Beaudoin & Graham, 2011), but other explanations for the caffeine-independent effects also exist (Tunnicliffe & Shearer, 2008).

There is, however, also strong evidence that adenosine can influence insulin secretion. We found that plasma insulin levels were significantly increased in A_1 (−/−) mice compared to their wild-type controls after a glucose challenge (Johansson et al., 2007a), but there was no difference in basal insulin levels. Using the perfused pancreas, a glucose infusion (raising levels from 4 to 16 mM) caused the expected rapid increase in insulin release, followed by a rapid return to a low steady level of insulin secretion in pancreata from wild-type mice (Johansson et al., 2007a). However, in pancreata from A_1 (−/−) mice, the insulin secretion during the second phase was markedly enhanced compared to the situation in the wild type (Fig. 3). The total AUC was more than doubled. The basal insulin secretion was also higher in the pancreas lacking A_1 receptors. There was also a tendency for an increase in the pulsatility in the late secretory response.

The role of A_1 receptors in regulating pulsatility was further explored in a follow-up study (Salehi et al., 2009). Whereas no pulsatility in insulin secretion was found in the perfused control pancreas, a pancreas from an A_1 knockout mouse exhibited a pulsatility with a period of $\sim$4 min. Already in the wild type, there were pulses of glucagon release and antisynchronous somatostatin pulses (Salehi et al., 2009). The loss of A_1 receptors resulted in a prolongation of these pulses and a loss of their antisynchronous behavior. The decrease in somatostatin release probably partly explains why the overall release of glucagon was increased. Indeed, the normal depression of glucagon release with increasing glucose is lost in the absence of adenosine A_1 receptors. Thus, A_1 receptors appear to play an important role in the regulation of the release of pancreatic hormones. Blocking the A_1-mediated inhibition of insulin release can explain why caffeine raises insulin levels, and the enhanced glucagon release can explain why this does not result in a major

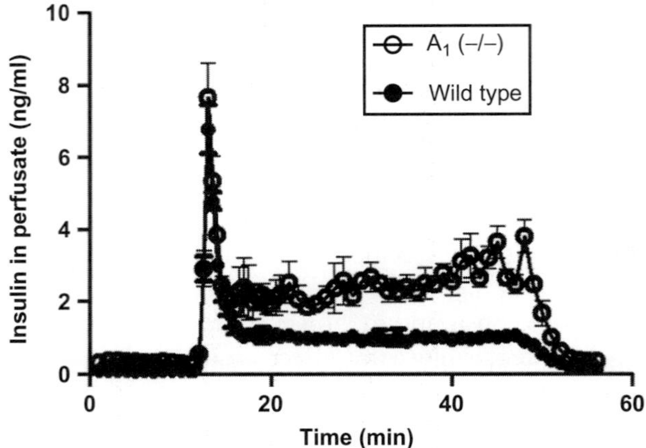

FIGURE 3 Endogenous adenosine reduces insulin release and controls its pulsatility. Pancreas from wild-type or A_1R knockout mice were perfused with low or high glucose medium. Results are mean and SEM from for separate experiments. Data from Johansson et al. (2007a) by permission.

fall in glucose. Together these findings could explain at least part of the altered glucose homeostasis. In addition, there are probably important effects mediated via other adenosine receptors, and there are important actions in the central nervous system, which we will briefly discuss below.

IV. Central Regulation of Metabolism, Sleep, and Wakefulness

It has been repeatedly emphasized that the regulation of sleep is intimately coupled to energy homeostasis (Horne, 2009; Saper, 2006). One of the reasons why we sleep is probably to conserve energy. Further, energy homeostasis and sleep/wakefulness show similarities in how they are regulated. As an example, hypocretins or orexins (Hcrt/Orx), neuropeptides that are synthesized by hypothalamic neurons, are involved in the regulation of feeding, thermoregulation, as well as the sleep–wakefulness cycle (Nuñez et al., 2009). Numerous other factors also regulate these processes in an apparently concerted manner, and one of these is adenosine.

A role for adenosine in the regulation of sleep–wakefulness has long been recognized as adenosine increases upon sleep deprivation and can induce sleep (Porkka-Heiskanen et al., 1997). In a recent study, it was reported that there are major (two- to fourfold) differences in ATP levels in the brain depending on time of the day (Dworak et al., 2010). As the highest levels were observed during sleep, this would be compatible with major energy differences and also with a critical role of adenosine and ATP. However, the study should be repeated using a different method as the levels of ATP were determined from slices cut after decapitation of the animal, and it has long been recognized that decapitation *per se* induces massive changes in ATP levels (Lowry et al., 1964), and slicing induces further falls (Fredholm et al., 1984). Recently, it was shown that overexpression of cytosolic adenosine deaminase, which slightly reduces adenosine levels, is sufficient to alter sleep physiology and to, for example, increase wakefulness and decrease REM (rapid eye movement) and nREM deep sleep without rapid eye movements (Palchykova et al., 2010). Via adenosine A_1 receptors, adenosine has been shown to reduce the activity of orexinergic neurons (Thakkar et al., 2008), basal forebrain cholinergic neurons (Basheer et al., 2004; Rainnie et al., 1994), tuberomammillary histaminergic neurons (Oishi et al., 2008), as well as locus coeruleus noradrenergic neurons and pontine serotoninergic neurons (studies showing reduced turnover or release of these transmitters; Harms et al., 1979; Reinhard et al., 1983; Shefner & Chiu, 1986). Since all these different neurons are involved in maintaining wakefulness, one would expect adenosine to be able to produce sleep by reducing neuronal activity at these different sites. Conversely, one will expect adenosine receptor antagonists, such as caffeine, to produce wakefulness by reducing these effects of adenosine at A_1 receptors.

With these results as a background, it was surprising that sleep was quite normal in A_1 receptor knockout mice (Stenberg et al., 2003). This surprising finding is best explained by assuming that the loss of adenosine A_1 receptors has been compensated for in the knockout animals. Circumstantial support for this is provided by the finding that a conditional A_1 receptor knockout mouse does show altered sleep and altered cognitive responses after sleep deprivation (Bjorness et al., 2009). Nevertheless, the results show that adenosine cannot be the only or even most important sleep-inducing substance in brain, neither can A_1 receptors be indispensable. Further, it has been shown that intracerebral injection of A_{2A} receptor agonist could induce sleep and affect sleep centers, and that the effect of the sleep-inducing substance prostaglandin D_2 was mediated through A_{2A} receptors (Scammell et al., 2001). Thus, it was not a complete surprise when it was shown that the alerting effect of caffeine was unaffected in A_1 receptor knockout mice, but completely eliminated in mice lacking the A_{2A} receptor (Huang et al., 2005). Further, the relevant A_{2A} receptors appear to be located in the basal ganglia, where they are known to coexist with dopamine D_2 receptors. Dopamine and D_2 receptors have been previously implicated in sleep control.

Dopamine is necessary for feeding as shown by the hypophagia in dopamine depleted animals, but too much dopamine signaling can also inhibit it (Palmiter, 2007). There is excellent evidence that adenosine and dopamine interact to regulate various motor behaviors and reward behaviors (Jenner et al., 2009; Kim & Palmiter, 2008; Salmi et al., 2005). The results mentioned above, together with the data showing that dopamine agonists can influence sleep, indicate that wakefulness is also controlled by A_{2A} and

D_2 receptors. Interestingly, one of the symptoms that precede the motor symptoms of Parkinson's disease is disordered sleep (Ferrer et al., 2010). Further, there is evidence that they interact in the regulation of feeding (Kim & Palmiter, 2003). Thus, the hypophagia that is known to occur after lesions of the dopaminergic pathways can be restored by blockade of adenosine receptors. Further, it is clear that goal-directed actions are sensitive to the work-related cost, and that these effort-related processes are regulated by adenosine A_{2A} receptors and the striatopallidal neurons (Mingote et al., 2008). The importance of food supply for the survival of the organism makes it imperative that food is a major reward. Thus, the brain circuitry involved in reward will interact closely with feeding behavior. Indeed, it is known that food deprivation is one of the most powerful influences on reward circuitry and dopamine-related behaviors (Palmiter, 2007).

One possible explanation is that factors such as leptin and insulin, which signal abundance of energy (circulating glucose and stored fat), act both on the neurons in the arcuate nucleus that regulate feeding (stimulating POMC/ART neurons; inhibiting NPY/AgRP neurons) and on the dopamine neurons in the ventral tegmental area (VTA), which are inhibited (Palmiter, 2007). By contrast, factors such as ghrelin and orexin that are released in fasting from peripheral and central sites have the opposite actions on both the arcuate nucleus and VTA. As mentioned earlier, adenosine can regulate insulin and orexin release. It is also known that activation of A_1 adenosine receptors can stimulate leptin release (Rice et al., 2000) and reduce ghrelin release (Yang et al., 2010a). In addition, the actions of dopamine in the basal ganglia are critically dependent upon the activity of adenosine acting predominantly on A_{2A} receptors. Thus, adenosine is one of the factors that ensures that the critically important balance between activity and energy balance is well maintained.

V. Regulation of Temperature

A very important factor in regulating energy consumption and in the sustained need for energy is the maintenance of body temperature. Heat is generated as a by-product of ATP production and ATP utilization. When temperature falls, thermogenic processes are turned on: increased muscle activity that may turn into shivering and heat production in brown adipose tissue (as well as in specialized white adipocytes) that depends on uncoupling of oxidative phosphorylation (Morrison et al., 2008). The process is regulated by heat and cold-sensitive neurons both peripherally and centrally. By contrast, when the supply of energy is very low (or when the prospects for future energy supply are bleak), animals may go into hibernation involving deep sleep and reduced body temperature. This may develop further into torpor (Drew et al., 2007). Torpor involves a remarkable drop in whole body metabolism down to 1–2% of basal values in awake animals and shows some deep similarities to sleep (Heller & Ruby, 2004). It has been suggested that torpor and sleep may have a common evolutionary origin with energy conservation as the major concern (Heller & Ruby, 2004). The process is common in small animals, but very rare in larger ones (Geiser, 2004). For example, mice may undergo daily bouts of torpor characterized by a metabolic rate that is one-fifth of basal and a body temperature of some 20 °C (Geiser, 2004). The mouse may be called a daily heterotherm rather than a homeotherm.

The body temperature may also fall in response to hypoxia. This so-called hypoxia-induced anapyrexia may have beneficial consequences as the fall in body temperature results in a decrease in the need for metabolizable energy and for oxygen (Steiner & Branco, 2002). Much of the conceptual framework rests on the idea of set points, that is, temperature values that are "defended." It has been pointed out that much of the experimental data are incompatible with the existence of a single set point and that hence the definitions of, for example, anapyrexia should be seriously reconsidered (Romanovsky, 2004), but this point of view is not uncontested (Cabanac, 2006). Be that as it may, the phenomenon may be very important: Given that a reduction in body temperature increases the area surviving an infarct, this could be an important endogenous mechanism to increase survival after a cerebral insult (Drew et al., 2007).

An important group of neurons are the hypocretin-containing neurons in the anterior preoptic area. Increasing temperature in this area by over pressing an uncoupling protein in hypocretin-containing neurons led to a reduced body temperature, and also to an increased life span (Tabarean et al., 2010). Given that decreased nutrient intake is a well-known means to extend life span, this suggests a link between temperature control and food intake as well. Even though this response may be particularly prominent in a small animal like mouse, a similar phenomenon may occur in man, and may also be used therapeutically. It is notable that these hypocretin neurons are inhibited by adenosine (section IV).

Indeed, adenosine is one possible mediator of this hypoxic response as shown by the fact that antagonism of A_1 effects by a local injection of DPCPX in the ventral preoptic area known to be important for temperature control reduced the hypoxia-induced fall in body temperature (Barros et al., 2006). The hypoxic ventilatory drive was also reduced in this study. It has been known for many years that adenosine levels rise in brain following hypoxia. Recently, focus has been on ATP release from glial cells that is triggered by a combination of hypoxia and hypercapnia (Erlichman et al., 2010; Gourine et al., 2010). Such ATP release has been shown to be a powerful stimulus for hypoxia-induced ventilation (Gourine et al., 2005).

It has long been known that systemic and intracerebroventricular administration of adenosine analogues produces a marked fall in body temperature. Pharmacological studies have mainly implicated A_1 receptors, and this has

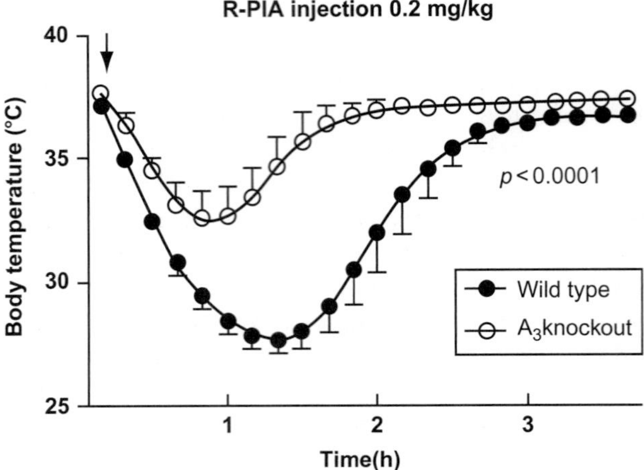

FIGURE 4 A_3 receptors and temperature. The temperature fall induced by the purportedly A_1 selective agonist R-PIA is substantially reduced in A_3 knockout mice. Results are mean and SEM. From Yang et al. (2010b) by permission.

been supported by the finding that the temperature drop can be substantially reduced in mice lacking A_1 receptors (Johansson et al., 2001; Yang et al., 2007). Further, body temperature was somewhat higher in A_1 (−/−) mice than in wild types of both sexes (Yang et al., 2007). By contrast, it tended to be slightly lower in A_{2A} (−/−) and A_1A_{2A} double (−/−) mice (Yang et al., 2009). Surprisingly, a major part of the adenosine-induced fall in body temperature was lost in mice lacking A_3 receptors (Yang et al., 2010b) (Fig. 4). Since these receptors are not abundant on hypothalamic neurons, this result indicates that other cells such as mast cells (Hua et al., 2008) or microglial cells (Hammarberg et al., 2003), which do possess functional A_3 receptors, may be involved in the temperature fall.

VI. Conclusion

It was realized from the very beginning of adenosine research that this nucleoside could regulate metabolism by affecting respiration and circulation. It also soon became clear that energy expenditure in, for example, the brain can be altered by reducing excitatory neurotransmission. This review briefly covers some additional ways in which adenosine will influence metabolism. In the periphery, both lipid and carbohydrate are directly regulated at the level of cells producing or consuming nutrients and also by altering the level of hormones that control lipid and glucose metabolism such as insulin, leptin, and ghrelin.

There is also increasing evidence, briefly highlighted above, that adenosine is one of the factors that regulates aspects of energy homeostasis centrally.

We know that energy needs are related to activity and that adenosine regulates activity, for example, by being involved in sleep–wakefulness. There is evidence that these processes are closely integrated with processes such as food intake and temperature control, and the evidence that adenosine is playing an important role here as well is mounting. Given that there is a deep link between energy homeostasis and the ability to cope with acute trauma and even with longevity, the possibility that adenosine has very wide ranging effects is intriguing. However, there are many open questions.

Acknowledgments

The authors are grateful to many collaborators and several granting agencies, notably the Swedish Science Research Council, Hjärnfonden and the Novo Nordisk fund.

Conflict of Interest: There is no conflict of interest to declare.

Abbreviations

AUC	area under the curve
AgRP	Agouti-related protein
CART	cocaine- and amphetamine-regulated transcript
DMH	dorsomedial hypothalamus
Hcrt	hypocretin (see also Orx)
LHA	lateral hypothalamus
NPY	neuropeptide Y
Orx	orexin (also called hypocretin)
POMC	proopiomelanocortin
VLPO	ventrolateral preoptic area
VTA	ventral tegmental area

References

Arch, J. R. S., & Newsholme, E. A. (1978). The control of the metabolism and the hormonal role of adenosine. *Essays in Biochemistry*, *14*, 88–123.

Astrup, A., & Toubro, S. (1993). Thermogenic, metabolic, and cardiovascular responses to ephedrine and caffeine in man. *International Journal of Obesity*, *17*(Suppl. 1), S41–S43.

Barraco, R. A., & Janusz, C. A. (1989). Respiratory effects of 5′-ethylcarboxamidoadenosine, an analog of adenosine, following microinjections into the nucleus tractus solitarius of rats. *Brain Research*, *480*, 360–364.

Barros, R. C., Branco, L. G., & Cárnio, E. C. (2006). Respiratory and body temperature modulation by adenosine A1 receptors in the anteroventral preoptic region during normoxia and hypoxia. *Respiratory Physiology & Neurobiology*, *153*, 115–125.

Basheer, R., Strecker, R. E., Thakkar, M. M., & McCarley, R. W. (2004). Adenosine and sleep-wake regulation. *Progress in Neurobiology*, *73*, 379–396.

Beaudoin, M. S., & Graham, T. E. (2011). Methylxanthines and human health: Epidemiological and experimental evidence. *Handbook of Experimental Pharmacology*, *200*, 509–548.

Berne, R. M. (1963). Cardiac nucleotides in hypoxia: Possible role in regulation of coronary blood flow. *The American Journal of Physiology*, *204*, 317–322.

Bjorness, T. E., Kelly, C. L., Gao, T., Poffenberger, V., & Greene, R. W. (2009). Control and function of the homeostatic sleep response by adenosine A1 receptors. *The Journal of Neuroscience*, *29*, 1267–1276.

Bracco, D., Ferrarra, J. M., Arnaud, M. J., Jéquier, E., & Schutz, Y. (1995). Effects of caffeine on energy metabolism, heart rate, and methylxanthine metabolism in lean and obese women. *The American Journal of Physiology*, *269*, E671–E678.

Buttigieg, J., & Nurse, C. A. (2004). Detection of hypoxia-evoked ATP release from chemoreceptor cells of the rat carotid body. *Biochemical and Biophysical Research Communications*, *322*, 82–87.

Cabanac, M. (2006). Adjustable set point: To honor Harold T. Hammel. *Journal of Applied Physiology*, *100*, 1338–1346.

Crist, G. H., Xu, B., Lanoue, K. F., & Lang, C. H. (1998). Tissue-specific effects of in vivo adenosine receptor blockade on glucose uptake in Zucker rats. *The FASEB Journal*, *12*, 1301–1308.

Darby, M., Kuzmiski, J. B., Panenka, W., Feighan, D., & MacVicar, B. A. (2003). ATP released from astrocytes during swelling activates chloride channels. *Journal of Neurophysiology*, *89*, 1870–1877.

Dhalla, A. K., Chisholm, J. W., Reaven, G. M., & Belardinelli, L. (2009). A1 adenosine receptor: Role in diabetes and obesity. *Handbook of Experimental Pharmacology*, *193*, 271–295.

Dhalla, A. K., Shryock, J. C., Shreeniwas, R., & Belardinelli, L. (2003). Pharmacology and therapeutic applications of A1 adenosine receptor ligands. *Current Topics in Medicinal Chemistry*, *3*, 369–385.

Dhalla, A. K., Wong, M. Y., Voshol, P. J., Belardinelli, L., & Reaven, G. M. (2007). A1 adenosine receptor partial agonist lowers plasma FFA and improves insulin resistance induced by high-fat diet in rodents. *American Journal of Physiology. Endocrinology and Metabolism*, *292*, E1358–E1363.

Dole, V. P. (1962). Insulin-like actions of ribonucleic acid, adenylic acid, and adenosine. *The Journal of Biological Chemistry*, *237*, 2758–2762.

Dong, Q., Ginsberg, H. N., & Erlanger, B. F. (2001). Overexpression of the A1 adenosine receptor in adipose tissue protects mice from obesity-related insulin resistance. *Diabetes, Obesity & Metabolism*, *3*, 360–366.

Drew, K. L., Buck, C. L., Barnes, B. M., Christian, S. L., Rasley, B. T., & Harris, M. B. (2007). Central nervous system regulation of mammalian hibernation: Implications for metabolic suppression and ischemia tolerance. *Journal of Neurochemistry*, *102*, 1713–1726.

Dulloo, A. G., Geissler, C. A., Horton, T., Collins, A., & Miller, D. S. (1989). Normal caffeine consumption: Influence on thermogenesis and daily energy expenditure in lean and postobese human volunteers. *The American Journal of Clinical Nutrition*, *49*, 44–50.

Dunwiddie, T. V., Hoffer, B. J., & Fredholm, B. B. (1981). Alkylxanthines elevate hippocampal excitability. Evidence for a role of endogenous adenosine. *Naunyn-Schmiedeberg's Archives of Pharmacology*, *316*, 326–330.

Dworak, M., McCarley, R. W., Kim, T., Kalinchuk, A. V., & Basheer, R. (2010). Sleep and brain energy levels: ATP changes during sleep. *The Journal of Neuroscience*, *30*, 9007–9016.

Eltzschig, H. K., Eckle, T., Mager, A., Küper, N., Karcher, C., Weissmüller, T., et al. (2006). ATP release from activated neutrophils occurs via connexin 43 and modulates adenosine-dependent endothelial cell function. *Circulation Research*, *99*, 1100–1108.

Enjyoji, K., et al. (2008). Deletion of cd39/entpd1 results in hepatic insulin resistance. *Diabetes*, *57*, 2311–2320.

Erlichman, J. S., Leiter, J. C., & Gourine, A. V. (2010). ATP, glia and central respiratory control. *Respiratory Physiology & Neurobiology*, *173*, 305–311.

Fain, J. N., Pointer, R. H., & Ward, W. F. (1972). Effects of adenosine nucleosides on adenylate cyclase, phosphodiesterase, cyclic adenosine monophosphate accumulation, and lipolysis in fat cells. *The Journal of Biological Chemistry*, *247*, 6866–6872.

Fain, J. N., & Wieser, P. B. (1975). Effects of adenosine deaminase on cyclic adenosine monophosphate accumulation, lipolysis, and glucose metabolism of fat cells. *The Journal of Biological Chemistry*, *250*, 1027–1034.

Faris, A., & Spence, D. M. (2008). Measuring the simultaneous effects of hypoxia and deformation on ATP release from erythrocytes. *The Analyst*, *133*, 678–682.

Ferrer, I., Martinez, A., Blanco, R., Dalfó, E., & Carmona, M. (2010). Neuropathology of sporadic Parkinson disease before the appearance of parkinsonism: Preclinical Parkinson disease. *Journal of Neural Transmission*, Sep 23, 2010. Epub ahead of print.

Fredholm, B. B., Dunwiddie, T. V., Bergman, B., & Lindström, K. (1984). Levels of adenosine and adenine nucleotides in slices of rat hippocampus. *Brain Research*, *295*, 127–136.

Geiser, F. (2004). Metabolic rate and body temperature reduction during hibernation and daily torpor. *Annual Review of Physiology*, *66*, 239–274.

Gourine, A. V., Llaudet, E., Dale, N., & Spyer, K. M. (2005). Release of ATP in the ventral medulla during hypoxia in rats: Role in hypoxic ventilatory response. *The Journal of Neuroscience*, *25*, 1211–1218.

Gourine, A. V., et al. (2010). Astrocytes control breathing through pH-dependent release of ATP. *Science*, *329*, 571–575.

Green, A., Milligan, G., & Dobias, S. B. (1992). Gi down-regulation as a mechanism for heterologous desensitization in adipocytes. *The Journal of Biological Chemistry*, *267*, 3223–3229.

Hammarberg, C., Schulte, G., & Fredholm, B. B. (2003). Evidence for functional adenosine A3 receptors in microglia cells. *Journal of Neurochemistry*, *86*, 1051–1054.

Harms, H. H., Wardeh, G., & Mulder, A. H. (1979). Effect of adenosine on depolarization-induced release of various radiolabelled neurotransmitters from slices of rat corpus striatum. *Neuropharmacology*, *18*, 577–580.

Hedqvist, P., & Fredholm, B. B. (1976). Effects of adenosine on adrenergic neurotransmission; prejunctional inhibition and postjunctional enhancement. *Naunyn-Schmiedeberg's Archives of Pharmacology*, *293*, 217–223.

Heller, H. C., & Ruby, N. F. (2004). Sleep and circadian rhythms in mammalian torpor. *Annual Review of Physiology*, *66*, 275–289.

Hjemdahl, P., & Fredholm, B. B. (1976). Cyclic AMP-dependent and independent inhibition of lipolysis by adenosine and decreased pH. *Acta Physiologica Scandinavica*, *96*, 170–179.

Horne, J. (2009). REM sleep, energy balance and "optimal foraging". *Neuroscience and Biobehavioral Reviews*, *33*, 466–474.

Hua, X., Chason, K. D., Fredholm, B. B., Deshpande, D. A., Penn, R. B., & Tilley, S. L. (2008). Adenosine induces airway hyperresponsiveness through activation of A3 receptors on mast cells. *The Journal of Allergy and Clinical Immunology*, *122*, 107–113.

Huang, Z.-L., Qu, W.-M., Eguchi, N., Chen, J.-F., Schwarzschild, M. A., Fredholm, B. B., et al. (2005). Adenosine A_{2A}, but not A_1, receptors mediate the arousal effect of caffeine. *Nature Neuroscience*, *8*, 858–859.

Jenner, P., Mori, A., Hauser, R., Morelli, M., Fredholm, B., & Chen, J.-F. (2009). Adenosine, adenosine A2A antagonists and Parkinson's disease. *Parkinsonism & Related Disorders*, *15*, 406–413.

Johansson, S., Lindgren, E., Yang, J.-N., Herling, A. W., & Fredholm, B. B. (2008). Adenosine A1 receptors regulate lipolysis and lipogenesis in mouse adipose tissue—Interactions with insulin. *European Journal of Pharmacology*, *597*, 92–101.

Johansson, S., Yang, J., Lindgren, E., & Fredholm, B. B. (2007b). Eliminating the antilipolytic adenosine A1 receptor does not lead to compensatory changes of PGE2 or nicotinic acid effects on lipolysis. *Acta Physiologica*, *190*, 87–96.

Johansson, B., Halldner, L., Dunwiddie, T. V., Masino, S. A., Poelchen, W., Giménez-Llort, L., et al. (2001). Hyperalgesia, anxiety, and decreased hypoxic neuroprotection in mice lacking the adenosine A1 receptor. *Proceedings of the National Academy of Sciences of the United States of America*, *98*, 9407–9412.

Johansson, S., Salehi, A., Sundström, M., Westerblad, H., Lundqvist, I., Carlsson, P.-O., et al. (2007a). A1 receptor deficiency causes increased insulin and glucagon secretion in mice. *Biochemical Pharmacology*, *74*, 1628–1635.

Kim, D. S., & Palmiter, R. D. (2003). Adenosine receptor blockade reverses hypophagia and enhances locomotor activity of dopamine-deficient mice. *Proceedings of the National Academy of Sciences of the United States of America*, *100*, 1346–1351.

Kim, D. S., & Palmiter, R. D. (2008). Interaction of dopamine and adenosine receptor function in behavior: Studies with dopamine-deficient mice. *Frontiers in Bioscience*, *13*, 2311–2318.

LaNoue, K. F., & Martin, L. F. (1994). Abnormal A1 adenosine receptor function in genetic obesity. *The FASEB Journal*, *8*, 72–80.

Lee, T. F., Li, D. J., Jacobson, K. A., & Wang, L. C. (1990). Improvement of cold tolerance by selective A1 adenosine receptor antagonists in rats. *Pharmacology, Biochemistry and Behavior*, *37*, 107–112.

Liang, H. X., Belardinelli, L., Ozeck, M. J., & Shryock, J. C. (2002). Tonic activity of the rat adipocyte A1-adenosine receptor. *British Journal of Pharmacology*, *135*, 1457–1466.

Londos, C., Cooper, D. M., & Wolff, J. (1980). Subclasses of external adenosine receptors. *Proceedings of the National Academy of Sciences of the United States of America*, *77*, 2551–2554.

Lopez-Garcia, E., van Dam, R. M., Rajpathak, S., Willett, W. C., Manson, J. E., & Hu, F. B. (2006). Changes in caffeine intake and long-term weight change in men and women. *The American Journal of Clinical Nutrition*, *83*, 674–680.

Lowry, O. H., Passonneau, J. V., Hasselberger, F. X., & Schulz, D. W. (1964). Effect of ischemia on known substrates and cofactors of the glycolytic pathway in brain. *The Journal of Biological Chemistry*, *239*, 18–30.

Mingote, S., et al. (2008). Nucleus accumbens adenosine A2A receptors regulate exertion of effort by acting on the ventral striatopallidal pathway. *The Journal of Neuroscience*, *28*, 9037–9046.

Monteiro, E. C., & Ribeiro, J. A. (1987). Ventilatory effects of adenosine mediated by carotid body chemoreceptors in the rat. *Naunyn-Schmiedebergs Archives of Pharmacology*, *335*, 143–148.

Morrison, S. F., Nakamura, K., & Madden, C. J. (2008). Central control of thermogenesis in mammals. *Experimental Physiology*, *93*, 773–797.

Musante, L., Zegarra-Moran, O., Montaldo, P. G., Ponzoni, M., & Galietta, L. J. (1999). Autocrine regulation of volume-sensitive anion channels in airway epithelial cells by adenosine. *The Journal of Biological Chemistry*, *274*, 11701–11707.

Nuñez, A., Rodrigo-Angulo, M. L., Andrés, I. D., & Garzón, M. (2009). Hypocretin/Orexin neuropeptides: Participation in the control of sleep-wakefulness cycle and energy homeostasis. *Current Neuropharmacology*, *7*, 50–59.

Oishi, Y., Huang, Z. L., Fredholm, B. B., Urade, Y., & Hayaishi, O. (2008). Adenosine in the tuberomammillary nucleus inhibits the histaminergic system via A1 receptors and promotes non-rapid eye movement sleep. *Proceedings of the National Academy of Sciences of the United States of America*, *105*, 19992–19997.

Palchykova, S., Winsky-Sommerer, R., Shen, H. Y., Boison, D., Gerling, A., & Tobler, I. (2010). Manipulation of adenosine kinase affects sleep regulation in mice. *The Journal of Neuroscience, 30*, 13157–13165.

Palmiter, R. D. (2007). Is dopamine a physiologically relevant mediator of feeding behavior? *Trends in Neurosciences, 30*, 375–381.

Porkka-Heiskanen, T., Strecker, R. E., Thakkar, M., Bjorkum, A. A., Greene, R. W., & McCarley, R. W. (1997). Adenosine: A mediator of the sleep-inducing effects of prolonged wakefulness. *Science, 276*, 1265–1268.

Rainnie, D. G., Grunze, H. C., McCarley, R. W., & Greene, R. W. (1994). Adenosine inhibition of mesopontine cholinergic neurons: Implications for EEG arousal. *Science, 263*, 689–692, published erratum appears in Science 1994 July 1, 265 (5168), 16.

Reinhard, J. F., Galloway, M. P., & Roth, R. H. (1983). Noradrenergic modulation of serotonin synthesis and metabolism. II. Stimulation by 3-isobutyl-1-methylxanthine. *The Journal of Pharmacology and Experimental Therapeutics, 226*, 764–769.

Rice, A. M., Fain, J. N., & Rivkees, S. A. (2000). A1 adenosine receptor activation increases adipocyte leptin secretion. *Endocrinology, 141*, 1442–1445.

Romanovsky, A. A. (2004). Do fever and anapyrexia exist? Analysis of set point-based definitions. *American Journal of Physiology: Regulatory, Integrative and Comparative Physiology, 287*, R992–R995.

Salehi, A., Parandah, F., Fredholm, B. B., Garpengiesser, E., & Hellman, B. (2009). Absence of adenosine A1 receptors unmasks pulses of insulin release and prolongs those of glucagon and somatostatin. *Life Science, 85*, 470–476.

Salmi, P., Chergui, K., & Fredholm, B. B. (2005). Adenosine-dopamine interactions revealed in knockout mice. *Journal of Molecular Neuroscience, 26*, 239–244.

Saper, C. B. (2006). Staying awake for dinner: Hypothalamic integration of sleep, feeding, and circadian rhythms. *Progress in Brain Research, 153*, 243–252.

Scammell, T. E., Gerashchenko, D. Y., Mochizuki, T., McCarthy, M. T., Estabrooke, I. V., Sears, C. A., et al. (2001). An adenosine A2a agonist increases sleep and induces Fos in ventrolateral preoptic neurons. *Neuroscience, 107*, 653–663.

Shearer, J., Sellars, E. A., Farah, A., Graham, T. E., & Wasserman, D. H. (2007). Effects of chronic coffee consumption on glucose kinetics in the conscious rat. *Canadian Journal of Physiology and Pharmacology, 85*, 823–830.

Shefner, S. A., & Chiu, T. H. (1986). Adenosine inhibits locus coeruleus neurons: An intracellular study in a rat brain slice preparation. *Brain Research, 366*, 364–368.

Steiner, A. A., & Branco, L. G. (2002). Hypoxia-induced anapyrexia: Implications and putative mediators. *Annual Review of Physiology, 64*, 263–288.

Stenberg, D., Litonius, E., Halldner, L., Johansson, B., Fredholm, B. B., & Porkka-Heiskanen, T. (2003). Sleep and its homeostatic regulation in mice lacking the adenosine A1 receptor. *Journal of Sleep Research, 12*, 283–290.

Tabarean, I., Morrison, B., Marcondes, M. C., Bartfai, T., & Conti, B. (2010). Hypothalamic and dietary control of temperature-mediated longevity. *Ageing Research Reviews, 9*, 41–50.

Thakkar, M. M., Engemann, S. C., Walsh, K. M., & Sahota, P. K. (2008). Adenosine and the homeostatic control of sleep: Effects of A1 receptor blockade in the perifornical lateral hypothalamus on sleep-wakefulness. *Neuroscience, 153*, 875–880.

Tu, J., Le, G., & Ballard, H. J. (2010). Involvement of the cystic fibrosis transmembrane conductance regulator in the acidosis-induced efflux of ATP from rat skeletal muscle. *The Journal of Physiology, 588*, 4563–4578.

Tunnicliffe, J. M., & Shearer, J. (2008). Coffee, glucose homeostasis, and insulin resistance: Physiological mechanisms and mediators. *Applied Physiology, Nutrition, and Metabolism, 33*, 1290–1300.

Vial, C., Owen, P., Opie, L. H., & Posel, D. (1987). Significance of release of adenosine triphosphate and adenosine induced by hypoxia or adrenaline in perfused rat heart. *Journal of Molecular and Cellular Cardiology*, *19*, 187–197.

Vosper, H. (2009). Niacin: A re-emerging pharmaceutical for the treatment of dyslipidaemia. *British Journal of Pharmacology*, *158*, 429–441.

Xu, B., Berkich, D. A., Crist, G. H., & LaNoue, K. F. (1998). A1 adenosine receptor antagonism improves glucose tolerance in Zucker rats. *The American Journal of Physiology*, *274*, E271–E279.

Yang, J., Chen, J., & Fredholm, B. B. (2009). Physiological roles of A1 and A2A adenosine receptors in regulating heart rate, body temperature, and locomotion as revealed using knockout mice and caffeine. *American Journal of Physiology. Heart and Circulatory Physiology*, *296*, H1141–H1149.

Yang, J.-N., Tiselius, C., Daré, E., Johansson, B., Valen, G., & Fredholm, B. B. (2007). Sex differences in mouse heart rate and body temperature and in their regulation by adenosine A1 receptors. *Acta Physiologica*, *190*, 63–75.

Yang, J., Wang, Y., Garcia-Roves, P., Björnholm, M., & Fredholm, B. B. (2010). Adenosine A3 receptors regulate heart rate, motor activity and body temperature. *Acta Physiologica*, *199*, 221–230.

Yang, G. K., Yip, L., Fredholm, B. B., Kieffer, T. J., & Kwok, Y. N. (2010). Involvement of adenosine signaling in controlling the release of ghrelin from the mouse stomach. *The Journal of Pharmacology and Experimental Therapeutics*, *336*, 77–86.

Joel Linden
Division of Inflammation Biology, La Jolla Institute of Allergy and Immunology,
La Jolla, California, USA

Regulation of Leukocyte Function by Adenosine Receptors

Abstract

The immune system responds to cues in the microenvironment to make acute and chronic adaptations in response to inflammation and injury. Locally produced purine nucleotides and adenosine provide receptor-mediated signaling to all bone-marrow derived cells of the immune system to modulate their responses. This review summarizes recent advances in our understanding of the effects of adenosine signaling through G protein-coupled adenosine receptors on cells of the immune system. Adenosine A_{2A} receptors (A_{2A}Rs) have a generally suppressive effect on the activation of immune cells. Moreover, their transcription is strongly induced by signals that activate macrophages or dendritic cells through toll-like receptors, or T cells through T cell receptors. A_{2A}R induction is responsible for producing a gradual dissipation of inflammatory responses. A_{2A}R activation is particularly effective in limiting the activation of invariant NKT (iNKT) cells that play a central role in acute reperfusion injury. A_{2A} agonists have clinical promise for the treatment of vaso-occlusive tissue injury. Blockade of A_{2A} receptors may be useful to enhance immune-mediated killing of cancer cells. A_{2B}R expression also is transcriptionally regulated by hypoxia, cytokines, and oxygen radicals. Acute A_{2B}R activation attenuates the production of proinflammatory cytokines from macrophages, but sustained activation facilitates macrophage and dendritic cell remodeling and the production of acute phase proteins and angiogenic factors that may participate in evoking insulin resistance and tissue fibrosis. A_{2B}R activation also influences macrophage and neutrophil function by influencing expression of the

Advances in Pharmacology, Volume 61

1054-3589/11 $35.00
10.1016/B978-0-12-385526-8.00004-7

anti-inflammatory netrin receptor, UNC5B. The therapeutic significance of adenosine-mediated effects on the immune system is discussed.

I. Introduction

Both innate and adaptive immunity are strongly influenced by purinergic signaling. Innate immunity is the most ancient system that protects multicellular hosts from infections and is comprised of immune cells that are activated in response to either pathogen-associated molecular patterns (PAMPs) or sterile host tissue injury resulting in inflammation in response to damage-associated molecular patterns (DAMPs; Pelegrin, 2008). The adaptive immune system evolved subsequent to the innate system and utilizes antigen presenting macrophages and DCs, MHC molecules, and TCRs to recognize specific pathogenic antigens or host autoantigens. All cells of the immune system express multiple purinergic receptors, and these receptors play a major role in their regulation. The reader is directed to previous reviews for background information about adenosine signaling in the immune system (Hasko et al., 2007; Kumar & Sharma, 2009). This review focuses on recent findings that have shed new light on the role that purinergic signaling plays in regulating both innate and adaptive immune responses. Of particular interest are recent discoveries demonstrating that adenosine receptor transcripts can be rapidly upregulated in response to local cues such as activation of excitatory receptors or tissue hypoxia. It has also become evident that the extracellular metabolism of adenine nucleotides by ectoenzymes such as CD39 and CD73 is a major source of adenosine, based on proinflammatory responses in mice upon deletion of these enzymes.

A diagram of the suppressive effects of A_{2A}Rs on adaptive and innate immunity is shown in Fig. 1. Conventional T cells are part of the adaptive immune system. Selective activation of highly variable T cell receptors results in the expansion of these cells and the release of cytokines such as INF-γ. A minor subset of T cells known as invariant NKT (iNKT) cells express invariant T cell receptors. In addition to responding to various pathogens, iNKT cells are activated by injury to host tissues and contribute to sterile inflammation. Since NKT cells possess T cell receptors than can be rapidly activated by innate signals from either pathogens or danger signals produced by the injured host, they bridge innate and adaptive immunity. Both systems are strongly influenced by inducible A_{2A}R signaling as well as other purinergic receptors. Suppression of the innate immune response due to adenosine signaling can be beneficial to limit tissue inflammation and injury. However, too much immunosuppression by adenosine can blunt the ability of the immune system to control infections (Hasko et al., 2008). Activation of adaptive immune responses can be beneficial, for example, by enhancing immune surveillance of tumors (Jin et al., 2010), or harmful, for example, by reducing immune sensitization to persistent viral infections (Alam et al.,

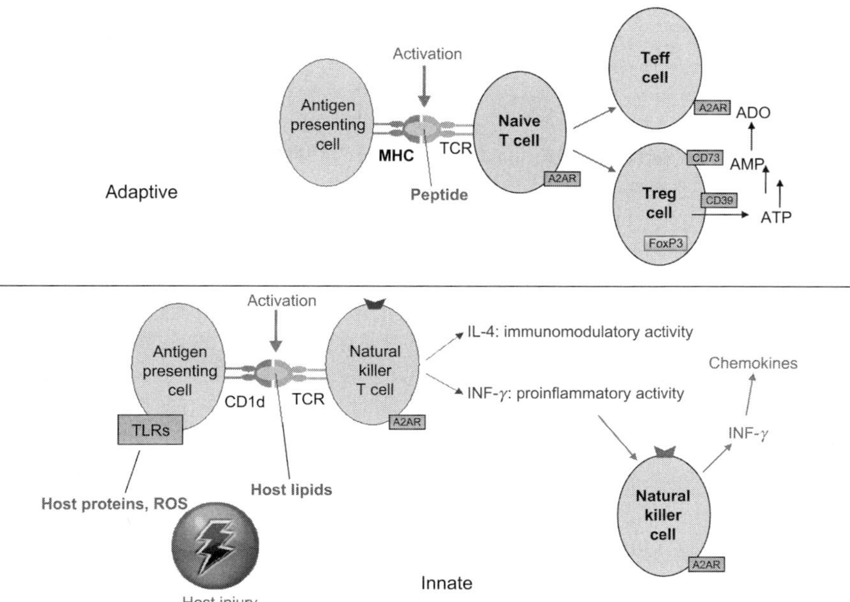

FIGURE 1 Comparison of $A_{2A}R$ effects on T cells and iNKT cells. The *top panel* illustrates that in the adaptive immune response, peptide antigens are processed by antigen presenting cells and presented on major histocompatibility complex (MHC) molecules to variable T cell receptors. Upon TCR activation, naive T cells expand and generate T effector (Teff) cells, T regulatory (Treg) cells, or other types of daughter T cells. $A_{2A}R$ activation on naive T cells during antigen presentation enhances the production of Treg cells and produces persistent anergy of Teff cells. Activation of $A_{2A}Rs$ on Teff cells during TCR activation suppresses their expansion and cytokine production. Among lymphocytes, only Treg cells express ectonucleotidases CD39 and CD73 that generate adenosine from the extracellular metabolism of adenosine nucleotides. The *bottom panel* illustrates the innate response of NKT cells. Glycolipid antigens can be derived from pathogens but also are thought to be generated from glycolipids derived from necrotic host cells and are presented by the MHC-like antigen presenting molecule, CD1d, to invariant TCRs on NKT cells. NKT cells usually express TCRs and NK cell markers such as NK1.1. Upon activation of their TCR, iNKT cells rapidly produce large quantities of several cytokines including IFN-γ and IL-4. NK cells are transactivated by cytokines released form NKT cells and produce additional IFN-γ which stimulates the production of IFN-γ inducible chemokines that recruit additional leukocytes into the inflamed tissues. $A_{2A}Rs$ are induced upon TCR activation of NKT and NK cells, and $A_{2A}R$ signaling strongly suppresses cytokine production by these cells.

2009). We discuss how recent developments may be useful to the goal of exploiting adenosine signaling for therapeutic uses such as treatment of reperfusion injury, chronic inflammatory diseases, and tumor killing.

II. Immune Responses to Adenosine Receptor Signaling

Activation of the immune system elicits immune cell-mediated killing of pathogens and the release of proinflammatory cytokines. The rapid induction of proinflammatory mediators by the immune system is accompanied by the initiation of transcriptional programs that limit inflammation. These include production of TGF-β, IL-10, vascular endothelial growth factor (VEGF), insulin-like growth factor-1, HO-1, and netrin-1. Adenosine and the A_{2A} and A_{2B} receptors are included among anti-inflammatory factors that are produced or induced during inflammation. A_1R signaling also is important in immune regulation, but it acts primarily by influencing the sympathetic nervous system. Prejunctional A_1 receptors inhibit the release of the sympathetic cotransmitters norepinephrine and ATP. All primary and secondary immune organs receive sympathetic innervation from sympathetic postganglionic neurons (Nance & Sanders, 2007). Innate immune cells express both α- and β-adrenergic receptor subtypes, while T and B lymphocytes express anti-inflammatory β2 adrenergic receptors exclusively. The A_3 receptor has been implicated in influencing neutrophil chemotaxis (Chen et al., 2006) and mast cell degranulation (Feoktistov et al., 2003) and may contribute to inhibiting reperfusion injury (Ge et al., 2010), but in general, the role of the A_3 receptor in immune regulation remains enigmatic (Gessi et al., 2008).

A. Platelets

Platelets are activated during sterile inflammation that occurs in response to tissue trauma or ischemia/reperfusion injury (IRI). Substantial platelet activation is associated with sickle cell disease that has been extensively studied as a model of simultaneous IRI in multiple tissues. Intravital microscopy analyses in mice with sickle cell disease indicate that sickle RBCs interact primarily with adherent platelets and leukocytes in postcapillary and collecting venules leading to vascular obstruction (Turhan et al., 2002). ATP and ADP released from activated or damaged cells activate platelets via two G protein-coupled ADP receptors ($P2Y_1$ and $P2Y_{12}$) and via ATP through the ligand-gated P2X1 receptor (Oury et al., 2006).

It is now appreciated that the metabolic flux of adenine nucleotides and adenosine in the extracellular space regulates platelet activation due to counterbalancing signaling through P2 and adenosine receptors (Iyu et al., 2011). Activation of A_{2A} receptors on platelets causes an increase in cyclic AMP accumulation and a decrease in platelet aggregation (Cooper et al., 1995; Table I). In A_{2A} receptor-knockout mice, platelet aggregation is increased, proving the importance of this receptor subtype in limiting platelet activation (Ledent et al., 1997). Platelet activation is not only important for regulation platelet aggregation and secretion but also because it stimulates the production of platelet heteroaggregates with other leukocytes including monocytes, eosinophils, and neutrophils (Polanowska-Grabowska et al., 2010). Blockade of P-selectin-mediated platelet–leukocyte aggregation is beneficial in the animal models of vascular injury (Merhi et al., 1999). Hence, platelet A_{2A}R activation may contribute to reduced sterile inflammation by direct effects on singlet platelets and platelet–leukocyte heteroaggregates. Although it was thought

TABLE I Summary of the Effects of $A_{2A}R$ and $A_{2B}R$ Signaling on Some Cells of the Immune System

	A_{2A}	*A_{2B}*	*Other*
Platelets	↑ Cyclic AMP ↓ Aggregation ↓ Secretion ↓ Leukocyte heteroaggregates	↓ $P2Y_1$ expression ↓ ADP-induced aggregation	ADP and ATP receptors
Neutrophils	↑ Cyclic AMP ↓ Oxidative burst ↓ α4/β1 integrin (VLA-4)		ATP release, pannexin channels
Macrophages	↑ M1 to M2-like switch ↓ TNF-α, IL-12 ↑ VEGF, IL-10 Induced by HO-1 Induced by endotoxin	↓ TNF, IL-12 ↑ VEGF, IL-10 ↑ IL-6 Induced by HIF Induced by IFN-γ Induced by diabetes Controls UNC5B expression	M1 inflammatory M2 angiogenic
T cells	↑ Cyclic AMP ↓ IFN-γ production ↓ CD-69 ↓ Proliferation, IL-2 ↑ Anergy Induced by TCR activation ↑ Treg production ↑ Treg function		CD73 and CD39 (Tregs only)
iNKT cells	↑ Cyclic AMP ↓ IFN-γ production ↓ TNF-α Induced by TCR activation		Activated by lipid antigens Coactivated by TIM-1 Coactivates NK cells

↓ decrease; ↑ increase.

that the only adenosine receptor on platelets was the $A_{2A}R$, Yang et al. (2010) recently showed that systemic inflammation induces the expression of $A_{2B}Rs$ on platelets and activation of these receptors inhibits the expression of the $P2Y_1$ receptor and ADP-induced platelet aggregation.

B. Neutrophils

Tissue trauma or IRI results in an inflammatory cascade that ultimately results in neutrophil infiltration into tissues (Lappas et al., 2006; McDonald et al., 2010). In the absence of infection, neutrophil accumulation in tissues can be very destructive. Platelet activation is associated with increased platelet adhesion to microvascular endothelium (Brittain et al., 1993), and formation of platelet heteroaggregates with erythrocytes (Inwald et al., 2000) and leukocytes including neutrophils, monocytes, and eosinophils. Oxidative burst in activated neutrophils and elevated expression of α4/β1 integrin (VLA-4, CD49d/CD29) are decreased as a result of $A_{2A}R$ activation (Fredholm et al., 1996; Revan et al., 1996; Sullivan et al., 2001, 2004b).

Neutrophils release ATP through pannexin-1 hemichannels in response to inflammatory mediators (Chen et al., 2010). Released ATP is necessary for maintaining neutrophil activation, but metabolism of ATP to adenosine inhibits neutrophil activation and adhesion to endothelial cells by direct effects on neutrophils (Sullivan et al., 2001) as well as indirect effects that reduced cytokine-mediated expression of P-selectin and ICAM-1 on endothelial cells (Okusa et al., 2000). Neutrophils are guided to sites of tissue injury by chemokines and formal peptides released from necrotic cells (McDonald et al., 2010). Thus, purinergic signaling is one of the several mechanisms required for regulation of neutrophil trafficking during inflammation. $A_{2B}Rs$ also indirectly influence neutrophil trafficking by effects on tissue production of cytokines that are chemotactic to neutrophils such as KC. For example, $A_{2B}R$ activation plays a role in mediating lung inflammation after ischemia–reperfusion by stimulating neutrophil chemotaxis (Anvari et al., 2010).

C. Macrophages and DCs

Macrophages are broadly classified into inflammatory M1 (NOS2+) and angiogenic M2 (arginase+). Toll-like receptor (TLR) 2, 4, 7, and 9 agonists, together with $A_{2A}R$ agonists, switch macrophages from an M1- to an M2-like phenotypes. This switch involves induction of $A_{2A}Rs$ by TLR agonists, diminished TNF-α and IL-12 production, and enhanced production of VEGF and IL-10 (Grinberg et al., 2009). LPS suppresses PLCβ1 and β2 expression in macrophages *in vitro* and in several tissues *in vivo*. Signaling through TLRs suppresses PLC-β2 and this switches M1 macrophages into an M2-like state (Grinberg et al., 2009). Recognition of apoptotic cells also polarizes macrophages toward the anti-inflammatory M2-like phenotype by a process involving macrophage production of sphingosine-1-phosphate and VEGF and the induction of the $A_{2A}R$ (Weis et al., 2009). These responses are mediated in part by the transcription factor HO-1. These findings suggest that HO-1, which is induced by apoptotic cell-derived S1P, is involved in macrophage polarization toward an M2 phenotype that includes $A_{2A}R$ induction (Weis et al., 2009).

The release of proinflammatory cytokines such as TNF-α and IL-12 can be inhibited by either $A_{2A}R$ or $A_{2B}R$ activation. $A_{2B}R$ receptors are induced in response to arterial injury or by IFN-γ. Stimulation of A_{2B}Rs inhibits the IFN-γ-induced expression of MHC class II genes, nitric oxide synthase, and proinflammatory cytokines (Xaus et al., 1999).

In addition to binding adenosine, the A_{2B}R has also been reported to bind another anti-inflammatory signaling molecule, netrin-1 (Corset et al., 2000). Netrin-1 mediates its functions through stimulation of the deleted in colorectal cancer (DCC) family receptors DCC and neogenin, and the UNC5 family receptors UNC5A, UNC5B, UNC5C, and UNC5D (Barallobre et al., 2005). Netrin-1 can act as chemoattractant or chemorepellent. The DCC family of receptors mediates attraction to netrin-1, whereas the UNC5 family of receptors forms a netrin-1-dependent complex with DCC and mediates repulsion (Hong et al., 1999). In addition to its function in neuronal development, netrin-1 expressed outside the nervous system inhibits migration of leukocytes *in vitro* and *in vivo* and attenuates inflammation-mediated tissue injury. The netrin-1 receptor UNC5B is highly expressed on human monocytes, granulocytes, and lymphocytes, and netrin-1 acting through UNC5B receptor inhibits migration of monocytes (Wang et al., 2009) *in vitro*. Activation of the A_{2B}R, originally proposed to contribute to netrin effects on axons, is not required for axon outgrowth or Xenopus spinal axon attraction to netrin-1. Thus, DCC plays a central role in netrin signaling of axon growth and guidance independent of A_{2B}R activation (Stein et al., 2001). Administration of recombinant netrin-1 before or after renal IRI reduced kidney injury, apoptosis, monocyte and neutrophil infiltration, and cytokine and chemokine production (Tadagavadi et al., 2010). Analysis of different netrin-1 receptors on leukocytes showed very high expression of UNC5B but little or no expression of UNC5A, UNC5C, UNC5D, neogenin, or DCC. These findings suggest that the A_{2B}R may in fact not be the netrin-1 receptor. Rather, A_{2B}R activation may influence the expression of the netrin receptor, UNC5B, on macrophages and other leukocytes. Neutralization of UNC5B receptor reduced netrin-1-mediated protection against renal IRI, and it increased monocyte and neutrophil infiltration, as well as serum and renal cytokine and chemokine production, with increased kidney injury. These studies suggest that netrin-1 acts through UNC5B receptors that are regulated by A_{2B}R signaling to reduce inflammation.

D. T Cells

Incubation of purified C57BL/6 murine CD4(+) T lymphocytes with anti-CD3 mAb serves as a model of TCR-mediated activation and results in increased IFN-γ production and cell surface expression of activation markers, CD25 and CD69. Signaling through the TCR causes a rapid fivefold increase in A_{2A}R mRNA, which is correlated with a significant increase in the efficacy of A_{2A}R-mediated cAMP accumulation in these cells (Lappas et al., 2005). A_{2A}R stimulation not only inhibits the generation of adaptive effector

T cells but also promotes the induction of adaptive regulatory T cells. *In vitro*, antigen recognition in the setting of $A_{2A}R$ engagement induces T-cell anergy, even in the presence of costimulation (Zarek et al., 2008). T cells initially stimulated in the presence of an $A_{2A}R$ agonist fail to proliferate and produce IL-2 and IFN-γ when rechallenged in the absence of $A_{2A}R$ stimulation.

$A_{2A}R$ stimulation inhibits interleukin-6 expression while enhancing the production of TGF-β. TGF-β favors the production of anti-inflammatory T regulatory cells, while IL-6, in conjunction with TGF-β, favors the production of inflammatory Th17 inflammatory cells. Consequently, treating mice with $A_{2A}R$ agonists not only inhibits Th1 and Th17 effector cell generation but also promotes the generation of Foxp3(+) T regulatory cells. Overall, the effect of $A_{2A}R$ activation on T cells is to promote long-term T-cell anergy and the generation of adaptive T regulatory cells.

$A_{2A}Rs$ also regulate the function of T regulatory cells. Although the transfer of T regulatory cells (CD45RB(low)) blocks colitis induced by pathogenic CD45RB(high) Th cells, CD45RB(low) cells from $A_{2A}R$-deficient mice do not prevent colitis (Naganuma et al., 2006). $A_{2A}R$ agonists suppress the production of proinflammatory cytokines by CD45RB(high) and CD45RB(low) T cells in association with a loss of mRNA stability. In contrast, anti-inflammatory cytokines, including IL-10 and TGF-β, are minimally affected. Oral administration of the $A_{2A}R$ agonist ATL313 attenuated colitis in mice receiving CD45RB(high) Th cells. These data suggest that $A_{2A}R$ activation controls T-cell-mediated colitis by suppressing the expression of proinflammatory cytokines while sparing anti-inflammatory activity mediated by IL-10 and TGF-β.

$A_{2B}R$ stimulation has not been reported to have strong direct effects on T-cell function. However, activation of $A_{2B}Rs$ may indirectly promote the development of tissue rejection by inhibiting $CD4^{+}/CD25^{+}/Foxp3^{+}$ regulatory T-cell infiltration (Zhao et al., 2010).

E. NKT Cells

A_{2A} agonists have also been found to reduce injury following ischemia or trauma in liver (Alchera et al., 2008; Ben-Ari et al., 2005; Cao et al., 2009; Day et al., 2004, 2005b; Harada et al., 2000), kidney (Day et al., 2003, 2005a; Okusa et al., 1999, 2001), skin (Peirce et al., 2001), lung (Gazoni et al., 2008; Rivo et al., 2007; Sharma et al., 2010), heart (Patel et al., 2009; Rork et al., 2008; Xi et al., 2009; Yang et al., 2006b), intestine (Di Paola et al., 2010), and spinal cord (Cassada et al., 2002; Li et al., 2006; Reece et al., 2008). The cellular targets of $A_{2A}Rs$ initially were not clear. As noted above, platelets, neutrophils, and macrophages express $A_{2A}Rs$

that, respectively, inhibit oxidative burst and adhesion molecule expression (Sullivan et al., 2004a) and cytokine production (Murphree et al., 2005). We introduced *loxp* sites flanking the first coding exon of the $A_{2A}R$ gene, *adora2a*, and crossed these mice to LysMCre mice. All lines were made congenic to C57BL/6J using marker-assisted selection. The resultant LysM-Cre × $A_{2A}R^{f/f}$ mice selectively lack $A_{2A}Rs$ in neutrophils and macrophages. Nevertheless, $A_{2A}R$ activation was still highly effective at reducing injury in response to liver or lung IRI (Reutershan et al., 2007). Adoptive transfer of $CD4^+$ (but not $CD8^+$ T cells) to $Rag1^{-/-}$ mice reconstituted injury from IRI (Zhai et al., 2006). The A_{2A} agonist, ATL146e, inhibited this injury if the transferred cells had $A_{2A}Rs$, but not if they lacked $A_{2A}Rs$ (Yang et al., 2006b). This result is striking because $Rag1^{-/-}$ mice reconstituted with $A_{2A}R^{-/-}$ $CD4^+$ T cells have a normal complement of $A_{2A}Rs$ in all cells except the reconstituted T cells. The results indicate that despite the widespread distribution of $A_{2A}Rs$ on platelets and leukocytes, A_{2A} agonists reduce IRI primarily by their effects on T cells.

In 2005, Shimamura et al. found that liver reperfusion injury was associated with an expansion and activation of CD1d-restricted NKT cells (Shimamura et al. (2005)). Subsequently, we found that depletion of NKT and NK cells with PK136, an antibody that binds to NK1.1 found only on NKT and NK cells, or an anti-CD1d antibody produces protection from liver IRI that is equivalent to and not additive to protection by ATL146e (Lappas et al., 2006). These studies indicate that the adenosine-sensitive T cells that mediate IRI are iNKT cells. The putative endogenous ligands that are responsible for activating iNKT following IRI have not been identified, but recent studies suggest that tissue injury may result in the formation of one or more galactose-containing glycolipids that can activate the invariant TCR (Darmoise et al., 2010). In addition, iNKT cell activation may be facilitated by the binding of phosphatidylserine on the surface of apoptotic cells to T cell Ig-like mucin-like-1 (TIM-1) receptors on NKT cells (Lee et al., 2010). Hepatic preconditioning produced by preactivating NKT cells protects the liver from IRI via an IL-13 response and induction of $A_{2A}Rs$ (Cao et al., 2009).

As sickle cell disease is characterized by persistent multiorgan microvascular IRI, we examined the role of iNKT cells in sickle cell disease. Deletion or blockade of iNKT cell activation was found to greatly attenuate pulmonary vaso-occlusive pathophysiology in sickle cell mice. In addition, sickle cell patients were found to have increased numbers of activated iNKT cells in their blood (Wallace et al., 2009). These findings suggest that iNKT cells orchestrate a leukocyte inflammatory cascade that triggers vaso-occlusive episodes. $A_{2A}R$ agonists produce substantial protection to mouse lungs in sickle cell disease, primarily by targeting A_{2A} receptors that are induced on iNKT cells and NK cells (Wallace & Linden, 2010).

III. Disease Relevance of Adenosine to Immune Signaling

A. Diabetes

Inflammation in diabetes may be triggered in part by elevated concentrations of free fatty acids that increase CD11c+ macrophage accumulation and activation in adipose tissue (Nguyen et al., 2007). Insulin resistance due to a high-fat diet causes macrophage accumulation in adipose tissue and M2-like remodeling (Shaul et al., 2010). Endothelial dysfunction is also a hallmark of diabetes because inflammatory mediators activate receptors and transcription factors such as nuclear factor-κB, TLRs, c-Jun amino terminal kinase, and the receptor for advanced glycation end products, which cause systemic endothelial dysfunction (Goldberg, 2009). Signaling through the $A_{2B}R$ also contributes to insulin resistance by altering the production of IL-6 and other cytokines. IL-6 is produced primarily by macrophages and adipocytes and drives the production of CRP.

Several studies have linked adenosine receptor blockade with reversal of insulin resistance. Challis and coworkers reported that adenosine receptor antagonists (Challis et al., 1984) or degradation of adenosine with adenosine deaminase (Budohoski et al., 1984) reverse insulin resistance in skeletal muscle isolated from diabetic animals. The orally active adenosine receptor antagonist BW-1433, was found to persistently reverse insulin resistance in obese Zucker rats (Crist et al., 1998, 2001; Xu et al., 1998). In mice rendered insulin resistant due to a high-fat diet, *ADORA2B* gene deletion was reported to reduce body fat, reduce liver glycogen, increase energy expenditure, and increase lean body mass (Treadway et al., 2006). It is notable that statins stimulate the induction of CD73 and have been shown to cause insulin resistance. Statins also enhance ischemia-mediated vasodilation in people, and this is blocked by caffeine, consistent with an effect to enhance adenosine production (Meijer et al., 2010). Enhanced adenosine production, by activating $A_{2B}Rs$, may contribute to the effect of statins to provoke insulin resistance.

Diabetes triggers induction of $A_{2B}R$ mRNA in macrophages and endothelial cells, resulting in increased IL-6 production in response to $A_{2B}R$ activation (Figler et al., 2011). Deletion of the mouse $A_{2B}R$ resulted acutely in a proinflammatory phenotype manifested as mild vascular inflammation at baseline and exacerbation of cytokine production in response to endotoxin (Yang et al., 2006a). Thus, in some settings, signaling by the $A_{2B}R$ reduces inflammation. However, persistant activation of $A_{2B}Rs$ increased IL-6 plasma levels in mice, and by several types of isolated cells (Linden, 2006), including macrophages (Ryzhov et al., 2008b) and dendritic cells (Novitskiy et al., 2008; Ryzhov et al., 2008b). IL-6 is directly involved in stimulating the production of transcription factors that enhance CRP production (Young et al., 2008). Analyses of the cloned human $A_{2B}R$ promoter identified a functional binding site for hypoxia-inducible factor (Kong et al., 2006) and identified TNF-α and the oxidative stress-promoting enzyme NAD(P)H oxidase as additional regulators of $A_{2B}R$ gene expression (Kolachala et al., 2005). Since elevated TNF-α and oxidative stress are associated with diabetes (Castoldi et al., 2007; Gokulakrishnan et al., 2009), it is reasonable to speculate that these factors contribute to induction of $A_{2B}R$ mRNA in diabetics. Hence, $A_{2B}R$-facilitated production of IL-6 and other adipokines by macrophages that accumulate in adipose tissue of obese animals and people may contribute to insulin resistance associated with type II diabetes (Figler et al., 2011). Chronic activation of $A_{2B}Rs$ has been implicated in other pathological processes, such as pulmonary fibrosis (Sun et al., 2006).

B. Cancer

Both agonists and antagonists of adenosine receptors have been evaluated in mouse models of cancer and, in some cases, have direct effects on tumor cells that sometimes express various adenosine receptor subtypes (Fishman et al., 2009; Merighi et al., 2007). Another approach has been to target adenosine receptors in immunocompetent hosts for blockade as a means of enhancing immune killing of tumors. Most tumors are thought to produce some degree of immune activation that might be exploited to facilitate tumor rejection. For example, in bladder cancer, activation of the immune system by the immune adjuvant bacillus Calmette–Guerin (BCG) has been shown to significantly reduce tumor progression (Demkow et al., 2008). Sequential activation of NKT cells and NK cells provides effective innate immunotherapy of cancer (Smyth et al., 2005). As discussed above, signaling through A_{2A} and A_{2B} receptors generally has a strong negative effect on T cell responses. Activation of the $A_{2A}R$ on T effector cells can reduce by 98% INF-γ release (Lappas et al., 2005). $A_{2A}R$ activation on CD1d-restricted NKT cells reduces the production of INF-γ, TNF-α, and IL-2 in response to glycolipid antigens (Lappas et al., 2006). Treating mice with synthetic A_{2A} agonists inhibits Th1 and Th17 effector cell generation and promotes the generation of $Foxp3^+$ regulatory T cells (Zarek et al., 2008). Given the suppressive effects of $A_{2A}Rs$ on T cells and other leukocytes, $A_{2A}R$ blockade or deletion has been investigated to enhance immune killing of tumors. These studies have met with some success in immunocompetent mouse models with syngeneic tumors (Lukashev et al., 2007; Ohta and Sitkovsky, 2011; Ohta et al., 2006). Ohta et al. (2006) showed that solid tumors produce high concentrations of adenosine and demonstrated that genetic deletion of the $A_{2A}R$ resulted in rejection of established immunogenic lung tumors in $\sim$60% of mice with no rejection observed in control WT mice. Caffeine, a weak nonselective adenosine receptor antagonist, also significantly increased tumor rejection.

In addition to conventional Foxp3+ T regulatory cells, adaptive regulatory T cells (Tr1) are induced in the periphery upon encountering cognate antigens. In cancer, their frequency is increased; however, Tr1-mediated suppression mechanisms have only recently begun to be studied. Both ecto-nucleotidases (CD39/CD73) and cyclooxygenase 2 (COX-2) are involved in Tr1-mediated suppression. The concomitant inhibition of prostaglandin E2 and adenosine receptors via their common intracellular cyclic AMP pathway has been suggested as an additional approach for improving results of immune therapies for cancer (Mandapathil et al., 2010).

In addition to their effects on the function of T cells, $A_{2A}R$ and $A_{2B}R$ blockade may have indirect effects on tumor angiogenesis. In addition to effects of A_{2B} signaling on macrophages and DCs, both A_{2B} and A_3 receptors have been shown to facilitate the release of angiogenic factors from mast cells (Feoktistov et al., 2003). $A_{2B}R$ blockade impairs production of IL-8 in a mouse melanoma model (Merighi et al., 2009). In a Lewis lung carcinoma isograft model, deletion of the host $A_{2B}R$ lowered tumor levels of VEGF and attenuated tumor growth (Ryzhov et al., 2008a). Since $A_{2A}R$ activation strongly suppresses the production of IFN-γ by both NKT and NK cells, blockade of these receptors increases the production of IFN-γ-inducible chemokines. CXC chemokines are important in controlling leukocyte trafficking, enhancing innate and adaptive immunity, and regulating angiogenesis. CXC chemokines behave as both potent promoters of Th1-dependent cell-mediated immunity and inhibitors of angiogenesis. These chemokines bind to a specific receptor known as CXCR3. This receptor has been found on Th1 T cells, B cells, NK cells, and endothelial cells. The CXCR3 ligands represent the major chemoattractants for the recruitment of Th1 cells during cell-mediated immunity. Recently, CXCR3 has been found to exist in two alternatively spliced mRNAs (CXCR3A and CXCR3B). CXCR3B is expressed on endothelial cells and mediates the angiostatic effects of CXCR3 ligands, whereas CXCR3A appears to be expressed on T cells, B cells, and NK cells (Struyf et al., 2010). IL-2 is the major agonist for triggering the expression of CXCR3A on these leukocytes. The regulation of the expression of CXCR3B on endothelial cells remains to be fully elucidated. In addition to their role in mediating Th1-mediated immunity, CXCR3 ligands are potent and efficacious cytokines for inhibiting angiogenesis induced by VEGF, bFGF, and ELR+ CXC chemokines. $A_{2A}R$ blockade enhances the production of interferon-inducible CXC chemokines to promote Th1 immunity and inhibit angiogenesis. Studies are ongoing in several laboratories to evaluate effects of $A_{2A}R$ and $A_{2B}R$ blockade on tumor progression.

IV. Conclusion

It is now clear that purineric signaling exerts major regulatory effects on the immune system. $A_{2A}R$ activation produces strong anti-inflammatory effects on multiple cell types. As A_{2A} agonists make their way toward the clinic, it may be possible to exploit their anti-inflammatory effects to inhibit tissue injury in response to acute insults such as tissue transplantation, myocardial infarction, and flares in autoimmune diseases or sickle cell anemia. $A_{2B}R$ signaling is more complex. Although $A_{2B}R$ activation seems to produce some of the acute anti-inflammatory effects on macrophages as are produced by A_{2A} agonists, acute $A_{2B}R$ activation may elevate blood glucose, and prolonged $A_{2B}R$ signaling results in tissue reparative programs, such as fibrosis, angiogenesis, and IL-6 production that may be detrimental in some instances. $A_{2B}R$ antagonists are currently in clinical development for the treatment of asthma (due in part to inhibition of mast cell deregulation). It will be of interest to determine if such antagonists prove to be useful for the treatment of chronic inflammatory states such as pulmonary fibrosis, type II diabetes, and others.

Disclosure Statement

The author is a paid consultant to Forest Laboratories which has A_{2A} agonists and A_{2B} antagonists in clinical development.

Abbreviations

DAMPs	damage-associated molecular patterns
DCC	deleted in colorectal cancer
DCs	dendritic cells
ECs	endothelial cells
HIF-α	hypoxia-inducible factor-α
HO-1	heme oxygenase-1
IL	interleukin
iNKT	invariant NKT
IRI	ischemia reperfusion injury
MHC	major histocompatibility complex
TCR	T cell receptor
TGF-β	transforming growth factor-β
TIM-1	T cell Ig-like mucin-like-1
VEGF	vascular endothelial growth factor

References

Alam, M. S., Kurtz, C. C., Wilson, J. M., Burnette, B. R., Wiznerowicz, E. B., Ross, W. G., Figler, R. A., Linden, J., Crowe, S. E., & Ernst, P. B. (2009). A2A adenosine receptor (AR) activation inhibits pro-inflammatory cytokine production by human CD4+ helper T cells and regulates Helicobacter-induced gastritis and bacterial persistence. *Mucosal Immunology*, 2, 232–242.

Alchera, E., Tacchini, L., Imarisio, C., Dal Ponte, C., De Ponti, C., Gammella, E., Cairo, G., Albano, E., & Carini, R. (2008). Adenosine-dependent activation of hypoxia-inducible factor-1 induces late preconditioning in liver cells. *Hepatology*, *48*, 230–239.

Anvari, F., Sharma, A. K., Fernandez, L. G., Hranjec, T., Ravid, K., Kron, I. L., & Laubach, V. E. (2010). Tissue-derived proinflammatory effect of adenosine A2B receptor in lung ischemia-reperfusion injury. *The Journal of Thoracic and Cardiovascular Surgery*, *140*, 871–877.

Barallobre, M. J., Pascual, M., Del Rio, J. A., & Soriano, E. (2005). The Netrin family of guidance factors: Emphasis on Netrin-1 signalling. *Brain Research. Brain Research Reviews*, *49*, 22–47.

Ben-Ari, Z., Pappo, O., Sulkes, J., Cheporko, Y., Vidne, B. A., & Hochhauser, E. (2005). Effect of adenosine A2A receptor agonist (CGS) on ischemia/reperfusion injury in isolated rat liver. *Apoptosis*, *10*, 955–962.

Brittain, H. A., Eckman, J. R., Swerlick, R. A., Howard, R. J., & Wick, T. M. (1993). Thrombospondin from activated platelets promotes sickle erythrocyte adherence to human microvascular endothelium under physiologic flow: A potential role for platelet activation in sickle cell vaso-occlusion. *Blood*, *81*, 2137–2143.

Budohoski, L., Challiss, R. A., Cooney, G. J., McManus, B., & Newsholme, E. A. (1984). Reversal of dietary-induced insulin resistance in muscle of the rat by adenosine deaminase and an adenosine-receptor antagonist. *The Biochemical Journal*, *224*, 327–330.

Cao, Z., Yuan, Y., Jeyabalan, G., Du, Q., Tsung, A., Geller, D. A., & Billiar, T. R. (2009). Preactivation of NKT cells with alpha-GalCer protects against hepatic ischemia-reperfusion injury in mouse by a mechanism involving IL-13 and adenosine A2A receptor. *American Journal of Physiology. Gastrointestinal and Liver Physiology*, *297*, G249–G258.

Cassada, D. C., Tribble, C. G., Young, J. S., Gangemi, J. J., Gohari, A. R., Butler, P. D., Rieger, J. M., Kron, I. L., & Linden, J. (2002). Adenosine A2A analogue improves neurologic outcome after spinal cord trauma in the rabbit. *The Journal of Trauma*, *53*, 225–229, discussion 229–231.

Castoldi, G., Galimberti, S., Riva, C., Papagna, R., Querci, F., Casati, M., Zerbini, G., Caccianiga, G., Ferrarese, C., Baldoni, M., Valsecchi, M. G., & Stella, A. (2007). Association between serum values of C-reactive protein and cytokine production in whole blood of patients with type 2 diabetes. *Clinical Science (London)*, *113*, 103–108.

Challis, R. A., Budohoski, L., McManus, B., & Newsholme, E. A. (1984). Effects of an adenosine-receptor antagonist on insulin-resistance in soleus muscle from obese Zucker rats. *The Biochemical Journal*, *221*, 915–917.

Chen, Y., Corriden, R., Inoue, Y., Yip, L., Hashiguchi, N., Zinkernagel, A., Nizet, V., Insel, P. A., & Junger, W. G. (2006). ATP release guides neutrophil chemotaxis via P2Y2 and A3 receptors. *Science*, *314*, 1792–1795.

Chen, Y., Yao, Y., Sumi, Y., Li, A., To, U. K., Elkhal, A., Inoue, Y., Woehrle, T., Zhang, Q., Hauser, C., & Junger, W. G. (2010). Purinergic signaling: A fundamental mechanism in neutrophil activation. *Science Signaling*, *3*, ra45.

Cooper, J. A., Hill, S. J., Alexander, S. P., Rubin, P. C., & Horn, E. H. (1995). Adenosine receptor-induced cyclic AMP generation and inhibition of 5-hydroxytryptamine release in human platelets. *British Journal of Clinical Pharmacology*, *40*, 43–50.

Corset, V., Nguyen-Ba-Charvet, K. T., Forcet, C., Moyse, E., Chedotal, A., & Mehlen, P. (2000). Netrin-1-mediated axon outgrowth and cAMP production requires interaction with adenosine A2b receptor. *Nature*, *407*, 747–750.

Crist, G. H., Xu, B., Berkich, D. A., & LaNoue, K. F. (2001). Effects of adenosine receptor antagonism on protein tyrosine phosphatase in rat skeletal muscle. *The International Journal of Biochemistry & Cell Biology*, *33*, 817–830.

Crist, G. H., Xu, B., Lanoue, K. F., & Lang, C. H. (1998). Tissue-specific effects of in vivo adenosine receptor blockade on glucose uptake in Zucker rats. *The FASEB Journal*, *12*, 1301–1308.

Darmoise, A., Teneberg, S., Bouzonville, L., Brady, R. O., Beck, M., Kaufmann, S. H., & Winau, F. (2010). Lysosomal alpha-galactosidase controls the generation of self lipid antigens for natural killer T cells. *Immunity*, *33*, 216–228.

Day, Y. J., Huang, L., McDuffie, M. J., Rosin, D. L., Ye, H., Chen, J. F., Schwarzschild, M. A., Fink, J. S., Linden, J., & Okusa, M. D. (2003). Renal protection from ischemia mediated by A2A adenosine receptors on bone marrow-derived cells. *Journal of Clinical Investigation*, *112*, 883–891.

Day, Y. J., Huang, L., Ye, H., Linden, J., & Okusa, M. D. (2005a). Renal ischemia-reperfusion injury and adenosine 2A receptor-mediated tissue protection: Role of macrophages. *American Journal of Physiology. Renal Physiology*, *288*, F722–F731.

Day, Y. J., Li, Y., Rieger, J. M., Ramos, S. I., Okusa, M. D., & Linden, J. (2005b). A2A adenosine receptors on bone marrow-derived cells protect liver from ischemia-reperfusion injury. *Journal of Immunology*, *174*, 5040–5046.

Day, Y. J., Marshall, M. A., Huang, L., McDuffie, M. J., Okusa, M. D., & Linden, J. (2004). Protection from ischemic liver injury by activation of A2A adenosine receptors during reperfusion: Inhibition of chemokine induction. *American Journal of Physiology. Gastrointestinal and Liver Physiology*, *286*, G285–G293.

Demkow, T., Alter, A., & Wiechno, P. (2008). Intravesical bacillus Calmette-Guerin therapy for T1 superficial bladder cancer. *Urologia Internationalis*, *80*, 74–79.

Di Paola, R., Melani, A., Esposito, E., Mazzon, E., Paterniti, I., Bramanti, P., Pedata, F., & Cuzzocrea, S. (2010). Adenosine A2A receptor-selective stimulation reduces signaling pathways involved in the development of intestine ischemia and reperfusion injury. *Shock*, *33*, 541–551.

Feoktistov, I., Ryzhov, S., Goldstein, A. E., & Biaggioni, I. (2003). Mast cell-mediated stimulation of angiogenesis: Cooperative interaction between A2B and A3 adenosine receptors. *Circulation Research*, *92*, 485–492.

Figler, R. A., Wang, G., Srinivasan, S., Jung, D. Y., Zhiyou, Z., Pankow, J. S., Ravid, K., Fredholm, B., Hedrick, C. C., Rich, S. S., Kim, J. K., LaNoue, K. F., & Linden, J. (2011). Links between insulin resistance, adenosine A2B receptors and inflammatory markers in mice and humans. *Diabetes*, *60*, 1–11.

Fishman, P., Bar-Yehuda, S., Synowitz, M., Powell, J. D., Klotz, K. N., Gessi, S., & Borea, P. A. (2009). Adenosine receptors and cancer. *Handbook of Experimental Pharmacology*, *193*, 399–441.

Fredholm, B. B., Zhang, Y., & van der Ploeg, I. (1996). Adenosine A2A receptors mediate the inhibitory effect of adenosine on formyl-Met-Leu-Phe-stimulated respiratory burst in neutrophil leucocytes. *Naunyn Schmiedebergs Archives of Pharmacology*, *354*, 262–267.

Gazoni, L. M., Laubach, V. E., Mulloy, D. P., Bellizzi, A., Unger, E. B., Linden, J., Ellman, P. I., Lisle, T. C., & Kron, I. L. (2008). Additive protection against lung ischemia-reperfusion injury by adenosine A2A receptor activation before procurement and during reperfusion. *The Journal of Thoracic and Cardiovascular Surgery*, *135*, 156–165.

Ge, Z. D., van der Hoeven, D., Maas, J. E., Wan, T. C., & Auchampach, J. A. (2010). A(3) adenosine receptor activation during reperfusion reduces infarct size through actions on bone marrow-derived cells. *Journal of Molecular and Cellular Cardiology*, *49*, 280–286.

Gessi, S., Merighi, S., Varani, K., Leung, E., Mac Lennan, S., & Borea, P. A. (2008). The A3 adenosine receptor: An enigmatic player in cell biology. *Pharmacology & Therapeutics*, *117*, 123–140.

Gokulakrishnan, K., Mohanavalli, K. T., Monickaraj, F., Mohan, V., & Balasubramanyam, M. (2009). Subclinical inflammation/oxidation as revealed by altered gene expression profiles in subjects with impaired glucose tolerance and Type 2 diabetes patients. *Molecular and Cellular Biochemistry*, *324*, 173–181.

Goldberg, R. B. (2009). Cytokine and cytokine-like inflammation markers, endothelial dysfunction, and imbalanced coagulation in development of diabetes and its complications. *The Journal of Clinical Endocrinology and Metabolism*, *94*, 3171–3182.

Grinberg, S., Hasko, G., Wu, D., & Leibovich, S. J. (2009). Suppression of PLCbeta2 by endotoxin plays a role in the adenosine A(2A) receptor-mediated switch of macrophages from an inflammatory to an angiogenic phenotype. *The American Journal of Pathology*, *175*, 2439–2453.

Harada, N., Okajima, K., Murakami, K., Usune, S., Sato, C., Ohshima, K., & Katsuragi, T. (2000). Adenosine and selective A(2A) receptor agonists reduce ischemia/reperfusion injury of rat liver mainly by inhibiting leukocyte activation. *The Journal of Pharmacology and Experimental Therapeutics*, *294*, 1034–1042.

Hasko, G., Linden, J., Cronstein, B., & Pacher, P. (2008). Adenosine receptors: Therapeutic aspects for inflammatory and immune diseases. *Nature Reviews. Drug Discovery*, *7*, 759–770.

Hasko, G., Pacher, P., Deitch, E. A., & Vizi, E. S. (2007). Shaping of monocyte and macrophage function by adenosine receptors. *Pharmacology & Therapeutics*, *113*, 264–275.

Hong, K., Hinck, L., Nishiyama, M., Poo, M. M., Tessier-Lavigne, M., & Stein, E. (1999). A ligand-gated association between cytoplasmic domains of UNC5 and DCC family receptors converts netrin-induced growth cone attraction to repulsion. *Cell*, *97*, 927–941.

Inwald, D. P., Kirkham, F. J., Peters, M. J., Lane, R., Wade, A., Evans, J. P., & Klein, N. J. (2000). Platelet and leucocyte activation in childhood sickle cell disease: Association with nocturnal hypoxaemia. *British Journal Haematology*, *111*, 474–481.

Iyu, D., Glenn, J. R., White, A. E., Fox, S. C., & Heptinstall, S. (2011). Adenosine derived from ADP can contribute to inhibition of platelet aggregation in the presence of a P2Y12 antagonist. *Arteriosclerosis, Thrombosis, and Vascular Biology*, *31*, 416–422.

Jin, D., Fan, J., Wang, L., Thompson, L. F., Liu, A., Daniel, B. J., Shin, T., Curiel, T. J., & Zhang, B. (2010). CD73 on tumor cells impairs antitumor T-cell responses: A novel mechanism of tumor-induced immune suppression. *Cancer Research*, *70*, 2245–2255.

Kolachala, V., Asamoah, V., Wang, L., Obertone, T. S., Ziegler, T. R., Merlin, D., & Sitaraman, S. V. (2005). TNF-alpha upregulates adenosine 2b (A2b) receptor expression and signaling in intestinal epithelial cells: A basis for A2bR overexpression in colitis. *Cellular and Molecular Life Sciences*, *62*, 2647–2657.

Kong, T., Westerman, K. A., Faigle, M., Eltzschig, H. K., & Colgan, S. P. (2006). HIF-dependent induction of adenosine A2B receptor in hypoxia. *The FASEB Journal*, *20*, 2242–2250.

Kumar, V., & Sharma, A. (2009). Adenosine: An endogenous modulator of innate immune system with therapeutic potential. *European Journal of Pharmacology*, *616*, 7–15.

Lappas, C. M., Day, Y. J., Marshall, M. A., Engelhard, V. H., & Linden, J. (2006). Adenosine A2A receptor activation reduces hepatic ischemia reperfusion injury by inhibiting CD1d-dependent NKT cell activation. *The Journal of Experimental Medicine*, *203*, 2639–2648.

Lappas, C. M., Rieger, J. M., & Linden, J. (2005). A2A adenosine receptor induction inhibits IFN-gamma production in murine CD4+ T cells. *Journal of Immunology*, *174*, 1073–1080.

Ledent, C., Vaugeois, J. M., Schiffmann, S. N., Pedrazzini, T., El Yacoubi, M., Vanderhaeghen, J. J., Costentin, J., Heath, J. K., Vassart, G., & Parmentier, M. (1997). Aggressiveness, hypoalgesia and high blood pressure in mice lacking the adenosine A2a receptor. *Nature*, *388*, 674–678.

Lee, H. H., Meyer, E. H., Goya, S., Pichavant, M., Kim, H. Y., Bu, X., Umetsu, S. E., Jones, J. C., Savage, P. B., Iwakura, Y., Casasnovas, J. M., Kaplan, G., Freeman, G. J., DeKruyff, R. H., & Umetsu, D. T. (2010). Apoptotic cells activate NKT cells through T cell Ig-like mucin-like-1 resulting in airway hyperreactivity. *Journal of Immunology*, *185*, 5225–5235.

Li, Y., Oskouian, R. J., Day, Y. J., Rieger, J. M., Liu, L., Kern, J. A., & Linden, J. (2006). Mouse spinal cord compression injury is reduced by either activation of the adenosine A2A receptor on bone marrow-derived cells or deletion of the A2A receptor on non-bone marrow-derived cells. *Neuroscience*, *141*, 2029–2039.

Linden, J. (2006). New insights into the regulation of inflammation by adenosine. *Journal of Clinical Investigation*, *116*, 1835–1837.

Lukashev, D., Sitkovsky, M., & Ohta, A. (2007). From "Hellstrom Paradox" to anti-adenosinergic cancer immunotherapy. *Purinergic Signalling*, *3*, 129–134.

Mandapathil, M., Szczepanski, M. J., Szajnik, M., Ren, J., Jackson, E. K., Johnson, J. T., Gorelik, E., Lang, S., & Whiteside, T. L. (2010). Adenosine and prostaglandin E2 cooperate in the suppression of immune responses mediated by adaptive regulatory T cells. *The Journal of Biological Chemistry*, *285*, 27571–27580.

McDonald, B., Pittman, K., Menezes, G. B., Hirota, S. A., Slaba, I., Waterhouse, C. C., Beck, P. L., Muruve, D. A., & Kubes, P. (2010). Intravascular danger signals guide neutrophils to sites of sterile inflammation. *Science*, *330*, 362–366.

Meijer, P., Wouters, C. W., van den Broek, P. H., de Rooij, M., Scheffer, G. J., Smits, P., & Rongen, G. A. (2010). Upregulation of ecto-5′-nucleotidase by rosuvastatin increases the vasodilator response to ischemia. *Hypertension*, *56*, 722–727.

Merhi, Y., Provost, P., Chauvet, P., Theoret, J. F., Phillips, M. L., & Latour, J. G. (1999). Selectin blockade reduces neutrophil interaction with platelets at the site of deep arterial injury by angioplasty in pigs. *Arteriosclerosis, Thrombosis, and Vascular Biology*, *19*, 372–377.

Merighi, S., Benini, A., Mirandola, P., Gessi, S., Varani, K., Simioni, C., Leung, E., Maclennan, S., Baraldi, P. G., & Borea, P. A. (2007). Caffeine inhibits adenosine-induced accumulation of hypoxia-inducible factor-1alpha, vascular endothelial growth factor, and interleukin-8 expression in hypoxic human colon cancer cells. *Molecular Pharmacology*, *72*, 395–406.

Merighi, S., Simioni, C., Gessi, S., Varani, K., Mirandola, P., Tabrizi, M. A., Baraldi, P. G., & Borea, P. A. (2009). A(2B) and A(3) adenosine receptors modulate vascular endothelial growth factor and interleukin-8 expression in human melanoma cells treated with etoposide and doxorubicin. *Neoplasia*, *11*, 1064–1073.

Murphree, L. J., Sullivan, G. W., Marshall, M. A., & Linden, J. (2005). Lipopolysaccharide rapidly modifies adenosine receptor transcripts in murine and human macrophages: Role of NF-kappaB in A(2A) adenosine receptor induction. *The Biochemical Journal*, *391*, 575–580.

Naganuma, M., Wiznerowicz, E. B., Lappas, C. M., Linden, J., Worthington, M. T., & Ernst, P. B. (2006). Cutting edge: Critical role for A2A adenosine receptors in the T cell-mediated regulation of colitis. *Journal of Immunology*, *177*, 2765–2769.

Nance, D. M., & Sanders, V. M. (2007). Autonomic innervation and regulation of the immune system (1987–2007). *Brain, Behavior, and Immunity*, *21*, 736–745.

Nguyen, M. T., Favelyukis, S., Nguyen, A. K., Reichart, D., Scott, P. A., Jenn, A., Liu-Bryan, R., Glass, C. K., Neels, J. G., & Olefsky, J. M. (2007). A subpopulation of macrophages infiltrates hypertrophic adipose tissue and is activated by free fatty acids via Toll-like receptors 2 and 4 and JNK-dependent pathways. *The Journal of Biological Chemistry*, *282*, 35279–35292.

Novitskiy, S. V., Ryzhov, S., Zaynagetdinov, R., Goldstein, A. E., Huang, Y., Tikhomirov, O. Y., Blackburn, M. R., Biaggioni, I., Carbone, D. P., Feoktistov, I., & Dikov, M. M. (2008). Adenosine receptors in regulation of dendritic cell differentiation and function. *Blood*, *112*, 1822–1831.

Ohta, A., Gorelik, E., Prasad, S. J., Ronchese, F., Lukashev, D., Wong, M. K., Huang, X., Caldwell, S., Liu, K., Smith, P., Chen, J. F., Jackson, E. K., Apasov, S., Abrams, S., & Sitkovsky, M. (2006). A2A adenosine receptor protects tumors from antitumor T cells. *Proceedings of the National Academy of Sciences of the United States of America*, *103*, 13132–13137.

Ohta, A., & Sitkovsky, M. (2011). Methylxanthines, inflammation, and cancer: Fundamental mechanisms. *Handbook of Experimental Pharmacology*, *200*, 469–481.

Okusa, M. D., Linden, J., Huang, L., Rieger, J. M., Macdonald, T. L., & Huynh, L. P. (2000). A(2A) adenosine receptor-mediated inhibition of renal injury and neutrophil adhesion. *American Journal of Physiology. Renal Physiology*, *279*, F809–F818.

Okusa, M. D., Linden, J., Huang, L., Rosin, D. L., Smith, D. F., & Sullivan, G. (2001). Enhanced protection from renal ischemia-reperfusion [correction of ischemia:reperfusion] injury with A(2A)-adenosine receptor activation and PDE 4 inhibition. *Kidney International*, *59*, 2114–2125.

Okusa, M. D., Linden, J., Macdonald, T., & Huang, L. (1999). Selective A2A adenosine receptor activation reduces ischemia-reperfusion injury in rat kidney. *The American Journal of Physiology*, *277*, F404–F412.

Oury, C., Toth-Zsamboki, E., Vermylen, J., & Hoylaerts, M. F. (2006). The platelet ATP and ADP receptors. *Current Pharmaceutical Design*, *12*, 859–875.

Patel, R. A., Glover, D. K., Broisat, A., Kabul, H. K., Ruiz, M., Goodman, N. C., Kramer, C. M., Meerdink, D. J., Linden, J., & Beller, G. A. (2009). Reduction in myocardial infarct size at 48 hours after brief intravenous infusion of ATL-146e, a highly selective adenosine A2A receptor agonist. *American Journal of Physiology. Heart and Circulatory Physiology*, *297*, H637–H642.

Peirce, S. M., Skalak, T. C., Rieger, J. M., Macdonald, T. L., & Linden, J. (2001). Selective A (2A) adenosine receptor activation reduces skin pressure ulcer formation and inflammation. *American Journal of Physiology. Heart and Circulatory Physiology*, *281*, H67–H74.

Pelegrin, P. (2008). Targeting interleukin-1 signaling in chronic inflammation: Focus on P2X(7) receptor and Pannexin-1. *Drug News & Perspectives*, *21*, 424–433.

Polanowska-Grabowska, R., Wallace, K., Field, J. J., Chen, L., Marshall, M. A., Figler, R., Gear, A. R., & Linden, J. (2010). P-selectin-mediated platelet-neutrophil aggregate formation activates neutrophils in mouse and human sickle cell disease. *Arteriosclerosis, Thrombosis, and Vascular Biology*, *30*, 2392–2399.

Reece, T. B., Tribble, C. G., Okonkwo, D. O., Davis, J. D., Maxey, T. S., Gazoni, L. M., Linden, J., Kron, I. L., & Kern, J. A. (2008). Early adenosine receptor activation ameliorates spinal cord reperfusion injury. *Journal of Cardiovascular Medicine (Hagerstown)*, *9*, 363–367.

Reutershan, J., Cagnina, R. E., Chang, D., Linden, J., & Ley, K. (2007). Therapeutic anti-inflammatory effects of myeloid cell adenosine receptor A2a stimulation in lipopolysaccharide-induced lung injury. *Journal of Immunology*, *179*, 1254–1263.

Revan, S., Montesinos, M. C., Naime, D., Landau, S., & Cronstein, B. N. (1996). Adenosine A2 receptor occupancy regulates stimulated neutrophil function via activation of a serine/threonine protein phosphatase. *The Journal of Biological Chemistry*, *271*, 17114–17118.

Rivo, J., Zeira, E., Galun, E., Einav, S., Linden, J., & Matot, I. (2007). Attenuation of reperfusion lung injury and apoptosis by A2A adenosine receptor activation is associated with modulation of Bcl-2 and Bax expression and activation of extracellular signal-regulated kinases. *Shock*, *27*, 266–273.

Rork, T. H., Wallace, K. L., Kennedy, D. P., Marshall, M. A., Lankford, A. R., & Linden, J. (2008). Adenosine A2A receptor activation reduces infarct size in the isolated, perfused mouse heart by inhibiting resident cardiac mast cell degranulation. *American Journal of Physiology. Heart and Circulatory Physiology*, *295*, H1825–H1833.

Ryzhov, S., Novitskiy, S. V., Zaynagetdinov, R., Goldstein, A. E., Carbone, D. P., Biaggioni, I., Dikov, M. M., & Feoktistov, I. (2008a). Host A(2B) adenosine receptors promote carcinoma growth. *Neoplasia*, *10*, 987–995.

Ryzhov, S., Zaynagetdinov, R., Goldstein, A. E., Novitskiy, S. V., Blackburn, M. R., Biaggioni, I., & Feoktistov, I. (2008b). Effect of A2B adenosine receptor gene ablation on adenosine-dependent regulation of proinflammatory cytokines. *The Journal of Pharmacology and Experimental Therapeutics*, *324*, 694–700.

Sharma, A. K., Laubach, V. E., Ramos, S. I., Zhao, Y., Stukenborg, G., Linden, J., Kron, I. L., & Yang, Z. (2010). Adenosine A2A receptor activation on CD4+ T lymphocytes and neutrophils attenuates lung ischemia-reperfusion injury. *The Journal of Thoracic and Cardiovascular Surgery*, *139*, 474–482.

Shaul, M. E., Bennett, G., Strissel, K. J., Greenberg, A. S., & Obin, M. S. (2010). Dynamic, M2-like remodeling phenotypes of CD11c+ adipose tissue macrophages during high-fat diet–induced obesity in mice. *Diabetes*, *59*, 1171–1181.

Shimamura, K., Kawamura, H., Nagura, T., Kato, T., Naito, T., Kameyama, H., Hatakeyama, K., & Abo, T. (2005). Association of NKT cells and granulocytes with liver injury after reperfusion of the portal vein. *Cellular Immunology*, *234*, 31–38.

Smyth, M. J., Wallace, M. E., Nutt, S. L., Yagita, H., Godfrey, D. I., & Hayakawa, Y. (2005). Sequential activation of NKT cells and NK cells provides effective innate immunotherapy of cancer. *The Journal of Experimental Medicine*, *201*, 1973–1985.

Stein, E., Zou, Y., Poo, M., & Tessier-Lavigne, M. (2001). Binding of DCC by netrin-1 to mediate axon guidance independent of adenosine A2B receptor activation. *Science*, *291*, 1976–1982.

Struyf, S., Salogni, L., Burdick, M. D., Vandercappellen, J., Gouwy, M., Noppen, S., Proost, P., Opdenakker, G., Parmentier, M., Gerard, C., Sozzani, S., Strieter, R. M., & Van Damme, J. (2010). Angiostatic and chemotactic activities of the CXC chemokine CXCL4L1 (platelet factor-4 variant) are mediated by CXCR3. *Blood*, *117*, 480–488.

Sullivan, G. W., Fang, G., Linden, J., & Scheld, W. M. (2004a). A2A adenosine receptor activation improves survival in mouse models of endotoxemia and sepsis. *The Journal of Infectious Diseases*, *189*, 1897–1904.

Sullivan, G. W., Lee, D. D., Ross, W. G., DiVietro, J. A., Lappas, C. M., Lawrence, M. B., & Linden, J. (2004b). Activation of A2A adenosine receptors inhibits expression of alpha 4/beta 1 integrin (very late antigen-4) on stimulated human neutrophils. *Journal of Leukocyte Biology*, *75*, 127–134.

Sullivan, G. W., Rieger, J. M., Scheld, W. M., Macdonald, T. L., & Linden, J. (2001). Cyclic AMP-dependent inhibition of human neutrophil oxidative activity by substituted 2-propynylcyclohexyl adenosine A(2A) receptor agonists. *British Journal of Pharmacology*, *132*, 1017–1026.

Sun, C. X., Zhong, H., Mohsenin, A., Morschl, E., Chunn, J. L., Molina, J. G., Belardinelli, L., Zeng, D., & Blackburn, M. R. (2006). Role of A2B adenosine receptor signaling in adenosine-dependent pulmonary inflammation and injury. *Journal of Clinical Investigation*, *116*, 2173–2182.

Tadagavadi, R. K., Wang, W., & Ramesh, G. (2010). Netrin-1 regulates Th1/Th2/Th17 cytokine production and inflammation through UNC5B receptor and protects kidney against ischemia-reperfusion injury. *Journal of Immunology*, *185*, 3750–3758.

Treadway, J. L., Sacca, R., & Jones, B. K. (2006). Adenosine A(2B) receptor knock-out mice display an improved metabolic phenotype. *Diabetologia*, *49*, 44–45.

Turhan, A., Weiss, L. A., Mohandas, N., Coller, B. S., & Frenette, P. S. (2002). Primary role for adherent leukocytes in sickle cell vascular occlusion: A new paradigm. *Proceedings of the National Academy of Sciences of the United States of America*, *99*, 3047–3051.

Wallace, K. L., & Linden, J. (2010). Adenosine A2A receptors induced on iNKT and NK cells reduce pulmonary inflammation and injury in mice with sickle cell disease. *Blood*, *116*, 5010–5020.

Wallace, K. L., Marshall, M. A., Ramos, S. I., Lannigan, J. A., Field, J. J., Strieter, R. M., & Linden, J. (2009). NKT cells mediate pulmonary inflammation and dysfunction in murine sickle cell disease through production of IFN-gamma and CXCR3 chemokines. *Blood*, *114*, 667–676.

Wang, W., Reeves, W. B., Pays, L., Mehlen, P., & Ramesh, G. (2009). Netrin-1 overexpression protects kidney from ischemia reperfusion injury by suppressing apoptosis. *The American Journal of Pathology*, *175*, 1010–1018.

Weis, N., Weigert, A., von Knethen, A., & Brune, B. (2009). Heme oxygenase-1 contributes to an alternative macrophage activation profile induced by apoptotic cell supernatants. *Molecular Biology of the Cell*, *20*, 1280–1288.

Xaus, J., Mirabet, M., Lloberas, J., Soler, C., Lluis, C., Franco, R., & Celada, A. (1999). IFN-gamma up-regulates the A2B adenosine receptor expression in macrophages: A mechanism of macrophage deactivation. *Journal of Immunology*, *162*, 3607–3614.

Xi, J., McIntosh, R., Shen, X., Lee, S., Chanoit, G., Criswell, H., Zvara, D. A., & Xu, Z. (2009). Adenosine A2A and A2B receptors work in concert to induce a strong protection against reperfusion injury in rat hearts. *Journal of Molecular and Cellular Cardiology*, *47*, 684–690.

Xu, B., Berkich, D. A., Crist, G. H., & LaNoue, K. F. (1998). A1 adenosine receptor antagonism improves glucose tolerance in Zucker rats. *The American Journal of Physiology*, *274*, E271–E279.

Yang, D., Chen, H., Koupenova, M., Carroll, S. H., Eliades, A., Freedman, J. E., Toselli, P., & Ravid, K. (2010). A new role for the A2b adenosine receptor in regulating platelet function. *Journal of Thrombosis and Haemostasis*, *8*, 817–827.

Yang, D., Zhang, Y., Nguyen, H. G., Koupenova, M., Chauhan, A. K., Makitalo, M., Jones, M. R., St Hilaire, C., Seldin, D. C., Toselli, P., Lamperti, E., Schreiber, B. M., Gavras, H., Wagner, D. D., & Ravid, K. (2006a). The A2B adenosine receptor protects against inflammation and excessive vascular adhesion. *Journal of Clinical Investigation*, *116*, 1913–1923.

Yang, Z., Day, Y. J., Toufektsian, M. C., Xu, Y., Ramos, S. I., Marshall, M. A., French, B. A., & Linden, J. (2006b). Myocardial infarct-sparing effect of adenosine A2A receptor activation is due to its action on CD4+ T lymphocytes. *Circulation*, *114*, 2056–2064.

Young, D. P., Kushner, I., & Samols, D. (2008). Binding of C/EBPbeta to the C-reactive protein (CRP) promoter in Hep3B cells is associated with transcription of CRP mRNA. *Journal of Immunology*, *181*, 2420–2427.

Zarek, P. E., Huang, C. T., Lutz, E. R., Kowalski, J., Horton, M. R., Linden, J., Drake, C. G., & Powell, J. D. (2008). A2A receptor signaling promotes peripheral tolerance by inducing T-cell anergy and the generation of adaptive regulatory T cells. *Blood*, *111*, 251–259.

Zhai, Y., Shen, X. D., Hancock, W. W., Gao, F., Qiao, B., Lassman, C., Belperio, J. A., Strieter, R. M., Busuttil, R. W., & Kupiec-Weglinski, J. W. (2006). CXCR3+CD4+ T cells mediate innate immune function in the pathophysiology of liver ischemia/reperfusion injury. *Journal of Immunology*, *176*, 6313–6322.

Zhao, Y., Lapar, D. J., Steidle, J., Emaminia, A., Kron, I. L., Ailawadi, G., Linden, J., & Lau, C. L. (2010). Adenosine signaling via the adenosine 2B receptor is involved in bronchiolitis obliterans development. *The Journal of Heart and Lung Transplantation*, *29*, 1405–1414.

Igor Feoktistov* **and Italo Biaggioni**[†]

*Departments of Medicine and Pharmacology, Vanderbilt University, Division of Cardiovascular Medicine, Nashville, Tennessee, USA

[†]Departments of Medicine and Pharmacology, Vanderbilt University, Division of Clinical Pharmacology, Nashville, Tennessee, USA

Role of Adenosine A_{2B} Receptors in Inflammation

Abstract

Recent progress in our understanding of the unique role of A_{2B} receptors in the regulation of inflammation, immunity, and tissue repair was considerably facilitated with the introduction of new pharmacological and genetic tools. However, it also led to seemingly conflicting conclusions on the role of A_{2B} adenosine receptors in inflammation with some publications indicating proinflammatory effects and others suggesting the opposite. This chapter reviews the functions of A_{2B} receptors in various cell types related to inflammation and integrated effects of A_{2B} receptor modulation in several animal models of inflammation. It is argued that translation of current findings into novel therapies would require a better understanding of A_{2B} receptor functions in diverse types of inflammatory responses in various tissues and at different points of their progression.

I. Introduction

The extracellular accumulation of adenosine contributes to the regulation of inflammation, immunity, and tissue repair. Adenosine exerts its action by interacting with four adenosine receptor subtypes, A_1, A_{2A}, A_{2B}, and A_3, that belong to the family of seven transmembrane G-protein-coupled receptors (Fredholm et al., 2001a).

Among adenosine receptor subtypes, the A_{2B} receptor has the lowest affinity to adenosine requiring micromolar concentrations to become functional, whereas the affinities of other adenosine receptor subtypes are

Advances in Pharmacology, Volume 61

1054-3589/11 $35.00
10.1016/B978-0-12-385526-8.00005-9

significantly higher, rendering them active well below micromolar concentrations of adenosine (Fredholm et al., 2001b). Therefore, it is likely that A_{2B} receptors remain silent under the resting conditions when extracellular adenosine concentrations are low, estimated between 10 and few hundred nanomolar range (Fredholm, 2007), but their role becomes more important in pathophysiological conditions when adenosine concentrations are the highest.

Because both A_{2A} and A_{2B} receptor subtypes stimulate adenylate cyclase, A_{2B} receptors were often viewed as being a redundant low-affinity version of A_{2A} receptors. However, there is also strong evidence that A_{2B} receptors play a nonredundant role distinct from and often opposite to that of A_{2A} receptors.

II. Tools to Study A_{2B} Adenosine Receptor Function

Recent progress in our understanding of the unique role of A_{2B} receptors was considerably facilitated with the introduction of new pharmacological and genetic tools. Results from using these complementary approaches, however, have not always agreed with some publications indicating proinflammatory effects and others suggesting the opposite. We will address these issues later in this chapter but first will discuss potential pitfalls in using genetic and pharmacological tools that might have contributed to inadvertent misinterpretation of results.

Generation of mice deficient in A_{2B} adenosine receptor provided a powerful tool for investigating its function *in vivo* and *in vitro*. Several studies have found significant differences between A_{2B}KO and wild-type (WT) mouse phenotypes at rest. These differences include increases in basal TNF-α secretion, leukocyte adhesion, and vascular permeability as well as changes in expression of E-selectin, P-selectin, ICAM-1, IκB, and $P2Y_1$ receptors (Eckle et al., 2008a; Yang et al., 2006, 2010). This is surprising because basal levels of adenosine are thought to be too low to activate A_{2B} receptors. As with many gene knockouts, there is always concern that potential compensatory changes in the expression of other proteins may confound the specific role of A_{2B} receptors in any given process. Therefore, it is important to keep in mind limitations of gene knockouts so that appropriate controls can be included and correct conclusions can be made.

A number of potent and selective A_{2B} receptor antagonists have been synthesized over the past decade (for review, see Kalla & Zablocki, 2009). They were used together with antagonists of other adenosine receptor subtypes to elucidate the role of A_{2B} receptors in inflammatory models. When interpreting these studies, however, attention should be paid to the use of antagonists at concentrations that remain selective for the receptor to be targeted. For example, DPCPX and ZM241385 are often used as selective A_1 and A_{2A} antagonists, respectively, but the fact that they can also bind to human and mouse A_{2B} receptors at low nanomolar concentrations (Kreckler et al., 2006; Linden et al., 1999) is often overlooked. The table in this chapter summarizes binding affinities of adenosine antagonists at human and mouse adenosine receptor subtypes commonly used to study the role of A_{2B} receptors.

The goal of attaining selectivity for A_{2B} agonists has been more elusive than for antagonists, but several potent A_{2B} agonists have been developed recently (for review, see Baraldi et al., 2009). BAY 60-6583 has been described as a high-affinity (EC_{50} of 3 nM) and specific A_{2B} agonist based on its effects on the activity of reporters coexpressed together with human adenosine receptors (Eckle et al., 2007). However, the reported affinity of BAY 60-6583 to A_{2B} receptors determined by radioligand binding was approximately two order magnitude lower (K_i of 330–750 nM; Auchampach et al., 2009). It is likely, therefore, that the affinity of BAY 60-6583 to A_{2B} receptors was overestimated in a functional reporter assay, possibly due to the significant receptor reserve in cells overexpressing A_{2B} receptors (Linden et al., 1999). Comparative binding studies at all four adenosine receptor subtypes would be necessary to validate the selectivity of this compound toward A_{2B} receptors.

In summary, the combination of novel genetic and pharmacological approaches provides powerful tools to dissect the role of adenosine signaling through A_{2B} receptors in physiological and pathological processes. As important as these advances are, we wanted to alert the reader to the importance of taking into account the pitfalls and limitations of these approaches for the correct interpretation of results of studies on the role of A_{2B} receptors in adenosine-dependent regulation of inflammatory responses.

III. A_{2B} Receptors on Immune Cells

A. Neutrophils, Lymphocytes, Platelets

A_{2B} receptors are ubiquitously expressed and, therefore, it is not surprising that they are present on various cells of hematopoietic origin. In most cases, A_{2B} receptors are coexpressed with A_{2A} receptors. A_{2B} receptor transcripts are found in neutrophils (Fredholm et al., 1996), lymphocytes (Gessi et al., 2005), and even platelets (Amisten et al., 2008). Little is known about their specific functions in these cells, and A_{2A} receptors appear to predominate (Csoka et al., 2008; Fredholm et al., 1996; Yang et al., 2010).

Recently, the neuronal guidance molecule netrin-1 was proposed to inhibit neutrophil migration through activation of A_{2B} receptors located on these cells (Rosenberger et al., 2009). Although intriguing, the issue of interactions between netrin-1 and A_{2B} receptors remains controversial. It has been previously hypothesized that A_{2B} receptors can also serve as receptors for netrin-1 (Corset et al., 2000). However, there is evidence against a direct effect of netrin-1 on A_{2B} receptors. First, binding assays in COS cells over-expressing A_{2B} receptors failed to demonstrate netrin-1 binding to these receptors (McKenna et al., 2008). Second, the lack of effects of 100 μM DPCPX and ZM241385, known to inhibit human A_{2B} receptors (see Table I), on the netrin-1-dependent inhibition of neutrophil migration (Rosenberger et al., 2009) argues against this hypothesis.

B. Mast Cells

1. Role of A_{2B} Receptors in Mast Cell Degranulation

Adenosine has distinct effects on mast cell degranulation of preformed mediators and on release of newly generated cytokine/growth factors. Adenosine is known to potentiate antigen-induced degranulation of mast cells. This effect of adenosine is mediated by A_3 adenosine receptors in rodent (Ramkumar et al., 1993; Salvatore et al., 2000), but perhaps not in human (Walker et al., 1997) or canine (Auchampach et al., 1997) mast cells.

The role of A_{2B} receptors in mast cell degranulation is less clear. A_{2B}, but not A_3, receptors both stimulated directly and potentiated the effects of the

TABLE I Affinity of Commonly Used Antagonists to Human (h) and Mouse (m) Adenosine Receptor Subtypes Determined in Radioligand Binding Assays (Inhibition Constants in Micromol/Liter)

	Subtypes			
Compounds	A_1	A_{2A}	A_{2B}	A_3
ALT-801	h 5.0[a]	h 0.7[a]	h 0.02[a]	h 6.3[a]
	m 5.2[a]	m 3.5[a]	m 0.2[a]	m > 10[a]
CVT 6883	h 1.9[b]	h 3.3[b]	h 0.02[b]	h 1.0[b]
MRS 1706	h 0.2[c]	h 0.1[c]	h 0.001[c]	h 0.2[c]
MRS 1754	h 0.4[c]	h 0.5[c]	h 0.002[c]	h 0.6[c]
	m 0.009[d]	m > 10[d]	m 0.003[d]	m > 10[d]
PSB1115	h > 10[b]	h > 10[b]	h 0.05[b]	h > 10[b]
DPCPX	h 0.004[e]	h 0.1[e]	h 0.05[f]	h 4.0[e]
	m 0.002[d]	m 0.6[d]	m 0.09[d]	m > 10[d]
ZM 241385	h 0.3[g]	h 0.0008[g]	h 0.03[f]	h > 10[g]
	m 0.2[d]	m 0.0007[d]	m 0.03[d]	m > 100[d]

[a] Kolachala et al. (2008a).
[b] Kalla and Zablocki (2009).
[c] Kim et al. (2000).
[d] Kreckler et al. (2006).
[e] Klotz et al. (1998).
[f] Linden et al. (1999).
[g] Ongini et al. (1999).

calcium ionophore A23187 on degranulation of canine BR mastocytoma cells (Auchampach et al., 1997). Conversely, A_{2B} receptors were proposed to play an inhibitory role in degranulation of mouse bone marrow-derived mast cells (BMMCs), based on the finding that A_{2B}KO mice show an exaggerated antigen-induced mast cell degranulation (Hua et al., 2007). Acknowledging the implausibility of explaining this phenomenon by tonic stimulation of the low-affinity A_{2B} receptor by endogenous adenosine, Hua et al. proposed the alternative explanation that A_{2B} receptors may be constitutively active in WT BMMCs even in the absence of an agonist. However, we determined that A_{2B} receptors expressed in WT BMMCs display no constitutive activity. Further, our work demonstrated that A_{2B} receptors do not inhibit the A_3 receptor-mediated potentiation of antigen-induced degranulation of BMMCs (Ryzhov et al., 2008d). It is likely therefore that exaggerated antigen-induced degranulation in BMMCs deficient of A_{2B} receptors is unrelated to the loss of adenosine signaling function of A_{2B} receptors. Compensatory developmental changes in mice or rearrangement of proteins normally coupled to the A_{2B} receptor as a result of the A_{2B} knockout may have contributed in this phenomenon.

In contrast, pharmacological studies suggested that A_{2A} but not A_{2B} receptors can attenuate antigen-induced degranulation in human cord blood-derived mast cells (Suzuki et al., 1998) and primary human lung mast cells (Duffy et al., 2007). This effect was attributed to the ability of A_{2A} receptors to close K^+ channel KCa3.1 by a cAMP-independent mechanism (Duffy et al., 2007).

2. Role of A_{2B} Receptors in Cytokine/Growth Factor Secretion from Mast Cells

The role of A_{2B} receptors in regulation of cytokine/growth factor secretion from mast cells is better understood than its effects on degranulation. Studies in HMC-1 cells showed that only A_{2B}, but not A_{2A} or A_3, receptors stimulate secretion of angiogenic factors IL-8 and vascular endothelial growth factor (VEGF), and the Th2 cytokines IL-13 and IL-4 (Feoktistov & Biaggioni, 1995; Feoktistov et al., 2003; Ryzhov et al., 2004, 2006). Conditioned media from A_{2B} receptor-activated mast cells stimulated human umbilical vein endothelial cell (HUVEC) proliferation and migration, and induced capillary tube formation. These proangiogenic effects of A_{2B} receptor-stimulated mast cells were attributed primarily to VEGF release because they were blocked by anti-VEGF antibody (Feoktistov et al., 2003). Coculturing B lymphocytes with A_{2B} receptor-stimulated mast cells induced IgE production by B lymphocytes, an effect that appeared to be secondary to increased secretion of Th2 cytokines IL-4 and IL-13 by the mast cells (Ryzhov et al., 2004).

Like human HMC-1 cells, mouse BMMCs express A_3, A_{2A}, and A_{2B}, but not A_1 receptors (Feoktistov et al., 2003; Meade et al., 2002; Ryzhov et al.,

2008d). Activation of adenosine receptors stimulated IL-13 and VEGF secretion only in WT but not in A_{2B}KO BMMCs. Contrary to adenosine action on degranulation, which is only apparent in antigen-stimulated BMMCs, these effects do not require activation of FcεRI receptors, as they are evident even in the absence of antigen (Ryzhov et al., 2008d).

The notion that only A_{2B} receptors, but not A_{2A} receptors coexpressed in the same cells, were able to stimulate cytokine/growth factor secretion could seem paradoxical because both receptor subtypes were thought to act by stimulation of adenylate cyclase through Gs proteins. However, studies in HMC-1 revealed that in contrast to A_{2A} receptors, A_{2B} receptors are also coupled to phospholipase C (PLC), as evidenced by increase in inositol phosphate production with consequent mobilization of intracellular calcium. These A_{2B} receptor-dependent pathways are stimulated through a cholera toxin- and pertussis toxin-insensitive G-protein, presumably of the Gq family (Feoktistov & Biaggioni, 1995). In addition, stimulation of A_{2B} receptors activates the small GTP-binding protein p21ras. This event triggers ERK signaling pathway with sequential stimulation of Raf, MEK1/2, and ERK1/2 protein kinase activities (Feoktistov et al., 1999). We have also demonstrated the coupling of adenosine receptors to JNK and p38 MAPK signaling pathways (Feoktistov et al., 1999). The fact that A_{2B} receptors are coupled to multiple intracellular signaling pathways in mast cells explains their ability to regulate the generation and secretion of diverse cytokines and growth factors.

A_{2B} receptors stimulate release of VEGF, IL-13, and IL-8 from mast cells by mechanisms that involve activation of ERK and p38 MAPK (Ryzhov et al., 2006, 2008d). Stimulation of the receptor tyrosine kinase c-kit with stem cell factor (Meade et al., 2002) or the receptor complex ST2/IL1RAP with IL-33 (Silver et al., 2010) synergized with A_{2B} receptors in the upregulation of IL-8 from HMC-1 cells. Functional analysis of cells transfected with full-length and truncated receptor constructs revealed that the A_{2B} receptor C-terminus is important for coupling to Gs and Gq proteins. However, the A_{2B} receptor C-terminus is not essential for upregulation of IL-8. Instead, integrity of the third intracellular loop of the A_{2B} receptor was crucial for IL-8 stimulation (Ryzhov et al., 2009).

Whereas A_{2B} receptor-mediated stimulation of IL-8 and IL-13 is cAMP-independent, stimulation of adenylate cyclase was required (but not sufficient) for upregulation of VEGF and IL-4 (Ryzhov et al., 2006, 2008d). The dual coupling of A_{2B} receptors to Gs/Gq proteins with concurrent stimulation of diverse intracellular pathways is necessary for adenosine-dependent regulation of IL-4 production in HMC-1. A_{2B} adenosine receptors induce IL-4 generation via Gq-mediated stimulation of PLC_{β}, inositol trisphosphate-mediated mobilization of intracellular Ca^{2+}, and activation of nuclear factor of activated T cells (NFAT) by calcineurin. This process is potentiated via Gs-mediated stimulation of adenylate cyclase and activation of protein kinase A (PKA) and may involve the increase in protein levels of NFATc1. Thus, the existence of cross talk between Gq-PLC$_{\beta}$ and Gs-adenylate cyclase signaling pathways in regulation of IL-4 secretion enables A_{2B} receptors, coupled to both Gq and Gs, to effectively stimulate IL-4 production in mast cells (Ryzhov et al., 2006).

C. Dendritic Cells

1. Dendritic Cell Functions

Dendritic cells play an important role in bridging innate and adaptive immunity. It is generally accepted that conventional dendritic cells arise from bone-marrow hematopoietic progenitors or peripheral blood monocytes that migrate into peripheral tissues and differentiate into immature dendritic cells. Immature dendritic cells in tissues are constantly sampling their microenvironment for the presence of antigens. Upon activation by pathogens and other inflammatory stimuli, dendritic cells undergo phenotypical maturation and migrate toward the secondary lymphoid organs. On reaching these organs, dendritic cells develop into mature cells capable to present antigens to naïve T lymphocytes, thus initiating the development of adaptive immune responses (Dominguez & Ardavin, 2010). The presence of A_{2B} receptors on monocytes and dendritic cells (Novitskiy et al., 2008) suggests that their activation may influence both differentiation and maturation of dendritic cells.

2. Role of A_{2B} Receptors in Dendritic Cell Differentiation

Using a combination of genetic and pharmacological approaches, we have recently shown that stimulation of A_{2B} receptors *in vitro* and *in vivo* induces generation of a phenotypically and functionally distinct subset of dendritic cells. These "adenosine-differentiated" cells are impaired in their ability to induce T-cell proliferation and IFN-γ production. These cells also produce high levels of immunomodulatory cytokines IL-6, IL-10, and TGF-β. It is possible that by upregulating IL-10 and TGF-β, adenosine-differentiated dendritic cells could affect Th1-mediated immune reactions, induce the generation of regulatory T cells, and polarize the immune response toward a Th2 type. Because these cells also express high levels of the tolerance-inducing enzymes indoleamine 2,3-dioxygenase (IDO) and arginase, they can impair T-cell signal transduction and function. Adenosine-differentiated dendritic cells also secrete high levels of angiogenic factors VEGF and IL-8. Both immunosuppressive and proangiogenic properties of these cells could be beneficial for tumor growth. Indeed, our studies *in vivo* demonstrated that the presence of "adenosine-differentiated" dendritic cells significantly promoted tumor growth in a mouse Lewis lung carcinoma model (Novitskiy et al., 2008).

3. Role of A_{2B} Receptors in Dendritic Cell Maturation

In addition to modulation of cell differentiation, A_{2B} receptors were shown also to affect maturation of dendritic cells. A_{2B} receptors inhibited Th1 immune response-promoting cytokines IL-12 p70 and IL-23 but enhanced IL-10 secretion by TLR-activated bone-marrow-derived dendritic cells. Moreover, stimulation of A_{2B} receptors during dendritic cell maturation increased expression and enzymatic activity of IDO and arginase in LPS-activated dendritic cells (Ben et al., 2008). A_{2B} receptors were shown to be responsible for formation of a dendritic cell fraction expressing low levels of MHC-II and costimulatory molecule CD86. These cells were characterized by an increased expression of A_{2B} receptor transcripts compared to the rest of dendritic cell population. Only this subset of "adenosine-matured" dendritic cells expressed lower levels of IL-12p40 but higher levels of IL-10 and had poor capacity to stimulate CD4+ T cells, compared to cells matured in the absence of adenosine stimulation (Wilson et al., 2009). Thus, stimulation of A_{2B} receptors, not only during dendritic cell differentiation but also during maturation, may lead to the formation of a cell population capable of impairing Th1 differentiation of CD4+ T cells and promoting immune tolerance. The role of A_{2B} receptors in adenosine actions on both differentiation and maturation of dendritic cells appears to be nonredundant because specific stimulation of other adenosine receptor subtypes did not produce similar effects. Furthermore, these effects were observed only in WT but not in A_{2B}KO cells and were inhibited by selective A_{2B} antagonists (Ben et al., 2008; Novitskiy et al., 2008; Wilson et al., 2009).

D. Monocytes/Macrophages

Adenosine has been recognized as an important regulator of monocyte/macrophage functions. Some responses to adenosine are mediated by cAMP-dependent mechanisms. These responses include the inhibition of LPS-induced TNF-α production (Kreckler et al., 2009), the potentiation of IL-10 production (Nemeth et al., 2005), and the inhibition of proliferation induced by monocyte colony stimulating factor (M-CSF; Xaus et al., 1999b). The role of A_{2B} receptors in adenosine-dependent regulation of these monocyte/macrophage functions is often masked by A_{2A} receptors coexpressed in the same cells. A_{2B} receptors can control cAMP-dependent functions only in those cell models where the expression of dominant A_{2A} receptors is negligible or absent. For example, macrophages generated from mouse bone marrow *in vitro* expressed predominantly A_{2B} adenosine receptors and only negligible levels of A_{2A} receptors. Stimulation of A_{2B}, but not A_{2A}, receptors increased cAMP levels in this cell preparation and inhibited M-CSF-induced macrophage proliferation by upregulating the cyclin-dependent kinase inhibitor p27Kip-1 via activation of cAMP–PKA pathway (Xaus et al., 1999b). IFN-γ further upregulated the expression of A_{2B} receptors on bone marrow-derived macrophages resulting in an increased cAMP production in response to stimulation with NECA, which in turn downregulated both MHC-II and iNOS expression (Xaus et al., 1999a). Due to negligible expression of A_{2A} receptors on RAW264.7 cells, A_{2B} receptors were capable of inhibiting TNF-α production and augmenting IL-10 production in this macrophage-like cell line following activation with LPS (Nemeth et al., 2005).

Similarly, A_{2B} receptors inhibited LPS-induced TNF-α release in peritoneal macrophages isolated from mice lacking A_{2A} receptors. However, selective inhibition of A_{2B} receptors had no effect on the adenosine-dependent inhibition of LPS-induced TNF-α secretion from WT mouse peritoneal macrophages due to the dominant role of A_{2A} receptors in regulation of this cAMP-dependent event (Kreckler et al., 2006). Indeed, NECA-induced cAMP accumulation was similar in peritoneal macrophages obtained from WT and A_{2B}KO animals, indicating the dominant role of A_{2A} receptors in this process. Furthermore, the absence of A_{2B} adenosine receptors did not affect adenosine receptor-dependent suppression of LPS-activated TNF-α release from peritoneal macrophages (Ryzhov et al., 2008c). Comprehensive pharmacological analysis of adenosine-dependent inhibition of LPS-induced TNF-α release from human primary monocytes (Zhang et al., 2005) and alveolar macrophages (Buenestado et al., 2010) revealed that this effect was exclusively mediated by A_{2A} receptors, corroborating the findings in mouse peritoneal macrophages.

However, A_{2B} receptors may have distinctive functions in macrophages even in the presence of otherwise dominant A_{2A} receptors. Pretreatment of mouse primary alveolar macrophages with the selective A_{2B} antagonist MRS 1706 or genetic ablation of A_{2B} receptors resulted in a loss of NECA-stimulated increases in osteopontin expression (Schneider et al., 2010). Likewise, pharmacological inhibition with selective A_{2B} antagonists or genetic ablation of A_{2B} receptors completely abrogated NECA-induced increase in IL-6 release from peritoneal macrophages (Ryzhov et al., 2008c). Of interest, elevation of cAMP in murine macrophages attenuates LPS-induced TNF-α secretion (Kreckler et al., 2009) but has no effect on basal IL-6 release (Tang et al., 1998). It is possible that the differential regulation of TNF-α and IL-6 secretion by A_{2A} and A_{2B} receptors in mouse peritoneal macrophages can be explained by coupling of these receptors to distinct intracellular pathways. Further studies are needed to delineate the signaling pathways linking activation of A_{2B} receptors to cytokine production in macrophages.

IV. A_{2B} Receptors on Endothelial Cells

A. Endothelial Cells in Inflammation

Vascular endothelium lines all vessels in the body and serves as a dynamic and selective barrier regulating the flow of nutrients, biologically active molecules, and cells across blood vessel walls. Endothelial cells, in close cooperation with other cell types, play an important role in inflammation and subsequent tissue remodeling. Activation of endothelial cells by inflammatory stimuli increases vascular permeability and the expression of adhesion molecules on the endothelial surface, which in turn promote edema, and leukocyte attachment and extravasation leading to initiation of an inflammatory cascade. Later in inflammation, endothelial cells play a central role in the expansion, regression, and remodeling of preexisting blood vessels, a process commonly known as angiogenesis. Adenosine has been implicated in modulation of all these events (Sands & Palmer, 2005).

Differential expression of cell-surface adenosine receptors is part of the phenotypic heterogeneity of endothelial cells, and endothelial responses to adenosine can differ depending on the relative expression of adenosine receptor subtypes. For example, HUVECs express predominantly A_{2A} adenosine receptors, whereas the human microvascular endothelial cells HMEC-1 express predominantly A_{2B} receptors (Feoktistov et al., 2002). Predominant expression of A_{2B} receptors has been demonstrated also in cardiac microvascular endothelial cells (Ryzhov et al., 2008b) and endothelial cells of high endothelial venules (Takedachi et al., 2008). Furthermore, there are multiple factors that can modify an endothelial phenotype, including mechanical forces, biologically active compounds, the composition of extracellular matrix, and contact with circulating and tissue-based cells. Conditions present in inflammatory processes provide powerful stimuli for such phenotypic changes. The Th1 cytokines IL-1 and TNF-α increase expression of both A_{2A} and A_{2B} adenosine receptors in human dermal microvascular endothelial cells. IFN-γ treatment increases the expression of A_{2B} receptors but decreases the expression of A_{2A} receptors (Khoa et al., 2003). Hypoxia, a condition often present in inflamed tissues, also selectively increases A_{2B} expression in endothelial cells (Eltzschig et al., 2003; Feoktistov et al., 2004) by a mechanism that involves the oxygen-sensitive hypoxia-inducible factor-1α (HIF-1α)-dependent transactivation of the A_{2B} receptor promoter (Kong et al., 2006). An increase in the A_{2B} receptor expression may not only increase the effects of adenosine but also affect their outcomes. For example, HUVECs express predominantly A_{2A} adenosine receptors and do not produce VEGF in response to adenosine (Feoktistov et al., 2002). Hypoxia decreased A_{2A} and increased A_{2B} receptor expression in these cells. Consistent with these changes in receptor expression, adenosine stimulated VEGF release under hypoxic but not normoxic conditions, indicating that hypoxia increased the expression of A_{2B} receptors that were functionally coupled to upregulation of VEGF (Feoktistov et al., 2004).

B. Role of A_{2B} Receptors in the Expression of Endothelial Adhesion Molecules

Adenosine has been shown to inhibit the expression of adhesion molecules and leukocyte recruitment by activated endothelial cells. Whereas A_{2A} receptors have been implicated in these effects both *in vitro* and *in vivo* (Palmer & Trevethick, 2008), the role of A_{2B} receptors is less understood.

Genetic ablation of A_{2B} receptor in mice produced a phenotype that was interpreted as supporting an inhibitory role of A_{2B} receptors in the expression of endothelial adhesion molecules and leukocyte recruitment. Intravital examination of mesenteric venules revealed an increased number of leukocytes rolling or adhered to the vascular wall of A_{2B}KO mice as compared to WT mice. Based on these findings and also on the increased levels of the adhesion molecules ICAM-1, P-selectin and E-selectin in protein extracts isolated from mesenteric arteries of A_{2B}KO mice, it has been suggested that A_{2B} receptors tonically downregulate the expression of endothelial adhesion molecules and leukocyte recruitment (Yang et al., 2006). However, these effects of A_{2B} receptor ablation *in vivo* are likely to be secondary to the increased basal plasma levels of TNF-α, a potent activator of endothelial adhesion molecules (Mackay et al., 1993). In contrast, our studies showed that stimulation of A_{2B} receptors in cardiac microvascular endothelial cells induced rapid cell-surface expression of P-selectin by a mechanism likely involving exocytosis of the content of Weibel–Palade bodies (Ryzhov et al., 2008b). Further studies are needed to elucidate the role of endothelial A_{2B} receptor activation on the expression of adhesion molecules involved in the recruitment of inflammatory cells by activated endothelium.

C. Role of A_{2B} Receptors in Regulation of Endothelial Barrier Function

Numerous studies *in vitro* have shown that adenosine acting on A_{2A} or/and A_{2B} receptors decreases endothelial permeability (for review, see Biaggioni & Feoktistov, 2005). *In vivo*, mice lacking either apyrase (CD39) or ecto-5′-nucleotidase (CD73), enzymes involved in the generation of extracellular adenosine, had a higher leakage of albumin through endothelium in various tissues, as measured by the Evans Blue technique (Eltzschig et al., 2003; Thompson et al., 2004). Remarkably, a similar phenotype was found in mice lacking A_{2B} receptors but not other adenosine receptor subtypes (Eckle et al., 2008a). The increased basal vascular permeability was even further increased in A_{2B}KO mice subjected to ambient hypoxia (Eckle et al., 2008a) supporting the previous evidence obtained *in vitro* on the role of A_{2B} receptors in regulation of endothelial barrier function. Studies in bone marrow chimeric mice suggested a predominant role of vascular A_{2B} receptors but not those located on bone marrow-derived cells in this response. Surprisingly, A_{2A}KO mice did not demonstrate loss of barrier function *in vivo* (Eckle et al., 2008a), as it would be expected from *in vitro* studies that implicated A_{2A} receptors in adenosine-dependent regulation of permeability of several endothelial cell types (Umapathy et al., 2010; Wang & Huxley, 2006). Although some of the differences in vascular permeability between A_{2B}KO and A_{2A}KO mice used in these studies may be attributed to their different genetic background (C57Bl6 and CD1, respectively), these results can be explained by the predominant role of A_{2B} receptors in hypoxic endothelial cells due to HIF-1α-dependent increase in A_{2B} receptor expression (Kong et al., 2006). HIF-1α was also shown to transactivate the promoter of CD73 (Synnestvedt et al., 2002) and repress the promoter of the equilibrative nucleoside transporter (Eltzschig et al., 2005). Taken together, these effects of hypoxia would upregulate A_{2B} receptors and produce high levels of adenosine at the endothelial surface, thus promoting A_{2B} receptor signaling. However, the reason for the augmented vascular permeability in A_{2B}KO mice at rest, when extracellular adenosine levels are low, is less clear. Explanation of this phenomenon will require further investigation.

Reduction of endothelial permeability by adenosine seems to be mediated by the cAMP–PKA-dependent pathway because this effect was mimicked by reagents elevating cAMP or stimulating PKA. These effects were associated with a rearrangement of the F-actin component of the cytoskeleton, enhanced cell-surface expression of cell–cell junctional protein VE-cadherin, and an involvement of myosin-light-chain phosphatase (Umapathy et al., 2010). However, the exact mechanism of adenosine-induced endothelial barrier enhancement remains largely unknown. It has been proposed that it can be explained in part by relaxation of actin cytoskeletal tension, as a result of phosphorylation by PKA of actin-associated, vasodilator-stimulated phosphoprotein (Comerford et al., 2002). Inhibition of the RhoA-dependent pathway has been implicated in adenosine- and cAMP-dependent regulation of endothelial barrier function (Harrington et al., 2004; Waschke et al., 2004). In addition, stimulation of ERK via A_{2B} receptors was also proposed to promote barrier function through dephosphorylation of the myosin II regulatory light chains (Srinivas et al., 2004).

D. Role of A_{2B} Receptors in Regulation of Endothelial Cell Proliferation

Depending on the endothelial cell studied, either A_{2A} or A_{2B} receptors have been implicated in stimulation of endothelial cell proliferation. Adenosine A_{2B} receptors have been shown to mediate the proliferative actions of adenosine in human retinal microvascular endothelial cells (Afzal et al., 2003; Grant et al., 1999, 2001; Mino et al., 2001), in porcine coronary artery, and in rat aortic endothelial cells (Dubey et al., 2002). The proliferative effects of adenosine on endothelial cells are mediated at least partly by stimulating the production of growth factors that facilitate new blood vessel formation. Adenosine increased VEGF production in pig cerebral microvascular endothelial cells (Fischer et al., 1995), but the adenosine receptor subtype involved in VEGF upregulation in these cells remains uncertain because of nonspecific concentrations of antagonists used in that study. Takagi and associates showed that adenosine upregulates VEGF mRNA in bovine retinal microvascular cells via A_{2A} receptors (Takagi et al., 1996). However, adenosine upregulated VEGF mRNA expression and protein secretion via A_{2B} receptors in human retinal endothelial cells (Grant et al., 1999).

VEGF is not the only angiogenic factor modulated by adenosine in endothelial cells. In human retinal microvascular endothelial cells, A_{2B} receptor activation upregulated also basic fibroblast growth factor (bFGF) and insulin-like factor-1 (Grant et al., 1999). In immortalized human dermal microvascular endothelial cells HMEC-1, stimulation of A_{2B} adenosine receptors upregulated bFGF and IL-8 in addition to VEGF (Feoktistov et al., 2002). The intracellular mechanisms mediating these effects are not clear. Although some data suggest that cAMP may play a role in the effects of adenosine on VEGF secretion (Takagi et al., 1996), other studies found that cAMP–PKA-independent stimulation of MAPK pathways was primarily responsible for proangiogenic effects of A_{2B} receptors on microvascular endothelial cells (Feoktistov et al., 2002; Grant et al., 1999, 2001).

It is commonly accepted that VEGF production is regulated by HIF-1α. However, A_{2B} receptors can upregulate VEGF production in microvascular endothelial cells by an HIF-1α-independent mechanism (Feoktistov et al., 2004). Because adenosine levels are increased in hypoxia, stimulation of A_{2B} receptors on endothelial cells can complement HIF-1α-dependent actions of hypoxia in the regulation of angiogenesis. Stimulation of A_{2B} receptors results in secretion of additional angiogenic factors (IL-8) not induced by hypoxia *per se* (when adenosine is scavenged by adenosine deaminase (ADA)) and in greater VEGF production from endothelial cells (Ryzhov et al., 2007). Thus, A_{2B} receptor-dependent release of angiogenic factors can contribute to the overall effect of hypoxia and provide an autocrine pathway regulating endothelial cell growth during chronic inflammation or in the resolution phase of acute inflammation.

V. A_{2B} Receptors on Epithelial Cells

A. Intestinal Epithelial Cells

Epithelial cells participate in inflammatory processes by maintaining mucosal integrity, producing biologically active mediators, and modulating local immune responses. Functional adenosine A_{2B} receptors are expressed on intestinal epithelial cells. Epithelial A_{2B} receptors were shown to upregulate chloride secretion through the activation of apical cystic fibrosis conductance regulator (CFTR; Strohmeier et al., 1995). The effect of A_{2B} receptors on epithelial secretion has received particular attention because of its potential relevance to intestinal inflammation. As part of the pathophysiology of these disorders, neutrophils are recruited into intestinal crypts, where they release AMP, which is then converted to adenosine at the epithelial cell surface by CD73. It is adenosine that then acts on epithelial A_{2B} receptors to stimulate chloride secretion (Madara et al., 1993). This pathway normally serves to hydrate the mucosal surface, thereby protecting the intestine by preventing the translocation of bacteria, bacterial products, and antigens to lamina propria. When stimulated during inflammation, chloride secretion facilitates fluid movement to lumen in order to flush pathogens from the mucosal surface, thus contributing to inflammation-associated diarrhea (Kolachala et al., 2008b).

Epithelial A_{2B} receptors were also shown to modulate the release of proteins involved in intestinal inflammation. Stimulation of A_{2B} receptors on colonic epithelial cells induced an increase in IL-6 secretion into the luminal compartment at levels sufficient to activate neutrophils (Sitaraman et al., 2001). Stimulation of A_{2B} receptors also induced fibronectin synthesis and secretion from the apical surface of intestinal epithelial cells (Walia et al., 2004). Apical fibronectin significantly enhanced the adherence and invasion of *Salmonella typhimurium* to cultured epithelial cells as well as consequent IL-8 secretion (Walia et al., 2004). Intracellular mechanisms mediating these effects were suggested to involve stimulation of IL-6 and fibronectin transcription via cAMP/PKA-mediated activation of nuclear cAMP-responsive element-binding protein (CREB; Sitaraman et al., 2001; Walia et al., 2004).

A_{2B} receptor functions on intestinal epithelium can be modulated by INF-γ and TNF-α, factors elevated at various stages of inflammation. TNF-α reportedly increases A_{2B} receptor expression in colonic epithelial cells by a posttranscriptional mechanism possibly involving downregulation of A_{2B} receptor-specific micro-RNA miR27b and miR128a (Kolachala et al., 2010), whereas INF-γ affects A_{2B} receptor signaling through the direct inhibition of adenylate cyclase expression (Kolachala et al., 2005).

B. Pulmonary Epithelial Cells

A_{2B} receptors have been suggested to play an important role in regulation of ion and water transport in airway epithelium. Airborne particles, including pathogens, are absorbed by the mucus layer lining the airways, where they are inactivated by the innate mucosal defense system and removed via mucociliary and cough clearance. This process largely depends on active transepithelial salt transport involving the CFTR and amiloride-sensitive epithelial Na^+ channel (ENaC). The role of A_{2B} receptors in the regulation of CFTR was demonstrated in the human airway epithelial Calu-3 cell line and primary bronchial epithelial cells (Huang et al., 2001; Lazarowski et al., 2004). Like in intestinal epithelium, A_{2B} receptors stimulate CFTR by a mechanism involving cAMP/PKA-dependent pathway (Lazarowski et al., 2004). In bronchial ciliated epithelium, CFTR can inhibit ENaC, and activation of CFTR in the presence of ENaC inhibition generates Cl^- secretion and liquid transport to maintain airway surface liquid volume (Donaldson & Boucher, 2007). In distal airspaces of the lung, sodium movement through ENaC and chloride movement through CFTR may be coupled (Reddy et al., 1999), and activation of CFTR-mediated chloride transport can result in fluid absorption (Fang et al., 2006). Therefore, adenosine-dependent, cAMP-mediated pathways could promote alveolar ion absorption and fluid clearance. Indeed, adenosine receptors were shown to regulate alveolar liquid clearance in rat type II pneumocytes (Factor et al., 2007), and high levels of A_{2B} receptor expression were described in murine type II alveolar epithelial cells (Cagnina et al., 2009).

In addition to stimulation of cAMP-dependent pathway, apical adenosine A_{2B} receptors in Calu-3 cells can regulate anion secretion through stimulation of basolateral K_{Ca} channels via PLC/Ca^{2+} signaling. This pathway synergizes with cAMP-dependent modulation of apical CFTR channels for transepithelial anion secretion and a mixed secretion of chloride and bicarbonate (Wang et al., 2008). The mechanism of A_{2B} receptor-mediated transepithelial liquid transport appears even more complicated in view of the recent report that A_{2B} receptor signaling regulates gap junctional intercellular communication between epithelial cells by the release of PGE_2 and subsequent activation of basolateral EP_4 receptors. It has been suggested that this mechanism may contribute to the spread of ions, second messenger and/or co-factor exchange between cells to fully activate CFTR and ensure efficient Cl^- secretion (Scheckenbach et al., 2010).

Physical stimulation of airway surfaces was suggested to evoke liquid secretion by producing ATP that then locally converted to adenosine and sensed by A_{2B} adenosine receptors (Huang et al., 2001). In nasal epithelial cells, A_{2B} receptor agonists elicited sustained responses in ciliary beat frequencies (Morse et al., 2001). Thus, A_{2B} receptors were proposed to stimulate mucociliary clearance in response to injurious stimuli to remove them from airway surfaces. Whether tonic stimulation of A_{2B} receptors is required at rest in order to maintain the mucus clearance is less clear, given the low affinity of this receptor subtype. A recent report by Rollins et al. appeared to support this hypothesis by demonstrating that the selective A_{2B} antagonist ATL801 inhibited autoregulation of airway surface liquid height in human bronchial epithelial cells (Rollins et al., 2008). However, the lack of effects of micromolar concentrations of DPCPX and ZM241385 (Rollins et al., 2008) known to inhibit human A_{2B} receptors (see Table I) adds uncertainty to the authors' conclusion.

A_{2B} receptors can also regulate cytokine secretion from pulmonary epithelium. Stimulation of A_{2B} receptors on human primary bronchial epithelial cells (HBECs) upregulated the expression of several cytokines including CXCL2, CXCL3, and IL-19 (Zhong et al., 2005). A_{2B} receptor-stimulated IL-19 released from HBECs was able to activate the monocytic cell line THP-1 and induce TNF-α secretion. In turn, TNF-α released from THP-1 upregulated the expression of A_{2B} receptors in HBECs, thus providing a positive feedback loop to facilitate the effects of adenosine (Zhong et al., 2005).

VI. Role of A_{2B} Receptors on Fibroblasts

The human fibroblast cell line VA13 was the first cell type where a low-affinity adenosine receptor was originally described back in 1980 (Bruns), and later designated as A_{2B} (Bruns et al., 1986). Fibroblasts represent a heterogeneous population of cells, which may differ in phenotype and function not only between anatomical sites but also even within a single tissue where immature cells (often called mesenchymal fibroblasts) exist with fibroblasts of various degree of differentiation. Fibroblasts play an important role in the progression of inflammation by secreting various factors that define the tissue microenvironment and modulate immune cell functions. They also contribute to tissue remodeling by increased proliferation, differentiation, and generation of various components of the extracellular matrix (for review, see Flavell et al., 2008).

Stimulation of A_{2B} receptors on mouse cardiac fibroblasts was shown to promote IL-6 release. This effect was independent of the Gs–cAMP–PKA pathway but required protein kinase Cδ and p38 MAPK activation (Feng et al., 2010). In contrast, A_{2B} receptor-mediated increase of IL-6 release from human gingival fibroblasts was attributed, at least in part, to the activation of Gs–cAMP–PKA pathway (Murakami et al., 2000). Stimulation of these cells with high concentrations (10–50 μM) of adenosine agonists potentiated IL-6 and IL-8 release induced by IL-1β and upregulated the expression of hyaluronate synthase mRNA (Murakami et al., 2001). In contrast, stimulation of A_{2B} receptors on synovial fibroblasts counteracted the effects of IL-1β by decreasing MMP1 mRNA expression (Boyle et al., 1996).

The role of A_{2B} receptors in regulation of IL-6 secretion under normoxic and hypoxic conditions was studied in human pulmonary fibroblasts. Stimulation of A_{2B} receptors under normoxic conditions increased the release of IL-6 and promoted the differentiation of human lung fibroblasts to myofibroblasts. Hypoxia amplified these effects by upregulating the expression of A_{2B} adenosine receptors. The findings that adenosine increases the release of IL-6, and this cytokine in turn induces differentiation of fibroblasts into myofibroblasts, suggest a mechanism whereby adenosine could participate in the remodeling process of chronic inflammatory diseases (Zhong et al., 2005).

Studies in corpus cavernosal fibroblast cells isolated from WT and A_{2B}KO mice suggested that A_{2B} receptors upregulate the expression of TGF-β1 and promote fibrosis. Pharmacological analysis in WT cells showed that NECA increased TGF-β1 mRNA expression and procollagen I mRNA levels, which was completely abolished by the A_{2B} receptor-specific antagonist, MRS1754. Furthermore, this effect was not seen in the A_{2B}KO cells (Wen et al., 2010).

In contrast, A_{2B} receptors diminished collagen production and proliferation of rat cardiac fibroblasts (Dubey et al., 2001; Epperson et al., 2009). The diverse effects of A_{2B} receptors on fibroblasts of different origins can be attributed to the known phenotypical and functional heterogeneity of these cells.

VII. Role of A_{2B} Receptors in Animal Models of Inflammation

The recent generation of A_{2B}KO mice and the development of selective A_{2B} antagonists have made it possible to test an integrated role of A_{2B} receptors in complex animal models of acute and chronic inflammation. Results have been interpreted as evidence for either "proinflammatory" or "anti-inflammatory" role of A_{2B} receptors, and results from genetic and pharmacological approaches have not always been in agreement.

A_{2B}KO mice appear to have a proinflammatory phenotype compared to WT controls, characterized by elevated TNF-α plasma concentrations (Yang et al., 2006) and increased vascular permeability for albumin in the colon, kidney, and lung (Eckle et al., 2008a). This is surprising because this is seen even at rest, when the low extracellular adenosine levels are not expected to stimulate A_{2B} receptors. Short-term exposure of mice to ambient hypoxia induced significantly higher vascular leak in all organs of A_{2B}KO mice compared to WT control (Eckle et al., 2008a). A_{2B}KO mice had also increased pulmonary albumin leakage after acute lung injury produced by either mechanical ventilation (Eckle et al., 2008b) or LPS inhalation (Eckle et al., 2008b; Schingnitz et al., 2010), compared to WT controls. In parallel to pulmonary edema, an increase in neutrophil infiltration and tissue levels of IL-1β, IL-6, and TNF-α was also higher in A_{2B}KO mice. Studies using bone marrow chimeric mice showed that A_{2B} receptors located on stromal cells rather than those on bone marrow-derived cells contribute to these differences in lung injury between A_{2B}KO and WT animals (Eckle et al., 2008b; Schingnitz et al., 2010).

Furthermore, it has been reported that genetic deficiency of A_{2B} receptors increased the mortality of mice suffering from cecal ligation and puncture-induced sepsis. The increased mortality of A_{2B} knockout mice was associated with increased inflammatory indices measured in the spleen, heart, and plasma in comparison with WT animals. Again, experiments using bone-marrow chimeras revealed that it is the lack of A_{2B} receptors on nonhematopoietic cells that is primarily responsible for the increased inflammation of septic A_{2B} receptor-deficient mice (Csoka et al., 2010). Another study employing bone-marrow chimeras suggested that vascular A_{2B} receptors also play an important role in protective effects of ischemic preconditioning in a mouse model of acute renal failure from ischemia (Grenz et al., 2008). Thus, it is possible that an increase in vascular permeability in mice lacking A_{2B} receptors may explain, at least in part, the observed exacerbation of acute tissue injury seen in these different disease models.

This evidence obtained in models of acute tissue injury seems to support an "anti-inflammatory" role of A_{2B} receptors. However, even similar models of acute injury may produce opposite results in different tissues. For example, A_{2B} receptors were proposed to play a protective role in gastrointestinal model of ischemia–reperfusion injury based on enhanced intestinal injury observed in A_{2B}KO mice (Hart et al., 2009). In contrast, the lungs of A_{2B}KO mice were significantly protected in a pulmonary ischemia–reperfusion model, as evidenced by reduced pulmonary artery pressure, increased lung compliance, decreased myeloperoxidase, and reduced levels of TNF-α, IL-6, CXCL-1, CCL2, and CCL5 (Anvari et al., 2010). These results would suggest a "proinflammatory" role of A_{2B} receptors in the lung, in contrast to their "anti-inflammatory" actions in the intestine. Experiments using bone-marrow chimeras also suggested that these effects were due to A_{2B} receptor activation primarily on resident pulmonary cells and not bone marrow-derived cells (Anvari et al., 2010).

It becomes increasingly clear that the final outcome of genetic ablation of A_{2B} receptors may be dependent not only on the tissue but also on the model of inflammation studied. Although genetic ablation of adenosine A_{2B} receptors in mice has been shown to facilitate acute inflammatory responses to antigen challenges in passively sensitized mice (Hua et al., 2007), this is not the case in chronic inflammation, a process dependent on the complex interplay between multiple cells and inflammatory factors. We studied the effects of A_{2B} receptor gene ablation in the context of ovalbumin-induced chronic pulmonary inflammation. We found that repetitive airway allergen challenge induced a significant increase in adenosine levels in fluid recovered by bronchoalveolar lavage (BAL). Genetic ablation of A_{2B} receptors significantly attenuated allergen-induced chronic pulmonary inflammation as evidenced by a reduction in the number of BAL eosinophils and in peribronchial eosinophilic infiltration. The most striking difference in the pulmonary inflammation induced in A_{2B}KO and WT mice was the lack of allergen-induced IL-4 release in the airways of A_{2B}KO animals, in line with a significant reduction in IL-4 protein and mRNA levels in lung tissue. In addition, attenuation of TGF-β1 release in airways of A_{2B}KO mice correlated with reduced airway smooth muscle and goblet cell hyperplasia/hypertrophy. It was concluded, therefore, that genetic removal of A_{2B} adenosine receptors in mice leads to inhibition of allergen-induced chronic pulmonary inflammation and airway remodeling (Zaynagetdinov et al., 2010). This conclusion was also supported by pharmacological evidence; antagonism of adenosine A_{2B} receptors by the selective A_{2B} antagonist CVT-6883 resulted in inhibition of airway inflammation induced by chronic exposure to allergen (Mustafa et al., 2007).

Inhibition of adenosine A_{2B} receptors by the selective A_{2B} antagonist CVT-6883 also reduced pulmonary inflammation and airway remodeling in ADA-deficient mice (Sun et al., 2006). These mice are characterized by elevated lung tissue levels of adenosine and exhibit a lung phenotype with features of lung inflammation, bronchial hyperresponsiveness, enhanced mucus secretion, increased IgE synthesis, and elevated levels of proinflammatory cytokines and angiogenic factors that could be reversed by lowering adenosine levels with exogenous ADA (Blackburn et al., 2000). Paradoxically, genetic removal of the A_{2B} receptors from ADA-deficient mice led to enhanced pulmonary inflammation and airway destruction. The authors suggested that loss of pulmonary barrier function in A_{2B}KO mice and excessive airway neutrophilia contributed to the enhanced tissue damage observed in this model (Zhou et al., 2009). In addition, TNF-α levels, known to be elevated in A_{2B}KO mice at rest (Ryzhov et al., 2008c; Yang et al., 2006), were also markedly increased in the lungs of ADA/A_{2B} double-knockout mice, which could activate pathways that influence the trafficking of neutrophils in the lung (Zhou et al., 2009). To explain the opposite effects of A_{2B} receptor antagonism and genetic ablation, the authors emphasized that pharmacological inhibition of A_{2B} receptors was introduced later in the disease process, 10 days after triggering pulmonary disease by withdrawing mice from ADA replacement therapy (Zhou et al., 2009). Thus, the differential effects of pharmacological antagonism and genetic deletion of A_{2B} receptors in ADA-deficient mice may indicate the importance of timing when elimination of A_{2B} receptor signaling could either promote or attenuate the development of pulmonary disease in ADA-deficient mouse model. The tonic increase in TNF-α in A_{2B}KO but not in WT mice may also contribute to this phenomenon.

The use of A_{2B}KO mice has also resulted in different outcomes of bleomycin-induced pulmonary fibrosis depending on the research model used. Bleomycin instilled directly into the respiratory tract produced acute damage resulting in extensive apoptosis of airway epithelial cells, followed by infiltration of granulocytes early after the challenge. These mice developed severe fibrosis due to failed wound healing. In contrast, in the intraperitoneal bleomycin model, lung injury and inflammation are not induced acutely, but mice do develop chronic pulmonary fibrosis. The fibrotic process was significantly reduced in A_{2B}KO mice in the intraperitoneal bleomycin model, but not in the intratracheal model, compared to corresponding WT controls (Zhou et al., 2011). In contrast, pharmacological antagonism of A_{2B} receptors with CVT-6883 attenuated pulmonary inflammation and fibrosis in WT mice subjected to lung injury induced by intratracheal instillation of bleomycin (Sun et al., 2006). These differences in the results obtained from genetic and pharmacological targeting of A_{2B} receptors could be explained by the loss of pulmonary vascular barrier seen in A_{2B}KO mice even at rest (Eckle et al., 2008a). This phenotype would be expected to amplify any model of acute pulmonary injury that involves an acute phase of inflammation with extensive edema and neutrophil infiltration.

The role of A_{2B} receptors in promoting intestinal inflammation was demonstrated in both acute and chronic mouse models of colitis. Genetic removal or pharmacological antagonism of A_{2B} receptors with ATL-801 attenuated clinical and histological features of intestinal inflammation in parallel with reduction of IL-6 and CXCL-1 secretion in several mouse models of experimental colitis (Kolachala et al., 2008a, 2008c). In contrast, both genetic ablation and pharmacological antagonism of A_{2B} with PSB1115 promoted intestinal inflammation in a mouse model of acute colitis as reflected by increased weight loss, colonic shortening, and disease activity indices (Frick et al., 2009). The reason for disparate results of these studies is unclear but may reflect the complexity of events orchestrated in these models when small differences in protocols or in composition of intestinal flora present in animals may greatly affect outcomes.

Adding more to controversy on the role of A_{2B} receptors in inflammatory responses, A_{2B} receptors were proposed to play a role in protective effects of ischemic preconditioning in models of myocardial ischemia–reperfusion injury (Eckle et al., 2007; Kuno et al., 2007). Interestingly, *in situ* ischemic preconditioning conferred cardioprotection in A_1KO, A_{2A}KO, or A_3KO mice but not in A_{2B}KO mice (Eckle et al., 2007). However, a recent comprehensive study also employing A_{2B}KO mice, and using the selective A_{2B} antagonist ATL-801 both in mouse and in rat models of myocardial ischemia–reperfusion injury, argued against contribution of A_{2B} receptors at least in the acute phase of ischemic preconditioning (Maas et al., 2010).

Stimulation of angiogenesis and modulation of immune response by A_{2B} receptors may play an important role in promotion of cancer growth, which can be considered in broad terms to share characteristics with chronic inflammatory processes. Indeed, Lewis lung carcinoma tumors grown in host animals lacking A_{2B} adenosine receptors contained significantly lower levels of VEGF and displayed lower intratumor vascular density compared to tumors grown in WT animals. This difference in neovascularization and tumor tissue VEGF levels was due to A_{2B} receptor-dependent VEGF production by host tumor-associated cells. Furthermore, analysis of host immune cells in tumors suggested that A_{2B} receptor signaling may favor the expansion of myeloid-derived suppressor cells. These observations raise the interesting possibility that host A_{2B} receptors on immune cells not only stimulate tumor angiogenesis but also suppress immune surveillance, thus engaging two distinct mechanisms to promote tumor survival and growth (Ryzhov et al., 2008a). It is also possible that not only tumors but also infectious agents may exploit host A_{2B} receptors for their advantage. Stimulation of host A_{2B} receptors was implicated in persistence of chlamidal infection (Pettengill et al., 2009) and establishing of Leishmania infection (de Almeida Marques-da-Silva et al., 2008). However, the role of other adenosine receptor subtypes in these infections is not known, and the mechanisms responsible for these effects need to be defined in detail.

VIII. Conclusions

To be beneficial, the inflammatory reaction must be acute, destroying an injurious agent within a short period of time and in a localized area, while inducing an immune response. This is achieved through a complex series of events involving multiple cell types and secreted factors. Given the actions of adenosine to return the host to a homeostatic state, it is not surprising that A_{2B} receptors may operate to control the extent of acute inflammatory responses. However, premature suppression of acute inflammation by adenosine may promote its progression to a chronic status to the detriment of the host.

We believe that labeling A_{2B} receptors as "anti-inflammatory" or "proinflammatory" may be overly simplistic and misleading because it fails to appreciate just how complex each of the different types of inflammation is and conveys little about precise mechanisms. For example, the inflammation in a model of ventilation-induced lung injury is clearly different from the inflammation in a model of allergen-induced chronic lung disease. Therefore, it is not surprising that inhibition of A_{2B} receptors produced opposite effects in these models. As we illustrated in this chapter, A_{2B} receptors may play different roles even in similar types of inflammation but occurring in different tissues. Furthermore, A_{2B} receptors can play different roles at different points in the progression of inflammation. In this respect, A_{2B} receptor functions may be reflective of pleiotropic effects of the secreted factors. As an example, IL-6 was shown to promote inflammation in models of chronic inflammatory diseases, whereas in models of acute inflammation, it can suppress local and systemic acute inflammatory responses (Gabay, 2006). Moreover, various combinations of cytokines released from A_{2B} receptor-activated cells can produce different outcomes. TGF-β1 produced by "adenosine-differentiated" dendritic cells (Novitskiy et al., 2008) favors the emergence of adaptive Tregs, but in combination with IL-4 or IL-6, it may stimulate Th9 and Th17 cells, respectively (Ernst et al., 2010).

Targeting the low-affinity A_{2B} receptor, as opposed to other adenosine receptor subtypes, in the development of novel therapeutic approaches for treatment of inflammatory diseases is especially appealing because these receptors are likely activated only in pathophysiological environment, while remaining silent in normal tissues. This characteristic could provide specificity in therapy of certain immune-related disorders, while decreasing likelihood of side effects. However, translation of current findings into novel therapies would require a better understanding of A_{2B} receptor functions in diverse types of inflammatory responses in various tissues and at different points of their progression.

Acknowledgments

This work was supported in part by National Institutes of Health Grants R01HL095787 and R01CA138923.

Conflict of Interest: The authors have been recipients of research grants from Gilead Palo Alto, Inc.

Abbreviations

ADA	adenosine deaminase
BAL	bronchoalveolar lavage
bFGF	basic fibroblast growth factor
BMMC	bone marrow-derived mast cell
CFTR	cystic fibrosis transmembrane conductance regulator
ENaC	epithelial sodium channel
ERK	extracellular signal-regulated kinase
HBEC	human bronchial epithelial cell
HIF-1α	hypoxia-inducible factor 1-alpha
HMC-1	human mast cell-1
HMEC-1	human microvascular endothelial cell-1
HUVEC	human umbilical vein endothelial cell
ICAM	intercellular adhesion molecule
IDO	indoleamine 2,3-dioxygenase
IL1RAP	interleukin-1 receptor accessory protein
INF-γ	interferon-gamma
JNK	c-Jun N-terminal kinase
KO	knockout
LPS	lipopolysaccharide
MAPK	mitogen-activated protein kinase
M-CSF	macrophage colony stimulating factor
MHC-II	major histocompatibility complex class II
MMP-1	matrix metalloproteinase-1
NFAT	nuclear factor of activated T cells
PKA	protein kinase A
PLC	phospholipase C
TGF-β1	transforming growth factor beta one
TLR	toll-like receptors
TNF-α	tumor necrosis factor alpha
VEGF	vascular endothelial growth factor
WT	wild type

References

Afzal, A., Shaw, L. C., Caballero, S., Spoerri, P. E., Lewin, A. S., Zeng, D., et al. (2003). Reduction in preretinal neovascularization by ribozymes that cleave the A_{2B} adenosine receptor mRNA. *Circulation Research*, *93*, 500–506.

Amisten, S., Braun, O. O., Bengtsson, A., & Erlinge, D. (2008). Gene expression profiling for the identification of G-protein coupled receptors in human platelets. *Thrombosis Research*, *122*, 47–57.

Anvari, F., Sharma, A. K., Fernandez, L. G., Hranjec, T., Ravid, K., Kron, I. L., et al. (2010). Tissue-derived proinflammatory effect of adenosine A_{2B} receptor in lung ischemia-reperfusion injury. *The Journal of Thoracic and Cardiovascular Surgery*, *140*, 871–877.

Auchampach, J. A., Jin, X., Wan, T. C., Caughey, G. H., & Linden, J. (1997). Canine mast cell adenosine receptors: Cloning and expression of the A_3 receptors and evidence that degranulation is mediated by the A_{2B} receptor. *Molecular Pharmacology*, *52*, 846–860.

Auchampach, J. A., Kreckler, L. M., Wan, T. C., Maas, J. E., van der Hoeven, D., Gizewski, E., et al. (2009). Characterization of the A_{2B} adenosine receptor from mouse, rabbit, and dog. *The Journal of Pharmacology and Experimental Therapeutics*, *329*, 2–13.

Baraldi, P. G., Tabrizi, M. A., Fruttarolo, F., Romagnoli, R., & Preti, D. (2009). Recent improvements in the development of A_{2B} adenosine receptor agonists. *Purinergic Signalling*, *5*, 3–19.

Ben, A. A., Lefort, A., Hua, X., Libert, F., Communi, D., Ledent, C., et al. (2008). Modulation of murine dendritic cell function by adenine nucleotides and adenosine: Involvement of the A_{2B} receptor. *European Journal of Immunology*, *38*, 1610–1620.

Biaggioni, I., & Feoktistov, I. (2005). Adenosine and endothelial cell function. In G. Hasko, B. N. Cronstein & C. Szabo (Eds.), *Adenosine receptors: Therapeutic aspects for inflammatory and immune diseases* (pp. 109–131). UK, USA: CRC Press.

Blackburn, M. R., Volmer, J. B., Thrasher, J. L., Zhong, H., Crosby, J. R., Lee, J. J., et al. (2000). Metabolic consequences of adenosine deaminase deficiency in mice are associated with defects in alveogenesis, pulmonary inflammation, and airway obstruction. *The Journal of Experimental Medicine*, *192*, 159–170.

Boyle, D. L., Sajjadi, F. G., & Firestein, G. S. (1996). Inhibition of synoviocyte collagenase gene expression by adenosine receptor stimulation. *Arthritis and Rheumatism*, *39*, 923–930.

Bruns, R. F. (1980). Adenosine receptor activation in human fibroblasts: Nucleoside agonists and antagonists. *Canadian Journal of Physiology and Pharmacology*, *58*, 673–691.

Bruns, R. F., Lu, G. H., & Pugsley, T. A. (1986). Characterization of the A_2 adenosine receptor labeled by [^{3}H]NECA in rat striatal membranes. *Molecular Pharmacology*, *29*, 331–346.

Buenestado, A., Grassin, D. S., Arnould, I., Besnard, F., Naline, E., Blouquit-Laye, S., et al. (2010). The role of adenosine receptors in regulating production of tumour necrosis factor-alpha and chemokines by human lung macrophages. *British Journal of Pharmacology*, *159*, 1304–1311.

Cagnina, R. E., Ramos, S. I., Marshall, M. A., Wang, G., Frazier, C. R., & Linden, J. (2009). Adenosine A_{2B} receptors are highly expressed on murine type II alveolar epithelial cells. *The American Journal of Physiology*, *297*, L467–L474.

Comerford, K. M., Lawrence, D. W., Synnestvedt, K., Levi, B. P., & Colgan, S. P. (2002). Role of vasodilator-stimulated phosphoprotein in PKA-induced changes in endothelial junctional permeability. *The FASEB Journal*, *16*, 583–585.

Corset, V., Nguyen-Ba-Charvet, K. T., Forcet, C., Moyse, E., Dotal, A., & Mehlen, P. (2000). Netrin-1-mediated axon outgrowth and cAMP production requires interaction with adenosine A_{2b} receptor. *Nature*, *407*, 747–750.

Csoka, B., Himer, L., Selmeczy, Z., Vizi, E. S., Pacher, P., Ledent, C., et al. (2008). Adenosine A_{2A} receptor activation inhibits T helper 1 and T helper 2 cell development and effector function. *The FASEB Journal*, *22*, 3491–3499.

Csoka, B., Nemeth, Z. H., Rosenberger, P., Eltzschig, H. K., Spolarics, Z., Pacher, P., et al. (2010). A_{2B} adenosine receptors protect against sepsis-induced mortality by dampening excessive inflammation. *Journal of Immunology*, *185*, 542–550.

de Almeida Marques-da-Silva, E., de Oliveira, J. C., Figueiredo, A. B., de Souza Lima, J. D., Carneiro, C. M., & Rangel Fietto, J. L. (2008). Extracellular nucleotide metabolism in Leishmania: Influence of adenosine in the establishment of infection. *Microbes and Infection*, *10*, 850–857.

Dominguez, P. M., & Ardavin, C. (2010). Differentiation and function of mouse monocyte-derived dendritic cells in steady state and inflammation. *Immunological Reviews*, *234*, 90–104.

Donaldson, S. H., & Boucher, R. C. (2007). Sodium channels and cystic fibrosis. *Chest*, *132*, 1631–1636.

Dubey, R. K., Gillespie, D. G., & Jackson, E. K. (2002). A_{2B} adenosine receptors stimulate growth of porcine and rat arterial endothelial cells. *Hypertension*, *39*, 530–535.

Dubey, R. K., Gillespie, D. G., Zacharia, L. C., Mi, Z., & Jackson, E. K. (2001). A_{2b} receptors mediate the antimitogenic effects of adenosine in cardiac fibroblasts. *Hypertension*, *37*, 716–721.

Duffy, S. M., Cruse, G., Brightling, C. E., & Bradding, P. (2007). Adenosine closes the K+ channel KCa3.1 in human lung mast cells and inhibits their migration via the adenosine A_{2A} receptor. *European Journal of Immunology*, *37*, 1653–1662.

Eckle, T., Faigle, M., Grenz, A., Laucher, S., Thompson, L. F., & Eltzschig, H. K. (2008a). A_{2B} adenosine receptor dampens hypoxia-induced vascular leak. *Blood*, *111*, 2024–2035.

Eckle, T., Grenz, A., Laucher, S., & Eltzschig, H. K. (2008b). A_{2B} adenosine receptor signaling attenuates acute lung injury by enhancing alveolar fluid clearance in mice. *The Journal of Clinical Investigation*, *118*, 3301–3315.

Eckle, T., Krahn, T., Grenz, A., Kohler, D., Mittelbronn, M., Ledent, C., et al. (2007). Cardioprotection by ecto-5′-nucleotidase (CD73) and A2B adenosine receptors. *Circulation*, *115*, 1581–1590.

Eltzschig, H. K., Abdulla, P., Hoffman, E., Hamilton, K. E., Daniels, D., Schonfeld, C., et al. (2005). HIF-1-dependent repression of equilibrative nucleoside transporter (ENT) in hypoxia. *The Journal of Experimental Medicine*, *202*, 1493–1505.

Eltzschig, H. K., Ibla, J. C., Furuta, G. T., Leonard, M. O., Jacobson, K. A., Enjyoji, K., et al. (2003). Coordinated adenine nucleotide phosphohydrolysis and nucleoside signaling in posthypoxic endothelium: Role of ectonucleotidases and adenosine A_{2B} receptors. *The Journal of Experimental Medicine*, *198*, 783–796.

Epperson, S. A., Brunton, L. L., Ramirez-Sanchez, I., & Villarreal, F. (2009). Adenosine receptors and second messenger signaling pathways in rat cardiac fibroblasts. *The American Journal of Physiology*, *296*, C1171–C1177.

Ernst, P. B., Garrison, J. C., & Thompson, L. F. (2010). Much ado about adenosine: Adenosine synthesis and function in regulatory T cell biology. *Journal of Immunology*, *185*, 1993–1998.

Factor, P., Mutlu, G. M., Chen, L., Mohameed, J., Akhmedov, A. T., Meng, F. J., et al. (2007). Adenosine regulation of alveolar fluid clearance. *Proceedings of the National Academy of Sciences of the United States of America*, *104*, 4083–4088.

Fang, X., Song, Y., Hirsch, J., Galietta, L. J., Pedemonte, N., Zemans, R. L., et al. (2006). Contribution of CFTR to apical-basolateral fluid transport in cultured human alveolar epithelial type II cells. *The American Journal of Physiology*, *290*, L242–L249.

Feng, W., Song, Y., Chen, C., Lu, Z. Z., & Zhang, Y. (2010). Stimulation of adenosine A_{2B} receptors induces interleukin-6 secretion in cardiac fibroblasts via the PKC-delta-P38 signalling pathway. *British Journal of Pharmacology*, *159*, 1598–1607.

Feoktistov, I., & Biaggioni, I. (1995). Adenosine A_{2b} receptors evoke interleukin-8 secretion in human mast cells. An enprofylline-sensitive mechanism with implications for asthma. *The Journal of Clinical Investigation*, *96*, 1979–1986.

Feoktistov, I., Goldstein, A. E., & Biaggioni, I. (1999). Role of p38 mitogen-activated protein kinase and extracellular signal-regulated protein kinase kinase in adenosine A_{2B} receptor-mediated interleukin-8 production in human mast cells. *Molecular Pharmacology*, *55*, 726–735.

Feoktistov, I., Goldstein, A. E., Ryzhov, S., Zeng, D., Belardinelli, L., Voyno-Yasenetskaya, T., et al. (2002). Differential expression of adenosine receptors in human endothelial cells: Role of A_{2B} receptors in angiogenic factor regulation. *Circulation Research*, *90*, 531–538.

Feoktistov, I., Ryzhov, S., Goldstein, A. E., & Biaggioni, I. (2003). Mast cell-mediated stimulation of angiogenesis: Cooperative interaction between A_{2B} and A_3 adenosine receptors. *Circulation Research*, *92*, 485–492.

Feoktistov, I., Ryzhov, S., Zhong, H., Goldstein, A. E., Matafonov, A., Zeng, D., et al. (2004). Hypoxia modulates adenosine receptors in human endothelial and smooth muscle cells toward an A2B angiogenic phenotype. *Hypertension*, *44*, 649–654.

Fischer, S., Sharma, H. S., Karliczek, G. F., & Schaper, W. (1995). Expression of vascular permeability factor/vascular endothelial growth factor in pig cerebral microvascular endothelial cells and its upregulation by adenosine. *Brain Research*, *28*, 141–148.

Flavell, S. J., Hou, T. Z., Lax, S., Filer, A. D., Salmon, M., & Buckley, C. D. (2008). Fibroblasts as novel therapeutic targets in chronic inflammation. *British Journal of Pharmacology*, *153*, S241–S246.

Fredholm, B. B. (2007). Adenosine, an endogenous distress signal, modulates tissue damage and repair. *Cell Death and Differentiation*, *14*, 1315–1323.

Fredholm, B. B., Ijzerman, A. P., Jacobson, K. A., Klotz, K. N., & Linden, J. (2001a). International Union of Pharmacology XXV. Nomenclature and classification of adenosine receptors. *Pharmacological Reviews*, *53*, 527–552.

Fredholm, B. B., Irenius, E., Kull, B., & Schulte, G. (2001b). Comparison of the potency of adenosine as an agonist at human adenosine receptors expressed in Chinese hamster ovary cells. *Biochemical Pharmacology*, *61*, 443–448.

Fredholm, B. B., Zhang, Y., & van der Ploeg, I. (1996). Adenosine A_{2a} receptors mediate the inhibitory effect of adenosine on formyl-met-leu-phe-stimulated respiratory burst in neutrophil leucocytes. *Naunyn-Schmiedebergs Archives of Pharmacology*, *354*, 262–267.

Frick, J. S., Macmanus, C. F., Scully, M., Glover, L. E., Eltzschig, H. K., & Colgan, S. P. (2009). Contribution of adenosine A_{2B} receptors to inflammatory parameters of experimental colitis. *Journal of Immunology*, *182*, 4957–4964.

Gabay, C. (2006). Interleukin-6 and chronic inflammation. *Arthritis Research & Therapy*, *8*, S3.

Gessi, S., Varani, K., Merighi, S., Cattabriga, E., Pancaldi, C., Szabadkai, Y., et al. (2005). Expression, pharmacological profile, and functional coupling of A_{2B} receptors in a recombinant system and in peripheral blood cells using a novel selective antagonist radioligand, [^{3}H]MRE 2029-F20. *Molecular Pharmacology*, *67*, 2137–2147.

Grant, M. B., Davis, M. I., Caballero, S., Feoktistov, I., Biaggioni, I., & Belardinelli, L. (2001). Proliferation, migration, and ERK activation in human retinal endothelial cells through A_{2B} adenosine receptor stimulation. *Investigative Ophthalmology and Visual Science*, *42*, 2068–2073.

Grant, M. B., Tarnuzzer, R. W., Caballero, S., Ozeck, M. J., Davis, M. I., Spoerri, P. E., et al. (1999). Adenosine receptor activation induces vascular endothelial growth factor in human retinal endothelial cells. *Circulation Research*, *85*, 699–706.

Grenz, A., Osswald, H., Eckle, T., Yang, D., Zhang, H., Tran, Z. V., et al. (2008). The reno-vascular A_{2B} adenosine receptor protects the kidney from ischemia. *PLoS Medicine*, *5*, e137.

Harrington, E. O., Newton, J., Morin, N., & Rounds, S. (2004). Barrier dysfunction and RhoA activation are blunted by homocysteine and adenosine in pulmonary endothelium. *The American Journal of Physiology*, *287*, L1091–L1097.

Hart, M. L., Jacobi, B., Schittenhelm, J., Henn, M., & Eltzschig, H. K. (2009). Cutting edge: A_{2B} adenosine receptor signaling provides potent protection during intestinal ischemia/reperfusion injury. *Journal of Immunology*, *182*, 3965–3968.

Hua, X., Kovarova, M., Chason, K. D., Nguyen, M., Koller, B. H., & Tilley, S. L. (2007). Enhanced mast cell activation in mice deficient in the A_{2b} adenosine receptor. *The Journal of Experimental Medicine*, *204*, 117–128.

Huang, P., Lazarowski, E. R., Tarran, R., Milgram, S. L., Boucher, R. C., & Stutts, M. J. (2001). Compartmentalized autocrine signaling to cystic fibrosis transmembrane conductance regulator at the apical membrane of airway epithelial cells. *Proceedings of the National Academy of Sciences of the United States of America*, *98*, 14120–14125.

Kalla, R. V., & Zablocki, J. (2009). Progress in the discovery of selective, high affinity A_{2B} adenosine receptor antagonists as clinical candidates. *Purinergic Signalling*, *5*, 21–29.

Khoa, N. D., Montesinos, M. C., Williams, A. J., Kelly, M., & Cronstein, B. N. (2003). Th1 cytokines regulate adenosine receptors and their downstream signaling elements in human microvascular endothelial cells. *Journal of Immunology*, *171*, 3991–3998.

Kim, Y. C., Ji, X., Melman, N., Linden, J., & Jacobson, K. A. (2000). Anilide derivatives of an 8-phenylxanthine carboxylic congener are highly potent and selective antagonists at human A_{2B} adenosine receptors. *Journal of Medicinal Chemistry*, *43*, 1165–1172.

Klotz, K. N., Hessling, J., Hegler, J., Owman, C., Kull, B., Fredholm, B. B., et al. (1998). Comparative pharmacology of human adenosine receptor subtypes—Characterization of stably transfected receptors in CHO cells. *Naunyn-Schmiedebergs Archives of Pharmacology*, *357*, 1–9.

Kolachala, V., Asamoah, V., Wang, L., Srinivasan, S., Merlin, D., & Sitaraman, S. V. (2005). Interferon-gamma down-regulates adenosine 2b receptor-mediated signaling and short circuit current in the intestinal epithelia by inhibiting the expression of adenylate cyclase. *The Journal of Biological Chemistry*, *280*, 4048–4057.

Kolachala, V. L., Bajaj, R., Chalasani, M., & Sitaraman, S. V. (2008a). Purinergic receptors in gastrointestinal inflammation. *The American Journal of Physiology*, *294*, G401–G410.

Kolachala, V., Ruble, B., Vijay-Kumar, M., Wang, L., Mwangi, S., Figler, H., et al. (2008b). Blockade of adenosine A_{2B} receptors ameliorates murine colitis. *British Journal of Pharmacology*, *155*, 127–137.

Kolachala, V. L., Vijay-Kumar, M., Dalmasso, G., Yang, D., Linden, J., Wang, L., et al. (2008c). A_{2B} adenosine receptor gene deletion attenuates murine colitis. *Gastroenterology*, *135*, 861–870.

Kolachala, V. L., Wang, L., Obertone, T. S., Prasad, M., Yan, Y., Dalmasso, G., et al. (2010). Adenosine 2B receptor expression is post-transcriptionally regulated by microRNA. *The Journal of Biological Chemistry*, *285*, 18184–18190.

Kong, T., Westerman, K. A., Faigle, M., Eltzschig, H. K., & Colgan, S. P. (2006). HIF-dependent induction of adenosine A_{2B} receptor in hypoxia. *The FASEB Journal*, *20*, 2242–2250.

Kreckler, L. M., Gizewski, E., Wan, T. C., & Auchampach, J. A. (2009). Adenosine suppresses lipopolysaccharide-induced tumor necrosis factor-alpha production by murine macrophages through a protein kinase A- and exchange protein activated by cAMP-independent signaling pathway. *The Journal of Pharmacology and Experimental Therapeutics*, *331*, 1051–1061.

Kreckler, L. M., Wan, T. C., Ge, Z. D., & Auchampach, J. A. (2006). Adenosine inhibits tumor necrosis factor-alpha release from mouse peritoneal macrophages via A_{2A} and A_{2B} but not the A_3 adenosine receptor. *The Journal of Pharmacology and Experimental Therapeutics*, *317*, 172–180.

Kuno, A., Critz, S. D., Cui, L., Solodushko, V., Yang, X. M., Krahn, T., et al. (2007). Protein kinase C protects preconditioned rabbit hearts by increasing sensitivity of adenosine A_{2b}-dependent signaling during early reperfusion. *Journal of Molecular and Cellular Cardiology*, *43*, 262–271.

Lazarowski, E. R., Tarran, R., Grubb, B. R., van Heusden, C. A., Okada, S., & Boucher, R. C. (2004). Nucleotide release provides a mechanism for airway surface liquid homeostasis. *The Journal of Biological Chemistry*, *279*, 36855–36864.

Linden, J., Thai, T., Figler, H., Jin, X., & Robeva, A. S. (1999). Characterization of human A_{2B} adenosine receptors: Radioligand binding, western blotting, and coupling to G_q in human embryonic kidney 293 cells and HMC-1 mast cells. *Molecular Pharmacology*, *56*, 705–713.

Maas, J. E., Wan, T. C., Figler, R. A., Gross, G. J., & Auchampach, J. A. (2010). Evidence that the acute phase of ischemic preconditioning does not require signaling by the A_{2B} adenosine receptor. *Journal of Molecular and Cellular Cardiology*, *49*, 886–893.

Mackay, F., Loetscher, H., Stueber, D., Gehr, G., & Lesslauer, W. (1993). Tumor necrosis factor alpha (TNF-α)-induced cell adhesion to human endothelial cells is under dominant control of one TNF receptor type, TNF-R55. *The Journal of Experimental Medicine*, *177*, 1277–1286.

Madara, J. L., Patapoff, T. W., Gillece-Castro, B., Colgan, S. P., Parkos, C. A., Delp, C., et al. (1993). 5′-adenosine monophosphate is the neutrophil-derived paracrine factor that elicits chloride secretion from T84 intestinal epithelial cell monolayers. *The Journal of Clinical Investigation*, *91*, 2320–2325.

McKenna, W. L., Wong-Staal, C., Kim, G. C., Macias, H., Hinck, L., & Bartoe, J. L. (2008). Netrin-1-independent adenosine A_{2b} receptor activation regulates the response of axons to netrin-1 by controlling cell surface levels of UNC5A receptors. *Journal of Neurochemistry*, *104*, 1081–1090.

Meade, C. J., Worrall, L., Hayes, D., & Protin, U. (2002). Induction of interleukin 8 release from the HMC-1 mast cell line: Synergy between stem cell factor and activators of the adenosine A_{2b} receptor. *Biochemical Pharmacology*, *64*, 317–325.

Mino, R. P., Spoerri, P. E., Caballero, S., Player, D., Belardinelli, L., Biaggioni, I., et al. (2001). Adenosine receptor antagonists and retinal neovascularization in vivo. *Investigative Ophthalmology & Visual Science*, *42*, 3320–3324.

Morse, D. M., Smullen, J. L., & Davis, C. W. (2001). Differential effects of UTP, ATP, and adenosine on ciliary activity of human nasal epithelial cells. *The American Journal of Physiology*, *280*, C1485–C1497.

Murakami, S., Hashikawa, T., Saho, T., Takedachi, M., Nozaki, T., Shimabukuro, Y., et al. (2001). Adenosine regulates the IL-1 beta-induced cellular functions of human gingival fibroblasts. *International Immunology*, *13*, 1533–1540.

Murakami, S., Terakura, M., Kamatani, T., Hashikawa, T., Saho, T., Shimabukuro, Y., et al. (2000). Adenosine regulates the production of interleukin-6 by human gingival fibroblasts via cyclic AMP/protein kinase A pathway. *Journal of Periodontal Research*, *35*, 93–101.

Mustafa, S. J., Nadeem, A., Fan, M., Zhong, H., Belardinelli, L., & Zeng, D. (2007). Effect of a specific and selective A_{2B} adenosine receptor antagonist on adenosine agonist AMP and allergen-induced airway responsiveness and cellular influx in a mouse model of asthma. *The Journal of Pharmacology and Experimental Therapeutics*, *320*, 1246–1251.

Nemeth, Z. H., Lutz, C. S., Csoka, B., Deitch, E. A., Leibovich, S. J., Gause, W. C., et al. (2005). Adenosine augments IL-10 production by macrophages through an A_{2B} receptor-mediated posttranscriptional mechanism. *Journal of Immunology*, *175*, 8260–8270.

Novitskiy, S. V., Ryzhov, S., Zaynagetdinov, R., Goldstein, A. E., Huang, Y., Tikhomirov, O. Y., et al. (2008). Adenosine receptors in regulation of dendritic cell differentiation and function. *Blood*, *112*, 1822–1831.

Ongini, E., Dionisotti, S., Gessi, S., Irenius, E., & Fredholm, B. B. (1999). Comparison of CGS 15943, ZM 241385 and SCH 58261 as antagonists at human adenosine receptors. *Naunyn-Schmiedebergs Archives of Pharmacology*, *359*, 7–10.

Palmer, T. M., & Trevethick, M. A. (2008). Suppression of inflammatory and immune responses by the A_{2A} adenosine receptor: An introduction. *British Journal of Pharmacology*, *153*, S27–S34.

Pettengill, M. A., Lam, V. W., & Ojcius, D. M. (2009). The danger signal adenosine induces persistence of chlamydial infection through stimulation of A_{2b} receptors. *PLoS ONE*, *4*, e8299.

Ramkumar, V., Stiles, G. L., Beaven, M. A., & Ali, H. (1993). The A_3 adenosine receptor is the unique adenosine receptor which facilitates release of allergic mediators in mast cells. *The Journal of Biological Chemistry*, *268*, 16887–16890.

Reddy, M. M., Light, M. J., & Quinton, P. M. (1999). Activation of the epithelial Na^+ channel (ENaC) requires CFTR Cl- channel function. *Nature*, *402*, 301–304.

Rollins, B. M., Burn, M., Coakley, R. D., Chambers, L. A., Hirsh, A. J., Clunes, M. T., et al. (2008). A_{2B} adenosine receptors regulate the mucus clearance component of the lung's innate defense system. *American Journal of Respiratory Cell and Molecular Biology*, *39*, 190–197.

Rosenberger, P., Schwab, J. M., Mirakaj, V., Masekowsky, E., Mager, A., Morote-Garcia, J. C., et al. (2009). Hypoxia-inducible factor-dependent induction of netrin-1 dampens inflammation caused by hypoxia. *Nature Immunology*, *10*, 195–202.

Ryzhov, S., Goldstein, A. E., Biaggioni, I., & Feoktistov, I. (2006). Cross-talk between G_s- and G_q-coupled pathways in regulation of interleukin-4 by A_{2B} adenosine receptors in human mast cells. *Molecular Pharmacology*, *70*, 727–735.

Ryzhov, S., Goldstein, A. E., Matafonov, A., Zeng, D., Biaggioni, I., & Feoktistov, I. (2004). Adenosine-activated mast cells induce IgE synthesis by B lymphocytes: An A_{2B}-mediated process involving Th2 cytokines IL-4 and IL-13 with implications for asthma. *Journal of Immunology*, *172*, 7726–7733.

Ryzhov, S., McCaleb, J. L., Goldstein, A. E., Biaggioni, I., & Feoktistov, I. (2007). Role of adenosine receptors in the regulation of angiogenic factors and neovascularization in hypoxia. *The Journal of Pharmacology and Experimental Therapeutics*, *382*, 565–572.

Ryzhov, S., Novitskiy, S. V., Zaynagetdinov, R., Goldstein, A. E., Carbone, D. P., Biaggioni, I., et al. (2008a). Host A_{2B} adenosine receptors promote carcinoma growth. *Neoplasia (New York)*, *10*, 987–995.

Ryzhov, S., Solenkova, N. V., Goldstein, A. E., Lamparter, M., Fleenor, T., Young, P. P., et al. (2008b). Adenosine receptor-mediated adhesion of endothelial progenitors to cardiac microvascular endothelial cells. *Circulation Research*, *102*, 356–363.

Ryzhov, S., Zaynagetdinov, R., Goldstein, A. E., Matafonov, A., Biaggioni, I., & Feoktistov, I. (2009). Differential role of the carboxy-terminus of the A_{2B} adenosine receptor in stimulation of adenylate cyclase, phospholipase Cβ, and interleukin-8. *Purinergic Signalling*, *5*, 289–298.

Ryzhov, S., Zaynagetdinov, R., Goldstein, A. E., Novitskiy, S. V., Blackburn, M. R., Biaggioni, I., et al. (2008c). Effect of A_{2B} adenosine receptor gene ablation on adenosine-dependent regulation of proinflammatory cytokines. *The Journal of Pharmacology and Experimental Therapeutics*, *324*, 694–700.

Ryzhov, S., Zaynagetdinov, R., Goldstein, A. E., Novitskiy, S. V., Dikov, M. M., Blackburn, M. R., et al. (2008d). Effect of A_{2B} adenosine receptor gene ablation on proinflammatory adenosine signaling in mast cells. *Journal of Immunology*, *180*, 7212–7220.

Salvatore, C. A., Tilley, S. L., Latour, A. M., Fletcher, D. S., Koller, B. H., & Jacobson, M. A. (2000). Disruption of the A_3 adenosine receptor gene in mice and its effect on stimulated inflammatory cells. *The Journal of Biological Chemistry*, *275*, 4429–4434.

Sands, W. A., & Palmer, T. M. (2005). Adenosine receptors and the control of endothelial cell function in inflammatory disease. *Immunology Letters*, *101*, 1–11.

Scheckenbach, K. L., Losa, D., Dudez, T., Bacchetta, M., O'Grady, S., Crespin, S., et al. (2010). PGE_2 regulation of CFTR activity and air surface liquid volume requires gap junctional communication. *American Journal of Respiratory Cell and Molecular Biology*.

Schingnitz, U., Hartmann, K., Macmanus, C. F., Eckle, T., Zug, S., Colgan, S. P., et al. (2010). Signaling through the A_{2B} adenosine receptor dampens endotoxin-induced acute lung injury. *Journal of Immunology*, *184*, 5271–5279.

Schneider, D. J., Lindsay, J. C., Zhou, Y., Molina, J. G., & Blackburn, M. R. (2010). Adenosine and osteopontin contribute to the development of chronic obstructive pulmonary disease. *The FASEB Journal*, *24*, 70–80.

Silver, M. R., Margulis, A., Wood, N., Goldman, S. J., Kasaian, M., & Chaudhary, D. (2010). IL-33 synergizes with IgE-dependent and IgE-independent agents to promote mast cell and basophil activation. *Inflammation Research*, *59*, 207–218.

Sitaraman, S. V., Merlin, D., Wang, L., Wong, M., Gewirtz, A. T., Si-Tahar, M., et al. (2001). Neutrophil-epithelial crosstalk at the intestinal lumenal surface mediated by reciprocal secretion of adenosine and IL-6. *The Journal of Clinical Investigation*, *107*, 861–869.

Srinivas, S. P., Satpathy, M., Gallagher, P., Lariviere, E., & Van Driessche, W. (2004). Adenosine induces dephosphorylation of myosin II regulatory light chain in cultured bovine corneal endothelial cells. *Experimental Eye Research*, *79*, 543–551.

Strohmeier, G. R., Reppert, S. M., Lencer, W. I., & Madara, J. L. (1995). The A_{2b} adenosine receptor mediates cAMP responses to adenosine receptor agonists in human intestinal epithelia. *The Journal of Biological Chemistry*, *270*, 2387–2394.

Sun, C. X., Zhong, H., Mohsenin, A., Morschl, E., Chunn, J. L., Molina, J. G., et al. (2006). Role of A_{2B} adenosine receptor signaling in adenosine-dependent pulmonary inflammation and injury. *The Journal of Clinical Investigation*, *116*, 2173–2182.

Suzuki, H., Takei, M., Nakahata, T., & Fukamachi, H. (1998). Inhibitory effect of adenosine on degranulation of human cultured mast cells upon cross-linking of Fc epsilon RI. *Biochemical and Biophysical Research Communications*, *242*, 697–702.

Synnestvedt, K., Furuta, G. T., Comerford, K. M., Louis, N., Karhausen, J., Eltzschig, H. K., et al. (2002). Ecto-5′-nucleotidase (CD73) regulation by hypoxia-inducible factor-1 mediates permeability changes in intestinal epithelia. *The Journal of Clinical Investigation*, *110*, 993–1002.

Takagi, H., King, G. L., Ferrara, N., & Aiello, L. P. (1996). Hypoxia regulates vascular endothelial growth factor receptor KDR/Flk gene expression through adenosine A_2 receptors in retinal capillary endothelial cells. *Investigative Ophthalmology & Visual Science*, *37*, 1311–1321.

Takedachi, M., Qu, D., Ebisuno, Y., Oohara, H., Joachims, M. L., McGee, S. T., et al. (2008). CD73-generated adenosine restricts lymphocyte migration into draining lymph nodes. *Journal of Immunology*, *180*, 6288–6296.

Tang, Y., Feng, Y., & Wang, X. (1998). Calcitonin gene-related peptide potentiates LPS-induced IL-6 release from mouse peritoneal macrophages. *Journal of Neuroimmunology*, *84*, 207–212.

Thompson, L. F., Eltzschig, H. K., Ibla, J. C., Van De Wiele, C. J., Resta, R., Morote-Garcia, J. C., et al. (2004). Crucial role for ecto-5′-nucleotidase (CD73) in vascular leakage during hypoxia. *The Journal of Experimental Medicine*, *200*, 1395–1405.

Umapathy, N. S., Fan, Z., Zemskov, E. A., Alieva, I. B., Black, S. M., & Verin, A. D. (2010). Molecular mechanisms involved in adenosine-induced endothelial cell barrier enhancement. *Vascular Pharmacology*, *52*, 199–206.

Walia, B., Castaneda, F. E., Wang, L., Kolachala, V. L., Bajaj, R., Roman, J., et al. (2004). Polarized fibronectin secretion induced by adenosine regulates bacterial-epithelial interaction in human intestinal epithelial cells. *The Biochemical Journal*, *382*, 589–596.

Walker, B. A., Jacobson, M. A., Knight, D. A., Salvatore, C. A., Weir, T., Zhou, D., et al. (1997). Adenosine A_3 receptor expression and function in eosinophils. *American Journal of Respiratory Cell and Molecular Biology*, *16*, 531–537.

Wang, J., & Huxley, V. H. (2006). Adenosine A_{2A} receptor modulation of juvenile female rat skeletal muscle microvessel permeability. *American Journal of Physiology. Heart and Circulatory Physiology*, *291*, H3094–H3105.

Wang, D., Sun, Y., Zhang, W., & Huang, P. (2008). Apical adenosine regulates basolateral Ca2+-activated potassium channels in human airway Calu-3 epithelial cells. *The American Journal of Physiology*, *294*, C1443–C1453.

Waschke, J., Drenckhahn, D., Adamson, R. H., Barth, H., & Curry, F. E. (2004). cAMP protects endothelial barrier functions by preventing Rac-1 inhibition. *The American Journal of Physiology*, *287*, H2427–H2433.

Wen, J., Jiang, X., Dai, Y., Zhang, Y., Tang, Y., Sun, H., et al. (2010). Increased adenosine contributes to penile fibrosis, a dangerous feature of priapism, via A_{2B} adenosine receptor signaling. *The FASEB Journal*, *24*, 740–749.

Wilson, J. M., Ross, W. G., Agbai, O. N., Frazier, R., Figler, R. A., Rieger, J., et al. (2009). The A_{2B} adenosine receptor impairs the maturation and immunogenicity of dendritic cells. *Journal of Immunology*, *182*, 4616–4623.

Xaus, J., Mirabet, M., Lloberas, J., Soler, C., Lluis, C., Franco, R., et al. (1999a). IFN-gamma up-regulates the A_{2B} adenosine receptor expression in macrophages: A mechanism of macrophage deactivation. *Journal of Immunology*, *162*, 3607–3614.

Xaus, J., Valledor, A. F., Cardo, M., Marques, L., Beleta, J., Palacios, J. M., et al. (1999b). Adenosine inhibits macrophage colony-stimulating factor-dependent proliferation of macrophages through the induction of p27kip-1 expression. *Journal of Immunology*, *163*, 4140–4149.

Yang, D., Chen, H., Koupenova, M., Carroll, S. H., Eliades, A., Freedman, J. E., et al. (2010). A new role for the A_{2b} adenosine receptor in regulating platelet function. *Journal of Thrombosis and Haemostasis*, *8*, 817–827.

Yang, D., Zhang, Y., Nguyen, H. G., Koupenova, M., Chauhan, A. K., Makitalo, M., et al. (2006). The A_{2B} adenosine receptor protects against inflammation and excessive vascular adhesion. *The Journal of Clinical Investigation*, *116*, 1913–1923.

Zaynagetdinov, R., Ryzhov, S., Goldstein, A. E., Yin, H., Novitskiy, S. V., Goleniewska, K., et al. (2010). Attenuation of chronic pulmonary inflammation in A_{2B} adenosine receptor knockout mice. *American Journal of Respiratory Cell and Molecular Biology*, *42*, 564–571.

Zhang, J. G., Hepburn, L., Cruz, G., Borman, R. A., & Clark, K. L. (2005). The role of adenosine A_{2A} and A_{2B} receptors in the regulation of TNF-alpha production by human monocytes. *Biochemical Pharmacology*, *69*, 883–889.

Zhong, H., Belardinelli, L., Maa, T., & Zeng, D. (2005). Synergy between A_{2B} adenosine receptors and hypoxia in activating human lung fibroblasts. *American Journal of Respiratory Cell and Molecular Biology*, *32*, 2–8.

Zhou, Y., Mohsenin, A., Morschl, E., Young, H. W., Molina, J. G., Ma, W., et al. (2009). Enhanced airway inflammation and remodeling in adenosine deaminase-deficient mice lacking the A_{2B} adenosine receptor. *Journal of Immunology*, *182*, 8037–8046.

Zhou, Y., Schneider, D. J., Morschl, E., Song, L., Pedroza, M., Karmouty-Quintana, H., et al. (2011). Distinct roles for the A_{2B} adenosine receptor in acute and chronic stages of bleomycin-induced lung injury. *Journal of Immunology*, *186*, 1097–1106.

Michael Koeppen, Tobias Eckle, and Holger K. Eltzschig

Mucosal Inflammation Program, Department of Anesthesiology,
University of Colorado Denver, Aurora, Colorado, USA

Interplay of Hypoxia and A_{2B} Adenosine Receptors in Tissue Protection

Abstract

That adenosine signaling can elicit adaptive tissue responses during conditions of limited oxygen availability (hypoxia) is a long-suspected notion that recently gained general acceptance from genetic and pharmacologic studies of the adenosine signaling pathway. As hypoxia and inflammation share an interdependent relationship, these studies have demonstrated that adenosine signaling events can be targeted to dampen hypoxia-induced inflammation. Here, we build on the hypothesis that particularly the A_{2B} adenosine receptor ($ADORA_{2B}$) plays a central role in tissue adaptation to hypoxia. In fact, the $ADORA_{2B}$ requires higher adenosine concentrations than any of the other adenosine receptors. However, during conditions of hypoxia or ischemia, the hypoxia-elicited rise in extracellular adenosine is sufficient to activate the $ADORA_{2B}$. Moreover, several studies have demonstrated very robust induction of the $ADORA_{2B}$ elicited by transcriptional mechanisms involving hypoxia-dependent signaling pathways and the transcription factor "hypoxia-induced factor" 1. In the present chapter, genetic and pharmacologic evidence is presented to support our hypothesis of a tissue protective role of $ADORA_{2B}$ signaling during hypoxic conditions, including hypoxia-elicited vascular leakage, organ ischemia, or acute lung injury. All these disease models are characterized by hypoxia-elicited tissue inflammation. As such, the $ADORA_{2B}$ has emerged as a therapeutic target for dampening hypoxia-induced inflammation and tissue adaptation to limited oxygen availability.

Advances in Pharmacology, Volume 61

1054-3589/11 $35.00
10.1016/B978-0-12-385526-8.00006-0

I. Introduction

In the extracellular compartment, adenosine has been strongly implicated as a signaling molecule (Aherne et al., 2010; Eltzschig, 2009; Jacobson & Gao, 2006). This notion goes back to research work from almost a century ago. In the 1920s, Drury and Szent-Gyorgyi from the University of Cambridge, UK, performed an experiment, where they injected extracts from cardiac tissues intravenously into a whole animal. Surprisingly, they noticed a transient disturbance of the cardiac rhythm and slowing of the heart rate (Drury & Szent-Gyorgyi, 1929). By utilizing chemical purification steps, the authors demonstrated that the biologically active compound of this extract from animal hearts was an "adenine compound" (Drury & Szent-Gyorgyi, 1929). Today, we know that signaling effects of adenosine involves four different adenosine receptors (ARs): the $ADORA_1$, $ADORA_{2A}$, $ADORA_{2B}$, and $ADORA_3$ (Fig. 1; Eckle et al., 2009; Fredholm, 2007; Fredholm et al., 2001; Gao & Jacobson, 2007; Hasko & Cronstein, 2004; Hasko et al., 2008, 2009; Jacobson & Gao, 2006). For instance, signaling events through the $ADORA_1$ have been implicated in the heart-rate slowing effects of extracellular adenosine, as was noticed by Drury and Szent-Gyorgyi (Drury & Szent-Gyorgyi, 1929; Matherne et al., 1997; Yang et al., 2009). In fact, intravascular treatment with an adenosine bolus will

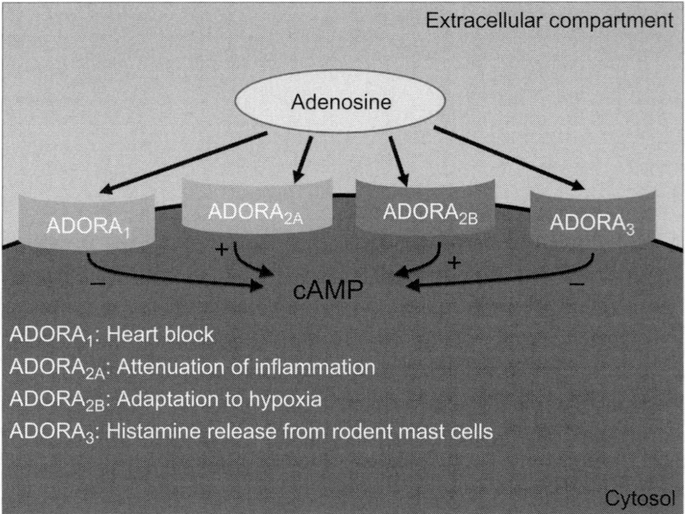

FIGURE 1 *Schematic of ARs.* At present, four different ARs have been described ($ADORA_1$, $ADORA_{2A}$, $ADORA_{2B}$, and $ADORA_3$). They are anchored in the cell membrane and represent G-protein-coupled receptors that are activated by extracellular adenosine. As part of the signal transduction, they are known to alter intracellular cyclic adenosine monophosphate (cAMP) levels. Some well-characterized functions characteristic of the individual receptors are displayed.

result in a temporary heart block in humans (Delacretaz, 2006) or in mice (Koeppen et al., 2009). While *Adora*$_{2a}$–/–, *Adora*$_{2b}$–/–, or *Adora*$_3$–/– mice will exhibit a transient heart block following adenosine injection, *Adora*$_1$–/– mice do not demonstrate any alteration of their heart rate when injected with adenosine (Koeppen et al., 2009). The ADORA$_{2A}$ has been implicated in regulation of the vascular tone (Morrison et al., 2002), or the attenuation of inflammatory responses—particularly through activation of ADORA$_{2A}$s expressed on inflammatory cells (Cronstein, 1994; Cronstein et al., 1990, 1992, 1993; Day et al., 2003, 2004; Lappas et al., 2006; Linden, 2005, 2006a; Yang et al., 2005). This was first described in pioneer research work from the laboratory of Bruce Cronstein who observed an anti-inflammatory effect of A2 receptors expressed on neutrophils (Cronstein, 1994; Cronstein et al., 1983, 1985, 1990, 1991, 1992). Genetic studies from the laboratory of Misha Sitkovsky confirmed such findings by identifying that *Adora*$_{2a}$–/– mice are more prone to develop tissue inflammation than wild-type mice (Lukashev et al., 2003; Ohta & Sitkovsky, 2001; Sitkovsky, 2009; Sitkovsky & Lukashev, 2005; Sitkovsky et al., 2004, 2008; Thiel et al., 2005). Studies from Joel Linden's research group revealed that ADORA$_{2A}$s expressed on T-cells can dampen inflammatory responses (Clark et al., 2007; Day et al., 2003, 2004, 2005; Gazoni et al., 2008; Lappas et al., 2006; Linden, 2005, 2006a; Naganuma et al., 2006; Odashima et al., 2005; Okusa et al., 2001; Reutershan et al., 2007; Sharma et al., 2010, 2009b; Yang et al., 2005). The ADORA$_3$ has been implicated in the release of histamine from rodent mast cells (Zhong et al., 2003) and in cardioprotection from ischemia (Jordan et al., 1999; Takano et al., 2001). Moreover, ADORA$_3$ agonists are presently being examined in clinical Phase II trials for treatments targeting diseases such as cancer, arthritis, and psoriasis (Jacobson et al., 2009). Of the four known ARs, the ADORA$_{2B}$ was the most recent to be characterized, and only during the past years, its role in different disease models has emerged (Aherne et al., 2010; Eltzschig et al., 2009b; Hasko et al., 2008). Its first pharmacological characterization dates back to 1980 (Bruns, 1980). While the ADORA$_{2B}$ was cloned at the same time as the ADORA$_3$, the development of selective tools for the ADORA$_{2B}$ has been lagging behind. This may in part be due to the fact that the ADORA$_{2B}$ is the most adenosine-insensitive AR among the four (Aherne et al., 2010; Eltzschig, 2009). However, during conditions of hypoxia, extracellular adenosine levels are elevated to levels that may be sufficient to cause activation of the ADORA$_{2B}$ (Eckle et al., 2008a; Eltzschig et al., 2003, 2004, 2005; Loffler et al., 2007; Morote-Garcia et al., 2008a, 2009). In fact, several more recent studies have implicated ADORA$_{2B}$ signaling events in the attenuation of hypoxia-induced inflammation (Aherne et al., 2010; Eckle et al., 2008a, 2008d; Eltzschig et al., 2003, 2004, 2009b; Grenz et al., 2007a, 2007b, 2007c, 2007d; Hart et al., 2009; Rosenberger et al., 2010; Schingnitz et al., 2010). While several studies have implicated the ADORA$_{2B}$ in vascular (Yang et al., 2006, 2008) or epithelial inflammation (Frick et al., 2009; Khoury et al.,

2007), we will specifically review here the relationship between hypoxia and inflammation and provide evidence for the hypothesis, that the $ADORA_{2B}$ plays an important role in tissue protection during conditions of limited oxygen availability.

II. Relationship Between Hypoxia and Inflammation

A. Acute Hypoxia Exposure Causes Tissue Inflammation

It had been suspected for some time that hypoxia can cause tissue inflammation (Colgan & Taylor, 2010). During high-altitude mountaineering, climbers are exposed to ambient hypoxia. This is associated with inflammatory tissue changes. For example, a study examined the plasma levels of interleukin 6 (IL-6), interleukin 1 receptor antagonist (IL-1ra), and C-reactive protein (CRP) in climbers who spend three nights at the Jungfraujoch, Switzerland, at an elevation of 3458 m. They found that circulating IL-6, IL-1ra, and CRP is upregulated in response to ambient hypoxic conditions at high altitude, and the systemic increase of these inflammatory markers reflects considerable tissue inflammation (Hartmann et al., 2000). Other inflammatory changes that occur in the setting of high-altitude mountain climbing include the development of cerebral or pulmonary edema. For example, a very elegant study examined arterial blood gas samples that were obtained from mountain climbers at different time points throughout their ascent to the top of Mount Everest (Grocott et al., 2009). The level of environmental hypobaric hypoxia that affects climbers at the summit of Mount Everest (8848 m [29,029 ft]) is close to the limit of tolerance by humans. In this study, samples of arterial blood from 10 climbers during their ascent to and descent from the summit of Mount Everest were obtained and partial pressures of arterial oxygen (PaO_2) and carbon dioxide ($PaCO_2$), pH, and hemoglobin and lactate concentrations were measured. These studies found that the PaO_2 fell with increasing altitude, whereas SaO_2 was relatively stable. The hemoglobin concentration increased such that the oxygen content of arterial blood was maintained at or above sea-level values until the climbers reached an elevation of 7100 m (23,294 ft). In four samples taken at 8400 m (27,559 ft)—at which altitude the barometric pressure was 272 mm Hg (36.3 kPa)—the mean PaO(2) in subjects breathing ambient air was 24.6 mm Hg (3.28 kPa), with a range of 19.1–29.5 mm Hg (2.55–3.93 kPa). The mean PaCO(2) was 13.3 mm Hg (1.77 kPa), with a range of 10.3–15.7 mm Hg (1.37–2.09 kPa). At 8400 m, the mean arterial oxygen content was 26% lower than it was at 7100 m (145.8 ml l^{-1} as compared with 197.1 ml l^{-1}). The mean calculated alveolar–arterial oxygen difference was 5.4 mm Hg (0.72 kPa). As such, these studies demonstrate that the elevated alveolar–arterial oxygen difference that is seen in subjects who are in conditions of extreme hypoxia represents subclinical high-altitude pulmonary edema (Grocott et al., 2009). Other studies in mountain climbers observed a high frequency of upper respiratory infections that were resistant to treatment with antibiotics—until the climbers would return to altitude levels with higher oxygen concentration (Sarnquist, 1983). *Together, such findings in humans exposed to ambient hypoxia demonstrate inflammatory changes characterized by the leakage of fluid into different organs, and a systemic inflammatory response.*

These findings in humans are consistent with experimental studies that demonstrate inflammatory alterations during conditions of limited oxygen availability. In fact, experimental studies demonstrate that the endothelial barrier becomes leaky upon hypoxia exposure (Ogawa et al., 1990, 1992; Yan et al., 1997). Studies of mice exposed to ambient hypoxia reveal attenuation of the epithelial (Synnestvedt et al., 2002) or vascular barrier function (Eltzschig et al., 2003; Thompson et al., 2004). Mice exposed to ambient hypoxia (8% oxygen for 4–8 h) show increases in lung water content and demonstrate fluid leakage into different organs, such as the lungs or the brain (Dieterich et al., 2006; Eckle et al., 2008a; Eltzschig et al., 2003, 2006b; Morote-Garcia et al., 2008a, 2009; Rosenberger et al., 2009). Moreover, circulating cytokine levels are elevated upon hypoxia exposure (Rosenberger et al., 2009). Likewise, ambient hypoxia exposure of mice is associated with the accumulation of inflammatory cells in different organs, such as the lungs, the intestine, or the kidneys (Eltzschig et al., 2004; Rosenberger et al., 2009). Together, such experimental studies demonstrate that ambient hypoxia exposure is associated with an inflammatory phenotype characterized by fluid leakage into different organs, elevated inflammatory markers, and the accumulation of inflammatory cells into multiple organs. It is important to point out that several clinical conditions are characterized by hypoxia-induced inflammation. For example, hypoxia during intestinal (Hart et al., 2008a, 2009), hepatic (Eltzschig et al., 2009b; Hart et al., 2008c, 2010), renal (Grenz et al., 2007b, 2007c, 2008), or myocardial ischemia (Eckle et al., 2008d; Eltzschig et al., 2009a; Kohler et al., 2007) drives inflammatory changes, and anti-inflammatory approaches to dampen hypoxia-induced inflammation have been proposed for these disorders (Eltzschig, 2009). Similarly, limited oxygen availability is the cause of graft inflammation during solid organ transplantation. It is well established that ischemia of an organ graft closely correlates with the inflammatory changes that inversely correlate with early organ function and immune tolerance (Andrade et al., 2006; De Perrot et al., 2002). For example, a recent study in patients undergoing kidney transplantation demonstrates that graft ischemia time correlates with the expressional levels of toll-like receptor (TLR)4 (Kruger et al., 2009). To assess the functional significance of TLR4 in human kidney transplantation, this study determined whether TLR4 mutations influence intragraft gene-expression profiles and immediate graft function. Compared with kidneys expressing wild-type alleles, kidneys with a TLR4 loss-of-function allele contained less tumor necrosis factor α, monocyte chemoattractant protein-1 (MCP-1), and more heme oxygenase 1 (HO1), and exhibited a higher rate of immediate graft function. These results represent evidence that donor TLR4 contributes to graft inflammation and sterile injury following cold preservation and transplantation in humans (Kruger et al., 2009).

Together, such studies demonstrate that hypoxia or ischemia promotes an inflammatory phenotype, and such inflammatory changes represent an important therapeutic target in human conditions, including acute organ ischemia, high-altitude climbing, or organ transplantation.

B. Inflammatory Diseases Are Characterized by Tissue Hypoxia

As outlined above, ambient hypoxia can cause an inflammatory phenotype. However, hypoxia and inflammation share an interdependent relationship. Along these lines, several studies have demonstrated that inflammatory diseases are associated with tissue hypoxia. A famous example for tissue hypoxia during inflammatory conditions comes from the laboratory of Sean Colgan. Here, Karhausen et al. (2004) used an *in vivo* staining technique for hypoxia (Hodgkiss et al., 1991). For this purpose, they injected mice with nitroimidazole compounds that are retained in hypoxic tissues and can be visualized histologically by antibody staining techniques. These studies demonstrated that already at baseline—without inflammatory stimulation—the intestinal mucosa displays tissue hypoxia, which the authors refer to as "physiological hypoxia" (Karhausen et al., 2003, 2004, 2005). This is not surprising when considering the fact that the intestinal lumen is anaerobic, and as such the enterocytes are exposed to a very steep oxygen gradient across the mucosa. However, upon exposure of mice to protocols of experimental colitis, the amount of hypoxia dramatically increases. While under physiologic conditions, hypoxia staining is limited to the apical aspects of the intestinal surface, the staining extends into the submucosal tissues and is much more intense in the inflamed colon (Karhausen et al., 2004). Active inflammation is characterized by dramatic shifts in tissue metabolism and perfusion. These changes include diminished availability of oxygen (Colgan & Taylor, 2010; Haddad, 2003; Saadi et al., 2002; Taylor & Colgan, 2007) with subsequent lactate accumulation and resultant metabolic acidosis. Such shifts in tissue metabolism result, at least in part, from profound recruitment of inflammatory cells, in particular myeloid cells such as neutrophils (polymorphonuclear cells) and monocytes (Colgan & Taylor, 2010; Taylor & Colgan, 2007). The vast majority of inflammatory cells are not resident cells but are recruited to inflammatory lesions (Lewis et al., 1999). As such, it is important to understand the interactions between microenvironmental metabolic changes (e.g., hypoxia) as they relate to molecular mechanisms of leukocyte recruitment and intestinal epithelial dysfunction during inflammation. Together, such studies demonstrate tissue hypoxia caused by inflammation (Colgan & Taylor, 2010; Taylor & Colgan, 2007). While these studies have clearly pointed out that tissue hypoxia can be caused by inflammatory diseases of the intestine or the lungs, we want to point out that tissue hypoxia is not simply a bystander of inflammatory diseases but can greatly impact the microenvironment, particularly by regulating changes in oxygen-dependent gene expression (Taylor, 2008). As hypoxia has been known for many decades to elicit adaptive and protective changes in gene expression (e.g., high-altitude adaptation by transcriptional regulation of erythropoietin expression and subsequent increases in oxygen carrying capacity), we think it will be of particular interest from a therapeutic perspective to identify endogenous anti-inflammatory pathways that are elicited by hypoxia signaling in an inflamed microenvironment (Aherne et al., 2010; Eckle et al., 2007b, 2008a, 2008c, 2008d; Eltzschig et al., 2003, 2004, 2009b; Frick et al., 2009; Grenz et al., 2007a, 2007b, 2007c, 2007d, 2008; Hart et al., 2009; Schingnitz et al., 2010).

C. Oxygen Sensing, Hypoxia-Inducible Factors, and Prolyl Hydroxylases

The survival of all metazoan organisms is highly dependent on the regulation of oxygen delivery and utilization of their tissues. Research work over the past 15 years discovered the transcription factor "hypoxia-inducible factor 1 (*HIF-1*)" as master regulator of mammalian oxygen homeostasis (Hirota & Semenza, 2005; Semenza, 1999a, 1999b, 2001, 2003, 2007, 2009). However, more recent evidence also implicates HIF in the transcriptional coordination of inflammatory events (Cramer et al., 2003; Haeberle et al., 2008; Karhausen et al., 2004, 2005; Kuhlicke et al., 2007; Morote-Garcia et al., 2009; Ratcliffe, 2007; Rius et al., 2008; Robinson et al., 2008; Rosenberger et al., 2009; Schofield & Ratcliffe, 2004; Sitkovsky & Lukashev, 2005; Zheng et al., 2009). HIF is an α/β heterodimer that binds hypoxia response elements (HREs) at target gene loci under hypoxic conditions (Fig. 2). In the presence of oxygen, HIF is inactivated by posttranslational hydroxylation of specific amino acid residues within its α subunits. Prolyl hydroxylation promotes interaction with the von Hippel–Lindau protein (pVHL; Kaelin, 2005, 2007, 2008; Kaelin & Ratcliffe, 2008), E3 ubiquitin ligase complex, and proteolytic inactivation by proteasomal degradation, while asparaginyl hydroxylation blocks coactivator recruitment. These hydroxylation steps are catalyzed by a set of nonheme Fe(II)- and 2-oxoglutarate-dependent dioxygenases (prolyl hydroxylases) whose absolute requirement for molecular oxygen confers sensitivity to hypoxia (Schofield & Ratcliffe, 2004). *HIF-1*α was the original HIF isoform identified by affinity purification using oligonucleotides from the EPO locus (Wang et al., 1995), while *HIF-2*α and HIF-3α were identified by homology searches

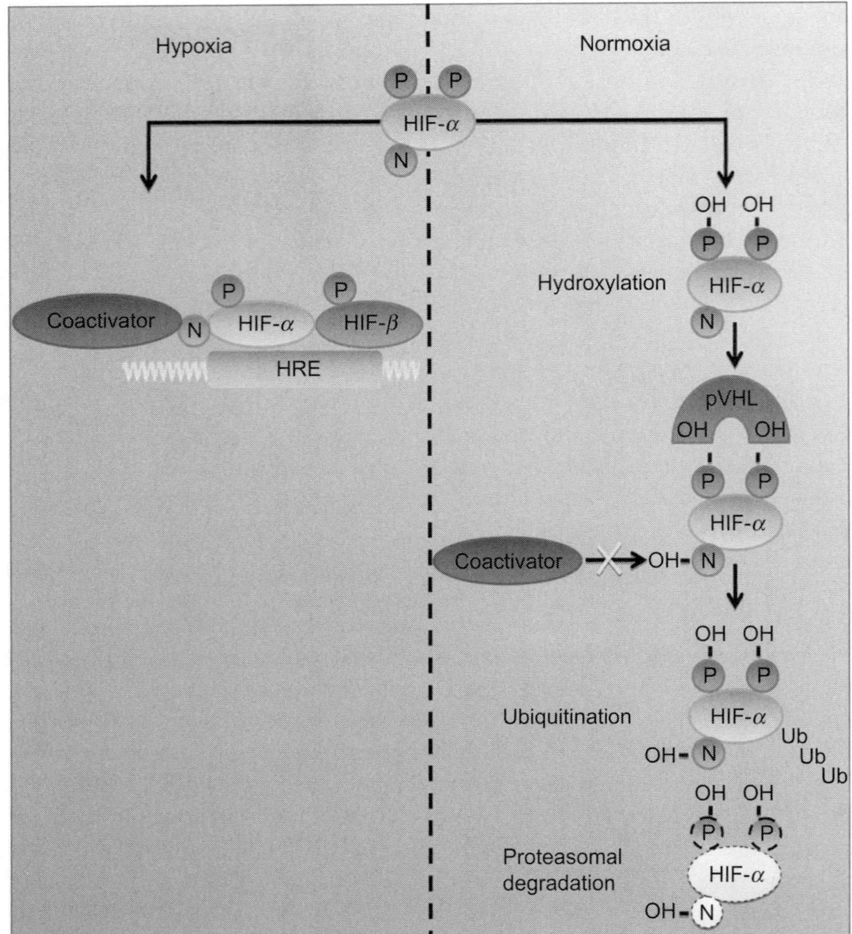

FIGURE 2 *Regulation of hypoxia-inducible factor (HIF) protein levels under normoxic or hypoxic conditions.* In normoxia, hydroxylation at two proline residues promotes HIF-α association with pVHL and HIF-α destruction via the ubiquitin/proteasome pathway, while hydroxylation of an asparagine residue blocks association with coactivators. In hypoxia, these processes are suppressed, allowing HIF-α subunits (both HIF-1α and HIF-2α) to escape proteolysis, dimerize with HIF-1β, recruit coactivators, and activate transcription via HREs. N, asparagine; P, proline; OH, hydroxyl group; Ub, ubiquitin [modified, from *The Journal of Clinical Investigation* with permission (Ratcliffe, 2007)].

or screens for interaction partners with *HIF-1*β. HIF-3α is the more distantly related isoform and, in certain splicing arrangements, encodes a polypeptide that antagonizes HRE-dependent gene expression. However, *HIF-1*α and *HIF-2*α are closely related and both activate HRE-dependent gene transcription (Wenger, 2002). Nevertheless, knockout studies in mice demonstrate that *HIF-1*α and *HIF-2*α play nonredundant roles, and inactivation of each one results in a distinctly different phenotype (Mastrogiannaki et al., 2009; Milkiewicz et al., 2007; Percy et al., 2008; Rankin et al., 2007; Ratcliffe, 2007; Rosenberger et al., 2002). This may result, in part, from differences in tissue-specific and temporal patterns of induction of each isoform (Holmquist-Mengelbier et al., 2006; Schofield & Ratcliffe, 2004), but, not uncommonly, both isoforms are expressed within a given cell type, and the results of several studies suggest that *HIF-1α* and *HIF-2α* may have distinct transcriptional targets (Rankin et al., 2007; Ratcliffe, 2007). For example, the transcription of genes encoding enzymes that operate in a coordinated way in the glycolytic pathway appears to be driven by *HIF-1α* and not by *HIF-2α* (Hu et al., 2003).

Under well-oxygenated conditions, HIFα becomes hydroxylated at one (or both) of two highly conserved prolyl residues by members of the prolyl hydroxylase domain (PHD) family (also called EgIN family; Aragones et al., 2008, 2009; Bishop et al., 2008; Fraisl et al., 2009; Kaelin, 2005; Mazzone et al., 2009; Schofield & Ratcliffe, 2004; Tambuwala et al., 2010). Hydroxylation of either of these prolyl residues generates a binding site for the pVHL tumor suppressor protein, which is a component of a ubiquitin ligase complex. As a result, HIFα is polyubiquitylated and subjected to proteasomal degradation when oxygen is available. The PHD proteins belong to the Fe(II)- and 2-oxoglutarate-dependent oxygenase superfamily, whose activity is absolutely dependent on oxygen. Accordingly, the rate of HIF hydroxylation is suppressed by hypoxia. Under low oxygen conditions, or in cells lacking functional pVHL, HIFα accumulates, dimerizes with an HIFβ family member, translocates to the nucleus, and transcriptionally activates different genes, including genes involved in erythropoiesis, angiogenesis, autophagy, and energy metabolism (Kaelin, 2005; Kaelin & Ratcliffe, 2008). Factor inhibiting HIF (FIH) 1 (Dayan et al., 2006; Mahon et al., 2001), like the PHD family members, is an Fe(II)- and 2-oxoglutarate-dependent dioxygenase. When oxygen is available, FIH1 hydroxylates a conserved asparaginyl residue within the HIF1α and HIF2α, leading to a steric clash that prevents the recruitment of the coactivators p300 and CBP (Kaelin, 2005; see Fig. 2). FIH1 remains active at lower oxygen concentrations than the PHDs and so might suppress the activity of HIFα proteins that escape destruction in moderate hypoxia (Dayan et al., 2006). HIFα stability is also regulated by other signaling pathways. For example, HSP90 inhibitors and histone acetylase inhibitors promote HIFα degradation in a pVHL-independent manner (Kong et al., 2006a). HIFα can also be sumoylated, but there are conflicting reports as to whether this leads to increased or decreased HIFα stability (Carbia-Nagashima et al., 2007; Cheng et al., 2007). Sumoylated HIFα can be recognized by pVHL in a hydroxylation-independent manner *in vitro*, and mice lacking Sumo-specific protease 1 (SENP1) develop anemia due to failure to stabilize HIFα (Cheng et al., 2007).

Both the PHDs and FIH1 are dioxygenases (enzymes that incorporate both atoms of molecular oxygen into their products). One oxygen atom is used in the oxidative decarboxylation of 2-oxoglutarate yielding succinate and CO_2, and the other is incorporated directly into the oxidized amino acid residue of HIFα (Kaelin & Ratcliffe, 2008). A number of studies have measured the relationship between HIF hydroxylase activity and oxygen concentration using HIFα peptide substrates. Initially reported values for the apparent K_M for oxygen (the concentration of oxygen that supports a half-maximal initial catalytic rate) were in the range of 230–250 μM for the PHDs and 90 μM for FIH1 (Kaelin & Ratcliffe, 2008). More recent studies using longer (and more physiological) HIFα polypeptides suggest that the PHD oxygen K_M values are closer to 100 μM (Kaelin, 2005; Kaelin & Ratcliffe, 2008). These values are nevertheless well above oxygen concentrations in tissues, which are typically in the range of 10–30 μM (though lower in certain locations, such as the renal medulla and many solid tumors). Importantly, because molecular oxygen is not produced by animal cells, it can be assumed that these measurements represent the maximum concentration available to the enzyme, irrespective of intracellular gradients. Thus intracellular pO_2 will essentially always be below the apparent K_M for oxygen, allowing enzyme activity to be modulated by molecular oxygen availability over the entire physiological range. Genetic, pharmacological, and biochemical studies indicate that hydroxylation is itself limiting for HIFα degradation and that HIFα is limiting for the HIF transcriptional response. Therefore, HIF hydroxylation directly links changes in molecular oxygen concentration to the regulation of a major transcription cascade, thus operating as an "oxygen sensor," that is, a specific biological interface with oxygen that has a prime function in control (Fig. 2).

D. Animal Models for Hypoxia Signaling

Gene-targeted mice for *HIF-1α* (*Hif-1α*$^{-/-}$ mice, complete *HIF-1α* deficiency) die at midgestation due to cardiovascular malformations and mesenchymal cell death (Fig. 3; Carmeliet et al., 1998; Yu et al., 1999). Heterozygote mice (*Hif-1α*$^{+/-}$) develop normally but show delayed adaptation to hypoxia, including delayed development of polycythemia, right ventricular hypertrophy, pulmonary hypertension, and pulmonary vascular remodeling (Yu et al., 1999). Moreover, hearts of *Hif-1α*$^{+/-}$mice are not protected by ischemic preconditioning (IP; Cai et al., 2007). Conditional deletion of *HIF-1α* in the colon leads to increased susceptibility to experimental colitis (Karhausen et al., 2004), and deletion of *HIF-1α* in the myeloid lineage is associated with attenuated neutrophil apoptosis during hypoxia (Walmsley et al., 2005), or impairment of myeloid cell aggregation, motility, invasiveness, and bacterial killing (Cramer et al., 2003). Moreover, conditional deletion of *HIF-1α* in the epidermis dampens epidermal oxygen sensing, and systemic responses to ambient hypoxia (Boutin et al., 2008). At present, human diseases caused by direct alterations of the HIF-1α gene

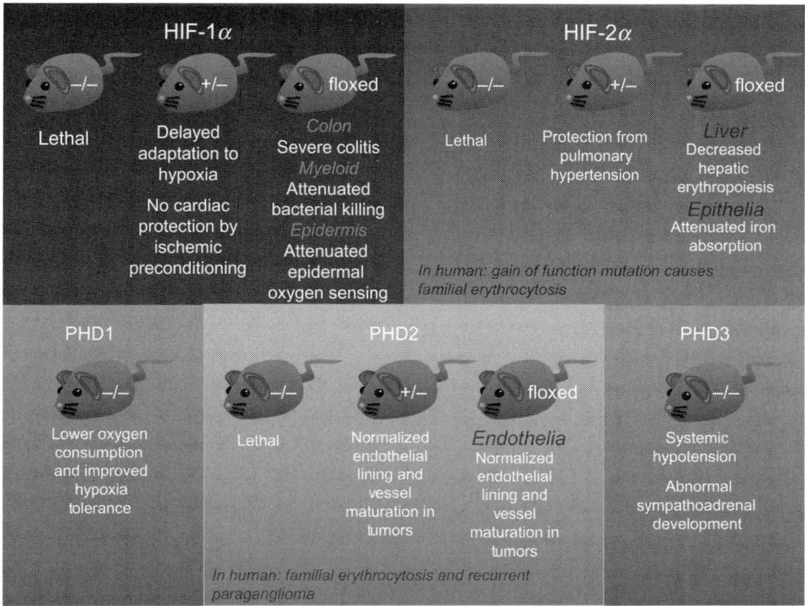

FIGURE 3 *Genetic models for the hypoxia-inducible factor (HIF) or prolyl hydroxylase pathway.* Over the past 15 years, genetic models for the different isoforms for HIFs and the oxygen-sensing prolyl hydroxylases have been generated. Some of the phenotypic characterizations are displayed.

are unknown. However, mutations in the von Hippel–Lindau gene are associated with elevated levels of HIF and tumors in vascular organs (Kaelin, 2007). Gene-targeted mice for *HIF-2α* (*Hif-2α*$^{-/-}$ mice) die shortly after birth from fatal respiratory distress syndrome due to insufficient surfactant production by alveolar type 2 cells (Compernolle et al., 2002). *Hif-2α*$^{+/-}$ mice develop normally and appear to be protected against pulmonary hypertension and right ventricular dysfunction during prolonged hypoxia exposure (Brusselmans et al., 2003). Conditional deletion of *HIF-2α* in the liver is associated with decreased hepatic erythropoietin production (Rankin et al., 2007) and targeted deletion in intestinal epithelia causes impaired iron absorption (Mastrogiannaki et al., 2009). Moreover, a recent study in patients demonstrated that a gain-of-function mutation in the HIF-2α gene is associated with familial erythrocytosis (Percy et al., 2008). Some of these phenotypes are summarized in Fig. 3. *Taken together, many studies demonstrate a conserved and highly regulated molecular pathway for cellular adaptation to hypoxia, and hypoxia-regulated gene expression. In the center of this pathway stands HIF as key transcription factor for regulating hypoxia-dependent transcriptional alterations. It is important to point out that a group of hydroxylases–PHDs and FIH—that regulate HIF stability and*

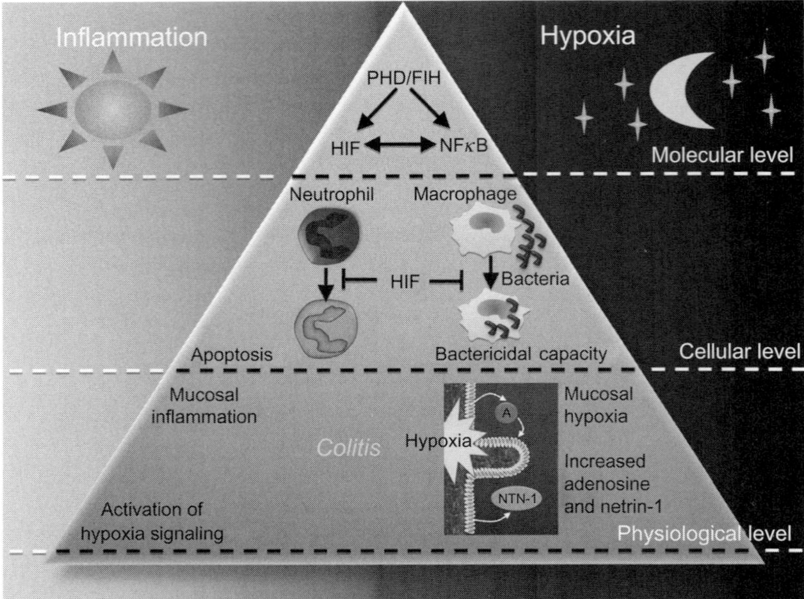

FIGURE 4 *Hypoxia and inflammation—two sides of the same coin.* Like day and night, inflammation and hypoxia represent two distinctively different, but very closely related events. On a molecular level, pathways that regulate gene transcription during inflammation or hypoxia are interdependently connected. As such, hypoxia-inducible factor (HIF) and nuclear factor κB (NFκB) conversely control each other's transcription. Moreover, the oxygen-sensing system consistent of prolyl hydroxylases and factor inhibiting HIF (FIH) control HIF and NFκB functions on a posttranslational level. On a cellular level, HIF is involved in inhibiting neutrophil apoptosis during hypoxia, or the bactericidal capacity of myeloid cells, such as macrophages. On a physiological level, HIF is stabilized during mucosal hypoxia, eliciting tissue protective and anti-inflammatory pathways, for example, via enhancing adenosine generation, and signaling, or the production of the anti-inflammatory guidance molecule netrin-1. Conversely, tissue hypoxia as occurs during mucosal inflammation in the setting of experimental colitis or inflammatory bowel disease is associated with activation hypoxia-dependent signaling pathways.

activity represents the oxygen-sensing element of this pathway. Moreover, hypoxia- and inflammatory-signaling events are interdependent on the molecular, cellular, and physiologic levels (see Fig. 4*).*

III. Effect of Hypoxia on the Extracellular Availability of Adenosine

In the previous sections, we have described the relationship between hypoxia and inflammation and how hypoxia signaling can affect the outcome of inflammatory diseases. In order to support our hypothesis that adenosine signaling through the $ADORA_{2B}$ represents an endogenous pathway to dampen hypoxia-induced inflammation, we will now go on to discuss the effect of hypoxia on extracellular adenosine accumulation. It has been a long-suspected notion that extracellular levels of adenosine are elevated during conditions of limited oxygen availability. For example, studies measuring interstitial adenosine concentrations in swine or perfused rabbit hearts via microdialysis noted a 6- or 12-fold increase in extracellular adenosine during conditions of limited oxygen availability (Headrick, 1996; Schulz et al., 1995). More recently, studies in genetic models have provided convincing evidence for specific metabolic pathways that are responsible for elevating extracellular adenosine levels during hypoxia or ischemia (Eckle et al., 2007b; Grenz et al., 2007b, 2007c; Kohler et al., 2007).

A. Extracellular Adenosine Production Is Increased During Hypoxia

In the extracellular compartment, adenosine stems mainly from the phosphohydrolysis of precursor nucleotides, ATP, ADP, and AMP. During conditions of hypoxia or inflammation, the release of precursor nucleotides is increased. Due to the fact that the intracellular ATP concentration is very high (between 4 and 7 mM), apoptosis or necrosis of cells is associated with the release of high levels of ATP. For instance, a very elegant study demonstrated that ATP released from apoptotic cells serves as a find-me signal for phagocytes (Elliott et al., 2009). Other studies demonstrate the controlled release of nucleotides under hypoxic or inflammatory conditions. Neutrophils can release ATP when activated by inflammatory stimuli in a controlled fashion (Eltzschig et al., 2003, 2006a). In initial studies using high-performance liquid chromatography and luminometric ATP detection assays, research from Sean Colgan's and our laboratory revealed that PMNs release ATP through activation-dependent pathways. *In vitro* models of endothelial barrier function and neutrophil/endothelial adhesion indicated that PMN-derived ATP signals through endothelial ARs, thereby promoting endothelial barrier function and attenuating PMN/endothelial adhesion. After excluding lytic ATP release, we used pharmacological strategies to reveal a potential mechanism involved in PMN-dependent ATP release (e.g., verapamil, dipyridamole, brefeldin A, 18-α-glycyrrhetinic acid, connexin-mimetic peptides). These studies showed that PMN ATP release occurs through connexin 43 (Cx43) hemichannels in a protein/phosphatase-A-dependent manner. Findings in human PMNs were confirmed in PMNs derived from induced $Cx43^{-/-}$ mice, whereby activated PMNs release less than 15% of ATP relative to littermate controls, whereas Cx43 heterozygote PMNs were intermediate in their capacity for ATP release ($P < 0.01$). Taken together, these results identified a role for Cx43 in activated PMN ATP release, therein contributing to the innate metabolic control of the inflammatory milieu. Again, other studies demonstrated that ATP released on the leading edge of PMN can serve as a signal for neutrophil chemotaxis (puringegic chemotaxis; Linden, 2006b; Chen et al., 2006). Moreover, directed nucleotide release can occur from platelets via granular release (Novak, 2003), or from bladder cells upon stretch stimulation (Wang et al., 2005). Other studies demonstrated endothelial cell-dependent ATP during hypoxia via connexin hemichannels (Faigle et al., 2008). Taken together, such studies demonstrate increased nucleotide release during inflammatory or hypoxic conditions.

Nucleotides released during hypoxia or inflammation are rapidly converted to adenosine in an enzymatically controlled fashion. The first step of the enzymatic breakdown of precursor nucleotides to adenosine involves ATP/ADP phosphohydrolysis. Extracellular ATP/ADP phosphohydrolysis is mainly achieved by ecto-nucleoside triphosphate diphosphohydrolases (E-NTPDases), a recently described family of ubiquitously expressed membrane-bound enzymes (Robson et al., 2005; Zimmermann, 2000). The catalytic sites of plasma membrane expressed E-NTPDases 1–3 and 8 are exposed to the extracellular milieu, the others are intracellular (Robson et al., 2005) The presumptive biological role of plasma membrane-bound E-NTPDases (E-NTPDases 1–3 and 8) is to fine-tune extracellular nucleotide levels. Particularly, E-NTPDase1 (CD39) has been implicated in the extracellular generation of adenosine in different models of tissue protection during hypoxia. CD39 plays an important role in vascular endothelial function by blocking platelet aggregation via the phosphohydrolysis of ATP and ADP from the blood to maintain vascular integrity (Marcus et al., 1997; Pinsky et al., 2002). At the same time, E-NTPDase1 is also important in the maintenance of platelet functionality by preventing platelet $P2Y_1$-receptor desensitization. As such, mice gene targeted for E-NTPDase1 ($cd39^{-/-}$ mice) show prolonged bleeding time with minimally perturbed coagulation parameters (Enjyoji et al., 1999).

Several studies have provided strong evidence that CD39 is transcriptionally induced during conditions of ambient hypoxia. First evidence comes from studies in intestinal epithelia that demonstrate induction of CD39 transcript levels upon hypoxia exposure (Synnestvedt et al., 2002). Similarly, the capacity of model endothelia (HMEC-1) to convert ATP to adenosine is dramatically increased when these cells are exposed to hypoxia (Eltzschig et al., 2003). Mechanistic studies revealed hypoxia-dependent increases in CD39 mRNA and immunoreactivity on endothelia. Examination of the human CD39 gene promoter identified a region important in hypoxia inducibility. Multiple levels of analysis, including site-directed mutagenesis, chromatin immunoprecipitation, and inhibition by antisense, revealed a critical role for transcription factor Sp1 in hypoxia induction of CD39. Sp1 is a human transcription factor involved in gene expression in the early development of an organism. It belongs to the Sp/KLF family of transcription factors. The protein is 785 amino acids long, with a molecular weight of 81 kDa. The SP1 transcription factor contains a zinc finger protein motif, by which it binds directly to DNA and enhances gene transcription (Chu & Ferro, 2005). Previous studies have implicated SP1 in the regulation of hypoxia-elicited changes in gene expression (Cummins & Taylor, 2005). Using a combination of $cd39^{-/-}$ mice and Sp1 small interfering RNA in *in vivo* cardiac ischemia models revealed Sp1-mediated induction of cardiac CD39 during myocardial ischemia. In summary, these results identify a novel Sp1-dependent regulatory pathway for CD39 and indicate the likelihood that CD39 is central to protective responses to hypoxia/ischemia (Eltzschig, et al., 2009a). Other studies demonstrated the existence of this pathway for intestinal epithelia exposed to ischemia–reperfusion (IR) injury (Hart et al., 2010). Together, these studies demonstrate that hypoxia enhances the enzymatically controlled conversion of ATP/ADP to AMP. While this process is under the transcriptional control of SP1, other components of the adenosine pathway (e.g., CD73, ENT1, ENT2, AK) are under the transcriptional control of HIF.

It is important to point out that AMP serves as the enzymatic substrate for the ecto-5′-nucleotidase (CD73)—the pacemaker enzyme for the extracellular production of adenosine (Colgan et al., 2006). CD73 is a glycosylphosphatidylinositol-anchored enzyme that is expressed on the extracellular surface of most cell types (Resta et al., 1998), with expressional levels highest in the colon, followed by the brain, the kidney, the lung, and the heart (Thompson et al., 2004). First evidence for a role of hypoxia in CD73 regulation comes from studies of intestinal epithelia exposed to ambient hypoxia (Synnestvedt et al., 2002). Microarray RNA analysis revealed an increase in CD73 in hypoxic epithelial cells. Metabolic studies of CD73 function in intact epithelia revealed that hypoxia enhances CD73 function as much as 6 ± 0.5-fold over normoxia. Examination of the CD73 gene promoter identified at least one binding site for *HIF-1* and inhibition of *HIF-1*α expression by antisense oligonucleotides resulted in significant inhibition of hypoxia-inducible CD73 expression. Studies using luciferase reporter constructs revealed a significant increase in activity in cells subjected to hypoxia, which was lost in truncated constructs lacking the *HIF-1* site. Mutagenesis of the *HIF-1*α binding site resulted in a nearly complete loss of hypoxia inducibility. *In vivo* studies in a murine hypoxia model revealed that hypoxia-induced CD73 may serve to protect the epithelial barrier, since the CD73 inhibitor α,β-methylene ADP promotes increased intestinal permeability. These results identified an *HIF-1*-dependent regulatory pathway for CD73 and indicate the likelihood that CD73 protects the epithelial barrier during hypoxia (Synnestvedt et al., 2002). Other studies in myocardial (Eckle et al., 2007b), intestinal (Hart et al., 2008a), or renal ischemia (Grenz et al., 2007b) confirmed induction of CD73 during conditions of limited oxygen availability and revealed a tissue protective role of enzymatic adenosine production via CD73-dependent conversion of AMP to adenosine in these models. *Together, these studies demonstrate that enzymatic production of extracellular adenosine is increased during conditions of inflammatory hypoxia.*

B. Uptake of Extracellular Adenosine Is Dampened During Hypoxia

Extracellular adenosine signaling events are rapidly terminated due to passive diffusion of adenosine through diffusion-limited channels that are located in the cell membrane. This mainly involves nucleoside transporters of the equilibrative type (equilibrative nucleoside transporters, ENTs). These transporters are integrated into the cell membrane and represent diffusion channels for adenosine, where adenosine can follow a diffusion gradient. As outlined above, extracellular adenosine production is significantly increased under hypoxic conditions (Eckle et al., 2009; Eltzschig, 2009). Therefore, the gradient across the cell membrane is directed from the extracellular compartment toward the intracellular space. Therefore, repression of ENT expression may serve to further enhance extracellular adenosine signaling during hypoxia. In fact, a recent study pursued the hypothesis that diminished uptake of adenosine effectively enhances extracellular adenosine signaling. Initial studies indicated that the half-life of extracellular adenosine was increased by as much as fivefold after exposure of endothelia to hypoxia. Examination of expressional levels of the ENT1 and ENT2 revealed a transcriptionally dependent decrease in mRNA, protein, and function in endothelia and epithelia. Examination of the ENT1 promoter identified a *HIF-1*-dependent repression of ENT1 during hypoxia. Using *in vitro* and *in vivo* models of adenosine signaling, the authors revealed that decreased adenosine uptake promotes vascular barrier and dampens neutrophil tissue accumulation during hypoxia. Moreover, epithelial HIF-1α mutant animals displayed increased epithelial ENT1 expression. Together, these results identify transcriptional repression of ENT as an innate mechanism to elevate extracellular adenosine during hypoxia (Eltzschig et al., 2005). Consistent with these findings, other studies confirmed attenuated expression of ENTs and enhanced adenosine signaling during hypoxia (Casanello et al., 2005; Chaudary et al., 2004; Loffler et al., 2007; Morote-Garcia et al., 2009, 2009; Rose et al., 2010). At present, the contribution of concentrative nucleoside transporters to adenosine transport under hypoxic conditions is unknown.

C. Alteration of Adenosine Metabolism with Hypoxia Exposure

Based on the observation that increases in intracellular adenosine can effectively elevate extracellular adenosine, a recent study addressed the contribution of adenosine kinase (AK, intracellular conversion of adenosine to adenosine monophosphate [AMP]) to vascular adenosine responses (Morote-Garcia et al., 2008a). Initial *in vitro* studies of ambient hypoxia revealed prominent repression of endothelial AK transcript ($85 \pm 2\%$ reduction), protein, and function. Transcription factor binding assays and *HIF-1α* loss- and gain-of-function studies suggested a role for *HIF-1* in transcriptional repression of AK. Moreover, repression of AK by ambient hypoxia was abolished in conditional HIF-1α mutant mice *in vivo*. Studies of endothelial barrier function revealed that inhibition or siRNA repression of AK is associated with enhanced adenosine-dependent barrier responses *in vitro*. Moreover, *in vivo* studies of vascular barrier function demonstrated that AK inhibition with 5′-iodotubericidin significantly attenuated hypoxia-induced vascular leakage in multiple organs and reduced hypoxia-associated increases in lung water. Taken together, these data reveal a critical role of AK in modulating vascular adenosine responses and suggest pharmacologic inhibitors of AK in the treatment of conditions associated with hypoxia-induced vascular leakage (e.g., sepsis or acute lung injury (ALI); Morote-Garcia et al., 2008a).

While acute increases in adenosine are important to counterbalance excessive inflammation or vascular leakage, chronically elevated adenosine levels may be toxic. Thus, a different study reasoned that clearance mechanisms might exist to offset deleterious influences of chronically elevated adenosine (Eltzschig et al., 2006b; Van Linden & Eltzschig, 2007). Guided by microarray results revealing induction of endothelial adenosine deaminase (ADA) mRNA in hypoxia, the authors used *in vitro* and *in vivo* models of adenosine signaling, confirming induction of ADA protein and activity. Further studies in human endothelia revealed that ADA-complexing protein CD26 is coordinately induced by hypoxia, effectively localizing ADA activity at the endothelial cell surface. Moreover, ADA surface binding was effectively blocked with glycoprotein 120 (gp120) treatment, a protein known to specifically compete for ADA–CD26 binding. Functional studies of murine hypoxia revealed that inhibition of ADA with deoxycoformycin enhances protective responses mediated by adenosine (vascular leak and neutrophil accumulation). Analysis of plasma ADA activity in pediatric patients with chronic hypoxia undergoing cardiac surgery demonstrated a 4.1 ± 0.6-fold increase in plasma ADA activity compared with controls. Taken together, these results reveal induction of ADA as innate metabolic adaptation to chronically elevated adenosine levels during hypoxia. In contrast, during acute hypoxia associated with vascular leakage and excessive inflammation, ADA inhibition may serve as therapeutic strategy (Eltzschig et al., 2006b). Similar to the observed induction of ADA with hypoxia, other studies have demonstrated that ADA is also induced in patients with diabetes (Kurtul et al., 2004).

The notion that ADA is induced with hypoxia stands somewhat in contrast to the studies discussed above. In fact, many studies provide strong evidence that extracellular adenosine levels are elevated during conditions of acute hypoxia. However, the time course of events may explain some of these differences. While the coordination of CD39 and CD73 induction and repression of ENTs and AK most likely represent acute responses to hypoxia that result in enhanced adenosine signaling events, induction of ADA most likely resembles a subsequent event to offset potentially toxic effects of chronic adenosine elevation (Blackburn, 2003; Blackburn et al., 2000, 2009; Chunn et al., 2005; Mi et al., 2008; Peng et al., 2009; Schneider et al., 2010; Zhou et al., 2009).

IV. Alteration of $ADORA_{2B}$ Signaling by Hypoxia

After having reviewed mechanism of how acute hypoxia will enhance extracellular adenosine concentrations, we next will discuss the effect of hypoxia on adenosine signaling events through different ARs. One of the first comparative studies on the effect of hypoxia on signaling events through ARs compared the functional influence of activated polymorphonuclear leukocytes (PMNs) on normoxic and posthypoxic endothelial cells. These studies indicated that activated PMNs preferentially promote endothelial barrier function in posthypoxic endothelial cells ($>60\%$ increase over normoxia). Extension of these findings identified at least one soluble mediator as extracellular ATP. Subsequent studies revealed that ATP is coordinately hydrolyzed to adenosine at the endothelial cell surface by hypoxia-induced CD39 and CD73. Studies *in vitro* and in cd39-null mice identified these surface ectoenzymes as critical control points for posthypoxia-associated protection of vascular permeability. Importantly, additional insight was gained through microarray analysis of the transcriptional responses of different ARs. These studies profiled the relative expression of ARs in normoxic and hypoxic (12 h exposure to pO_2 20 Torr) endothelial cells by microarray analysis. For these experiments, nonimmortalized human dermal microvascular endothelial cells (HMVECs) were used. Interestingly, these experiments demonstrated that the $ADORA_{2B}$ was selectively induced by hypoxia and that other isoforms were either not changed ($ADORA_3$) or significantly downregulated. These microarray results were verified in HMEC-1 by real-time PCR in RNA derived from endothelial cells exposed to a time course of hypoxia and consistently revealed that the $ADORA_{2B}$ was increased by as much as 4 ± 0.5-fold ($P < 0.01$ compared with normoxia). Identical results were obtained with HMVEC as an endothelial source (Eltzschig, unpublished data). Extensions of these findings revealed that surface protein levels of the $ADORA_{2B}$ were similarly increased by exposure of HMEC-1 to hypoxia. To determine whether such hypoxia-induced $AdoRA_{2B}$ expression was functional, endothelial permeability assays were employed using the selective $ADORA_{2B}$ antagonist MRS 1754 (Ji et al., 2001). MRS 1754 (100 nM) significantly shifted the adenosine dose–response curve to the right but had little influence on adenosine responses in normoxic endothelia. These were among the first studies to demonstrate a functional role of $ADORA_{2B}$ transcription and protein induction upon hypoxia exposure (Eltzschig et al., 2003).

Additional studies again from the laboratory of Sean Colgan provided mechanistic insight into the transcriptional mechanism governing $ADORA_{2B}$ expression during hypoxia, and provide another level of evidence on the role of HIFs in enhancing extracellular adenosine signaling events (Kong et al., 2006b). Here, the authors examined AR control and amplification of signaling through transcriptional regulation of endothelial and epithelial ARs. Initial studies confirmed previous findings indicating selective induction of human $ADORA_{2B}$ by hypoxia. Analysis of the cloned human $ADORA_{2B}$ promoter identified a previously unknown hypoxia-responsive region, including a functional binding site for HIF within the $ADORA_{2B}$ promoter. Further studies examining *HIF-1*α DNA binding and *HIF-1*α gain and loss of function confirmed strong dependence of $ADORA_{2B}$ induction on *HIF-1 in vitro* and *in vivo* mouse models. Additional studies in endothelia overexpressing full-length $ADORA_{2B}$ revealed functional phenotypes of increased barrier function and enhanced angiogenesis. Taken together, these results demonstrate transcriptional coordination of $ADORA_{2B}$ by HIF and amplified adenosine signaling during hypoxia. These findings may provide an important link between hypoxia and metabolic conditions associated with inflammation and angiogenesis (Kong et al., 2006b). Moreover, studies from the team of Igor Feoktistov, Italo Biaggioni, and colleagues provide convincing evidence that hypoxia modulates the expression of ARs in human endothelial and smooth muscle cells toward an A_{2B} "angiogenic" phenotype (Feoktistov et al., 2004). While many of the studies mentioned above were carried out in an immortalized human endothelial cell line 1 (HMEC-1; Eltzschig et al., 2003), their studies were performed in human umbilical vein endothelial cells (HUVECs) or bronchial smooth muscle cells (BSMCs; Feoktistov et al., 2004).

Taken together, these studies demonstrate that hypoxia-dependent signaling pathways under the transcriptional coordination of HIF result in the robust induction of the $ADORA_{2B}$ *and set the stage for functional studies of* $ADORA_{2B}$ *signaling in models linking inflammation and hypoxia.*

V. Examples of $ADORA_{2B}$ Signaling in Tissue Protection from Hypoxia

A. Vascular Leakage During Hypoxia

Previous studies have implicated extracellular nucleotide metabolites, predominantly adenosine, as triggers of endogenous protective mechanisms in a number of acute injury models (Baxter, 2002; Fredholm, 2007; Linden, 2001; Mubagwa & Flameng, 2001; Ohta & Sitkovsky, 2001; Sitkovsky & Lukashev, 2005; Sitkovsky et al., 2004). As outlined above, extracellular adenosine is derived primarily through phosphohydrolysis of AMP, and

CD73, a ubiquitously expressed ectoenzyme, is the pacemaker of this reaction (Thompson et al., 2004). Studies on the role of CD73 in tissue injury showed that $CD73^{-/-}$ mice develop profound vascular leakage and pulmonary edema upon hypoxia exposure (Thompson et al., 2004). Once generated into the extracellular space, adenosine can signal through any of four G-protein-coupled ARs ($ADORA_1/ADORA_{2A}/ADORA_{2B}/ADORA_3$). All these receptors are expressed on vascular endothelia (Eltzschig et al., 2003) and have been implicated in tissue protection in different models of injury (Eckle et al., 2007a, 2007b, 2008a; Eltzschig et al., 2005; Fredholm, 2007; Lankford et al., 2006; Lappas et al., 2006; Linden, 2005; Ohta & Sitkovsky, 2001; Salvatore et al., 2000; Sitkovsky & Lukashev, 2005; Sitkovsky et al., 2004; Yang et al., 2005, 2006).

Changes in vascular barrier function closely coincide with tissue injury of many etiologies and result in fluid loss, edema, and organ dysfunction (Michel & Curry, 1999; Stevens et al., 2000; Webb, 2000). The predominant barrier (~90%) to movement of macromolecules across a blood vessel wall is presented by the vascular endothelium (Stevens et al., 1999, 2000). Under physiological conditions, macromolecules such as albumin (molecular weight ~70 kDa) can cross the endothelial monolayer via a paracellular route (e.g., by passing between adjacent endothelia) with some contribution of transcellular passage (Michel, 1998; Stan, 2002). Endothelial barrier function correlates inversely with the size of molecules that can gain entry into tissues and differs between tissues of different origins. Endothelial permeability is highly regulated and may increase markedly upon exposure to inflammatory stimuli (e.g., lipopolysaccharide, bacteria, bacterial compounds, prostaglandins, reactive oxygen species, leukotrienes) or adverse conditions such as ischemia or hypoxia (Dejana et al., 2001; Eltzschig et al., 2005; Kempf et al., 2005; Luscinskas et al., 2002a, 2002b; Riess et al., 2004; Stevens et al., 2000).

Given that activation of ARs can lead to an elevation of intracellular cAMP, and that elevated cAMP in endothelia promotes barrier function (Moore et al., 1998; Stevens et al., 2000), a recent study considered the possibility of endothelial AR signaling to regulate vascular permeability (Eckle et al., 2008a). For this purpose, the authors examined the contribution of individual ARs ($ADORA_1/ADORA_{2A}/ADORA_{2B}/ADORA_3$) to vascular leak induced by hypoxia. Initial profiling studies revealed that siRNA-mediated repression of the $ADORA_{2B}$ selectively increased endothelial leak in response to hypoxia *in vitro*. In parallel, vascular permeability was significantly increased in vascular organs of $ADORA_{2B}$*−/−* mice subjected to ambient hypoxia (8% oxygen, 4 h; e.g., lung: 2.1 ± 0.12-fold increase). By contrast, hypoxia-induced vascular leak was not accentuated in $ADORA_1$*−/−*, $ADORA_{2A}$*−/−*, or $ADORA_3$*−/−* deficient mice, suggesting a degree of specificity for the $ADORA_{2B}$. Further studies in wild-type mice revealed that the selective $ADORA_{2B}$ antagonist PSB1115 resulted in profound increases in hypoxia-associated vascular leakage, while $ADORA_{2B}$

agonist (BAY 60-6583 [2-[6-amino-3,5-dicyano-4-[4-(cyclopropylmethoxy)-phenyl]pyridin-2-ylsulfanyl]acetamide]) treatment was associated with almost complete reversal of hypoxia-induced vascular leakage. Studies in bone marrow chimeric $ADORA_{2B}$ mice suggested a predominant role of vascular $ADORA_{2B}$s in this response, while hypoxia-associated increases in tissue neutrophils were, at least in part, mediated by $ADORA_{2B}$ expressing hematopoietic cells. Taken together, these studies provide pharmacologic and genetic evidence for vascular $ADORA_{2B}$ signaling as central control point of hypoxia-associated vascular leak (Eckle et al., 2008a).

B. Myocardial Ischemia

Myocardial ischemia represents a major health problem of western countries. Current therapeutic interventions focus mainly on early and persistent coronary reperfusion and additional pharmacological strategies to increase resistance to myocardial ischemia are currently areas of intense investigation (Frangogiannis et al., 2005; Herrmann, 2006; Walsh et al., 2005). A powerful strategy for cardioprotection would be to recapitulate the consequences of IP (Murry et al., 1986), where short and repeated episodes of IR prior to myocardial infarction result in attenuation of infarct size. Despite multiple attempts to identify the underlying molecular mechanisms, pharmacological strategies utilizing such pathways have yet to be further defined and introduced into clinical practice (Kloner & Rezkalla, 2006). Recent studies have implicated extracellular adenosine in the modulation of acute inflammation and tissue protection, particularly during conditions of limited oxygen availability (Sitkovsky & Lukashev, 2005; Sitkovsky et al., 2004). CD73, a ubiquitously expressed glycosyl phosphatidylinositol-anchored ectoenzyme, is the pacemaker of this reaction (Thompson et al., 2004). Due to its transcriptional induction by hypoxia (Synnestvedt et al., 2002; Thompson et al., 2004), CD73-dependent adenosine generation is particularly prominent during conditions of limited oxygen availability, as may occur during myocardial ischemia (Eltzschig et al., 2003). Nevertheless, pharmacological studies on the role of CD73-dependent adenosine generation in cardioprotection during IR have yielded conflicting results (Kitakaze et al., 1994; Miki et al., 1998). Extracellular adenosine produced by CD73 can signal through any of the four ARs, and all four ARs have been associated with tissue protection in a variety of physiological settings (Linden, 2005; Sitkovsky et al., 2004; Weissmuller et al., 2005). While mRNA coding for all four AR is present in cardiac tissue (Morrison et al., 2006), the contribution of individual receptors to cardioprotection from IR remains controversial (Auchampach et al., 2004; Lankford et al., 2006), and may—in part—be related to a lack of studies in which all four AR gene-targeted mice are subjected to the same IP protocol in parallel.

In order to elucidate the contribution of CD73-dependent adenosine generation and to clarify the role of individual ARs in cardioprotection during IP, a recent study adopted a previously described model of chronic cardiomyopathy utilizing a hanging-weight system for intermittent coronary artery occlusion (Dewald et al., 2004). Systematic evaluation of this model revealed highly reproducible infarct sizes and cardiac protection by IP, thus minimizing the variability associated with knot-based coronary occlusion systems (Eckle et al., 2006). In the present study, this model was applied in mice gene targeted for *cd73* or each individual AR. In addition, specific pharmacological adenosine therapeutics were used to confirm the findings from gene-targeted mice (Eckle et al., 2007b). Transcriptional profiling of preconditioned cardiac tissue revealed a prominent induction of CD73 and $ADORA_{2B}$ mRNA. Pharmacological inhibition or targeted gene deletion of *cd73* abolished the cardioprotective effects of *in situ* IP. Similarly, IP was abrogated in mice gene-targeted for the $ADORA_{2B}$, while mice deficient in each of the other ARs showed reduced infarct sizes following IP. Moreover, soluble 5′-nucleotidase or $ADORA_{2B}$ agonist treatment mimicked cardioprotection by IP as it was associated with significant attenuation of myocardial infarct sizes following ischemia. Taken together, these studies suggest manipulation of CD73 enzyme activity to increase extracellular adenosine concentrations and signaling through $ADORA_{2B}$ as therapeutic strategies for the treatment of coronary artery disease (Eckle et al., 2007b).

To gain further inside into mechanisms of how the $ADORA_{2B}$ is regulated during conditions of myocardial ischemia, a subsequent study addressed the functional role of HIF during myocardial ischemia or IP and the relationship of hypoxia-dependent signaling pathways to the $ADORA_{2B}$ (Eckle et al., 2008d). For this purpose, the authors pursued the contribution of *HIF-1* to cardiac IP utilizing loss- and gain-of-function studies of *HIF-1α*. They employed a novel technique of cardiac *in vivo* siRNA repression of selective genes via intraventricular infusion of specific siRNA followed by studies of murine *in situ* preconditioning and myocardial infarction. The authors gained first insight from Western blot analysis for *HIF-1α* from preconditioned myocardium showing activation of *HIF-1α* following IP treatment. While cardioprotective effects of IP were abolished following siRNA repression of *HIF-1α*, pharmacological or genetic activation of *HIF-1α* was associated with a similar degree of cardioprotection as IP treatment itself. Additional studies of endpoint signaling following HIF-activation suggested a critical role of HIF-dependent activation of purinergic signaling pathways through the $ADORA_{2B}$. Taken together, these studies reveal a central role of *HIF-1* in myocardial IP via transcriptional activation of purinergic signaling pathways to increase myocardial resistance to subsequent ischemic tissue injury and suggest pharmacological strategies of *HIF-1* activation in the treatment of acute myocardial ischemia.

C. Acute Kidney Injury

Acute kidney injury (AKI) is characterized by a decrease in the glomerular filtration rate (GFR), occurring over a period of minutes to days (Abuelo, 2007). In hospitalized patients, over 50% of cases of AKI are caused by renal ischemia, or more than 80% in the critical care setting (Abuelo, 2007; Schrier & Wang, 2004). A recent study of hospitalized patients revealed that only a mild increase in the serum creatinine level (0.3–0.4 mg/dl) is associated with a 70% greater risk of death than in persons without any increase (Abuelo, 2007). Along these lines, surgical procedures requiring cross-clamping of the aorta and renal vessels are associated with renal failure rates of up to 30%. Similarly, AKI after cardiac surgery occurs in over 10% of patients under normal circumstances and is associated with dramatic increases in mortality. AKI and chronic kidney disease are also common complications after liver transplantation. For example, the incidence of AKI following liver transplantation is at least 50%, and 8–17% of patients end up requiring renal replacement therapy (Yalavarthy et al., 2007). Moreover, delayed graft function due to tubule cell injury during kidney transplantation is frequently related to ischemia-associated AKI (Parikh et al., 2006). In addition, AKI occurs in approximately 20% of patients suffering from sepsis (Schrier & Wang, 2004). The occurrence of sepsis-associated AKI is associated with dramatic increases of morbidity and mortality (Abuelo, 2007; Schrier & Wang, 2004). Therefore, additional therapeutic modalities to prevent renal injury from ischemia are urgently needed.

Previous studies had demonstrated increased production of extracellular adenosine via CD39 and CD73 during AKI modeled by renal ischemia (Grenz et al., 2007b, 2007c). To further study individual ARs in AKI, we developed murine models of renal ischemia and renal IP (Grenz et al., 2007a). IP represents a powerful strategy for kidney protection, and recent advances in transgenic mice may help elucidate its molecular mechanisms. However, murine IP is technically challenging, and experimental details significantly influence results. Thus we developed a novel model for renal IP using a hanging-weight system for isolated renal artery occlusion. In contrast to previous models, this technique eliminates the need for clamping the vascular pedicle (artery/vein). In fact, assessment of renal injury after different time periods of ischemia (10–60 min) revealed highly reproducible increases in plasma creatinine and potassium levels, while creatinine clearance, urinary flow, and potassium/sodium excretion were significantly attenuated. Using different numbers of IP cycles, we found maximal protection with four cycles of 4 min of IR. In contrast, no significant renal protection was observed with IP of the vascular pedicle. To assess transcriptional responses in this model, we isolated RNA from preconditioned kidneys and found time-dependent induction of erythropoietin mRNA and plasma levels with IP. Taken together, this model provides highly reproducible renal injury and protection by IP, thus minimizing variability associated with previous techniques based on clamping of the renal pedicle (Grenz et al., 2007a).

Utilizing this model, we identified a novel pathway of renal protection from ischemia using IP as model (Grenz et al., 2008). Studies in gene-targeted mice for each individual AR confirmed renal protection by IP in *ADORA*$_1$*−/−*, *ADORA*$_{2A}$*−/−*, or *ADORA*$_3$*−/−* mice. In contrast, protection from ischemia was abolished in *ADORA*$_{2B}$*−/−* mice. This was associated with corresponding changes in tissue inflammation. In accordance, the ADORA$_{2B}$-antagonist PSB1115 blocked renal protection by IP, while treatment with the selective ADORA$_{2B}$-agonist BAY 60-6583 dramatically improved renal function and histology following ischemia alone. Using an ADORA$_{2B}$-reporter model, we found exclusive expression of ADORA$_{2B}$s within the reno-vasculature. Studies using ADORA$_{2B}$ bone marrow chimera conferred kidney protection selectively to renal ADORA$_{2B}$s. These results identify the ADORA$_{2B}$ as a novel therapeutic target for providing potent protection from renal ischemia (Grenz et al., 2008).

D. Acute Lung Injury

An additional example for ADORA$_{2B}$-elicited protection from hypoxia-induced inflammation is ALI. ALI and acute respiratory distress syndrome (ARDS) are life-threatening disorders that can develop in the course of different clinical conditions such as pneumonia, acid aspiration, major trauma, or prolonged mechanical ventilation, and contribute significantly to critical illness (Ware & Matthay, 2000). Recent epidemiological studies showed that each year 75,000 patients in the United States alone die from ARDS (Rubenfeld et al., 2005). The pathogenesis of ALI is characterized by influx of a protein-rich edema fluid into the interstitial and intraalveolar spaces as a consequence of increased permeability of the alveolar–capillary barrier (Ware & Matthay, 2000) in conjunction with excessive invasion of inflammatory cells—particularly PMNs (Belperio et al., 2002; Martin, 2002; Reutershan et al., 2006, 2007). At present, only little is known about how to target the alveolar–capillary barrier function or leukocyte trafficking therapeutically during ALI. In fact, no such strategies have been translated into clinical practice, and of today, there is no specific therapy available for ALI beyond mechanical ventilation and other supportive measures (Ware & Matthay, 2000).

Despite the large impact of ALI on morbidity and mortality in critically ill patients (Ware & Matthay, 2000), many episodes are self-limiting and resolve spontaneously through unknown mechanisms. For example, patients undergoing major surgery requiring prolonged mechanical ventilation have an overall incidence of ALI between 0.2% and 5%, depending on the kind of surgery (Licker et al., 2003; Milot et al., 2001; Shorr et al., 2003). In a recent study to identify endogenous pathways to attenuate ventilator-induced lung injury (VILI), we found that extracellular adenosine accumulates in the supernatant of pulmonary epithelia exposed to cyclic mechanical stretch *in vitro* (Eckle et al., 2007a). Similarly, pulmonary adenosine levels were elevated during mechanical ventilation *in vivo* (Eckle et al., 2007a). In fact, mice deficient in extracellular adenosine production showed dramatic increases in pulmonary edema and pulmonary inflammation when exposed to VILI (Eckle et al., 2007a). However, adenosine-dependent signaling pathways of lung protection during VILI remain unknown. As such, extracellular adenosine can signal through any of the four G-protein-coupled ARs ($ADORA_1$, $ADORA_{2A}$, $ADORA_{2B}$, $ADORA_3$), which have all been implicated in tissue protection in different models of injury or inflammation (Eckle et al., 2007a, 2007b; Eltzschig et al., 2005; Fredholm, 2007; Lankford et al., 2006; Lappas et al., 2006; Linden, 2005; Ohta & Sitkovsky, 2001; Salvatore et al., 2000; Sitkovsky & Lukashev, 2005; Sitkovsky et al., 2004; Yang et al., 2005, 2006). Therefore, a recent study was designed to test the hypothesis that AR signaling plays an important role in tissue protection from VILI. For this purpose, we assessed VILI in genetic models for each individual AR utilizing an *in vivo* model of VILI (Eckle et al., 2008b). As these studies pointed toward a pivotal role of $ADORA_{2B}$ signaling, we confirmed these genetic studies using pharmacological approaches with specific $ADORA_{2B}$ agonists and antagonists. Finally, we created bone marrow chimeras to study $ADORA_{2B}$ effects on hemopoietic versus nonhematopoietic cells. The results from this study point toward a dual role of $ADORA_{2B}$ in VILI, as pulmonary $ADORA_{2B}$ signaling dampen capillary–alveolar leakage, while pulmonary inflammation during VILI is attenuated by a combination of hematopoietic and pulmonary $ADORA_{2B}$ signaling. Taken together, these studies are an important contribution of adenosine signaling through the $ADORA_{2B}$ in attenuating ALI and suggest $ADORA_{2B}$ agonists as potential therapeutic for VILI (Eckle et al., 2008c).

E. Gastrointestinal Ischemia and Inflammation

Transient abdominal ischemia caused by surgery, organ transplantation, and spontaneous ischemia leads to profound functional and structural alterations of the gastrointestinal tract. Although restoration of blood flow to an ischemic organ is essential to prevent irreversible tissue injury, reperfusion augments injury by causing destruction of vascular integrity, tissue edema, and disturbances in cellular energy balance (Carden & Granger, 2000; Eltzschig et al., 2009b). Clinically, IR injury of the intestine is a significant problem during surgery for abdominal aortic aneurysm, small bowel transplantation, cardiopulmonary bypass, strangulated hernias, and neonatal necrotizing enterocolitis (Mallick et al., 2004). Intestinal IR can proceed to a systemic response and may result in bacterial translocation, endotoxemia, ARDS, or acute hepatic injury (Eltzschig & Collard, 2004; Eltzschig et al., 2009b; Turnage et al., 1994).

Similar to ischemic episodes, intestinal inflammation is associated with tissue hypoxia. Indeed, sites of intestinal inflammation are characterized by significant changes in metabolic activity. Shifts in energy supply and demand can result in diminished delivery and/or availability of oxygen, leading to inflammation-associated tissue hypoxia, termed "inflammatory hypoxia" (Colgan & Taylor, 2010; Cummins et al., 2008; Eltzschig et al., 2009b; Frick et al., 2009; Karhausen et al., 2004, 2005; Robinson et al., 2008; Taylor & Colgan, 2007). As an example, models of murine colitis have provided compelling evidence that particularly the mucosal surface (especially intestinal epithelium) is prone to significant drops in pO_2 and resulting inflammatory hypoxia (Eltzschig et al., 2009b; Karhausen et al., 2004, 2005). As a result, studies of hypoxia signaling and pharmacologic targeting hypoxia-dependent signaling pathways have become an area of intense investigation in diseases such as inflammatory bowel disease (IBD).

Previous studies have shown that the endogenous signaling molecule adenosine plays a critical role in attenuating inflammatory hypoxia of mucosal organs (Hart et al., 2008a; Louis et al., 2008; Naganuma et al., 2006; Ohta & Sitkovsky, 2001; Sitkovsky & Lukashev, 2005; Sitkovsky et al., 2004). During conditions of intestinal ischemia or inflammation, extracellular adenosine signaling is enhanced and mainly stems from enzymatic phosphohydrolysis of its extracellular precursor molecules (ATP, ADP, or AMP) (Eltzschig et al., 2003, 2004; Hart et al., 2008a, 2008b, 2008c; Kohler et al., 2007; Thompson et al., 2004). Functional studies of extracellular adenosine signaling during inflammatory hypoxia have demonstrated attenuation of vascular leakage (Eckle et al., 2008a; Eltzschig et al., 2003; Morote-Garcia et al., 2008a; Thompson et al., 2004), inflammatory cell accumulation (Eltzschig et al., 2004), myocardial infarction (Eckle et al., 2007b, 2008d, 2008e; Kohler et al., 2007), ALI (Khoury et al., 2007; Eckle et al., 2007a), intestinal inflammation (Frick et al., 2009; Louis et al., 2008), liver (Hart et al., 2008c), or gut ischemia (Hart et al., 2008a, 2009). Of these four ARs, particularly the A_{2B} adenosine receptor is induced during conditions of limited oxygen availability (Eckle et al., 2008a; Eltzschig et al., 2003, 2004; Grenz et al., 2008; Hart et al., 2009; Kong et al., 2006b) or acute inflammation (Eckle et al., 2008b). Studies utilizing a combination of pharmacological approaches with $ADORA_{2B}$ agonists and antagonist, in conjunction with studies of gene-targeted mice for individual ARs, revealed a protective role of $ADORA_{2B}$ signaling in intestinal ischemia (Hart et al., 2009) or inflammation (Eltzschig et al., 2009b; Frick et al., 2009).

Recent studies from our laboratory performed profiling of mucosal scrapings following murine IR (Hart et al., 2009). These studies demonstrated selective induction of $ADORA_{2B}$ transcript levels. Moreover, gene-targeted mice for the *$ADORA_{2B}$* showed more profound intestinal IR injury compared with controls. In contrast, *$ADORA_{2A}$*−/− mice exhibited no differences in intestinal injury compared with littermate controls. In addition, selective inhibition of the $ADORA_{2B}$ resulted in enhanced intestinal inflammation and injury during IR. Further, $ADORA_{2B}$ agonist treatment (BAY 60-6583; Eckle et al., 2007b) protected from intestinal injury, inflammation, and permeability dysfunction in wild-type mice, whereas the therapeutic effects of BAY 60-6583 were abolished following targeted $ADORA_{2B}$ gene deletion. Taken together, these studies demonstrate the $ADORA_{2B}$ as a novel therapeutic target for protection during gastrointestinal IR injury (Hart et al., 2009).

VI. Other ARs than the $ADORA_{2B}$ During Hypoxia

While the current chapter is focused on the role of the $ADORA_{2B}$ during conditions of hypoxia, it is important to point out that all four ARs have been implicated in hypoxia responses. For example, previous studies from the laboratory of John Headrick have demonstrated an important role of the $ADORA_1$ in cardioprotection mediated by IP (Headrick, 1996; Headrick et al., 2003; Matherne et al., 1997; Reichelt et al., 2005). Similarly, other studies have implicated the $ADORA_3$ in cardioprotection from ischemia (Armstrong & Ganote, 1994). For example, very elegant studies from the research laboratory of John A. Auchampach provide strong evidence for the $ADORA_3$ in cardioprotection by IP (Takano et al., 2001) or during myocardial IR injury (Jordan et al., 1999). Moreover, a study from the laboratories of Carl White provides strong evidence that the $ADORA_{2A}$ is transcriptionally induced by hypoxia (Ahmad et al., 2009). Similar to previous studies of the $ADORA_{2B}$ (Kong et al., 2006b), this study demonstrates a role of HIF in this response. Interestingly, this response does not involve *HIF-1*α, but instead is coordinated by *HIF-2*α (Ahmad et al., 2009). Moreover, exciting research work from the laboratory of Pier Andrea Borea, Ferrara, Italy, demonstrate that adenosine signaling in turn can enhance the stabilization of *HIF-1*α protein levels (Gessi et al., 2010; Merighi et al., 2005). *Together these studies highlight the many levels of interdependence between hypoxia and adenosine signaling responses, and that A2BAR signaling only represents one aspect of this field.*

VII. Conclusion

This chapter highlights the role of extracellular adenosine signaling through the $ADORA_{2B}$ in endogenous tissue protection from conditions of limited oxygen availability. As such, many of these studies indicate that targeting extracellular adenosine signaling via the $ADORA_{2B}$ may represent a novel therapeutic approach for the treatment of medical conditions that require tissue adaptation to hypoxia, such as organ ischemia, solid organ transplantation, or ALI. It is important to point out that these models are all highly acute disease models, and there are data suggesting that prolonged elevation of extracellular adenosine levels and $ADORA_{2B}$ signaling could potentially also have detrimental consequences (Blackburn, 2003; Blackburn et al., 2009). Along the same lines, potentially unwanted side effects of pharmacological elevations of extracellular adenosine levels or specific AR agonists have to be addressed. For example, such unwanted side effects could include alterations in blood pressure, heart rate, or sleep–awake cycle (Yang et al., 2009), fatty liver disease (Peng et al., 2009), chronic forms of lung disease (Sun et al., 2006), or involve platelet function, thromboregulation, or bleeding (Enjyoji et al., 1999; Hart et al., 2008b; Pinsky et al., 2002). Moreover, it is important to point out that most of the studies described in the present chapter were carried out in animal models, in fact most of them in mice. Information about adenosine generation and signaling is still very limited. While many of the *in vitro* studies discussed in the present chapter were carried out in human cell lines (Eltzschig et al., 2003, 2004; Khoury et al., 2007), only very few involve patient- derived material (e.g., plasma levels of ADA in hypoxic patients; Eltzschig et al., 2006b). Therefore, it will be a critical challenge for the future to translate these findings from mice to man and from disease models into the treatment of human disease.

Acknowledgment

The authors want to acknowledge Shelley A. Eltzschig for the artwork during manuscript preparation. This chapter is supported by U.S. National Institutes of Health Grants R01-HL0921, R01-DK083385, and R01HL098294 to H. K. E.; K08 HL102267 to T. E.; and German National Science Foundation (DFG) grant to M. K.

Conflict of Interest: None of the authors has a conflict of interest.

Abbreviations

$ADORA_1$/$ADORA_{2A}$/$ADORA_{2B}$/ $ADORA_3$	AR subtype 1/2A/2B/3
ADA	adenosine deaminase
ADP	adenosine diphosphate
AMP	adenosine monophosphate
AR	adenosine receptor
AK	adenosine kinase
ATP	adenosine triphosphate
cAMP	cyclic adenosine monophosphate
CD39	ecto-apyrase, ENTPDase 1
CD73	ecto-5′-nucleotidase

DMOG	dimethyloxaloylglycine
ENT	equilibrative nucleoside transporter
EPO	erythropoietin
HO1	heme oxygenase 1
IKK	IκB kinase (enzyme complex)
IKKβ	one of three subunits of IKK
HIF	hypoxia-inducible factor
NFκB	nuclear factor kappa B
PaO_2	partial arterial oxygen pressure
$PaCO_2$	partial arterial carbon dioxide pressure
PMN	polymorphonuclear leukocyte (neutrophil)
SaO_2	oxygen saturation
VILI	ventilator-induced lung injury

References

Abuelo, J. G. (2007). Normotensive ischemic acute renal failure. *The New England Journal of Medicine*, *357*(8), 797–805.

Aherne, C. M., Kewley, E. M., & Eltzschig, H. K. (2010). The resurgence of A2B adenosine receptor signaling. *Biochimica et Biophysica Acta*, (in press).

Ahmad, A., Ahmad, S., Glover, L., Miller, S. M., Shannon, J. M., Guo, X., et al. (2009). Adenosine A2A receptor is a unique angiogenic target of *HIF*-2alpha in pulmonary endothelial cells. *Proceedings of the National Academy of Sciences of the United States of America*, *106*(26), 10684–10689.

Andrade, C. F., Kaneda, H., Der, S., Tsang, M., Lodyga, M., Chimisso Dos Santos, C., et al. (2006). Toll-like receptor and cytokine gene expression in the early phase of human lung transplantation. *The Journal of Heart and Lung Transplantation*, *25*(11), 1317–1323.

Aragones, J., Fraisl, P., Baes, M., & Carmeliet, P. (2009). Oxygen sensors at the crossroad of metabolism. *Cell Metabolism*, *9*(1), 11–22.

Aragones, J., Schneider, M., Van Geyte, K., Fraisl, P., Dresselaers, T., Mazzone, M., et al. (2008). Deficiency or inhibition of oxygen sensor Phd1 induces hypoxia tolerance by reprogramming basal metabolism. *Nature Genetics*, *40*(2), 170–180.

Armstrong, S., & Ganote, C. E. (1994). Adenosine receptor specificity in preconditioning of isolated rabbit cardiomyocytes: Evidence of A3 receptor involvement. *Cardiovascular Research*, *28*(7), 1049–1056.

Auchampach, J. A., Jin, X., Moore, J., Wan, T. C., Kreckler, L. M., Ge, Z. D., et al. (2004). Comparison of three different A1 adenosine receptor antagonists on infarct size and multiple cycle ischemic preconditioning in anesthetized dogs. *The Journal of Pharmacology and Experimental Therapeutics*, *308*(3), 846–856.

Baxter, G. F. (2002). Role of adenosine in delayed preconditioning of myocardium. *Cardiovascular Research*, *55*, 483–494.

Belperio, J. A., Keane, M. P., Burdick, M. D., Londhe, V., Xue, Y. Y., Li, K., et al. (2002). Critical role for CXCR2 and CXCR2 ligands during the pathogenesis of ventilator-induced lung injury. *The Journal of Clinical Investigation*, *110*(11), 1703–1716.

Bishop, T., Gallagher, D., Pascual, A., Lygate, C. A., de Bono, J. P., Nicholls, L. G., et al. (2008). Abnormal sympathoadrenal development and systemic hypotension in PHD3−/− mice. *Molecular and Cellular Biology*, *28*(10), 3386–3400.
Blackburn, M. R. (2003). Too much of a good thing: Adenosine overload in adenosine-deaminase-deficient mice. *Trends in Pharmacological Sciences*, *24*(2), 66–70.
Blackburn, M. R., Vance, C. O., Morschl, E., & Wilson, C. N. (2009). Adenosine receptors and inflammation. *Handbook of Experimental Pharmacology*, (193), 215–269.
Blackburn, M. R., Volmer, J. B., Thrasher, J. L., Zhong, H., Crosby, J. R., Lee, J. J., et al. (2000). Metabolic consequences of adenosine deaminase deficiency in mice are associated with defects in alveogenesis, pulmonary inflammation, and airway obstruction. *The Journal of Experimental Medicine*, *192*(2), 159–170.
Boutin, A. T., Weidemann, A., Fu, Z., Mesropian, L., Gradin, K., Jamora, C., et al. (2008). Epidermal sensing of oxygen is essential for systemic hypoxic response. *Cell*, *133*(2), 223–234.
Bruns, R. F. (1980). Adenosine receptor activation in human fibroblasts: Nucleoside agonists and antagonists. *Canadian Journal of Physiology and Pharmacology*, *58*(6), 673–691.
Brusselmans, K., Compernolle, V., Tjwa, M., Wiesener, M. S., Maxwell, P. H., Collen, D., et al. (2003). Heterozygous deficiency of hypoxia-inducible factor-2{alpha} protects mice against pulmonary hypertension and right ventricular dysfunction during prolonged hypoxia. *The Journal of Clinical Investigation*, *111*(10), 1519–1527.
Cai, Z., Zhong, H., Bosch-Marce, M., Fox-Talbot, K., Wang, L., Wei, C., et al. (2007). Complete loss of ischaemic preconditioning-induced cardioprotection in mice with partial deficiency of *HIF-1*{alpha}. *Cardiovascular Research*, *77*(3), 463–470.
Carbia-Nagashima, A., Gerez, J., Perez-Castro, C., Paez-Pereda, M., Silberstein, S., Stalla, G. K., et al. (2007). RSUME, a small RWD-containing protein, enhances SUMO conjugation and stabilizes *HIF-1*alpha during hypoxia. *Cell*, *131*(2), 309–323.
Carden, D. L., & Granger, D. N. (2000). Pathophysiology of ischaemia-reperfusion injury. *The Journal of Pathology*, *190*(3), 255–266.
Carmeliet, P., Dor, Y., Herbert, J. M., Fukumura, D., Brusselmans, K., Dewerchin, M., et al. (1998). Role of *HIF-1*alpha in hypoxia-mediated apoptosis, cell proliferation and tumour angiogenesis. *Nature*, *394*(6692), 485–490.
Casanello, P., Torres, A., Sanhueza, F., Gonzalez, M., Farias, M., Gallardo, V., et al. (2005). Equilibrative nucleoside transporter 1 expression is downregulated by hypoxia in human umbilical vein endothelium. *Circulation Research*, *97*(1), 16–24.
Chaudary, N., Naydenova, Z., Shuralyova, I., & Coe, I. R. (2004). Hypoxia regulates the adenosine transporter, mENT1, in the murine cardiomyocyte cell line, HL-1. *Cardiovascular Research*, *61*(4), 780–788.
Chen, Y., Corriden, R., Inoue, Y., Yip, L., Hashiguchi, N., Zinkernagel, A., et al. (2006). ATP release guides neutrophil chemotaxis via P2Y2 and A3 receptors. *Science*, *314*(5806), 1792–1795.
Cheng, J., Kang, X., Zhang, S., & Yeh, E. T. (2007). SUMO-specific protease 1 is essential for stabilization of HIF1alpha during hypoxia. *Cell*, *131*(3), 584–595.
Chu, S., & Ferro, T. J. (2005). Sp1: Regulation of gene expression by phosphorylation. *Gene*, *348*, 1–11.
Chunn, J. L., Molina, J. G., Mi, T., Xia, Y., Kellems, R. E., & Blackburn, M. R. (2005). Adenosine-dependent pulmonary fibrosis in adenosine deaminase-deficient mice. *Journal of Immunology*, *175*(3), 1937–1946.
Clark, A. N., Youkey, R., Liu, X., Jia, L., Blatt, R., Day, Y. J., et al. (2007). A1 adenosine receptor activation promotes angiogenesis and release of VEGF from monocytes. *Circulation Research*, *101*(11), 1130–1138.
Colgan, S. P., Eltzschig, H. K., Eckle, T., & Thompson, L. F. (2006). Physiological roles of 5′-ectonucleotidase (CD73). *Purinergic Signalling*, *2*(2), 351–360.

Colgan, S. P., & Taylor, C. T. (2010). Hypoxia: An alarm signal during intestinal inflammation. *Nature Reviews. Gastroenterology & Hepatology*, *7*(5), 281–287.

Compernolle, V., Brusselmans, K., Acker, T., Hoet, P., Tjwa, M., Beck, H., et al. (2002). Loss of *HIF-2*alpha and inhibition of VEGF impair fetal lung maturation, whereas treatment with VEGF prevents fatal respiratory distress in premature mice. *Natural Medicines*, *8*(7), 702–710.

Cramer, T., Yamanishi, Y., Clausen, B. E., Forster, I., Pawlinski, R., Mackman, N., et al. (2003). *HIF-1*alpha is essential for myeloid cell-mediated inflammation. *Cell*, *112*(5), 645–657.

Cronstein, B. N. (1994). Adenosine, an endogenous anti-inflammatory agent. *Journal of Applied Physiology*, *76*(1), 5–13.

Cronstein, B. N., Daguma, L., Nichols, D., Hutchison, A. J., & Williams, M. (1990). The adenosine/neutrophil paradox resolved: Human neutrophils possess both A1 and A2 receptors that promote chemotaxis and inhibit O_2 generation, respectively. *The Journal of Clinical Investigation*, *85*(4), 1150–1157.

Cronstein, B. N., Eberle, M. A., Gruber, H. E., & Levin, R. I. (1991). Methotrexate inhibits neutrophil function by stimulating adenosine release from connective tissue cells. *Proceedings of the National Academy of Sciences of the United States of America*, *88*(6), 2441–2445.

Cronstein, B. N., Kramer, S. B., Weissmann, G., & Hirschhorn, R. (1983). Adenosine: A physiological modulator of superoxide anion generation by human neutrophils. *The Journal of Experimental Medicine*, *158*(4), 1160–1177.

Cronstein, B. N., Levin, R. I., Philips, M., Hirschhorn, R., Abramson, S. B., & Weissmann, G. (1992). Neutrophil adherence to endothelium is enhanced via adenosine A1 receptors and inhibited via adenosine A2 receptors. *Journal of Immunology*, *148*(7), 2201–2206.

Cronstein, B. N., Naime, D., & Ostad, E. (1993). The antiinflammatory mechanism of methotrexate. Increased adenosine release at inflamed sites diminishes leukocyte accumulation in an in vivo model of inflammation. *The Journal of Clinical Investigation*, *92*(6), 2675–2682.

Cronstein, B., Rosenstein, E., Kramer, S., Weissmann, G., & Hirschhorn, R. (1985). Adenosine: A physiologic modulator of superoxide anion generation by human neutrophils. Adenosine acts via an A2 receptor on human neutrophils. *Journal of Immunology*, *135* (2), 1366–1371.

Cummins, E. P., Seeballuck, F., Keely, S. J., Mangan, N. E., Callanan, J. J., Fallon, P. G., et al. (2008). The hydroxylase inhibitor dimethyloxalylglycine is protective in a murine model of colitis. *Gastroenterology*, *134*(1), 156–165.

Cummins, E. P., & Taylor, C. T. (2005). Hypoxia-responsive transcription factors. *Pflugers Archiv*, *450*(6), 363–371, Epub 2005 Jul 9.

Day, Y. J., Huang, L., McDuffie, M. J., Rosin, D. L., Ye, H., Chen, J. F., et al. (2003). Renal protection from ischemia mediated by A2A adenosine receptors on bone marrow-derived cells. *The Journal of Clinical Investigation*, *112*(6), 883–891.

Day, Y. J., Li, Y., Rieger, J. M., Ramos, S. I., Okusa, M. D., & Linden, J. (2005). A2A adenosine receptors on bone marrow-derived cells protect liver from ischemia-reperfusion injury. *Journal of Immunology*, *174*(8), 5040–5046.

Day, Y. J., Marshall, M. A., Huang, L., McDuffie, M. J., Okusa, M. D., & Linden, J. (2004). Protection from ischemic liver injury by activation of A2A adenosine receptors during reperfusion: Inhibition of chemokine induction. *American Journal of Physiology. Gastrointestinal and Liver Physiology*, *286*(2), G285–293.

Dayan, F., Roux, D., Brahimi-Horn, M. C., Pouyssegur, J., & Mazure, N. M. (2006). The oxygen sensor factor-inhibiting hypoxia-inducible factor-1 controls expression of distinct genes through the bifunctional transcriptional character of hypoxia-inducible factor-1alpha. *Cancer Research*, *66*(7), 3688–3698.

De Perrot, M., Sekine, Y., Fischer, S., Waddell, T. K., McRae, K., Liu, M., et al. (2002). Interleukin-8 release during early reperfusion predicts graft function in human lung transplantation. *American Journal of Respiratory and Critical Care Medicine*, *165*(2), 211–215.

Dejana, E., Spagnuolo, R., & Bazzoni, G. (2001). Interendothelial junctions and their role in the control of angiogenesis, vascular permeability and leukocyte transmigration. *Thrombosis and Haemostasis*, *86*(1), 308–315.

Delacretaz, E. (2006). Clinical practice. Supraventricular tachycardia. *New England Journal of Medicine*, *354*(10), 1039–1051.

Dewald, O., Frangogiannis, N. G., Zoerlein, M. P., Duerr, G. D., Taffet, G., Michael, L. H., et al. (2004). A murine model of ischemic cardiomyopathy induced by repetitive ischemia and reperfusion. *Thoracic and Cardiovascular Surgeon*, *52*(5), 305–311.

Dieterich, H. J., Weissmuller, T., Rosenberger, P., & Eltzschig, H. K. (2006). Effect of hydroxyethyl starch on vascular leak syndrome and neutrophil accumulation during hypoxia. *Critical Care Medicine*, *34*(6), 1775–1782.

Drury, A. N., & Szent-Gyorgyi, A. (1929). The physiological activity of adenine compounds with especial reference to their action upon the mammalian heart. *The Journal of Physiology*, *68*(3), 213–237.

Eckle, T., Faigle, M., Grenz, A., Laucher, S., Thompson, L. F., & Eltzschig, H. K. (2008a). A2B adenosine receptor dampens hypoxia-induced vascular leak. *Blood*, *111*(4), 2024–2035.

Eckle, T., Fullbier, L., Grenz, A., & Eltzschig, H. K. (2008b). Usefulness of pressure-controlled ventilation at high inspiratory pressures to induce acute lung injury in mice. *American Journal of Physiology. Lung Cellular and Molecular Physiology*, *295*(4), L718–L724.

Eckle, T., Fullbier, L., Wehrmann, M., Khoury, J., Mittelbronn, M., Ibla, J., et al. (2007a). Identification of ectonucleotidases CD39 and CD73 in innate protection during acute lung injury. *Journal of Immunology*, *178*(12), 8127–8137.

Eckle, T., Grenz, A., Kohler, D., Redel, A., Falk, M., Rolauffs, B., et al. (2006). Systematic evaluation of a novel model for cardiac ischemic preconditioning in mice. *American Journal of Physiology. Heart and Circulatory Physiology*, *291*(5), H2533–H2540.

Eckle, T., Grenz, A., Laucher, S., & Eltzschig, H. K. (2008c). A2B adenosine receptor signaling attenuates acute lung injury by enhancing alveolar fluid clearance in mice. *The Journal of Clinical Investigation*, *118*(10), 3301–3315.

Eckle, T., Koeppen, M., & Eltzschig, H. K. (2009). Role of extracellular adenosine in acute lung injury. *Physiology (Bethesda)*, *24*(5), 298–306.

Eckle, T., Kohler, D., Lehmann, R., El Kasmi, K. C., & Eltzschig, H. K. (2008d). Hypoxia-inducible factor-1 is central to cardioprotection: A new paradigm for ischemic preconditioning. *Circulation*, *118*(2), 166–175.

Eckle, T., Krahn, T., Grenz, A., Kohler, D., Mittelbronn, M., Ledent, C., et al. (2007b). Cardioprotection by ecto-5′-nucleotidase (CD73) and A2B adenosine receptors. *Circulation*, *115*(12), 1581–1590.

Elliott, M. R., Chekeni, F. B., Trampont, P. C., Lazarowski, E. R., Kadl, A., Walk, S. F., et al. (2009). Nucleotides released by apoptotic cells act as a find-me signal to promote phagocytic clearance. *Nature*, *461*(7261), 282–286.

Eltzschig, H. K. (2009). Adenosine: An old drug newly discovered. *Anesthesiology*, *111*(4), 904–915.

Eltzschig, H. K., Abdulla, P., Hoffman, E., Hamilton, K. E., Daniels, D., Schonfeld, C., et al. (2005). *HIF-1*-dependent repression of equilibrative nucleoside transporter (ENT) in hypoxia. *The Journal of Experimental Medicine*, *202*(11), 1493–1505.

Eltzschig, H. K., & Collard, C. D. (2004). Vascular ischaemia and reperfusion injury. *British Medical Bulletin*, *70*(1), 71–86.

Eltzschig, H. K., Eckle, T., Mager, A., Kuper, N., Karcher, C., Weissmuller, T., et al. (2006a). ATP release from activated neutrophils occurs via connexin 43 and modulates adenosine-dependent endothelial cell function. *Circulation Research*, *99*(10), 1100–1108.

Eltzschig, H. K., Faigle, M., Knapp, S., Karhausen, J., Ibla, J., Rosenberger, P., et al. (2006b). Endothelial catabolism of extracellular adenosine during hypoxia: The role of surface adenosine deaminase and CD26. *Blood*, *108*(5), 1602–1610.

Eltzschig, H. K., Ibla, J. C., Furuta, G. T., Leonard, M. O., Jacobson, K. A., Enjyoji, K., et al. (2003). Coordinated adenine nucleotide phosphohydrolysis and nucleoside signaling in posthypoxic endothelium: Role of ectonucleotidases and adenosine A2B receptors. *The Journal of Experimental Medicine*, *198*(5), 783–796.

Eltzschig, H. K., Kohler, D., Eckle, T., Kong, T., Robson, S. C., & Colgan, S. P. (2009a). Central role of Sp1-regulated CD39 in hypoxia/ischemia protection. *Blood*, *113*(1), 224–232.

Eltzschig, H. K., Rivera-Nieves, J., & Colgan, S. P. (2009b). Targeting the A2B adenosine receptor during gastrointestinal ischemia and inflammation. *Expert Opinion on Therapeutic Targets*, *13*(11), 1267–1277.

Eltzschig, H. K., Thompson, L. F., Karhausen, J., Cotta, R. J., Ibla, J. C., Robson, S. C., et al. (2004). Endogenous adenosine produced during hypoxia attenuates neutrophil accumulation: Coordination by extracellular nucleotide metabolism. *Blood*, *104*(13), 3986–3992.

Enjyoji, K., Sevigny, J., Lin, Y., Frenette, P. S., Christie, P. D., Esch, J. S., 2nd, et al. (1999). Targeted disruption of cd39/ATP diphosphohydrolase results in disordered hemostasis and thromboregulation. *Natural Medicines*, *5*(9), 1010–1017.

Faigle, M., Seessle, J., Zug, S., El Kasmi, K. C., & Eltzschig, H. K. (2008). ATP release from vascular endothelia occurs across Cx43 hemichannels and is attenuated during hypoxia. *PLoS One*, *3*(7), e2801.

Feoktistov, I., Ryzhov, S., Zhong, H., Goldstein, A. E., Matafonov, A., Zeng, D., et al. (2004). Hypoxia modulates adenosine receptors in human endothelial and smooth muscle cells toward an A2B angiogenic phenotype. *Hypertension*, *44*(5), 649–654.

Fraisl, P., Aragones, J., & Carmeliet, P. (2009). Inhibition of oxygen sensors as a therapeutic strategy for ischaemic and inflammatory disease. *Nature Reviews. Drug Discovery*, *8*(2), 139–152.

Frangogiannis, N. G., Ren, G., Dewald, O., Zymek, P., Haudek, S., Koerting, A., et al. (2005). Critical role of endogenous thrombospondin-1 in preventing expansion of healing myocardial infarcts. *Circulation*, *111*(22), 2935–2942.

Fredholm, B. B. (2007). Adenosine, an endogenous distress signal, modulates tissue damage and repair. *Cell Death and Differentiation*, *14*(7), 1315–1323.

Fredholm, B. B., IJzerman, A. P., Jacobson, K. A., Klotz, K. N., & Linden, J. (2001). International Union of Pharmacology. XXV. Nomenclature and classification of adenosine receptors. *Pharmacological Reviews*, *53*(4), 527–552.

Frick, J. S., MacManus, C. F., Scully, M., Glover, L. E., Eltzschig, H. K., & Colgan, S. P. (2009). Contribution of adenosine A2B receptors to inflammatory parameters of experimental colitis. *Journal of Immunology*, *182*(8), 4957–4964.

Gao, Z. G., & Jacobson, K. A. (2007). Emerging adenosine receptor agonists. *Expert Opinion on Emerging Drugs*, *12*(3), 479–492.

Gazoni, L. M., Laubach, V. E., Mulloy, D. P., Bellizzi, A., Unger, E. B., Linden, J., et al. (2008). Additive protection against lung ischemia-reperfusion injury by adenosine A2A receptor activation before procurement and during reperfusion. *The Journal of Thoracic and Cardiovascular Surgery*, *135*(1), 156–165.

Gessi, S., Fogli, E., Sacchetto, V., Merighi, S., Varani, K., Preti, D., et al. (2010). Adenosine modulates *HIF-1*{alpha}, VEGF, IL-8, and foam cell formation in a human model of hypoxic foam cells. *Arteriosclerosis, Thrombosis, and Vascular Biology*, *30*(1), 90–97.

Grenz, A., Eckle, T., Zhang, H., Huang, D. Y., Wehrmann, M., Kohle, C., et al. (2007a). Use of a hanging-weight system for isolated renal artery occlusion during ischemic preconditioning in mice. *American Journal of Physiology—Renal Physiology*, *292*(1), F475–F485.

Grenz, A., Osswald, H., Eckle, T., Yang, D., Zhang, H., Tran, Z. V., et al. (2008). The renovascular A2B adenosine receptor protects the kidney from ischemia. *PLoS Medicine*, *5*(6), e137.

Grenz, A., Zhang, H., Eckle, T., Mittelbronn, M., Wehrmann, M., Kohle, C., et al. (2007b). Protective role of ecto-5′-nucleotidase (CD73) in renal ischemia. *Journal of the American Society of Nephrology*, *18*(3), 833–845.

Grenz, A., Zhang, H., Hermes, M., Eckle, T., Klingel, K., Huang, D. Y., et al. (2007c). Contribution of E-NTPDase1 (CD39) to renal protection from ischemia-reperfusion injury. *The FASEB Journal*, *21*(11), 2863–2873.

Grenz, A., Zhang, H., Weingart, J., von Wietersheim, S., Eckle, T., Schnermann, J. B., et al. (2007d). Lack of effect of extracellular adenosine generation and signalling on renal erythropoietin secretion during hypoxia. *American Journal of Physiology—Renal Physiology*, 293.

Grocott, M. P., Martin, D. S., Levett, D. Z., McMorrow, R., Windsor, J., & Montgomery, H. E. (2009). Arterial blood gases and oxygen content in climbers on Mount Everest. *The New England Journal of Medicine*, *360*(2), 140–149.

Haddad, J. J. (2003). Science review: Redox and oxygen-sensitive transcription factors in the regulation of oxidant-mediated lung injury: Role for hypoxia-inducible factor-1alpha. *Critical Care*, *7*(1), 47–54.

Haeberle, H. A., Durrstein, C., Rosenberger, P., Hosakote, Y. M., Kuhlicke, J., Kempf, V. A., et al. (2008). Oxygen-independent stabilization of hypoxia inducible factor (HIF)-1 during RSV infection. *PLoS One*, *3*(10), e3352.

Hart, M. L., Gorzolla, I. C., Schittenhelm, J., Robson, S. C., & Eltzschig, H. K. (2010). SP1-dependent induction of CD39 facilitates hepatic ischemic preconditioning. *Journal of Immunology*, *184*(7), 4017–4024.

Hart, M. L., Henn, M., Kohler, D., Kloor, D., Mittelbronn, M., Gorzolla, I. C., et al. (2008a). Role of extracellular nucleotide phosphohydrolysis in intestinal ischemia-reperfusion injury. *The FASEB Journal*, *22*(8), 2784–2797.

Hart, M. L., Jacobi, B., Schittenhelm, J., Henn, M., & Eltzschig, H. K. (2009). Cutting edge: A2B adenosine receptor signaling provides potent protection during intestinal ischemia/reperfusion injury. *Journal of Immunology*, *182*(7), 3965–3968.

Hart, M. L., Kohler, D., Eckle, T., Kloor, D., Stahl, G. L., & Eltzschig, H. K. (2008b). Direct treatment of mouse or human blood with soluble 5′-nucleotidase inhibits platelet aggregation. *Arteriosclerosis, Thrombosis, and Vascular Biology*, *28*(8), 1477–1483.

Hart, M. L., Much, C., Gorzolla, I. C., Schittenhelm, J., Kloor, D., Stahl, G. L., et al. (2008c). Extracellular adenosine production by ecto-5′-nucleotidase protects during murine hepatic ischemic preconditioning. *Gastroenterology*, *135*(5), 1739–1750 e1733.

Hartmann, G., Tschop, M., Fischer, R., Bidlingmaier, C., Riepl, R., Tschop, K., et al. (2000). High altitude increases circulating interleukin-6, interleukin-1 receptor antagonist and C-reactive protein. *Cytokine*, *12*(3), 246–252.

Hasko, G., & Cronstein, B. N. (2004). Adenosine: An endogenous regulator of innate immunity. *Trends in Immunology*, *25*(1), 33–39.

Hasko, G., Csoka, B., Nemeth, Z. H., Vizi, E. S., & Pacher, P. (2009). A(2B) adenosine receptors in immunity and inflammation. *Trends in Immunology*, *30*, 263–270, Epub 2009, May 7.

Hasko, G., Linden, J., Cronstein, B., & Pacher, P. (2008). Adenosine receptors: Therapeutic aspects for inflammatory and immune diseases. *Nature Reviews. Drug Discovery*, *7*(9), 759–770.

Headrick, J. P. (1996). Ischemic preconditioning: Bioenergetic and metabolic changes and the role of endogenous adenosine. *Journal of Molecular and Cellular Cardiology*, *28*(6), 1227–1240.

Headrick, J. P., Hack, B., & Ashton, K. J. (2003). Acute adenosinergic cardioprotection in ischemic-reperfused hearts. *American Journal of Physiology. Heart and Circulatory Physiology*, *285*(5), H1797–1818.

Herrmann, H. C. (2006). Update and rationale for ongoing acute myocardial infarction trials: Combination therapy, facilitation, and myocardial preservation. *American Heart Journal*, *151*(Suppl. 6), S30–39.

Hirota, K., & Semenza, G. L. (2005). Regulation of hypoxia-inducible factor 1 by prolyl and asparaginyl hydroxylases. *Biochemical and Biophysical Research Communications*, *338* (1), 610–616.

Hodgkiss, R. J., Jones, G., Long, A., Parrick, J., Smith, K. A., Stratford, M. R., et al. (1991). Flow cytometric evaluation of hypoxic cells in solid experimental tumours using fluorescence immunodetection. *British Journal of Cancer*, *63*(1), 119–125.

Holmquist-Mengelbier, L., Fredlund, E., Lofstedt, T., Noguera, R., Navarro, S., Nilsson, H., et al. (2006). Recruitment of *HIF-1*alpha and *HIF-2*alpha to common target genes is differentially regulated in neuroblastoma: *HIF-2*alpha promotes an aggressive phenotype. *Cancer Cell*, *10*(5), 413–423.

Hu, C. J., Wang, L. Y., Chodosh, L. A., Keith, B., & Simon, M. C. (2003). Differential roles of hypoxia-inducible factor 1alpha (*HIF-1*alpha) and *HIF-2*alpha in hypoxic gene regulation. *Molecular and Cellular Biology*, *23*(24), 9361–9374.

Jacobson, K. A., & Gao, Z. G. (2006). Adenosine receptors as therapeutic targets. *Nature Reviews. Drug Discovery*, *5*(3), 247–264.

Jacobson, K. A., Klutz, A. M., Tosh, D. K., Ivanov, A. A., Preti, D., & Baraldi, P. G. (2009). Medicinal chemistry of the A3 adenosine receptor: Agonists, antagonists, and receptor engineering. *Handbook of Experimental Pharmacology*, (193), 123–159.

Ji, X., Kim, Y. C., Ahern, D. G., Linden, J., & Jacobson, K. A. (2001). [3H]MRS 1754, a selective antagonist radioligand for A(2B) adenosine receptors. *Biochemical Pharmacology*, *61*(6), 657–663.

Jordan, J. E., Thourani, V. H., Auchampach, J. A., Robinson, J. A., Wang, N. P., & Vinten-Johansen, J. (1999). A(3) adenosine receptor activation attenuates neutrophil function and neutrophil-mediated reperfusion injury. *The American Journal of Physiology*, *277*(5 Pt. 2), H1895–1905.

Kaelin, W. G. (2005). Proline hydroxylation and gene expression. *Annual Review of Biochemistry*, *74*, 115–128.

Kaelin, W. G. (2007). Von Hippel-Lindau disease. *Annual Review of Pathology*, *2*, 145–173.

Kaelin, W. G., Jr. (2008). The von Hippel-Lindau tumour suppressor protein: O_2 sensing and cancer. *Nature Reviews. Cancer*, *8*(11), 865–873.

Kaelin, W. G., Jr., & Ratcliffe, P. J. (2008). Oxygen sensing by metazoans: The central role of the HIF hydroxylase pathway. *Molecular Cell*, *30*(4), 393–402.

Karhausen, J., Furuta, G. T., Tomaszewski, J. E., Johnson, R. S., Colgan, S. P., & Haase, V. H. (2004). Epithelial hypoxia-inducible factor-1 is protective in murine experimental colitis. *The Journal of Clinical Investigation*, *114*(8), 1098–1106.

Karhausen, J., Haase, V. H., & Colgan, S. P. (2005). Inflammatory hypoxia: Role of hypoxia-inducible factor. *Cell Cycle*, *4*(2), 256–258.

Karhausen, J., Ibla, J. C., & Colgan, S. P. (2003). Implications of hypoxia on mucosal barrier function. *Cellular and Molecular Biology (Noisy-le-grand, France)*, *49*(1), 77–87.

Kempf, V. A. J., Lebiedziejewski, M., Alitalo, K., Walzlein, J.-H., Ehehalt, U., Fiebig, J., et al. (2005). Activation of hypoxia-inducible factor-1 in bacillary angiomatosis: Evidence for a role of hypoxia-inducible factor-1 in bacterial infections. *Circulation*, *111*(8), 1054–1062.

Khoury, J., Ibla, J. C., Neish, A. S., & Colgan, S. P. (2007). Antiinflammatory adaptation to hypoxia through adenosine-mediated cullin-1 deneddylation. *The Journal of Clinical Investigation*, *117*(3), 703–711.

Kitakaze, M., Hori, M., Morioka, T., Minamino, T., Takashima, S., Sato, H., et al. (1994). Infarct size-limiting effect of ischemic preconditioning is blunted by inhibition of 5′-nucleotidase activity and attenuation of adenosine release. *Circulation*, *89*(3), 1237–1246.

Kloner, R. A., & Rezkalla, S. H. (2006). Preconditioning, postconditioning and their application to clinical cardiology. *Cardiovascular Research*, *70*(2), 297–307.

Koeppen, M., Eckle, T., & Eltzschig, H. K. (2009). Selective deletion of the A1 adenosine receptor abolishes heart-rate slowing effects of intravascular adenosine in vivo. *PLoS One*, *4*(8), e6784.

Kohler, D., Eckle, T., Faigle, M., Grenz, A., Mittelbronn, M., Laucher, S., et al. (2007). CD39/ectonucleoside triphosphate diphosphohydrolase 1 provides myocardial protection during cardiac ischemia/reperfusion injury. *Circulation*, *116*(16), 1784–1794.

Kong, X., Lin, Z., Liang, D., Fath, D., Sang, N., & Caro, J. (2006a). Histone deacetylase inhibitors induce VHL and ubiquitin-independent proteasomal degradation of hypoxia-inducible factor 1alpha. *Molecular and Cellular Biology*, *26*(6), 2019–2028.

Kong, T., Westerman, K. A., Faigle, M., Eltzschig, H. K., & Colgan, S. P. (2006b). HIF-dependent induction of adenosine A2B receptor in hypoxia. *The FASEB Journal*, *20*(13), 2242–2250.

Kruger, B., Krick, S., Dhillon, N., Lerner, S. M., Ames, S., Bromberg, J. S., et al. (2009). Donor toll-like receptor 4 contributes to ischemia and reperfusion injury following human kidney transplantation. *Proceedings of the National Academy of Sciences of the United States of America*, *106*(9), 3390–3395.

Kuhlicke, J., Frick, J. S., Morote-Garcia, J. C., Rosenberger, P., & Eltzschig, H. K. (2007). Hypoxia inducible factor (HIF)-1 coordinates induction of toll-like receptors TLR2 and TLR6 during hypoxia. *PLoS One*, *2*(12), e1364.

Kurtul, N., Pence, S., Akarsu, E., Kocoglu, H., Aksoy, Y., & Aksoy, H. (2004). Adenosine deaminase activity in the serum of type 2 diabetic patients. *Acta Medica (Hradec Kralove)*, *47*(1), 33–35.

Lankford, A. R., Yang, J. N., Rose'Meyer, R., French, B. A., Matherne, G. P., Fredholm, B. B., et al. (2006). Effect of modulating cardiac A1 adenosine receptor expression on protection with ischemic preconditioning. *American Journal of Physiology. Heart and Circulatory Physiology*, *290*(4), H1469–H1473.

Lappas, C. M., Day, Y. J., Marshall, M. A., Engelhard, V. H., & Linden, J. (2006). Adenosine A2A receptor activation reduces hepatic ischemia reperfusion injury by inhibiting CD1d-dependent NKT cell activation. *The Journal of Experimental Medicine*, *203*(12), 2639–2648.

Lewis, J. S., Lee, J. A., Underwood, J. C., Harris, A. L., & Lewis, C. E. (1999). Macrophage responses to hypoxia: Relevance to disease mechanisms. *Journal of Leukocyte Biology*, *66* (6), 889–900.

Licker, M., de Perrot, M., Spiliopoulos, A., Robert, J., Diaper, J., Chevalley, C., et al. (2003). Risk factors for acute lung injury after thoracic surgery for lung cancer. *Anesthesia and Analgesia*, *97*(6), 1558–1565.

Linden, J. (2001). Molecular approach to adenosine receptors: Receptor-mediated mechanisms of tissue protection. *Annual Review of Pharmacology and Toxicology*, *41*, 775–787.

Linden, J. (2005). Adenosine in tissue protection and tissue regeneration. *Molecular Pharmacology*, *67*(5), 1385–1387.

Linden, J. (2006a). New insights into the regulation of inflammation by adenosine. *The Journal of Clinical Investigation*, *116*(7), 1835–1837.

Linden, J. (2006b). Purinergic chemotaxis. *Science*, *314*(5806), 1689–1690.

Loffler, M., Morote-Garcia, J. C., Eltzschig, S. A., Coe, I. R., & Eltzschig, H. K. (2007). Physiological roles of vascular nucleoside transporters. *Arteriosclerosis, Thrombosis, and Vascular Biology*, *27*(5), 1004–1013.

Louis, N. A., Robinson, A. M., MacManus, C. F., Karhausen, J., Scully, M., & Colgan, S. P. (2008). Control of IFN-alphaA by CD73: Implications for mucosal inflammation. *Journal of Immunology*, *180*(6), 4246–4255.

Lukashev, D. E., Smith, P. T., Caldwell, C. C., Ohta, A., Apasov, S. G., & Sitkovsky, M. V. (2003). Analysis of A2a receptor-deficient mice reveals no significant compensatory increases in the expression of A2b, A1, and A3 adenosine receptors in lymphoid organs. *Biochemical Pharmacology*, *65*(12), 2081–2090.

Luscinskas, F. W., Ma, S., Nusrat, A., Parkos, C. A., & Shaw, S. K. (2002a). Leukocyte transendothelial migration: A junctional affair. *Seminars in Immunology*, *14*(2), 105–113.

Luscinskas, F. W., Ma, S., Nusrat, A., Parkos, C. A., & Shaw, S. K. (2002b). The role of endothelial cell lateral junctions during leukocyte trafficking. *Immunological Reviews*, *186*, 57–67.

Mahon, P. C., Hirota, K., & Semenza, G. L. (2001). FIH-1: A novel protein that interacts with *HIF-1*alpha and VHL to mediate repression of *HIF-1* transcriptional activity. *Genes & Development*, *15*(20), 2675–2686.

Mallick, I. H., Yang, W., Winslet, M. C., & Seifalian, A. M. (2004). Ischemia-reperfusion injury of the intestine and protective strategies against injury. *Digestive Diseases and Sciences*, *49*(9), 1359–1377.

Marcus, A. J., Broekman, M. J., Drosopoulos, J. H. F., Islam, N., Alyonycheva, T. N., Safier, L. B., et al. (1997). The endothelial cell ecto-ADPase responsible for inhibition of platelet function is CD39. *The Journal of Clinical Investigation*, *99*(6), 1351–1360.

Martin, T. R. (2002). Neutrophils and lung injury: Getting it right. *The Journal of Clinical Investigation*, *110*(11), 1603–1605.

Mastrogiannaki, M., Matak, P., Keith, B., Simon, M. C., Vaulont, S., & Peyssonnaux, C. (2009). *HIF-2*alpha, but not *HIF-1*alpha, promotes iron absorption in mice. *The Journal of Clinical Investigation*, *119*(5), 1159–1166.

Matherne, G. P., Linden, J., Byford, A. M., Gauthier, N. S., & Headrick, J. P. (1997). Transgenic A1 adenosine receptor overexpression increases myocardial resistance to ischemia. *Proceedings of the National Academy of Sciences of the United States of America*, *94*(12), 6541–6546.

Mazzone, M., Dettori, D., Leite de Oliveira, R., Loges, S., Schmidt, T., Jonckx, B., et al. (2009). Heterozygous deficiency of PHD2 restores tumor oxygenation and inhibits metastasis via endothelial normalization. *Cell*, *136*(5), 839–851.

Merighi, S., Benini, A., Mirandola, P., Gessi, S., Varani, K., Leung, E., et al. (2005). A3 adenosine receptors modulate hypoxia-inducible factor-1alpha expression in human A375 melanoma cells. *Neoplasia*, *7*(10), 894–903.

Mi, T., Abbasi, S., Zhang, H., Uray, K., Chunn, J. L., Xia, L. W., et al. (2008). Excess adenosine in murine penile erectile tissues contributes to priapism via A2B adenosine receptor signaling. *The Journal of Clinical Investigation*, *118*(4), 1491–1501.

Michel, C. C. (1998). Capillaries, caveolae, calcium and cyclic nucleotides: A new look at microvascular permeability. *Journal of Molecular and Cellular Cardiology*, *30*(12), 2541–2546.

Michel, C. C., & Curry, F. E. (1999). Microvascular permeability. *Physiological Reviews*, *79* (3), 703–761.

Miki, T., Miura, T., Bunger, R., Suzuki, K., Sakamoto, J., & Shimamoto, K. (1998). Ecto-5''-nucleotidase is not required for ischemic preconditioning in rabbit myocardium in situ. *The American Journal of Physiology*, *275*(4 Pt. 2), H1329–1337.

Milkiewicz, M., Doyle, J. L., Fudalewski, T., Ispanovic, E., Aghasi, M., & Haas, T. L. (2007). *HIF-1*alpha and *HIF-2*alpha play a central role in stretch-induced but not shear-stress-induced angiogenesis in rat skeletal muscle. *The Journal of Physiology*, *583*(Pt. 2), 753–766.

Milot, J., Perron, J., Lacasse, Y., Letourneau, L., Cartier, P. C., & Maltais, F. (2001). Incidence and predictors of ARDS after cardiac surgery. *Chest*, *119*(3), 884–888.

Moore, T. M., Chetham, P. M., Kelly, J. J., & Stevens, T. (1998). Signal transduction and regulation of lung endothelial cell permeability. Interaction between calcium and cAMP. *American Journal of Physiology*, *275*, L203–222.

Morote-Garcia, J. C., Rosenberger, P., Kuhlicke, J., & Eltzschig, H. K. (2008a). *HIF-1-*dependent repression of adenosine kinase attenuates hypoxia-induced vascular leak. *Blood*, *111*, 5571–5580.

Morote-Garcia, J. C., Rosenberger, P., Nivillac, N. M., Coe, I. R., & Eltzschig, H. K. (2009). Hypoxia-inducible factor-dependent repression of equilibrative nucleoside transporter 2 attenuates mucosal inflammation during intestinal hypoxia. *Gastroenterology*, *136*(2), 607–618.

Morrison, R. R., Talukder, M. A., Ledent, C., & Mustafa, S. J. (2002). Cardiac effects of adenosine in A(2A) receptor knockout hearts: Uncovering A(2B) receptors. *American Journal of Physiology. Heart and Circulatory Physiology*, *282*(2), H437–444.

Morrison, R. R., Teng, B., Oldenburg, P. J., Katwa, L. C., Schnermann, J. B., & Mustafa, S. J. (2006). Effects of targeted deletion of A1 adenosine receptors on post-ischemic cardiac function and expression of adenosine receptor subtypes. *American Journal of Physiology. Heart and Circulatory Physiology*, *291*(4), H1875–H1882.

Mubagwa, K., & Flameng, W. (2001). Adenosine, adenosine receptors and myocardial protection: An updated overview. *Cardiovascular Research*, *52*, 25–39.

Murry, C. E., Jennings, R. B., & Reimer, K. A. (1986). Preconditioning with ischemia: A delay of lethal cell injury in ischemic myocardium. *Circulation*, *74*(5), 1124–1136.

Naganuma, M., Wiznerowicz, E. B., Lappas, C. M., Linden, J., Worthington, M. T., & Ernst, P. B. (2006). Cutting edge: Critical role for A2A adenosine receptors in the T cell-mediated regulation of colitis. *Journal of Immunology*, *177*(5), 2765–2769.

Novak, I. (2003). ATP as a signaling molecule: The exocrine focus. *News in Physiological Sciences*, *18*(1), 12–17.

Odashima, M., Bamias, G., Rivera-Nieves, J., Linden, J., Nast, C. C., Moskaluk, C. A., et al. (2005). Activation of A2A adenosine receptor attenuates intestinal inflammation in animal models of inflammatory bowel disease. *Gastroenterology*, *129*(1), 26–33.

Ogawa, S., Gerlach, H., Esposito, C., Pasagian-Macaulay, A., Brett, J., & Stern, D. (1990). Hypoxia modulates the barrier and coagulant function of cultured bovine endothelium. Increased monolayer permeability and induction of procoagulant properties. *The Journal of Clinical Investigation*, *85*(4), 1090–1098.

Ogawa, S., Koga, S., Kuwabara, K., Brett, J., Morrow, B., Morris, S. A., et al. (1992). Hypoxia-induced increased permeability of endothelial monolayers occurs through lowering of cellular cAMP levels. *The American Journal of Physiology*, *262*(3 Pt. 1), C546–554.

Ohta, A., & Sitkovsky, M. (2001). Role of G-protein-coupled adenosine receptors in downregulation of inflammation and protection from tissue damage. *Nature*, *414* (6866), 916–920.

Okusa, M. D., Linden, J., Huang, L., Rosin, D. L., Smith, D. F., & Sullivan, G. (2001). Enhanced protection from renal ischemia-reperfusion [correction of ischemia:reperfusion] injury with A(2A)-adenosine receptor activation and PDE 4 inhibition. *Kidney International*, *59*(6), 2114–2125.

Parikh, C. R., Jani, A., Mishra, J., Ma, Q., Kelly, C., Barasch, J., et al. (2006). Urine NGAL and IL-18 are predictive biomarkers for delayed graft function following kidney transplantation. *American Journal of Transplantation*, *6*(7), 1639–1645.

Peng, Z., Borea, P. A., Wilder, T., Yee, H., Chiriboga, L., Blackburn, M. R., et al. (2009). Adenosine signaling contributes to ethanol-induced fatty liver in mice. *The Journal of Clinical Investigation*, *119*(3), 582–594.

Percy, M. J., Furlow, P. W., Lucas, G. S., Li, X., Lappin, T. R., McMullin, M. F., et al. (2008). A gain-of-function mutation in the HIF2A gene in familial erythrocytosis. *The New England Journal of Medicine*, *358*(2), 162–168.

Pinsky, D. J., Broekman, M. J., Peschon, J. J., Stocking, K. L., Fujita, T., Ramasamy, R., et al. (2002). Elucidation of the thromboregulatory role of CD39/ectoapyrase in the ischemic brain. *The Journal of Clinical Investigation*, *109*(8), 1031–1040.

Rankin, E. B., Biju, M. P., Liu, Q., Unger, T. L., Rha, J., Johnson, R. S., et al. (2007). Hypoxia-inducible factor-2 (*HIF-2*) regulates hepatic erythropoietin in vivo. *The Journal of Clinical Investigation*, *117*(4), 1068–1077.

Ratcliffe, P. J. (2007). *HIF-1* and *HIF-2*: Working alone or together in hypoxia? *The Journal of Clinical Investigation*, *117*(4), 862–865.

Reichelt, M. E., Willems, L., Molina, J. G., Sun, C. X., Noble, J. C., Ashton, K. J., et al. (2005). Genetic deletion of the A1 adenosine receptor limits myocardial ischemic tolerance. *Circulation Research*, *96*(3), 363–367.

Resta, R., Yamashita, Y., & Thompson, L. F. (1998). Ecto-enzyme and signaling functions of lymphocyte CD73. *Immunological Reviews*, *161*, 95–109.

Reutershan, J., Cagnina, R. E., Chang, D., Linden, J., & Ley, K. (2007). Therapeutic anti-inflammatory effects of myeloid cell adenosine receptor A2a stimulation in lipopolysaccharide-induced lung injury. *Journal of Immunology*, *179*(2), 1254–1263.

Reutershan, J., Morris, M. A., Burcin, T. L., Smith, D. F., Chang, D., Saprito, M. S., et al. (2006). Critical role of endothelial CXCR2 in LPS-induced neutrophil migration into the lung. *The Journal of Clinical Investigation*, *116*(3), 695–702.

Riess, T., Andersson, S. G. E., Lupas, A., Schaller, M., Schafer, A., Kyme, P., et al. (2004). Bartonella adhesin A mediates a proangiogenic host cell response. *The Journal of Experimental Medicine*, *200*(10), 1267–1278.

Rius, J., Guma, M., Schachtrup, C., Akassoglou, K., Zinkernagel, A. S., Nizet, V., et al. (2008). NF-kappaB links innate immunity to the hypoxic response through transcriptional regulation of *HIF-1*alpha. *Nature*, *153*(7196), 807–811.

Robinson, A., Keely, S., Karhausen, J., Gerich, M. E., Furuta, G. T., & Colgan, S. P. (2008). Mucosal protection by hypoxia-inducible factor prolyl hydroxylase inhibition. *Gastroenterology*, *134*(1), 145–155.

Robson, S. C., Wu, Y., Sun, X., Knosalla, C., Dwyer, K., & Enjyoji, K. (2005). Ectonucleotidases of CD39 family modulate vascular inflammation and thrombosis in transplantation. *Seminars in Thrombosis and Hemostasis*, *31*(2), 217–233.

Rose, J. B., Naydenova, Z., Bang, A., Eguchi, M., Sweeney, G., Choi, D. S., et al. (2010). Equilibrative nucleoside transporter 1 plays an essential role in cardioprotection. *American Journal of Physiology - Heart and Circulatory Physiology*, *298*(3), H771–H777.

Rosenberger, C., Mandriota, S., Jurgensen, J. S., Wiesener, M. S., Horstrup, J. H., Frei, U., et al. (2002). Expression of hypoxia-inducible factor-1alpha and -2alpha in hypoxic and ischemic rat kidneys. *Journal of the American Society of Nephrology*, *13*(7), 1721–1732.

Rosenberger, P., Schwab, J. M., Mirakaj, V., Masekowsky, E., Mager, A., Morote-Garcia, J. C., et al. (2009). Hypoxia-inducible factor-dependent induction of netrin-1 dampens inflammation caused by hypoxia. *Nature Immunology*, *10*(2), 195–202.

Rubenfeld, G. D., Caldwell, E., Peabody, E., Weaver, J., Martin, D. P., Neff, M., et al. (2005). Incidence and outcomes of acute lung injury. *The New England Journal of Medicine*, *353* (16), 1685–1693.

Saadi, S., Wrenshall, L. E., & Platt, J. L. (2002). Regional manifestations and control of the immune system. *The FASEB Journal*, *16*(8), 849–856.

Salvatore, C. A., Tilley, S. L., Latour, A. M., Fletcher, D. S., Koller, B. H., & Jacobson, M. A. (2000). Disruption of the A(3) adenosine receptor gene in mice and its effect on stimulated inflammatory cells. *The Journal of Biological Chemistry*, *275*(6), 4429–4434.

Sarnquist, F. H. (1983). Physicians on Mount Everest. A clinical account of the 1981 American Medical Research Expedition to Everest. *The Western Journal of Medicine*, *139*(4), 480–485.

Schingnitz, U., Hartmann, K., Macmanus, C. F., Eckle, T., Zug, S., Colgan, S. P., et al. (2010). Signaling through the A2B adenosine receptor dampens endotoxin-induced acute lung injury. *Journal of Immunology*, *184*(9), 5271–5279.

Schneider, D. J., Lindsay, J. C., Zhou, Y., Molina, J. G., & Blackburn, M. R. (2010). Adenosine and osteopontin contribute to the development of chronic obstructive pulmonary disease. *The FASEB Journal*, *24*(1), 70–80.

Schofield, C. J., & Ratcliffe, P. J. (2004). Oxygen sensing by HIF hydroxylases. *Nature Reviews. Molecular Cell Biology*, *5*(5), 343–354.

Schrier, R. W., & Wang, W. (2004). Acute renal failure and sepsis. *The New England Journal of Medicine*, *351*(2), 159–169.

Schulz, R., Rose, J., Post, H., & Heusch, G. (1995). Involvement of endogenous adenosine in ischaemic preconditioning in swine. *Pflugers Archiv*, *430*(2), 273–282.

Semenza, G. L. (1999a). Perspectives on oxygen sensing. *Cell*, *98*(3), 281–284.

Semenza, G. L. (1999b). Regulation of mammalian O_2 homeostasis by hypoxia-inducible factor 1. *Annual Review of Cell and Developmental Biology*, *15*, 551–578.

Semenza, G. L. (2001). *HIF-1*, O(2), and the 3 PHDs: How animal cells signal hypoxia to the nucleus. *Cell*, *107*(1), 1–3.

Semenza, G. L. (2003). Targeting *HIF-1* for cancer therapy. *Nature Reviews. Cancer*, *3*(10), 721–732.

Semenza, G. L. (2007). Life with oxygen. *Science*, *318*(5847), 62–64.

Semenza, G. L. (2009). Defining the role of hypoxia-inducible factor 1 in cancer biology and therapeutics. *Oncogene*, *29*(5), 625–634.

Sharma, A. K., Laubach, V. E., Ramos, S. I., Zhao, Y., Stukenborg, G., Linden, J., et al. (2010). Adenosine A2A receptor activation on CD4+ T lymphocytes and neutrophils attenuates lung ischemia-reperfusion injury. *The Journal of Thoracic and Cardiovascular Surgery*, *139*(2), 474–482.

Sharma, A. K., Linden, J., Kron, I. L., & Laubach, V. E. (2009b). Protection from pulmonary ischemia-reperfusion injury by adenosine A2A receptor activation. *Respiratory Research*, *10*, 58.

Shorr, A. F., Abbott, K. C., & Agadoa, L. Y. (2003). Acute respiratory distress syndrome after kidney transplantation: Epidemiology, risk factors, and outcomes. *Critical Care Medicine*, *31*(5), 1325–1330.

Sitkovsky, M. V. (2009). T regulatory cells: Hypoxia-adenosinergic suppression and re-direction of the immune response. *Trends in Immunology*, *30*(3), 102–108.

Sitkovsky, M. V., Kjaergaard, J., Lukashev, D., & Ohta, A. (2008). Hypoxia-adenosinergic immunosuppression: Tumor protection by T regulatory cells and cancerous tissue hypoxia. *Clinical Cancer Research*, *14*(19), 5947–5952.

Sitkovsky, M., & Lukashev, D. (2005). Regulation of immune cells by local-tissue oxygen tension: HIF1 alpha and adenosine receptors. *Nature Reviews. Immunology*, *5*(9), 712–721.

Sitkovsky, M. V., Lukashev, D., Apasov, S., Kojima, H., Koshiba, M., Caldwell, C., et al. (2004). Physiological control of immune response and inflammatory tissue damage by hypoxia-inducible factors and adenosine A2A receptors. *Annual Review of Immunology*, *22*(1), 657–682.

Stan, R. V. (2002). Structure and function of endothelial caveolae. *Microscopy Research and Technique*, *57*, 350–364.

Stevens, T., Creighton, J., & Thompson, W. J. (1999). Control of cAMP in lung endothelial cell phenotypes. Implications for control of barrier function. *American Journal of Physiology*, *277*, L119–L126.

Stevens, T., Garcia, J. G. N., Shasby, D. M., Bhattacharya, J., & Malik, A. B. (2000). Mechanisms regulating endothelial cell barrier function. *American Journal of Physiology. Lung Cellular and Molecular Physiology*, *279*, L419–L422.

Sun, C. X., Zhong, H., Mohsenin, A., Morschl, E., Chunn, J. L., Molina, J. G., et al. (2006). Role of A2B adenosine receptor signaling in adenosine-dependent pulmonary inflammation and injury. *The Journal of Clinical Investigation*, *116*(8), 2173–2182.

Synnestvedt, K., Furuta, G. T., Comerford, K. M., Louis, N., Karhausen, J., Eltzschig, H. K., et al. (2002). Ecto-5′-nucleotidase (CD73) regulation by hypoxia-inducible factor-1 mediates permeability changes in intestinal epithelia. *The Journal of Clinical Investigation*, *110*(7), 993–1002.

Takano, H., Bolli, R., Black, R. G., Jr., Kodani, E., Tang, X. L., Yang, Z., et al. (2001). A(1) or A(3) adenosine receptors induce late preconditioning against infarction in conscious rabbits by different mechanisms. *Circulation Research*, *88*(5), 520–528.

Tambuwala, M. M., Cummins, E. P., Lenihan, C. R., Kiss, J., Stauch, M., Scholz, C. C., et al. (2010). Loss of prolyl hydroxylase-1 protects against colitis through reduced epithelial cell apoptosis and increased barrier function. *Gastroenterology*, *139*(6), 2093–2101.

Taylor, C. T. (2008). Interdependent roles for hypoxia inducible factor and nuclear factor-kappaB in hypoxic inflammation. *The Journal of Physiology*, *586*(Pt. 17), 4055–4059.

Taylor, C. T., & Colgan, S. P. (2007). Hypoxia and gastrointestinal disease. *Journal of Molecular Medicine*, *85*(12), 1295–1300.

Thiel, M., Chouker, A., Ohta, A., Jackson, E., Caldwell, C., Smith, P., et al. (2005). Oxygenation inhibits the physiological tissue-protecting mechanism and thereby exacerbates acute inflammatory lung injury. *PLoS Biology*, *3*(6), e174.

Thompson, L. F., Eltzschig, H. K., Ibla, J. C., Van De Wiele, C. J., Resta, R., Morote-Garcia, J. C., et al. (2004). Crucial role for ecto-5′-nucleotidase (CD73) in vascular leakage during hypoxia. *The Journal of Experimental Medicine*, *200*(11), 1395–1405.

Turnage, R. H., Guice, K. S., & Oldham, K. T. (1994). The effects of hypovolemia on multiple organ injury following intestinal reperfusion. *Shock*, *1*(6), 408–412.

Van Linden, A., & Eltzschig, H. K. (2007). Role of pulmonary adenosine during hypoxia: Extracellular generation, signaling and metabolism by surface adenosine deaminase/CD26. *Expert Opinion on Biological Therapy*, *7*(9), 1437–1447.

Walmsley, S. R., Print, C., Farahi, N., Peyssonnaux, C., Johnson, R. S., Cramer, T., et al. (2005). Hypoxia-induced neutrophil survival is mediated by *HIF-1*{alpha}-dependent NF-{kappa}B activity. *The Journal of Experimental Medicine*, *201*(1), 105–115.

Walsh, M. C., Bourcier, T., Takahashi, K., Shi, L., Busche, M. N., Rother, R. P., et al. (2005). Mannose-binding lectin is a regulator of inflammation that accompanies myocardial ischemia and reperfusion injury. *Journal of Immunology*, *175*(1), 541–546.

Wang, G., Jiang, B., Rue, E., & Semenza, G. (1995). Hypoxia-inducible factor 1 is a basic-helix-loop-helix-PAS heterodimer regulated by cellular O_2 tension. *Proceedings of the National Academy of Sciences of the United States of America*, *92*(12), 5510–5514.

Wang, E. C. Y., Lee, J.-M., Ruiz, W. G., Balestreire, E. M., von Bodungen, M., Barrick, S., et al. (2005). ATP and purinergic receptor-dependent membrane traffic in bladder umbrella cells. *The Journal of Clinical Investigation*, *115*(9), 2412–2422.

Ware, L. B., & Matthay, M. A. (2000). The acute respiratory distress syndrome. *The New England Journal of Medicine*, *342*(18), 1334–1349.

Webb, A. R. (2000). Capillary leak. Pathogenesis and treatment. *Minerva Anestesiologica*, *66*(5), 255–263.

Weissmuller, T., Eltzschig, H. K., & Colgan, S. P. (2005). Dynamic purine signaling and metabolism during neutrophil-endothelial interactions. *Purinergic Signalling*, *1*(3), 229–239.

Wenger, R. H. (2002). Cellular adaptation to hypoxia: O_2-sensing protein hydroxylases, hypoxia-inducible transcription factors, and O_2-regulated gene expression. *The FASEB Journal*, *16*(10), 1151–1162.

Yalavarthy, R., Edelstein, C. L., & Teitelbaum, I. (2007). Acute renal failure and chronic kidney disease following liver transplantation. *Hemodialysis International*, *11*(Suppl. 3), S7–S12.

Yan, S. F., Ogawa, S., Stern, D. M., & Pinsky, D. J. (1997). Hypoxia-induced modulation of endothelial cell properties: Regulation of barrier function and expression of interleukin-6. *Kidney International*, *51*(2), 419–425.

Yang, J. N., Chen, J. F., & Fredholm, B. B. (2009). Physiological roles of A1 and A2A adenosine receptors in regulating heart rate, body temperature, and locomotion as revealed using knockout mice and caffeine. *American Journal of Physiology. Heart and Circulatory Physiology*, *296*(4), H1141–H1149.

Yang, Z., Day, Y. J., Toufektsian, M. C., Ramos, S. I., Marshall, M., Wang, X. Q., et al. (2005). Infarct-sparing effect of A2A-adenosine receptor activation is due primarily to its action on lymphocytes. *Circulation*, *111*(17), 2190–2197.

Yang, D., Koupenova, M., McCrann, D. J., Kopeikina, K. J., Kagan, H. M., Schreiber, B. M., et al. (2008). The A2b adenosine receptor protects against vascular injury. *Proceedings of the National Academy of Sciences of the United States of America*, *105*(2), 792–796.

Yang, D., Zhang, Y., Nguyen, H. G., Koupenova, M., Chauhan, A. K., Makitalo, M., et al. (2006). The A2B adenosine receptor protects against inflammation and excessive vascular adhesion. *The Journal of Clinical Investigation*, *116*(7), 1913–1923.

Yu, A. Y., Shimoda, L. A., Iyer, N. V., Huso, D. L., Sun, X., McWilliams, R., et al. (1999). Impaired physiological responses to chronic hypoxia in mice partially deficient for hypoxia-inducible factor 1{alpha}. *The Journal of Clinical Investigation*, *103*(5), 691–696.

Zheng, W., Kuhlicke, J., Jackel, K., Eltzschig, H. K., Singh, A., Sjoblom, M., et al. (2009). Hypoxia inducible factor-1 (*HIF-1*)-mediated repression of cystic fibrosis transmembrane conductance regulator (CFTR) in the intestinal epithelium. *The FASEB Journal*, *23*(1), 204–213.

Zhong, H., Shlykov, S. G., Molina, J. G., Sanborn, B. M., Jacobson, M. A., Tilley, S. L., et al. (2003). Activation of murine lung mast cells by the adenosine A3 receptor. *Journal of Immunology*, *171*(1), 338–345.

Zhou, Y., Mohsenin, A., Morschl, E., Young, H. W., Molina, J. G., Ma, W., et al. (2009). Enhanced airway inflammation and remodeling in adenosine deaminase-deficient mice lacking the A2B adenosine receptor. *Journal of Immunology*, *182*(12), 8037–8046.

Zimmermann, H. (2000). Extracellular metabolism of ATP and other nucleotides. *Naunyn-Schmiedeberg's Archives of Pharmacology*, *362*(4–5), 299–309.

Kenneth A. Jacobson*, Zhan-Guo Gao*, Anikó Göblyös[†], and Adriaan P. IJzerman[†]

*Molecular Recognition Section, Laboratory of Bioorganic Chemistry, National Institute of Diabetes and Digestive and Kidney Diseases, National Institutes of Health, Bethesda, Maryland, USA

[†]Division of Medicinal Chemistry, Leiden/Amsterdam Center for Drug Research, Leiden University, Leiden, The Netherlands

Allosteric Modulation of Purine and Pyrimidine Receptors

Abstract

Among the purine and pyrimidine receptors, the discovery of small molecular allosteric modulators has been most highly advanced for the A_1 and A_3 adenosine receptors (ARs). These AR modulators have allosteric effects that are structurally separated from the orthosteric effects in SAR studies. The benzoylthiophene derivatives tend to act as allosteric agonists as well as selective positive allosteric modulators (PAMs) of the A_1 AR. A 2-amino-3-aroylthiophene derivative T-62 has been under development as a PAM of the A_1 AR for the treatment of chronic pain. Several structurally distinct classes of allosteric modulators of the human A_3 AR have been reported: 3-(2-pyridinyl)isoquinolines, 2,4-disubstituted quinolines, 1H-imidazo-[4,5-*c*]quinolin-4-amines, endocannabinoid 2-arachidonylglycerol, and the food dye Brilliant Black BN. Site-directed mutagenesis of A_1 and A_3 ARs has identified residues associated with the allosteric effect, distinct from those that affect orthosteric binding. A few small molecular allosteric modulators have been reported for several of the P2X ligand-gated ion channels and the G protein-coupled P2Y receptor nucleotides. Metal ion modulation of the P2X receptors has been extensively explored. The allosteric approach to modulation of purine and pyrimidine receptors looks promising for development of drugs that are event and site specific in action.

Advances in Pharmacology, Volume 61

1054-3589/11 $35.00
10.1016/B978-0-12-385526-8.00007-2

I. Introduction

Drug design for cell surface receptors has largely focused on either competitive agonist or antagonist ligands that occupy the principal (orthosteric) binding sites of these receptors, that is, the sites at which the native ligands for these receptors act. Recently, interest has grown in the modulation of clinically validated receptors by small molecules that act at allosteric (from the Greek *allos*, "other," and *stereos*, "space") sites, that is, those sites for binding on the receptor protein that are not identical with the orthosteric binding sites of the native ligands. Changeux and colleagues first introduced the concept of allosteric modulation of receptor action for the nicotinic cholinergic receptors, that is, channels for cations that are activated by the neurotransmitter acetylcholine (Changeux, 2010). Now, the approach of allosteric modulation of the action of a native agonist has grown in importance for the ligand design and pharmacology of both G protein-coupled receptors (GPCRs) and ligand-gated ion channels (LGICs). For example, two such agents that are already in clinical use for GPCR modulation are Cinacalcet (Harrington & Fotsch, 2007) and Maraviroc (Yang & Rotstein, 2010), which act as an allosteric agonist of the calcium sensing receptor and inhibitor of chemokine coreceptors required for HIV entry, respectively. The widely used benzodiazepines allosterically enhance the activation of $GABA_A$ chloride channels.

In the area of GPCRs, in particular, many therapeutic agents currently in use act as orthosteric agonists and antagonists, but there is a need to expand the ways in which GPCRs and other cell surface receptors may be modulated. Two major types of allosteric modulators for GPCRs have been defined: positive allosteric modulators (PAMs), which increase the affinity, potency and/or efficacy of the agonist, and negative allosteric modulators (NAMs), which may decrease the above parameters (Christopoulos, 2002). Further divisions based on pharmacological parameters are also relevant. For example, a PAM may only modulate the action of the native agonists, that is, magnify or enhance the effect of an endogenous molecular signal. Alternatively, it might have its own agonist action by binding to a site different from the binding site of the native agonist and would therefore be classified as an allosteric agonist.

There are several points of justification for studying allosteric modulation of cell surface receptors. First of all, the receptors are often widely distributed throughout the body—leading to problems of side effects when an orthosteric agonist is administered therapeutically. The action of a PAM would be expected to be more tissue- and event specific than the action of a stable, exogenously administered orthosteric agonist, which would circulate throughout the body (Conn et al., 2009). Second, allosteric modulators may have an inherently greater chance of achieving subtype selectivity than the orthosteric ligands. In the case of some GPCRs, such as muscarinic acetylcholine receptors, the design of subtype-selective orthosteric agonists and antagonists has progressed very slowly until recently, largely because the amino acid residues within the orthosteric binding site are highly conserved if not identical across the subtypes. It is thought that greater subtype selectivity could be obtained by targeting other regions of the receptors, such as the extracellular loops (ELs) in Class A GPCRs where there is more structural variation than in transmembrane domains (TMs). In fact, this approach has resulted in muscarinic receptor modulators of great selectivity (Conn et al., 2009). Another possible advantage of PAMs over orthosteric agonists is the possibility to alter the spectrum of second messenger effects or produce a bias toward a particular pathway based on conformational variation of the receptor in its activated state (Stewart et al., 2010).

Finally, an additional potential advantage of allosteric modulators is the preferential activation of receptors in areas of low receptor density or low receptor reserve. A full agonist or a partial agonist will always activate areas where receptor density is highest. In contrast, it may be possible to have preferential action on areas of lower receptor density with PAMs, as was illustrated for the A_1 adenosine receptor (AR) (Childers et al., 2005).

Assay methods used to identify allosteric modulators of the ARs have included both radioligand binding and functional assays. Initially, screening typically has consisted of detecting an increase in the level of binding of radioligand to membranes expressing a given receptor subtype. Functional assays using an EC_{50} or EC_{80} concentration of an orthosteric ligand are used to screen for modulators, particularly in industry.

One commonly used screen is a single point dissociation assay with the goal to detect a change in k_{off} as indicated by a change in the remaining radioligand after a fixed time of dissociation. This assay differs from a full dissociation kinetic experiment in the number of time-points included. Thus, the index of modulator activity known as the allosteric enhancer (AE) score, also called the *K*-score (Ferguson et al., 2008), is related to the percentage of specifically bound agonist remaining after a fix time of dissociation (e.g., 10 min). A more labor-intensive binding method has been to look for alteration of the dissociation rate of a radioligand. Thus, a decreased off-rate of an agonist ligand in the presence of a fixed concentration of the candidate modulator may indicate a PAM, and conversely, an increased off-rate may indicate a NAM. Although these preliminary indicators are often validated in subsequent analysis, especially for the A_1 AR, the feature of PAM or NAM depends on the overall effect of the modulator on the equilibrium binding of the probe, which is a ratio of the effect of the modulator on the dissociation (k_{off}) and association (k_{on}) rates of the tracer. Thus, the phenomenon of allosteric modulation of a receptor cannot be conclusively determined by binding alone, but rather at least one functional assay needs to be carried out in general. Such functional assays may consist of the enhancement (for a PAM) or reduction (for a NAM) of binding of a radiolabeled guanine nucleotide ([^{35}S]GTPγS) in response to a known receptor agonist, or effects on agonist-induced changes in adenylate cyclase or other second messenger systems. The [^{35}S]GTPγS assay measures receptor-mediated G protein activation.

For allosteric modulation, in general, the experimental conditions may have a far greater influence on the outcome and the conclusions than is normally encountered in routine screening. Different conclusions may be reached for the same PAM in different binding assays, for example, membrane versus whole cell studies, guanine nucleotide versus agonist radioligand, and cloned receptors (and their expression system) versus endogenous receptors. It is also worth noting that *in vivo* effects of PAMs are not necessarily predicted by cell-based assays. For example, tissue selectivities *in vitro* for AEs have been described (Leung et al., 1995). However, the source of these differences, in receptor coupling or the tissue environment, remains to be determined. Also, to note is that many PAMs are poorly soluble in water and require DMSO for a stock solution, of which the stability and solubility when diluted may affect the results.

Another important concept is probe dependency, in which the allosteric properties of a particular PAM or NAM may be highly variable depending on which orthosteric ligand is used in the experiment. Slight differences in the region or mode of binding of orthosteric ligands within the same receptor are likely responsible for this phenomenon.

This review focuses on the purine and pyrimidine receptors: four subtypes of GPCRs that respond to extracellular adenosine (ARs), eight subtypes of GPCRs that respond to extracellular purine and pyrimidine nucleotides (P2Y receptors), and seven subtypes of cation-permeable LGICs that respond to extracellular adenine nucleotides (P2X receptors).

II. AR Modulators

The subtypes of ARs are numbered A_1, A_{2A}, A_{2B}, and A_3. Activation of the A_1 and A_3 ARs leads to the inhibition of adenylate cyclase, while the other two subtypes are stimulatory (Fredholm et al., 2011). Endogenous adenosine acts as a mediator in numerous organs and tissues to protect against damaging effects of stress, such as in ischemia. Many novel drug concepts have been proposed based on administration of selective AR agonists and antagonists (Jacobson & Gao, 2006). Fortunately, the lack of selective ligands that has plagued the muscarinic acetylcholine receptor field is not a limitation for the ARs because both agonist and antagonist ligands that are thousands of fold selective have been reported for most of the subtypes. However, the ubiquity of the ARs throughout the body does present a problem of lack of selectivity for even highly selective agonists. Native adenosine is rapidly degraded and does not migrate beyond the target site. In stress situations, the extracellular concentration of adenosine is elevated locally, which avoids side effects in other organs. The action of a stable synthetic PAM of the ARs may be more selective than an orthosteric agonist because it boosts the effect of local adenosine elevation that occurs in response to a physiological need (i.e., stress) to an organ or tissue. Thus, allosteric enhancement of native adenosine in activating the ARs is a particularly attractive option for therapeutics—leading to site- and event-specific action. A similar rationale could be presented for allosteric modulation of receptors for nucleotides.

A second advantage of allosteric modulators of ARs concerns pharmacokinetics, giving such modulators a clear benefit in the central nervous system. Adenosine agonists, which are nearly always nucleoside derivatives and thus have a highly hydrophilic (i.e., ribose) region, tend not to readily penetrate the blood–brain barrier (BBB). The brain entry of such nucleoside derivatives is typically only 1–2% of free passage across the BBB. Thus, for induction of AR activation in the brain, where adenosine levels can be greatly elevated in response to stress or hypoxia, a freely penetrating PAM (i.e., belonging to a different chemical class that might pass the BBB more easily) would be more effective than a nucleoside agonist.

The structure–activity relationship (SAR) of small organic molecules as allosteric modulators is well explored for A_1 and A_3 ARs, with isolated reports of examples for allosteric or "noncompetitive" (i.e., potentially allosteric) ligands for other AR subtypes (Gao et al., 2005; Göblyös & IJzerman, 2009). Most of the examples of allosteric modulators of ARs are PAMs. The effects of metal ions as allosteric modulators of the ARs have also been reported (Gao & IJzerman, 2000).

Some classes of GPCR allosteric modulators, for example, amiloride analogues, affect multiple members of the AR family (Gao et al., 2003b). Amiloride analogues have been characterized as allosteric modulators of the A_{2A} AR, although it is recognized that they also interact with other GPCRs, such as dopamine receptors. Thus, the amiloride analogues lack specificity due to interaction with many other protein sites, including other AR subtypes (Gao & IJzerman, 2000). At the A_{2A} AR, 5-(*N*,*N*-dialkyl)amiloride derivatives containing a cyclic 5-(*N*,*N*-hexamethylene)amiloride group such as **12** (HMA, Fig. 1B) increase the dissociation rate of antagonist radioligand. Such amiloride analogues also allosterically modulate action of ligands at both A_1 and A_3 ARs (Gao et al., 2003b). At the A_1 AR, their behavior is similar to the A_{2A} AR. At the A_3 AR, they additionally decrease the dissociation rate of agonist radioligand. They also compete for orthosteric binding at these three subtypes. Thus, amiloride analogues are not useful as selective allosteric pharmacological probes of specific AR subtypes.

Other nonselective AR allosteric modulators include SCH-202676 (*N*-(2,3-diphenyl-1,2,4-thiadiazol-5-(2H)-ylidene)methanamine), which affects a wide range of structurally unrelated GPCRs and has highly divergent effects on purine receptors (Gao et al., 2004b; van den Nieuwendijk et al., 2004). However, later studies suggested that SCH-202676 is a chemical modifying agent, rather than a true allosteric modulator (Göblyös et al., 2005; Lewandowicz et al., 2006). Another nonselective agent, Brilliant black G, was shown to decrease antagonist affinity at both A_1 and A_3 ARs (May et al., 2010a).

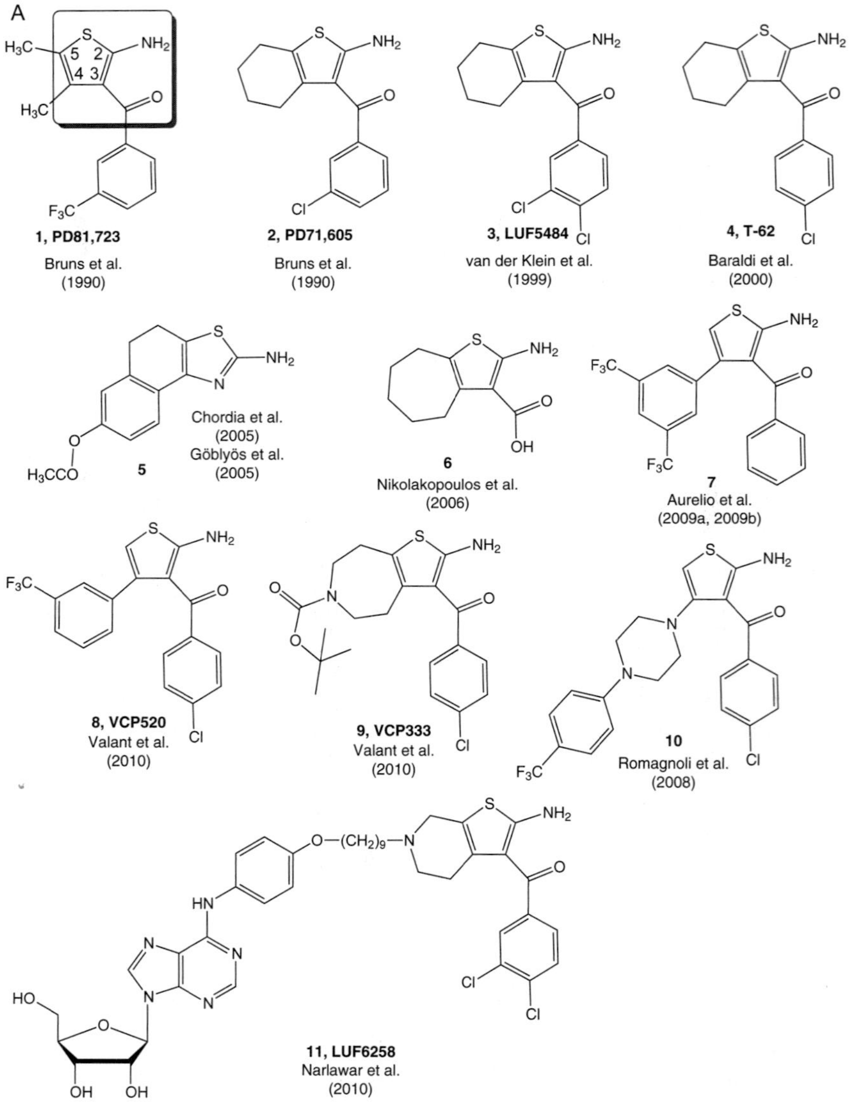

FIGURE 1 Continued

B

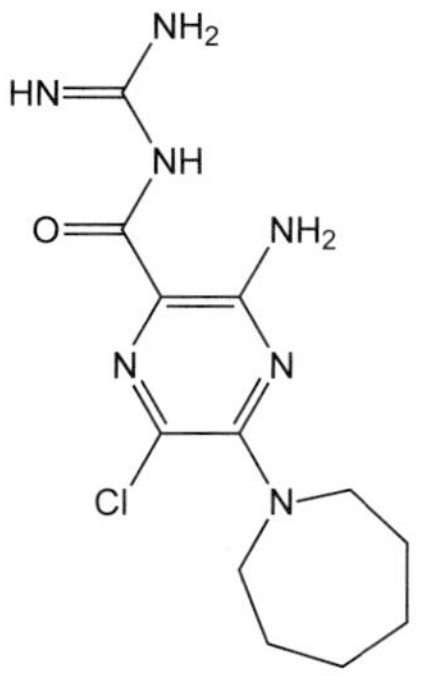

12, 5-(*N*,*N*-Hexamethylene)amiloride
Gao et al. (2003a,b)

C

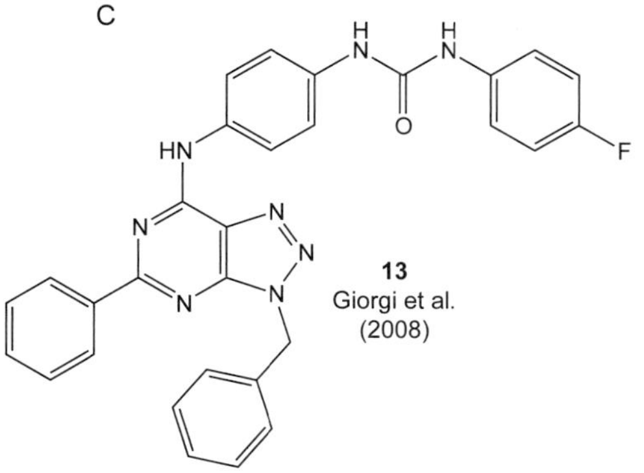

13
Giorgi et al.
(2008)

D

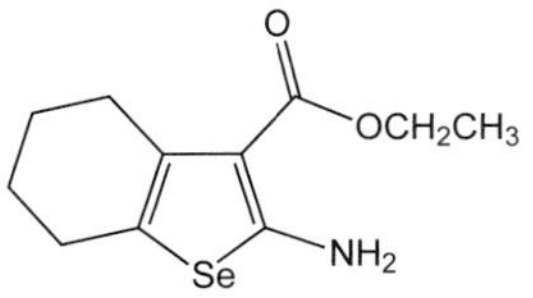

14
Aumann et al.
(2007)

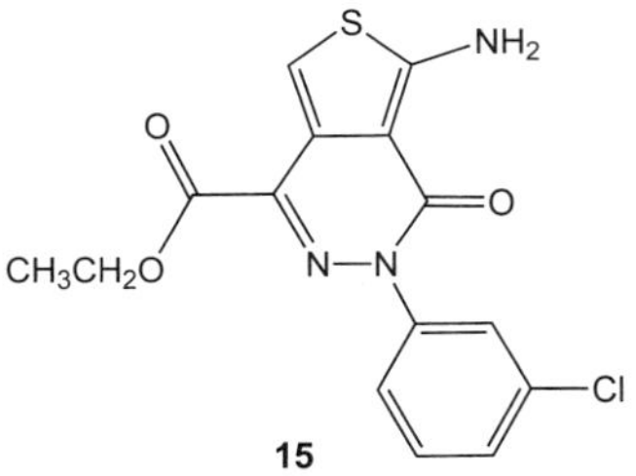

15
Ferguson et al.
(2008)

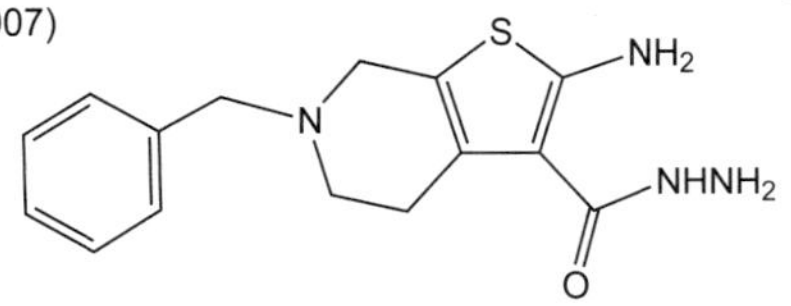

16
Aurelio et al.
(2009a,b)

FIGURE 1 Allosteric modulators of the A_1 AR, including benzoylthiophene derivatives (A) and an amiloride derivative lacking subtype selectivity (B), and of the A_{2A} AR (C). The minimal, principal pharmacophore of the class of aroylthiophenes as A_1 AR PAMs is shown in a box on PD81,723 **2**. Compound **11** is a bitopic modulator, containing an agonist (6-phenyl substituted adenosine) moiety. In (D), there are recently published 2-amino-3-substituted thiophene derivatives as allosteric modulators for A_1 adenosine receptors. Numbering of ring substitutions is shown for compound **1**.

A. SAR of A_1 AR Allosteric Modulators

1. Benzoylthiophenes and Related Allosteric Modulators of the A_1 AR

The benzoylthiophenes were the first A_1 AR allosteric modulators to be identified (Bruns and Fergus, 1990). They were identified as PAMs of the A_1 AR initially by increasing the level of agonist radioligand ([^{3}H]N^6-cyclohexyladenosine (CHA)) bound to the receptor in rat brain membranes. Among the various analogues coming out of a chemical library screen, the prototypical benzoylthiophene determined to act as an A_1 AR allosteric modulator is PD81,723 **1** (Fig. 1A). This compound is still used extensively as a pharmacological standard.

The structure of **1** has been extensively modified in subsequent studies, and the SAR of benzoylthiophene derivatives as PAMs has been documented (Romagnoli et al., 2010). Many analogues have been prepared and found to have comparable or more favorable activity as allosteric modulators (Baraldi et al., 2003, 2004, 2007; Gao et al., 2005; Kourounakis et al., 2000; Romagnoli, 2006). The 2-amino-3-carbonyl thiophene moiety is required as a minimal pharmacophore (shown in box). The aroyl group can be substituted with other phenyl and heteroaromatic groups. The 4,5-alkyl substituents of the thiophene ring may be cyclized, with cycloalkyl chains. Cyclohexyl groups as in **2–4** have been incorporated in various analogues (van der Klein et al., 1999), although cycloheptyl rings as in **6** are also tolerated (Nikolakopoulos et al., 2006). In certain cases, the aroyl group can be simplified in the form of a carboxylic acid as in **6**. The synthesis of novel benzoylthiophene analogues VCP520 **8** and VCP333 **9** has been described (Valant et al., 2010).

An atypical structural class, 2-aminothiazoles including **5**, was reported as PAMs of the A_1 AR, but their allosteric activity is not always evident and appears to be limited to specific salt forms (Chordia et al., 2005, Göblyös et al., 2005). The 3-piperazinyl derivative **10** was found to act as a PAM of the A_1 AR (Romagnoli et al., 2008).

An allosteric modulator of the A_1 AR, the benzoylthiophene analogue T-62 **4** (Baraldi et al., 2000), has been in clinical trials for treating chronic pain (Kiesman et al., 2009). T-62 is active in the central nervous system and produces a beneficial effect in several *in vivo* pain models. An antinociceptive effect has been studied following the intrathecal administration of T-62 in carrageenan-inflamed rats. This allosteric AR modulation reduced hypersensitivity following peripheral inflammation by a central mechanism (Li et al., 2003). In investigation of the mode of action of T-62 in brain slices, it was found to selectively enhance the G_i protein coupling of the A_1 AR (Childers et al., 2005). T-62 also raises the basal levels of [^{35}S]GTPγS binding in brain slices and therefore acts as an allosteric agonist. T-62 has been radiolabeled, and its binding properties are indicative of allosteric binding (Romagnoli et al., 2006).

A new 3,5-di(trifluoromethyl)benzoylthiophene derivative **7** from the Scammells lab was shown to act as an allo agonist of the A_1AR (Aurelio et al., 2009a,b). Activation of the extracellular-signal-regulated kinase (ERK) phosphorylation pathway required higher concentrations of the derivative than for G protein modulation (based on [^{35}S]GTPγS binding), suggesting the possibility of signaling bias, pending clarification in additional studies.

Valant et al. (2010) used a combination of membrane-based and intact-cell radioligand binding, multiple signaling assays, and a native tissue bioassay to characterize the allosteric interaction between benzoylthiophenes and various radiolabeled agonists and antagonists of the A_1 AR. The findings were consistent with a ternary complex model involving binding of the benzoylthiophene modulator at a single extracellular allosteric site. As noted previously, the benzoylthiophenes can also serve as allosteric agonists, and the consequent signaling pathways are biased with respect to signaling from a standard orthosteric agonist. However, when allowed access to the intracellular milieu, the benzoylthiophenes have a secondary action as direct G protein inhibitors, which was also seen after stimulation of another GPCR. Thus, there are multiple modes of interaction with the A_1 AR, which should be taken into account in pharmacological experiments.

Interestingly, 2-aminoselenophene-3-carboxylates also proved to be PAMs of the A_1 AR. Compound **14** (Fig. 1D) had an AE score of 64%, and it was significantly more potent as a PAM of the A_1 AR than its thiophene analogue. However, it is noteworthy that this compound is not stable under mildly acidic conditions (Aumann et al., 2007).

In a pharmacophore-based library screen, ethyl 5-amino-3-(4-*tert*-butylphenyl)-4-oxo-3,4-dihydrothieno[3,4-*d*]pyridazine-1-carboxylate was identified as a new allosteric modulator of the A_1 AR. On the basis of this lead compound, various derivatives were prepared and evaluated for activity at the human (h) A_1 AR. However, these compounds turned out to be a new class of hA_1 AR antagonists that can also recognize the receptor's allosteric site with lower potency. Among them, compound **15** proved to be the most potent (Ferguson et al., 2008). Similar results were found for a series of 2-amino-4,5,6,7-tetrahydrothieno[2,3-*c*]pyridines, for example, compound **16** with an AE score of 83% (Aurelio et al., 2009a, 2009b). These findings emphasized the caveat that changes in orthosteric ligand dissociation kinetics induced by a test compound cannot guarantee that the predominant pharmacological effect will be allosteric. Compounds that act allosterically and/or orthosterically at the hA_1 AR often have close structural resemblance, which suggests that the allosteric site on the A_1 AR is closer or more similar to the orthosteric site on the A_1 AR than in other Class A GPCRs (see Fig. 1D). Novel conformationally rigid analogues of the benzoylthiophenes were screened at the hA_1 AR, and (2-aminoindeno[2,1-*b*]thiophen-3-yl)(phenyl)methanones with *para*-chloro substitution displayed considerable PAM activity (Aurelio et al., 2010).

2. Bitopic Allosteric Modulators of the A_1 AR

The concept of bitopic allosteric modulators, which bridge orthosteric and allosteric binding regions on a given GPCR protein, has been introduced (Mohr et al., 2010; Valant et al., 2009). IJzerman and coworkers recently designed such a bitopic ligand for the A_1 AR by tethering pharmacophores using spacers of varying lengths (Narlawar et al., 2010). The bivalent ligand N^6-[2-amino-3-(3,4-dichlorobenzoyl)-4,5,6,7-tetrahydrothieno[2,3-*c*]pyridin-6-yl-9-nonyloxy-4-phenyl]-adenosine **11** (LUF6258) with a 9 carbon atom spacer did not show significant changes in affinity or potency in the presence of PD81,723, indicating that this ligand bridged both sites on the A_1 AR. Further, this bitopic ligand displayed an increase in efficacy, but not potency, compared to the parent, monovalent agonist. Molecular modeling and docking of this ligand suggested that the allosteric site could be located both in proximity to the orthosteric site and in the vicinity of EL2.

B. A_{2A} AR Allosteric Modulators

Allosteric modulation of the G_s-coupled A_{2A} AR is much less well advanced than allosteric modulation of the A_1 AR. In an early abstract, Bruns et al. reported that a benzopyran-2-one derivative named PD120,918 (not shown) was an enhancer of agonist radioligand ([^{3}H]NECA in the presence of an A_1 AR agonist N^6-cyclopentyladenosine (CPA)—prior to the development of A_{2A} AR-selective radioligands) binding at the A_{2A} AR in rat striatal membranes (Bruns and Lu, 1989; Gao et al., 2005), but this lead was not subsequently explored.

Amiloride and its analogues were demonstrated to be allosteric inhibitors for the A_{2A} AR too. Among the derivatives tested HMA proved to be the most potent allosteric inhibitor (Fig. 1B). Amiloride analogues increased the dissociation rate of the antagonist [^{3}H]ZM 241385 from the A_{2A} AR; however, they did not show any effect on the dissociation rate of the agonist [^{3}H]CGS21680 (Gao & IJzerman, 2000). Sodium ions (high concentrations of NaCl) rather decreased the dissociation rate of the antagonist [^{3}H] ZM241385 from the A_{2A} AR in a concentration-dependent manner.

Recently, a 2-phenyl-9-benzyl-8-azaadenine derivative **13** (Fig. 1C) was reported to be a PAM of both agonist and antagonist radioligand binding at the A_{2A} AR, and it increased the potency of an A_{2A} AR agonist to induced relaxation of rat aortic rings (Giorgi et al., 2008).

Addex Pharmaceuticals is developing PAMs of the A_{2A} AR for treatment of inflammatory diseases, such as psoriasis and osteoarthritis, but so far the compounds remain in the preclinical phase (http://www.addexpharma.com/press-releases/press-release-details/article/addex-rd-day-highlights-broadened-therapeutic-potential-of-allosteric-modulation-platform-to-includ/).

A cholesterol-sequestering cyclodextrin molecule, enhanced adenosine A_{2A} AR-activated transepithelial short circuit current from the basolateral side of colonic epithelial cells (Lam et al., 2009). Thus, cholesterol content modulates agonist-selective signaling at this receptor.

C. A_{2B} AR Allosteric Modulators

Allosteric modulators for the A_{2B} AR have not yet been reported. Until recently, there was no suitable radioligand for this receptor subtype that could be obtained commercially. However, with the advent of the radiolabeled antagonist [^{3}H]MRS1754 and other radioligands such as [^{3}H] PSB603 (Borrmann et al., 2009), these studies are now feasible. However, a radiolabeled agonist is not available, and therefore, identification of PAMs of agonist binding is still hampered.

D. SAR of A_3 AR Allosteric Modulators

Lead compounds for allosteric modulators of the A_3 AR were discovered in the course of screening structurally diverse chemical libraries in binding at this subtype (Gao et al., 2001, 2002). Just as in the initial discovery of PD81,723 **1** as a PAM at the A_1 AR, certain lead molecules were found to increase the level of agonist radioligand ([^{125}I]AB-MECA) binding at this other G_i-coupled subtype. Classes of heterocyclic ligands that became prototypical PAMs of the A_3 AR include 3-(2-pyridinyl)isoquinolines (e.g., VUF5455 **17**) and 1H-imidazo-[4,5-*c*]quinolin-4-amines (e.g., DU124183 **18** and LUF6000 **21**).

1. 3-(2-Pyridinyl)Isoquinoline Derivatives as Allosteric Modulators of the A_3 AR

It is to be noted that some members of the same chemical classes identified as PAMs are pure antagonists of radioligand binding at the A_3 AR, and only certain members of these groups of heterocycles were found to decrease the rate of agonist dissociation in addition to displacing the radioligand. Thus, the interaction of these compound classes with the A_3 AR is complex. For example, IJzerman and coworkers reported the pyridinylisoquinoline derivative VUF5455 **17** (*N*-(2-methoxyphenyl)-*N′*-[2-(3-pyrindinyl)-4-quinazolinyl]-urea) to be a selective antagonist of the A_3AR with a potent K_i value of 4 nM. Other members of this chemical series were found to be less potent antagonists of the A_3 AR, leading to a distinct SAR in the inhibition of radioligand binding at the orthosteric site.

The effects of the reference A_3 AR agonist Cl-IB-MECA on forskolin-induced cAMP formation were significantly enhanced by several 3-(2-pyridinyl)isoquinoline derivatives, previously identified as potential antagonists for the hA_3 AR (van Muijlwijk-Koezen et al., 1998). VUF5455 was shown to be selective for the agonistic state of the A_3 AR. In competitive binding studies on cloned hA_3 AR, VUF5455 displayed modest affinity as an orthosteric antagonist (K_i=1680 nM). Replacement of the 7-methyl group of VUF5455 by H (VUF8504) had no influence on the allosteric activity but increased the A_3 AR affinity nearly 100-fold (K_i=17.3 nM). Exchanging the 4′-methoxy group of VUF8504 by methyl (VUF8502) or H (VUF8507) lowered the A_3 AR affinity (K_i values of 96 and 204 nM) without affecting the allosteric activity. The corresponding imino instead of carboxamido analogues displayed moderate A_3 AR affinity (K_i values 300–700 nM) but were devoid of allosteric properties. Thus, although VUF5455 is not devoid of A_3 AR antagonistic activity, the compound might be used as a lead for the design of pure PAMs of the A_3 AR.

Subsequently, various pyridinylisoquinolines were also found to exhibit allosteric properties, mainly enhancement but in some cases inhibition, with respect to the binding of radiolabeled A_3 AR agonist (Fig. 2). The SAR of pyridinylisoquinolines in orthosteric binding to the A_3 AR is distinct from SAR in allosteric enhancement. It was proposed that the displacement of orthosteric radioligand at the A_3 AR was competitive, but this assumption has not been conclusively established. As many structural homologues of the pyridinylisoquinolines were already available, it was feasible to characterize their profile as PAMs of A_3 AR.

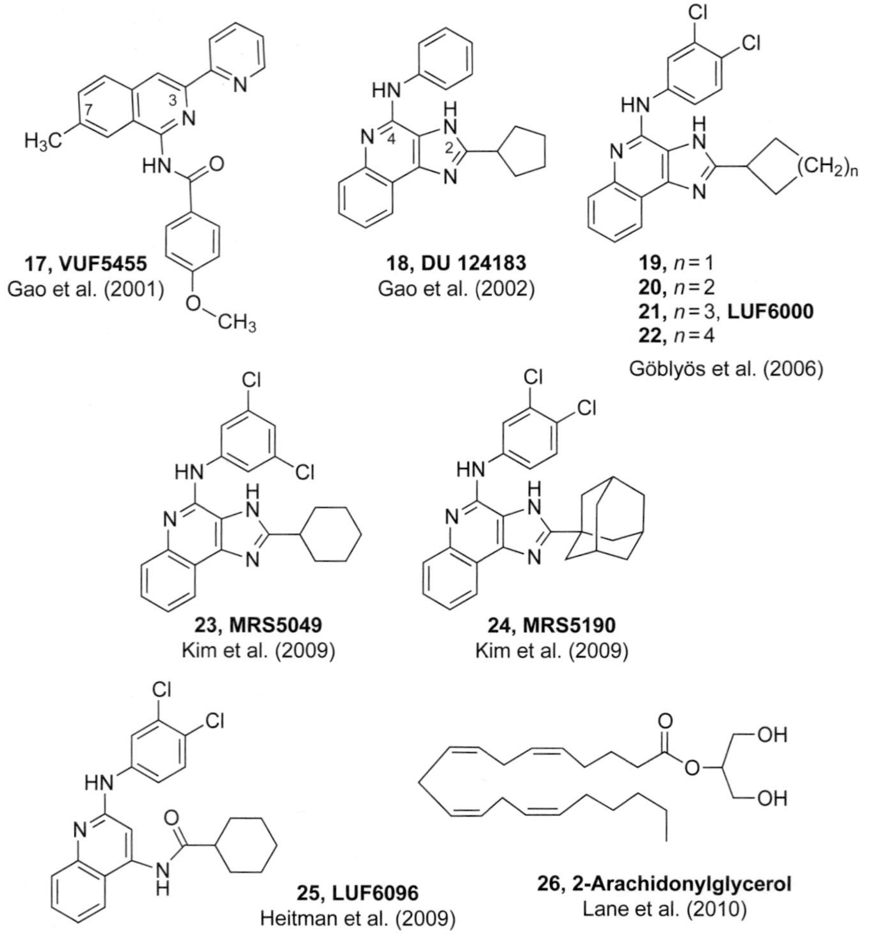

17, VUF5455
Gao et al. (2001)

18, DU 124183
Gao et al. (2002)

19, $n=1$
20, $n=2$
21, $n=3$, **LUF6000**
22, $n=4$
Göblyös et al. (2006)

23, MRS5049
Kim et al. (2009)

24, MRS5190
Kim et al. (2009)

25, LUF6096
Heitman et al. (2009)

26, 2-Arachidonylglycerol
Lane et al. (2010)

FIGURE 2 Structures of pyridinylisoquinoline and imidazoquinolinamine derivatives that act as PAMs of the human A_3 AR and structure of 2-AG, which acts as a NAM of the A_3 AR. Numbering of ring substitutions is shown for compounds **17** and **18**.

In a recent study with single living cells, it was shown that both the A_3 AR enhancer VUF5455 and the A_1 AR enhancer PD81,723 increased the dissociation rate of a fluorescent-tethered nucleoside agonist ABA-X-BY630 from the A_3 AR and decreased it at the A_1 AR (May et al., 2010b). The latter finding was surprising and may have to do with the probe dependency of allosteric modulators. These results in whole cells are in contrast to previous findings in membrane preparations and emphasize the complexity of the interactions involved in allosteric modulation.

2. 1H-Imidazo-[4,5-c]Quinolin-4-Amine Derivatives as Allosteric Modulators of the A_3 AR

Another structural class of AR antagonists that was subsequently found to include PAMs of the A_3 AR is the imidazoquinolinamines (Gao et al., 2002). IJzerman and coworkers originally introduced the imidazoquinolinamines as A_1 AR antagonists (van Galen et al., 1991). In addition to inhibiting the binding of competitive radioligands, the compound DU124183 was found to be a PAM of agonist binding at the A_3 AR. This action was initially shown by a reduction in the off-rate of bound agonist radioligand and then conclusively by a functional enhancement at the A_3 AR.

The SAR of a further series of imidazoquinolinamines as PAMs of the A_3 AR has been explored in detail (Gao et al., 2008b; Göblyös et al., 2006). Allosteric enhancement of the A_3 AR was demonstrated by reducing the dissociation rate of agonist radioligand and by enhancement of maximal guanine nucleotide binding in the presence of the reference agonist Cl-IB-MECA. There was clearly a divergence of the structural requirements for allosteric action and the inhibition of binding of orthosteric radioligands. Modification of the 2 and 4 positions was most useful for demonstrating the difference in SAR. The most favorable groups for allosteric enhancement of the A_3 AR were found to be 2-cyclohexyl and 4-phenylamino. By structural modification in the series, that is, altering the size of the 2-cycloalkyl ring (**19–22**) and by substitution of the 4-phenylamino group, the allosteric effects were enhanced without increasing the orthosteric inhibition. The combination of favorable structural modifications resulted in the prototypical A_3 AR PAM of the imidazoquinolinamine class, LUF6000 **21** (*N*-(3,4-dichlorophenyl)-2-cyclohexyl-1*H*-imidazo[4,5-*c*]quinolin-4-amine). The potency of Cl-IB-MECA in binding to the hA_3 AR was not affected by LUF6000, but the maximal functional effect of the agonist was increased.

Further exploration of steric and electronic effects of substitution at the 2 and 4 positions of a series of imidazoquinolinamines as PAMs of the A_3AR was reported (Kim et al., 2009). The enhancing ability with minimal inhibition of orthosteric radioligand binding was maintained with 3,5-dichloro substitution in MRS5049 **23**. Enhancement was observed by two bridged derivatives containing 2-bicycloalkyl groups as in the adamantyl derivative MRS5190 **24**, indicating that rigid steric bulk is tolerated at the 2 position. Hydrophobicity is also a requirement at that position. Introduction of nitrogen atoms in the 2-cycloalkyl substituents abolished the allosteric enhancement.

Scission of the imidazole ring in the structure of LUF6000 led to a series of 2,4-disubstituted quinolines as PAMs of the A_3 AR (Heitman et al., 2009). The same substitution pattern as in LUF6000 led to the most potent allosteric modulator in the series, LUF6096 **25**, with negligible orthosteric activity on A_1 and A_3 ARs, even less than observed for LUF6000.

It has been shown that adenosine derivatives are highly structure dependent in activating as well as binding to the A_3 AR. It is even possible to convert nucleoside agonists to A_3 AR antagonists. The PAM LUF6000 was found to convert the nucleoside antagonist of the A_3 AR, MRS542, into an A_3 AR agonist, which is an example of a complete reversal of the nature of the action of an antagonist by a PAM of a GPCR. The experimental result is consistent with the prediction from a mathematical model (Gao et al., 2008b). This phenomenon was not observed with nonnucleoside antagonists of the A_3 AR such as MRS1220.

Lane and coworkers recently reported that some endogenous cannabinoid ligands also modulate the A_3 AR (Lane et al., 2010). 2-Arachidonylglycerol **26** (2-AG) was able to inhibit agonist [^{125}I]AB-MECA binding and increase the rate of [^{125}I]AB-MECA dissociation, suggesting that 2-AG acts as a NAM. The presence of A_3 AR in astrocytes and microglia suggests that this finding may be relevant to cerebral ischemia, a pathological condition in which levels of 2-AG are raised.

The synthetic azo food dye Brilliant Black BN (500 μM, tetrasodium (6Z)-4-acetamido-5-oxo-6-[[7-sulfonato-4-(4-sulfonatophenyl)azo-1-naphthyl] hydrazono]naphthalene-1,7-disulfonate) decreased the affinity of certain antagonists acting at the A_1 AR and at the A_3 AR but had no effect on calcium mobilization stimulated by the nonselective AR agonist NECA (May et al., 2010a). This allosteric effect was ascribed to a significant increase in dissociation rate of the antagonist.

E. Hypotheses for Binding Modes of AR Allosteric Modulators

The potential binding modes of AR allosteric modulators have been probed in site-directed mutagenesis studies (Barbhaiya et al., 1996; De Ligt et al., 2005; Gao et al., 2003a; Heitman et al., 2006; Kourounakis et al., 2001). However, the residues studied only yield a partial view of the amino acids involved in the allosteric regulation of ARs. Chimeras of the A_1 AR having A_{2A} AR substitutions (Bhattacharya et al., 2006) have also provided some insight into the structural basis of allosteric modulation. Asp55(2.50) (using in parentheses the Ballesteros numbering for each TM; Ballesteros & Weinstein, 1995) is probably responsible for allosteric regulation of ligand binding by sodium ions and amilorides, but in G14T(1.37) and T277A(7.42) mutant A_1 ARs, PD81,723 loses its enhancing activity with respect to CPA. As the potency of CPA alone is also drastically diminished by these mutations, it is not clear yet whether these two amino acids are also part of the PD 81,723 binding site or not.

Gao and colleagues studied the E13Q(1.39) and H278Y(7.43) mutations of the A_{2A} AR. The authors concluded the two residues Glu13 and His278, which are closely linked spatially, are the most important for agonist recognition and partly responsible for the allosteric regulation by sodium ions (Gao & IJzerman, 2000; Gao et al., 2003a,b).

Mutagenesis of the hA_3 AR has shown that the PAMs imidazoquinolinamine DU124183 and pyridinylisoquinoline VUF5455 lost their allosteric effects upon F182A(5.43) and N274A(7.45) mutation. The D107N(3.49) mutation eliminated the effects of DU124183, but not of VUF5455. Other residues, such as Trp243(6.48) and Asn30(1.50), were modulatory. Asn274 in TM7 was required for allosteric binding of the imidazoquinolinamine but not for maintaining the orthosteric binding site. His95(3.37) and Phe182 (5.43) are important for orthosteric binding of the DU124183 **18** (Gao et al., 2003a). A conserved His272(7.43) residue in TM7 is required for A_3 AR radioligand binding, and therefore, it was not possible to establish the effect of its mutation to Ala on allosteric enhancement. A docking study for a PAM, VUF5455 **17**, in the agonist-occupied hA_3 AR molecular model was reported based on an energetically favorable interaction of this heterocyclic derivative with the outer portions of the receptor—near the ELs (Gao et al., 2003a). This would allow the simultaneous binding of both agonist and PAM to different regions of the receptor protein, as has been shown for allosteric modulation of muscarinic receptors (Conn et al., 2009). The determination of the A_{2A} AR crystal structure (Jaakola et al., 2008) has not yet had an impact on the determination of the allosteric binding site(s) on the ARs. It has also been hypothesized that agonists and PAMs might bind on opposite protomers of homodimeric receptor pairs (Schwartz & Holst, 2006). However, it is not clear if this hypothesis is applicable to the binding of PAMs at the ARs.

III. "Translational" Assessment of a Prototypical PAM of A_1 AR

The characteristics of an allosteric effect depend on the nature of both receptor and orthosteric ligand, on the cellular context and on the pharmacological read-out (Christopoulos & Kenakin, 2002). Interestingly, the PAM PD81,723 has been evaluated in a great number of pharmacological assays, and hence, a translational assessment of its modulatory potency is now possible for the first time. This analysis follows its effect on the signaling cascade of receptor, G protein and second messengers, down to various organ and tissue systems, and finally *in vivo*.

A. Receptor Effects

The modulatory effects of PD81,723 were first discovered in radioligand binding assays by Bruns and colleagues. When three different tritiated agonists, CHA, NECA, and R-PIA, were used, PD81,723 enhanced their binding to the A_1 AR from rat brain tissue in a highly similar way (~40% at 10–30 μM). The binding of the tritiated inverse agonist DPCPX was inhibited, approximately 30% at 10 μM PD81,723 (Bruns et al., 1990). This behavior was replicated in brain tissue from dog and guinea pig (Jarvis et al., 1999). However, adipocyte membranes from all these species, which also express the A_1 AR, appeared insensitive to PD81,723.

Not all of the data in the early reports on benzoylthiophenes were internally consistent. For example, certain data were explained by the antagonist properties for the prototypical PAMs, which were not confirmed in later studies. For example, Childers et al. (2005) did not observe antagonist properties of T-62 in autoradiographic studies. This may relate to the fact that a steady-state increase in agonist binding in the presence of a PAM is not necessarily associated with a measured decrease in k_{off} for the agonist. Similarly, compounds that decreased k_{off} may not necessary show an increase in agonist binding at steady-state or equilibrium conditions.

Allosteric enhancement by PD81,723 was also noticed when radioligand agonist ([^{125}I]ABA) binding to the hA_1 AR stably expressed in Chinese hamster ovary (CHO) cells was studied (Bhattacharya & Linden, 1995; see also Figler et al., 2003). PD81,723 tested at 20 μM caused a threefold increase in the fraction of receptors found in a high-affinity G protein-coupled conformation. When [^{3}H]2-chloro-N^6-cyclopentyladenosine (CCPA), another agonist radioligand, was studied, no significant changes in K_{D} values were noted in the absence or presence of 10 μM PD81,723, not only for stably expressed receptors in CHO cells (Kourounakis et al., 2001) but also in human and rat brain membranes (Baraldi et al., 2004). B_{max} values, however, appeared somewhat increased in the presence of PD81,723 (Baraldi et al., 2003). This latter finding had been reported previously by Kollias-Baker et al. (1997) using [^{3}H]CHA, yet another radiolabeled agonist. In radioligand displacement studies with [^{3}H]DPCPX (Bhattacharya & Linden, 1995), a 2.4-fold increase in the potency of the agonist R-PIA was observed, confirmed in later observations (Kourounakis et al., 2001). This enhanced affinity was easily reconciled with the observation that PD81,723 caused a 1.5-fold increase of the dissociation half-life of [^{125}I]ABA from the receptor; a similar twofold increase was found for COS-7 cell membranes expressing the canine A_1 AR (Mizumura et al., 1996). Bhattacharya and Linden were among the first to suggest that PD81,723 binds to the A_1 AR at a site distinct from the agonist (i.e., orthosteric) binding site and stabilizes agonist-R–G_i complexes. This applied also to studies of an A_1/A_{2A} AR chimera having the third intracellular loop of the A_1 AR replaced by that of the A_{2A} AR (Bhattacharya et al., 2006). On this receptor construct, PD81,723 caused a comparable increase of the dissociation half-life of [^{125}I]ABA from the receptor. Heitman et al. (2006) also studied the effects of PD81,723 on the hA_1 AR. The authors first examined its influence on equilibrium saturation experiments with the radiolabeled inverse agonist [^{3}H]DPCPX. The K_D value of [^{3}H]DPCPX was increased 5.5 times in the presence of 10 μM PD81,723, suggesting that PD81,723 promotes a receptor state that is unfavorable for this inverse agonist. In equilibrium displacement experiments with the agonist CPA, PD81,723 increased the affinity of CPA for the low-affinity state of the receptor by 3.6-fold, while the affinity for the high-affinity state was not altered. A similar finding was observed in another cellular background, that is, COS cells expressing the hA_1 AR (De Ligt et al., 2005). These effects were somewhat different from the findings by Linden and coworkers described above, in which an increase in the fraction of receptors in the high-affinity state was noticed. The effect in the Heitman study was probe dependent; PD81,723 slightly decreased the affinity of LUF5831, a nonribose agonist for the receptor, whereas another ribose-containing agonist (N^6-4-methoxyphenyladenosine) again showed a fourfold increase in affinity (Narlawar et al., 2010). This enhancing effect vanished in the latter study when hybrid (bitopic) agonist ligands were studied, in which the classic adenosine-like agonist was linked and coupled to a PD81,723-like pharmacophore. Using lower concentrations of PD81,723, like 3 μM in a study by Musser et al. (1999), did not cause significant changes in B_{max} or K_D values of radiolabeled agonists and antagonists on rat brain membranes or CHO cells expressing the rat A_1 AR.

B. G Protein Effects

The G protein dependency of some of the receptor interactions of PD81,723 was revealed by a number of research groups. It was first found that the effects of (stable derivatives of) GTP on agonist binding to the A_1 AR were influenced by PD81,723. Bhattacharya and Linden (1995) demonstrated that the PAM caused a 2.2-fold increase in the concentration of GTPγS required to half-maximally uncouple receptor-G protein complexes. In a similar experimental setup (Kollias-Baker et al., 1994a), the IC_{50} values for GppNHp to reduce specific binding of [^{3}H]CHA to guinea pig cardiac membranes increased from 1.5 μM in the absence of PD81,723 (30 μM) to 10 μM in its presence. The effects of PD81,723 were also examined in direct [^{35}S]GTPγS binding experiments. In a recent study, the potency of the agonist N^6-4-methoxyphenyladenosine to stimulate GTPγS binding on cell membranes expressing the hA_1 AR was increased 4.9-fold by 10 μM PD81,723 (Narlawar et al., 2010). Remarkably, Kollias-Baker et al. (1997) showed that PD81,723 also caused a direct stimulation of GTPγS binding to cell membranes expressing the hA_1 AR, that is, in the absence of an added agonist.

This increase in binding was abrogated in the presence of DPCPX and not influenced by the addition of adenosine deaminase, excluding a role for endogenous adenosine in these observations. In yet another approach, Klaasse et al. (2004) studied the behavior of PD81,723 on receptor-$G\alpha_i$ protein fusion products. The effects of 10 μM PD81,723 on ligand binding were rather similar for both the unfused A_1 AR (expressed in G protein-poor COS cells) and the different fusion proteins (mutations in the α-subunit and expressed in the same cell line). In the presence of PD81,723, CPA's affinity for the unfused receptor increased fourfold. The allosteric modulator increased the affinity of CPA for the various fusion proteins to a varying but overall quite similar extent, namely two- to sixfold. The binding of the inverse agonist DPCPX to the fusion products was not much affected by PD81,723. In another "fusion" study, Bhattacharya et al. (2006) examined chimeric constructs between A_1 and A_{2A} ARs. The authors showed that the allosteric effect of PD81,723 was maintained in a construct in which the third intracellular loop of the G_i-coupled A_1 AR was replaced with the analogous sequence of the G_s-coupled A_{2A} AR. PD81,723 increased the potency of CPA to *increase* cAMP accumulation in cells expressing this chimeric receptor with or without pretreatment with pertussis toxin. The results suggest that the recognition site for PD81,723 is on the A_1 AR, and that the enhancer directly stabilizes the receptor in a conformation capable of coupling to G_i or G_s.

C. Second Messengers and Intracellular Pathways

1. cAMP

In intact CHO cells expressing the hA_1 AR, PD81,723 increased the potency of the reference agonist R-PIA to decrease forskolin-stimulated cAMP accumulation by 3.3-fold (Bhattacharya & Linden, 1995). Musser et al. (1999) further studied its effects on adenylate cyclase and cAMP production. The compound inhibited basal adenylyl cyclase (AC) activity as well as forskolin-, cholera toxin-, and pertussis toxin-stimulated AC activity in "empty" CHO cells and CHO cells carrying the rat A_1 AR gene. For instance, basal AC activity was significantly inhibited in both cell lines by a high concentration of PD81,723 (30 μM). In CHO-A_1 cells, half-maximal inhibition of forskolin-stimulated AC occurred at 5 μM PD81,723 compared to 10 μM in CHO cells. Some of these effects may actually occur at the level of adenylate cyclase itself because [^{3}H]forskolin was displaced from purified enzyme from rat liver by PD81,723 with an IC_{50} of 96 μM. Apparently, two mechanisms appear to contribute to the observed effects of PD81,723: allosteric enhancement of A_1 AR function and direct effect on adenylate cyclase. Kollias-Baker et al. (1997) studied CHO cells stably expressing the hA_1 AR in which PD81,723 acted in synergism with the agonist (*R*)-PIA to inhibit forskolin-stimulated cAMP formation. In a more extensive study by Kourounakis et al. (2001), PD81,723 (10 μM) alone inhibited cAMP production to approximately 70–80% of forskolin-stimulated levels. The agonist CPA decreased cAMP production to 70% at a concentration of 1 nM, while PD81,723 further decreased the cAMP production in combination with CPA. In this cell line, DPCPX increased the cAMP production by approximately 30%. In the additional presence of PD81,723, cAMP production decreased to forskolin-only levels. For the agonist CPA, a dose–response curve was recorded in the absence (EC_{50} 4.2 nM) and presence (EC_{50} 0.6 nM) of 10 μM PD81,723.

2. MAP Kinase

PD81,723 (10 μM) enhanced the potency of an A_1 AR agonist, N^6-4-methoxyphenyladenosine, to stimulate ERK1/2 phosphorylation in CHO cells expressing the hA_1 AR by fourfold. It also enhanced the efficacy of the agonist by 30% (Narlawar et al., 2010).

3. DNA Synthesis

PD81,723 (3 μM) enhanced the potency of the selective A_1 AR agonist CCPA (1 μM) to stimulate DNA synthesis in pig coronary artery smooth muscle cells. A similar effect was seen with adenosine as the agonist (100 μM). Control experiments showed that treatment of the cells with DPCPX or with pertussis toxin abolished the stimulatory effects on DNA synthesis (Shen et al., 2005).

4. Receptor Desensitization and Internalization

Bhattacharya and Linden (1996) studied the desensitization and downregulation of the hA_1 AR in CHO cells. Pretreatment with 20 μM PD81,723 or 10 μM CPA caused a 1.5- and 4.0-fold desensitization measured as a reduced potency of CPA to lower cAMP levels in the cells. Pretreatment with these agents did not modify the acute effect of PD81,723 to increase the potency of CPA fivefold. Radioligand binding experiments were performed to measure receptor downregulation in cell membranes and in intact cells. Pretreatment of the cells with PD81,723 had no significant effect on the number of receptors. Pretreatment of cells with CPA produced large reductions in the binding of agonist and antagonist radioligands to both membranes and intact cells. The authors speculated that the relatively small degree of functional desensitization and downregulation of receptors caused by long-term exposure of cells to PD81,723 is encouraging in terms of the therapeutic potential of such PAMs. Klaasse et al. (2005) studied the long-term effect of PD81,723 on receptor internalization. To visualize this process, the receptor was engineered to contain a C-terminal YFP tag. The introduction of this marker did not affect the radioligand binding properties of the receptor. CHO cells stably expressing this receptor were subjected during 16 h to varying concentrations of the agonist CPA in the absence or presence of 10 μM of PD81,723. CPA itself was able to internalize 25% and 40% of the receptors at a concentration of 400 nM or 4 μM, respectively. Addition of PD81,723 alone had no effect on internalization. However, a slight amount of internalization induced by with PD81,723 was obtained already at 40 nM of CPA, and 59% of the receptors internalized at 400 nM CPA.

D. Effects on Tissues

I. Heart

One of the first studies of cardiac effects of PD81,723 was performed by Mudumbi et al. (1993). The authors investigated the effects of PD81,723 in spontaneously contracting right atria and electrically stimulated left atria isolated from Sprague–Dawley rats. The reference A_1 AR agonist CPA produced a concentration-dependent inhibition of heart rate in right atria (chronotropy) and contractile parameters in left atria (inotropy). In both right and left atrium PD81,723 (5 μM) significantly left-shifted the concentration–response curves for CPA. In the same year, Amoah-Apraku et al. (1993) studied PD81,723 as an enhancer of the negative dromotropic effect of exogenous adenosine in guinea pig isolated and *in situ* hearts. In isolated hearts, PD81,723 alone produced only a small stimulus-to-His bundle (S–H) interval prolongation of 1.5–4 ms, which could be blocked by the A_1-selective antagonist 8-cyclopentyltheophylline (CPT). Under hypoxia, leading to an increase of interstitial adenosine levels, the S–H interval was also prolonged, and this was increased twofold in the presence of 5 μM PD81,723 (Kollias-Baker et al., 1994b). PD81,723 (5 μM) significantly increased the potency of adenosine for prolongation of the S–H interval from 7 to 4 μM. This potentiation by PD81,723 was also dose dependent (Kollias-Baker et al., 1994a). The effect was A_1 AR dependent; PD81,723 had no effect when, for example, carbachol was used. In *in situ* hearts, PD81,723 (2 μmol/kg, i.v.) caused a significant leftward and upward shift of the adenosine dose–response curve for inducing atrium-to-His bundle (A–H) interval prolongation. As a consequence, the degree of atrioventricular block caused by adenosine was also increased. Martynyuk et al. (2002) studied the molecular mechanisms underlying the adenosine-induced slowing of atrioventricular nodal conduction in guinea pig isolated hearts and in single atrial myocytes. This is a rate-dependent process, of which the authors analyzed the A–H interval (heart) or patch-clamp recordings (myocytes). A decrease in atrial cycle length from 300 to 190 ms decreased the concentration of adenosine needed to cause atrioventricular nodal block from 8 to 3 μM. PD81,723 (5 μM) potentiated the negative dromotropic effect of adenosine. In atrial myocytes, adenosine augmented a time- and voltage-dependent K^+ current, which was also potentiated by PD81,723. It should be mentioned though that PD81,723 appears to act as a direct inhibitor of some K^+ channels in guinea pig atrial myocytes too, most notably inward rectifying ones (Brandts et al., 1997). Mizumura et al. (1996) performed *in vivo* studies in dogs and examined the phenomenon of cardiac preconditioning. To determine if PD81,723 lowers the threshold for ischemic preconditioning, anesthetized dogs were subjected to coronary artery occlusion and subsequent reperfusion. Myocardial infarct size was significantly decreased by a combination of PD81,723 and preconditioning (a short period of artery occlusion preceding the main event), which beneficial effect could be blocked by the intravenous administration of DPCPX.

2. Brain

Janusz et al. (1991) were the first to evaluate the actions of PD81,723 in brain slices. PD81,723 dose dependently enhanced the inhibitory effects of exogenously applied adenosine in hippocampal brain slices as indicated by two parameters, the amplitude of the population spike and paired-pulse facilitation. PD81,723 had no effect when administered alone but required the presence of adenosine. In a different experimental setup on the same preparation, adenosine reduced the duration of epileptiform bursting in a dose-dependent manner. Application of PD81,723 at concentrations as high as 100 μM also resulted in a dose-dependent reduction in the duration of the triggered burst (Janusz & Berman, 1993). Phillis et al. (1994) studied the effects of PD81,723 in ischemia-evoked amino acid transmitter release from rat cerebral cortex. When administered at 10 mg/kg, i.p., PD81,723 significantly depressed glutamate efflux but not of GABA. However, in a gerbil model of forebrain ischemia PD81,723, studied at three dosages, failed to protect against ischemia/reperfusion-evoked cerebral injury (Cao & Phillis, 1995). Bueters et al. (2002) characterized the effects of PD81,723 on striatal acetylcholine release. Upon local administration in conscious rats via a microdialysis probe, the compound (0.1–100 μM) caused a concentration-dependent increase of extracellular acetylcholine levels of approximately 40%, which was similar to that obtained by the selective A_1 AR antagonist CPT. In competition experiments, PD81,723 did not change the inhibition of acetylcholine release by CPA, whereas CPT caused an eightfold rightward shift of the CPA dose–response curve. Apparently, the putative antagonistic action of PD81,723 in this animal model appeared to counteract its allosteric action. Meno et al. (2003) investigated the effects of PD81,723 (3 or 10 mg/kg, i.p.) on hippocampal injury and Morris water maze performance following hyperglycemic cerebral ischemia and reperfusion in the rat. Only at the lower dose, a significant reduction of hippocampal injury was observed, in line with an improved water maze performance suggesting that “reinforcement” of endogenously produced adenosine provides neuroprotection in this animal model.

Therefore, PD81,723 enhances A_1 AR agonist binding and function in many but not all tissues (e.g., adipocytes; see Jarvis et al., 1999) examined. Its effects are relatively modest, that is, agonist potencies tend to be increased by a factor of 2–5 with PD81,723 at micromolar (3–100 μM) concentrations. Probe dependency has been established; for instance, nonribose agonists are less sensitive, if at all, than adenosine-like derivatives to the influence of PD81,723. In *in vivo* and *ex vivo* experimental results vary, as a differentiation between enhancing and antagonistic effects of PD81,723 is not always easily observed.

IV. Allosteric Modulators of P2Y and P2X Receptors for Nucleotides

A. P2Y Receptor Modulation

P2Y receptors respond to various extracellular nucleotides, including ATP, UTP, ADP, UDP, and UDP-glucose. The subtypes of P2Y receptors are numbered $P2Y_1$, $P2Y_2$, $P2Y_4$, $P2Y_6$, $P2Y_{11}$, $P2Y_{12}$, $P2Y_{13}$, and $P2Y_{14}$. The three $P2Y_{12}$–$P2Y_{14}$ subtypes inhibit adenylate cyclase through G_i protein, and the other five subtypes activate phospholipase $C\beta$ through G_q protein.

The $P2Y_1$ receptor is activated by endogenous ADP to induce platelet aggregation, muscle relaxation, and vasodilation. The inhibitory effect of 2,2′-pyridylisatogen tosylate **27** (PIT, Fig. 3) on the $hP2Y_1$ receptor is allosteric (Gao et al., 2004a; Spedding et al., 2000). PIT blocked the accumulation of inositol phosphates induced by the potent synthetic agonist 2-methylthio-ADP (2-MeSADP) and by ADP in 1321N1 astrocytoma cells stably expressing the $hP2Y_1$ receptor. The antagonism occurred in a concentration-dependent manner but was noncompetitive, and it did not inhibit the binding of a selective $P2Y_1$ receptor antagonist radioligand. PIT had no significant effect on agonist activation of other P2Y receptors examined. Thus, PIT selectively and noncompetitively blocked $P2Y_1$ receptor signaling without affecting nucleotide binding.

SCH-202676 was shown to inhibit ATP-induced Na^+–K^+ pump activity mediated via the $P2Y_1$ receptor in depolarized skeletal muscle (Broch-Lips et al., 2010), although it did not inhibit the binding of the selective radioligand [^{3}H]MRS2279 to the $P2Y_1$ receptor (Gao et al., 2004a, 2004b). It has been reported that SCH-202676 affects a number of GPCRs (Fawzi et al., 2001). However, later studies suggested SCH-202676 modulates some GPCRs via thiol modification rather than via true allosteric mechanisms (Göblyös et al., 2005; Lewandowicz et al., 2006).

A uridine 5′-methylene-phosphonate derivative **28** was a relatively potent ($EC_{50} = 1.6 \pm 0.4$ μM) agonist at the $P2Y_2$ receptor and had no effect on the $P2Y_4$ and $P2Y_6$ receptors (Cosyn et al., 2009). However, the maximal agonist effect observed was $<50\%$ of that observed with the native agonist UTP. UMP itself was inactive at this receptor. High concentrations of **28** failed to antagonize activation of the $P2Y_2$ receptor by UTP, suggesting that it potentially activates the $P2Y_2$ receptor through an allosteric mechanism. At the $P2Y_4$ receptor, diadenosine polyphosphates potentiated the UTP agonist response, which was not observed at other P2Y subtypes (Patel et al., 2001).

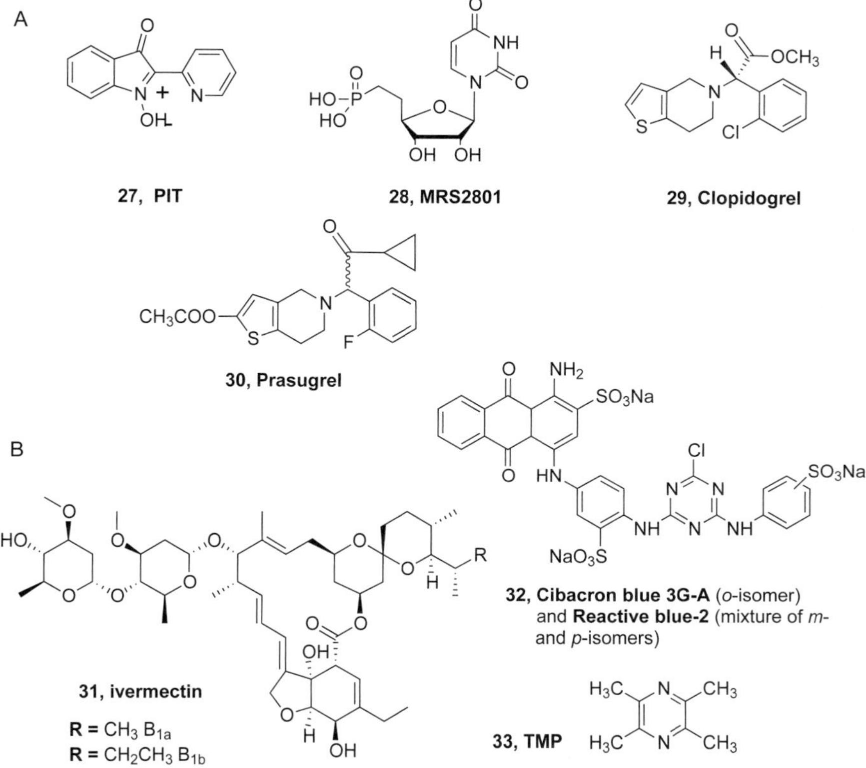

FIGURE 3 Structures of compounds found to act as allosteric modulators of P2Y (A) and P2X (B) receptors.

Antagonists of the $P2Y_{12}$ receptor are clinically used as antithrombotic agents. Several of these agents, clopidogrel **29** and prasugrel **30**, are thienopyridine derivatives that contain masked thiol groups that are liberated *in vivo*. The reactive metabolites act as irreversible inhibitors of the binding of ADP at this receptor, and their site of action might prove to be allosteric. They have been shown to bind covalently to form disulfides with Cys residues that normally form a bridge between EL2 and TM3 (Algaier et al., 2008).

Other modulators of P2Y receptor signaling that are possibly allosteric modulators are cysteinyl leukotriene antagonists (Mamedova et al., 2005) and bicyclic diketopiperazines (Besada et al., 2005).

B. P2X Receptor Modulation

The subtypes of P2X receptors are numbered P2X1 through P2X7. These subunits combine to form trimeric LGICs that are activated by extracellular ATP at various concentrations. The recent determination of the zebrafish P2X4 crystal structure (Kawate et al., 2009) has provided insights into the structure of this family of ion channels.

Zn^{2+} ions, Cu^{2+} ions, and pH can act as allosteric modulators of action at the P2X receptors (Evans, 2009). Other metals, such as lanthanides, may also modulate P2X receptors, and ethanol has been shown to reduce the potency of ATP at various P2X receptors (reviewed in Coddou et al., 2011). The sedative propofol and various lipids also modulate action at P2X receptors.

The differential sensitivity of various P2X receptor subtypes to metal ions and to protons has been probed using mutagenesis. Zn^{2+} at low micromolar concentrations increased the channel activity of P2X2 and P2X4 receptors (Brake et al., 1994; Seguela et al., 1996; Soto et al., 1996; Xiong et al., 1999). The same treatment decreased the activity of P2X1 and P2X7 receptors (Virginio et al., 1997; Wildman et al., 1999). The effects of Cu^{2+} ions (Xiong et al., 1999) and protons (King et al., 1996; Li et al., 1996; Stoop et al., 1997; Wildman et al., 1998, 1999) of various P2X receptor subtypes have been extensively explored. Various His and Asp residues have been found to be involved in these effects using site-directed mutagenesis. The potentiation of the P2X4 receptor activity by Zn^{2+} is dependent on Cys132 and to a lesser extent on Thr133 (Coddou et al., 2007). Neither of these residues affects inhibition by Cu^{2+}, which is dependent on Asp138 and His140. Thus, this receptor region contains a pocket for trace metal coordination with two distinct and separate sites for dications as a PAM or NAM. For the P2X2 receptor, His120 and His213 were identified as part of an intersubunit binding site that accounts for Zn^{2+} potentiation (Nagaya et al., 2005); these residues are also involved in Cu^{2+} potentiation (Lorca et al., 2005).

The antiparasitic drug ivermectin **31** (a mixture of macrocyclic lactone disaccharides 22,23-dihydroavermectin B_{1a} + 22,23-dihydroavermectin B_{1b}) has been reported to enhance currents at the P2X4 receptor but not at other P2X receptors (Khakh et al., 1999). Its mechanism of action and specific amino acid residues involved in the effect have been investigated (Coddou et al., 2011, Priel & Silberberg, 2004; Silberberg et al., 2007; Toulmé et al., 2006; Zemkova et al., 2010).

Cibacron blue **30** allosterically modulates the rat P2X4 receptor (Miller et al., 1998). It was also found to be a PAM of the P2X3 receptor that also restored the ATP responsiveness to acutely desensitized receptors (Alexander et al., 1999). There has been confusion about which isomer of this dye is designated Cibacron blue. Cibacron blue 3GA refers to the *ortho*-isomer of **32**, but many studies have assumed Reactive blue-2 (mixture of *meta*- and *para*-isomers) to be synonymous with Cibacron blue.

Tetramethylpyrazine (TMP, **33**), an alkaloid in traditional Chinese medicine, inhibits the effects of nucleotides at the P2X3 receptor in primary afferent transmission in neuropathic pain states (Gao et al., 2008a). It has been proposed to bind at an allosteric site on the large extracellular region of the P2X receptor.

V. Conclusion

Allosteric modulators have been most highly developed for the A_1 and A_3 ARs among the purine and pyrimidine receptors. In fact, a 2-amino-3-aroylthiophene derivative T-62 has been under development as a PAM of the A_1 AR for the treatment of chronic pain. The prototypic PAM of the A_1 AR, PD81,723, has been evaluated in a great number of pharmacological assays, which makes possible a translational assessment of its modulatory potency. The benzoylthiophene derivatives tend to act as allosteric agonists as well as pure PAMs of this subtype and lack action at other AR subtypes. Two classes of A_3 AR allosteric modulators have been explored: 3-(2-pyridinyl)isoquinolines (e.g., VUF5455) and 1H-imidazo-[4,5-*c*]quinolin-4-amines (e.g., DU124183 and LUF6000), which selectively decreased the agonist dissociation rate at the hA_3ARs, but not at A_1 and A_{2A} ARs. These A_3 AR modulators have allosteric effects that can be structurally separated from the orthosteric effects in SAR studies. Nucleoside derivatives that are A_3 selective antagonists and low efficacy agonists can be converted into full agonists by coadministration of the PAM LUF6000. Site-directed mutagenesis of A_1 and A_3 receptors has identified residues associated with the allosteric effect. Distinct amino acid residues affect orthosteric versus allosteric binding. Thus, there are clear advantages to the design of allosteric modulators of ARs. Small molecular allosteric modulators have been reported for several of the P2Y nucleotide receptors and P2X receptors, but there is much room for exploration of this approach into the nucleotide receptor field. Allosteric modulation of the P2X receptors by metal ions and protons has been extensively studied by site-directed mutagenesis. In conclusion, allosteric modulation of purine and pyrimidine receptors looks promising for development of drugs that are event- and site specific in action.

Acknowledgments

K. A. J. acknowledges support from the Intramural Research Program of the NIH, National Institute of Diabetes & Digestive & Kidney Diseases. A. P. I. J. thanks Dutch Top Institute Pharma (project The GPCR forum, D1.105) for generous financial support.

Conflict of Interest: The authors are inventors on patents for LUF6000 and LUF6096.

Abbreviations

AR	adenosine receptor
BBB	blood–brain barrier
CCPA	2-chloro-N^6-cyclopentyladenosine
CGS21680	2-*p*-(2-carboxyethyl)phenethylamino-5′-*N*-ethylcarboxamidoadenosine hydrochloride

CHA	N^6-cyclohexyladenosine
Cl-IB-MECA	2-chloro-N^6-(3-iodobenzyl)-adenosine-5′-*N*-methyluronamide
CPA	N^6-cyclopentyladenosine
CPT	8-cyclopentyltheophylline
DU124183	*N*-phenyl-2-cyclopentyl-1*H*-imidazo[4,5-*c*]quinolin-4-amine
EL	extracellular loop
ERK	extracellular-signal-regulated kinase
GPCR	G protein-coupled receptor
HMA	5-(*N*,*N*-hexamethylene)amiloride
LGIC	ligand-gated ion channel
LUF5484	(2-amino-4,5,6,7-tetrahydrobenzo[*b*]thiophen-3-yl)(3,4-dichlorophenyl)methanone
LUF6000	*N*-(3,4-dichlorophenyl)-2-cyclohexyl-1*H*-imidazo[4,5-*c*] quinolin-4-amine
LUF6258	*N*(6)-[2-amino-3-(3,4-dichlorobenzoyl)-4,5,6,7-tetrahydrothieno[2,3-*c*]pyridin-6-yl-9-nonyloxy-4-phenyl]-adenosine
MRS542	2-chloro-N^6-(3-iodobenzyl)-adenosine
MRS1220	*N*-[9-chloro-2-(2-furanyl)[1,2,4]triazolo[1,5-*c*]quinazolin-5-yl]benzeneacetamide
MRS5049	2-cyclohexyl-*N*-(3,5-dichlorophenyl)-1*H*-imidazo[4,5-*c*] quinolin-4-amine
MRS5190	2-(1-adamantanyl)-*N*-(3,4-dichlorophenyl)-1*H*-imidazo[4,5-*c*] quinolin-4-amine
NECA	5′-(*N*-ethylcarboxamido)adenosine
PAM	positive allosteric modulator
PD120,918	4-methyl-2-oxo-2*H*-chromen-7-yl methylcarbamate
PD71,605	(2-amino-4,5,6,7-tetrahydrobenzo[*b*]thiophen-3-yl)-(2-chlorophenyl)-methanone
PD81,723	(2-amino-4,5-dimethyl-3-thienyl)-[3-(trifluoromethyl)phenyl] methanone
PIT	pyridyl isatogen tosylate
SAR	structure–activity relationship
SCH-202676	*N*-(2,3-diphenyl-1,2,4-thiadiazol-5-(2H)-ylidene) methanamine
T62	2-amino-3-(4-chlorobenzoyl)-5,6,7,8-tetrahydrobenzothiophene
TM	transmembrane domain
VCP333	*tert*-butyl 2-amino-3-(4-chlorobenzoyl)-7,8-dihydro-4*H*-thieno[2,3-*d*]azepine-6(5*H*)-carboxylate
VCP520	2-amino-4-(3,5-bis(trifluoromethyl)phenyl)thiophen-3-yl) (phenyl)methanone

VUF5455	(*N*-(2-methoxyphenyl)-*N′*-[2-(3-pyrindinyl)-4-quinazolinyl]-urea
VUF8502	4-methyl-*N*-[3-(2-pyridinyl)-1-isoquinolinyl]benzamide
VUF8504	methoxy-*N*-[3-(2-pyridinyl)-1-isoquinolinyl]benzamide
VUF8507	*N*-[3-(2-pyridinyl)-1-isoquinolinyl]benzamide
ZM241385	4-{2-[7-amino-2-(2-furyl)-1,2,4-triazolo[1,5-a]1,3,5]triazin-5-yl-amino]ethyl}phenol

References

Alexander, K., Niforatos, W., Bianchi, B., Burgard, E. C., Lynch, K. J., Kowaluk, E. A., Jarvis, M. F., & van Biesen, T. (1999). Allosteric modulation and accelerated resensitization of human $P2X_3$ receptors by Cibacron blue. *The Journal of Pharmacology and Experimental Therapeutics*, *291*, 1135–1142.

Algaier, I., Jakubowski, J. A., Asai, F., & von Kügelgen, I. (2008). Interaction of the active metabolite of prasugrel, R-138727, with cysteine 97 and cysteine 175 of the human $P2Y_{12}$ receptor. *Journal of Thrombosis and Haemostasis*, *6*, 1908–1914.

Amoah-Apraku, B., Xu, J., Lu, J. Y., Pelleg, A., Bruns, R. F., & Belardinelli, L. (1993). Selective potentiation by an A1 adenosine receptor enhancer of the negative dromotropic action of adenosine in the guinea pig heart. *The Journal of Pharmacology and Experimental Therapeutics*, *266*, 611–617.

Aumann, K. M., Scammells, P. J., White, J. M., & Schiesser, C. H. (2007). On the stability of 2-aminoselenophene-3-carboxylates: Potential dual-acting selenium-containing allosteric enhancers of A_1 adenosine receptor binding. *Organic & Biomolecular Chemistry*, *5*, 1276–1281.

Aurelio, L., Valant, C., Figler, H., Flynn, B. L., Linden, J., Sexton, P. M., Christopoulos, A., & Scammells, P. J. (2009a). 3- and 6-Substituted 2-amino-4,5,6,7-tetrahydrothieno[2,3-c] pyridines as A1 adenosine receptor allosteric modulators and antagonists. *Bioorganic & Medicinal Chemistry*, *17*, 7353–7361.

Aurelio, L., Valant, C., Flynn, B. L., Sexton, P. M., Christopoulos, A., & Scammells, P. J. (2009b). Allosteric modulators of the adenosine A_1 receptor: Synthesis and pharmacological evaluation of 4-substituted 2-amino-3-benzoylthiophenes. *Journal of Medicinal Chemistry*, *52*, 4543–4547.

Aurelio, L., Valant, C., Flynn, B. L., Sexton, P. M., White, J. M., Christopoulos, A., & Scammells, P. J. (2010). Effects of conformational restriction of 2-amino-3-benzoylthiophenes on A_1 adenosine receptor modulation. *Journal of Medicinal Chemistry*, *53*, 6550–6559.

Ballesteros, J. A., & Weinstein, H. (1995). Integrated methods for the construction of three-dimensional models and computational probing of structure-function relations in G protein-coupled receptors. *Methods in Neurosciences*, *25*, 366–428.

Baraldi, P. G., Iaconinoto, M. A., Moorman, A. R., Carrion, M. D., Cara, C. L., Preti, D., Lopez, O. C., Fruttarolo, F., Tabrizi, M. A., & Romagnoli, R. (2007). Allosteric enhancers for A_1 adenosine receptor. *Mini Reviews in Medicinal Chemistry*, *7*, 559–569.

Baraldi, P. G., Pavani, M. G., Shryock, J. C., Moorman, A. R., Iannatta, V., Borea, P. A., & Romagnoli, R. (2004). Synthesis of 2-amino-3-heteroaroylthiophenes and evaluation of their activity as potential allosteric enhancers at the human A_1 receptor. *European Journal of Medicinal Chemistry*, *39*, 855–865.

Baraldi, P. G., Romagnoli, R., Pavani, M. G., Nunez, M. C., Tabrizi, M. A., Shryock, J. C., Leung, E., Moorman, A. R., Uluoglu, C., Iannotta, V., Merighi, S., & Borea, P. A. (2003). Synthesis and biological effects of novel 2-amino-3-naphthoylthiophenes as allosteric enhancers of the A_1 adenosine receptor. *Journal of Medicinal Chemistry*, *46*, 794–809.

Baraldi, P. G., Zaid, A. N., Lampronti, I., Fruttarolo, F., Pavani, M. G., Tabrizi, M. A., Shryock, J. C., Leung, E., & Romagnoli, R. (2000). Synthesis and biological effects of a new series of 2-amino-3-benzoylthiophenes as allosteric enhancers of A_1 adenosine receptor. *Bioorganic & Medicinal Chemistry Letters*, *10*, 1953–1957.

Barbhaiya, H., McClain, R., IJzerman, A., & Rivkees, S. A. (1996). Site-directed mutagenesis of the human A_1 adenosine receptor: Influences of acidic and hydroxy residues in the first four transmembrane domains on ligand binding. *Molecular Pharmacology*, *50*, 1635–1642.

Besada, P., Mamedova, L., Thomas, C. J., Costanzi, S., & Jacobson, K. A. (2005). Design and synthesis of new bicyclic diketopiperazines as scaffolds for receptor probes of structurally diverse functionality. *Organic & Biomolecular Chemistry*, *3*, 2016–2025.

Bhattacharya, S., & Linden, J. (1995). The allosteric enhancer, PD 81,723, stabilizes human A1 adenosine receptor coupling to G proteins. *Biochimica et Biophysica Acta*, *1265*, 15–21.

Bhattacharya, S., & Linden, J. (1996). Effects of long-term treatment with the allosteric enhancer, PD81,723, on Chinese hamster ovary cells expressing recombinant human A1 adenosine receptors. *Molecular Pharmacology*, *50*, 104–111.

Bhattacharya, S., Youkey, R. L., Ghartey, K., Leonard, M., Linden, J., & Tucker, A. L. (2006). The allosteric enhancer PD 81,723 increases chimaeric A1/A2A adenosine receptor coupling with Gs. *The Biochemical Journal*, *396*, 139–146.

Borrmann, T., Hinz, S., Bertarelli, D. C., Li, W., Florin, N. C., Scheiff, A. B., & Müller, C. E. (2009). 1-Alkyl-8-(piperazine-1-sulfonyl)phenylxanthines: Development and characterization of adenosine A_{2B} receptor antagonists and a new radioligand with subnanomolar affinity and subtype specificity. *Journal of Medicinal Chemistry*, *52*, 3994–4006.

Brake, A. J., Wagenbach, M. J., & Julius, D. (1994). New structural motif for ligand-gated ion channels defined by an ionotropic ATP receptor. *Nature*, *371*, 519–523.

Brandts, B., Bünemann, M., Hluchy, J., Sabin, G. V., & Pott, L. (1997). Inhibition of muscarinic K+ current in guinea-pig atrial myocytes by PD 81,723, an allosteric enhancer of adenosine binding to A1 receptors. *British Journal of Pharmacology*, *121*, 1217–1223.

Broch-Lips, M., Pedersen, T. H., & Nielsen, O. B. (2010). Effect of purinergic receptor activation on Na+–K+ pump activity, excitability, and function in depolarized skeletal muscle. *American Journal of Physiology. Cell Physiology*, *298*, C1438–C1444.

Bruns, R. F., & Fergus, J. H. (1990). Allosteric enhancement of adenosine A_1 receptor binding and function by 2-amino-3-benzoylthiophenes. *Molecular Pharmacology*, *38*, 939–949.

Bruns, R. F., Fergus, J. H., Coughenour, L. L., Courtland, G. G., Pugsley, T. A., Dodd, J. H., & Tinney, F. J. (1990). Structure-activity relationships for enhancement of adenosine A_1 receptor binding by 2-amino-3-benzoylthiophenes. *Molecular Pharmacology*, *38*, 950–958.

Bruns, R. F., & Lu, G. H. (1989). Enhancement of adenosine agonist binding by a substituted 1-benzopyran-2-one. In: J. A. Ribeiro (Ed.), *Adenosine receptors in the nervous system*. London: Taylor and Francis, p. 192.

Bueters, T. J., van Helden, H. P., Danhof, M., & IJzerman, A. P. (2002). Effects of the adenosine A_1 receptor allosteric modulators PD 81,723 and LUF 5484 on the striatal acetylcholine release. *European Journal of Pharmacology*, *454*, 177–182.

Cao, X., & Phillis, J. W. (1995). Adenosine A1 receptor enhancer, PD 81,723, and cerebral ischemia/reperfusion injury in the gerbil. *General Pharmacology*, *26*, 1545–1548.

Changeux, J. P. (2010). Allosteric receptors: From electric organ to cognition. *Annual Review of Pharmacology and Toxicology*, *50*, 1–38.

Childers, S. R., Li, X. L., Xiao, R., & Eisenach, J. C. (2005). Allosteric modulation of adenosine A_1 receptor coupling to G-proteins in brain. *Journal of Neurochemistry*, *93*, 715–723.

Chordia, M. D., Zigler, M., Murphree, L. J., Figler, H., Macdonald, T. L., Olsson, R. A., & Linden, J. (2005). 6-aryl-8H-indeno[1,2-d]thiazol-2- ylamines: A1 adenosine receptor agonist allosteric enhancers having improved potency. *Journal of Medicinal Chemistry*, *48*, 5131–5139.

Christopoulos, A. (2002). Allosteric binding sites on cell-surface receptors: Novel targets for drug discovery. *Nature Reviews. Drug Discovery*, *1*, 198–210.

Christopoulos, A., & Kenakin, T. (2002). G protein-coupled receptor allosterism and complexing. *Pharmacological Reviews*, *54*, 323–374.

Coddou, C., Acuna-Castillo, C., Bull, P., & Huidobro-Toro, J. P. (2007). Dissecting the facilitator and inhibitor allosteric metal sites of the $P2X_4$ receptor channel: Critical roles of Cys-132 for zinc-potentiation and Asp-138 for copper-inhibition. *The Journal of Biological Chemistry*, *282*, 36879–36886.

Coddou, C., Yan, Z., Obsil, T., Huidobro-Toro, J. P., & Stojilkovic, S. S. (2011). Activation and regulation of purinergic P2X receptor channels. *Pharmacological Reviews* (submitted for publication).

Conn, P. J., Christopoulos, A., & Lindsley, C. W. (2009). Allosteric modulators of GPCRs: A novel approach for the treatment of CNS disorders. *Nature Reviews. Drug Discovery*, *8*, 41–54.

Cosyn, L., Van Calenbergh, S., Joshi, B. V., Ko, H., Carter, R. L., Harden, T. K., & Jacobson, K. A. (2009). Synthesis and P2Y receptor activity of nucleotide 5′-phosphonate derivatives. *Bioorganic & Medicinal Chemistry Letters*, *19*, 3002–3005.

De Ligt, R. A. F., Rivkees, S. A., Lorenzen, A., Leurs, R., & IJzerman, A. P. (2005). A "locked-on", constitutively active mutant of the adenosine A1 receptor. *European Journal of Pharmacology*, *510*, 1–8.

Evans, R. J. (2009). Orthosteric and allosteric binding sites of P2X receptors. *European Biophysics Journal*, *38*, 319–327.

Fawzi, A. B., Macdonald, D., Benbow, L. L., Smith-Torhan, A., Zhang, H., Weig, B. C., Ho, G., Tulshian, D., Linder, M. E., & Graziano, M. P. (2001). SCH-202676: An allosteric modulator of both agonist and antagonist binding to G protein-coupled receptors. *Molecular Pharmacology*, *59*, 30–37.

Ferguson, G. N., Valant, C., Horne, J., Figler, H., Flynn, B. L., Linden, J., Chalmers, D. K., Sexton, P. M., Christopoulos, A., & Scammells, P. J. (2008). 2-Aminothienopyridazines as novel adenosine A_1 receptor modulators and antagonists. *Journal of Medicinal Chemistry*, *51*, 6165–6172.

Figler, H., Olsson, R. A., & Linden, J. (2003). Allosteric enhancers of A1 adenosine receptors increase receptor-G protein coupling and counteract guanine nucleotide effects on agonist binding. *Molecular Pharmacology*, *64*, 1557–1564.

Fredholm, B. B., IJzerman, A. P., Jacobson, K. A., Linden, J., & Müller, C. (2011). Nomenclature and classification of adenosine receptors—An update. *Pharmacological Reviews*, *63*, 1–34.

Gao, Z. G., Gross, A. S., & Jacobson, K. A. (2004b). Effects of SCH-202676 on adenosine and P2Y receptors. *Life Sciences*, *74*, 3173–3180.

Gao, Z. G., & IJzerman, A. P. (2000). Allosteric modulation of A_{2A} adenosine receptors by amiloride analogues and sodium ions. *Biochemical Pharmacology*, *60*, 669–676.

Gao, Z. G., Kim, S. K., Gross, A. S., Chen, A., Blaustein, J., & Jacobson, K. A. (2003a). Identification of essential residues involved in the allosteric modulation of the human A_3 adenosine receptor. *Molecular Pharmacology*, *63*, 1021–1031.

Gao, Z. G., Kim, S. K., IJzerman, A. P., & Jacobson, K. A. (2005). Allosteric modulation of the adenosine family of receptor. *Mini Reviews in Medicinal Chemistry*, *5*, 545–553.

Gao, Z. G., Kim, S. G., Soltysiak, K. A., Melman, N., IJzerman, A. P., & Jacobson, K. A. (2002). Selective allosteric enhancement of agonist binding and function at human A_3 adenosine receptors by a series of imidazoquinoline derivatives. *Molecular Pharmacology*, *62*, 81–89.

Gao, Z. G., Mamedova, L., Tchilibon, S., Gross, A. S., & Jacobson, K. A. (2004a). 2, 2′-Pyridylisatogen tosylate antagonizes P2Y1 receptor signaling without affecting nucleotide binding. *Biochemical Pharmacology*, *68*, 231–237.

Gao, Z. G., Melman, N., Erdmann, A., Kim, S. G., Müller, C. E., & IJzerman, A. P. (2003b). Differential allosteric modulation by amiloride analogues of agonist and antagonist binding at A_1 and A_3 adenosine receptors. *Biochemical Pharmacology*, *65*, 525–534.

Gao, Z. G., van Muijlwijk-Koezen, J. E., Chen, A., Müller, C. E., IJzerman, A. P., & Jacobson, K. A. (2001). Allosteric modulation of A_3 adenosine receptors by a series of 3-(2-pyridinyl)isoquinoline derivatives. *Molecular Pharmacology*, *60*, 1057–1063.

Gao, Y., Xu, C., Liang, S., Zhang, A., Mu, S., Wang, Y., & Wan, F. (2008a). Effect of tetramethylpyrazine on primary afferent transmission mediated by P2X3 receptor in neuropathic pain states. *Brain Research Bulletin*, *77*, 27–32.

Gao, Z. G., Ye, K., Göblyös, A., IJzerman, A. P., & Jacobson, K. A. (2008b). Flexible modulation of agonist efficacy at the human A_3 adenosine receptor by an imidazoquinoline allosteric enhancer LUF6000 and its analogues. *BMC Pharmacology*, *8*, 20.

Giorgi, I., Biagi, G., Bianucci, A. M., Borghini, A., Livi, O., Leonardi, M., Pietra, D., Calderone, V., & Martelli, A. (2008). N^6-1,3-diphenylurea derivatives of 2-phenyl-9-benzyladenines and 8-azaadenines: Synthesis and biological evaluation as allosteric modulators of A2A adenosine receptors. *European Journal of Medicinal Chemistry*, *43*, 1639–1647.

Göblyös, A., Gao, Z. G., Brussee, J., Connestari, R., Neves Santiago, S., Ye, K., IJzerman, A. P., & Jacobson, K. A. (2006). Structure activity relationships of 1*H*-imidazo[4,5-*c*]quinolin-4-amine derivatives new as allosteric enhancers of the A_3 adenosine receptor. *Journal of Medicinal Chemistry*, *49*, 3354–3361.

Göblyös, A., & IJzerman, A. P. (2009). Allosteric modulation of adenosine receptors. *Purinergic Signalling*, *5*, 51–61.

Göblyös, A., Santiago, S. N., Pietra, D., Mulder-Krieger, T., von Frijtag Drabbe Künzel, J., Brussee, J., & IJzerman, A. P. (2005). Synthesis and biological evaluation of 2-aminothiazoles and their amide derivatives on human adenosine receptors. Lack of effect of 2-aminothiazoles as allosteric enhancers. *Bioorganic & Medicinal Chemistry*, *13*, 2079–2087.

Harrington, P. E., & Fotsch, C. (2007). Calcium sensing receptor activators: Calcimimetics. *Current Medicinal Chemistry*, *14*, 3027–3034.

Heitman, L. H., Göblyös, A., Zweemer, A. M., Bakker, R., Mulder-Krieger, T., van Veldhoven, J. P., de Vries, H., Brussee, J., & IJzerman, A. P. (2009). A series of 2,4-disubstituted quinolines as a new class of allosteric enhancers of the adenosine A_3 receptor. *Journal of Medicinal Chemistry*, *52*, 926–931.

Heitman, L. H., Mulder-Krieger, T., Spanjersberg, R. F., von Frijtag Drabbe Künzel, J. K., Dalpiaz, A., & IJzerman, A. P. (2006). Allosteric modulation, thermodynamics and binding to wild-type and mutant (T277A) adenosine A_1 receptors of LUF5831, a novel nonadenosine-like agonist. *British Journal of Pharmacology*, *147*, 533–541.

Jaakola, V. P., Griffith, M. T., Hanson, M. A., Cherezov, V., Chien, E. Y., Lane, J. R., IJzerman, A. P., & Stevens, R. C. (2008). The 2.6 angstrom crystal structure of a human A_{2A} adenosine receptor bound to an antagonist. *Science*, *322*, 1211–1217.

Jacobson, K. A., & Gao, Z. G. (2006). Adenosine receptors as therapeutic targets. *Nature Reviews. Drug Discovery*, *5*, 247–264.

Janusz, C. A., & Berman, R. F. (1993). The adenosine binding enhancer, PD 81,723, inhibits epileptiform bursting in the hippocampal brain slice. *Brain Research*, *619*, 131–136.

Janusz, C. A., Bruns, R. F., & Berman, R. F. (1991). Functional activity of the adenosine binding enhancer, PD 81,723, in the in vitro hippocampal slice. *Brain Research*, *567*, 181–187.

Jarvis, M. F., Gessner, G., Shapiro, G., Merkel, L., Myers, M., Cox, B. F., & Martin, G. E. (1999). Differential effects of the adenosine A1 receptor allosteric enhancer PD 81,723 on agonist binding to brain and adipocyte membranes. *Brain Research*, *840*, 75–83.

Kawate, T., Michel, J. C., Birdsong, W. T., & Gouaux, E. (2009). Crystal structure of the ATP-gated P2X4 ion channel in the closed state. *Nature*, *460*, 592–599.

Khakh, B. S., Proctor, W. R., Dunwiddie, T. V., Labarca, C., & Lester, H. A. (1999). Allosteric control of gating and kinetics at $P2X_4$ receptor channels. *The Journal of Neuroscience*, *19*, 7289–7299.

Kiesman, W. F., Elzein, E., & Zablocki, J. (2009). A_1 adenosine receptor antagonists, agonists and allosteric enhancers. In C. N. Wilson & S. J. Mustafa (Eds.), *Adenosine receptors in health and disease, Handbook of experimental pharmacology* (pp. 25–58). Berlin, Heidelberg: Springer-Verlag.

Kim, Y., de Castro, S., Gao, Z. G., IJzerman, A. P., & Jacobson, K. A. (2009). Novel 2- and 4-substituted 1H-imidazo[4,5-c]quinolin-4-amine derivatives as allosteric modulators of the A_3 adenosine receptor. *Journal of Medicinal Chemistry*, *52*, 2098–2108.

King, B. F., Ziganshin, L. E., Pintor, J., & Burnstock, G. (1996). Full sensitivity of $P2X_2$ purinoceptor revealed by changing extracellular pH. *British Journal of Pharmacology*, *117*, 1371–1373.

Klaasse, E., de Ligt, R. A., Roerink, S. F., Lorenzen, A., Milligan, G., Leurs, R., & IJzerman, A. P. (2004). Allosteric modulation and constitutive activity of fusion proteins between the adenosine A_1 receptor and different 351Cys-mutated Gi alpha-subunits. *European Journal of Pharmacology*, *499*, 91–98.

Klaasse, E. C., van den Hout, G., Roerink, S. F., de Grip, W. J., IJzerman, A. P., & Beukers, M. W. (2005). Allosteric modulators affect the internalization of human adenosine A_1 receptors. *European Journal of Pharmacology*, *522*, 1–8.

Kollias-Baker, C., Ruble, J., Dennis, D., Bruns, R. F., Linden, J., & Belardinelli, L. (1994a). Allosteric enhancer PD 81,723 acts by novel mechanism to potentiate cardiac actions of adenosine. *Circulation Research*, *75*, 961–971.

Kollias-Baker, C. A., Ruble, J., Jacobson, M., Harrison, J. K., Ozeck, M., Shryock, J. C., & Belardinelli, L. (1997). Agonist-independent effect of an allosteric enhancer of the A_1 adenosine receptor in CHO cells stably expressing the recombinant human A_1 receptor. *The Journal of Pharmacology and Experimental Therapeutics*, *281*, 761–768.

Kollias-Baker, C., Xu, J., Pelleg, A., & Belardinelli, L. (1994b). Novel approach for enhancing atrioventricular nodal conduction delay mediated by endogenous adenosine. *Circulation Research*, *75*, 972–980.

Kourounakis, A. P., van de Klein, P. A. M., & IJzerman, A. P. (2000). Elucidation of stucture-activity relationships of 2-amino-3-benzoylthiophenes: Study of their allosteric enhanceing vs. antagonistic activity on adenosine A1 receptors. *Drug Development Research*, *49*, 227–237.

Kourounakis, A., Visser, C., de Groote, M., & IJzerman, A. P. (2001). Differential effects of the allosteric enhancer (2-amino-4,5-dimethyltrienyl)[3-(trifluoromethyl)phenyl] methanone (PD 81,723) on agonist and antagonist binding and function at the human wild-type and a mutant (T277A) adenosine A1 receptor. *Biochemical Pharmacology*, *61*, 137–144.

Lam, R. S., Nahirny, D., & Duszyk, M. (2009). Cholesterol-dependent regulation of adenosine A_{2A} receptor-mediated anion secretion in colon epithelial cells. *Experimental Cell Research*, *315*, 3028–3035.

Lane, J. R., Beukers, M. W., Mulder-Krieger, T., & IJzerman, A. P. (2010). The endocannabinoid 2-arachidonylglycerol is a negative allosteric modulator of the human A_3 adenosine receptor. *Biochemical Pharmacology*, *79*, 48–56.

Leung, E., Walsh, L. K., Kim, E. J., Lazar, D. A., Seran, C. S., Wong, E. H., & Eglen, R. M. (1995). Enhancement of the adenosine A receptor functions by benzoylthiophenes in guinea pig tissues in vitro. *Naunyn-Schmiedeberg's Archives of Pharmacology*, *352*, 206–212.

Lewandowicz, A. M., Vepsäläinen, J., & Laitinen, J. T. (2006). The 'allosteric modulator' SCH-202676 disrupts G protein-coupled receptor function via sulphydryl-sensitive mechanisms. *British Journal of Pharmacology*, *147*, 422–429.

Li, X., Conklin, D., Pan, H.-L., & Eisenach, J. C. (2003). Allosteric adenosine receptor modulation reduces hypersensitivity following peripheral inflammation by a central mechanism. *The Journal of Pharmacology and Experimental Therapeutics*, *305*, 950–955.

Li, C., Peoples, R. W., & Weight, F. F. (1996). Proton potentiation of ATP-gated ion channel responses to ATP and Zn^{2+} in rat nodose ganglion neurons. *Journal of Neurophysiology*, *76*, 3048–3058.

Lorca, R. A., Coddou, C., Gazitua, M. C., Bull, P., Arredondo, C., & Huidobro-Toro, J. P. (2005). Extracellular histidine residues identify common structural determinants in the copper/zinc P2X2 receptor modulation. *Journal of Neurochemistry*, *95*, 499–512.

Mamedova, L., Capra, V., Accomazzo, M. R., Gao, Z. G., Ferrario, S., Fumigalli, M., Abbracchio, M. P., Rovati, G. E., & Jacobson, K. A. (2005). CysLT1 leukotriene receptor antagonists inhibit the effects of nucleotides acting at P2Y receptors. *Biochemical Pharmacology*, *71*, 115–125.

Martynyuk, A. E., Zima, A., Seubert, C. N., Morey, T. E., Belardinelli, L., & Dennis, D. M. (2002). Potentiation of the negative dromotropic effect of adenosine by rapid heart rates: Possible ionic mechanism. *Basic Research in Cardiology*, *97*, 295–304.

May, L. T., Briddon, S. J., & Hill, S. J. (2010a). Antagonist selective modulation of adenosine A_1 and A_3 receptor pharmacology by the food dye brilliant black BN: Evidence for allosteric interactions. *Molecular Pharmacology*, *77*, 678–686.

May, L. T., Self, T. J., Briddon, S. J., & Hill, S. J. (2010b). The effect of allosteric modulators on the kinetics of agonist-G protein-coupled receptor interactions in single living cells. *Molecular Pharmacology*, *78*, 511–523.

Meno, J. R., Higashi, H., Cambray, A. J., Zhou, J., D'Ambrosio, R., & Winn, H. R. (2003). Hippocampal injury and neurobehavioral deficits are improved by PD 81,723 following hyperglycemic cerebral ischemia. *Experimental Neurology*, *183*, 188–196.

Miller, K. J., Michel, A. D., Chessell, I. P., & Humphrey, P. P. (1998). Cibacron blue allosterically modulates the rat $P2X_4$ receptor. *Neuropharmacology*, *37*, 1579–1586.

Mizumura, T., Auchampach, J. A., Linden, J., Bruns, R. F., & Gross, G. J. (1996). PD 81,723, an allosteric enhancer of the A1 adenosine receptor, lowers the threshold for ischemic preconditioning in dogs. *Circulation Research*, *79*, 415–423.

Mohr, K., Tränkle, C., Kostenis, E., Barocelli, E., De Amici, M., & Holzgrabe, U. (2010). Rational design of dualsteric GPCR ligands: Quests and promise. *British Journal of Pharmacology*, *159*, 997–1008.

Mudumbi, R. V., Montamat, S. C., Bruns, R. F., & Vestal, R. E. (1993). Cardiac functional responses to adenosine by PD 81,723, an allosteric enhancer of the adenosine A1 receptor. *The American Journal of Physiology*, *264*, H1017–H1022.

Musser, B., Mudumbi, R. V., Liu, J., Olson, R. D., & Vestal, R. E. (1999). Adenosine A1 receptor-dependent and -independent effects of the allosteric enhancer PD 81,723. *The Journal of Pharmacology and Experimental Therapeutics*, *288*, 446–454.

Nagaya, N., Tittle, R. K., Saar, N., Dellal, S. S., & Hume, R. I. (2005). An intersubunit zinc binding site in rat $P2X_2$ receptors. *The Journal of Biological Chemistry*, *280*, 25982–25993.

Narlawar, R., Lane, J. R., Doddareddy, M., Lin, J., Brussee, J., & IJzerman, A. P. (2010). Hybrid ortho/allosteric ligands for the adenosine A_1 receptor. *Journal of Medicinal Chemistry*, *53*, 3028–3037.

Nikolakopoulos, G., Figler, H., Linden, J., & Scammels, P. J. (2006). 2-Aminothiophene-3-carboxylates and carboxamides as adenosine A_1 receptor allosteric enhancers. *Bioorganic & Medicinal Chemistry*, *14*, 2358–2365.

Patel, K., Barnes, A., Camacho, J., Paterson, C., Boughtflower, R., Cousens, D., & Marshall, F. (2001). Activity of diadenosine polyphosphates at P2Y receptors stably expressed in 1321N1 cells. *European Journal of Pharmacology*, *430*, 203–210.

Phillis, J. W., Smith-Barbour, M., Perkins, L. M., & O'Regan, M. H. (1994). Characterization of glutamate, aspartate, and GABA release from ischemic rat cerebral cortex. *Brain Research Bulletin*, *34*, 457–466.

Priel, A., & Silberberg, S. D. (2004). Mechanism of ivermectin facilitation of human $P2X_4$ receptor channels. *The Journal of General Physiology*, *123*, 281–293.

Romagnoli, R. (2006). Synthesis and biological characterization of [3H](2-amino-4,5,6,7-tetrahydrobenzo[b]thiophen-3-yl)-(4-chlorophenyl)-methanone, the first radiolabelled adenosine A1 allosteric enhancer. *Bioorganic & Medicinal Chemistry*, *16*, 1402–1404.

Romagnoli, R., Baraldi, P. G., Carrion, M. D., Cara, C. L., Cruz-Lopez, O., Iaconinoto, M. A., Preti, D., Shryock, J. C., Moorman, A. R., Vincenzi, F., Varani, K., & Borea, P. A. (2008). Synthesis and biological evaluation of 2-amino-3-(4-chlorobenzoyl)-4-[*N*-(substituted) piperazin-1-yl]thiophenes as potent allosteric enhancers of the A_1 adenosine receptor. *Journal of Medicinal Chemistry*, *51*, 5875–5879.

Romagnoli, R., Baraldi, P. G., Moorman, A. R., Iaconinoto, M. A., Carrion, M. D., Cara, C. L., Tabrizi, M. A., Preti, D., Fruttarolo, F., Baker, S. P., Varani, K., & Borea, P. A. (2006). Microwave-assisted synthesis of thieno[2,3-c]pyridine derivatives as a new series of allosteric enhancers at the adenosine A1 receptor. *Bioorganic & Medicinal Chemistry*, *16*, 5530–5533.

Romagnoli, R., Baraldi, P. G., Tabrizi, M. A., Gessi, S., Borea, P. A., & Merighi, S. (2010). Allosteric enhancers of A_1 adenosine receptors: State of the art and new horizons for drug development. *Current Medicinal Chemistry*, *17*, 3488–3502.

Schwartz, T. W., & Holst, B. (2006). Ago-allosteric modulation and other types of allostery in dimeric 7TM receptors. *Journal of Receptors and Signal Transduction*, *26*, 107–128.

Seguela, P., Haghighi, A., Soghomonian, J.-J., & Cooper, E. (1996). A novel neuronal P2X ATP receptor ion channel with widespread distribution in the brain. *The Journal of Neuroscience*, *16*, 448–455.

Shen, J., Halenda, S. P., Sturek, M., & Wilden, P. A. (2005). Novel mitogenic effect of adenosine on coronary artery smooth muscle cells: Role for the A1 adenosine receptor. *Circulation Research*, *96*, 982–990.

Silberberg, S. D., Li, M., & Swartz, K. J. (2007). Ivermectin interaction with transmembrane helices reveals widespread rearrangements during opening of P2X receptor channels. *Neuron*, *54*, 263–274.

Soto, F., Garcia-Guzman, M., Gomez-Hernandez, J. M., Hollmann, M., Karschin, C., & Stuhmer, W. (1996). $P2X_4$: An ATP-activated ionotropic receptor cloned from rat brain. *Proceedings of the National Academy of Sciences of the United States of America*, *93*, 3684–3688.

Spedding, M., Menton, K., Markham, A., & Weetman, D. F. (2000). Antagonists and the purinergic nerve hypothesis: 2,2'-pyridylisatogen tosylate (PIT), an allosteric modulator of P2Y receptors. A retrospective on a quarter century of progress. *Journal of the Autonomic Nervous System*, *81*, 225–227.

Stewart, G. D., Sexton, P. M., & Christopoulos, A. (2010). Prediction of functionally selective allosteric interactions at an M3 muscarinic acetylcholine receptor mutant using Saccharomyces cerevisiae. *Molecular Pharmacology*, *78*, 205–214.

Stoop, R., Surprenant, A., & North, R. A. (1997). Different sensitivities to pH of ATP-induced currents at four cloned P2X receptors. *Journal of Neurophysiology*, *78*, 1837–1840.

Toulmé, E., Soto, F., Garret, M., & Boué-Grabot, E. (2006). Functional properties of internalization-deficient $P2X_4$ receptors reveal a novel mechanism of ligand-gated channel facilitation by ivermectin. *Molecular Pharmacology*, *69*, 576–587.

Valant, C., Aurelio, L., Urmaliya, V. B., White, P., Scammells, P. J., Sexton, P. M., & Christopoulos, A. (2010). Delineating the mode of action of adenosine A_1 receptor allosteric modulators. *Molecular Pharmacology*, *78*, 444–455.

Valant, C., Sexton, P. M., & Christopoulos, A. (2009). Orthosteric/allosteric bitopic ligands: Going hybrid at GPCRs. *Molecular Interventions*, *9*, 125–135.

van den Nieuwendijk, A. M., Pietra, D., Heitman, L., Göblyös, A., & IJzerman, A. P. (2004). Synthesis and biological evaluation of 2,3,5-substituted [1,2,4]thiadiazoles as allosteric modulators of adenosine receptors. *Journal of Medicinal Chemistry*, *47*, 663–672.

van der Klein, P. A. M., Kourounakis, A. P., & IJzerman, A. P. (1999). Allosteric modulation of the adenosine A_1 receptor: Synthesis and biological evaluation of novel 2-amino-3-benzoylthiophenes as allosteric enhancers of agonist binding. *Journal of Medicinal Chemistry*, *42*, 3629–3635.

van Galen, P. J., Nissen, P., van Wijngaarden, I., IJzerman, A. P., & Soudijn, W. (1991). 1H-imidazo[4,5-c]quinolin-4-amines: Novel non-xanthine adenosine antagonists. *Journal of Medicinal Chemistry*, *34*, 1202–1206.

van Muijlwijk-Koezen, J. E., Timmerman, H., Link, R., van der Goot, H., & IJzerman, A. P. (1998). A novel class of adenosine A_3 receptor ligands. 1. 3-(2-Pyridinyl)isoquinoline derivatives. *Journal of Medicinal Chemistry*, *41*, 3987–3993.

Virginio, C., Church, D., North, R. A., & Surprenant, A. (1997). Effects of divalent cations, protons and calmidazolium at the rat P2X7 receptor. *Neuropharmacology*, *36*, 1285–1294.

Wildman, S. S., King, B. F., & Burnstock, G. (1998). Zn^{2+} modulation of ATP responses at recombinant $P2X_2$ receptors and its dependence on extracellular pH. *British Journal of Pharmacology*, *123*, 1214–1220.

Wildman, S. S., King, B. F., & Burnstock, G. (1999). Modulatory activity of extracellular H^+ and Zn^{2+} on ATP-responses at $rP2X_1$ and $rP2X_3$ receptors. *British Journal of Pharmacology*, *128*, 486–492.

Xiong, K., Peoples, R. W., Montgomery, J. P., Chiang, Y., Stewart, R. R., Weight, F. F., & Li, C. (1999). Differential modulation by copper and zinc of $P2X_2$ and $P2X_4$ receptor function. *Journal of Neurophysiology*, *81*, 2088–2094.

Yang, H., & Rotstein, D. M. (2010). Novel CCR5 antagonists for the treatment of HIV infection: A review of compounds patented in 2006–2008. *Expert Opinion on Therapeutic Patents*, *20*, 325–354.

Zemkova, H., Kucka, M., Li, S., Gonzalez-Iglesias, A. E., Tomic, M., & Stojilkovic, S. S. (2010). Characterization of purinergic $P2X_4$ receptor channels expressed in anterior pituitary cells. *American Journal of Physiology. Endocrinology and Metabolism*, *298*, E644–E651.

Eduardo R. Lazarowski, Juliana I. Sesma, Lucia Seminario-Vidal, and Silvia M. Kreda

Cystic Fibrosis/Pulmonary Research & Treatment Center, School of Medicine, University of North Carolina, Chapel Hill, North Carolina, USA

Molecular Mechanisms of Purine and Pyrimidine Nucleotide Release

Abstract

Given the widespread importance of purinergic receptor-evoked signaling, understanding how ATP and other nucleotides are released from cells in a regulated manner is an essential physiological question. Nonlytic release of ATP, UTP, UDP-glucose, and other nucleotides occurs in all cell types and tissues via both constitutive mechanisms, that is, in the absence of external stimuli, and to a greater extent in response to biochemical or mechanical/physical stimuli. However, a molecular understanding of the processes regulating nucleotide release has only recently begun to emerge. It is generally accepted that nucleotide release occurs in two different scenarios, exocytotic release from the secretory pathway or via conductive/transport mechanisms, and a critical review of our current understanding of these mechanisms is presented in this chapter.

I. Introduction

Nearly 70 years have elapsed since Fritz Lipman and Herman Kalckar established that ATP is the major source of energy for metabolic processes. In addition, ATP and other nucleotides participate in vital cellular functions, e.g., as cofactors of enzymatic reactions, second messenger cascades, protein glycosylation, and DNA and RNA synthesis. It was long assumed that control mechanisms should be in place to ensure that cellular nucleotides *are not released* to the extracellular milieu. However, we would not be writing these lines, let alone this book, had that concept remained

Advances in Pharmacology, Volume 61

1054-3589/11 $35.00
10.1016/B978-0-12-385526-8.00008-4

unchallenged. Thanks to the vision of Geoff Burnstock, we know now that control mechanisms are in place to ensure that cellular nucleotides *are released* to the extracellular milieu *in a regulated manner*. What we know and do not know about these mechanisms is what this article is about.

The notion that nucleotides are released from cells to accomplish extracellular signaling originated in a series of observations indicating that (i) stimulation of vagal nonadrenergic innervations resulted in accumulation of adenosine and inosine in vascular perfusates from the stomachs of guinea-pigs and toads; (ii) ATP perfused through the toad stomach vasculature was broken down to adenosine and other species; (iii) stimulation of portions of Auerbach's plexus isolated from turkey gizzard caused the release of ATP, ADP, and AMP; and (iv) ATP was the most potent nucleotide causing relaxation of guinea pig gut preparations that contained nonadrenergic inhibitory nerves. Based on these findings, it was suggested that ATP or a related nucleotide is the transmitter substance released by nonadrenergic inhibitory nerves in the gut (Burnstock et al., 1970).

Despite the skepticism with which this proposal was initially greeted, the concept that adenine and uridine nucleotides as well as UDP-sugars function as extracellular signaling molecules has been verified in countless biological, biochemical, physiological, and genetic observations. At least 15 nucleotide-activated cell surface receptors exist in the human genome, and remarkably broad and varied physiological responses are known to occur downstream of nucleotide receptor activation (reviewed in Burnstock, 2006; see Chapters 11–13). Activation of platelets by nucleotides released from red cells was an early observation in the field (Gaarder et al., 1961). Observation of responses to nucleotides in all peripheral tissues, including those not significantly innervated by the autonomic nervous system, indicates that, in addition to purinergic neurotransmission, nucleotides released from nonneuronal sources underlie many important physiological processes (Burnstock, 1972, 1978, 2006; Burnstock & Kennedy, 1985; Burnstock et al., 1970). The significance of nucleotides as extracellular signaling molecules also is underscored by the ubiquitous distribution of several classes of cell surface and secreted ectoenzymes that control nucleotide actions, by catalyzing their breakdown and interconversion (recently reviewed in Yegutkin, 2008; see Chapters 9 and 10).

II. Nucleotide Release Is a Feature of Living Cells

In addition to ATP release from nerve terminals, extracellular ATP has been shown to be released from epithelial and endothelial cells, smooth muscles and fibroblasts, astrocytes, glial cells, macrophages, circulating neutrophils, lymphocytes, monocytes, platelets, red cells, endocrine and exocrine pancreatic cells, hepatocytes and cholangiocytes, and in innumerable transformed cell lines. Virtually, all tissues and cell types, including yeasts and plants, exhibit regulated release of ATP (Corriden & Insel, 2010; Demidchik et al., 2003; Esther et al., 2008; Jeter et al., 2004; Koshlukova et al., 1999; Lazarowski et al., 2003a; Praetorius & Leipziger, 2009; Weerasinghe et al., 2009; Yegutkin, 2008; Zhong et al., 2003).

Given the widespread importance of purinergic receptor-evoked signaling, understanding how ATP and other nucleotides are released from cells in a regulated manner is an essential physiological question. However, a molecular understanding of the processes regulating nucleotide release has only recently begun to emerge. A critical rather than descriptive review of the progress toward this understanding is attempted in the present chapter. Many excellent studies have been published on various aspects of nucleotide release but, due to space concerns, we have narrowed the discussion to what we consider the most recent, mechanistically relevant findings. While neuronal/excitatory ATP release is discussed to some extent, an emphasis is put on mechanisms of nucleotide release from peripheral, nonneuronal tissues.

The availability of the luciferin–luciferase assay, which allows the selective quantification of ATP with picomolar sensitivity, greatly contributed to the widespread interest in studying cellular ATP release. In addition, high-performance liquid chromatography (HPLC)-based techniques quantify fluorescent derivatives of ATP, ADP, AMP, and adenosine with nanomolar sensitivity (Lazarowski et al., 2004; Levitt et al., 1984). Adenosine dinucleoside polyphosphates ($Ap_{(n)}A$) that display relative selectivity on various P2X and P2Y receptors (Table I) have been also identified in extracellular solutions (Miras-Portugal et al., 1999; Pintor et al., 2000). The realization that a subset of P2Y receptors is activated by uridine nucleotides and nucleotide-sugars (Table I; see Chapter 12) further stirred an interest in assessing the occurrence of extracellular uridine nucleotides. Consequently, highly sensitive assays for the quantification of UTP, UDP, and UDP-sugars were developed to illustrate that these molecules are released from cells, in addition to ATP. Thus, all nucleotides/

TABLE I Naturally Occurring Purinergic Receptor Agonists[a]

Nucleotide/nucleoside	*Receptor*
ATP	P2X1–P2X7, $P2Y_2$, $P2Y_{11}$
ADP	$P2Y_1$, $P2Y_{12}$, $P2Y_{13}$
UTP	$P2Y_2$, $P2Y_4$
UDP	$P2Y_6$, $P2Y_{14}$
UDP-glucose (and other UDP-sugars)	$P2Y_{14}$
Adenosine	A_1, A_{2A}, A_{2B}, A_3
Ap_3A, Ap_4A	$P2Y_1$, $P2Y_2$, P2X1–P2X4

[a] The agonist–receptor selectivity corresponds to human receptors.

nucleosides known to activate purinergic receptors have been detected in extracellular solutions bathing living cells (Table I).

A. ATP Release from Resting Cells

The occurrence of basal ATP release was first suggested by studies indicating that resting levels of extracellular ATP conferred tonic activity to the $P2Y_2$ and $P2Y_{11}$ receptors endogenously expressed on Madin–Darby canine kidney (MDCK) cells. That is, resting MDCK cells exhibited basal inositol phosphate and cyclic AMP formation activities that were reduced by the addition of the nucleotidase apyrase or P2Y receptor antagonists (Ostrom et al., 2000). It was further established that steady-state levels of extracellular ATP on resting cells reflected a balance between rates of constitutive ATP release (calculated as 20–200 fmol/min per million cells) and ATP hydrolysis (Lazarowski et al., 2000). The physiological relevance of basal or constitutive ATP release may reside in determining the "set point" for P2Y/P2X receptor-mediated signaling (Corriden & Insel, 2010; Ostrom et al., 2000). Although constitutive ATP release may not suffice to promote P2Y/P2X receptor-mediated responses in many cells due to rapid hydrolysis, it provides a pathway for activation of adenosine receptors. For examples, in the airways, where adenosine is markedly more stable than ATP, constitutive release of ATP (and conversion to adenosine) confers tonic A_{2B} receptor-promoted CFTR (cystic fibrosis transmembrane conductance regulator) Cl^- channel activity that is crucial for ion/water transport in the airways (Huang et al., 2001; Lazarowski et al., 2004).

B. Stimulated ATP Release

1. Agonist-Promoted ATP Release

Release of ATP from adrenaline-, ADP-, or thrombin-stimulated platelets has been known since the early 1960s and 1970s (Holmsen et al., 1972; O'Brien, 1963). Subsequently, aortic endothelial and smooth muscle cells were shown to exhibit thrombin-elicited ATP release (Pearson & Gordon, 1979), and these studies were followed by demonstration of G protein-coupled receptor (GPCR)-evoked ATP release in several nonexcitatory cells, including endothelial cells, pancreatic acini, MDCK, COS-7, and HEK-293 cells (Buxton et al., 2001; Ostrom et al., 2000; Sorensen & Novak, 2001; Yang et al., 1994). However, signaling cascades and release pathways involved in agonist-elicited ATP release were not revealed until recently. As discussed in Section III, GPCR-promoted ATP release has been associated with various mechanistic scenarios. For example, protease-activated receptor (PAR) promoted Ca^{2+}- and Rho-dependent release of ATP and UDP-glucose from vesicles in astrocytes (PAR1) and goblet epithelial cells (PAR1 and PAR2) (Blum et al., 2008; Joseph et al., 2003; Kreda et al.,

2008, 2010); PAR3 evoked Ca^{2+}- and Rho-mediated conductive ATP release from nonmucous lung epithelial cells (Seminario-Vidal et al., 2009); chemotactic peptide induced ATP release via plasma membrane pannexin/connexin hemichannels in neutrophils (Chen et al., 2006, 2010; Eltzschig et al., 2006, 2008); complement C5a evoked pannexin 1 (Px1)-independent ATP release from macrophages; gustative stimulus induced ATP release from taste buds via either pannexin channels or vesicle exocytosis (Huang et al., 2007; Iwatsuki et al., 2009); and prostaglandin and β-adrenergic receptor promoted cyclic AMP-dependent conductive ATP release from erythrocytes (Sprague et al., 2008). In addition to GPCRs, ligation of the T cell receptor (TCR) complex with anti-CD3/CD28 antibodies resulted in enhanced ATP release from splenocytes and Jurkat cells, and both conductive and exocytotic pathways have been proposed to mediate such release (Schenk et al., 2008; Tokunaga et al., 2010; Yip et al., 2009). Forskolin and other reagents known to elevate cellular cyclic AMP levels promoted ATP release in retinal epithelial (Reigada and Mitchell, 2005) and alveolar type II cells (Bove et al., 2010). Elevation of intracellular Ca^{2+} is a commonly used approach to induce exocytotic nucleotide release from neuroendocrine cells (Kreda et al., 2007; Obermuller et al., 2005; Pintor et al., 2000).

2. *Stress-Induced ATP Release*

A striking observation from the early days of purinergic signaling research was the widespread occurrence of robust ATP release in response to physical stresses. Mechanical release of ATP was first reported during sustained exercise of human forearm muscle (Forrester, 1972a, 1972b), but nearly two decades elapsed until the physiological relevance of mechanical ATP release became apparent. In a series of studies during the early 1990s, Burnstock and colleagues illustrated that endothelial cells released ATP in response to increased flow rate, suggesting that flow-promoted ATP release results in P2Y (now $P2Y_1$) receptor-mediated vasodilatation (Bodin et al., 1991; Milner et al., 1990, 1992). Upon availability of molecular biology techniques that facilitated the cloning and expression GPCRs in null systems, robust P2Y-receptor expression-dependent formation of second messengers was noted in cultured cells subjected to mechanical stress, for example, medium displacement or cell wash (Filtz et al., 1994; Lazarowski et al., 1995; Parr et al., 1994). A vast number of studies followed illustrating that nonlytic release of ATP occurred in practically every cell type subjected to physical stresses such as shear stress, hydrostatic pressure, compressive stress, mechanical loading, plasma membrane stretch, hypoxia, and cell swelling (reviewed in Button and Boucher, 2008; Fitz, 2007; Lazarowski et al., 2003a; Sprague et al., 2007; Yegutkin, 2008). Epithelial cells are particularly sensitive to mechanical forces. Compelling evidence for a physiological role of mechanically induced ATP release derived from studies indicating that distention of the urinary bladder resulted in enhanced ATP release from the urothelium. ATP-mediated activation of P2X3 receptors on subepithelial sensory nerve terminals controls bladder volume reflexes of the autonomic nervous system (Cockayne et al., 2000; Ferguson et al., 1997; Vlaskovska et al., 2001). It has been proposed that purinergic mechanosensory transduction effects neural reflexes in distended tubes and sacs, including ureter, vagina, salivary and bile ducts, gut, and urinary and gall bladders (Burnstock, 2009).

Cell volume homeostasis is crucial for normal cell function. Under physiological conditions, cell volume is continuously compromised by the presence of nonpermeable, charged macromolecules and ion imbalance. Perturbations in cell volume occur, for example, as a consequence of transport of nutrients (e.g., active solute uptake by enterocytes, renal tubular cells, or hepatocytes), secretion of fluid and electrolytes by glandular cells, and cell migration and proliferation (Franco et al., 2008). Any physiological or pathological osmotic perturbation will induce a transmembrane flow of water that rapidly restores this equilibrium and gives rise to a concurrent swelling or shrinkage of the cell. The process during which swollen cells decrease their volume is known as regulatory volume decrease (RVD) and involves an efflux of K^+, Cl^-, and organic osmolytes. The Cl^- channel crucially involved in this process is known as the volume-regulated chloride channel (VRAC) or volume-sensitive organic osmolyte and anion channel (VSOAC). Regulatory mechanisms that compensate for cell volume perturbations include ATP release-promoted RVD, as discussed further below.

As acute cell swelling is one of the most robust means to promote nonlytic ATP release, hypotonicity-induced cell swelling is commonly used in studies of regulatory mechanisms and pathways involved in nucleotide release. The vast literature describing hypotonic stress-promoting ATP release includes studies with hepatocytes and cholangiocytes; endothelial cells; airway, alveolar, retinal pigment, and intestine epithelial cells; glial cells and astrocytes; medullary thick ascending limb and kidney epithelial cell lines; and erythrocytes (see Section III and Blum et al., 2010; Boudreault & Grygorczyk, 2002, 2004; Darby et al., 2003; De Vuyst et al., 2005; Gatof et al., 2004; Guyot & Hanrahan, 2002; Hisadome et al., 2002; Jans et al., 2002; Okada et al., 2006; Pafundo et al., 2008, 2010; Ransford et al., 2009; Reigada & Mitchell, 2005; Roman et al., 1997; Sabirov et al., 2001; Shinozuka et al., 2001; Silva & Garvin, 2008; Wang et al., 1996; Watt et al., 1998; Wijk et al., 2003).

C. Release of Uridine Nucleotides and Nucleotide Sugars

1. Release of UTP

Cellular release of UTP was first reported in studies indicating that perfused [^{3}H]uridine-labeled endothelial cells released [^{3}H] species in response to changes in flow rates (Saïag et al., 1995). Taking advantage of the high selectivity that UDP-glucose pyrophosphorylase exhibits for the conversion of UTP to UDP-glucose in the presence of glucose-1 phosphate, an assay that quantifies the UTP-dependent conversion of [^{14}C]glucose-1P to [^{14}C]UDP-glucose with nanomolar sensitivity was developed (Lazarowski & Harden, 1999). Using this assay, the mass of UTP released from various cell types and tissues was quantified under various physiological conditions. It was thus established that, like ATP, UTP is released from cells constitutively and that stimuli promoting ATP release also result in enhanced release of UTP. In most (but not all) cases where UTP release has been measured, the extracellular UTP:ATP mass accumulation ratio (~1:3) resembles the intracellular UTP:ATP ratio, suggesting that both nucleotides were released via common mechanisms and from a common subcellular nucleotide pool (Lazarowski & Harden, 1999). Enhanced release of UTP has been observed with various cultured cell types subjected to mechanical stresses (Lazarowski & Harden, 1999; Lazarowski et al., 1997; Tatur et al., 2008), as well as in apoptotic T lymphocytes (Elliott et al., 2009), lung secretions from RSV-infected mice (Davis et al., 2006), and the blood following heart ischemia (Erlinge et al., 2005). Davis et al. illustrated that removal of UTP from the bronchoalveolar liquid of RSV-infected mice *in vivo* markedly improved alveolar fluid clearance (Davis et al., 2004, 2006). These observations suggest that UTP secreted into the alveolar space plays a role in the pathogenesis of lung edema associated with viral infection. The recent observation that UTP was more efficient than ATP in promoting Cl^- secretion from primary cultures of alveolar type II cells further supports the involvement of UTP release in alveolar epithelial cell homeostasis (Bove et al., 2010).

2. Assessing Extracellular UDP

An enzymatic, HPLC-coupled assay that quantities the phosphorylation of UDP by nucleoside diphosphokinase (NDPK) in the presence of [^{32}P]ATP has been developed to measure UDP in extracellular solutions. Using this assay, Tatur et al. (2008) demonstrated that extracellular UDP levels sharply increased, simultaneously with UTP, ATP, and ADP, during hypotonic cell swelling of lung epithelial A549 cells. The fact that cells were continuously perfused to minimize nucleotide (e.g., UTP) hydrolysis by ecto-NTPDases suggests that the presence of UDP in the perfused solution reflected cellular release of this molecule. As discussed below, observation of UDP release would be consistent with contribution of a vesicular nucleotide pool.

3. Release of UDP-Glucose, UDP-N-Acetylglucosamine, and UDP-Galactose

Expression of the human $P2Y_{14}$ receptor in COS-7 cells resulted in G protein-promoted inositol phosphate accumulation that was reversed by enzymatic removal of UDP-sugars from the medium (Fricks et al., 2008; Lazarowski et al., 2003b). Direct demonstration of UDP-glucose release was achieved by quantifying this nucleotide-sugar in the extracellular medium of COS-7 and other cells. Using a principle similar to that applied to the UTP assay, UDP-glucose pyrophosphorylase was used to drive the pyrophosphorolysis of UDP-glucose in the presence of radiolabeled pyrophosphate (^{32}PPi); the UDP-glucose-dependent formation of [^{32}P]UTP is quantified by HPLC (Lazarowski et al., 2003b). GPCR-promoted Ca^{2+}-dependent UDP-glucose release was observed with thrombin-stimulated 1321N1 astrocytoma cells (Kreda et al., 2008). Given the abundant expression of the $P2Y_{14}$ receptor in astrocytes and glial cells (Bianco et al., 2005; Fumagalli et al., 2003; Moore et al., 2003), release of UDP-sugars from astrocytes potentially has important autocrine/paracrine functions in the brain.

Increased UDP-glucose release rates from primary cultures of human bronchial epithelial (HBE) cells exhibiting goblet cell hyperplasia suggested a link between UDP-glucose release and mucin secretion (Okada et al., 2010). Studies illustrating that Ca^{2+}-regulated mucin secretion from goblet-like Calu-3 airway epithelial cells was accompanied by enhanced release of UDP-glucose (and ATP) provide further evidence that UDP-glucose release from goblet cells is associated with mucin secretion (Kreda et al., 2007). As UDP-sugars are not substrates of ecto-NTPDases, release of UDP-glucose potentially leads to nucleotide-sugar accumulation in the extracellular milieu at concentrations capable of promoting $P2Y_{14}$ receptor activation (EC_{50} = 80–100 nM; Harden et al., 2010). Indeed, high nanomolar/submicromolar levels of UDP-glucose (and UDP-*N*-acetylglucosamine, see below) were found in lung secretions from patients with cystic fibrosis (CF), a chronic lung disease characterized by progressive goblet cell metaplasia, mucus plugs, and neutrophil inflammation (Sesma et al., 2009). As the $P2Y_{14}$ receptor is functionally expressed in neutrophils, lymphocytes, and other inflammatory cells (Moore et al., 2003; Scrivens & Dickenson, 2006), UDP-sugar release from airway epithelial cells has potential pathophysiological implications (i.e., as local modulators of leukocyte function in inflamed lungs).

A recently developed assay for the quantification of UDP-galactose selectively converts this UDP-sugar to UDP and lactose (in the presence of 1-4-β-galactosyltransferase), and the resulting UDP is quantified as above. Enhanced release of UDP-galactose accompanies UDP-glucose release from thrombin-stimulated 1321N1 astrocytoma cells (Lazarowski, 2010).

UDP-*N*-acetylglucosamine is the second most abundant cellular nucleotide after ATP and is a full agonist at the $P2Y_{14}$ receptor (Harden et al., 2010), albeit less potent that UDP-glucose and UDP-galactose. Based on a UDP-*N*-acetylglucosamine pyrophosphorylase (AGX2)-catalyzed reaction, the release of this UDP-sugar was recently reported (Sesma et al., 2009). UDP-*N*-acetylglucosamine release was demonstrated in airway epithelial cells and in yeast and was similar in magnitude to UDP-glucose release. Like UDP-glucose, pharmacologically relevant concentrations of UDP-*N*-acetylglucosamine were found in lung secretions from CF patients (Sesma et al., 2009).

III. Mechanisms of Nucleotide Release

The initial observation that ATP is stored within and released from secretory granules in neuroexcitatory tissues suggested that exocytotic ATP release may be a commonly occurring phenomenon. However, in most nonexcitatory tissues in which ATP release has been documented, *bona fide* ATP-containing granules have not been identified. Thus, it was generally believed that conductive/transport mechanisms mediate ATP release in nonexcitable tissues. This concept has been changing over the past few years. It is now accepted that, to some extent, exocytotic and nonexocytotic nucleotide release exists in both excitatory and nonexcitatory cells.

A. Exocytotic Release of Nucleotides

1. Ca^{2+}-Regulated Exocytosis in Excitatory/Endocrine Tissues

Neurons, chromaffin cells, platelets, and other tissues regulate the release of ATP, neurotransmitters, and other extracellular mediators after packaging them in specialized granules called synaptic vesicles, chromaffin granules, or electron dense-core granules (Anderson et al., 1974; Burnstock, 1997; Dean et al., 1984; Evans & Surprenant, 1992; Evans et al., 1992; Gualix et al., 1996, 1999; Kanner & Schuldiner, 1987). Tightly regulated machinery controls the plasma membrane insertion of specialized granules in secretory tissues in response to a relevant signal. For example, stimulation of chromaffin cells by preganglionic sympathetic neurons results in granule transport along the filaments of the cytoskeleton network to the subplasma membrane compartment, fusion of the granule with the plasma membrane in a Ca^{2+}-dependent manner, and release of contents into the extracellular space, a process commonly referred to as regulated exocytosis (Burgoyne & Morgan, 2003).

a. Chromaffin Cells and Neurons Our current understanding of the mechanism of ATP storage in secretory granules derives mostly from studies with chromaffin cells, due in part to the fact that chromaffin granules are large and relatively easy to isolate without disturbing their luminal content. It has been recognized as early as the 1970s that catecholamines, serotonin, and ATP are transported and costored in chromaffin granules using the electrochemical force provided by the V-type H^+-ATPase of the granule membrane (Aberer et al., 1978; Bankston & Guidotti, 1996; Gualix et al., 1996, 1999; Hanada et al., 1990; Winkler, 1976). The V-type H^+-ATPase pumps H^+ ions into granules, generating an electrical potential ($\Delta\psi$) and a pH gradient that drive neurotransmitters and ATP into vesicles. Bafilomycin A_1 is a potent and highly selective inhibitor of the V-type H^+-ATPase (Bankston & Guidotti, 1996) and is widely used to assess the involvement of secretory granules in the storage/release of ATP. However, the molecular mechanism responsible for the uptake of cytosolic ATP into secretory granules has remained unknown, until recently.

By examining the substrate selectivity and transport properties of orphan members of the SLC17 family of ion transporters (Fredriksson et al., 2008; Reimer & Edwards, 2004; Sreedharan et al., 2010), Moriyama and coworkers provided evidence that SLC17A9 is a vesicular nucleotide transporter (thereafter named VNUT) that displays pharmacological and biochemical features of the ATP transporter endogenously expressed in chromaffin cells (Sawada et al., 2008; Fig. 1). That is, reconstitution of purified recombinant SLC17A9 into liposomes resulted in $\Delta\psi$-driven Cl^--dependent ATP transport activity that was inhibited by 4,4′-diisothiocyanostilbene-2,2′-disulfonate (DIDS). Strong SLC17A9 immunoreactivity that colocalized with synaptotagmin (a secretory granule marker) was observed in chromaffin-like PC12 cells. Knocking down SLC17A9 by siRNA decreased SLC17A9 immunoreactivity and KCl-triggered ATP release in PC12 cells (Sawada et al., 2008). The data suggest that SLC17A9 may be the elusive nucleotide transporter of synaptic vesicles and nerve terminals. Supporting this concept, a recent *in situ hybridization* study indicated that SLC17A9 mRNA is widely expressed in the mouse brain, with particularly high levels of expression in regions of the hippocampus known to display purinergic neurotransmission (Sreedharan et al., 2010). Interestingly, the presence of SLC17A9 immunoreactivity in astrocytes in brain slices supports the notion that astrocytes release ATP via regulated exocytosis (Coco et al., 2003; Pangrsic et al., 2007; Pascual et al., 2005; Zhang et al., 2007).

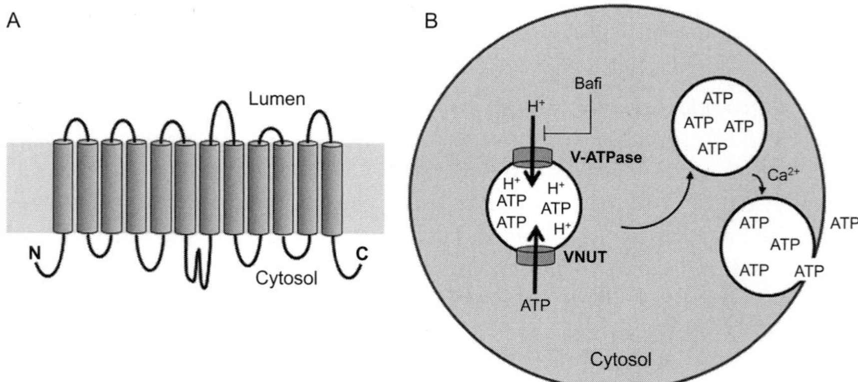

FIGURE 1 *Vesicular nucleotide transporter.* Schematic representation of SLC17A9/VNUT displaying its predicted 12 transmembrane-spanning domains (A). VNUT transports ATP to the lumen of secretory granules, using the electrochemical gradient provided by the bafilomycin A_1 (Bafi)-sensitive proton pump V-ATPase (B).

b. Pancreatic β Cells The existence of an adenine nucleotide-rich compartment associated with insulin granules was first reported in 1975 by Leitner et al., who discovered (i) a correlation between glucose-induced insulin secretion and ATP levels in the effluent of perfused rat islets and (ii) that the granular fraction of islets stored adenine nucleotides (Leitner et al., 1975). Those findings were subsequently confirmed by independent groups (Detimary et al., 1996; Hutton et al., 1983). The physiological implication of these observations is highlighted by the fact that P2Y, P2X, and adenosine receptors promote Ca^{2+} and cyclic AMP-mediated responses in insulin-secreting β-cells (reviewed in Novak, 2008). However, it was only recently established that insulin-secreting cells use the same secretory compartment to release small molecules, including ATP, either alone or together with insulin, via Ca^{2+}-regulated exocytosis. Combining capacitance measurements, electrophysiological detection of ATP release, and single-granule imaging, Obermuller et al. (2005) illustrated that rat Ins1 cells exhibited exocytotic events following voltage clamp depolarization or after dialyzing the cell interior with a solution containing 2 μM Ca^{2+}. Intriguingly, while ATP release was detected ~0.1 ms after membrane fusion, peptide release lagged ~2 s relative to ATP release. These authors also observed that 70% of the exocytotic events were accompanied by ATP release, but not by insulin secretion. Premature closure of the fusion pore may have prevented the release of peptides, but not small molecules (e.g., ATP), to the extracellular milieu, a phenomenon called "transient fusion" or "kiss-and-run" exocytosis. However, an alternative explanation to these observations was that ATP was released from a separate vesicular compartment closely associated with insulin granules. The dilemma was addressed using β-cells isolated from rat pancreatic islets preloaded with serotonin, a well-known component of insulin secretory granules (Aspinwall et al., 1999). Stimulation of exocytosis via intracellular dialysis with Ca^{2+} or by photoliberation of caged Ca^{2+} resulted in closely parallel recordings of ATP and serotonin release, strongly suggesting that ATP release originated from the same compartment as serotonin, that is, the insulin granule (Karanauskaite et al., 2009).

2. Exocytotic Nucleotide Release from Exocrine Tissues/Secretory Cells

Regulated exocytosis of secretory granules has been extensively characterized in pancreatic acinar cells, mast cells, neutrophils, goblet epithelial cells, alveolar type II cells, sperm, and other tissues (reviewed in Burgoyne & Morgan, 2003). Conceivably, ATP could be costored in granules together with pancreatic digestive enzymes, mucins, lung surfactant, inflammatory proteases, histamine, and other molecules susceptible to regulated exocytotic release. The high expression levels of SLC17A9 transcripts found in the stomach, intestine, liver, lung, skeletal muscle, thyroid, spleen, and blood cells suggest that nucleotide storage and release from secretory granules is not restricted to excitable tissues (Sawada et al., 2008; Sreedharan et al., 2010). Two recent examples of nucleotide release from secretory granules of exocrine tissues are discussed here.

a. Pancreatic Acini Early studies with pancreatic acinar cells by Novak and coworkers indicated that zymogen granules avidly take up and store the fluorescent ATP analogue MANT-ATP. As ATP was released concomitantly with digestive enzymes during cholinergic stimulation of the pancreatic acini, zymogen granules may well be the source of released ATP (Sorensen & Novak, 2001). To test this hypothesis, the Novak's group assessed the ability of isolated pancreatic zymogen granules to uptake and store ATP. ATP uptake in zymogen granules was markedly inhibited by bafiloimycin A_1, was stimulated by Cl^-, and exhibited K_m values, pH dependence, and susceptibility to inhibitors similar to VNUT (Haanes & Novak, 2010). Western blot analysis indicated strong VNUT immunoreactivity in isolated zymogen granules. Altogether, the data suggest that the exocrine pancreas utilizes a mechanism similar to neuroendocrine cells for the storage and release of ATP. The physiological relevance of ATP secretion from acini granules is supported by the presence of several P2Y, P2X, and adenosine receptors in the pancreatic acini and distal ducts. Based on the ATP and digestive enzyme content within the zymogen granule and assuming that ATP release parallels enzyme secretion, zymogen granule discharge would contribute with $>8\ \mu M$ ATP to the pancreatic juice, high enough to promote P2Y/P2X receptor activation. Cholinergic stimulation of acini causes release of ATP onto the lumen, potentially leading to upregulation of secretin-evoked bicarbonate and fluid secretion of pancreatic ducts (Novak, 2008).

b. Goblet Epithelial Cells Goblet cells represent a subpopulation of cells within the airways that produce and secrete gel-forming mucins, essential components of the mucociliary clearance machinery that traps and removes microorganisms from the lung (Davis & Dickey, 2008). Gel-forming mucins, the major components of mucus, are extraordinarily large glycoproteins (up to several thousand kDa) and highly water adsorbent. Proper hydration of released mucins is essential for airway homeostasis (Boucher, 2007a; Evans & Koo, 2009; Livraghi et al., 2009; Mall et al., 2004). Puzzlingly, goblet cells lack ion transport activities necessary for the proper hydration of released mucins. Instead, these activities reside in neighboring ciliated cells, the most abundant cell type in normal airways. Electrolyte transport activities at the apical membrane of ciliated cells consist of two major components: Cl^- secretion and Na^+ absorption, which promote or reduce the rate/volume of water secretion into the airways, respectively. Cl^- secretion is primarily driven by the cyclic AMP-regulated Cl^- channel CFTR and to a lesser extent by the calcium-activated Cl^- channel (CaCC), recently identified as ANO/TMEM16 (Caputo et al., 2008; Schroeder et al., 2008; Yang et al.,

2008). Na^+ absorption occurs through the epithelial sodium channel ENaC (Boucher, 2007b). CFTR activation leads to inhibition of ENaC, in addition to enhanced Cl^- secretion. The adenosine A_{2B} and $P2Y_2$ receptors are the major regulators of CFTR and CaCC activities in airway epithelia, respectively (Lazarowski & Boucher, 2009; Fig. 2).

A long-standing fundamental question in lung biology is "how do mucin-secreting goblet cells signal toward ciliated cells to promote water transport into the airway lumen?" That is, what intercellular signaling mechanism relevant to Cl^- and Na^+ channel activities on ciliated cells is triggered by the goblet cell mucin discharge? Kreda et al. (2007) hypothesized that ATP release from goblet cells (and subsequent ATP hydrolysis and adenosine formation) provides a mechanism for mucin hydration during mucin granule secretion (Kreda et al., 2007). Using goblet-like Calu-3 cells, the authors first examined the potential correlation between Ca^{2+}-promoted mucin secretion and ATP release. When polarized cultures of Calu-3 cell monolayers were briefly exposed to the Ca^{2+} ionophore ionomycin, a robust mucin granule

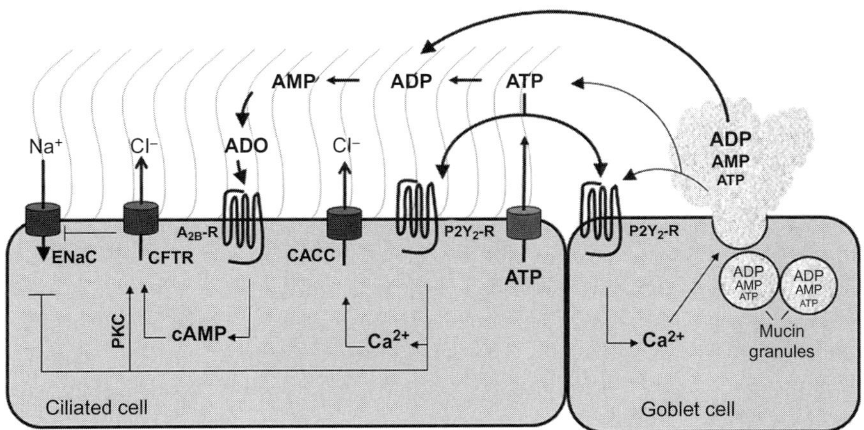

FIGURE 2 *Purinergic regulation of lung mucociliary clearance functions.* The schematic diagram represents airway epithelial ciliated and goblet cells. Mucin exocytosis from agonist-stimulated goblet cells is accompanied by release of adenyl nucleotides present in mucin granules. ADP is the prevalent species within mucin granules, followed by AMP and ATP. Pannexin 1 hemichannels (Px1) potentially mediate ATP release from agonist/stress-stimulated ciliated cells, The constitutive plasma membrane insertion of cargo vesicles (not shown but see Fig. 3) likely provides a pathway for the tonic (constitute) release of nucleotides from both ciliated and goblet cells. Airway surface liquid ATP promotes $P2Y_2$ receptor-mediated Ca^{2+}-activated Cl^- channel (CaCC) activity, and inhibition of the epithelial Na^+ channel ENaC (via PIP_2 depletion, not shown). Potentially, the $P2Y_2$ receptor promotes protein kinase C (PKC)-mediated CFTR activation. ATP/ADP/AMP hydrolysis results in adenosine accumulation (ADO), which in turns activates the A_{2B} receptor, leading to cyclic AMP (cAMP) and protein kinase A-mediated CFTR activation. CFTR inhibits ENaC by mechanisms that are not well understood. The $P2Y_2$ receptor expressed on goblet cells promotes Ca^{2+}-dependent mucin secretion.

discharge occurred, with concomitant accumulation of mucins in the apical bath. Ca^{2+}-promoted mucin secretion was accompanied by enhanced ATP release to the apical bath. Real-time confocal microscopic analyses revealed that mucin granules labeled with the fluorophore FM 1–43 or quinacrine underwent Ca^{2+}-regulated exocytosis, exhibiting kinetics overlapping with that of ATP release. Pharmacological agents that inhibit mucin secretion (e. g., cytoskeleton inhibitors) also inhibited ATP release. Ca^{2+}-promoted ATP release was markedly reduced by bafilomycin A_1, suggesting that a V-type H^+-ATPase-dependent ATP storage compartment was an important source of released ATP (Kreda et al., 2007). These observations were expanded to primary cultures of HBE cells, which were induced to develop goblet cell metaplasia by infection with respiratory syncytial virus or treatment with interleukin-13. Goblet cell metaplastic cultures displayed enhanced mucin secretion, which was accompanied by increased rates of ATP release. Intracellular calcium chelation (BAPTA-AM) or disruption of the secretory pathways (nocodazole, brefeldin A, and *N*-ethylmaleimide) decreased both mucin secretion and ATP release in goblet cell metaplastic HBE cultures (Okada et al., 2010). Altogether, the data suggested that the mucin granule itself stores and releases ATP concomitantly with mucins.

In a follow-up study, mucin granules isolated via density gradients were shown to be enriched with ATP, but ADP and AMP levels in granules were up to 10-fold more abundant than ATP (Kreda et al., 2010). The relatively high levels of ADP and AMP in mucin granules suggested that these nucleotide species are coreleased with ATP and mucins. Indeed, a marked increase of extracellular ADP/AMP accumulation accompanied that of ATP on Calu-3 cells stimulated with the mucin secretagogue thrombin. As ATP release and hydrolysis could not account for the net increase in ADP/AMP levels in Calu-3 cell secretions, the most likely interpretation of these data is that ADP/AMP were released directly from mucin granules (Kreda et al., 2010). By releasing predominantly ADP and AMP, mucin granules have the capacity to minimize autocrine stimulation of mucin release, while favoring adenosine formation, selectively activating ion/water secretion from ciliated cells (Fig. 2).

As Ca^{2+}-promoted ATP release from airway epithelial goblet cells was sensitive to bafilomycin A_1, these cells likely utilize a mechanism of ATP storage in mucin granules similar to that described above for zymosan and chromaffin granules. The presence of SLC17A9 transcripts in the rat lung (Sreedharan et al., 2010) and in Calu-3 cells (Sesma, J. and Lazarowski, E., unpublished) supports this hypothesis.

3. Regulated Nucleotide Release from Orphan Vesicles

Aside from the above-discussed examples, strong, although rather indirect, evidence of exocytotic ATP release has been reported with a number of cells, including lymphocytes (Tsukimoto et al., 2009), liver cells (Dolovcak et al., 2009; Feranchak et al., 2010; Gatof et al., 2004), lung and intestinal epithelial cells (Boudreault & Grygorczyk, 2004; Groulx et al., 2006; Tatur et al., 2007; Wijk et al., 2003), and osteoblasts (Orriss et al., 2009). However, in most cases, the existence of biochemically well-defined secretory granules has not been unambiguously demonstrated.

a. TCR-Stimulated Lymphocytes Activation of cell surface TCRs results in rapid and robust release of ATP from lymphocytes, leading to P2X receptor-evoked calcium-dependent responses, for example, activation of nuclear factor of activated T cells and production of interleukin IL-2, and proliferation (Schenk et al., 2008; Smith-Garvin et al., 2009; Tsukimoto et al., 2009; Yip et al., 2009). Although initial studies suggested that TRC-promoted ATP release encompasses a conductive pathway (Schenk et al., 2008; Yip et al., 2009; see Section III.B.2), a recent report by Tokunaga et al. provided evidence for an exocytotic pathway contributing to ATP release from activated T cells. These authors showed that TCR-activated T lymphocytes and T cell-derived Jurkat lymphoma cells displayed robust Ca^{2+}-dependent ATP release that was inhibited not only by channel blockers but also by inhibitors of vesicle trafficking/exocytosis (e.g., brefeldin A) and bafilomycin A_1. LY294002, an inhibitor of phosphoinositide-3 kinase (PI-3 kinase; a modulator of vesicular trafficking and cytoskeletal organization) (Lindmo & Stenmark, 2006), also reduced ATP release in TCR-stimulated lymphocytes. Incubation of murine $CD4^+$ T cells or Jurkat cells with MANT-ATP resulted in strong vesicular-associated fluorescence, suggesting the presence of an ATP-rich intracellular compartment. Knocking down SLC17A9 in Jurkat cells (via shRNA) reduced TCR-evoked ATP release from these cells (Tokunaga et al., 2010). The data suggest that T cells store and release ATP by Ca^{2+}-regulated exocytotic mechanisms similar to those described with secretory cells. The identity of the putative SLC17A9-expressing ATP-rich compartment remains unknown.

b. ATP Release from Swollen Hepatocytes and Cholangiocytes In the liver, ATP release from hepatocytes and cholangiocytes contributes to bile formation and stimulation of biliary secretion. During the past several years, a series of studies by Fritz and colleagues provided compelling evidence for the existence of an exocytotic pathway for ATP release in liver cells. This group built their case based on the initial observation that hypotonic cell swelling of Mz-ChA-1 cholangiocarcinoma cells resulted in abrupt increase (10-fold) in exocytosis (as measured by the plasma membrane fluorescence probe FM 1–43) and ATP release rates (35-fold) (Gatof et al., 2004). These findings were recently expanded to HTC rat hepatoma cells. Using a highly sensitive CCD camera to record light burst from luciferin/luciferase superfused cells, ATP release in real time was assessed. Following hypotonic shock-induced cell swelling, luminescence increased dramatically in the form of discreet abrupt point source bursts. After loading the cells with quinacrine (reported to fluorescently -load ATP-rich compartments; Chander et al., 1986), fluorescent vesicles were visualized within the 100-nm submembrane space of resting cells, via total internal reflection fluorescence (TIRF) imaging. Real-time visualization of quinacrine granules in nonstimulated cells indicated discrete, focal regions of high fluorescence intensity. Fluorescent vesicles occasionally moved toward the plasma membrane, displaying a transient increase in intensity followed by rapid fading, thus, suggesting spontaneous exocytotic events. Further, the number of exocytosis-like events increased abruptly after exposure of the cells to hypotonic swelling and was accompanied by robust ATP release. Three-dimensional reconstruction of fluorescent confocal microscopy images confirmed that quinacrine-labeled granules rapidly disappeared in response to hypotonicity. Hypotonicity-evoked exocytosis was further verified using the plasma membrane fluorescence probe FM 1–43, that is, fusion of vesicles with the plasma membrane resulted in incorporation FM 1–43 from the extracellular medium to the lipid–liquid interface, resulting in increased fluorescence. Both quinacrine granule labeling and swelling-induced ATP release were inhibited by bafilomycin A_1 as well as by brefeldin A and nocodazole. Exocytotic events and ATP release were impaired by pharmacological inhibitors of protein kinase C (PKC) and PI-3 kinase. Altogether, the data suggest that swollen hepatocytes release ATP from a vesicular, not yet identified quinacrine-loadable compartment susceptible of PKC- and PI-3 kinase regulation. Of note, quinacrine is a weak base that accumulates in its protonated form in acidic compartments, including lysosomes and phagosomes, in addition to a variety of secretory vesicles whose ATP content has not been otherwise assessed (Costa et al., 1984; Di et al., 2002; Goren et al., 1984; Kolber & Henkart, 1988).

4. Constitutive Release of ATP and UDP-Sugars from the Secretory Pathway

Studies of ATP release in the yeast *Saccharomyces cerevisiae* provided initial evidence for the involvement of the secretory pathway in the constitutive release of nucleotides from nonexcitatory cells (Esther et al., 2008; Zhong et al., 2003). Yeast overexpressing Mcd4p (a Golgi-resident transporter postulated to transport ATP to the secretory pathway) displayed increased ATP release, relative to control cells (Esther et al., 2008; Zhong et al., 2003). As ATP release from yeast dramatically decreased when glucose was omitted from the extracellular medium, it was hypothesized that ATP release reflected an exocytotic mechanism initiated by activation of a glucose-sensing cell surface receptor (Zhong et al., 2003). However, HPLC analysis of yeast secretions indicated that, in the absence of glucose, extracellular AMP levels increased robustly, mirroring ATP decrease, and that the net mass of AMP/ADP/ATP released from cells was not affected by extracellular glucose (Esther et al., 2008). Moreover, short-term removal of glucose from the extracellular medium did not affect the ability of yeast to secrete invertase, a marker of exocytosis (Esther et al., 2008). The simplest interpretation of these data is that yeast releases nucleotides constitutively, via vesicle exocytosis, and that the energy balance of the cell determines the relative levels of ATP and AMP within the releasable vesicular pool.

Additional support for the involvement of the secretory pathway in the release of nucleotides was generated from the observation that, in most cells, ATP release is accompanied by the release of UDP-sugars. Constitutive trafficking of secretory vesicles to the plasma membrane, followed by vesicle–plasma membrane fusion events, is required for the insertion of new cell membrane. This pathway also is essential for the export of cell surface proteins and for the secretion of growth factors, antibodies, cytokines and chemokines, and a variety of soluble proteins. The fact that UDP-sugars participate in glycosylation reactions within the secretory pathway suggested that these molecules are released as cargo during the export of glycoconjugates to the plasma membrane (Fig. 3). To test this hypothesis, Sesma et al. examined the effect of removing/overexpressing endoplasmic reticulum (ER)/Golgi-resident UDP-sugar transporters on the cellular release of their cognate substrate. UDP-sugars are synthesized in the cytosol and concentrated in the lumen of the ER and Golgi apparatus to serve as sugar donor substrates for

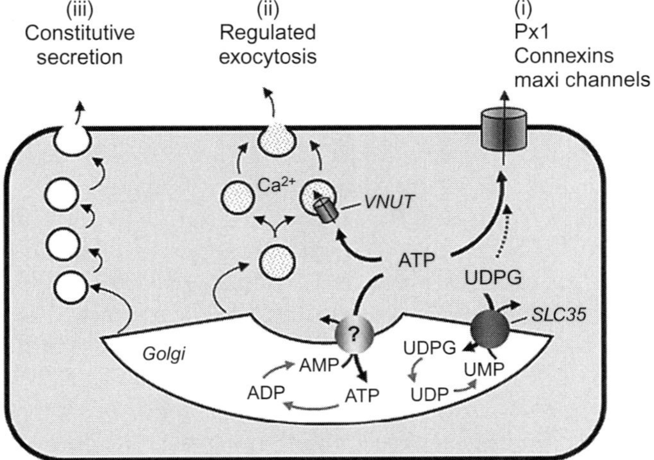

FIGURE 3 *Potential pathways for nucleotide and UDP-sugar release.* Several scenarios possibly account for the basal and stimulated release of nucleotides. (i) Several candidate ATP conductances, including Px1, connexins, and maxi anion channels efflux cytosolic ATP and UTP out of the cells. (ii) VNUT transports ATP into dense-core granules and vesicles competent for Ca^{2+} regulated exocytosis. (iii) UDP-glucose and other UDP-sugars (UDPG), and hypothetically ATP, enter the secretory pathway via ER/Golgi-resident SLC35-like transporters, using UMP and AMP as antiporter substrates, respectively. ER/Golgi nucleotides serving in glycosylation and energy-driven reactions and their luminal metabolites are released as residual cargo molecules of the constitutive secretory pathway.

glycosyltransferase reactions (Berninsone & Hirschberg, 2000; Parodi, 2000; Perez & Hirschberg, 1986). UDP, a by-product of this process, is in turn hydrolyzed to UMP (Berninsone & Hirschberg, 2000). The entry of UDP-sugars to the ER/Golgi is mediated by UDP-sugar transporters, which use luminal UMP as antiporter substrate (Fig. 3). UDP-sugar/UMP translocators belong to the family of solute carrier SLC35 ER/Golgi nucleotide-sugar transporters (Ishida & Kawakita, 2004). The identity of the Golgi UDP-glucose/UMP translocator is not known, but UDP-*N*-acetylglucosamine/UMP translocators have been cloned and characterized (Ishida & Kawakita, 2004). Deletion of the *YEA4* gene, which encodes the UDP-*N*-acetylglucosamine transporter in yeast, resulted in decreased release of this UDP-sugar, which was accompanied by reduced synthesis of chitin, a glucosamine-rich cell wall component (Sesma et al., 2009). The reduced UDP-*N*-acetylglucosamine release and chitin synthesis and export in *yea4*Δ cells were reversed by complementing the mutant strain with the WT *YEA4* gene (Sesma et al., 2009). In humans, three gene products, *SLC35A3*, *SLC35B4*, and *SLC35D2* (also known as *HFRC1*), have been characterized as Golgi-resident UDP-*N*-acetylglucosamine/UMP translocators (Ishida & Kawakita, 2004; Ishida et al., 2005; Suda et al., 2004). Human bronchial epithelial 16HBE14o$^-$ cells stably overexpressing SLC35D2 displayed enhanced cell surface immunoreactivity toward heparan sulfate (a glycosaminoglycan rich in *N*-acetylglucosamine) and increased apical surface binding of WGA (a lectin that recognizes *N*-acetylglucosamine and sialic acids in airway epithelia). Overexpression of SLC35D2 also resulted in an increased rate of brefeldin A-sensitive mucosal release of UDP-*N*-acetylglucosamine, relative to control cells (Sesma et al., 2009). Altogether, these results indicate that, by regulating the entry of nucleotides to the ER/Golgi, nucleotide transporters contribute to the cellular release of their cognate substrates.

Similar to UDP-sugars, ATP is translocated to the ER and Golgi, via not yet identified ATP/AMP antiporters, where it serves as an energy source for protein folding reactions (Guillen & Hirschberg, 1995; Hirschberg et al., 1998). As ATP and UDP-sugars entering the ER/Golgi lumen, as well as their luminal metabolic products (ADP, AMP, UDP, UMP), do not diffuse back to the cytosol, tonic release of ATP, UDP-sugars, as well as ADP and UDP, likely reflects the continuous (constitutive) recycling of proteins and glycoconjugates on the apical plasma membrane and exocytotic release of cocargo solutes, for example, nucleotides (Fig. 3).

B. Conductive Nucleotide Release

The fact that many cells release ATP in the absence of apparent vesicle/granule secretion events suggests that conductive or transport mechanisms provide alternative nucleotide release pathways. Given the high concentration of ATP in the cytosol (1–10 mM), relative to the extracellular milieu

(0.001–10 μM), rapid and robust efflux of cytosolic ATP would be facilitated by a transient opening of a nucleotide-permissible plasma membrane channel or pore, without compromising cellular viability. Several candidate proteins have been postulated to behave as plasma membrane ATP channels or transporters. In many cases, however, evidence supporting the potential cause–effect relationship between channel activity and ATP efflux is circumstantial, relying almost entirely on the use of broadly selective channel/transporter inhibitors.

In spite of initial evidence suggesting that members of the ABC-family of plasma membrane transporters (e.g., the multidrug resistance (MDR)-1 protein and the CFTR Cl^- channel) may provide a pathway for the release of cytosolic ATP (Schwiebert, 1999), the putative ATP channel/transport function attributed to these proteins has not been corroborated. Similarly, a splice variant of the murine mitochondrial voltage-dependent anion channel-1 (VDAC-1) was postulated to function as a plasma membrane ATP pore, based on the observation that cells isolated from VDAC-1 knockout mice exhibited decreased ATP release (Okada et al., 2004). However, plasma membrane insertion of mitochondrial VDAC has not been demonstrated. Bioinformatics sequence analysis of the human VDAC-1 failed to identify a splicing signal compatible with a putative plasma membrane isoform of human VDAC-1 (Okada et al., 2004). It has been postulated that P2X7 receptor channel activation facilitates the release of ATP (Pellegatti et al., 2005; Suadicani et al., 2006), but as the P2X7 receptor is associated with the pore-forming protein Px1 (Pelegrin & Surprenant, 2006), pannexin (see below) may have been the actual conduit for ATP release in response to P2X7 receptor activation.

Recently, two classes of plasma membrane channels have attracted major attention as potential ATP conduits: ATP conducting Cl^- channels and connexin/pannexin hemichannels.

1. ATP Conducting Cl^- Channels

a. Maxi Anion Channels Excellent, comprehensive reviews on the properties of maxi anion channels have been recently published (Sabirov & Okada, 2005, 2009), and, therefore, only the most salient observations linking these channels with an ATP release function are discussed here. An ATP-conductive large-conductance anion channel was first described by Sabirov et al. in murine mammary C127i cells (Sabirov et al., 2001) and subsequently found in a variety of tissues and cell types, including cardiomyocytes, astrocytes, and kidney macula densa (Bell et al., 2003; Dutta et al., 2004, 2008; Liu et al., 2008a, 2008b). Electrophysiological and biophysical studies indicate that maxi anion channels exhibit large single-channel conductance of 300–400 pS, have a wide pore (effective radius of ~1.3 nm), discriminate between Na^+ and halides, and allow the passage of small organic anions, including signaling molecules such as glutamate and ATP^{4-} (Sabirov &

Okada, 2004, 2009; Sabirov et al., 2001). The maxi anion channel is also permeable to ATP^{2-}, ADP^{3-}, and UTP^{4-} (Sabirov & Okada, 2005). Maxi anion channels are activated by osmotic cell swelling, ischemia/hypoxia, and in response to excision of the membrane patch, and are inhibited by Gd^{3+} and by a number of anion channel blockers of relative selectivity, such as DIDS, NPPB (5-nitro-2-(3-phenylpropylamino)benzoic acid), SITS (4-acetamido-4′-isothiocyanatostilben-2,2′-disulfonate), and DPC (diphenylamine-2-carboxylate), but not by glybenclamide (Sabirov & Okada, 2009). ATP release from cells exhibiting maxi anion channel conductance is also inhibited by SITS, NPPB, and Gd^{3+} (Sabirov & Okada, 2009). By measuring ATP release via a P2X2 receptor-based biosensor inserted into the patch pipette, maxi anion channel activity was shown to coincide, spatially, with P2X2 receptor-dependent currents. Altogether, these studies strongly suggest that the maxi anion channel is a physiologically relevant ATP conduit.

Osmotic cell swelling and hypoxia/ischemia are well-recognized maxi anion channel stimuli, but the signaling mechanism transducing osmotic or hypoxic stresses into channel activation are incompletely characterized. Previous studies on cortical collecting duct cells indicated that a large single-channel Cl^- conductance reminiscent of the maxi anion channels described above was inhibited by removal of cytosolic Ca^{2+} and activated by Ca^{2+} ionophores (Light et al., 1990). Similarly, large-conductance anion channels present in Swiss 3T3 fibroblasts and peritoneal macrophages were activated by Ca^{2+} ionophores (Kawahara & Takuwa, 1991; Kolb & Ubl, 1987). Whether Ca^{2+}-promoted activation of these channels resulted in enhanced ATP conductance was not reported. However, lowering the Ca^{2+} concentration in the cytosolic bath of C127i cell-excised patches failed to affect excision-stimulated maxi anion channel activity (Toychiev et al., 2009), suggesting that ATP-conductive maxi channels are not regulated by changes in cytosolic Ca^{2+}. It was recently reported that maxi anion channel activity (in response to patch excision) was enhanced by protein tyrosine kinase inhibitors, but not by PKC-selective inhibitors or protein serine/threonine kinase inhibitors. Further, a cocktail of tyrosine phosphatase inhibitors has a robust inhibitory effect on the C127i cell maxi anion channel activity (Toychiev et al., 2009). The maxi anion channel-like activity present in mouse skin fibroblasts was reduced in cells isolated from receptor protein tyrosine phosphatase (RPTP)ζ-deficient mice. These observations suggest that maxi anion channels (or a closely associated regulatory protein) are maintained in an inactive state by tyrosine phosphorylation. Relevant to ATP release, it would now be important to determine whether the above-reported tyrosine phosphatase regulation applies to the ATP conductance that has been associated with the maxi anion channel and, if so, how intact cells transduce osmotic/hypoxic stress into tyrosine phosphatase activation-promoted maxi anion channel-mediated ATP release.

The molecular identity of the maxi anion channel is not known. Contrary to initial reports suggesting that mitochondrial VDAC and maxi anion channels share a number of bioelectrical properties, genetic evidence indicates that these two activities are not related. That is, the maxi anion channel activity observed in wild-type murine fibroblasts remained essentially unchanged in fibroblasts from mice deficient in VDAC-1, VDAC-2, or VDAC-3, and in VDAC-1/VDAC-3 double knockouts in which VDAC-2 was knocked down by siRNA (Sabirov & Okada, 2005, 2009; Sabirov et al., 2006).

b. Tweety The three human homologues to the *Drosophila flightless* gene *tweety* (TTYH1, TTYH2, and TTYH3) are predicted to encode large-conductance Cl^- channels with properties similar to the maxi anion channel described above (Suzuki & Mizuno, 2004). However, expression of two splice variants of TTYH1 (TTYH1-SV and TTYH1-E) in null cells failed to produce maxi anion channel activity in excised patches comparable with that in C127i cells (Sabirov & Okada, 2009). Interestingly, TTYH3, a predicted 523 amino acid-long protein with six transmembrane-spanning domains and broad tissue distribution, exhibited CaCC activity (Suzuki & Mizuno, 2004) and has been recently implicated in the release of ATP. Knocking down TTHY3 via shRNA markedly decreased ATP release from HL60 cells stimulated with the formyl-bacterial peptide fMLP (Chen et al., 2010).

c. Volume-Regulated Anion Channels Based on the estimated pore radius ($\sim$0.7 nm) and other considerations (see below), the ubiquitously distributed volume-regulated anion channels (VRACs) are predicted to be permeable to organic anions, including amino acids and ATP (Hisadome et al., 2002; Nilius & Droogmans, 2003; Okada, 1997; Okada et al., 2009; Sabirov & Okada, 2005). VRAC, also known as the volume-sensitive outwardly rectifying (VSOR) Cl^- channel, is rapidly activated in response to osmotic cell swelling. VRAC Cl^- channel activity is inhibited by a broad spectrum of channel, transporter, and enzyme inhibitors, including glibenclamide, verapamil, tamoxifen, fluoxetine, nordihydroguaiaretic acid (NDGA), 1,9-dideoxyforskolin, niflumic acid, quinine, NPPB, DIDS, and SITS (Blum et al., 2010; Nilius & Droogmans, 2003). VRAC activity also is suppressed by high concentrations (1 mM) of extracellular ATP and other nucleotides. The voltage dependency of VRAC inhibition by ATP is consistent with a binding site for the blocker within the open pore, supporting the concept that ATP can physically pass through the channel. VRAC inhibitors suppressed ATP release in aortic endothelial cells (Hisadome et al., 2002) and in 1321N1 astrocytoma cells (Blum et al., 2010) exposed to hypotonic solutions, but not in other cells (reviewed in Sabirov & Okada, 2005). The rather low selectivity of the reagents used to inhibit VRAC activity has been a problem in assigning an ATP conduit role to this Cl^- channel. A recent report illustrated that bradykinin promotes Cl^- channel activity in mouse astrocytes with electrophysiological and pharmacological properties similar, if not identical, to VRAC. Bradykinin-elicited VRAC-like activation was mirrored by the release of glutamate, which was inhibited by DIDS and other VRAC blockers but, surprisingly, was not accompanied by ATP release (Liu et al., 2009). Thus, until the molecular identity of VRAC is established and protocols to selectively knockdown/out VRAC developed, the potential role of VRAC as an ATP channel remains unknown.

2. Connexin and Pannexin Hemichannels, General Considerations

Connexin and pannexin hemichannels have recently emerged as candidate conduits for ATP release in a broad range of tissues and cell types. Connexin subunits are the building blocks of vertebrate gap junction channels, formed at sites of direct cell-to-cell contact. Twenty-one human connexin isoforms have been identified, with predicted size of individual species ranging from 23 to 62 kDa and, accordingly, referred as to Cx*n*, where *n* denotes the molecular weight (e.g., Cx30, Cx32, Cx43). A separate family of mammalian proteins (pannexins) exhibits approximately 20% primary sequence similarity to that of innexins, the components of gap junctions in invertebrates. Connexins and pannexins share a similar predicted structure of four transmembrane domains, with N and C termini located in the cytosol (reviewed in Scemes et al., 2009). Unlike connexins, pannexins display consensus sequences for glycosylation. Hemichannel assemblies composed of connexin subunits are known as connexons, whereas those composed of pannexins are called pannexons. Six subunits form a hemichannel (Scemes et al., 2009).

Connexons concentrate at gap junction domains of the cell membrane. Connexons from two adjacent cells align and bind to each other, forming an intercellular gap that allows the passage of small cytosolic molecules between cells. In addition, some connexons may localize to nonjunctional regions of the plasma membrane, thus forming functional plasma membrane channels that are not involved in cell contact (Scemes et al., 2009). In contrast to connexons, plasma membrane pannexons do not readily assemble into the plaque-like ensembles that typify gap junctions; pannexin glycosylation likely prevents the docking between pannexons on adjacent cells (Boassa et al., 2008; Penuela et al., 2007). Most hemichannels exhibit permeability to ATP and small dyes (e.g., propidium and ethidium iodide, carboxyfluorescein, YoPro, and Lucifer Yellow). Opening of connexins and pannexins can be induced by membrane depolarization, typically in the 40–60 mV range for connexins and 20–40 mV for pannexins. Some connexin (but not pannexin) hemichannels are activated by lowering the extracellular Ca^{2+} concentration, likely due to Ca^{2+}-induced conformational changes leading to pore closure (Muller et al., 2002). The cell biology, pharmacological, and electrophysiological properties of connexins and pannexins have been extensively reviewed (Dahl & Locovei, 2006; Evans et al., 2006; Ma et al., 2009; MacVicar & Thompson, 2010; Scemes et al., 2009; Thompson & MacVicar, 2008).

a. Connexin Hemichannel-Mediated ATP Release Nedergaard and coworkers pioneered the concept that connexin hemichannels conduct ATP, by illustrating that expression of Cx43 or Cx32 in gap junction-deficient C6 rat glioma cells increased ATP release in response to extracellular Ca^{2+} removal. A subsequent study indicated that lowering the extracellular Ca^{2+} concentrations in astrocytes, endothelial cells, bronchial epithelial cells, and Cx43-overexpressing C6 (Cx43-C6) cells resulted in increased uptake of the hemichannel probe propidium iodide, in addition to enhanced ATP release (Arcuino et al., 2002). More recently, by combining real-time bioluminescence imaging of ATP release from Cx43-C6 cells with electrophysiological measurement of hemichannel opening, this group elegantly illustrated that ATP efflux coincided with channel openings, and that excised Cx43-expressing patches exhibited ATP conductance (Kang et al., 2008).

In line with the above-described observations, Stout et al. reported that mechanical stimulation (with glass microBeads) of Cx43-C6 cells and mouse astrocytes resulted in increased ATP release and dye uptake, which were inhibited by the connexin hemichannel inhibitor flufenamic acid (Stout et al., 2002). In addition, De Vuyst et al. reported that Cx43-C6 and Cx32-C6 cells, but not wild-type (WT) C6 cells, displayed enhanced ATP release and propidium iodide uptake following a controlled increase of intracellular calcium (500 nM Ca^{2+}) or exposure of the cells to divalent cation-free solutions (De Vuyst et al., 2005, 2009). Based on the effect of a range of pharmacological inhibitors (on ATP release in Cx43-C6 cells), these authors proposed that elevation of intracellular Ca^{2+} triggers hemichannel opening via multiple signaling steps, including Ca^{2+}/calmodulin-dependent pathways, arachidonic acid metabolism, MAP kinase activation, formation of reactive oxygen species and nitric oxide, and depolarization (De Vuyst et al., 2009). How these pathways/activities potentially promote Cx43 opening remains unclear.

Physiological roles for connexin-mediated ATP release have been recently suggested. Release of ATP in specialized areas of the ventral surface of the medulla oblongata (VMS) is an important signaling step in the central chemosensory transduction of hypoxia (CO_2 stimulus). Released ATP may excite respiratory neurons via P2X and P2Y receptors on their ventrally directed dendrites that approach the VMS. Using an *in vitro* model of the VMS, Huckstepp et al. illustrated that CO_2-triggered ATP release was accompanied by the uptake of hemichannel dyes and inhibited by agents that reduce connexin (but not pannexin) hemichannel activity. Cx26 is preferentially expressed in GFAP-positive cells in areas of VMS displaying CO_2-dependent dye uptake, suggesting that Cx26-mediated release of ATP in response to changes in CO_2 levels contributes to central respiratory chemosensitivity (Huckstepp et al., 2010). The potential contribution of hemichannels in the release of ATP from astrocytes during ischemia and epilepsy has been recently reviewed (Dale & Frenguelli, 2009).

In the inner ear, ATP-evoked Ca^{2+} wave propagation has been linked to noise-induced hearing loss and development of hair cell-afferent synapses. Ca^{2+} wave propagation in cochlear organotypic cells (e.g., triggered by photostimulation of caged IP_3) was reduced by apyrase, suramin, carbenoxolone, niflumic acid, flufenamic acid, and lanthanum, suggesting that Ca^{2+} waves reflected hemichannel opening-mediated ATP release (Anselmi et al., 2008). Cx26 and Cx30 are the major connexins expressed in the cochlea. Cochlear organotypic cultures from Cx26 (tissue-targeted)- or Cx30-deficient mice, but not Px1-deficient mice, exhibited reduced ATP release, impaired Ca^{2+} wave propagation, and reduced dye efflux (in calcein-loaded cultures), suggesting that Cx26 and Cx30 operate as ATP conduits in the inner ear (Anselmi et al., 2008).

ATP release in response to inflammatory mediators is a fundamental mechanism required for neutrophil activation and immune defense (Chen et al., 2006, 2010; Eltzschig et al., 2006, 2008). Nonspecific inhibitors of gap junction channels and a Cx43-blocking peptide reduced ATP release from fMLP-activated human neutrophils. fMLP-promoted ATP release from neutrophils was paralleled by Cx43 dephosphorylation (as visualized with an antiphospho-Cx43 antibody) and inhibited by the protein phosphatase-A inhibitor okadaic acid. Neutrophils isolated from Cx43 knockout mice exhibited impaired ATP release (Eltzschig et al., 2006). These findings suggested that neutrophil ATP release occurs through Cx43 hemichannels in a protein phosphatase-A-dependent manner (Eltzschig et al., 2006). However, the involvement of Cx43 in ATP release from neutrophils in humans has been questioned by a recent study indicating no detectable levels of Cx43 in these cells or neutrophil-like HL-60 cells (Chen et al., 2010 and see below).

Phenotypes associated with connexin gene deletion or RNA silencing should be interpreted with caution. For example, Cx43 may exhibit functions that are independent of channel activity. Cx43 may form complexes with a variety of proteins, including transcriptional proteins (e.g., β-catenin and members of src family of oncogenes), membrane receptors, cell signaling molecules and scaffolding proteins (e.g., ZO1), tubulin, and caveolin-1 (Gilleron et al., 2008; Langlois et al., 2010; Singh & Lampe, 2003; Toyofuku et al., 1998). Astrocytes from Cx43 knockout mice exhibited decreased $P2Y_1$ (and increased $P2Y_2$) receptor expression (Scemes, 2008). Further, deletion of genes encoding Cx32, Cx36, or Cx43 affected the transcription of a large number of unrelated genes (Spray & Iacobas, 2007). Array analysis of Cx43 knockout and knockdown astrocytes indicated transcriptional changes affecting genes encoding proteins involved in a wide diversity of cell functions (Iacobas et al., 2008). Last, the physiological relevance of Cx-mediated ATP release in response to low extracellular Ca^{2+}

and/or highly depolarizing protocols is not well understood (Scemes et al., 2007, 2009).

b. Px1 as an ATP Release Conduit The fact that Px1 does not form gap junctions in mammalian tissues (Penuela et al., 2007), can be activated at physiological membrane depolarization conditions, and is not affected by extracellular Ca^{2+} concentrations (Bruzzone et al., 2003, 2005) suggests that this protein may be a relevant conductive pore for ATP release. Initial support for this hypothesis was provided by studies by Dahl and coworkers indicating that (i) expression of human Px1 in *Xenopus oocytes* resulted in the appearance of a membrane channel activity that exhibited ATP conductance, (ii) conditions favoring hemichannel opening (high K^+) enhanced ATP release from Px1-expressing oocytes, and (iii) mechanical stimulation (membrane stretch) or elevation of intracellular Ca^{2+} promoted Px1 activation (Bao et al., 2004; Locovei et al., 2006b). Further, human erythrocytes exhibited mechanosensitive large-conductance channel activity consistent with Px1, and both high potassium and hypotonic stress enhanced ATP release and dye uptake from erythrocytes in a carbenoxolone-sensitive manner (Locovei et al., 2006a). More recently, this group identified the anti-inflammatory drug probenecid as an effective inhibitor of Px1 (but not connexin) activity (Silverman et al., 2008).

In addition to probenecid, low micromolar concentrations of the licorice root component carbenoxolone and a peptide mimicking a segment of the first extracellular loop of Px1 (WRQAAFVDSY, ^{10}Panx1) are relatively selective inhibitors of Px1 (Ma et al., 2009; Pelegrin & Surprenant, 2006; Wang et al., 2007) and have been shown to reduce ATP release in a broad range of tissues. For example, probenecid, carbenoxolone, and/or ^{10}Panx1 inhibited ATP release from hypoxic red cells (Sridharan et al., 2010), pressured retina (Reigada et al., 2008), electrically stimulated skeletal muscle (Buvinic et al., 2009), TRC-stimulated lymphocytes (Yip et al., 2009), and chemoattractant-activated neutrophils (Chen et al., 2010). Applying a gustatory stimulus (cycloheximide/denatonium/saccharin) to isolated taste bud receptor cells resulted in carbenoxolone-sensitive ATP release (RT-PCR and *in situ* hybridization analysis indicated abundant expression of Px1 in these cells) (Huang et al., 2007). Consistent with Px1 activation, elevation of intracellular Ca^{2+} in taste bud cells (via $P2Y_2$ receptor activation) resulted in enhanced carbenoxolone-sensitive uptake of the dye carboxyfluorescein (Huang et al., 2007). However, Romanov et al. reported that depolarization of taste cells resulted in ATP release and dye uptake that were independent of Ca^{2+} and not affected by carbenoxolone. Instead, these responses were reduced by connexin inhibitors (e.g., the connexin-blocking peptide ^{43}GAP26 and octanol), suggesting that connexins rather than Px1 may be responsible for the ATP release observed in these preparations (Romanov et al., 2007). Whether these findings reflect stimulus selectivity (taste receptor agonists vs. voltage gating) and/or different contribution of connexins and Px1 within taste cell subpopulations would require further scrutiny with cell type/tissue-specific disruption of the relevant connexin gene, as well as Px1 gene deletion (Romanov et al., 2008).

Recently, Px1-knockdown and Px1 overexpression have been reported to decrease and enhance, respectively, ATP release in hypotonic stress-stimulated epithelial cells. In the eye, the hydrostatic (intraocular) pressure is regulated in part by the rate of aqueous humor formation, which in turn is regulated by adenosine and ATP acting, presumably, via A_3 and $P2Y_2$ receptors, respectively. ATP release from hypotonic stress-stimulated cultures of bovine ciliary epithelial cells was reduced ($\sim$40%) by carbenoxolone and probenecid, while overexpression of bovine Px1 in HEK293T cells resulted in increased release of ATP (Li et al., 2011). Using primary cultures of HBE cells (dominated by ciliated cells), Ransford et al. (2009) illustrated that hypotonic shock-stimulated ATP release was reduced by Px1 inhibitors as well as by knocking down Px1 via shRNA. In line with this report, studies in our lab indicated that agonist (thrombin)- and hypotonic stress-stimulated HBE and A549 lung epithelial cells displayed enhanced ATP release and propidium iodide uptake, which were reduced by carbenoxolone and/or Px1 siRNA (Lazarowski et al., 2010; Seminario-Vidal et al., 2009).

It could be argued that opening of such as a large, nonselective channel as Px1 may result in ion gradient collapse and cell death. Indeed, Px1 activation has been associated with cell death in ischemic (O_2/glucose-deprived) hippocampal and cortical neurons (Thompson et al., 2006), aberrant bursting in *N*-methyl-D-aspartate receptor (NMDAR)-stimulated pyramidal neurons (Thompson et al., 2008), and apoptotic T lymphocytes (Chekeni et al., 2010). It has been also suggested that Px1 forms a large pore contributing to cell death during prolonged activation of the P2X7 receptor (Locovei et al., 2007; Pelegrin & Surprenant, 2006). Thus, strict control mechanisms should regulate Px1 opening/closing under physiologic (i.e., nonlethal) conditions. Based on the observation that extracellular ATP exhibits Px1 channel blocker activity (Ma et al., 2009; Qiu & Dahl, 2009), it was suggested that ATP release may act as negative feedback on Px1, preventing deleterious long-lasting opening of the channel (Qiu & Dahl, 2009). However, ATP IC_{50} values for inhibiting Px1 activity are in the high ($>$700) micromolar range (Ma et al., 2009; Qiu & Dahl, 2009) and, therefore, unlikely to occur under conditions in which physiological ATP release has been observed. Alternatively, cell surface localization and activity of Px1 may be regulated by its interaction with cytoskeletal components, as suggested by the fact that Px1 coimmunoprecipitated and cosedimented with F-actin (Bhalla-Gehi et al., 2010). Interestingly, disrupting Rho signaling (a known upstream effector of cytoskeleton rearrangements) markedly decreased the carbenoxolone-sensitive propidium iodide uptake and ATP release in lung epithelial cells (Lazarowski et al., 2010; Seminario-Vidal et al., 2009).

In addition to its involvement in the physiological release of ATP, a mechanism for the irreversible activation of Px1 during apoptotic cell death was recently unveiled. Using a model of anti-Fas-exposed T cells, Ravichandran and colleagues demonstrated that apoptotic lymphocytes release the "find me" signals ATP and UTP at the earliest stages of death to recruit phagocytes, and that this release was inhibited by a cell permeable pan-caspase inhibitor (Elliott et al., 2009). They subsequently demonstrated that carbenoxolone, probenecid, and Px1 knockdown (via siRNA) reduced ATP and UTP release from apoptotic T cells. Conversely, Px1 overexpression enhanced the release of nucleotides from apoptotic Jurkat cells (Chekeni et al., 2010). Whole cell patch-clamp recording of single-cell channel activity indicated the presence Px1-dependent carbenoxolone-sensitive currents in apoptotic (but not control, "living") cells. A caspase-cleavage site (site B) within the C terminus of Px1 (DVVD at residues 376–379) was crucial for the induction of Px1-associated responses during apoptosis. That is, mutating aspartic acid to either alanine or glutamic acid rendered Px1 resistant to caspase, resulting in the loss of currents, impaired dye uptake, and reduced ATP release. Consistent with these observations, a truncation mutant that mimics the cleavage of Px1 at site B exhibited enhanced dye uptake and current-voltage relationship (in nonapoptotic cells) that resembles that of activated wild-type Px1, suggesting that removal of site B resulted in constitutively active Px1 (Chekeni et al., 2010). Thus, site B may function as a dominant-negative domain within Px1.

IV. Conclusion

There is little doubt that exocytotic and conductive/transport mechanisms operate to a variable extent in practically all cell types. In many cases, secretory cells display conductive nucleotide release in addition to exocytotic pathways, and conversely, vesicular nucleotide release has been documented with cells in which the presence of secretory granules could not be morphologically demonstrated. However, as recently noted by Blum et al. (2010), a common feature of most stimuli that promote nucleotide release is the activation of cytoskeleton rearrangements, which in turn facilitates the translocation, reorganization, or fusion to the plasma membrane of components of the nucleotide release machinery. Thus, demonstrating that vesicles are the source of released nucleotides, as opposed to a pathway for the plasma membrane insertion of a nucleotide channel or its regulator, has been challenging.

The occurrence of exocytotic nucleotide release from exocrine/endocrine tissues is firmly supported by the demonstration that isolated secretory granules store ATP and other nucleotides. The recent identification of the vesicular nucleotide transporter SLC17A9/VNUT has provided a valuable tool for assessing the contribution of secretory pathways to the release of ATP from chromaffin cells, nerve terminals, pancreatic acinar cells, and potentially other tissues in which exocytotic nucleotide release has been postulated. The observation that purified VNUT conferred ATP, ADP, and UTP transport activity to reconstituted vesicles suggests that the secretory pathway could be a source of released ADP and uridine nucleotides, in addition to ATP. However, the apparent lack of expression of SLC17A9 in discrete brain regions known to exhibit purinergic neurotransmission suggests that additional vesicular nucleotide transporters exist. Predictably, ATP transporters other than SLC17A9 may be also expressed in nonneuronal tissues. Elucidating whether orphan members of the broadly expressed α-group of SLC transporter (Fredriksson et al., 2008; Sreedharan et al., 2010) function as vesicular nucleotide transporters would greatly expand our understanding of the contribution of the secretory pathways to nucleotide release.

A separate class of SLC transporters is expressed in the ER and Golgi apparatus and translocates UDP-sugars from the cytosol to these organelles, using UMP as luminal antiporter substrate (Berninsone & Hirschberg, 1998). ER/Golgi-resident UDP-*N*-acetylglucosamine/UMP translocators (human SLC35D2 (HFRC1) and yeast Yea4) contribute to the release of this UDP-sugar from nonmucous epithelial cells and from yeast, strongly suggesting that the ER/Golgi lumen represents a major source of UDP-sugar release in "nonsecretory" cells (Sesma et al., 2009). Identifying the ER/Golgi-resident UDP-glucose/UMP translocator at the molecular level would greatly advance our knowledge of the autocrine/paracrine mechanisms leading to $P2Y_{14}$ receptor activation.

Extensively investigated, although less completely characterized, are the pathways leading to cytosolic nucleotide release. Candidate plasma membrane ATP channels include members of the connexin family of gap junction proteins and the nonjunctional channel Px1. Assessing the effect of connexin- and pannexin-selective deletion on the temporal and spatial correlation between channel activity and ATP release would strengthen the case for these proteins as plasma membrane ATP conduits. The realization that the C terminus of Px1 contains a proteolysis-sensitive inhibitory domain suggests an additional degree of complexity in the understanding of mechanisms regulating the activation/inactivation of this channel in living cells. Addressing these issues, identifying the molecular components of maxi anion channels and VRACs and elucidating the signaling mechanisms transducing osmotic/mechanical stresses into nucleotide release are pending challenges to the field.

Acknowledgments

We thank Lisa Brown for editorial assistance of the chapter. Supported by National Institute of Health grant P01-HL034322.

Conflict of Interest : The authors have no potential conflict of interest.

Abbreviations

CFTR	cystic fibrosis transmembrane conductance regulator
Cx	connexin
DIDS	4,4′-diisothiocyanostilbene-2,2′-disulfonate
fMLP	formyl-Met-Leu-Phe
GFAP	glial fibrillary acidic protein
GPCR	G protein-coupled receptors
NBPP	5-nitro-2-(3-phenylpropylamino)benzoic acid
Px	pannexin
shRNA	short hairpin RNA
siRNA	small interfering RNA
SLC	solute carrier
TCR	T cell receptor
VDAC	voltage-dependent anion channel
VRAC	volume-regulated anion channel
VSOR	volume-sensitive outwardly rectifying

References

Aberer, W., Kostron, H., Huber, E., & Winkler, H. (1978). A characterization of the nucleotide uptake of chromaffin granules of bovine adrenal medulla. *The Biochemical Journal*, *172*, 353–360.

Anderson, P., Rohlich, P., Slorach, S. A., & Uvnas, B. (1974). Morphology and storage properties of rat mast cell granules isolated by different methods. *Acta Physiologica Scandinavica*, *91*, 145–153.

Anselmi, F., Hernandez, V. H., Crispino, G., Seydel, A., Ortolano, S., Roper, S. D., Kessaris, N., Richardson, W., Rickheit, G., Filippov, M. A., Monyer, H., & Mammano, F. (2008). ATP release through connexin hemichannels and gap junction transfer of second messengers propagate Ca2+ signals across the inner ear. *Proceedings of the National Academy of Sciences of the United States of America*, *105*, 18770–18775.

Arcuino, G., Lin, J. H., Takano, T., Liu, C., Jiang, L., Gao, Q., Kang, J., & Nedergaard, M. (2002). Intercellular calcium signaling mediated by point-source burst release of ATP. *Proceedings of the National Academy of Sciences of the United States of America*, *99*, 9840–9845.

Aspinwall, C. A., Huang, L., Lakey, J. R., & Kennedy, R. T. (1999). Comparison of amperometric methods for detection of exocytosis from single pancreatic beta-cells of different species. *Analytical Chemistry*, *71*, 5551–5556.

Bankston, L. A., & Guidotti, G. (1996). Characterization of ATP transport into chromaffin granule ghosts—Synergy of ATP and serotonin accumulation in chromaffin granule ghosts. *The Journal of Biological Chemistry*, *271*, 17132–17138.

Bao, L., Locovei, S., & Dahl, G. (2004). Pannexin membrane channels are mechanosensitive conduits for ATP. *FEBS Letters*, *572*, 65–68.

Bell, P. D., Lapointe, J. Y., Sabirov, R., Hayashi, S., Peti-Peterdi, J., Manabe, K., Kovacs, G., & Okada, Y. (2003). Macula densa cell signaling involves ATP release through a maxi anion channel. *Proceedings of the National Academy of Sciences of the United States of America*, *100*, 4322–4327.

Berninsone, P., & Hirschberg, C. B. (1998). Nucleotide sugars, nucleotide sulfate, and ATP transporters of the endoplasmic reticulum and Golgi apparatus. *Annals of the New York Academy of Sciences*, *842*, 91–99.

Berninsone, P. M., & Hirschberg, C. B. (2000). Nucleotide sugar transporters of the Golgi apparatus. *Current Opinion in Structural Biology*, *10*, 542–547.

Bhalla-Gehi, R., Penuela, S., Churko, J. M., Shao, Q., & Laird, D. W. (2010). Pannexin1 and pannexin3 delivery, cell surface dynamics, and cytoskeletal interactions. *The Journal of Biological Chemistry*, *285*, 9147–9160.

Bianco, F., Fumagalli, M., Pravettoni, E., D'Ambrosi, N., Volonte, C., Matteoli, M., Abbracchio, M. P., & Verderio, C. (2005). Pathophysiological roles of extracellular nucleotides in glial cells: Differential expression of purinergic receptors in resting and activated microglia. *Brain Research. Brain Research Reviews*, *48*, 144–156.

Blum, A. E., Joseph, S. M., Przybylski, R. J., & Dubyak, G. R. (2008). Rho-family GTPases modulate Ca(2+)-dependent ATP release from astrocytes. *American Journal of Physiology. Cell Physiology*, *295*, C231–C241.

Blum, A. E., Walsh, B. C., & Dubyak, G. R. (2010). Extracellular osmolarity modulates g protein-coupled receptor-dependent ATP release from 1321N1 astrocytoma cells. *American Journal of Physiology. Cell Physiology*, *298*, C386–C396.

Boassa, D., Qiu, F., Dahl, G., & Sosinsky, G. (2008). Trafficking dynamics of glycosylated pannexin 1 proteins. *Cell Communication & Adhesion*, *15*, 119–132.

Bodin, P., Bailey, D., & Burnstock, G. (1991). Increased flow-induced ATP release from isolated vascular endothelial cells but not smooth muscle cells. *British Journal of Pharmacology*, *103*, 1203–1205.

Boucher, R. C. (2007a). Airway surface dehydration in cystic fibrosis: Pathogenesis and therapy. *Annual Review of Medicine*, *58*, 157–170.

Boucher, R. C. (2007b). Evidence for airway surface dehydration as the initiating event in CF airway disease. *Journal of Internal Medicine*, *261*, 5–16.

Boudreault, F., & Grygorczyk, R. (2002). Cell swelling-induced ATP release and gadolinium-sensitive channels. *The American Journal of Physiology*, *282*, C219–C226.

Boudreault, F., & Grygorczyk, R. (2004). Cell swelling-induced ATP release is tightly dependent on intracellular calcium elevations. *The Journal of Physiology*, *561*, 499–513.

Bove, P. F., Grubb, B. R., Okada, S. F., Ribeiro, C. M., Rogers, T. D., Randell, S. H., O'Neal, W. K., & Boucher, R. C. (2010). Human alveolar type II cells secrete and absorb liquid in response to local nucleotide signaling. *The Journal of Biological Chemistry*, *285*, 34939–34949.

Bruzzone, R., Barbe, M. T., Jakob, N. J., & Monyer, H. (2005). Pharmacological properties of homomeric and heteromeric pannexin hemichannels expressed in Xenopus oocytes. *Journal of Neurochemistry*, 92, 1033–1043.

Bruzzone, R., Hormuzdi, S. G., Barbe, M. T., Herb, A., & Monyer, H. (2003). Pannexins, a family of gap junction proteins expressed in brain. *Proceedings of the National Academy of Sciences of the United States of America*, *100*, 13644–13649.

Burgoyne, R. D., & Morgan, A. (2003). Secretory granule exocytosis. *Physiological Reviews*, *83*, 581–632.

Burnstock, G. (1972). Purinergic nerves. *Pharmacological Reviews*, *24*, 509–581.

Burnstock, G. (1978). A basis for distinguishing two types of purinergic receptor. In L. B. Straub (Ed.), *Cell membrane receptors for drugs and hormones: A multidisciplinary approach* (pp. 107–118). New York: Raven Press.

Burnstock, G. (1997). The past, present and future of purine nucleotides as signalling molecules. *Neuropharmacology*, *36*, 1127–1139.

Burnstock, G. (2006). Purinergic signalling. *British Journal of Pharmacology*, *147*(Suppl. 1), S172–S181.

Burnstock, G. (2009). Purinergic mechanosensory transduction and visceral pain. *Molecular Pain*, *5*, 69.

Burnstock, G., Campbell, G., Satchell, D., & Smythe, A. (1970). Evidence that adenosine triphosphate or a related nucleotide is the transmitter substance released by non-adrenergic inhibitory nerves in the gut. *British Journal of Pharmacology*, *120*, 337–357.

Burnstock, G., & Kennedy, C. (1985). Is there a basis for distinguishing two types of P_2-purinoceptor? *General Pharmacology*, *16*, 433–440.

Button, B., & Boucher, R. C. (2008). Role of mechanical stress in regulating airway surface hydration and mucus clearance rates. *Respiratory Physiology & Neurobiology*, *163*, 189–201.

Buvinic, S., Almarza, G., Bustamante, M., Casas, M., Lopez, J., Riquelme, M., Saez, J. C., Huidobro-Toro, J. P., & Jaimovich, E. (2009). ATP released by electrical stimuli elicits calcium transients and gene expression in skeletal muscle. *The Journal of Biological Chemistry*, *284*, 34490–34505.

Buxton, I. L., Kaiser, R. A., Oxhorn, B. C., & Cheek, D. J. (2001). Evidence supporting the nucleotide axis hypothesis: ATP release and metabolism by coronary endothelium. *American Journal of Physiology. Heart and Circulatory Physiology*, *281*, H1657–H1666.

Caputo, A., Caci, E., Ferrera, L., Pedemonte, N., Barsanti, C., Sondo, E., Pfeffer, U., Ravazzolo, R., Zegarra-Moran, O., & Galietta, L. J. (2008). TMEM16A, a membrane protein associated with calcium-dependent chloride channel activity. *Science*, *322*, 590–594.

Chander, A., Johnson, R. G., Reicherter, J., & Fisher, A. B. (1986). Lung lamellar bodies maintain an acidic internal pH. *The Journal of Biological Chemistry*, *261*, 6126–6131.

Chekeni, F. B., Elliott, M. R., Sandilos, J. K., Walk, S. F., Kinchen, J. M., Lazarowski, E. R., Armstrong, A. J., Penuela, S., Laird, D. W., Salvesen, G. S., Isakson, B. E., Bayliss, D. A., & Ravichandran, K. S. (2010). Pannexin 1 channels mediate 'find-me' signal release and membrane permeability during apoptosis. *Nature*, *467*, 863–867.

Chen, Y., Corriden, R., Inoue, Y., Yip, L., Hashiguchi, N., Zinkernagel, A., Nizet, V., Insel, P. A., & Junger, W. G. (2006). ATP release guides neutrophil chemotaxis Via P2Y2 and A3 receptors. *Science*, *314*, 1792–1795.

Chen, Y., Yao, Y., Sumi, Y., Li, A., To, U. K., Elkhal, A., Inoue, Y., Woehrle, T., Zhang, Q., Hauser, C., & Junger, W. G. (2010). Purinergic signaling: A fundamental mechanism in neutrophil activation. *Science Signaling*, *3*, ra45.

Cockayne, D. A., Hamilton, S. G., Zhu, Q. M., Dunn, P. M., Zhong, Y., Novakovic, S., Malmberg, A. B., Cain, G., Berson, A., Kassotakis, L., Hedley, L., Lachnit, W. G., Burnstock, G., McMahon, S. B., & Ford, A. P. (2000). Urinary bladder hyporeflexia and reduced pain-related behaviour in P2X3-deficient mice. *Nature*, *407*, 1011–1015.

Coco, S., Calegari, F., Pravettoni, E., Pozzi, D., Taverna, E., Rosa, P., Matteoli, M., & Verderio, C. (2003). Storage and release of ATP from astrocytes in culture. *The Journal of Biological Chemistry*, *278*, 1354–1362.

Corriden, R., & Insel, P. A. (2010). Basal release of ATP: An autocrine-paracrine mechanism for cell regulation. *Science Signaling*, *3*, re1.

Costa, J. L., Fay, D. D., & Kirk, K. L. (1984). Quinacrine and basic amines in human platelets: Subcellular compartmentation and effects on serotonin. *Research Communications in Chemical Pathology and Pharmacology*, *43*, 25–42.

Dahl, G., & Locovei, S. (2006). Pannexin: To gap or not to gap, is that a question? *IUBMB Life*, *58*, 409–419.

Dale, N., & Frenguelli, B. G. (2009). Release of adenosine and ATP during ischemia and epilepsy. *Current Neuropharmacology*, *7*, 160–179.

Darby, M., Kuzmiski, J. B., Panenka, W., Feighan, D., & MacVicar, B. A. (2003). ATP released from astrocytes during swelling activates chloride channels. *Journal of Neurophysiology*, *89*, 1870–1877.

Davis, C. W., & Dickey, B. F. (2008). Regulated airway goblet cell mucin secretion. *Annual Review of Physiology*, *70*, 487–512.

Davis, I. C., Lazarowski, E. R., Hickman-Davis, J. M., Fortenberry, J. A., Chen, F. P., Zhao, X., Sorscher, E., Graves, L. M., Sullender, W. M., & Matalon, S. (2006). Leflunomide prevents alveolar fluid clearance inhibition by respiratory syncytial virus. *American Journal of Respiratory and Critical Care Medicine*, *173*, 673–682.

Davis, I. C., Sullender, W. M., Hickman-Davis, J. M., Lindsey, J. R., & Matalon, S. (2004). Nucleotide-mediated inhibition of alveolar fluid clearance in BALB/c mice after respiratory syncytial virus infection. *American Journal of Physiology. Lung Cellular and Molecular Physiology*, *286*, L112–L120.

De Vuyst, E., Decrock, E., Cabooter, L., Dubyak, G. R., Naus, C. C., Evans, W. H., & Leybaert, L. (2005). Intracellular calcium changes trigger connexin 32 hemichannel opening. *The EMBO Journal*, *25*, 34–44.

De Vuyst, E., Wang, N., Decrock, E., De Bock, M., Vinken, M., Van Moorhem, M., Lai, C., Culot, M., Rogiers, V., Cecchelli, R., Naus, C. C., Evans, W. H., & Leybaert, L. (2009). Ca(2+) regulation of connexin 43 hemichannels in c6 glioma and glial cells. *Cell Calcium*, *46*, 176–187.

Dean, G. E., Fishkes, H., Nelson, P. J., & Rudnick, G. (1984). The hydrogen ion-pumping adenosine triphosphatase of platelet dense granule membrane. Differences from F1F0- and phosphoenzyme-type ATPases. *The Journal of Biological Chemistry*, *259*, 9569–9574.

Demidchik, V., Nichols, C., Oliynyk, M., Dark, A., Glover, B. J., & Davies, J. M. (2003). Is ATP a signaling agent in plants? *Plant Physiology*, *133*, 456–461.

Detimary, P., Jonas, J. C., & Henquin, J. C. (1996). Stable and diffusible pools of nucleotides in pancreatic islet cells. *Endocrinology*, *137*, 4671–4676.

Di, A., Krupa, B., Bindokas, V. P., Chen, Y., Brown, M. E., Palfrey, H. C., Naren, A. P., Kirk, K. L., & Nelson, D. J. (2002). Quantal release of free radicals during exocytosis of phagosomes. *Nature Cell Biology*, *4*, 279–285.

Dolovcak, S., Waldrop, S. L., Fitz, J. G., & Kilic, G. (2009). 5-Nitro-2-(3-phenylpropylamino) benzoic acid (NPPB) stimulates cellular ATP release through exocytosis of ATP-enriched vesicles. *The Journal of Biological Chemistry*, *284*, 33894–33903.

Dutta, A. K., Korchev, Y. E., Shevchuk, A. I., Hayashi, S., Okada, Y., & Sabirov, R. Z. (2008). Spatial distribution of maxi-anion channel on cardiomyocytes detected by smart-patch technique. *Biophysical Journal*, *94*, 1646–1655.

Dutta, A. K., Sabirov, R. Z., Uramoto, H., & Okada, Y. (2004). Role of ATP-conductive anion channel in ATP release from neonatal rat cardiomyocytes in ischaemic or hypoxic conditions. *The Journal of Physiology*, *559*, 799–812.

Elliott, M. R., Chekeni, F. B., Trampont, P. C., Lazarowski, E. R., Kadl, A., Walk, S. F., Park, D., Woodson, R. I., Ostankovich, M., Sharma, P., Lysiak, J. J., Harden, T. K., Leitinger, N., & Ravichandran, K. S. (2009). Nucleotides released by apoptotic cells act as a find-me signal to promote phagocytic clearance. *Nature*, *461*, 282–286.

Eltzschig, H. K., Eckle, T., Mager, A., Kuper, N., Karcher, C., Weissmuller, T., Boengler, K., Schulz, R., Robson, S. C., & Colgan, S. P. (2006). ATP release from activated neutrophils occurs via connexin 43 and modulates adenosine-dependent endothelial cell function. *Circulation Research*, *99*, 1100–1108.

Eltzschig, H. K., Macmanus, C. F., & Colgan, S. P. (2008). Neutrophils As sources of extracellular nucleotides: Functional consequences at the vascular interface. *Trends in Cardiovascular Medicine*, *18*, 103–107.

Erlinge, D., Harnek, J., van Heusden, C., Olivecrona, G., Jern, S., & Lazarowski, E. (2005). Uridine triphosphate (UTP) is released during cardiac ischemia. *International Journal of Cardiology*, *100*, 427–433.

Esther, C. R., Jr., Sesma, J. I., Dohlman, H. G., Ault, A. D., Clas, M. L., Lazarowski, E. R., & Boucher, R. C. (2008). Similarities between UDP-glucose and adenine nucleotide release in yeast: Involvement of the secretory pathway. *Biochemistry*, *47*, 9269–9278.

Evans, C. M., & Koo, J. S. (2009). Airway mucus: The good, the bad, the sticky. *Pharmacology & Therapeutics*, *121*, 332–348.

Evans, R. J., Derkach, V., & Surprenant, A. (1992). ATP mediates fast synaptic transmission in mammalian neurons. *Nature*, *357*, 503–505.

Evans, R. J., & Surprenant, A. (1992). Vasoconstriction of guinea-pig submucosal arterioles following sympathetic nerve stimulation is mediated by the release of ATP. *British Journal of Pharmacology*, *106*, 242–249.

Evans, W. H., De Vuyst, E., & Leybaert, L. (2006). The gap junction cellular internet: Connexin hemichannels enter the signalling limelight. *The Biochemical Journal*, *397*, 1–14.

Feranchak, A. P., Lewis, M. A., Kresge, C., Sathe, M., Bugde, A., Luby-Phelps, K., Antich, P. P., & Fitz, J. G. (2010). Initiation of purinergic signaling by exocytosis of ATP-containing vesicles in liver epithelium. *The Journal of Biological Chemistry*, *285*, 8138–8147.

Ferguson, D. R., Kennedy, I., & Burton, T. J. (1997). ATP is released from rabbit urinary bladder epithelial cells by hydrostatic pressure changes—A possible sensory mechanism? *The Journal of Physiology*, *505*, 503–511.

Filtz, T. M., Li, Q., Boyer, J. L., Nicholas, R. A., & Harden, T. K. (1994). Expression of a cloned P2Y purinergic receptor that couples to phospholipase C. *Molecular Pharmacology*, *46*, 8–14.

Fitz, J. G. (2007). Regulation of cellular ATP release. *Transactions of the American Clinical and Climatological Association*, *118*, 199–208.

Forrester, T. (1972a). A quantitative estimation of adenosine triphosphate released from human forearm muscle during sustained exercise. *The Journal of Physiology*, *221*, 25P–26P.

Forrester, T. (1972b). An estimate of adenosine triphosphate release into the venous effluent from exercising human forearm muscle. *The Journal of Physiology*, *224*, 611–628.

Franco, R., Panayiotidis, M. I., & de la Paz, L. D. (2008). Autocrine signaling involved in cell volume regulation: The role of released transmitters and plasma membrane receptors. *Journal of Cellular Physiology*, *216*, 14–28.

Fredriksson, R., Nordstrom, K. J., Stephansson, O., Hagglund, M. G., & Schioth, H. B. (2008). The solute carrier (SLC) complement of the human genome: Phylogenetic classification reveals four major families. *FEBS Letters*, *582*, 3811–3816.

Fricks, I. P., Maddileti, S., Carter, R. L., Lazarowski, E. R., Nicholas, R. A., Jacobson, K. A., & Harden, T. K. (2008). UDP is a competitive antagonist at the human P2Y14 receptor. *The Journal of Pharmacology and Experimental Therapeutics*, *325*, 588–594.

Fumagalli, M., Brambilla, R., D'Ambrosi, N., Volonte, C., Matteoli, M., Verderio, C., & Abbracchio, M. P. (2003). Nucleotide-mediated calcium signaling in rat cortical astrocytes: Role of P2X and P2Y receptors. *Glia*, *43*, 218–230.

Gaarder, A., Jonsen, J., Laland, S., Hellem, A., & Owren, P. A. (1961). Adenosine diphosphate in red cells as a factor in the adhesiveness of human blood platelets. *Nature*, *192*, 531–532.

Gatof, D., Kilic, G., & Fitz, J. G. (2004). Vesicular exocytosis contributes to volume-sensitive ATP release in biliary cells. *American Journal of Physiology. Gastrointestinal and Liver Physiology*, *286*, G538–G546.

Gilleron, J., Fiorini, C., Carette, D., Avondet, C., Falk, M. M., Segretain, D., & Pointis, G. (2008). Molecular reorganization of Cx43, Zo-1 and Src complexes during the endocytosis of gap junction plaques in response to a non-genomic carcinogen. *Journal of Cell Science*, *121*, 4069–4078.

Goren, M. B., Swendsen, C. L., Fiscus, J., & Miranti, C. (1984). Fluorescent markers for studying phagosome-lysosome fusion. *Journal of Leukocyte Biology*, *36*, 273–292.

Groulx, N., Boudreault, F., Orlov, S. N., & Grygorczyk, R. (2006). Membrane reserves and hypotonic cell swelling. *The Journal of Membrane Biology*, *214*, 43–56.

Gualix, J., Abal, M., Pintor, J., Garcia-Carmona, F., & Miras-Portugal, M. T. (1996). Nucleotide vesicular transporter of bovine chromaffin granules. Evidence for a mnemonic regulation. *The Journal of Biological Chemistry*, *271*, 1957–1965.

Gualix, J., Pintor, J., & Miras-Portugal, M. T. (1999). Characterization of nucleotide transport into rat brain synaptic vesicles. *Journal of Neurochemistry*, *73*, 1098–1104.

Guillen, E., & Hirschberg, C. B. (1995). Transport of adenosine triphosphate into endoplasmic reticulum proteoliposomes. *Biochemistry*, *34*, 5472–5476.

Guyot, A., & Hanrahan, J. W. (2002). ATP release from human airway epithelial cells studied using a capillary cell culture system. *The Journal of Physiology*, *545*, 199–206.

Haanes, K. A., & Novak, I. (2010). ATP storage and uptake by isolated pancreatic zymogen granules. *The Biochemical Journal*, *429*, 303–311.

Hanada, H., Moriyama, Y., Maeda, M., & Futai, M. (1990). Kinetic studies of chromaffin granule H+-ATPase and effects of bafilomycin A1. *Biochemical and Biophysical Research Communications*, *170*, 873–878.

Harden, T. K., Sesma, J. I., Fricks, I. P., & Lazarowski, E. R. (2010). Signalling and pharmacological properties of the P2Y14 receptor. *Acta Physiologica (Oxford, England)*, *199*, 149–160.

Hirschberg, C. B., Robbins, P. W., & Abeijon, C. (1998). Transporters of nucleotide sugars, ATP, and nucleotide sulfate in the endoplasmic reticulum and Golgi apparatus. *Annual Review of Biochemistry*, *67*, 49–69.

Hisadome, K., Koyama, T., Kimura, C., Droogmans, G., Ito, Y., & Oike, M. (2002). Volume-regulated anion channels serve as an auto/paracrine nucleotide release pathway in aortic endothelial cells. *The Journal of General Physiology*, *119*, 511–520.

Holmsen, H., Day, H. J., & Setkowsky, C. A. (1972). Secretory mechanisms. Behaviour of adenine nucleotides during the platelet release reaction induced by adenosine diphosphate and adrenaline. *Biochemical Journal*, *129*, 67–82.

Huang, P. B., Lazarowski, E. R., Tarran, R., Milgram, S. L., Boucher, R. C., & Stutts, M. J. (2001). Compartmentalized autocrine signaling to cystic fibrosis transmembrane conductance regulator at the apical membrane of airway epithelial cells. *Proceedings of the National Academy of Sciences of the United States of America*, *98*, 14120–14125.

Huang, Y. J., Maruyama, Y., Dvoryanchikov, G., Pereira, E., Chaudhari, N., & Roper, S. D. (2007). The role of pannexin 1 hemichannels in ATP release and cell-cell communication in mouse taste buds. *Proceedings of the National Academy of Sciences of the United States of America*, *104*, 6436–6441.

Huckstepp, R. T., Id Bihi, R., Eason, R., Spyer, K. M., Dicke, N., Willecke, K., Marina, N., Gourine, A. V., & Dale, N. (2010). Connexin hemichannel-mediated CO2-dependent release of ATP in the medulla oblongata contributes to central respiratory chemosensitivity. *The Journal of Physiology*, *588*, 3901–3920.

Hutton, J. C., Penn, E. J., & Peshavaria, M. (1983). Low-molecular-weight constituents of isolated insulin-secretory granules. bivalent cations, adenine nucleotides and inorganic phosphate. *Biochemical Journal*, *210*, 297–305.

Iacobas, D. A., Iacobas, S., Urban-Maldonado, M., Scemes, E., & Spray, D. C. (2008). Similar transcriptomic alterations in Cx43 knockdown and knockout astrocytes. *Cell Communication & Adhesion*, *15*, 195–206.

Ishida, N., & Kawakita, M. (2004). Molecular physiology and pathology of the nucleotide sugar transporter family (SLC35). *Pflugers Archiv*, *447*, 768–775.

Ishida, N., Kuba, T., Aoki, K., Miyatake, S., Kawakita, M., & Sanai, Y. (2005). Identification and characterization of human Golgi nucleotide sugar transporter SLC35D2, a novel member of the SLC35 nucleotide sugar transporter family. *Genomics*, *85*, 106–116.

Iwatsuki, K., Ichikawa, R., Hiasa, M., Moriyama, Y., Torii, K., & Uneyama, H. (2009). Identification of the vesicular nucleotide transporter (VNUT) in taste cells. *Biochemical and Biophysical Research Communications*, *388*, 1–5.

Jans, D., Srinivas, S. P., Waelkens, E., Segal, A., Lariviere, E., Simaels, J., & Van Driessche, W. (2002). Hypotonic treatment evokes biphasic ATP release across the basolateral membrane of cultured renal epithelia (A6). *The Journal of Physiology*, *545*, 543–555.

Jeter, C. R., Tang, W., Henaff, E., Butterfield, T., & Roux, S. J. (2004). Evidence of a novel cell signaling role for extracellular adenosine triphosphates and diphosphates in arabidopsis. *The Plant Cell*, *16*, 2652–2664.

Joseph, S. M., Buchakjian, M. R., & Dubyak, G. R. (2003). Colocalization of ATP release sites and ecto-ATPase activity at the extracellular surface of human astrocytes. *The Journal of Biological Chemistry*, *278*, 23342.

Kang, J., Kang, N., Lovatt, D., Torres, A., Zhao, Z., Lin, J., & Nedergaard, M. (2008). Connexin 43 hemichannels are permeable to ATP. *The Journal of Neuroscience*, *28*, 4702–4711.

Kanner, B. I., & Schuldiner, S. (1987). Mechanism of transport and storage of neurotransmitters. *CRC Critical Reviews in Biochemistry*, *22*, 1–38.

Karanauskaite, J., Hoppa, M. B., Braun, M., Galvanovskis, J., & Rorsman, P. (2009). Quantal ATP release in rat beta-cells by exocytosis of insulin-containing LDCVs. *Pflugers Archiv*, *458*, 389–401.

Kawahara, K., & Takuwa, N. (1991). Bombesin activates large-conductance chloride channels in Swiss 3T3 fibroblasts. *Biochemical and Biophysical Research Communications*, *177*, 292–298.

Kolb, H. A., & Ubl, J. (1987). Activation of anion channels by zymosan particles in membranes of peritoneal macrophages. *Biochimica et Biophysica Acta*, *899*, 239–246.

Kolber, M. A., & Henkart, P. A. (1988). Quantitation of secretion by rat basophilic leukemia cells by measurements of quinacrine uptake. *Biochimica et Biophysica Acta*, *939*, 459–466.

Koshlukova, S. E., Lloyd, T. L., Araujo, M. W., & Edgerton, M. (1999). Salivary histatin 5 induces non-lytic release of ATP from Candida albicans leading to cell death. *The Journal of Biological Chemistry*, *274*, 18872–18879.

Kreda, S. M., Okada, S. F., van Heusden, C. A., O'Neal, W., Gabriel, S., Abdullah, L., Davis, C. W., Boucher, R. C., & Lazarowski, E. R. (2007). Coordinated release of nucleotides and mucin from human airway epithelial Calu-3 cells. *The Journal of Physiology*, *584*, 245–259.

Kreda, S. M., Seminario-Vidal, L., Heusden, C., & Lazarowski, E. R. (2008). Thrombin-promoted release of UDP-glucose from human astrocytoma cells. *British Journal of Pharmacology*, *153*, 1528–1537.

Kreda, S. M., Seminario-Vidal, L., van Heusden, C. A., O'Neal, W., Jones, L., Boucher, R. C., & Lazarowski, E. R. (2010). Receptor-promoted exocytosis of airway epithelial mucin granules containing a spectrum of adenine nucleotides. *The Journal of Physiology*, *588*, 2255–2267.

Langlois, S., Cowan, K. N., Shao, Q., Cowan, B. J., & Laird, D. W. (2010). The tumor-suppressive function of Connexin43 in keratinocytes is mediated in part via interaction with caveolin-1. *Cancer Research*, *70*, 4222–4232.

Lazarowski, E., Seminario-VIdal, L. S., Kreda, S., Sesma, J., Okada, S., & Boucher, R. (2010). Rho-dependent pannexin 1-mediated ATP release from airway epithelia. *Purinergic Signalling*, *6*(Suppl. 1), S71.

Lazarowski, E. R. (2010). Quantification of extracellular UDP-galactose. *Analytical Biochemistry*, *396*, 23–29.

Lazarowski, E. R., & Boucher, R. C. (2009). Purinergic receptors in airway epithelia. *Current Opinion in Pharmacology*, *9*, 262–267.

Lazarowski, E. R., Boucher, R. C., & Harden, T. K. (2000). Constitutive release of ATP and evidence for major contribution of ecto-nucleotide pyrophosphatase and nucleoside diphosphokinase to extracellular nucleotide concentrations. *The Journal of Biological Chemistry*, *275*, 31061–31068.

Lazarowski, E. R., Boucher, R. C., & Harden, T. K. (2003a). Mechanisms of release of nucleotides and integration of their action as P2X- and P2Y-receptor activating molecules. *Molecular Pharmacology*, *64*, 785–795.

Lazarowski, E. R., & Harden, T. K. (1999). Quantitation of extracellular UTP using a sensitive enzymatic assay. *British Journal of Pharmacology*, *127*, 1272–1278.

Lazarowski, E. R., Homolya, L., Boucher, R. C., & Harden, T. K. (1997). Direct demonstration of mechanically induced release of cellular UTP and its implication for uridine nucleotide receptor activation. *The Journal of Biological Chemistry*, *272*, 24348–24354.

Lazarowski, E. R., Shea, D. A., Boucher, R. C., & Harden, T. K. (2003b). Release of cellular UDP-glucose as a potential extracellular signaling molecule. *Molecular Pharmacology*, *63*, 1190–1197.

Lazarowski, E. R., Tarran, R., Grubb, B. R., van Heusden, C. A., Okada, S., & Boucher, R. C. (2004). Nucleotide release provides a mechanism for airway surface liquid homeostasis. *The Journal of Biological Chemistry*, *279*, 36855–36864.

Lazarowski, E. R., Watt, W. C., Stutts, M. J., Boucher, R. C., & Harden, T. K. (1995). Pharmacological selectivity of the cloned human P-2U-purinoceptor—Potent activation by diadenosine tetraphosphate. *British Journal of Pharmacology*, *116*, 1619–1627.

Leitner, J. W., Sussman, K. E., Vatter, A. E., & Schneider, F. H. (1975). Adenine nucleotides in the secretory granule fraction of rat islets. *Endocrinology*, *96*, 662–677.

Levitt, B., Head, R. J., & Westfall, D. P. (1984). High-pressure liquid chromatographic-fluorometric detection of adenosine and adenine nucleotides: Application to endogenous content and electrically induced release of adenyl purines in guinea pig vas deferens. *Analytical Biochemistry*, *137*, 93–100.

Li, A., Leung, C. T., Peterson-Yantorno, K., Mitchell, C. H., & Civan, M. M. (2011). Pathways for ATP release by bovine ciliary epithelial cells, the initial step in purinergic regulation of aqueous humor inflow. *American Journal of Physiology. Cell Physiology*, in press.

Light, D. B., Schwiebert, E. M., Fejes-Toth, G., Naray-Fejes-Toth, A., Karlson, K. H., McCann, F. V., & Stanton, B. A. (1990). Chloride channels in the apical membrane of cortical collecting duct cells. *The American Journal of Physiology*, *258*, F273–F280.

Lindmo, K., & Stenmark, H. (2006). Regulation of membrane traffic by phosphoinositide 3-kinases. *Journal of Cell Science*, *119*, 605–614.

Liu, H. T., Akita, T., Shimizu, T., Sabirov, R. Z., & Okada, Y. (2009). Bradykinin-induced astrocyte-neuron signalling: Glutamate release is mediated by ROS-activated volume-sensitive outwardly rectifying anion channels. *The Journal of Physiology*, *587*, 2197–2209.

Liu, H. T., Sabirov, R. Z., & Okada, Y. (2008a). Oxygen-glucose deprivation induces ATP release via maxi-anion channels in astrocytes. *Purinergic Signalling*, *4*, 147–154.

Liu, H. T., Toychiev, A. H., Takahashi, N., Sabirov, R. Z., & Okada, Y. (2008b). Maxi-anion channel as a candidate pathway for osmosensitive ATP release from mouse astrocytes in primary culture. *Cell Research*, *18*, 558–565.

Livraghi, A., Grubb, B. R., Hudson, E. J., Wilkinson, K. J., Sheehan, J. K., Mall, M. A., O'Neal, W. K., Boucher, R. C., & Randell, S. H. (2009). Airway and lung pathology due to mucosal surface dehydration in {beta}-epithelial Na+ channel-overexpressing mice: Role of TNF-{alpha and IL-4R{alpha} signaling, influence of neonatal development, and limited efficacy of glucocorticoid treatment. *Journal of Immunology*, *182*, 4357–4367.

Locovei, S., Bao, L., & Dahl, G. (2006a). Pannexin 1 in erythrocytes: Function without a gap. *Proceedings of the National Academy of Sciences of the United States of America*, *103*, 7655–7659.

Locovei, S., Scemes, E., Qiu, F., Spray, D. C., & Dahl, G. (2007). Pannexin1 is part of the pore forming unit of the P2X(7) receptor death complex. *FEBS Letters*, *581*, 483–488.

Locovei, S., Wang, J., & Dahl, G. (2006b). Activation of pannexin 1 channels by ATP through P2Y receptors and by cytoplasmic calcium. *FEBS Letters*, *580*, 239–244.

Ma, W., Hui, H., Pelegrin, P., & Surprenant, A. (2009). Pharmacological characterization of pannexin-1 currents expressed in mammalian cells. *The Journal of Pharmacology and Experimental Therapeutics*, *328*, 409–418.

MacVicar, B. A., & Thompson, R. J. (2010). Non-junction functions of pannexin-1 channels. *Trends in Neurosciences*, *33*, 93–102.

Mall, M., Grubb, B. R., Harkema, J. R., O'Neal, W. K., & Boucher, R. C. (2004). Increased airway epithelial Na+ absorption produces cystic fibrosis-like lung disease in mice. *Nature Medicines*, *10*, 487–493.

Milner, P., Bodin, P., Loesch, A., & Burnstock, G. (1990). Rapid release of endothelin and ATP from isolated aortic endothelial cells exposed to increased flow. *Biochemical and Biophysical Research Communications*, *170*, 649–656.

Milner, P., Bodin, P., Loesch, A., & Burnstock, G. (1992). Increased shear stress leads to differential release of endothelin and ATP from isolated endothelial cells from 4- and 12-month-old male rabbit aorta. *Journal of Vascular Research*, *29*, 420–425.

Miras-Portugal, M. T., Gualix, J., Mateo, J., Diaz-Hernandez, M., Gomez-Villafuertes, R., Castro, E., & Pintor, J. (1999). Diadenosine polyphosphates, extracellular function and catabolism. *Progress in Brain Research*, *120*, 397–409.

Moore, D. J., Murdock, P. R., Watson, J. M., Faull, R. L., Waldvogel, H. J., Szekeres, P. G., Wilson, S., Freeman, K. B., & Emson, P. C. (2003). GPR105, a novel Gi/o-coupled UDP-glucose receptor expressed on brain glia and peripheral immune cells, is regulated by immunologic challenge: Possible role in neuroimmune function. *Brain Research. Molecular Brain Research*, *118*, 10–23.

Muller, D. J., Hand, G. M., Engel, A., & Sosinsky, G. E. (2002). Conformational changes in surface structures of isolated connexin 26 gap junctions. *The EMBO Journal*, *21*, 3598–3607.

Nilius, B., & Droogmans, G. (2003). Amazing chloride channels: An overview. *Acta Physiologica Scandinavica*, *177*, 119–147.

Novak, I. (2008). Purinergic receptors in the endocrine and exocrine pancreas. *Purinergic Signalling*, *4*, 237–253.

O'Brien, J. R. (1963). Some effects of adrenaline and anti-adrenaline compounds in vitro and in vivo. *Nature*, *200*, 763–764.

Obermuller, S., Lindqvist, A., Karanauskaite, J., Galvanovskis, J., Rorsman, P., & Barg, S. (2005). Selective nucleotide-release from dense-core granules in insulin-secreting cells. *Journal of Cell Science*, *118*, 4271–4282.

Okada, S. F., Nicholas, R. A., Kreda, S. M., Lazarowski, E. R., & Boucher, R. C. (2006). Physiological regulation of ATP release at the apical surface of human airway epithelia. *The Journal of Biological Chemistry*, *281*, 22992–23002.

Okada, S. F., O'Neal, W. K., Huang, P., Nicholas, R. A., Ostrowski, L. E., Craigen, W. J., Lazarowski, E. R., & Boucher, R. C. (2004). Voltage-dependent anion channel-1 (VDAC-1) contributes to ATP release and cell volume regulation in murine cells. *The Journal of General Physiology*, *124*, 513–526.

Okada, S. F., Zhang, L., Kreda, S. M., Abdullah, L. H., Davis, C. W., Pickles, R. J., Lazarowski, E. R., & Boucher, R. C. (2010). Coupled nucleotide and mucin hypersecretion from goblet cell metaplastic human airway epithelium. *American Journal of Respiratory Cell and Molecular Biology*, in press.

Okada, Y. (1997). Volume expansion-sensing outward-rectifier Cl-channel: Fresh start to the molecular identity and volume sensor. *The American Journal of Physiology*, *273*, C755–C789.

Okada, Y., Sato, K., & Numata, T. (2009). Pathophysiology and puzzles of the volume-sensitive outwardly rectifying anion channel. *The Journal of Physiology*, *587*, 2141–2149.

Orriss, I. R., Knight, G. E., Utting, J. C., Taylor, S. E., Burnstock, G., & Arnett, T. R. (2009). Hypoxia stimulates vesicular ATP release from rat osteoblasts. *Journal of Cellular Physiology*, *220*, 155–162.

Ostrom, R. S., Gregorian, C., & Insel, P. A. (2000). Cellular release of and response to ATP as key determinants of the set-point of signal transduction pathways. *The Journal of Biological Chemistry*, *275*, 11735–11739.

Pafundo, D. E., Alvarez, C. L., Krumschnabel, G., & Schwarzbaum, P. J. (2010). A volume regulatory response can be triggered by nucleosides in human erythrocytes, a perfect osmometer no longer. *The Journal of Biological Chemistry*, *285*, 6134–6144.

Pafundo, D. E., Chara, O., Faillace, M. P., Krumschnabel, G., & Schwarzbaum, P. J. (2008). Kinetics of ATP release and cell volume regulation of hyposmotically challenged goldfish hepatocytes. *American Journal of Physiology. Regulatory, Integrative and Comparative Physiology*, *294*, R220–R233.

Pangrsic, T., Potokar, M., Stenovec, M., Kreft, M., Fabbretti, E., Nistri, A., Pryazhnikov, E., Khiroug, L., Giniatullin, R., & Zorec, R. (2007). Exocytotic release of ATP from cultured astrocytes. *The Journal of Biological Chemistry*, *282*, 28749–28758.

Parodi, A. J. (2000). Protein glucosylation and its role in protein folding. *Annual Review of Biochemistry*, *69*, 69–93.

Parr, C. E., Sullivan, D. M., Paradiso, A. M., Lazarowski, E. R., Burch, L. H., Olsen, J. C., Erb, L., Weisman, G. A., Boucher, R. C., & Turner, J. T. (1994). Cloning and expression of a human-P(2U) nucleotide receptor, a target for cystic-fibrosis pharmacotherapy. *Proceedings of the National Academy of Sciences of the United States of America*, *91*, 3275–3279.

Pascual, O., Casper, K. B., Kubera, C., Zhang, J., Revilla-Sanchez, R., Sul, J. Y., Takano, H., Moss, S. J., McCarthy, K., & Haydon, P. G. (2005). Astrocytic purinergic signaling coordinates synaptic networks. *Science*, *310*, 113–116.

Pearson, J. D., & Gordon, J. L. (1979). Vascular endothelial and smooth muscle cells in culture selectively release adenine nucleotides. *Nature*, *281*, 384–386.

Pelegrin, P., & Surprenant, A. (2006). Pannexin-1 mediates large pore formation and interleukin-1beta release by the ATP-gated P2X7 receptor. *The EMBO Journal*, *25*, 5071–5082.

Pellegatti, P., Falzoni, S., Pinton, P., Rizzuto, R., & Di Virgilio, F. (2005). A novel recombinant plasma membrane-targeted luciferase reveals a new pathway for ATP secretion. *Molecular Biology of the Cell*, *16*, 3659–3665.

Penuela, S., Bhalla, R., Gong, X. Q., Cowan, K. N., Celetti, S. J., Cowan, B. J., Bai, D., Shao, Q., & Laird, D. W. (2007). Pannexin 1 and pannexin 3 are glycoproteins that exhibit many distinct characteristics from the connexin family of gap junction proteins. *Journal of Cell Science*, *120*, 3772–3783.

Perez, M., & Hirschberg, C. B. (1986). Topography of glycosylation reactions in the rough endoplasmic reticulum membrane. *The Journal of Biological Chemistry*, *261*, 6822–6830.

Pintor, J., Diaz-Hernandez, M., Gualix, J., Gomez-Villafuertes, R., Hernando, F., & Miras-Portugal, M. T. (2000). Diadenosine polyphosphate receptors. From rat and guinea-pig brain to human nervous system. *Pharmacology & Therapeutics*, *87*, 103–115.

Praetorius, H. A., & Leipziger, J. (2009). ATP release from non-excitable cells. *Purinergic Signalling*, *5*, 433–446.

Qiu, F., & Dahl, G. (2009). A permeant regulating its permeation pore: Inhibition of pannexin 1 channels by ATP. *American Journal of Physiology. Cell Physiology*, *296*, C250–C255.

Ransford, G. A., Fregien, N., Qiu, F., Dahl, G., Conner, G. E., & Salathe, M. (2009). Pannexin 1 contributes to ATP release in airway epithelia. *American Journal of Respiratory Cell and Molecular Biology*, *41*, 525–534.

Reigada, D., Lu, W., Zhang, M., & Mitchell, C. H. (2008). Elevated pressure triggers a physiological release of ATP from the retina: Possible role for pannexin hemichannels. *Neuroscience*, *157*, 396–404.

Reigada, D., & Mitchell, C. H. (2005). Release of ATP from retinal pigment epithelial cells involves both CFTR and vesicular transport. *American Journal of Physiology. Cell Physiology*, *288*, C132–C140.

Reimer, R. J., & Edwards, R. H. (2004). Organic anion transport is the primary function of the SLC17/type I phosphate transporter family. *Pflugers Archiv*, *447*, 629–635.

Roman, R. M., Wang, Y., Lidofsky, S. D., Feranchak, A. P., Lomri, N., Scharschmidt, B. F., & Fitz, J. G. (1997). Hepatocellular ATP-binding cassette protein expression enhances ATP release and autocrine regulation of cell volume. *The Journal of Biological Chemistry, 272*, 21970–21976.

Romanov, R. A., Rogachevskaja, O. A., Bystrova, M. F., Jiang, P., Margolskee, R. F., & Kolesnikov, S. S. (2007). Afferent neurotransmission mediated by hemichannels in mammalian taste cells. *The EMBO Journal, 26*, 657–667.

Romanov, R. A., Rogachevskaja, O. A., Khokhlov, A. A., & Kolesnikov, S. S. (2008). Voltage dependence of ATP secretion in mammalian taste cells. *The Journal of General Physiology, 132*, 731–744.

Sabirov, R., & Okada, Y. (2005). ATP release via anion channels. *Purinergic Signalling, 1*, 311–328.

Sabirov, R. Z., Dutta, A. K., & Okada, Y. (2001). Volume-dependent ATP-conductive large-conductance anion channel as a pathway for swelling-induced ATP release. *The Journal of General Physiology, 118*, 251–266.

Sabirov, R. Z., & Okada, Y. (2004). Wide nanoscopic pore of maxi-anion channel suits its function as an ATP-conductive pathway. *Biophysical Journal, 87*, 1672–1685.

Sabirov, R. Z., & Okada, Y. (2009). The maxi-anion channel: A classical channel playing novel roles through an unidentified molecular entity. *The Journal of Physiological Sciences, 59*, 3–21.

Sabirov, R. Z., Sheiko, T., Liu, H., Deng, D., Okada, Y., & Craigen, W. J. (2006). Genetic demonstration that the plasma membrane maxi-anion channel and voltage-dependent anion channels (VDACs) are unrelated proteins. *The Journal of Biological Chemistry, 281*, 1897–1904.

Saïag, B., Bodin, P., Schacoori, V., Catheline, M., Rault, B., & Burnstock, G. (1995). Uptake and flow-induced release of uridine nucleotides from isolated vascular endothelial cells. *Endothelium, 2*, 279–285 [Abstract].

Sawada, K., Echigo, N., Juge, N., Miyaji, T., Otsuka, M., Omote, H., Yamamoto, A., & Moriyama, Y. (2008). Identification of a vesicular nucleotide transporter. *Proceedings of the National Academy of Sciences of the United States of America, 105*, 5683–5686.

Scemes, E. (2008). Modulation of astrocyte P2Y1 receptors by the carboxyl terminal domain of the gap junction protein Cx43. *Glia, 56*, 145–153.

Scemes, E., Spray, D. C., & Meda, P. (2009). Connexins, pannexins, innexins: Novel roles of "hemi-channels" *Pflugers Archiv, 457*, 1207–1226.

Scemes, E., Suadicani, S. O., Dahl, G., & Spray, D. C. (2007). Connexin and pannexin mediated cell-cell communication. *Neuron Glia Biology, 3*, 199–208.

Schenk, U., Westendorf, A. M., Radaelli, E., Casati, A., Ferro, M., Fumagalli, M., Verderio, C., Buer, J., Scanziani, E., & Grassi, F. (2008). Purinergic control of T cell activation by ATP released through pannexin-1 hemichannels. *Science Signaling, 1*, ra6.

Schroeder, B. C., Cheng, T., Jan, Y. N., & Jan, L. Y. (2008). Expression cloning of TMEM16A as a calcium-activated chloride channel subunit. *Cell, 134*, 1019–1029.

Schwiebert, E. M. (1999). ABC transporter-facilitated ATP conductive transport. *The American Journal of Physiology, 276*, C1–C8.

Scrivens, M., & Dickenson, J. M. (2006). Functional expression of the P2Y(14) receptor in human neutrophils. *European Journal of Pharmacology, 543*, 166–173.

Seminario-Vidal, L., Kreda, S., Jones, L., O'Neal, W., Trejo, J., Boucher, R. C., & Lazarowski, E. R. (2009). Thrombin promotes release of ATP from lung epithelial cells through coordinated activation of Rho- and Ca2+-dependent signaling pathways. *The Journal of Biological Chemistry, 284*, 20638–20648.

Sesma, J. I., Esther, C. R., Jr., Kreda, S. M., Jones, L., O'Neal, W., Nishihara, S., Nicholas, R. A., & Lazarowski, E. R. (2009). ER/Golgi nucelotide sugar transporters contribute to the cellular release of UDP-sugar signaling molecules. *The Journal of Biological Chemistry, 284*, 12572–12583.

Shinozuka, K., Tanaka, N., Kawasaki, K., Mizuno, H., Kubota, Y., Nakamura, K., Hashimoto, M., & Kunitomo, M. (2001). Participation of ATP in cell volume regulation in the endothelium after hypotonic stress. *Clinical and Experimental Pharmacology & Physiology*, *28*, 799–803.

Silva, G. B., & Garvin, J. L. (2008). TRPV4 mediates hypotonicity-induced ATP release by the thick ascending limb. *American Journal of Physiology. Renal Physiology*, *295*, F1090–F1095.

Silverman, W., Locovei, S., & Dahl, G. (2008). Probenecid, a gout remedy, inhibits pannexin 1 channels. *American Journal of Physiology. Cell Physiology*, *295*, C761–C767.

Singh, D., & Lampe, P. D. (2003). Identification of connexin-43 interacting proteins. *Cell Communication & Adhesion*, *10*, 215–220.

Smith-Garvin, J. E., Koretzky, G. A., & Jordan, M. S. (2009). T cell activation. *Annual Review of Immunology*, *27*, 591–619.

Sorensen, C. E., & Novak, I. (2001). Visualization of ATP release in pancreatic acini in response to cholinergic stimulus. Use of fluorescent probes and confocal microscopy. *The Journal of Biological Chemistry*, *276*, 32925–32932.

Sprague, R. S., Bowles, E. A., Hanson, M. S., DuFaux, E. A., Sridharan, M., Adderley, S., Ellsworth, M. L., & Stephenson, A. H. (2008). Prostacyclin analogs stimulate receptor-mediated CAMP synthesis and ATP release from rabbit and human erythrocytes. *Microcirculation*, *15*, 461–471.

Sprague, R. S., Stephenson, A. H., & Ellsworth, M. L. (2007). Red not dead: Signaling in and from erythrocytes. *Trends in Endocrinology and Metabolism*, *18*, 350–355.

Spray, D. C., & Iacobas, D. A. (2007). Organizational principles of the connexin-related brain transcriptome. *The Journal of Membrane Biology*, *218*, 39–47.

Sreedharan, S., Shaik, J. H., Olszewski, P. K., Levine, A. S., Schioth, H. B., & Fredriksson, R. (2010). Glutamate, aspartate and nucleotide transporters in the SLC17 family form four main phylogenetic clusters: Evolution and tissue expression. *BMC Genomics*, *11*, 17.

Sridharan, M., Adderley, S. P., Bowles, E. A., Egan, T. M., Stephenson, A. H., Ellsworth, M. L., & Sprague, R. S. (2010). Pannexin 1 is the conduit for low oxygen tension-induced ATP release from human erythrocytes. *American Journal of Physiology. Heart and Circulatory Physiology*, *299*, H1146–H1152.

Stout, C. E., Costantin, J. L., Naus, C. C., & Charles, A. C. (2002). Intercellular calcium signaling in astrocytes via ATP release through connexin hemichannels. *The Journal of Biological Chemistry*, *277*, 10482–10488.

Suadicani, S. O., Brosnan, C. F., & Scemes, E. (2006). P2X7 receptors mediate ATP release and amplification of astrocytic intercellular Ca2+ signaling. *The Journal of Neuroscience*, *26*, 1378–1385.

Suda, T., Kamiyama, S., Suzuki, M., Kikuchi, N., Nakayama, K., Narimatsu, H., Jigami, Y., Aoki, T., & Nishihara, S. (2004). Molecular cloning and characterization of a human multisubstrate specific nucleotide-sugar transporter homologous to drosophila fringe connection. *The Journal of Biological Chemistry*, *279*, 26469–26474.

Suzuki, M., & Mizuno, A. (2004). A novel human Cl(−) channel family related to drosophila flightless locus. *The Journal of Biological Chemistry*, *279*, 22461–22468.

Tatur, S., Groulx, N., Orlov, S. N., & Grygorczyk, R. (2007). Ca2+-dependent ATP release from A549 cells involves synergistic autocrine stimulation by coreleased uridine nucleotides. *The Journal of Physiology*, *584*, 419–435.

Tatur, S., Kreda, S., Lazarowski, E., & Grygorczyk, R. (2008). Calcium-dependent release of adenosine and uridine nucleotides from A549 cells. *Purinergic Signalling*, *4*, 139–146.

Thompson, R. J., Jackson, M. F., Olah, M. E., Rungta, R. L., Hines, D. J., Beazely, M. A., MacDonald, J. F., & MacVicar, B. A. (2008). Activation of pannexin-1 hemichannels augments aberrant bursting in the hippocampus. *Science*, *322*, 1555–1559.

Thompson, R. J., & MacVicar, B. A. (2008). Connexin and pannexin hemichannels of neurons and astrocytes. *Channels (Austin)*, 2, 81–86.

Thompson, R. J., Zhou, N., & MacVicar, B. A. (2006). Ischemia opens neuronal gap junction hemichannels. *Science*, *312*, 924–927.

Tokunaga, A., Tsukimoto, M., Harada, H., Moriyama, Y., & Kojima, S. (2010). Involvement of SLC17A9-dependent vesicular exocytosis in the mechanism of ATP release during T cell activation. *The Journal of Biological Chemistry*, *285*, 17406–17416.

Toychiev, A. H., Sabirov, R. Z., Takahashi, N., Ando-Akatsuka, Y., Liu, H., Shintani, T., Noda, M., & Okada, Y. (2009). Activation of maxi-anion channel by protein tyrosine dephosphorylation. *American Journal of Physiology. Cell Physiology*, *297*, C990–C1000.

Toyofuku, T., Yabuki, M., Otsu, K., Kuzuya, T., Hori, M., & Tada, M. (1998). Direct association of the gap junction protein connexin-43 with ZO-1 in cardiac myocytes. *The Journal of Biological Chemistry*, *273*, 12725–12731.

Tsukimoto, M., Tokunaga, A., Harada, H., & Kojima, S. (2009). Blockade of murine T cell activation by antagonists of P2Y6 and P2X7 receptors. *Biochemical and Biophysical Research Communications*, *384*, 512–518.

Vlaskovska, M., Kasakov, L., Rong, W., Bodin, P., Bardini, M., Cockayne, D. A., Ford, A. P., & Burnstock, G. (2001). P2X3 knock-out mice reveal a major sensory role for urothelially released ATP. *The Journal of Neuroscience*, *21*, 5670–5677.

Wang, J., Ma, M., Locovei, S., Keane, R. W., & Dahl, G. (2007). Modulation of membrane channel currents by gap junction protein mimetic peptides: Size matters. *American Journal of Physiology. Cell Physiology*, *293*, C1112–C1119.

Wang, Y., Roman, R., Lidofsky, S. D., & Fitz, J. G. (1996). Autocrine signaling through ATP release represents a novel mechanism for cell volume regulation. *Proceedings of the National Academy of Sciences of the United States of America*, *93*, 12020–12025.

Watt, W. C., Lazarowski, E. R., & Boucher, R. C. (1998). Cystic fibrosis transmembrane regulator-independent release of ATP—Its implications for the regulation of P2Y(2) receptors in airway epithelia. *The Journal of Biological Chemistry*, *273*, 14053–14058.

Weerasinghe, R. R., Swanson, S. J., Okada, S. F., Garrett, M. B., Kim, S. Y., Stacey, G., Boucher, R. C., Gilroy, S., & Jones, A. M. (2009). Touch induces ATP release in arabidopsis roots that is modulated by the heterotrimeric G-protein complex. *FEBS Letters*, *583*, 2521–2526.

Wijk, v.d.T., Tomassen, S. F., Houtsmuller, A. B., de Jonge, H. R., & Tilly, B. C. (2003). Increased vesicle recycling in response to osmotic cell swelling. Cause and consequence of hypotonicity-provoked ATP release. *The Journal of Biological Chemistry*, *278*, 40020–40025.

Winkler, H. (1976). The composition of adrenal chromaffin granules: An assessment of controversial results. *Neuroscience*, *1*, 65–80.

Yang, S., Cheek, D. J., Westfall, D. P., & Buxton, I. L. (1994). Purinergic axis in cardiac blood vessels. Agonist-mediated release of ATP from cardiac endothelial cells. *Circulation Research*, *74*, 401–407.

Yang, Y. D., Cho, H., Koo, J. Y., Tak, M. H., Cho, Y., Shim, W. S., Park, S. P., Lee, J., Lee, B., Kim, B. M., Raouf, R., Shin, Y. K., & Oh, U. (2008). TMEM16A confers receptor-activated calcium-dependent chloride conductance. *Nature*, *455*, 1210–1215.

Yegutkin, G. G. (2008). Nucleotide- and nucleoside-converting ectoenzymes: Important modulators of purinergic signalling cascade. *Biochimica et Biophysica Acta*, *1783*, 673–694.

Yip, L., Woehrle, T., Corriden, R., Hirsh, M., Chen, Y., Inoue, Y., Ferrari, V., Insel, P. A., & Junger, W. G. (2009). Autocrine regulation of T-cell activation by ATP release and P2X7 receptors. *The FASEB Journal*, *23*, 1685–1693.

Zhang, Z., Chen, G., Zhou, W., Song, A., Xu, T., Luo, Q., Wang, W., Gu, X. S., & Duan, S. (2007). Regulated ATP release from astrocytes through lysosome exocytosis. *Nature Cell Biology*, *9*, 945–953.

Zhong, X., Malhotra, R., & Guidotti, G. (2003). ATP uptake in the Golgi and extracellular release require Mcd4 protein and the vacuolar H+-ATPase. *The Journal of Biological Chemistry*, *278*, 33436–33444.

Filip Kukulski*,†,1, Sébastien A. Lévesque*,†,1, and Jean Sévigny*,†

*Centre de Recherche en Rhumatologie et Immunologie, Centre Hospitalier Universitaire de Québec (pavillon CHUL), Québec, Québec, Canada

†Département de Microbiologie-Infectiologie et d'Immunologie, Faculté de Médecine, Université Laval, Québec, Québec, Canada

1These authors have contributed equally to this work

Impact of Ectoenzymes on P2 and P1 Receptor Signaling

Abstract

P2 receptors that are activated by extracellular nucleotides (e.g., ATP, ADP, UTP, UDP, Ap$_n$A) and P1 receptors activated by adenosine control a diversity of biological processes. The activation of these receptors is tightly regulated by ectoenzymes that metabolize their ligands. This review presents these enzymes as well as their roles in the regulation of P2 and P1 receptor activation. We focus specifically on the role of ectoenzymes in processes of our interest, that is, inflammation, vascular tone, and neurotransmission. An update on the development of ectonucleotidase inhibitors is also presented.

I. Introduction

Extracellular nucleotides via P2 receptors and extracellular adenosine via P1 receptors are involved in a number of physiological processes (Abbracchio & Burnstock, 1998; Abbracchio et al., 2009; Di Virgilio et al., 2009; Hasko et al., 2008). The presence of nucleotides in the extracellular space not only arises as a result of cellular damage but also occurs in a controlled fashion by their secretion from activated cells (reviewed in Chapter 8). The information encrypted in nucleotide release is delivered into the cells through plasma membrane-bound ionotropic P2X (P2X1–7) and metabotropic P2Y (P2Y$_{1,2,4,6,11-14}$) receptors. P2 receptor subtypes differ in respect to their selectivity toward nucleotides and are coupled to different intracellular signaling pathways. All P2X and P2Y$_{11}$ receptors are

Advances in Pharmacology, Volume 61

1054-3589/11 $35.00
10.1016/B978-0-12-385526-8.00009-6

activated by ATP; $P2Y_2$ by ATP and UTP; $P2Y_1$, $P2Y_{12}$, and $P2Y_{13}$ by ADP; $P2Y_4$ by UTP; $P2Y_6$ by UDP (and UTP in mouse; Kauffenstein et al., 2010a); and $P2Y_{14}$ by UDP-glucose (Abbracchio et al., 2006). In addition, nucleotides can also activate other G protein-coupled receptors such as cysteinyl leukotriene receptor-1 and -2 (CysLT1R and CysLT2R), and GPR17 (Ciana et al., 2006; Mamedova et al., 2005; von Kugelgen, 2006).

Extracellular adenosine originates either from the catabolism of nucleotides by ectonucleotidases or from transport through SLC29 transporter formerly known as equilibrative nucleoside transporters (ENT; Colgan et al., 2006; Conde & Monteiro, 2004; Parkinson et al., 2005; Sowa et al., 2009; Zylka et al., 2008). Once outside the cell, adenosine activates G protein-coupled P1 receptors (A_1, A_{2A}, A_{2B}, and A_3) that exert physiological responses (Jacobson & Gao, 2006) often opposite to those activated by extracellular nucleotides through P2 receptors.

The activation of P2 and P1 receptors is regulated by ectoenzymes that either eliminate or produce extracellular nucleotides and adenosine. This review presents these enzymes and recent development about their described functions in three systems of our interest: inflammation, vascular tone, and neurotransmission. In addition, recent progress in the development of specific inhibitors for some of these enzymes is also presented.

II. Enzymes Metabolizing Extracellular Nucleotides

Extracellular nucleotides are generally metabolized by plasma membrane-bound enzymes whose active site faces the extracellular environment. These enzymes are also called with the more general term of ectoenzymes that reflects this property. Nucleotides can also be metabolized by exoenzymes that are secreted from cells or originate from ectoenzymes shedded from the plasma membrane either by a proteolitic cleavage by proteases or by the cleavage of their glycosyl phosphatidylinositol (GPI) anchors by phospholipases. More specifically, the ectoenzymes involved in nucleotide hydrolysis are referred to as ectonucleotidases and those responsible for their phosphorylation are called ectokinases.

A. Ectonucleotidases Regulate P2 and P1 Signaling

Ectonucleotidases are involved in multiple aspects of P2 and P1 receptor signaling that include (i) termination of P2 receptor activation, (ii) protection of susceptible P2 receptors from desensitization, (iii) generation of ligands for P2 receptors by hydrolyzing either molecules containing a nucleotide moiety (i.e., Np_nN, NAD^+, UDP-glu, etc.) to nucleotides or triphosphonucleosides to diphosphonucleosides, and (iv) production of extracellular adenosine that activates P1 receptors.

There are four families of ectonucleotidases: nucleoside triphosphate diphosphohydrolases (NTPDases), nucleotide pyrophosphatases/phosphodiesterases (NPPs), alkaline and acid phosphatases (ALP and ACP, respectively), and ecto-5′-nucleotidase (a.k.a. CD73; Robson et al., 2006; Stefan et al., 2006; Yegutkin, 2008; Zimmermann, 2000, 2009). These enzymes differ in substrate specificity and hydrolysis, as well as in cellular localization. In agreement with the biological importance of ectonucleotidases, neurological, inflammatory, and autoimmune diseases have been associated with an alteration of the activity and/or expression of these enzymes, which, in some cases, is linked to genetic polymorphisms. These associated diseases include neurological condition-like stress (Fontella et al., 2005; Torres et al., 2002), epilepsy (Bonan et al., 2000), colitis (Friedman et al., 2009), type II diabetes (Friedman et al., 2008), multiple sclerosis (Spanevello et al., 2010a,b), and rheumatoid arthritis (Becker et al., 2010; Kehlen et al., 2001).

1. Nucleoside Triphosphate Diphosphohydrolases (EC 3.6.1.5)

The NTPDase family consists of eight members, that is, NTPDase1–8 (Robson et al., 2006). NTPDase1, -2, -3, and -8 are type II membrane proteins that efficiently hydrolyze all nucleotides activating P2 receptors. In contrast, NTPDase4–7 are localized in intracellular organelles, and it is generally assumed that these enzymes do not participate in the metabolism of extracellular nucleotides (Robson et al., 2006). However, NTPDase5 and -6 may appear at the cell surface or be released in the extracellular space as exoenzymes as demonstrated by heterologous expression (Braun et al., 2000; Hicks-Berger et al., 2000; Mulero et al., 1999). As these enzymes hydrolyze diphosphonucleosides and UTP, they could possibly regulate the activation of ADP-, UDP-, and UTP-dependent P2 receptors. However, the relatively high K_m and low specific activity of NTPDase5 and -6 suggest that the contribution of these enzymes to the hydrolysis of extracellular nucleotides is limited.

Plasma membrane-bound NTPDases hydrolyze adenine and uracil tri- and diphosphonucleosides as well as other nucleotides not demonstrated to activate P2 receptors. The hydrolysis of nucleotides by these enzymes requires the presence of calcium (Ca^{2+}) or magnesium (Mg^{2+}) ions (Kukulski et al., 2005; Zimmermann, 2001). NTPDases exhibit different pH optima but are all fairly active in the pH range of 7.0–8.5 (Kukulski et al., 2005). The K_m values for nucleotide hydrolysis by plasma membrane NTPDases, which are in a low micromolar range, predispose these enzymes to control P2 receptor activation. As all these enzymes efficiently hydrolyze ATP and UTP, they are expected to terminate the activation of P2X1–7 and $P2Y_{2,4,11}$ receptors (Table I; Failer et al., 2003; Kukulski et al., 2005). However, due to differences in ATP and UTP hydrolysis pattern, individual plasma membrane NTPDases could differentially affect the activation of ADP- and UDP-dependent P2 receptors. NTPDase1 (also known as CD39)

TABLE I Summary of Ectoenzymes Potentially Involved in the Modulation of P2 and P1 Receptor Signaling

Ectoenzyme	*Gene*[a]	*Other alias*	*Enzyme class*	*Enzymatic reaction*	*Substrates*	*Inhibitors*	*Functions or potential functions*[b]
Nucleoside triphosphate diphosphohydrolases (NTPDases)							
NTPDase1	*ENTPD1*	CD39, apyrase, ATPDase			NTP, NDP	NaN_3, 8-BuS-ATP, ARL 67156, POM-1, ticlopidine, clopidogrel	• Prevent $P2Y_1$ and P2X1 desensitization • Terminate P2 signaling ($P2Y_2$, $P2Y_6$, P2X7, etc.) • Favor adenosine generation
NTPDase2	*ENTPD2*	CD39L1, ecto-ATPase			NTP	PSB-6426, POM-1	• *Termination of P2 signaling* • *Switch P2 activation*
NTPDase3	*ENTPD3*	CD39L3, HB6	EC 3.6.1.5	NTP → NDP + P_i NDP → NMP + P_i	NTP, NDP	ARL 67156, hN3-H10s	• *Termination of P2 signaling* • *Transient switch of P2 activation*
NTPDase5	*ENTPD5*	CD39L4, PCPH			NDP		• *Termination of P2 signaling*
NTPDase6	*ENTPD6*	CD39L2			UDP, ADP, UTP		• *Termination of P2 signaling*
NTPDase8	*ENTPD8*	Hepatic ATPDase			NTP, NDP		• *Termination of P2 signaling* • *Transient switch of P2 activation*
Nucleotide pyrophosphatases/phosphodiesterases (NPPs)							
NPP1	*ENPP1*	CD203a, PC-1		NTP → NMP + PP_i NDP → NMP + P_i Np_nN → NMP + $Np_{(n-1)}$	ATP, Np_nN, pNP-TMP, NAD^+	ARL 67156, Me-Ap_5A-Me, Me-dAp_5dA-Me	• *Termination of P2 signaling*
NPP2	*ENPP2*	Autotaxin	EC 3.1.4.1 EC 3.6.1.9	NAD^+ → AMP + nicotinamide ribosyl-P	ATP, Np_nN, pNP-TMP, LPC, SPC	Me-Ap_5A-Me, Me-dAp_5dA-Me	• *Favor P2 activation (from dinucleotides' hydrolysis)*
NPP3	*ENPP3*	CD203c, B10 GP130[RB13-6]		3′-5′-NMPc → NMP	ATP, Np_nN, pNP-TMP		• *Favor adenosine generation*

Ecto-5′-nucleotidase							
Ecto-5′-nucleotidase	*NT5E*	CD73	EC 3.1.3.5	NMP → nucleoside + P_i	NMP	ADP, α,β-MeADP	• Activate P1 receptors
Phosphatases (ACP and ALP)							
PAP	*ACPP*	ACP3		NMP → nucleoside + P_i	NMP, pNP-P, various phosphorylated molecules (alkaloid, lipid, protein, sugar)	L-(+)-tartrate, NaF, α-BzNBzPhosphonic acid	• Activate P1 receptors
OcAP/TrAP	*ACP5*		EC 3.6.1.2	NTP → NDP + P_i NDP → NMP + P_i NMP → nucleoside + P_i Molecule-P → molecule + P_i	NTP, NDP, NMP, pNP-P, various phosphorylated molecules (alkaloid, lipid, protein, sugar)		• *Termination of P2 signaling* • *Activation of P1 receptors*
TNAP	*ALPL*	akp-2	EC 3.6.1.1	NTP → NDP + P_i NDP → NMP + P_i NMP → nucleoside + P_i Molecule-P → molecule + P_i	NTP, NDP, NMP, pNP-P, cAMP, various phosphorylated molecules (alkaloid, lipid, protein, sugar)	Levamisole	• *Termination of P2 signaling* • Activate P1 receptors
Ectokinases (NDPK and AK)							
NDPK	*NM23-H1* and/or *-H2*		EC 2.7.4.6	NTP + N′DP ↔ NDP + N′TP	NTP, NDP		• Switch P2 activation
AK	*AK1β*		EC 2.7.4.3	2ADP ↔ AMP + ATP	ADP	Ap_5A	• Favor ADP-activated receptors ($P2Y_1$, $P2Y_{12}$, $P2Y_{13}$) • Switch P2 activation

(*continued*)

TABLE I *(continued)*

Ectoenzyme	*Gene*[a]	*Other alias*	*Enzyme class*	*Enzymatic reaction*	*Substrates*	*Inhibitors*	*Functions or potential functions*[b]
Other ectoenzymes							
ART2.1, ART2.2	*Art2a, Art2b*	Ly92a Rt6, Ly92b	EC 2.4.2.31	NAD^+ + protein → ADP-ribolysated protein + nicotinamide	NAD^+	Rec. Ab s+16a-Fc	• Activate P2X7 receptor
ADA	*ADA*	Adenosine aminohydrolase	EC 3.5.4.4	Adenosine + H_2O → inosine + NH_4^+	Adenosine	EHNA, dCF coformycin	• Terminate/reduce P1 signaling • Favor nucleobase uptake
PNP	*PNP*	Inosine phosphorylase	EC 2.4.2.1	Inosine + H_2O → hypoxanthine + ribofuranose	Inosine, guanosine	8-aminoguanine, 5′-I-9-deazaIno, Acyclovir-2P	• Terminate P1 signaling • Favor nucleobase uptake
e-PD	–			3′-5′-AMPc → AMP	cAMP		• Activation of P1 signaling

[a] Gene names according to the HUGO nomenclature.
[b] Normal font is for known functions and italic for potential functions.

should terminate the activation of $P2Y_{1,12,13}$ receptors as it does not allow the accumulation of ADP during ATP hydrolysis. The capacity of this enzyme to hydrolyze ATP and ADP rapidly to AMP also protects P2X1 (Faria et al., 2008; Schaefer et al., 2007) and $P2Y_1$ receptors from desensitization (Enjyoji et al., 1999; Kauffenstein et al., 2010b). In contrast to NTPDase1, NTPDase2, -3, and -8 are expected to promote the activation of ADP-specific receptors due to sustained (NTPDase2) or transient (NTPDase3 and -8) accumulation of ADP in the presence of ATP. The latter enzymes also provide a substrate, ADP, for adenylate kinase (AK; Table I and Sections II.B.2 and III.A.1). Interestingly, all plasma membrane NTPDases dephosphorylate UTP with a significant accumulation of UDP, a $P2Y_6$ receptor ligand. The production of UDP by NTPDases may promote $P2Y_6$-dependent responses such as cytokine production including IL-6, IL-8, TNF-α, MCP-1, IP-10, MIP-2, and MIP-3α (Bar et al., 2008; Ben Yebdri et al., 2009; Kukulski et al., 2007; Warny et al., 2001). Distinct NTPDases also differentially affect the production of adenosine from extracellular ATP suggesting that they have the capacity to fine-tune the activation of adenosine receptors (see Section II.A.4).

Fluorescence resonance energy transfer (FRET) studies showed that NTPDase1 is in close proximity to a number of P2 and P1 receptors (Schicker et al., 2009). After heterologous expression of rat NTPDase1 with several P1 and P2 receptors (A_1, A_{2A}, P2X2, $P2Y_1$, $P2Y_2$, $P2Y_{12}$, and $P2Y_{13}$), this enzyme was detected in proximity of these receptors except for P2X2. In the same experiment, NTPDase2 was not located directly nearby receptors. These studies further support the importance of NTPDase1 in the control of P2 and P1 receptor activation. In agreement, NTPDase1 controls the activation of $P2Y_2$ receptor in endothelial cells and neutrophils (Kauffenstein et al., 2010b; Kukulski et al., submitted for publication), $P2Y_6$ in vascular smooth muscle cells (Kauffenstein et al., 2010a), and P2X7 in monocytes/macrophages (Hyman et al., 2009; Lévesque et al., 2010) and prevent desensitization of certain P2 receptors (Enjyoji et al., 1999; Kauffenstein et al., 2010b; Schaefer et al., 2007). Some of these studies are further discussed in Section III.

Plasma membrane NTPDases are ubiquitously expressed. NTPDase1 is expressed in endothelial (Enjyoji et al., 1999; Kaczmarek et al., 1996) and smooth muscle cells (Kauffenstein et al., 2010a; Sévigny et al., 1997) as well as in leukocytes including neutrophils (Hyman et al., 2009; Kukulski et al., submitted for publication), monocytes/macrophages (Hyman et al., 2009; Lévesque et al., 2010; Martín-Satué et al., 2009), lymphocytes (e.g., in Tregs, memory lymphocytes, B lymphocytes, and natural killer T cells; Beldi et al., 2008; Kansas et al., 1991; Moncrieffe et al., 2010; Nigam et al., 2010), natural killer cells (Beldi et al., 2010), Langerhans, and dendritic cells (Kansas et al., 1991; Mizumoto et al., 2002; Pizzirani et al., 2007). NTPDase1 was also detected in some epithelial cells (Fausther et al., 2010;

Kittel et al., 2004; Martín-Satué et al., 2009; Sorensen et al., 2003). NTPDase2 is present on the adventitial surface of blood vessels (Sévigny et al., 2002), type I cells of taste buds (Bartel et al., 2006), and different types of glial cells (Braun et al., 2004; Wink et al., 2006). So far, the localization of NTPDase3 was addressed in the brain, kidney, airways, and reproductive and digestive systems. These studies demonstrated that NTPDase3 is expressed in neurons from the brain (Belcher et al., 2006) and along the bowel (Lavoie et al., 2011), and also in some epithelial cells of the digestive, reproductive, renal, and respiratory systems (Fausther et al., 2010; Lavoie et al., 2011; Martín-Satué et al., 2009). More specifically in the kidney, NTPDase3 is expressed on thick ascending limb, distal tubule, and on cortical and outer medullary collecting ducts (Vekaria et al., 2006). In the pancreas, it is expressed in all Langerhan's islet cells (Lavoie et al., 2010), and in the gastric antrum in some enteroendocrine cells (Lavoie et al., 2011). The expression of NTPDase8 appears specific to a few tissues. It was detected by both Western (Fausther et al., 2007; Sévigny et al., 2000) and Northern blot (Bigonnesse et al., 2004) in liver, kidney, and intestine. In liver, NTPDase8 is predominantly expressed by hepatocytes in bile canaliculi (Fausther et al., 2007; Sévigny et al., 2000). In porcine kidney, NTPDase8 was immunolocalized in tubules on brush border membranes (therefore presumably on proximal tubules; Sévigny et al., 2000). It is also noteworthy that the expression of some NTPDases can be affected by inflammation (see Section III.A.3).

2. Nucleotide Pyrophosphatases/Diphosphodiesterases (EC 3.1.4.1, EC 3.6.1.9)

NPP family consists of seven members (NPP1–7) but only NPP1 (PC-1, CD203a), NPP2 (autotaxin), and NPP3 ($gp130^{RB13\text{-}6}$, B10, CD203c) have the capacity to hydrolyze nucleotides (Stefan et al., 2006). NPP6 and -7, as well as NPP2, hydrolyze the phosphodiester bonds of lysophospholipids or choline phosphodiesters (Stefan et al., 2006), whereas the biological activity of NPP4 and -5 is not yet known. NPP1 and NPP3 are type II membrane proteins, while NPP2 is produced as a zymogen (pre-pro-enzyme) that is secreted after proteolytic cleavage by a furine-like protease (Jansen et al., 2005). Soluble forms of NPP1 (Belli et al., 1993; Hosoda et al., 1999) and NPP3 (Meerson et al., 1998) have also been described.

For optimal activity, NPP1–3 require alkaline pH and divalent ions, that is, zinc (Zn^{2+}) and magnesium (Mg^{2+}; Gijsbers et al., 2001). By hydrolyzing ATP to AMP and pyrophosphate (PP_i) with K_m values of approximately 10 μM, these enzymes could terminate P2 receptor signaling (Table I). In addition to ATP hydrolysis, NPP1–3 can also hydrolyze phosphodiester bonds of cyclic nucleotides (i.e., cNMP) and several molecules containing a nucleotide moiety, that is, dinucleotide polyphosphates, nucleotide-sugars such as UDP-glucose, NAD^+, $NADP^+$, etc. (Table I; Canales et al., 1995;

Vollmayer et al., 2003; Stefan et al., 2006). As some of these molecules can activate P2 receptors, for example, dinucleotide polyphosphates, UDP-glucose, and NAD^+, their hydrolysis by NPPs would be expected to terminate this activation. At the same time, the hydrolysis of some of these molecules, for example, diadenosine pentaphosphates (Ap_5A), results in the removal of AK inhibitor (see Section III.A.1). NPPs can also generate nucleotides that activate P2 receptors. For example, the hydrolysis of dinucleotide polyphosphates to AMP and nucleoside (n-1)phosphates with K_m values in the low micromolar range (2–4 μM) can result in the production of nucleotides such as ADP and ATP (Table I; Grobben et al., 1999; Vollmayer et al., 2003). If these intermediary products accumulate before being further hydrolyzed by NPPs themselves, they could activate P2 receptors. In favor of this possibility, we observed that NPPs hydrolyzed preferentially dinucleotide polyphosphates over ATP (Lévesque et al., 2007; S. A. Lévesque & J. Sévigny, unpublished observations). Therefore, it is plausible that the hydrolysis of high concentrations of dinucleotide polyphosphates by NPPs may result in the generation of ATP and P2 receptor activation.

NPPs can also affect P1 receptor signaling by hydrolyzing ATP directly to the ecto-5′-nucleotidase substrate AMP, without any production of the ecto-5′-nucleotidase inhibitor ADP, contrasting with the effect produced by NTPDase2, -3, and -8. Therefore, NPPs would be expected to facilitate the production of adenosine and P1 receptor activation. Noteworthy, NPPs can produce AMP also from some dinucleotide polyphosphates and other molecules such as NAD^+ (Table I). Eventually, NPPs can also affect P2X7-mediated responses in murine macrophages by producing PP_i from ATP hydrolysis. Indeed, PP_i inhibits the P2X7-induced activation of NLRP3-inflammasome and thus IL-1β maturation and release (Pelegrin & Surprenant, 2009).

NPPs are widely distributed (Bollen et al., 2000; Goding et al., 2003). NPP1 is found in adipose tissue, urinary bladder, heart, kidney, liver, lung, and thymus (Petersen et al., 2007) as well as in airway epithelia and hepatocytes (Stefan et al., 2006). NPP1 is the major PP_i-generating enzyme on osteoblasts and chondrocytes where it plays a critical role in bone mineralization which is governed by the tight balance between phosphate and pyrophosphate concentration (Goding et al., 2003; Harmey et al., 2004). Among leukocytes, NPP1 is present in alveolar macrophages and subsets of B and T lymphocytes (Goding et al., 1998; Petersen et al., 2007; Takahashi et al., 1970). NPP3 is present on the apical side of airway and choroid-plexus epithelial cells, hepatocytes, cholangiocytes, and human basophiles (Buhring et al., 2001). NPP1 and -3 are expressed in rat C6 glioma cells where they are responsible for the hydrolysis of extracellular ATP (Grobben et al., 1999; Joseph et al., 2004). NPPase activity was also found in vascular smooth muscle cells where it is involved in cell growth

(Prosdocimo et al., 2009). Interestingly, the expression of NPP1-3 is affected by inflammation and other conditions (see Section III.A.3).

3. Alkaline and Acid Phosphatases (ALP; EC 3.1.3.1 and ALP; EC 3.1.3.2, Respectively)

The ALP and ACP families consist of several ectoenzymes that dephosphorylate various substrates including nucleotides, phosphorylated proteins, polysaccharides, and alkaloids.

Several members of mammalian ALP family consist of two identical subunits covalently bond by two disulphide bridges and attached to the plasma membrane via a GPI anchor at the C-terminus (Millan, 2006a). Four genes encode for ectoalkaline phosphatase isozymes in human, that is, *ALPL*, *ALPP*, *ALPP2*, and *ALP1* and three in mice, that is, *akp2*, *akp3*, and *akp5*. In human, the products of *ALPP*, *ALPP2*, and *ALP1* genes are highly homologous (90–98% identity) and have specific expression in placenta, germ cells, and intestine, respectively. In contrast, the product of *ALPL* or *akp2* gene called tissue nonspecific alkaline phosphatase (TNAP) has a wide distribution as it is abundantly expressed in bone, liver, and kidney, and to a lower extent in some other tissues (Hoshi et al., 1997; Millan, 2006a, 2006b).

In the presence of Zn^{2+} and Mg^{2+} ions, ALP hydrolyze a broad range of substrates including 5′-nucleotides (e.g., ATP, ADP, and AMP), pyridoxal-5′-phosphate (a phosphorylated form of vitamin B_6), 6′-phosphorylated sugars (e.g., glucose-6-P, fructose-6-P), 3′–5′-cyclic AMP (cAMP; released to extracellular space via certain ABC transporters, e.g., MRP4, MRP5, and MRP8), and PP_i (Table I; Hessle et al., 2002; Millan, 2006a; Say et al., 1991). ALP could terminate P2 receptor activation because they can hydrolyze nucleotides, but this function has yet to be demonstrated. In contrast, TNAP has a low capacity to hydrolyze ATP and UTP at physiological pH (Say et al., 1991; S. A. Lévesque & J. Sévigny, unpublished observation) and generally colocalizes with other ectonucleotidases (Langer et al., 2007). ALP can also contribute to adenosine production from AMP as a substrate, as demonstrated for the neuroblastoma × glioma hybrid NG108-15 cells and for the mucosal surface of airway epithelia (Ohkubo et al., 2000; Picher et al., 2003). Interestingly, the latter cells also express ecto-5′-nucleotidase, which is more effective than TNAP to hydrolyze AMP at a low concentration; however, at high concentration, the contribution of both enzymes to the hydrolysis of AMP was equivalent (Picher et al., 2003). These data suggest that TNAP may support ecto-5′-nucleotidase in the conditions that require more effective AMP hydrolysis and adenosine production.

The ACP family comprises five members named after a specific tissue, cell, or cell structure from which they were first discovered, that is, EAP (erythrocytes), LAP (lysosomes), PAP (prostate), MAP (macrophages), and OcAP/TrAP (osteoclasts, tartrate-resistant form). With the exception of OcAP/TrAP (see below), ACP does not hydrolyze ATP and is therefore not expected to regulate P2 receptor activation. However, ACP may affect P1 receptor activation by generating adenosine (Table I) as demonstrated for PAP in the dorsal root ganglia. Indeed, $PAP^{-/-}$ mice exhibit adenosine deficit in these nerve structures and are therefore more susceptible to thermal hyperalgesia and mechanical allodynia (Zylka et al., 2008). In addition to the spinal cord, PAP is also expressed in prostate, salivary glands, thymus, lung, kidney, brain, spleen, and thyroid. Interestingly, an active splicing variant of this enzyme can be secreted to the extracellular space (Quintero et al., 2007).

OcAP/TrAP, the only acid phosphatase capable of ATP and ADP hydrolysis, is generally present in lysosomal membranes of osteoclasts, macrophages, and dendritic cells (Hayman, 2008). This enzyme may potentially affect P2 receptor activation in the acidic resorptive space of the bone (Kaunitz & Yamaguchi, 2008).

4. *Ecto-5′-Nucleotidase (EC 3.1.3.5)*

The human family of 5′-nucleotidases has seven members, six cytosolic and one ectoenzyme, that is, ecto-5′-nucleotidase which is anchored in the plasma membrane via a GPI anchor at the C-terminus (Airas et al., 1997; Yegutkin, 2008). In mammals, ecto-5′-nucleotidase (also known as CD73) is a glycoprotein consisting of two 60–74 kDa subunits associated through noncovalent bonds (Zimmermann, 1992). This enzyme binds Zn^{2+} and other divalent cations, which are required for enzymatic activity, and has a broad specificity toward nucleoside monophosphates with a K_m in the range of 3–50 μM (Zimmermann, 1992). The main function of this enzyme resides in the generation of extracellular adenosine from AMP, the product of NTPDases, NPPs, and AK (Table I). This role of ecto-5′-nucleotidase was unequivocally confirmed by the generation of $Cd73^{-/-}$ mice. These animals exhibit markedly impaired production of extracellular adenosine that is not compensated *in vivo* by other ectoenzymes such as ALP and ACP or by the release of adenosine via ENT. As a result, $Cd73^{-/-}$ mice lose adenosine-mediated protection in inflammatory conditions such as hypoxia or lung injury (Eckle et al., 2007a, 2007b; Reutershan et al., 2009). The inflammatory mechanisms affected in these diseases by the deficiency of ecto-5′-nucleotidase are discussed in Section III. In contrast to its generally beneficial role, ecto-5′-nucleotidase was recently shown to stimulate breast cancer development (Stagg et al., 2010). This suggests that the activity of CD73 may become a therapeutic target for treatments against cancers that take an advantage of extracellular adenosine production for immunological escape. In addition to its role in inflammation and cancerogenesis, ecto-5′-nucleotidase is also involved in trophic effects such as the differentiation of neurons (Braun et al., 1995, 1998; Heilbronn & Zimmermann, 1995).

As alluded earlier, ecto-5′-nucleotidase accepts all nucleoside monophosphates as substrates but it does not hydrolyze nucleoside tri- and diphosphates. On the contrary, the substrates of NTPDases, that is, ATP and especially ADP, are potent inhibitors of this enzyme (Zimmermann, 1992). Thanks to this property, ecto-5′-nucleotidase can be expected to favor a delay between the activation of P2 receptors by released ATP and/or ADP and the activation of P1 receptors by adenosine, the hydrolysis product of these nucleotides. In agreement, an *in vitro* experiment reflecting extracellular ATP catabolism by each individual plasma membrane NTPDase in combination with ecto-5′-nucleotidase showed that a significant decrease in ATP and ADP concentration was required for the effective production of adenosine. As expected, adenosine generation was fastest in the presence of NTPDase1 that rapidly hydrolyzed both ATP and ADP. In contrast, in the presence of NTPDase2, adenosine production was very low due to the minimal hydrolysis of ADP by this enzyme. The results obtained with NTPDase3 and -8 were somehow intermediate in respect to those for NTPDase1 and -2 at high ATP level (500 μM), which correlates with a transient production of ADP by these enzymes (Fausther, Lecka et al., unpublished observation; Kukulski et al., 2005; Vorhoff et al., 2005), adenosine being produced only when ATP and ADP levels have fallen below 50–100 μM, even in the presence of high AMP levels (200 μM). Interestingly, the situation was different during the metabolism of low ATP concentration (1 μM) when the levels of ATP and ADP were insufficient to inhibit ecto-5′-nucleotidase and therefore adenosine generation was limited only by the rate of AMP production by NTPDases. Noteworthy, in these conditions, rat NTPDase8 acted like NTPDase2 as it hydrolyzed ATP to ADP with only minimal production of AMP (Fausther, Lecka et al., unpublished observation; Martín-Satué et al., 2010) which greatly limited adenosine generation by ecto-5′-nucleotidase. This was in agreement with the K_m value of this enzyme for ADP as a substrate that is much higher than those of NTPDase1 and -3 (Kukulski et al., 2005).

Ecto-5′-nucleotidase is abundantly expressed in colon, kidney, brain, liver, heart, and lung and along mouse reproductive tract (Martín-Satué et al., 2010; Yegutkin, 2008; Zimmermann, 1992). At the cellular level, this enzyme is present in neurons, endothelial cells, and $CD4^+$/$CD25^+$/$Foxp3^+$ Treg cells, while neutrophils, erythrocytes, platelets, and other blood cells express little or no CD73 (Yegutkin, 2008). The expression of ecto-5′-nucleotidase is often upregulated in inflammatory conditions (see Section III.A.3).

B. Ectokinases in P2 Receptor Activation

P2 receptor signaling can also be influenced by the ectokinases, that is, ectonucleotide diphosphokinases and AKs, that interconvert and regenerate nucleotides, respectively (Yegutkin, 2008; Yegutkin et al., 2002).

1. Nucleoside Diphosphate Kinases (NDPK; EC 2.7.4.6)

The human NDPK family consists of 10 members (NM23-H1 to H10) which are the products of the genes belonging to NM23 tumor metastasis suppressor (Boissan et al., 2009; Yegutkin, 2008). NDPKs are oligomers composed of 17–20 kDa subunits and are generally found inside the cells, in mitochondria, cytosol, and nuclei. However, two members of this family, that is, NM23-H1 and -H2, exist as ectoenzymes and therefore could be responsible for NDPK activities detected at the surface of endothelial cells, smooth muscle cells, astrocytoma and glioma cells, hepatocytes, keratinocytes, lymphocytes, and erythrocytes (Boissan et al., 2009; Yegutkin, 2008). NDPKs catalyze the transfer of γ-phosphate from one nucleoside 5′-triphosphate to another nucleoside 5′-diphosphate (NTP + N′DP → NDP + N′TP) at neutral pH in the presence of Mg^{2+} ions (Table I). These reactions may affect P2 receptor signaling. For example, using ATP and UDP as substrates, NDPK can generate UTP and ADP and thus transform the ligands of all P2X and $P2Y_{2,6,11}$ receptors to the ligands of $P2Y_{2,4,12,13}$ receptors. However, as these reactions are reversible, the real impact of NDPK on P2 receptors would depend on substrate availability and concentration.

2. Adenylate Kinase (AK; EC 2.7.4.3)

Only one member of the AK family, that is, ecto-AK or AK1β, is present at the cell surface. Other members are located intracellularly (AK1α and AK2–6), or have unknown localization (AK7). Ecto-AK processes exclusively adenine nucleotides, requires the presence of Mg^{2+} ions, and is inhibited by substrates of NPPs such as Ap_5A (Yegutkin et al., 2008). This enzyme catalyzes the reversible transphosphorylation reaction leading to ATP and AMP production from two molecules of ADP as a substrate (Table I). The rates of forward and reverse reactions depend on the concentration of ADP, ATP, and AMP. Thus, at given concentrations of these nucleotides, ecto-AK can either favor/prolong the activation of ATP-dependent P2 receptors and terminate the activation of ADP-dependent receptors, or do the opposite. In agreement with the former function, in human colon adenocarcinoma cells, ecto-AK potentiated $P2Y_{11}$-dependent IL-8 secretion through the production of ATP from ADP as a substrate (see Section III.A.1), whereas, in liver cells, it decreased the HDL transport induced by the ADP-dependent $P2Y_{13}$ receptor (Fabre et al., 2006). In the presence of [ADP] > [ATP], ecto-AK may also favor adenosine production and P1 activation by producing the ecto-5′-nucleotidase's and phosphatases' substrate AMP. However, the impact of ecto-AK on adenosine production appears to depend on the overall network of ectoenzymes expressed in a particular cell type. For example, in the lung epithelial cells expressing AK, the production of adenosine from ATP as a substrate was delayed compared to the production of uridine from UTP

(Picher & Boucher, 2003) while in colon adenocarcinoma cell line HT29, this enzyme actually facilitated adenosine generation during extracellular ATP metabolism by transforming ADP produced by NTPDase2, also expressed in these cells, to AMP. In agreement, the inhibition of AK during the incubation of HT29 cells with ATP markedly decreased adenosine production (F. Bahrami, F. Kukulski & J. Sévigny, unpublished observation). However, the exact effect of AK on P1 signaling requires to be rigorously demonstrated in a physiological context.

Based on activity assays, the presence of ecto-AK has been reported in hepatocytes, hepatic cell lines, epithelial and endothelial cells, keratinocytes, lymphocytes, and leukemic cell lines (Dzeja & Terzic, 2009; Lecka et al., 2010a; Yegutkin, 2008). Further studies confirming the presence and function of AK protein in these cells are now required.

C. Other Ectoenzymes Potentially Involved in Purinergic Signaling

1. Ecto-ADP-Ribosyltransferases (ARTs; EC 2.4.2.31)

In addition to ectonucleotidases and ectokinases, P2 receptor signaling can also be affected by ARTs. These enzymes catalyze mono-ADP ribosylation of certain membrane proteins, like P2X7 receptor, using extracellular NAD^+ as a substrate (Table I). The ADP ribosylation catalyzed by ART2.2 (also known as ART2b and Rt6) was shown to activate P2X7 in T lymphocytes (Adriouch et al., 2008; Hubert et al., 2010; Seman et al., 2003). Interestingly, ART2.1, which is highly similar to ART2.2 but active only in reducing conditions, is present on several types of leukocytes where its expression can be induced by IFN-β and -γ (Hong et al., 2007, 2009).

ARTs are expected to play a role in conditions associated with high levels of extracellular NAD^+, for example, inflammation or ischemia. Noteworthy, four ectoenzymes can possibly compete with ARTs for NAD^+: NPP1 and NPP3 that can metabolize NAD^+ to AMP and nicotinamine ribosyl phosphate (Fig. 1 and Table I), as well as CD38 and CD157 that produce nicotinamide and either ADP-ribose or cyclic ADP-ribose (see Chapter 10 for more details). Further studies are necessary to determine the importance of ARTs in P2 receptor signaling and to define how these enzymes interplay with other ectoenzymes engaged in extracellular nucleotide metabolism.

2. Adenosine Deaminase (ADA; EC 3.5.4.4) and Purine Nucleoside Phosphorylase (PNP; EC 2.4.2.1)

These enzymes can terminate the activation of P1 receptors by breaking down adenosine to inosine (adenosine deaminase (ADA)) and inosine to hypoxanthine (purine nucleoside phosphorylase (PNP); Fig. 1 and Table I). In human, two isoenzymes, ADA1 (a.k.a. ADA) and ADA2 (a.k.a. CECR1),

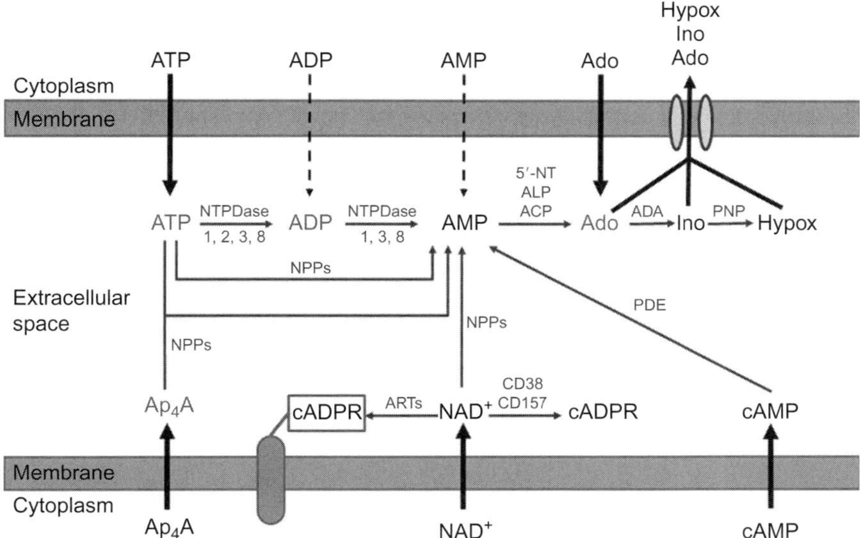

FIGURE 1 Extracellular metabolism of adenine nucleotides and related molecules. The enzymes (in red) correspond to 5′-NT, ecto-5′-nucleotidase; ADA, adenosine deaminase; ACP, acid phosphatases; ALP, alkaline phosphatases; ART, adenosine diphosphate ribosyltransferase; NPPs, nucleotide pyrophosphatases/phosphodiesterases; NTPDases, nucleoside triphosphate diphosphohydrolases; PDE, phosphodiesterases; and PNP, purine nucleoside phosphorylase. For sake of clarity, adenylate kinase (AK; 2ADP$\leftrightarrow$AMP+ATP) and nucleoside diphosphate kinase (NDPK; ADP+A′TP$\rightarrow$ATP+A′DP) are not represented in this figure although they are expected to participate in adenine nucleotide metabolism. (For interpretation of the references to color in this figure legend, the reader is referred to the Web version of this chapter.)

appear to be responsible for ADA activity (Wiginton et al., 1986; Zavialov & Engstrom, 2005). Both enzymes are present in the cytosol of cells but only ADA was linked to the ecto-ADA activity. Indeed, the membrane topology of ADA has not been completely elucidated, but interaction of ecto-ADA with the type II membrane protein dipeptidyl peptidase-4 (a.k.a. CD26; Weihofen et al., 2004) supports the view that this enzyme is present at the plasma membrane. Importantly, a commercial form of ADA is often used as an adenosine scavenger in assays aiming to verify the implication of adenosine receptors in different systems. This is an alternative approach to the utilization of adenosine receptor antagonists. However, it should be applied with caution as the product of the enzyme inosine was reported to have some physiological functions, possibly via the activation of A_1 and/or A_3 receptors, for which inosine has lower affinity than adenosine (Fredholm et al., 2001; Gomez & Sitkovsky, 2003; Jin et al., 1997), or via the activation of a yet undescribed receptor.

Inosine produced by ADA can further be hydrolyzed to hypoxanthine. Two genes *PNP1* and *PNP2* encode proteins with a capacity to catalyze this reaction as well as hydrolysis of guanosine to xanthine. Hitherto, PNP1 seems principally found in the cytosol, little information is available for PNP2. Finally, high PNP activity was detected at the surface of leukocytes such as lymphocytes, neutrophils, and dendritic cells (Pacheco et al., 2005; van Waeg & Van den Berghe, 1991; Yegutkin, 2008).

3. Ectophosphodiesterase (ecto-PD; EC 3.1.4.17)

Last, as ecto-PD hydrolyzes cAMP to AMP providing the substrate for ecto-5′-nucleotidase and phosphatases, it can therefore contribute to adenosine generation and P1 receptor activation (Fig. 1). Significant ecto-PD-like activity is present in energy-rich tissues including liver, kidney, fat tissue, and skeletal muscles (Chiavegatti et al., 2008; Jackson et al., 2007; Smoake et al., 1981; Zacher & Carey, 1999), but further studies are needed to confirm whether these activities belong to ecto-PD.

III. Examples of Physiological Roles of Ectoenzymes in P2 and P1 Signaling

A. Ectoenzymes in Inflammation

1. Ectonucleotidases and Ectokinases Control Cytokine Production

Cytokines are small soluble proteins produced by various cell types to mount immune responses. IL-8 is a key human cytokine involved in inflammatory recruitment of neutrophils (thus it is called a chemokine), cell proliferation and activation, and angiogenesis. In many cell types, the production of IL-8 is controlled by extracellular nucleotides (Ben Yebdri et al., 2009; Kukulski et al., 2007, 2009; Warny et al., 2001). A recent study has demonstrated that the endogenous activity of NTPDase1 tightly regulates the level of ATP in the media of human neutrophils, and in extension, its autocrine effect on $P2Y_2$-dependent IL-8 release (Kukulski et al., submitted for publication). Indeed, the inhibition of NTPDase1 in neutrophils resulted in the accumulation of ATP in the media of these cells and an increase of the basal level of IL-8 secretion, that is normally minimal (Kukulski et al., submitted for publication). The control of IL-8 production by NTPDase1 in neutrophils may have an important impact on the development of immune responses as the amount of IL-8 produced at inflamed sites by neutrophils already recruited may be critical to determine the magnitude of further leukocyte infiltration and activation. Thus, factors decreasing NTPDase1 activity (e.g., natural inhibitors, reactive oxygen species, mutations affecting the expression of this enzyme, etc.) may exacerbate inflammatory responses. In agreement, humans with decreased NTPDase1 expression are more susceptible to inflammatory bowel disease characterized by excessive neutrophil migration and IL-8 production (Friedman et al., 2009). Noteworthy, IL-8 release from human primary monocytes is also controlled by P2 receptors and can be markedly reduced by the addition of exogenous NTPDase1 activity (i.e., apyrase) to these cells to increase the endogenous level of this enzyme also expressed in these cells (Ben Yebdri et al., 2009; Kukulski et al., 2007).

New data from our laboratory show that IL-8 production can also be regulated by ecto-AK and NTPDase2 (Bahrami et al., submitted for publication). In HT29 colon carcinoma cell line, IL-8 release is mediated by $P2Y_{11}$ receptor, for which ATP is a better agonist than ADP. However, both nucleotides induce similar IL-8 release from HT29 cells due to ADP conversion to ATP by ecto-AK. The presence of ecto-AK is of particular importance in these cells as they also express NTPDase2 that rapidly converts ATP to ADP. Therefore, in these cells, ecto-AK helps keeping the activation of an ATP-dependent receptor(s) that would normally be rapidly diminished by NTPDase2. Ecto-AK in HT29 cells also makes possible the generation of adenosine by producing AMP in the same time as ATP (from ADP as the substrate). The question whether the expression of ecto-AK is peculiar to HT29, a cancer cell line, or is also present in normal intestinal epithelial cells will be further investigated.

NTPDase1 also controls the release of cytokines of the IL-1 family. In mouse macrophages, the endogenous activity of this enzyme significantly decreased P2X7-induced IL-1β release. Indeed, macrophages from NTPDase1-deficient mice (*Entpd1*$^{-/-}$) primed with lipopolysaccharide (LPS) or Pam_3CSK_4, to stimulate pro-IL-1β production, and then stimulated with extracellular ATP (2 mM) exhibited a twofold increase in IL-1β release compared to *Entpd1*$^{+/+}$ macrophages (Lévesque et al., 2010). Interestingly, NTPDase1 activity present on *Entpd1*$^{+/+}$ murine macrophages also preserved these cells from ATP-induced death while *Entpd1*$^{-/-}$ cells were very susceptible to this effect (Lévesque et al., 2010). In human umbilical vein endothelial cells (HUVECs), the release of IL-1α was markedly reduced by exogenous NTPDase1 activity (i.e., apyrase) or the augmentation of NTPDase1 activity by means of adenoviral overexpression (Imai et al., 2000).

2. Ectoenzymes Control Inflammatory Leukocyte Trafficking

Leukocytes play key roles in inflammation. Under physiological conditions, the majority of these cells circulate in the blood stream patrolling vascular endothelium for the presence of inflammatory cues. Infection and/or injury initiate rapid leukocyte recruitment through a stepwise migration process including rolling and adhesion to endothelium, transendothelial migration, and chemotaxis (Wagner & Roth, 2000). This last step of migration plays a key role in attracting extravasated immune cells at sites of inflammation where they kill and eliminate invading pathogens, model further immune responses, or initiate tissue repair and healing processes (Nathan, 2006; Nauseef, 2007). Mounting evidence shows that ectonucleotidases and ectokinases participate in all stages of leukocyte migration.

Neutrophils are the most abundant type of leukocytes in human blood whose infiltration is a hallmark of an acute inflammatory response. Flow cytometry analysis with specific anti-NTPDase1 antibodies revealed that more than 90% of neutrophils express NTPDase1 protein at the cell surface (Pulte et al., 2007). In agreement, these cells very efficiently hydrolyze extracellular ATP (Corriden et al., 2008; Kobayashi et al., 1997; Kukulski et al., submitted for publication; Pulte et al., 2007). During transendothelial migration, activated neutrophils release ATP that is subsequently rapidly dephosphorylated by NTPDase1 expressed on these cells as well as NTPDase1 and ecto-5′-nucleotidase present at the surface of the endothelium. This process generates extracellular adenosine that, through the activation of P1 receptors, inhibits the migration of neutrophils (Eltzschig et al., 2004, 2006, 2008; Weissmuller et al., 2005). In agreement with a lack of mRNA for ecto-5′-nucleotidase, neutrophils exhibit very low ectoAMPase activity and therefore have limited capacity to produce adenosine by themselves (Corriden et al., 2008; Kukulski et al., submitted for publication). Thus, the production of adenosine during the interaction of neutrophils with endothelial cells appears to be mainly due to endothelial ecto-5′-nucleotidase (Eltzschig et al., 2003). In agreement, endothelial ecto-5′-nucleotidase also regulates inflammatory migration of monocytes. Indeed, the knockdown of its expression with siRNA increased the adhesion of monocytes to the endothelium through the upregulation of intercellular adhesion molecule-1 (ICAM-1), vascular cell adhesion molecule-1 (VCAM-1), and E-selectin expression at the endothelial surface (Grunewald & Ridley, 2010). This study also demonstrated that ecto-5′-nucleotidase-deficient HUVECs have a more elongated shape, higher level of stress fiber formation, and increased permeability compared to normal HUVECs (Grunewald & Ridley, 2010).

In agreement with the role of NTPDase1 and ecto-5′-nucleotidase in neutrophil transendothelial migration, mice deficient in both enzymes exhibit increased neutrophil infiltration in hypoxia and LPS-induced lung injury (Eckle et al., 2007a; Eltzschig et al., 2004; Reutershan et al., 2009). Although the exacerbated leukocyte recruitment in these animals may be explained by a deficit in extracellular adenosine production, there is also a significant body of evidence demonstrating a key role of P2 receptors in inflammatory leukocyte trafficking in asthma (Idzko et al., 2007), lung inflammation (Cicko et al., 2010), cystic fibrosis (Boucher, 2007), and others. Moreover, NTPDase1 was shown to control *N*-formyl-Met-Leu-Phe (fMLF, also known as fMLP)-induced and P2X7-mediated expression of CD11b (or $\alpha_M\beta_2$, a subunit of MAC-1) in human neutrophils (Hyman et al., 2007). This protein is important in neutrophil adhesion to endothelium and its expression was markedly increased in *Entpd1*$^{-/-}$ neutrophils (deficient in NTPDase1; Hyman et al., 2007). Further, *Entpd1*$^{-/-}$ mice had more severe cerebral ischemia due to increased monocyte and neutrophil infiltration (Hyman et al., 2009). Thus, NTPDase1 is likely to control leukocyte migration via the regulation of both P2 and P1 receptor activation.

In addition to the regulation of neutrophil adhesion and transendothelial migration, ectonucleotidases also take part in the control of chemotaxis of these cells, that is, their movement toward a chemotactic agent. The latter includes the bacterial component fMLF and factors produced by an inflamed tissue or immune cells such as chemokine IL-8, leukotriene B4 (LTB4), and complement protein 5a (C5a). All these chemoattractants induce the release of ATP from neutrophils through maxi-anion channels and pannexin-1 (panx1) hemichannels (Chen et al., 2010). In their seminal work, Dr. Junger's group showed that neutrophil chemotaxis involves both the released ATP and its degradation product adenosine (Chen et al., 2006). Specifically, the authors proposed that ATP released from the leading edge of neutrophils activates $P2Y_2$ receptors to orient cell migration toward the chemoattractant and is subsequently broken down to adenosine that via A_3 receptors stimulates neutrophil movement (Chen et al., 2006). Although these opposite effects of adenosine on neutrophil trafficking, that is, the inhibition of transendothelial migration (see above) and the stimulation of chemotaxis, are not incompatible, the mechanism that segregates them remains to be elucidated. The impaired neutrophil recruitment in $P2Y_2$ and A_3 receptor knockout mice appears to support the *in vitro* model of neutrophil chemotaxis proposed by Dr. Junger (Chen et al., 2006; Inoue et al., 2008). However, we recently demonstrated that endothelial $P2Y_2$ receptor also plays a key role in neutrophil transendothelial migration (Kukulski et al., 2010). Therefore, further *in vivo* studies are necessary to determine the contribution of neutrophil versus endothelial $P2Y_2$ receptors in the recruitment of the former cells.

Interestingly, in addition to ectonucleotidases, the inflammatory migration of lymphocytes is also regulated by nucleoside kinases. Specifically, it was demonstrated that, while endothelial ectonucleotidases NTPDase1 and ecto-5′-nucleotidase decrease T lymphocyte transendothelial migration via hydrolysis of ATP to adenosine, ectokinases expressed at the surface of lymphocytes resynthesize ATP and thus promote the migration of these cells (Henttinen et al., 2003).

3. *Inflammatory Diseases and Cytokines Remodel Extracellular Nucleotide Metabolism*

The expression of ectonucleotidases can be markedly altered by pathological conditions which can modulate P2 and P1 receptor signaling. One of the best characterized conditions that upregulates the expression of NTPDase1 and ecto-5′-nucleotidase is hypoxia. The increase in the expression of these enzymes helps increase the production of extracellular adenosine that protects against hypoxia-induced injury by limiting leukocyte infiltration and epithelial barrier permeability (Eltzschig et al., 2009; Synnestvedt et al., 2002). NTPDase1 and ecto-5′-nucleotidase expression in hypoxia is regulated by transcription factors, specificity protein-1 (Sp1) and hypoxia-inducible factor-1 (HIF-1), respectively (Eckle et al., 2008; Eltzschig et al., 2009; Hart et al., 2010; Synnestvedt et al., 2002). In contrast, in RAW macrophages, the transcriptional regulation of NTPDase1 involves cAMP response element binding (CREB; Liao et al., 2010).

Cystic fibrosis affects not only the expression level of ectonucleotidases but also their cellular localization (Fausther et al., 2010). In aseptic cultures of bronchial epithelial cells, NTPDase3 is present at both the apical and basolateral side, while NTPDase1 is present exclusively at the apical surface. Following the treatment with mucopurulent material from cystic fibrosis patients, NTPDase distribution was markedly altered in these cells as NTPDase1 was mobilized to the basolateral side and NTPDase3 to the apical surface. In addition to the changes in NTPDase distribution, this treatment also decreased expression of NTPDase1 by half and increased by threefolds the expression of NTPDase3. Similar changes in the pattern of ectonucleotidase localization and expression were found in the biopsies from airways of cystic fibrosis patients compared to those of healthy people (Fausther et al., 2010). The remodeling of ectonucleotidase expression by cystic fibrosis may have an important impact on inflammatory response. For example, the relocation of NTPDase1 to the basolateral membrane containing ecto-5′-nucleotidase may facilitate adenosine formation and thus limit leukocyte infiltration. The mobilization of NTPDase3 to the apical surface may, in turn, facilitate the transient activation of ADP- and UDP-specific receptors ($P2Y_{1,12,13}$ and $P2Y_6$, respectively), which would be difficult in the presence of NTPDase1. Moreover, the increase in the expression of this enzyme may help reduce extracellular ATP level and thus decrease $P2Y_2$ receptor-mediated mucus production and airways obstruction.

NTPDase expression can also be affected by inflammatory cytokines. In the liver, the expression of NTPDase2 in portal fibroblasts is downregulated at the transcriptional level by IL-6 (Yu et al., 2008). Interestingly, this may represent a mechanism by which the aberrant proliferation of bile duct cells occurs in biliary cirrhosis, where IL-6 is markedly upregulated. Indeed, ductular proliferation is prevented by NTPDase2 expressed in portal fibroblasts but increased when NTPDase2 expression is downregulated (Jhandier et al., 2005).

Interestingly, the expression of NPPs also shows some plasticity and can be increased by diverse factors such as cytokines TGF-α and IL-3, glucocorticoids, and vigorous physical exercise (Hauswirth et al., 2007; Kehlen et al., 2001) and decreased by certain cytokines such as IFN-γ, IL-1, or IL-4 (Kehlen et al., 2001).

B. NTPDase1 Regulates Vascular Tone

We have demonstrated that NTPDase1 is responsible for the major ectonucleotidase activity present at the surface of both endothelial cells and smooth muscle cells (Enjyoji et al., 1999; Kaczmarek et al., 1996; Kauffenstein et al., 2010a, 2010b; Sévigny et al., 1997). More recently, we observed that this enzyme expressed on these cells regulates the vascular tone, that is, the state of contractile tension in the vessel walls that controls local and systemic blood flow.

1. Effect on Vasorelaxation

Extracellular ADP and ATP can induce vasorelaxation by acting on endothelial $P2Y_1$ and $P2Y_2$ receptors, respectively (Boarder & Hourani, 1998). *Entpd1*$^{-/-}$ mice injected with ATP or UTP showed a rapid decrease in blood pressure compared to *Entpd1*$^{+/+}$ animals (Kauffenstein et al., 2010b), supporting the view that NTPDase1 is the dominant ectonucleotidase in the blood vessel lumen. In agreement, lower amounts of ATP and ADP were sufficient to induce relaxation in *Entpd1*$^{-/-}$ aortic rings than in *Entpd1*$^{+/+}$ controls, confirming that both $P2Y_1$ and $P2Y_2$ receptors could be overactivated in NTPDase1-deficient animals. Interestingly, the regulation of vasorelaxation by NTPDase1 also revealed that this enzyme regulates the reactivity of $P2Y_1$ but not the one of $P2Y_2$. Indeed, while endothelial $P2Y_1$ receptor was more easily desensitized in *Entp1*$^{-/-}$ aortic rings, which was in agreement with previously reported desensitization of this receptor in *Entp1*$^{-/-}$ platelets (Enjyoji et al., 1999), the desensitization level of endothelial $P2Y_2$ was not changed in *Entp1*$^{-/-}$ aortic rings at physiological concentrations of ATP (Kauffenstein et al., 2010b). A difference in $P2Y_2$ receptor desensitization between *Entp1*$^{-/-}$ and *Entpd1*$^{+/+}$ aortic rings could be observed only at pharmacological concentrations (high mM) of ATP (Kauffenstein et al., 2010b).

2. Effect on Vasoconstriction

Either extracellular UDP or UTP induces a contractile response via $P2Y_6$ receptor in mouse aortic rings denuded from its endothelium. In agreement with an essential role of NTPDase1 in the control of $P2Y_6$-mediated vasoconstriction, the weak response in wild-type animals was markedly enhanced in aortic rings from *Entpd1*$^{-/-}$ animals treated with UDP or UTP (Kauffenstein et al., 2010a). Accordingly, UDP infusion *in vivo* increased blood pressure more importantly in NTPDase1-deficient mice. In addition, *Entpd1*$^{-/-}$ mesenteric arteries displayed an enhanced myogenic response. This suggests that, in the absence of hydrolysis by NTPDase1, endogenous nucleotides released after mechanical forces, such as stretching in this case, can induce a more potent contraction in *Entpd1*$^{-/-}$ vessels.

C. Role of Ectonucleotidases in Neurotransmission and Neuroprotection

In the central nervous system, low micromolar ATP concentrations are involved in neurotransmission, while higher millimolar concentrations have cytotoxic effects (Franke et al., 2006). Interestingly, two ectonucleotidases were purified to homogeneity from porcine brain synaptosomes and identified as NTPDase1 and NTPDase2 based on their ATP hydrolysis pattern, sensitivity to inhibitors, and immunoblot. Moreover, the kinetic properties of these enzymes suggest that they have the capacity to fine control the effects of ATP mentioned earlier. Indeed, these enzymes differed significantly in respect to K_m for ATP hydrolysis which corresponded to 97 and 270 μM, respectively, but had similar molecular activity, that is, hydrolysis efficiency (Kukulski & Komoszynski, 2003). This suggests that when the active sites of both enzymes are fully saturated with ATP, they would hydrolyze this substrate with a similar velocity. However, NTPDase1 K_{cat}/K_m coefficient is threefold higher than NTPDase2 (9×10^6 vs. 3×10^6 $M^{-1}s^{-1}$) which indicates that lower concentrations of ATP would be hydrolyzed faster by NTPDase1 than NTPDase2. These catalytic properties of brain NTPDase1 and NTPDase2 suggest that the importance of each enzyme in neurotransmission would depend on ATP concentration. Under physiological conditions, even after neuronal stimulation, ATP concentration in synapses does not exceed 20 μM (Zimmermann & Braun, 1996) and would be therefore expected to be hydrolyzed predominantly by NTPDase1 that has a K_m threefold lower than the K_m of NTPDase2. In agreement with this role of NTPDase1, its deletion in mice causes desensitization of pre- and postsynaptic P2 receptors (Schaefer et al., 2007). In addition to the removal of excitatory ATP, NTPDase1 would also facilitate the production of inhibitory adenosine. Indeed, brain ischemia increased the expression of this enzyme together with ecto-5′-nucleotidase that would be expected to enhance adenosine production (Braun et al., 1998). In comparison to NTPDase1, the contribution of NTPDase2 to the hydrolysis of extracellular ATP would markedly increase under pathological conditions, for example, during epilepsy seizure, migraine, hypoxia, or ischemia, that usually significantly raise the level of this nucleotide in synapses. In addition to ATP hydrolysis, the role of this enzyme in neuroprotection would also include ADP production that via $P2Y_1$ receptors inhibits the release of excitatory neurotransmitters, including those stimulating cell metabolism (Franke et al., 2006).

In rodents, immunohistochemistry did not show the presence of NTPDase1 in nerve structures but rather show NTPDase3 in neurons and NTPDase2 in glial cells (Braun et al., 1998). Due to its localization in the brain, it was proposed that NTPDase3 could be involved in the regulation of feeding and sleep–wake behavior (Belcher et al., 2006). Although the identity of the ectonucleotidases expressed in neuronal synapse of different species requires further confirmations, the presence of NTPDase3, that has also a low K_m for ATP, would be expected to play a similar function to NTPDase1 as presented earlier.

Ectonucleotidases can also regulate neurotransmission in the autonomic nervous system. For example, the effect of ATP metabolism on acetylcholine exocytosis in rat ileum myenteric plexus (Duarte-Araujo et al., 2009) concurs with the roles of NTPDase1 and NTPDase2 proposed for the regulation of ATP-induced neurotransmission in brain synaptosomes. The stimulation of these nerve structures with exogenous ATP triggered acetylcholine release, which was significantly increased in the presence of either ADA or P2Y$_1$ receptor antagonist MRS2179. The analysis of ATP metabolism by myenteric plexus revealed that a low concentration of exogenous ATP (10 μM) was rapidly hydrolyzed to AMP, which was then further dephosphorylated to adenosine that reduced acetylcholine exocytosis via P1 receptor activation. In contrast, the hydrolysis of higher ATP concentration (100 μM) was associated with a significant production of ADP that was not seen during the hydrolysis of the low ATP concentration. Similar to adenosine, ADP inhibited acetylcholine release via activation of P2Y$_1$ receptors. The ATP hydrolysis patterns in rat myenteric plexus appear to correspond to NTPDase1 and NTPDase2. In agreement, these enzymes colocalize in the neural retina (Ricatti et al., 2009). NTPDase1 was also localized in neurons of human pancreas and human and porcine heart (Kittel et al., 2002; Machida et al., 2005). In the latter organ, this enzyme plays a cardioprotective role in ischemia by regulating the release of norepinephrine. However, as for the central nervous system, other studies suggest that rodent enteric neurons do not express NTPDase1 but NTPDase3 (Lavoie et al., 2011). In the light of these results, NTPDase3 could be responsible for the regulation of acetylcholine exocytosis in rat myenteric plexus described earlier, while NTPDase1 is expressed mainly in blood vessels and laminae propria, and NTPDase2 in glial cells (Lavoie et al., 2011). Moreover, in the gut, ectonucleotidases modulate visceral hyperalgesia via adenosine production from extracellular ATP (Zahn et al., 2007). Ectonucleotidases also appear to control P2X3- and P2X2/3-mediated nociceptive neurotransmission in urinary bladder, ureter, gut, tooth pulp, etc. (Burnstock, 2009).

IV. Note on Ectonucleotidase Inhibitor Development

The main challenge in designing ectonucleotidase inhibitors is to obtain molecules that would be isoform selective and would not affect P2 receptors. ARL-67156 (6-*N*,*N*-diethyl-β,γ-dibromomethylene-D-ATP) is one of the most used general ectonucleotidase inhibitors that inhibits either NTPDases or NPPs but does not affect P2 receptors. However, it must be used with caution as it is a weak competitive inhibitor of NTPDase1 and NTPDase3 that does not affect

NTPDase2 (Iqbal et al., 2005; Lévesque et al., 2007). ARL-67156 also inhibits NPP1 but not very well NPP3 (Lévesque et al., 2007). Due to the fact that ARL-67156 is a weak competitive inhibitor, high concentrations of this product are necessary, but this may bring other problems since concentration above 500 μM can affect various P2 receptors, for example, $P2Y_2$, $P2Y_4$, $P2Y_{12}$, and P2X1 (Benham & Tsien, 1987; Crack et al., 1995).

Recently, several new ectonucleotidase inhibitors have been developed. For example, PSB-6426 synthesized from uridine-5′-carboxamide is a fairly good and selective NTPDase2 inhibitor with a K_i of 8.2 μM (Brunschweiger et al., 2008). Other nucleotide analogs with modification at position 8 of the adenine ring such as 8-BuS-ATP and 8-BuS-AMP were reported as strong inhibitors of ectoATPase activity from bovine spleen which is mainly due to NTPDase1 (Gendron et al., 2000; Halbfinger et al., 2003). A more extensive characterization with recombinant NTPDases suggests that these ATP analogs may be the most selective NTPDase1 inhibitors yet (J. Lecka & J. Sévigny, unpublished observations). Some non-nucleotide scaffolds have also been reported as ectonucleotidase inhibitors. For example, Dr. Muller's group has identified several polyoxometalate anionic complexes as potent NTPDase inhibitors (Muller et al., 2006). The most potent compound was $K_6H_2[TiW_{11}CoO_{40}]$, exhibiting K_i values of 0.14 μM for NTPDase1, 0.91 μM for NTPDase2, and 0.56 μM for NTPDase3, while another compound, $(NH_4)_{18}[NaSb_9W_{21}O_{86}]$, was selective for rat NTPDase2 and -3 versus NTPDase1. Another polyoxometalate compound called POM-1 ($2Na_2WO_4 \cdot 9WO_3 \cdot H_2O$) was reported to display minor selectivity for NTPDase1 and -2 over NTPDase3 (K_i values of 2.58, 3.26, > 10 μM, respectively). This molecule has been used in many recent studies. For example, it was shown to significantly inhibit ectoATPase activity on intestinal epithelial cells and NTPDase1 activity on Tregs (Sun et al., 2010; Weissmuller et al., 2008). It also abolished renal protection induced by ischemic preconditioning. This effect of POM-1 was attributed to a decrease in adenosine production due to NTPDase1 inhibition (Grenz et al., 2007). However, other studies suggest that POM-1 should be used with caution as it may have some effects that are independent of NTPDase inhibition (Wall et al., 2008). Our recent study suggests that the approach aiming at engineering of antibodies neutralizing ectonucleotidase activity may provide very selective ectonucleotidase inhibitors. Indeed, an anti-human NTPDase3 antibody generated in our laboratory markedly inhibits the activity of this enzyme but does not affect other NTPDases or NTPDase3 from other species (Munkonda et al., 2009). More recently, we also obtained a mouse antibody specifically inhibiting human NTPDase2 (J. Pelletier & J. Sévigny, unpublished observation). No potent and specific inhibitors of NTPDase8 have been reported so far. In fact, NTPDase8 appears as the most resistant NTPDase to most compounds tested so far (Munkonda et al., 2009; Sévigny et al., 2000).

A few NPP inhibitors have recently been identified. For example, [3-(*t*-butyldimethylsilyloxy)-phenyl]-1,3,3-oxadiazole-2 (3H)-thione was reported as an NPP1 inhibitor with K_i of 100 μM (Khan et al., 2009). Likewise, biscoumarin derivatives were identified as noncompetitive inhibitors of human NPP1 with K_i and IC_{50} values of 50 and 164 μM, respectively (Choudhary et al., 2006). More recently, diadenosine polyphosphonate derivatives, that is, diadenosine α,β-δ,ε-dimethylene-pentaphosphate (Me-Ap_5A-Me) and di-2′-deoxyadenosine α,β-δ,ε-dimethylene-pentaphosphate (Me-dAp_5dA-Me), were identified as potent inhibitors of NPP1-3 (Eliahu et al., 2010). Both compounds dramatically reduced the hydrolysis of pNP-TMP by intact osteocarcinoma and colon cancer cells that express some of these enzymes (Eliahu et al., 2010). Importantly, these diadenosine polyphosphonate derivatives had minor effect on plasma membrane NTPDases, ecto-5′-nucleotidase, and $P2Y_1$ and $P2Y_2$ receptors.

Finally, it is well known that most commonly used P2 receptor antagonists markedly inhibit ectonucleotidase activity, which should be taken into account in studies using these chemicals. The general P2 receptor antagonists such as PPADS, reactive blue 2 (RB2), and suramin as well as the selective P2X1 antagonist NF279 markedly affect the activity of all plasma membrane NTPDases (Iqbal et al., 2005; Munkonda et al., 2007). Suramin was also reported to inhibit NPPs (Rucker et al., 2007). In addition, we found that the thienopyridine antithrombotic drugs ticlopidine (Tyklid) and clopidogrel (Plavix), whose metabolites produced by the liver are $P2Y_{12}$ receptor antagonists, are selective and potent inhibitors of human NTPDase1 with a K_i apparent in the order of 10 μM with ADP as a substrate (Lecka et al., 2010b). Interestingly, these thienopyridines inhibited the ADPase activity of human NTPDase1 much more potently than the ATPase activity of this enzyme. It is noteworthy that these thienopyridines affected more modestly murine NTPDase1 (J. Lecka and J. Sévigny, unpublished observation). The available inhibitors for these and other ectoenzymes described in this review are included in Table I.

V. Conclusion

This brief review shows that P2 and P1 receptor activation can be controlled by a complex network of ectoenzymes. There are some enzymes that hydrolyze nucleotides and others that resynthesize them. The product of one enzyme can be either a substrate or an inhibitor of another ectoenzyme and distinct ectoenzymes can compete for the same substrate which may lead to different products. However, numerous studies suggest that the majority of biological processes are regulated by few dominant ectoenzymes and that the identities of these enzymes appear to correlate with the actual demand for P2 and P1 regulation in the particular cell type or tissue. For example, NTPDase1 (the most characterized NTPDase) and ecto-5′-nucleotidase dominate in the vascular system over other ectoenzymes in the production of extracellular adenosine, while ADA appears to be the only enzyme that terminates the activation of P1 receptors. NTPDase1 is also a dominant enzyme that prevents $P2Y_1$ receptor desensitization in the cells in contact with the blood and regulates the activation of $P2Y_1$, $P2Y_2$, $P2Y_6$, and P2X7 in immune cells and in the cells of the vascular wall. Nevertheless, it is now time to perform more extensive studies that will systematically determine the expression and activity of all ectoenzymes which have the potential to affect P2 and P1 receptor signaling in different tissues and cells.

Acknowledgments

This work was supported by grants from the Canadian Institutes of Health Research (CIHR) to J. S. who was a recipient of a Junior 2 Scholarship from the *Fonds de Recherche en Santé du Québec* (FRSQ).

Conflict of Interest: The authors declare no conflict of interest.

Abbreviations

ACP	acid phosphatase
ADA	adenosine deaminase
AK	adenylate kinase
ALP	alkaline phosphatase
Ap_4A	P^1P^4-di(adenosine-5′)tretraphosphate
Ap_5A	P^1P^5-di(adenosine-5′)pentaphosphate
ART	ecto-ADP-ribosyltransferase
CysLT1R and CysLT2R	cysteinyl leukotriene receptor-1 and -2
ecto-PD	ectophosphodiesterase
fMLP or fMLF	*N*-formyl-Met-Leu-Phe
GPI	glycosyl phosphatidylinositol
HUVECs	human umbilical vein endothelial cells
LPS	lipopolysaccharide
NDPK	nucleoside diphosphate kinase
NTPDase	nucleoside triphosphate diphosphohydrolase
NPP	nucleotide pyrophosphatase/phosphodiesterase
PNP	purine nucleoside phosphorylase
pNP-TMP	paranitrophenol-thymidine monophosphate

References

Abbracchio, M. P., & Burnstock, G. (1998). Purinergic signalling: Pathophysiological roles. *Japanese Journal of Pharmacology*, *78*(2), 113–145.

Abbracchio, M. P., Burnstock, G., Boeynaems, J. M., et al. (2006). International Union of Pharmacology LVIII: Update on the P2Y G protein-coupled nucleotide receptors: From molecular mechanisms and pathophysiology to therapy. *Pharmacological Reviews*, *58*(3), 281–341.

Abbracchio, M. P., Burnstock, G., Verkhratsky, A., et al. (2009). Purinergic signalling in the nervous system: An overview. *Trends in Neurosciences*, *32*(1), 19–29.

Adriouch, S., Bannas, P., Schwarz, N., et al. (2008). ADP-ribosylation at R125 gates the P2X7 ion channel by presenting a covalent ligand to its nucleotide binding site. *The FASEB Journal*, *22*(3), 861–869.

Airas, L., Niemela, J., Salmi, M., et al. (1997). Differential regulation and function of CD73, a glycosyl-phosphatidylinositol-linked 70-kD adhesion molecule, on lymphocytes and endothelial cells. *The Journal of Cell Biology*, *136*(2), 421–431.

Bahrami, F., Kukulski, F., Lecka, J., et al. P2Y11 receptor and ectoenzymes control Toll- like receptor 3-induced IL-8 production in HT-29 cells. Submitted.

Bar, I., Guns, P. J., Metallo, J., et al. (2008). Knockout mice reveal a role for P2Y6 receptor in macrophages, endothelial cells, and vascular smooth muscle cells. *Molecular Pharmacology*, *74*(3), 777–784.

Bartel, D. L., Sullivan, S. L., Lavoie, E. G., et al. (2006). Nucleoside triphosphate diphosphohydrolase-2 is the ecto-ATPase of type I cells in taste buds. *The Journal of Comparative Neurology*, *497*(1), 1–12.

Becker, L. V., Rosa, C. S., Souza Vdo, C., et al. (2010). Activities of enzymes that hydrolyze adenine nucleotides in platelets from patients with rheumatoid arthritis. *Clinical Biochemistry*, *43*(13–14), 1096–1100.

Belcher, S. M., Zsarnovszky, A., Crawford, P. A., et al. (2006). Immunolocalization of ecto-nucleoside triphosphate diphosphohydrolase 3 in rat brain: Implications for modulation of multiple homeostatic systems including feeding and sleep-wake behaviors. *Neuroscience*, *137*(4), 1331–1346.

Beldi, G., Banz, Y., Kroemer, A., et al. (2010). Deletion of CD39 on natural killer cells attenuates hepatic ischemia/reperfusion injury in mice. *Hepatology*, *51*(5), 1702–1711.

Beldi, G., Wu, Y., Banz, Y., et al. (2008). Natural killer T cell dysfunction in CD39-null mice protects against concanavalin A-induced hepatitis. *Hepatology*, *48*(3), 841–852.

Belli, S. I., van Driel, I. R., & Goding, J. W. (1993). Identification and characterization of a soluble form of the plasma cell membrane glycoprotein PC-1 (5′-nucleotide phosphodiesterase). *European Journal of Biochemistry*, *217*(1), 421–428.

Ben Yebdri, F., Kukulski, F., Tremblay, A., et al. (2009). Concomitant activation of P2Y(2) and P2Y(6) receptors on monocytes is required for TLR1/2-induced neutrophil migration by regulating IL-8 secretion. *European Journal of Immunology*, *39*(10), 2885–2894.

Benham, C. D., & Tsien, R. W. (1987). A novel receptor-operated Ca2+-permeable channel activated by ATP in smooth muscle. *Nature*, *328*(6127), 275–278.

Bigonnesse, F., Lévesque, S. A., Kukulski, F., et al. (2004). Cloning and characterization of mouse nucleoside triphosphate diphosphohydrolase-8. *Biochemistry*, *43*(18), 5511–5519.

Boarder, M. R., & Hourani, S. M. (1998). The regulation of vascular function by P2 receptors: Multiple sites and multiple receptors. *Trends in Pharmacological Sciences*, *19*(3), 99–107.

Boissan, M., Dabernat, S., Peuchant, E., et al. (2009). The mammalian Nm23/NDPK family: From metastasis control to cilia movement. *Molecular and Cellular Biochemistry*, *329*(1–2), 51–62.

Bollen, M., Gijsbers, R., Ceulemans, H., et al. (2000). Nucleotide pyrophosphatases/ phosphodiesterases on the move. *Critical Reviews in Biochemistry and Molecular Biology*, *35*(6), 393–432.

Bonan, C. D., Walz, R., Pereira, G. S., et al. (2000). Changes in synaptosomal ectonucleotidase activities in two rat models of temporal lobe epilepsy. *Epilepsy Research*, *39*(3), 229–238.

Boucher, R. C. (2007). Airway surface dehydration in cystic fibrosis: Pathogenesis and therapy. *Annual Review of Medicine*, *58*, 157–170.
Braun, N., Brendel, P., & Zimmermann, H. (1995). Distribution of 5′-nucleotidase in the developing mouse retina. *Brain Research. Developmental Brain Research*, *88*(1), 79–86.
Braun, N., Fengler, S., Ebeling, C., et al. (2000). Sequencing, functional expression and characterization of rat NTPDase6, a nucleoside diphosphatase and novel member of the ecto-nucleoside triphosphate diphosphohydrolase family. *The Biochemical Journal*, *351* (Pt 3), 639–647.
Braun, N., Sévigny, J., Robson, S. C., et al. (2004). Association of the ecto-ATPase NTPDase2 with glial cells of the peripheral nervous system. *Glia*, *45*(2), 124–132.
Braun, N., Zhu, Y., Krieglstein, J., et al. (1998). Upregulation of the enzyme chain hydrolyzing extracellular ATP after transient forebrain ischemia in the rat. *The Journal of Neuroscience*, *18*(13), 4891–4900.
Brunschweiger, A., Iqbal, J., Umbach, F., et al. (2008). Selective nucleoside triphosphate diphosphohydrolase-2 (NTPDase2) inhibitors: Nucleotide mimetics derived from uridine-5′-carboxamide. *Journal of Medicinal Chemistry*, *51*(15), 4518–4528.
Buhring, H. J., Seiffert, M., Giesert, C., et al. (2001). The basophil activation marker defined by antibody 97A6 is identical to the ectonucleotide pyrophosphatase/phosphodiesterase 3. *Blood*, *97*(10), 3303–3305.
Burnstock, G. (2009). Purinergic mechanosensory transduction and visceral pain. *Molecular Pain*, *5*, 69.
Canales, J., Pinto, R. M., Costas, M. J., et al. (1995). Rat liver nucleoside diphosphosugar or diphosphoalcohol pyrophosphatases different from nucleotide pyrophosphatase or phosphodiesterase I: Substrate specificities of Mg(2+)-and/or Mn(2+)-dependent hydrolases acting on ADP-ribose. *Biochimica et Biophysica Acta*, *1246*(2), 167–177.
Chen, Y., Corriden, R., Inoue, Y., et al. (2006). ATP release guides neutrophil chemotaxis via P2Y2 and A3 receptors. *Science*, *314*(5806), 1792–1795.
Chen, Y., Yao, Y., Sumi, Y., et al. (2010). Purinergic signaling: A fundamental mechanism in neutrophil activation. *Science Signaling*, *3*(125), ra45.
Chiavegatti, T., Costa, V. L., Jr., Araujo, M. S., et al. (2008). Skeletal muscle expresses the extracellular cyclic AMP-adenosine pathway. *British Journal of Pharmacology*, *153*(6), 1331–1340.
Choudhary, M. I., Fatima, N., Khan, K. M., et al. (2006). New biscoumarin derivatives-cytotoxicity and enzyme inhibitory activities. *Bioorganic & Medicinal Chemistry*, *14*(23), 8066–8072.
Ciana, P., Fumagalli, M., Trincavelli, M. L., et al. (2006). The orphan receptor GPR17 identified as a new dual uracil nucleotides/cysteinyl-leukotrienes receptor. *The EMBO Journal*, *25*(19), 4615–4627.
Cicko, S., Lucattelli, M., Muller, T., et al. (2010). Purinergic receptor inhibition prevents the development of smoke-induced lung injury and emphysema. *Journal of Immunology*, *185* (1), 688–697.
Colgan, S. P., Eltzschig, H. K., Eckle, T., et al. (2006). Physiological roles for ecto-5′-nucleotidase (CD73). *Purinergic Signalling*, *2*(2), 351–360.
Conde, S. V., & Monteiro, E. C. (2004). Hypoxia induces adenosine release from the rat carotid body. *Journal of Neurochemistry*, *89*(5), 1148–1156.
Corriden, R., Chen, Y., Inoue, Y., et al. (2008). Ecto-nucleoside triphosphate diphosphohydrolase 1 (E-NTPDase1/CD39) regulates neutrophil chemotaxis by hydrolyzing released ATP to adenosine. *The Journal of Biological Chemistry*, *283*(42), 28480–28486.
Crack, B. E., Pollard, C. E., Beukers, M. W., et al. (1995). Pharmacological and biochemical analysis of FPL 67156, a novel, selective inhibitor of ecto-ATPase. *British Journal of Pharmacology*, *114*(2), 475–481.

Di Virgilio, F., Ceruti, S., Bramanti, P., et al. (2009). Purinergic signalling in inflammation of the central nervous system. *Trends in Neurosciences*, *32*(2), 79–87.

Duarte-Araujo, M., Nascimento, C., Timoteo, M. A., et al. (2009). Relative contribution of ecto-ATPase and ecto-ATPDase pathways to the biphasic effect of ATP on acetylcholine release from myenteric motoneurons. *British Journal of Pharmacology*, *156*(3), 519–533.

Dzeja, P., & Terzic, A. (2009). Adenylate kinase and AMP signaling networks: Metabolic monitoring, signal communication and body energy sensing. *International Journal of Molecular Sciences*, *10*(4), 1729–1772.

Eckle, T., Fullbier, L., Wehrmann, M., et al. (2007). Identification of ectonucleotidases CD39 and CD73 in innate protection during acute lung injury. *Journal of Immunology*, *178*(12), 8127–8137.

Eckle, T., Kohler, D., Lehmann, R., et al. (2008). Hypoxia-inducible factor-1 is central to cardioprotection: A new paradigm for ischemic preconditioning. *Circulation*, *118*(2), 166–175.

Eckle, T., Krahn, T., Grenz, A., et al. (2007). Cardioprotection by ecto-5′-nucleotidase (CD73) and A2B adenosine receptors. *Circulation*, *115*(12), 1581–1590.

Eliahu, S., Lecka, J., Reiser, G., et al. (2010). Diadenosine-5′-5″-(boranated)polyphosphonate analogues as selective nucleotide pyrophosphatase/phosphodiesterase (NPP) inhibitors. *Journal of Medicinal Chemistry*, *53*(24), 8485–8497.

Eltzschig, H. K., Eckle, T., Mager, A., et al. (2006). ATP release from activated neutrophils occurs via connexin 43 and modulates adenosine-dependent endothelial cell function. *Circulation Research*, *99*(10), 1100–1108.

Eltzschig, H. K., Ibla, J. C., Furuta, G. T., et al. (2003). Coordinated adenine nucleotide phosphohydrolysis and nucleoside signaling in posthypoxic endothelium: Role of ectonucleotidases and adenosine A2B receptors. *The Journal of Experimental Medicine*, *198*(5), 783–796.

Eltzschig, H. K., Kohler, D., Eckle, T., et al. (2009). Central role of Sp1-regulated CD39 in hypoxia/ischemia protection. *Blood*, *113*(1), 224–232.

Eltzschig, H. K., Macmanus, C. F., & Colgan, S. P. (2008). Neutrophils as sources of extracellular nucleotides: Functional consequences at the vascular interface. *Trends in Cardiovascular Medicine*, *18*(3), 103–107.

Eltzschig, H. K., Thompson, L. F., Karhausen, J., et al. (2004). Endogenous adenosine produced during hypoxia attenuates neutrophil accumulation: Coordination by extracellular nucleotide metabolism. *Blood*, *104*(13), 3986–3992.

Enjyoji, K., Sévigny, J., Lin, Y., et al. (1999). Targeted disruption of CD39/ATP diphosphohydrolase results in disordered hemostasis and thromboregulation. *Nature Medicine*, *5*(9), 1010–1017.

Fabre, A. C., Vantourout, P., Champagne, E., et al. (2006). Cell surface adenylate kinase activity regulates the F(1)-ATPase/P2Y (13)-mediated HDL endocytosis pathway on human hepatocytes. *Cellular and Molecular Life Sciences*, *63*(23), 2829–2837.

Failer, B. U., Aschrafi, A., Schmalzing, G., et al. (2003). Determination of native oligomeric state and substrate specificity of rat NTPDase1 and NTPDase2 after heterologous expression in Xenopus oocytes. *European Journal of Biochemistry*, *270*(8), 1802–1809.

Faria, M., Magalhaes-Cardoso, T., Lafuente-de-Carvalho, J. M., et al. (2008). Decreased ecto-NTPDase1/CD39 activity leads to desensitization of P2 purinoceptors regulating tonus of corpora cavernosa in impotent men with endothelial dysfunction. *Nucleosides, Nucleotides & Nucleic Acids*, *27*(6), 761–768.

Fausther, M., Lecka, J., Kukulski, F., et al. (2007). Cloning, purification, and identification of the liver canalicular ecto-ATPase as NTPDase8. *American Journal of Physiology. Gastrointestinal and Liver Physiology*, *292*(3), G785–G795.

Fausther, M., Pelletier, J., Ribeiro, C. M., et al. (2010). Cystic fibrosis remodels the regulation of purinergic signaling by NTPDase1 (CD39) and NTPDase3. *American Journal of Physiology. Lung Cellular and Molecular Physiology*, *298*(6), L804–L818.

Fontella, F. U., Bruno, A. N., Balk, R. S., et al. (2005). Repeated stress effects on nociception and on ectonucleotidase activities in spinal cord synaptosomes of female rats. *Physiology & Behavior*, *85*(2), 213–219.

Franke, H., Krugel, U., & Illes, P. (2006). P2 receptors and neuronal injury. *Pflugers Archiv: European Journal of Physiology*, *452*(5), 622–644.

Fredholm, B. B., Irenius, E., Kull, B., et al. (2001). Comparison of the potency of adenosine as an agonist at human adenosine receptors expressed in Chinese hamster ovary cells. *Biochemical Pharmacology*, *61*(4), 443–448.

Friedman, D. J., Kunzli, B. M., Yi, A. R., et al. (2009). From the Cover: CD39 deletion exacerbates experimental murine colitis and human polymorphisms increase susceptibility to inflammatory bowel disease. *Proceedings of the National Academy of Sciences of the United States of America*, *106*(39), 16788–16793.

Friedman, D. J., Talbert, M. E., Bowden, D. W., et al. (2008). Functional ENTPD1 Polymorphisms in African-Americans with Diabetes and End-Stage Renal Disease. *Diabetes*, *58*(4), 999–1006.

Gendron, F. P., Halbfinger, E., Fischer, B., et al. (2000). Novel inhibitors of nucleoside triphosphate diphosphohydrolases: Chemical synthesis and biochemical and pharmacological characterizations. *Journal of Medicinal Chemistry*, *43*(11), 2239–2247.

Gijsbers, R., Ceulemans, H., Stalmans, W., et al. (2001). Structural and catalytic similarities between nucleotide pyrophosphatases/phosphodiesterases and alkaline phosphatases. *The Journal of Biological Chemistry*, *276*(2), 1361–1368.

Goding, J. W., Grobben, B., & Slegers, H. (2003). Physiological and pathophysiological functions of the ecto-nucleotide pyrophosphatase/phosphodiesterase family. *Biochimica et Biophysica Acta*, *1638*(1), 1–19.

Goding, J. W., Terkeltaub, R., Maurice, M., et al. (1998). Ecto-phosphodiesterase/pyrophosphatase of lymphocytes and non-lymphoid cells: Structure and function of the PC-1 family. *Immunological Reviews*, *161*, 11–26.

Gomez, G., & Sitkovsky, M. V. (2003). Differential requirement for A2a and A3 adenosine receptors for the protective effect of inosine in vivo. *Blood*, *102*(13), 4472–4478.

Grenz, A., Zhang, H., Hermes, M., et al. (2007). Contribution of E-NTPDase1 (CD39) to renal protection from ischemia-reperfusion injury. *The FASEB Journal*, *21*(11), 2863–2873.

Grobben, B., Anciaux, K., Roymans, D., et al. (1999). An ecto-nucleotide pyrophosphatase is one of the main enzymes involved in the extracellular metabolism of ATP in rat C6 glioma. *Journal of Neurochemistry*, *72*(2), 826–834.

Grunewald, J. K., & Ridley, A. J. (2010). CD73 represses pro-inflammatory responses in human endothelial cells. *Journal of Inflammation (London)*, *7*(1), 10.

Halbfinger, E., Gorochesky, K., Lévesque, S. A., et al. (2003). Photoaffinity labeling on magnetic microspheres (PALMm) methodology for topographic mapping: Preparation of PALMm reagents and demonstration of biochemical relevance. *Organic & Biomolecular Chemistry*, *1*(16), 2821–2832.

Harmey, D., Hessle, L., Narisawa, S., et al. (2004). Concerted regulation of inorganic pyrophosphate and osteopontin by akp2, enpp 1, and ank: An integrated model of the pathogenesis of mineralization disorders. *The American Journal of Pathology*, *164*(4), 1199–1209.

Hart, M. L., Gorzolla, I. C., Schittenhelm, J., et al. (2010). SP1-dependent induction of CD39 facilitates hepatic ischemic preconditioning. *Journal of Immunology*, *184*(7), 4017–4024.

Hasko, G., Linden, J., Cronstein, B., et al. (2008). Adenosine receptors: Therapeutic aspects for inflammatory and immune diseases. *Nature Reviews. Drug Discovery*, *7*(9), 759–770.

Hauswirth, A. W., Sonneck, K., Florian, S., et al. (2007). Interleukin-3 promotes the expression of E-NPP3/CD203C on human blood basophils in healthy subjects and in patients with birch pollen allergy. *International Journal of Immunopathology and Pharmacology*, *20* (2), 267–278.

Hayman, A. R. (2008). Tartrate-resistant acid phosphatase (TRAP) and the osteoclast/immune cell dichotomy. *Autoimmunity*, *41*(3), 218–223.

Heilbronn, A., & Zimmermann, H. (1995). 5′-nucleotidase activates and an inhibitory antibody prevents neuritic differentiation of PC12 cells. *The European Journal of Neuroscience*, *7* (6), 1172–1179.

Henttinen, T., Jalkanen, S., & Yegutkin, G. G. (2003). Adherent leukocytes prevent adenosine formation and impair endothelial barrier function by ecto-5′-nucleotidase/CD73-dependent mechanism. *The Journal of Biological Chemistry*, *278*(27), 24888–24895.

Hessle, L., Johnson, K. A., Anderson, H. C., et al. (2002). Tissue-nonspecific alkaline phosphatase and plasma cell membrane glycoprotein-1 are central antagonistic regulators of bone mineralization. *Proceedings of the National Academy of Sciences of the United States of America*, *99*(14), 9445–9449.

Hicks-Berger, C. A., Chadwick, B. P., Frischauf, A. M., et al. (2000). Expression and characterization of soluble and membrane-bound human nucleoside triphosphate diphosphohydrolase 6 (CD39L2). *The Journal of Biological Chemistry*, *275*(44), 34041–34045.

Hong, S., Brass, A., Seman, M., et al. (2007). Lipopolysaccharide, IFN-gamma, and IFN-beta induce expression of the thiol-sensitive ART2.1 ecto-ADP-ribosyltransferase in murine macrophages. *Journal of Immunology*, *179*(9), 6215–6227.

Hong, S., Brass, A., Seman, M., et al. (2009). Basal and inducible expression of the thiol-sensitive ART2.1 ecto-ADP-ribosyltransferase in myeloid and lymphoid leukocytes. *Purinergic Signalling*, *5*(3), 369–383.

Hoshi, K., Amizuka, N., Oda, K., et al. (1997). Immunolocalization of tissue non-specific alkaline phosphatase in mice. *Histochemistry and Cell Biology*, *107*(3), 183–191.

Hosoda, N., Hoshino, S. I., Kanda, Y., et al. (1999). Inhibition of phosphodiesterase/pyrophosphatase activity of PC-1 by its association with glycosaminoglycans. *European Journal of Biochemistry*, *265*(2), 763–770.

Hubert, S., Rissiek, B., Klages, K., et al. (2010). Extracellular NAD+ shapes the Foxp3+ regulatory T cell compartment through the ART2-P2X7 pathway. *The Journal of Experimental Medicine*, *207*(12), 2561–2568.

Hyman, M. C., Petrovic-Djergovic, D., Mazer, S. P., et al. (2007). ENTPDase (CD39) downregulates monocyte and neutrophil MAC-1 expression. *Circulation*, *116*, II_208.

Hyman, M. C., Petrovic-Djergovic, D., Visovatti, S. H., et al. (2009). Self-regulation of inflammatory cell trafficking in mice by the leukocyte surface apyrase CD39. *Journal of Clinical Investigation*, *119*(5), 1136–1149.

Idzko, M., Hammad, H., van Nimwegen, M., et al. (2007). Extracellular ATP triggers and maintains asthmatic airway inflammation by activating dendritic cells. *Nature Medicine*, *13*(8), 913–919.

Imai, M., Goepfert, C., Kaczmarek, E., et al. (2000). CD39 modulates IL-1 release from activated endothelial cells. *Biochemical and Biophysical Research Communications*, *270* (1), 272–278.

Inoue, Y., Chen, Y., Hirsh, M. I., et al. (2008). A3 and P2Y2 receptors control the recruitment of neutrophils to the lungs in a mouse model of sepsis. *Shock*, *30*(2), 173–177.

Iqbal, J., Vollmayer, P., Braun, N., et al. (2005). A capillary electrophoresis method for the characterization of ecto-nucleoside triphosphate diphosphohydrolases (NTPDases) and the analysis of inhibitors by in-capillary enzymatic microreaction. *Purinergic Signalling*, *1*(4), 349–358.

Jackson, E. K., Ren, J., Zacharia, L. C., et al. (2007). Characterization of renal ecto-phosphodiesterase. *The Journal of Pharmacology and Experimental Therapeutics*, *321*(2), 810–815.

Jacobson, K. A., & Gao, Z. G. (2006). Adenosine receptors as therapeutic targets. *Nature Reviews. Drug Discovery*, *5*(3), 247–264.

Jansen, S., Stefan, C., Creemers, J. W., et al. (2005). Proteolytic maturation and activation of autotaxin (NPP2), a secreted metastasis-enhancing lysophospholipase D. *Journal of Cell Science*, *118*(Pt 14), 3081–3089.

Jhandier, M. N., Kruglov, E. A., Lavoie, E. G., et al. (2005). Portal fibroblasts regulate the proliferation of bile duct epithelia via expression of NTPDase2. *The Journal of Biological Chemistry*, *280*(24), 22986–22992.

Jin, X., Shepherd, R. K., Duling, B. R., et al. (1997). Inosine binds to A3 adenosine receptors and stimulates mast cell degranulation. *Journal of Clinical Investigation*, *100*(11), 2849–2857.

Joseph, S. M., Pifer, M. A., Przybylski, R. J., et al. (2004). Methylene ATP analogs as modulators of extracellular ATP metabolism and accumulation. *British Journal of Pharmacology*, *142*(6), 1002–1014.

Kaczmarek, E., Koziak, K., Sévigny, J., et al. (1996). Identification and characterization of CD39/vascular ATP diphosphohydrolase. *The Journal of Biological Chemistry*, *271*(51), 33116–33122.

Kansas, G. S., Wood, G. S., & Tedder, T. F. (1991). Expression, distribution, and biochemistry of human CD39. Role in activation-associated homotypic adhesion of lymphocytes. *J Immunol*, *146*(7), 2235–2244.

Kauffenstein, G., Drouin, A., Thorin-Trescases, N., et al. (2010). NTPDase1 (CD39) controls nucleotide-dependent vasoconstriction in mouse. *Cardiovascular Research*, *85*(1), 204–213.

Kauffenstein, G., Furstenau, C. R., D'Orleans-Juste, P., et al. (2010). The ecto-nucleotidase NTPDase1 differentially regulates P2Y1 and P2Y2 receptor-dependent vasorelaxation. *British Journal of Pharmacology*, *159*(3), 576–585.

Kaunitz, J. D., & Yamaguchi, D. T. (2008). TNAP, TrAP, ecto-purinergic signaling, and bone remodeling. *Journal of Cellular Biochemistry*, *105*(3), 655–662.

Kehlen, A., Lauterbach, R., Santos, A. N., et al. (2001). IL-1 beta- and IL-4-induced down-regulation of autotaxin mRNA and PC-1 in fibroblast-like synoviocytes of patients with rheumatoid arthritis (RA). *Clinical and Experimental Immunology*, *123*(1), 147–154.

Khan, K. M., Fatima, N., Rasheed, M., et al. (2009). 1,3,4-Oxadiazole-2(3H)-thione and its analogues: A new class of non-competitive nucleotide pyrophosphatases/phosphodiesterases 1 inhibitors. *Bioorganic & Medicinal Chemistry*, *17*(22), 7816–7822.

Kittel, A., Garrido, M., & Varga, G. (2002). Localization of NTPDase1/CD39 in normal and transformed human pancreas. *The Journal of Histochemistry and Cytochemistry*, *50*(4), 549–556.

Kittel, A., Pelletier, J., Bigonnesse, F., et al. (2004). Localization of nucleoside triphosphate diphosphohydrolase-1 (NTPDase1) and NTPDase2 in pancreas and salivary gland. *The Journal of Histochemistry and Cytochemistry*, *52*(7), 861–871.

Kobayashi, T., Okada, T., Garcia del Saz, E., et al. (1997). Internalization of ecto-ATPase activity in human neutrophils upon stimulation with phorbol ester or formyl peptide. *Histochemistry and Cell Biology*, *107*(5), 353–363.

Kukulski, F., Bahrami, F., Ben Yebdri, F., et al. The ectonucleotidase NTPDase1 controls IL-8 production by human neutrophils. Submitted.

Kukulski, F., Ben Yebdri, F., Bahrami, F., et al. (2010). Endothelial P2Y2 receptor regulates LPS-induced neutrophil transendothelial migration in vitro. *Molecular Immunology*, *47* (5), 991–999.

Kukulski, F., Ben Yebdri, F., Lecka, J., et al. (2009). Extracellular ATP and P2 receptors are required for IL-8 to induce neutrophil migration. *Cytokine*, *46*(2), 166–170.

Kukulski, F., Ben Yebdri, F., Lefebvre, J., et al. (2007). Extracellular nucleotides mediate LPS-induced neutrophil migration in vitro and in vivo. *Journal of Leukocyte Biology*, *81*(5), 1269–1275.

Kukulski, F., & Komoszynski, M. (2003). Purification and characterization of NTPDase1 (ecto-apyrase) and NTPDase2 (ecto-ATPase) from porcine brain cortex synaptosomes. *European Journal of Biochemistry*, *270*(16), 3447–3454.

Kukulski, F., Lévesque, S. A., Lavoie, E. G., et al. (2005). Comparative hydrolysis of P2 receptor agonists by NTPDases 1,2,3 and 8. *Purinergic Signalling*, *1*(2), 193–204.

Langer, D., Ikehara, Y., Takebayashi, H., et al. (2007). The ectonucleotidases alkaline phosphatase and nucleoside triphosphate diphosphohydrolase 2 are associated with subsets of progenitor cell populations in the mouse embryonic, postnatal and adult neurogenic zones. *Neuroscience*, *150*(4), 863–879.

Lavoie, E. G., Fausther, M., Kauffenstein, G., et al. (2010). Identification of the ectonucleotidases expressed in mouse, rat and human Langerhans islets: Potential role of NTPDase3 in insulin secretion. *American Journal of Physiology. Endocrinology and Metabolism*, *299*(4), E647–E656.

Lavoie, E. G., Gulbransen, B. D., Martin-Satué, M., et al. (2011). Ectonucleotidase in mouse digestive system. Focus on NTPDase3 localization. *American Journal of Physiology. Gastrointestinal and Liver Physiology*, in press.

Lecka, J., Bloch-Boguslawska, E., Molski, S., et al. (2010). Extracellular purine metabolism in blood vessels (Part II): Activity of ecto-enzymes in blood vessels of patients with abdominal aortic aneurysm. *Clinical and Applied Thrombosis/Hemostasis*, *16*(6), 650–657.

Lecka, J., Rana, M. S., & Sévigny, J. (2010). Inhibition of vascular ectonucleotidase activities by the pro-drugs ticlopidine and clopidogrel favours platelet aggregation. *British Journal of Pharmacology*, *161*(5), 1150–1160.

Lévesque, S. A., Kukulski, F., Enjyoji, K., et al. (2010). NTPDase1 governs P2X(7)-dependent functions in murine macrophages. *European Journal of Immunology*, *40*(5), 1473–1485.

Lévesque, S. A., Lavoie, E. G., Lecka, J., et al. (2007). Specificity of the ecto-ATPase inhibitor ARL 67156 on human and mouse ectonucleotidases. *British Journal of Pharmacology*, *152*(1), 141–150.

Liao, H., Hyman, M. C., Baek, A. E., et al. (2010). cAMP/CREB-mediated transcriptional regulation of ectonucleoside triphosphate diphosphohydrolase 1 (CD39) expression. *The Journal of Biological Chemistry*, *285*(19), 14791–14805.

Machida, T., Heerdt, P. M., Reid, A. C., et al. (2005). Ectonucleoside triphosphate diphosphohydrolase 1/CD39, localized in neurons of human and porcine heart, modulates ATP-induced norepinephrine exocytosis. *The Journal of Pharmacology and Experimental Therapeutics*, *313*(2), 570–577.

Mamedova, L., Capra, V., Accomazzo, M. R., et al. (2005). CysLT1 leukotriene receptor antagonists inhibit the effects of nucleotides acting at P2Y receptors. *Biochemical Pharmacology*, *71*(1–2), 115–125.

Martín-Satué, M., Lavoie, E. G., Fausther, M., et al. (2010). High expression and activity of ecto-5′-nucleotidase/CD73 in the male murine reproductive tract. *Histochemistry and Cell Biology*, *133*(6), 659–668.

Martín-Satué, M., Lavoie, E. G., Pelletier, J., et al. (2009). Localization of plasma membrane bound NTPDases in the murine reproductive tract. *Histochemistry and Cell Biology*, *131* (5), 615–628.

Meerson, N. R., Delautier, D., Durand-Schneider, A. M., et al. (1998). Identification of B10, an alkaline phosphodiesterase of the apical plasma membrane of hepatocytes and biliary cells, in rat serum: Increased levels following bile duct ligation and during the development of cholangiocarcinoma. *Hepatology*, *27*(2), 563–568.

Millan, J. L. (2006a). Alkaline phosphatases : Structure, substrate specificity and functional relatedness to other members of a large superfamily of enzymes. *Purinergic Signalling*, *2*(2), 335–341.

Millan, J. L. (2006b). *Mammalian alkaline phosphatases: From biology to applications in medicine and biotechnology.* " Weinheim: WILEY-VCH Verlag GmbH & Co. KGaA.

Mizumoto, N., Kumamoto, T., Robson, S. C., et al. (2002). CD39 is the dominant Langerhans cell-associated ecto-NTPDase: Modulatory roles in inflammation and immune responsiveness. *Nature Medicine*, *8*(4), 358–365.

Moncrieffe, H., Nistala, K., Kamhieh, Y., et al. (2010). High expression of the ectonucleotidase CD39 on T cells from the inflamed site identifies two distinct populations, one regulatory and one memory T cell population. *Journal of Immunology*, *185*(1), 134–143.

Mulero, J. J., Yeung, G., Nelken, S. T., et al. (1999). CD39-L4 is a secreted human apyrase, specific for the hydrolysis of nucleoside diphosphates. *The Journal of Biological Chemistry*, *274*(29), 20064–20067.

Muller, C. E., Iqbal, J., Baqi, Y., et al. (2006). Polyoxometalates—A new class of potent ecto-nucleoside triphosphate diphosphohydrolase (NTPDase) inhibitors. *Bioorganic & Medicinal Chemistry Letters*, *16*(23), 5943–5947.

Munkonda, M. N., Kauffenstein, G., Kukulski, F., et al. (2007). Inhibition of human and mouse plasma membrane bound NTPDases by P2 receptor antagonists. *Biochemical Pharmacology*, *74*(10), 1524–1534.

Munkonda, M. N., Pelletier, J., Ivanenkov, V. V., et al. (2009). Characterization of a monoclonal antibody as the first specific inhibitor of human NTP diphosphohydrolase-3: Partial characterization of the inhibitory epitope and potential applications. *The FEBS Journal*, *276*(2), 479–496.

Nathan, C. (2006). Neutrophils and immunity: Challenges and opportunities. *Nature Reviews. Immunology*, *6*(3), 173–182.

Nauseef, W. M. (2007). How human neutrophils kill and degrade microbes: An integrated view. *Immunological Reviews*, *219*, 88–102.

Nigam, P., Velu, V., Kannanganat, S., et al. (2010). Expansion of FOXP3+ CD8 T cells with suppressive potential in colorectal mucosa following a pathogenic simian immunodeficiency virus infection correlates with diminished antiviral T cell response and viral control. *Journal of Immunology*, *184*(4), 1690–1701.

Ohkubo, S., Kimura, J., & Matsuoka, I. (2000). Ecto-alkaline phosphatase in NG108-15 cells: A key enzyme mediating P1 antagonist-sensitive ATP response. *British Journal of Pharmacology*, *131*(8), 1667–1672.

Pacheco, R., Martinez-Navio, J. M., Lejeune, M., et al. (2005). CD26, adenosine deaminase, and adenosine receptors mediate costimulatory signals in the immunological synapse. *Proceedings of the National Academy of Sciences of the United States of America*, *102* (27), 9583–9588.

Parkinson, F. E., Xiong, W., & Zamzow, C. R. (2005). Astrocytes and neurons: Different roles in regulating adenosine levels. *Neurological Research*, *27*(2), 153–160.

Pelegrin, P., & Surprenant, A. (2009). Dynamics of macrophage polarization reveal new mechanism to inhibit IL-1beta release through pyrophosphates. *The EMBO Journal*, *28* (14), 2114–2127.

Petersen, C. B., Nygard, A. B., Viuff, B., et al. (2007). Porcine ecto-nucleotide pyrophosphatase/phosphodiesterase 1 (NPP1/CD203a): Cloning, transcription, expression, mapping, and identification of an NPP1/CD203a epitope for swine workshop cluster 9 (SWC9) monoclonal antibodies. *Developmental and Comparative Immunology*, *31*(6), 618–631.

Picher, M., & Boucher, R. C. (2003). Human airway ecto-adenylate kinase. A mechanism to propagate ATP signaling on airway surfaces. *The Journal of Biological Chemistry*, *278* (13), 11256–11264.

Picher, M., Burch, L. H., Hirsh, A. J., et al. (2003). Ecto 5′-nucleotidase and nonspecific alkaline phosphatase. Two AMP-hydrolyzing ectoenzymes with distinct roles in human airways. *The Journal of Biological Chemistry*, *278*(15), 13468–13479.

Pizzirani, C., Ferrari, D., Chiozzi, P., et al. (2007). Stimulation of P2 receptors causes release of IL-1beta-loaded microvesicles from human dendritic cells. *Blood*, *109*(9), 3856–3864.

Prosdocimo, D. A., Douglas, D. C., Romani, A. M., et al. (2009). Autocrine ATP release coupled to extracellular pyrophosphate accumulation in vascular smooth muscle cells. *American Journal of Physiology. Cell Physiology*, *296*(4), C828–C839.

Pulte, E. D., Broekman, M. J., Olson, K. E., et al. (2007). CD39/NTPDase-1 activity and expression in normal leukocytes. *Thrombosis Research*, *121*(3), 309–317.

Quintero, I. B., Araujo, C. L., Pulkka, A. E., et al. (2007). Prostatic acid phosphatase is not a prostate specific target. *Cancer Research*, *67*(14), 6549–6554.

Reutershan, J., Vollmer, I., Stark, S., et al. (2009). Adenosine and inflammation: CD39 and CD73 are critical mediators in LPS-induced PMN trafficking into the lungs. *The FASEB Journal*, *23*(2), 473–482.

Ricatti, M. J., Alfie, L. D., Lavoie, E. G., et al. (2009). Immunocytochemical localization of NTPDases1 and 2 in the neural retina of mouse and zebrafish. *Synapse*, *63*(4), 291–307.

Robson, S. C., Sévigny, J., & Zimmermann, H. (2006). The E-NTPDase family of ectonucleotidases: Structure function relationships and pathophysiological significance. *Purinergic Signalling*, *2*(2), 409–430.

Rucker, B., Almeida, M. E., Libermann, T. A., et al. (2007). Biochemical characterization of ecto-nucleotide pyrophosphatase/phosphodiesterase (E-NPP, E.C. 3.1.4.1) from rat heart left ventricle. *Molecular and Cellular Biochemistry*, *306*(1–2), 247–254.

Say, J. C., Ciuffi, K., Furriel, R. P., et al. (1991). Alkaline phosphatase from rat osseous plates: Purification and biochemical characterization of a soluble form. *Biochimica et Biophysica Acta*, *1074*(2), 256–262.

Schaefer, U., Machida, T., Broekman, M. J., et al. (2007). Targeted deletion of ectonucleoside triphosphate diphosphohydrolase 1/CD39 leads to desensitization of pre- and postsynaptic purinergic P2 receptors. *The Journal of Pharmacology and Experimental Therapeutics*, *322*(3), 1269–1277.

Schicker, K., Hussl, S., Chandaka, G. K., et al. (2009). A membrane network of receptors and enzymes for adenine nucleotides and nucleosides. *Biochimica et Biophysica Acta*, *1793*(2), 325–334.

Seman, M., Adriouch, S., Scheuplein, F., et al. (2003). NAD-induced T cell death: ADP-ribosylation of cell surface proteins by ART2 activates the cytolytic P2X7 purinoceptor. *Immunity*, *19*(4), 571–582.

Sévigny, J., Lévesque, F. P., Grondin, G., et al. (1997). Purification of the blood vessel ATP diphosphohydrolase, identification and localisation by immunological techniques. *Biochimica et Biophysica Acta*, *1334*(1), 73–88.

Sévigny, J., Robson, S. C., Waelkens, E., et al. (2000). Identification and characterization of a novel hepatic canalicular ATP diphosphohydrolase. *The Journal of Biological Chemistry*, *275*(8), 5640–5647.

Sévigny, J., Sundberg, C., Braun, N., et al. (2002). Differential catalytic properties and vascular topography of murine nucleoside triphosphate diphosphohydrolase 1 (NTPDase1) and NTPDase2 have implications for thromboregulation. *Blood*, *99*(8), 2801–2809.

Smoake, J. A., McMahon, K. L., Wright, R. K., et al. (1981). Hormonally sensitive cyclic AMP phosphodiesterase in liver cells. An ecto-enzyme. *The Journal of Biological Chemistry*, *256* (16), 8531–8535.

Sorensen, C. E., Amstrup, J., Rasmussen, H. N., et al. (2003). Rat pancreas secretes particulate ecto-nucleotidase CD39. *The Journal of Physiology*, *551*(Pt 3), 881–892.

Sowa, N. A., Vadakkan, K. I., & Zylka, M. J. (2009). Recombinant mouse PAP has pH-dependent ectonucleotidase activity and acts through A(1)-adenosine receptors to mediate antinociception. *PLoS One*, *4*(1), e4248.

Spanevello, R. M., Mazzanti, C. M., Bagatini, M., et al. (2010). Activities of the enzymes that hydrolyze adenine nucleotides in platelets from multiple sclerosis patients. *Journal of Neurology*, *257*(1), 24–30.

Spanevello, R. M., Mazzanti, C. M., Schmatz, R., et al. (2010). The activity and expression of NTPDase is altered in lymphocytes of multiple sclerosis patients. *Clinica Chimica Acta*, *411*(3–4), 210–214.

Stagg, J., Divisekera, U., McLaughlin, N., et al. (2010). Anti-CD73 antibody therapy inhibits breast tumor growth and metastasis. *Proceedings of the National Academy of Sciences of the United States of America*, *107*(4), 1547–1552.

Stefan, C., Jansen, S., & Bollen, M. (2006). Modulation of purinergic signaling by NPP-type ectophosphodiesterases. *Purinergic Signalling*, *2*(2), 361–370.

Sun, X., Wu, Y., Gao, W., et al. (2010). CD39/ENTPD1 expression by CD4+Foxp3+ regulatory T cells promotes hepatic metastatic tumor growth in mice. *Gastroenterology*, *139*(3), 1030–1040.

Synnestvedt, K., Furuta, G. T., Comerford, K. M., et al. (2002). Ecto-5′-nucleotidase (CD73) regulation by hypoxia-inducible factor-1 mediates permeability changes in intestinal epithelia. *Journal of Clinical Investigation*, *110*(7), 993–1002.

Takahashi, T., Old, L. J., & Boyse, E. A. (1970). Surface alloantigens of plasma cells. *The Journal of Experimental Medicine*, *131*(6), 1325–1341.

Torres, I. L., Buffon, A., Silveira, P. P., et al. (2002). Effect of chronic and acute stress on ectonucleotidase activities in spinal cord. *Physiology & Behavior*, *75*(1–2), 1–5.

van Waeg, G., & Van den Berghe, G. (1991). Purine catabolism in polymorphonuclear neutrophils. Phorbol myristate acetate-induced accumulation of adenosine owing to inactivation of extracellularly released adenosine deaminase. *The Journal of Clinical Investigation*, *87*(1), 305–312.

Vekaria, R. M., Shirley, D. G., Sévigny, J., et al. (2006). Immunolocalization of ectonucleotidases along the rat nephron. *American Journal of Physiology—Renal Physiology*, *290*(2), F550–F560.

Vollmayer, P., Clair, T., Goding, J. W., et al. (2003). Hydrolysis of diadenosine polyphosphates by nucleotide pyrophosphatases/phosphodiesterases. *European Journal of Biochemistry*, *270*(14), 2971–2978.

von Kugelgen, I. (2006). Pharmacological profiles of cloned mammalian P2Y-receptor subtypes. *Pharmacology & Therapeutics*, *110*(3), 415–432.

Vorhoff, T., Zimmermann, H., Pelletier, J., et al. (2005). Cloning and characterization of the ecto-nucleotidase NTPDase3 from rat brain: Predicted secondary structure and relation to other members of the E-NTPDase family and actin. *Purinergic Signalling*, *1*(3), 259–270.

Wagner, J. G., & Roth, R. A. (2000). Neutrophil migration mechanisms, with an emphasis on the pulmonary vasculature. *Pharmacological Reviews*, *52*(3), 349–374.

Wall, M. J., Wigmore, G., Lopatar, J., et al. (2008). The novel NTPDase inhibitor sodium polyoxotungstate (POM-1) inhibits ATP breakdown but also blocks central synaptic transmission, an action independent of NTPDase inhibition. *Neuropharmacology*, *55*(7), 1251–1258.

Warny, M., Aboudola, S., Robson, S. C., et al. (2001). P2Y(6) nucleotide receptor mediates monocyte interleukin-8 production in response to UDP or lipopolysaccharide. *The Journal of Biological Chemistry*, *276*(28), 26051–26056.

Weihofen, W. A., Liu, J., Reutter, W., et al. (2004). Crystal structure of CD26/dipeptidyl-peptidase IV in complex with adenosine deaminase reveals a highly amphiphilic interface. *The Journal of Biological Chemistry*, *279*(41), 43330–43335.

Weissmuller, T., Campbell, E. L., Rosenberger, P., et al. (2008). PMNs facilitate translocation of platelets across human and mouse epithelium and together alter fluid homeostasis via epithelial cell-expressed ecto-NTPDases. *Journal of Clinical Investigation*, *118*(11), 3682–3692.

Weissmuller, T., Eltzschig, H. K., & Colgan, S. P. (2005). Dynamic purine signaling and metabolism during neutrophil–endothelial interactions. *Purinergic Signalling*, *1*(3), 229–239.

Wiginton, D. A., Kaplan, D. J., States, J. C., et al. (1986). Complete sequence and structure of the gene for human adenosine deaminase. *Biochemistry*, *25*(25), 8234–8244.

Wink, M. R., Braganhol, E., Tamajusuku, A. S., et al. (2006). Nucleoside triphosphate diphosphohydrolase-2 (NTPDase2/CD39L1) is the dominant ectonucleotidase expressed by rat astrocytes. *Neuroscience*, *138*(2), 421–432.

Yegutkin, G. G. (2008). Nucleotide- and nucleoside-converting ectoenzymes: Important modulators of purinergic signalling cascade. *Biochimica et Biophysica Acta*, *1783*(5), 673–694.

Yegutkin, G. G., Henttinen, T., Samburski, S. S., et al. (2002). The evidence for two opposite, ATP-generating and ATP-consuming, extracellular pathways on endothelial and lymphoid cells. *The Biochemical Journal*, *367*(Pt 1), 121–128.

Yegutkin, G. G., Jankowski, J., Jalkanen, S., et al. (2008). Dinucleotide polyphosphates contribute to purinergic signaling via inhibition of adenylate kinase activity. *Bioscience Reports*, *28*(4), 189–194.

Yu, J., Lavoie, E. G., Sheung, N., et al. (2008). IL-6 downregulates transcription of NTPDase2 via specific promoter elements. *American Journal of Physiology. Gastrointestinal and Liver Physiology*, *294*(3), G748–G756.

Zacher, L. A., & Carey, G. B. (1999). Cyclic AMP metabolism by swine adipocyte microsomal and plasma membranes. *Comparative Biochemistry and Physiology. Part B: Biochemistry & Molecular Biology*, *124*(1), 61–71.

Zahn, P. K., Straub, H., Wenk, M., et al. (2007). Adenosine A1 but not A2a receptor agonist reduces hyperalgesia caused by a surgical incision in rats: A pertussis toxin-sensitive G protein-dependent process. *Anesthesiology*, *107*(5), 797–806.

Zavialov, A. V., & Engstrom, A. (2005). Human ADA2 belongs to a new family of growth factors with adenosine deaminase activity. *The Biochemical Journal*, *391*(Pt 1), 51–57.

Zimmermann, H. (1992). 5′-nucleotidase: Molecular structure and functional aspects. *The Biochemical Journal*, *285*, 345–365.

Zimmermann, H. (2000). Extracellular metabolism of ATP and other nucleotides. *Naunyn Schmiedeberg's Archives of Pharmacology*, *362*(4–5), 299–309.

Zimmermann, H. (2001). Ectonucleotidases: Some recent developments and note on nomenclature. *Drug Development Research*, *52*, 44–56.

Zimmermann, H. (2009). Prostatic acid phosphatase, a neglected ectonucleotidase. *Purinergic Signalling*, *5*(3), 273–275.

Zimmermann, H., & Braun, N. (1996). Extracellular metabolism of nucleotides in the nervous system. *Journal of Autonomic Pharmacology*, *16*(6), 397–400.

Zylka, M. J., Sowa, N. A., Taylor-Blake, B., et al. (2008). Prostatic acid phosphatase is an ectonucleotidase and suppresses pain by generating adenosine. *Neuron*, *60*(1), 111–122.

Silvia Deaglio* and Simon C. Robson†

*Department of Genetics, Biology, and Biochemistry, University of Turin & Human Genetics Foundation, Turin, Italy

†Department of Medicine, Transplant Institute, Beth Israel Deaconess Medical Center, Harvard Medical School, Boston, Massachusetts, USA

Ectonucleotidases as Regulators of Purinergic Signaling in Thrombosis, Inflammation, and Immunity

Abstract

Evolving studies in models of transplant rejection, inflammatory bowel disease, and cancer, among others, have implicated purinergic signaling in clinical manifestations of vascular injury and thrombophilia, inflammation, and immune disturbance.

Within the vasculature, spatial and temporal expression of CD39 nucleoside triphosphate diphosphohydrolase (NTPDase) family members together with CD73 ecto-5′-nucleotidase control platelet activation, thrombus size, and stability. This is achieved by closely regulated phosphohydrolytic activities to scavenge extracellular nucleotides, maintain P2-receptor integrity, and coordinate adenosinergic signaling responses. The CD38/CD157 family of extracellular NADases degrades NAD^+ and generates Ca^{2+}-active metabolites, including cyclic ADP ribose and ADP ribose. These mediators regulate leukocyte adhesion and chemotaxis. These mechanisms are crucial in vascular homeostasis, hemostasis, thrombogenesis, and during inflammation.

There has been recent interest in ectonucleotidase expression by immune cells. CD39 expression identifies Langerhans-type dendritic cells and efficiently distinguishes T regulatory cells from other resting or activated T cells. CD39, together with CD73 in mice, serves as an integral component of the suppressive machinery of T cells. Purinergic responses also impact

Advances in Pharmacology, Volume 61

1054-3589/11 $35.00
10.1016/B978-0-12-385526-8.00010-2

generation of T helper-type 17 cells. Further, CD38 and changes in NAD^+ availability modulate ADP ribosylation of the cytolytic P2X7 receptor that deletes T regulatory cells.

Expression of CD39, CD73, and CD38 ectonucleotidases on either endothelial or immune cells allows for homeostatic integration and control of vascular inflammatory and immune cell reactions at sites of injury. Ongoing development of therapeutic strategies targeting these and other ectonucleotidases offers promise for the management of vascular thrombosis, disordered inflammation, and aberrant immune reactivity.

I. Introduction

This review addresses novel mechanisms concerning the role of purinergic/pyrimidinergic signaling in hemostasis, vascular thrombosis, and cellular immunity in inflammatory disorders such as in transplantation, sepsis, autoimmune diseases, and cancer (Robson et al., 2005, 2006).

Nucleosides are glycosylamines comprising the nucleobase attached to a pentose sugar ring. Examples of these include cytidine, uridine, adenosine, guanosine, thymidine, and inosine. Nucleosides can be phosphorylated by specific kinases (chiefly within the cell but also feasibly outside). These reactions generate nucleotides, the monomeric structural unit of nucleotide chains that form nucleic acids (RNA and DNA). Nucleotides also play important roles in cellular energy transport and transformation (nucleoside triphosphates such as ATP are the energy-rich end products of the majority of biochemical energy releasing pathways), in enzyme regulation, and as intracellular second messengers.

Clearly, cellular injury processes are detected by release of intracellular constituents that, given the parsimony of nature, could also be used as extracellular "danger" or signaling mediators. Hence, it is possible that specific extracellular messenger systems could have developed to facilitate recognition of extracellular nucleotides that would have served as sensors for environmental stresses. One can also propose that such novel modular systems have been generated and refined with the development of multicellular organisms, different organ systems, and the requirement for a circulatory system and mobile cells that need to leave and reenter the blood stream.

The organization of the purinergic/pyrimidinergic system in health and disease might include requirements for new gene families, with conservation or modification of adaptive genes to facilitate specificity of nucleotide metabolism and responses that we address in the context of the critical host defense mechanisms of hemostasis and blood coagulation, inflammation, and immunity.

II. Purinergic Signaling: A Paradigm Linking Coagulation, Inflammation, and Immunity

Extracellular nucleotides are involved in vascular cell and platelet activation that enhance the generation of fibrin after vascular injury: to cause hemostasis under physiological conditions and provoke thrombosis in pathological situations. These mediators, in a manner analogous to coagulation proteases such as thrombin, have cellular activation effects that are fibrin independent and dictated by the specific purinergic receptors.

Purinergic signaling mechanisms mediated by nucleotides and nucleosides have a complex influence on inflammatory processes. These are characterized by obligatory blood and vessel components that cause extravascular responses.

How purinergic responses dictate and/or modulate inflammation is still incompletely understood. A shared P2-mediated mechanism in several different experimental models involves the boosting of innate immune "danger" signals that are initially generated by bacterial infections with endotoxin or, in the context of transplantation, ischemia–reperfusion, or rejection. The pathways ultimately are switched over to promote adenosinergic-type responses, a process that at least in part involves upregulation of ectonucleotidases and ATP scavenging.

There is increasing evidence that purinergic signaling has dramatic positive and negative effects on cells involved in adaptive immunity, for example, T lymphocytes, and on inflammatory cells that drive chronic inflammation and fibrosis, for example, dendritic cells (DCs), stellate cells, and myofibroblasts. This review will illustrate key paradigms linking coagulation, inflammation, and immunity as integrated purinergic responses. Several inflammatory diseases with thrombophilia have been linked genetically and mechanistically to ectonucleotidases of the CD39 and CD38 families, among others. These ectonucleotidases, as well as principles of purinergic signaling, will be addressed here. We will also describe the importance of dysregulated nucleotide-mediated signaling in cancer and inflammatory bowel disease (IBD) and touch on how pathogens might also subvert these defenses.

III. Extracellular Nucleotides, Nucleosides, and Ectonucleotidases

Extracellular nucleotides (e.g., ATP, UTP, ADP, NAD), and the derivative nucleosides (e.g., adenosine from ATP), are released in a regulated manner by most cells to provide the initiators and primary components for purinergic responses (Luthje, 1989).

These nucleotide/nucleoside mediators bind specific purinergic receptors that after mediator release comprise the second requirement for such a signaling network. Almost all cells express multiple type-2 purinergic/pyrimidinergic (P2) receptors for nucleotides and adenosine or type-1 purinergic (P1) receptors (Burnstock & Knight, 2004). There are seven ionotropic (P2X1–7), at least eight metabotropic ($P2Y_{1,2,4,6,11-14}$) and four adenosine receptor subtypes (A_1, A_{2A}, A_{2B}, A_3) that have been identified and characterized to date (Burnstock, 2002). In short, P2X and $P2Y_{11}$ receptors are chiefly activated by extracellular ATP; $P2Y_2$ by ATP and UTP; $P2Y_1$ and $P2Y_{13}$ by ADP; $P2Y_4$ by UTP; $P2Y_6$ by UDP; and $P2Y_{14}$ by UDP-glucose.

The final regulatory component of purinergic systems comprises ectonucleotidases (Robson et al., 2001, 2006) that are the focus of this review. Such ectoenzymes hydrolyze extracellular nucleotides to generate other nucleotides and nucleosides that in turn differentially activate other P2 and ultimately adenosine receptors with often opposing effects to those seen with the initial P2-mediated effects (Beldi et al., 2008a). As an alternative mechanism, these enzymes can also act indirectly by limiting nucleotide (i.e., substrate) availability for other cell surface enzymes, which act on purinergic receptors, thereby indirectly modulating P2 receptor activities (Malavasi et al., 2008).

Within the past decade, ectonucleotidases belonging to several enzyme families have been discovered, cloned, and functionally characterized. In this review, we provide a brief overview of the vascular and immune ectonucleotidases and then focus on CD39, CD73, and CD38 to dissect out their pathophysiological roles (Deaglio & Malavasi, 2006; Di Virgilio et al., 2009; Dwyer et al., 2007; Malavasi et al., 2008; Sitkovsky et al., 2008).

CD39 is the prototype of the ecto-nucleoside triphosphate diphosphohydrolase (E-NTPDase) family (EC 3.6.1.5). These proteins comprise a group of ectoenzymes that hydrolyze extracellular nucleoside tri- and diphosphates. The ecto-nucleotidase chain or cascade, as initiated by NTPDases, is terminated by ecto-5′-nucleotidase (CD73; EC 3.1.3.5; Resta et al., 1998; Robson et al., 2006).

Together, ecto-5′-nucleotidase and adenosine deaminase (ADA; EC 3.5.4.4), another ectoenzyme that is involved in purine salvage pathways by converting adenosine to inosine, closely regulate local and pericellular extracellular concentrations of adenosine (Robson et al., 2006).

Most notably, however, in many tissues and cells, there are complex cell surface-located nucleotide hydrolyzing and interconverting machinery (Fig. 1). These mutiple ensembles include the ecto-nucleotide pyrophosphatase phosphodiesterases (E-NPPs; EC 3.1.4.1, EC 3.6.1.9), CD38, NAD-glycohydrolases, alkaline and acid phosphatases, diadenosine polyphosphate hydrolases, adenylate kinases, nucleoside diphosphate kinase, and potentially ecto-F1-Fo ATP synthases (Robson et al., 2006).

The *CD38/CD157* gene family encodes two extracellular enzymes characterized by a widespread and often complementary cell distribution

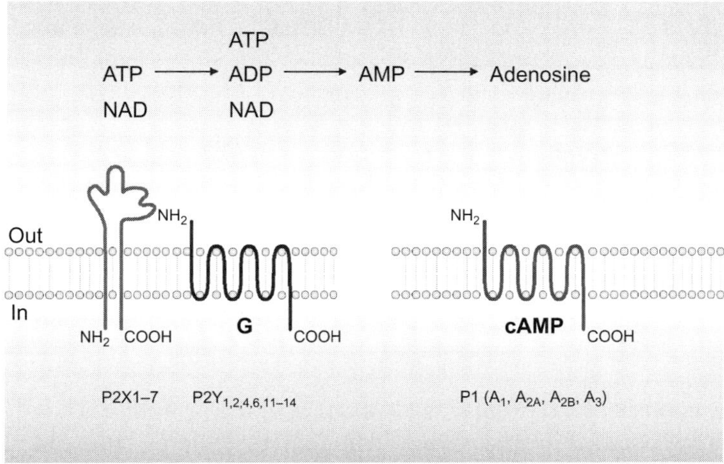

FIGURE 1 Purinergic mediators that are metabolized by CD39, CD73, and CD38 and respective receptors. Schematic depiction of extracellular nucleotides, for example, ATP, ADP, and NAD with derived adenosine nucleotide products as ligands for the corresponding purinergic receptors—see text for details of NAD catalysis by CD38 and impacts on purinergic signaling.

(Malavasi et al., 2008). Notwithstanding a high structural homology in the extracellular domain, they differ in membrane anchorage in that CD38 is a type II molecule, while CD157 is GPI anchored. These enzymes account for >95% of all NADase activities in mammals and were originally defined as cell surface activation molecules of leukocytes; they became classified as enzymes only later as a consequence of the identification of a sequence similarity between the human lymphocyte antigen CD38 (and later CD157) and the Aplysia ADP ribosyl cyclase (States et al., 1992). The catabolism of NAD+ and NAD(P) mediated by CD38 leads to the generation of potent intracellular Ca^{2+}-mobilizing compounds, including cyclic ADP ribose (cADPR) and ADP-ribose (ADPR); CD38 activity may be also impacted by extracellular ATP (Malavasi et al., 2008). As predicted by the structural homology, soluble CD157 incubated with NAD^+ produces cADPR and subsequently ADPR, indicating that this molecule is endowed with both ADPR cyclase and cADPR hydrolase activities. However, the catalytic efficiency of CD157 is several 100-folds lower than that of CD38 (Malavasi et al., 2008).

In addition to binding the TRPM2 membrane Ca^{2+} channels, ADPR, the main product of the reaction, can be covalently attached to proteins by ADP ribosyl transferases (ARTs). This posttranslational modification of target proteins can have dramatic effects on their functions. Thus, the CD38/CD157 family can modulate purinergic signaling by limiting NAD^+ availability to P2Y receptors, primarily $P2Y_1$ and $P2Y_{11}$ (Malavasi et al., 2008;

Moreschi et al., 2008). Alternatively, it is feasible that the enzymatic activity of CD38 may generate products that modulate purinergic signaling responses and also directly impacts on the activity of ARTs, affecting the amount of ADP ribosylated P2X7 (see Fig. 1).

To prevent redundancy and repetition in this issue, extracellular elements of nucleotide interconversion will not be addressed here. The interested reader is referred to Yegutkin et al. (2001, 2003, 2006) and to Chapter 8.

IV. Purinergic Signaling Responses in the Vasculature and Immune Systems

In the blood, extracellular nucleotides (e.g., ATP, ADP, UTP, UDP, and NAD^+) are released by leukocytes, lymphocytes, endothelium, and platelets by constituent mechanisms and in an enhanced manner with cell activation inclusive of degranulation and cell death. It is important to note here that many processes of arterial vascular injury are associated with the release of adenine nucleotides that exert a variety of inflammatory effects on the endothelium, platelets, and leukocytes (reviewed in Dubyak & Elmoatassim, 1993; Luthje, 1989). These events result in extracellular signals that are modulated by environmental factors (Luthje, 1989).

As alluded to above, these mediators bind the multiple P2Y and P2X receptors on platelets, endothelium, vascular smooth muscle cells, leukocytes, and immunocytes (Abbracchio & Burnstock, 1994). P2 receptors trigger and mediate short-term (acute) processes affecting cellular metabolism, nitric oxide (NO) release, adhesion, activation, and migration (Dubyak & Elmoatassim, 1993; Luthje, 1989). As an example in the vasculature, ATP released from endothelium during changes in flow (shear stress) or following exposure to hypoxic conditions activates P2Y receptors expressed by these cells and by vascular smooth muscle cells in an autocrine and paracrine manner to release NO, resulting in vessel relaxation as a purinergic event.

The intracellular NAD+ concentration is in the range of 1 mM, whereas the plasma concentration is in the range of 0.1 μM. The latter concentration is below the K_m of ARTs (Krebs et al., 2003). It is, however, feasible that high amounts of ecto-NAD^+ can be released during tissue injury as a consequence of cell lysis. Further, NAD^+ also seems to be released by nonlytic processes under various physiological conditions including hypoxia, inflammation, and mechanical or chemical activation (Bruzzone et al., 2001). Connexin 43 hemichannels, for instance, mediate transmembrane NAD+ fluxes in intact cells (Bruzzone et al., 2001). Hence, high local NAD^+ concentrations may be reached under various physiological and pathophysiological situations, which would permit ADP ribosylation of membrane proteins on ART-expressing neighboring cells. The fate of such generated NAD^+ is controlled by ecto-NADases, and cellular cross-talk that includes CD38 and select forms of ART, such as ART2.2 involved in immune regulation, has been demonstrated (Adriouch et al., 2008; Haag et al., 2007). Later, responses include those more protracted developmental responses, inclusive of cell proliferation, differentiation, and apoptosis (Burnstock & Knight, 2004; Harden et al., 1997; Weisman et al., 1998).

The dominant ectoenzymes or ectonucleotidases of the vasculature and lymphatics are now more fully characterized as E-NTPDases or E-NTPDases of the CD39 family, CD73/ecto-5′-nucleotidase, and potentially CD157. These important biological properties expressed by the endothelium, and associated cells, are responsible for the regulation of extracellular levels of nucleotides (Kaczmarek et al., 1996; Marcus et al., 1991, 2003; Robson et al., 1997, 2005).

NTPDases, in particular, might be able to serve as molecular switches at the membrane with differing functions according to the context of the signal. Colocalization of NTPDases with P2 and adenosine receptors has been demonstrated by fluorescence resonance energy transfer (FRET) techniques (Schicker et al., 2009); such structural aspects likely correlate with functional integration of responses and may confer added specificity. Such complexities must have been dictated by the ubiquitous nature of the signaling molecules, for example, ATP as mentioned above. However, extracellular nucleotide signaling should still be contrasted with the unique specificity of peptide hormones or vasoactive factors for often single, defined receptors (Goding & Howard, 1998; Sasamura et al., 1994).

Because of its high affinity for NAD^+ and its efficient hydrolase activity, CD38 represents the main extracellular NAD^+-metabolizing enzyme. When coexpressed with the ARTs, it competes for the substrate and limits their functional activities (Malavasi et al., 2008). This balance directly affects the amount of ADP-ribosylated P2X7 on arginine 125, a process mediated by ART2.2 (Adriouch et al., 2008), which catalyzes the covalent transfer of the ADPR group from NAD+ onto arginine residues of membrane target proteins. Consequently, NAD^+ released during inflammation participates in the regulation of T cell homeostasis *in vivo* through an ART2.2-dependent P2X7-mediated mechanism involving ADP ribosylation (Adriouch et al., 2008; Haag et al., 2007), a process, which is now referred to as NAD-induced cell death (NICD; Grahnert et al., 2010; Hubert et al., 2010).

Besides modulating extracellular NAD^+ levels, CD38/CD157 also catalyze the production of Ca^{2+}-regulated metabolites, including cADPR, ADPR, etc. These metabolites bind intracellular receptors and channels that lead to a transient increase in cytosolic Ca^{2+} concentrations (Malavasi et al., 2008). This activity is believed to potentiate the functions of chemokine receptors, as well as of antigen receptors in B and T lymphocytes, as observed by studying CD38 knockout mice and human disease models (Deaglio & Malavasi, 2006; Deaglio et al., 2001; Morabito et al., 2006).

Many aspects of these ectonucleotidase families and detailed expositions of structure/function relationships have been previously reviewed by us (Beldi et al., 2008a; Deaglio et al., 2006; Di Virgilio et al., 2009; Malavasi et al., 2008; Robson et al., 2006). These topics are fully addressed elsewhere in this issue and are not discussed further here (see Chapter 9).

V. Hemostasis and Thrombosis

A. Pathophysiology

Hemostasis must remain inactive under basal conditions but ready to immediately close off defects, shut off blood loss, and thereby minimize tissue injury. The pathophysiology of coagulation and platelet-activating mechanisms in hemostasis is integral to an understanding of thrombosis and inflammation in the vasculature.

The normal vascular endothelium provides a barrier that separates blood cells and plasma factors from highly reactive elements of the deeper layer of vessel wall and maintains blood fluidity and flow by inhibiting coagulation, platelet activation, and promoting fibrinolysis (Ross, 1995). These properties are governed by important thromboregulatory mechanisms that include the release of prostacyclin (Sinzinger et al., 1991), the generation of NO (Cooke & Dzau, 1997), and heparan sulfate expression (Ihrcke et al., 1993). Other key biological activities of the vasculature include ecto-nucleotide catalysis that generates nucleosides by phosphohydrolysis of the respective nucleotides (Robson et al., 1997b, 2001).

The initiating event of plasma coagulation is the exposure of abluminal or circulating tissue factor (TF) within microparticles to circulating Factor VII. Thus, any disruption in the endothelial barrier between these TF-expressing cells and circulating blood is an initiating event in plasma coagulation. The TF–Factor VIIa complex initiates a chain reaction by activating other zymogen coagulation factors in the blood and initiating feedback loops to enhance clotting activity. TF–Factor VIIa activates two zymogens: Factor IX and Factor X. Factor IX (complexed with Factor VIII which is in turn stabilized by von Willebrand factor (vWF)) serves to activate more Factor X. Activated Factor X in complex with Factor V activates thrombin (Factor II), the major protease responsible for cleavage of fibrinogen to form fibrin. Finally, thrombin accentuates the clotting process with positive feedback loops activating more Factor V, Factor VIII, and Factor XI (increases activated Factor IX) and Factor XIII, which covalently links fibrin molecules in a transglutaminase reaction to form an insoluble mesh. This cascade of activated zymogens and positive feedback loops propagates fibrin formation to impede local hemorrhage. The interested reader is referred to an excellent recent review of this topic by Furie and Furie (2008).

The hemostatic process is also initiated by damage to the endothelium with exposure of circulating platelets to subendothelial surfaces and any associated serine proteases of the coagulation cascade. There appear to be two separate and independent pathways for platelet activation: these involve vascular collagen-dependent activation and thrombin-dependent pathways. Vascular injury exposes circulating platelets to collagen-bound vWF, which binds via the glycoprotein receptor GP1bα and integrins such as GPIIbIIIa (which can also interact with fibrin(ogen)). Thrombin cleaves protease-activated receptors (PAR1 in humans) to initiate platelet activation. This process, in turn, activates other platelets by release of serotonin, thromboxane A2, and nucleotides, which operate in auto- and paracrine manners to amplify platelet activation (Furie & Furie, 2008).

Simultaneous activation of the coagulation cascade and platelets is thought to synergize to propagate thrombus formation (Furie & Furie, 2005). An alternative mechanism is that a form of TF that is initially inactive (or "encrypted") can be derived from cell-derived circulating microparticles and activated isomerization of a mixed disulfide and a free thiol to an intramolecular disulfide at sites of thrombus formation (Reinhardt et al., 2008). Thrombogenesis is blocked when the extracellular protein disulfide isomerase is inhibited (Reinhardt et al., 2008), perhaps preventing the activation of critical functions in platelet receptors and TF (Cho et al., 2008; Furie, 2009). The hemostatic process can be therefore further facilitated by expression of functionally active TF from cells and/or that expressed by cell-derived microparticles. It is of interest that such thrombus formation promoted by circulating microparticles can develop without implicating direct endothelial damage (Furie, 2009; Furie & Furie, 2008)

B. Impact of Extracellular Nucleotides, Nucleosides, and Ectonucleotidases on Hemostasis and Thrombosis

Over the past decade, extracellular nucleotides have been recognized as important mediators of a variety of processes including vascular inflammation and thrombosis with varying impacts in different systems (Robson et al., 2001). These cellular processes and nucleotide-triggered events are further modulated during angiogenesis and influence the development of atherosclerosis and restenosis following angioplasty (Burnstock & Knight, 2004; Goepfert et al., 2000, 2001; Wang et al., 2003; Wihlborg et al., 2004)

Extracellular nucleotides are continuously released from cells associated with the blood, for example, following the exocytosis of ATP/UTP-containing vesicles, facilitated diffusion by putative ABC transporters or by poorly understood electrodiffusional movements through ATP/nucleotide channels. It has been shown that rates of increase are higher in injured or stressed cells (Abraham et al., 1993; Franceschi et al., 1996; Grierson & Meldolesi, 1995; Luthje, 1989; Tsujimoto, 1997).

Several mechanisms account for the presence of nucleotides or nucleosides in plasma (Traut, 1994). As alluded to above, these include aggregating platelets, degranulating macrophages, excitatory neurons, injured cells, and cells undergoing mechanical or oxidative stress resulting in lysis, selective permeabilization of cellular membranes, and exocytosis of secretory vesicles, such as from platelet dense bodies (Fijnheer et al., 1992; Luthje, 1989). It is important to note here that many processes of arterial vascular injury are associated with the release of adenine nucleotides that exert a variety of inflammatory effects on endothelium, platelets, and leukocytes (reviewed in Dubyak & Elmoatassim, 1993; Luthje, 1989).

Adenosine is recognized as a bioactive agent in vascular inflammatory states, with effects mediated on both vascular cells and leukocytes (Ogura et al., 2006). In addition, adenosine has known antithrombotic effects by blocking induction of TF (Deguchi et al., 1998) via A_{2A} and A_3 receptors, particularly during ischemic or atherosclerotic processes, modulates the expression of antiapoptotic genes, and is immunosuppressive (Sitkovsky et al., 2004). Adenosine is constitutively present in the extracellular space at low concentrations, but under metabolically stressful and hypoxic conditions, the levels rise dramatically (Sitkovsky & Lukashev, 2005). Primary release of the mediator could occur *ab initio*, or this might follow conversion of released nucleotides to adenosine via ectonucleotidases.

The important function of ectonucleotidases in the vasculature is the modulation of P2-receptor-mediated signaling by the removal of extracellular ATP and ADP and related nucleotides. The ultimate generation of extracellular adenosine will abrogate and terminate nucleotide-mediated effects but will also activate adenosine receptors, with often opposing pathophysiological effects. Ectonucleotidases also produce the key molecules for purine salvage and consequent replenishment of ATP stores within multiple cell types. Indeed, although nucleotides appear not to be taken up by cells, their dephosphorylated nucleoside derivatives interact with several specific transporters to enable membrane passage. The regulated dephosphorylation of extracellular nucleotides by ectonucleotidases may be critical for appropriate purinergic/pyrimidinergic signaling and metabolic homeostasis (Enjyoji et al., 1999; Plesner, 1995; Robson et al., 2001, 2005).

The endothelial membrane expressed CD39/NTPDase-1 is the major ecto-nucleotidase in the vasculature (Enjyoji et al., 1999). Other NTPDases associated with the vasculature are the cell-associated ecto-ATPases (CD39L1 or NTPDase-2) and a soluble ecto-ADPases (CD39L2 or NTPDase-6; akin to the monocyte expressed CD39L4 or NTPDase-5; Chadwick & Frischauf, 1998; Mulero et al., 1999; Zimmermann, 1999).

The ectoenzyme CD39/NTPDase1 can be shown to efficiently bind and hydrolyze extracellular ADP (and ATP) to AMP. This phosphohydrolytic reaction limits the platelet activation response that is dependent upon the paracrine release of ADP and activation of specific purinergic receptors

(Marcus et al., 1991; Robson et al., 1997b, 2000). CD39L1/NTPDase2, a preferential nucleoside triphosphatase, activates platelets by converting the competitive antagonist (ATP) of platelet ADP receptors to the specific agonist of the $P2Y_1$ and $P2Y_{12}$ receptors.

In keeping with these biochemical properties, endothelial cells and vascular smooth muscle chiefly express CD39, where it serves as a thromboregulatory factor. In contrast, CD39L1 is associated with the adventitial surfaces of the muscularized vessels, microvascular pericytes, and the stromal cells and would potentially serve as a hemostatic factor (Sevigny et al., 2002).

The ectonucleotidase chain initiated by NTPDases is terminated by ecto-5′-nucleotidase (EC 3.1.3.5; Zimmermann, 1992). The ecto-5′-nucleotidase (CD73) is a glycoprotein member of the family of 5′-nucleotidases that operates in tandem with CD39 in the vasculature. Together, CD73 and ADA (EC 3.5.4.4), another ectoenzyme involved in purine salvage pathways that degrades adenosine to inosine, closely regulate local and pericellular extracellular and plasma concentrations of adenosine (Goding & Howard, 1998; Resta et al., 1998). CD73 exhibits high specificity toward nucleoside monophosphates and is inhibited by nucleoside diphosphates in a manner described as "feed-forward inhibition" that has the effect of promoting a platelet plug and limiting adenosine generation (Gordon et al., 1986) to enhance the process of thrombogenesis. This mechanism also explains, at least in part, why Cd39 deletion might have such pronounced effects on adenosine production and is the dominant functional ectonucleotidase within the vasculature (Gordon et al., 1986; Meghji et al., 1995).

The CD38 family of ectoenzymes is involved in recycling of extracellular nucleotides by metabolizing NAD through the generation of cADPR and ADPR (Howard et al., 1993). The surface expression levels of CD38 are tightly regulated *in vivo* in response to inflammatory signals and may also impact vascular responsiveness in this process (Malavasi et al., 1994). Specifically, a clear role for CD38 in the interactions taking place between circulating lymphocytes and the vessel wall has been recognized several years ago (Deaglio et al., 1996). The latter finding was the starting point for the search of an endothelial cell-bound ligand, which culminated in the identification of CD31 (platelet endothelial cell adhesion molecule) as a counter receptor for CD38 (Deaglio et al., 1998).

It is plausible that CD31 binding can affect the enzymatic activities of CD38, by influencing the dimerization of the molecule and the opening of the catalytic site (Liu et al., 2005). CD38/CD31 signals have been studied in humans in several models, by mimicking *in vitro* the conditions that might occur *in vivo*. The results indicate that this interaction is the starting point of a cascade of events that profoundly affects the gene signature of the cell. The modulated pathways are chiefly those leading to proliferation and chemotaxis (Deaglio & Malavasi, 2006; Deaglio et al., 2001; Morabito et al., 2006).

The CD38 paradigm is also representative of a more general trend involving several nucleotide-metabolizing enzymes (Deaglio et al., 2001). CD38 and CD157 may also serve as receptors, controlling intracellular signaling pathways apparently independently of the ectoenzymatic activity. The two functional activities might represent two separate evolutionary developments resulting in the final multifunctional molecule. Phylogenetic studies support the view that the enzymatic activity is the "older" conserved function, whereas at a later point, CD38 may have acquired membrane anchorage and the ability to function as a receptor (Ferrero & Malavasi, 1999).

These common traits also involve localization in membrane lipid microdomains, and the physical and functional association with partners specialized in signal transduction (Deaglio et al., 2002; Koziak et al., 2000; Pacheco et al., 2005).

C. Purinergic Events in Platelet Thrombus Formation

In vivo, activated platelets appear to contribute to thrombin generation through the exposure of phosphatidylserine, forming a procoagulant catalytic surface, and through platelet–leukocyte–microparticle interactions that result in exposure of TF to blood, thereby promoting clot formation (Falati et al., 2002, 2003). Platelet activation and integrin ligation in response to multiple agonists are known to be dependent upon the release of extracellular nucleotides and therefore regulated by specific antagonists of the $P2Y_1$, $P2Y_{12}$, and P2X1 receptors (Baurand et al., 2000; Hechler et al., 2003; von Kugelgen & Wetter, 2000).

How platelets and the coagulation systems coordinate the process of thrombogenesis can be directly visualized by *in vivo* confocal and widefield videomicroscopy visualizing platelet deposition, microparticle accumulation, and fibrin generation after direct, oxidant, or laser-induced vascular injury. Recent work has kinetically plotted the formation of platelet arterial thrombi in the living mouse. This pattern is characterized by rapid accumulation of platelets (accumulation phase), followed by a peak at about 100 s followed by a 50–75% decrease in platelet mass leading to the plateau or stabilization phase (Chou et al., 2004; Falati et al., 2002, 2003), for reasons that are not understood.

Intuitively, one would think that a thrombus would grow indefinitely as platelets accumulate, release ADP, and activate more platelets. But clearly this is not the case as the visualized thrombus decreases in size as bound platelets deaggregate and return to the circulation (Chou et al., 2004).

Plasma microparticles (<1.5 μm) originate from platelet and cell membrane lipid rafts and possibly regulate inflammatory responses and thrombogenesis (Falati et al., 2003; Furie, 2009; Furie & Furie, 2005). These actions are mediated through their phospholipid-rich surfaces and associated cell-derived surface molecules (Falati et al., 2003). Constitutively circulating microparticles are also associated with functional CD39, and accumulation of these at sites of vascular injury appears to influence local thrombus formation and evolution (Banz et al., 2008).

Purified, unstimulated platelets lack or have minimal CD39 functional activity but released CD39 present on microparticles may be a key modulatory influence of thrombosis. In this regard, monocyte-derived microparticles can bind to activated platelets in the developing thrombus in an interaction mediated by platelet P-selectin and microparticle P-selectin glycoprotein ligand 1(PSGL-1; Falati et al., 2002, 2003). We might infer that as the platelet thrombus forms, in parallel, microparticles accumulate in the growing aggregate, which deliver the associated membrane CD39. Hence, CD39, and associated NTPDase activity, is not initially a substantive part of the thrombus but only accumulates after platelet thrombus formation matures. The spatial and temporal expression of NTPDases in the accumulating microparticles may therefore regulate thrombus size by regulating the hydrolysis and hence inactivation of the platelet agonist ADP, ultimately resulting in the observed deaggregation responses (Atkinson et al., 2006).

There is, however, a complicating variable. Although CD39 is a key determinant of extracellular nucleotide fluxes in the vasculature, genetic deletion paradoxically results in disordered hemostasis, secondary to platelet dysfunction. When studied in the CD39-mutant mouse, platelet thrombi were limited after oxidant injury *in vivo* and purified Cd39-null platelets failed to aggregate to standard agonists *in vitro*. This pattern of platelet hypofunction was wholly reversible and associated with purinergic type $P2Y_1$ receptor desensitization and membrane integrin dysfunction (Enjyoji et al., 1999).

Cd39 deletion in mice produces a proinflammatory phenotype associated with quantitative and qualitative differences in the microparticle populations, as determined by 2D-gel, Western blot, and flow cytometry. *Cd39*-null microparticles are also more abundant, have significantly higher proportions of platelet- and endothelial-derived elements and decreased levels of interleukin (IL)-10, tumor necrosis factor receptor-I (TNF-RI), and matrix metalloprotease (MMP-2). Consequently, *Cd39*-null microparticles boost endothelial activation, as determined by inflammatory cytokine release and upregulation of adhesion molecules *in vitro*. These findings by Banz and our colleagues suggest a modulatory role for CD39 within microparticles in the exchange of regulatory signals between vascular cells (Banz et al., 2008).

VI. Inflammation

A. Platelets and Leukocytes

Inflammation potentiates thrombosis by both downregulating vascular natural anticoagulant mechanisms and provoking disordered purinergic mechanisms, as alluded to above. Inflammation and thrombosis are closely integrated by close interactions between platelets, circulating leukocytes, endothelial, and other vascular cells.

Platelets are a key element at the interface between thrombosis and inflammation. Activated platelets release extracellular nucleotides and soluble factors that may have both local and systemic effects on blood and vascular cells that compound inflammatory responses.

Polymorphonuclear neutrophils (PMN) are the cardinal cellular component of an acute inflammatory response, and purinergic signaling dictates their functions. Specifically, ATP appears to be released from the leading edge of neutrophils in response to stimulation by chemoattractants, such as the peptide *N*-formyl-met-leu-phe (fMLP). ATP released by cells among which are PMN activates $P2Y_2$ receptors and consequently forms adenosine that via A_3 receptors stimulates neutrophil movement (Chen et al., 2006). CD39 is the critical ectonucleotidase dictating hydrolysis of released ATP and directing cell migration by PMN. Upon stimulation of human PMN or differentiated HL-60 cells in a chemotactic gradient, CD39 tightly associates with the leading edge of polarized cells during their migration in a chemotactic gradient. Inhibition or genetic deletion of CD39 reduces the migration speed of PMN but not the chemotactic ability both *in vitro* and *in vivo* (Corriden et al., 2008).

Further, CD39 is the dominant ectonucleotidase expressed by monocyte-macrophages. Upregulation of TF expression by monocytes and major defects in their entry and migration into the substance of Matrigel plugs injected into the subcutaneous tissue of Cd39-null mice have been observed. In parallel, we also evaluated parameters of monocyte transendothelial migration influenced by ATP *in vitro* and noted failure of Cd39-null cells to migrate in response to exogenous nucleotides. This defect could be overcome by costimulation with serotonin, suggesting a degree of P2Y-receptor desensitization in Cd39-null monocytes, as previously noted in platelet functions (Goepfert et al., 2001; Robson et al., 2001). Ectoenzymes, specifically ecto-nucleotidases, play a key role in leukocyte trafficking but this complex area will not be dealt with here (for an excellent review on this topic, see Salmi & Jalkanen, 2005).

B. Inflammasome

This recently coined term refers to those multiprotein complexes comprising caspase family members that assemble in the cytoplasm following injury to cells, or exposure to "danger signals." Activated caspase-1 within the complex cleaves pro-IL-1, and others, leading to cytokine activation and secretion (Latz, 2010). ATP released from damaged cells may serve as a "danger signal" and binds P2X7 which is known to activate the NLRP3/ASC/caspase-1 inflammasome boosting release of active IL-1 (reviewed in Mariathasan et al., 2006; Ogura et al., 2006; and more recently in Aymeric et al., 2010; Stagg & Smyth, 2010).

The autocrine/paracrine release of ATP that generates the inflammasome activation via P2X7 receptor activation can be impacted upon by CD39 in endothelium or by myeloid cells. Crucially, overexpression of CD39 in endothelium efficiently abrogated initial phases of ATP secretion in response to lipopolysaccharide endotoxin and markedly inhibited IL-1 alpha release; comparable results were obtained with soluble NTPDase. These earlier data suggested that CD39 could modulate IL-1 release from endothelium (Imai et al., 2000) and might also impact the inflammasome.

In a more recent study, CD39 was shown to be the dominant ectonucleotidase expressed by murine peritoneal macrophages and to also regulate P2X7-dependent IL-1 secretion responses in these cells (Levesque et al., 2010).

For further details with respect to the impact of ectonucleotidases on inflammation and cytokine elaboration, please refer to Chapter 9.

C. Plasticity of Purinergic Responses in Inflammation

Inflammatory and vascular cells exhibit a certain degree of plasticity to respond to stress, as do adaptive input–output devices. The putative input arises from the extracellular milieu and includes biochemical signals triggered by extracellular nucleotides, and the resulting biomechanical responses are transduced by adhesion receptors in response to P2-mediated signals, such as the affinity changes in integrins and cell adhesion responses. The output is further manifested as alterations in cellular phenotype, such as with activation responses, and includes a number of structural and functional changes implicated in diverse pathophysiological processes.

As an example, acute inflammation and associated oxidative stress, for example, as in graft ischemia–reperfusion injury, have an immediate and direct impact on extracellular nucleotide metabolism secondary to substantive and immediate decreases in the biological activity of CD39. Palmitoylation status of CD39 and cholesterol content of membrane lipid rafts appear to impact changes on NTPDase activity under such episodes of stress. Loss of NTPDase activity can be ameliorated by statins (Kaneider et al., 2002) or by incorporation of saturated fatty acids into cell membranes and limitation of lipid peroxidation responses by antioxidants (Kaczmarek et al., 1996; Robson et al., 1997a).

Recent work has suggested that upregulation of CD39 can be influenced by the select transcription factors CREB (Liao et al., 2010), Sp1, and hypoxia-inducible factor-1 (HIF-1; Eltzschig et al., 2009). CD39 and CD73 as well as certain adenosine receptors are clearly upregulated in the setting of hypoxia (Eltzschig et al., 2003, 2004; Synnestvedt et al., 2002). Downregulation of adenosine transporter expression under such conditions further decreases adenosine uptake (Morote-Garcia et al., 2009). Such hypoxia–adenosinergic mechanisms boost production of extracellular adenosine and receptor signaling. These pathways have different effects in select experimental models when examining leukocyte and lymphocyte infiltration into tissues or peritoneum (Corriden et al., 2008) versus altering egress of cells across epithelial barriers, for example, as in the lung or gut (Eltzschig et al., 2003, 2004; Reutershan et al., 2009).

CD38 expression can be likewise readily modulated, following activation of B and T cells as well as in the context of contacts between lymphocytes and stromal cells (Deaglio et al., 2006). The gene includes a very long first intron that contains regulatory elements for vitamins (A and D) and for interleukins (interferons and IL-6), among others (Malavasi et al., 2008). The presence of a large CpG island suggests epigenetic regulation, as very recently demonstrated in the chronic lymphocytic leukemia (CLL) model. Lastly, CD38 has a well-characterized single-nucleotide polymorphism (SNP) located at the 5′ end of this intron (184 C→G), which leads to the presence (or absence) of a *Pvu*II restriction site. This SNP is located within an E-box and conditions binding of the E2A transcription factor in human B lymphocytes (Saborit-Villarroya et al., 2011).

VII. Immunity

High levels of ATP may be released by $CD4^+$ and $CD8^+$ T cells upon mitogenic or antigenic/TCR stimulation and serve to amplify activation of cells (la Sala et al., 2003). Acute P2 receptor-mediated stimulation of monocytes, lymphocytes, and endothelium causes largely proinflammatory responses, such as the release of IL-1 (or IL-8), as described above (Imai et al., 2000; la Sala et al., 2003; Warny et al., 2001). On DCs, exposure to extracellular ATP induces migration and differentiation to drive cellular immune responses (la Sala et al., 2001).

Multiple P2X and P2Y receptor subtypes are expressed by monocytes and DCs, whereas lymphocytes express only P2Y receptors (Burnstock & Knight, 2004). These various receptors operate in both auto- and paracrine loops and are considered to play a complex, important role in the regulation of vascular and immune cell-mediated responses. We will focus on how CD39 and CD38 impact these functions.

A. CD39

CD39 was first described as a B lymphocyte activation marker (Maliszewski et al., 1994) and has been also shown to be expressed on natural killer (NK) cells, monocytes, DC, and subsets of activated T cells (Koziak et al., 1999). The relevance of the expression of CD39 by these cells is still being explored but there have been several recent advances in understanding brought about, at least in part, by study of mutant mice.

Langerhans DCs have the capacity to recruit, activate, and polarize naive T cells and express high levels of CD39 (Mizumoto et al., 2002). Cd39-null DCs are unresponsive to ATP albeit these are susceptible to cell death, but only after prolonged exposure to nucleotides (Mizumoto et al., 2002). Mutant mice null for Cd39 have amplified inflammatory responses to irritant chemicals as a consequence of the lack of CD39 suppressive properties. However, there are major defects in DC formation, antigen presentation, and T cell responses to haptens in Cd39-null mice. These result in markedly attenuated responses to contact allergens in type IV hypersensitivity cutaneous responses that are also seen in IBD models following haptenic stimulation (Kunzli et al., 2010). These data suggest that Cd39 expression is required for optimal stimulation of hapten-reactive T cells in mice (Mizumoto et al., 2002). Cd39-null DCs are fully functional with respect to homing and phenotypic maturation but are less able to stimulate T cells. These immune findings are relevant to allograft rejection processes (Robson et al., 2001). Sequelae of these putative immune abnormalities include the relative failure of Cd39-null mice to reject allografts under limited costimulation blockade (Li et al., 2003). These data indicate a previously unrecognized role of CD39 and the effects on nucleotide-mediated signaling in immunological responses (Mizumoto et al., 2002) that has been extended in lymphocyte studies.

We have recently shown that absence of CD39 and consequent changes in P2 receptor signaling paradoxically limit interferon gamma (IFNγ) release by NK cells in inflammatory situations. Adoptive transfers of populations of mutant and wild-type immune cells into Rag2/common gamma null mice (deficient in T cells, B cells, and NK/NKT cells) suggest that CD39 deletion on NK cells provides end-organ protection in limited warm ischemia, decreasing tissue damage mediated by these innate immune cells (Beldi et al., 2010).

With Dr. Marilia Cascalho and colleagues, we have noted abnormalities in B cell memory responses to T-dependent antigens in Cd39-null mice with demonstrable abnormalities in postgerminal center terminal B cell differentiation (not shown here).

With Karen Dwyer, Wenda Gao, and colleagues, we have shown that CD39 and CD73 are surface markers of Treg cells (in the mouse) that impart a specific biochemical signature characterized by adenosine generation that has functional relevance for cellular immunoregulation mediated by effector T cell A_{2A} stimulation (Deaglio et al., 2007). CD39, albeit not consistently with CD73, is associated with human CD4–Treg, as defined by high expression of Foxp3 and low levels of CD127 (Dwyer et al., 2010). Comparable work has been also published in human systems (Borsellino et al., 2007; Mandapathil et al., 2010).

In other collaborative work recently published by Dwyer et al., CD39 has been used to characterize blood T cell populations to allow tracking of these cells in health and disease (Dwyer et al., 2010). Differential expression of CD25 and CD39 on circulating $CD4^+$ T cells differentiates between Treg and pathogenic cellular populations associated with secretion of inflammatory cytokines (IFNγ and IL-17). These latter cell populations are increased, with decreases in CD39-expressing Tregs, in patients with renal allograft rejection (Dwyer et al., 2010).

Lastly, work done by Guido Beldi and colleagues has shown an important NKT cell protective phenotype in Cd39-null mice following Concavalin A-induced immune hepatitis. These data indicate a role for modulated purinergic signaling to impact NKT-mediated mechanisms resulting in liver immune injury (Beldi et al., 2008b).

Nucleotides are potent inflammatory factors when suddenly released into the extracellular environment in high concentration, as with platelet degranulation. Hence, depending on the P2 or adenosine receptor subtype, the cell types, and signaling pathway involved, these receptors might preferentially trigger and mediate short-term (acute) processes that affect metabolism, adhesion, activation, or migration. Moreover, purinergic signaling also has profound impacts upon other more protracted reactions, including cell proliferation, differentiation, and apoptosis, such as seen in several chronic inflammatory states (Burnstock, 2002). These mechanisms could be also implicated in immune memory (Koshiba et al., 1997; Sitkovsky et al., 2004). Immune outcomes appear to differ somewhat when there are slow, gradual, and persistent increases in nucleotide fluxes where P2 receptor desensitization might occur. Under these conditions, purinergic effects deviate immune responses more toward nonresponsiveness (tolerance) rather than heightened reactivity (reviewed in Di Virgilio & Robson, 2009; Di Virgilio et al., 2009).

B. CD38

Expression of CD38 by activated T and B lymphocytes was the starting point for functional studies aimed at dissecting the role of the molecule in the immune compartment. The emerging picture is highly complex and mostly dependent on the lineage considered and the activation status of the cell. However, a few common trends emerge that are expanded upon below (reviewed in Malavasi et al., 2008).

It seems apparent that CD38 engagement by agonistic mAbs is followed by a signaling cascade typical of canonical receptors, including tyrosine phosphorylation of a sequential number of intracellular enzymes, nuclear events, and more long-term effects dependent on active protein synthesis. An increase of the cytoplasmic levels of calcium ions is also a common theme upon activation of CD38. The calcium wave is typically slow in rising as compared to the spikes obtained after signaling through the antigen receptors in T and B lymphocytes. Unlike the antigen receptors, a CD38-induced calcium wave may last several minutes before declining, indicating that the molecular mechanisms responsible for calcium mobilization might be different and possibly relying on the enzymatic activities of the molecule, which generate several Ca^{2+} active compounds.

It is still not clear what initiates the signal. A hypothesis is that this commences with provision of substrate, that is, NAD^{+}, and might be further regulated through protein–protein interactions. In the human, these would include CD31, the only nonsubstrate ligand so far identified that might focus biochemical activity at vascular sites, as alluded to above. The structural requirements for signaling include localization in critical areas of the plasma membrane, in close physical proximity with more typical signaling receptors, such as the B cell receptor complex in B cells (Deaglio et al., 2006), CD16 in NK cells (Deaglio et al., 2002), and also inclusive of the T cell receptor in T lymphocytes, MHC Class II and CD9 in monocytes, and the CCR7 and CXCR4 chemokine receptors, CD83 and CD11b in mature DCs (reviewed in Deaglio & Malavasi, 2006; Malavasi et al., 2008).

The long-term events described as a consequence of the interaction of CD38 with the CD31 ligand, as mimicked by agonistic mAbs, vary according to cell lineage and differentiation status. Two common themes as to the regulation of cellular responses by CD38 are evident. The first is linked to the amplification of the signal mediated by the antigen receptors in T and B lymphocytes and by CD16 in NK cells. The second is associated with the amplification of chemotaxis signals mediated by different chemokine receptors. Both avenues have the potential to modulate immune responsiveness and inflammation in health and disease, as addressed next.

VIII. Disease Processes

A. Cancer

Inflammation and coagulation have important roles in the biology of cancer, and effects mediated by nucleotides and other mediators are often dualistic.

First, patients with cancer might exhibit thrombophilia with coagulation activated toward a prothrombotic state and heightened activation of inflammation. Second, such a procoagulant environment may promote growth and dissemination by several mechanisms involving fibrin deposition, angiogenesis, and platelet activation. Dysregulation of coagulation with heightened levels of inflammatory mediators, such as extracellular nucleotides in cancer, impacts clotting and vascular homeostasis to provoke venous thrombosis. These mediators also stimulate cell signaling in the extra vascular compartment, through interaction with cell receptors like PARs and P2 receptors, to drive tumor growth. Platelet activation and associations with leukocytes may further promote inflammatory reactions to tumors.

Proteases like thrombin may drive malignant cell proliferation and are opposed by anticoagulant and anti-inflammatory actions, inclusive of the protein C-thrombomodulin mechanism. Likewise, nucleotide-mediated effects are abrogated by ectonucleotidases that generate adenosine. In such situations, cancer may promote growth and protect against host responses by subverting protective effects of inflammatory responses.

Experimental approaches that promote ATP-mediated activation of immune responses and tumor cytotoxicity, or inhibit generation of extracellular adenosine have been tested. We have demonstrated that metastatic tumors from a melanoma cell line and other cancers were strongly inhibited in mice with Cd39-null vasculature as a consequence of disordered angiogenesis (Beldi et al., 2008c; Jackson et al., 2007). Tumors likewise were decreased in size in wild-type mice with circulating Cd39-null bone marrow-derived cells (Sun et al., 2010). Our models suggest that functional CD39 expression on CD4(+)Foxp3(+) Tregs suppressed antitumor immunity mediated by NK cells *in vitro* and *in vivo*. Inhibition of CD39 activity by polyoxometalate-1, a pharmacologic inhibitor of NTPDase activity, significantly inhibited tumor growth in the experimental system validated by us (Sun et al., 2010).

Several primary tumors contain CD39-expressing Treg. There is another intriguing possibility in that tumor cells express ectonucleotidases that may primarily generate high levels of extracellular adenosine; cancer cells may, therefore, blunt host responses while promoting tumor growth without any need for Treg recruitment (Buffon et al., 2007; Stagg & Smyth 2010; Stagg et al., 2010). To address these possibilities, we and others have proposed that pharmacologic or targeted inhibition of CD39 and/or CD73 enzymatic activity may find utility as an adjunct therapy for primary and secondary malignancies (Hilchey et al., 2009; Mandapathil et al., 2010; Sun et al., 2010; Zhang, 2010).

The role of CD38 in human tumor models is intimately linked to CLL, a relatively common leukemia of adults (Chiorazzi et al., 2005a, 2005b). The disease is invariably characterized by the expansion of a population of mature B cells expressing CD5 and accumulating in the blood and lymphoid organs. The clinical behavior is highly heterogeneous with patients who survive decades without any need for therapy, along with patients who need to be treated in the first 12 months after diagnosis, become rapidly resistant to therapy, and ultimately die because of the leukemia. CD38 is expressed by the aggressive variant of the disease and has become a widely used and trusted marker (Deaglio et al., 2006).

The current view is that CD38 is intimately linked to disease progression and that expression conditions tumor growth and expansion (Deaglio et al., 2010a, 2010b). This view is linked to the role of CD38 in collaborating with the B cell receptor complex in delivering proliferation signals to the neoplastic cell. More recently, CD38 has also been recognized as critical in determining chemotactic responses to the CXCL12 chemokine, by working in physical and functional association with the CXCR4 receptor (Vaisitti et al., 2010).

These overall observations suggest that CD38 is a potential therapeutic target in CLL. It is reasonable to surmise that CD38-blocking reagents of antibody origin or based on modified substrate ligands will perturb growth circuits of the neoplastic clone, probably rendering it more susceptible to conventional chemotherapy. Alternatively, nonsubstrate ligands can be used to block CD38 and consequent biochemical interactions with the surrounding deleterious receptors (Deaglio & Malavasi, 2006; Malavasi et al., 2008). Human antihuman-CD38 mAbs are already being tested in clinical trials, providing further support to this proposal.

B. Inflammatory Bowel Disease

IBD is an often devastating disease and is associated with excessive inflammation in the bowel and extra intestinal tissues in genetically susceptible individuals (Podolsky, 2002). Dynamic balances of Treg to Th17 cells in the gut, alterations in bacterial flora, and other environmental factors may regulate inflammatory responses both locally and systematically (Boden & Snapper, 2008; Paust & Cantor, 2005). Associations between IBD and thrombophilia have been recognized for decades and are of unclear etiology lacking a clear molecular basis (Yoshida & Granger, 2009).

As noted above, purinergic signaling pathways are involved in thrombosis and vascular inflammation (Robson et al., 2005), as well as immune suppressive cellular responses (Deaglio et al., 2007). Importantly, genetic polymorphisms of CD39 are linked to Crohn's disease, and mice null for *Cd39* exhibit severe colitis in several experimental models (Friedman et al., 2009; Kunzli et al., 2010). The purinergic linkages between inflammation and clotting listed above might also indicate, at least in part, why this disease has such a high incidence of venous thrombosis, blood clotting issues, and platelet activation (Yoshida & Granger, 2009).

It may seem relevant to ask here as to the functions CD39 could be serving when expressed by Treg and/or DCs that are relevant to IBD. A population of $CD4^+$ lymphocytes noted to selectively express CD39 suggests a functional role for ectonucleotidases. Without such biochemical activity, Treg and vascular endothelium are not able to rapidly convert nucleotides into adenosine (Deaglio et al., 2007; Robson et al., 2005). Therefore, Cd39-null cells lack an endogenous "stop" or "brake" mechanism that might allow unfettered intestinal inflammation and correlate with the tendency of low expression of CD39 to be genetically associated with Crohn's disease (Friedman et al., 2009). In a reciprocal manner, heightened levels of CD39 likely promote generation of adenosine to effectively block Th17 functions (Dwyer et al., 2010).

Recent work by Atarashi et al. (2008) indicates that commensal bacteria release ATP that specifically generates Th17 cells following the activation of a unique subset of lamina propria cells. Importantly, the administration of exogenous ATP driving purinergic responses exacerbates an adoptive T transfer colitis model and boosts Th17 differentiation (Atarashi et al., 2008).

C. Infections and Pathogens

Certain parasitic pathogens of humans express surface-located NTPDases that have been linked to virulence (chiefly protozoans and helminths). For those parasites that are purine auxotrophs, these enzymes may play an important role in purine scavenging, although they may also influence the host response to infection.

Although E-NTPDases are rare in bacteria, expression of a secreted NTPDase in *Legionella pneumophila* has been noted (Sansom et al., 2008a, 2008b). This ectoenzyme enhances intracellular growth of the bacterium and potentially affects virulence.

Staphylococcus aureus and other bacterial pathogens exploit the immunomodulatory attributes of adenosine to escape host immune responses (Thammavongsa et al., 2009). Purine metabolism and ectoenzymes (CD73) influence the host response to infection to Lyme borreliosis, a common arthropod-borne infection caused by *Borrelia burgdorferi* (Yegutkin et al., 2010).

Those interested in reading more on possible effects of microbial ectonucleotidases on host–pathogen interactions should examine the review (Sansom et al., 2008b).

IX. Conclusion

Extracellular nucleotide-mediated vascular endothelial and accessory cell stimulation have clear consequences for platelet activation, thrombogenesis, angiogenesis, vascular remodeling, and the metabolic milieu of the vasculature, in response to inflammatory stress and/or immune reactions. These mechanisms not only impact hemostasis but also dictate processes of vascular injury, thromboregulatory disturbances, and defective angiogenesis with vascular remodeling. This review has attempted to summarize those specific components of purinergic signaling as pertaining to vascular injury, inflammation, and immunity with a focus on the CD39 and CD38 family of ectonucleotidases (Fig. 1).

We have suggested that modulated, distinctive ectonucleotidase expression (by endothelium, leukocytes, and platelets (and potentially by derived microparticles; Banz et al., 2008)) regulates nucleotide-mediated signaling in the vasculature in a temporal and spatial manner with consequences for thrombogenesis and platelet activation. We have also addressed components of extracellular nucleotide-mediated signaling pathway, chiefly in T cells that are impacted upon largely by CD39, the prototypic member of the E-NTPDase family of ectonucleotidases and those effects on T cell homeostasis pertinent to CD38 (as shown in Fig. 2). Modulated, distinct NTPDase expression appears to regulate nucleotide- and nucleoside-mediated signaling within the vasculature and in the immune system. Hence, expression of ectonucleotidases on either endothelial or immune cells integrates vascular inflammatory and immune cell reactions at sites of injury.

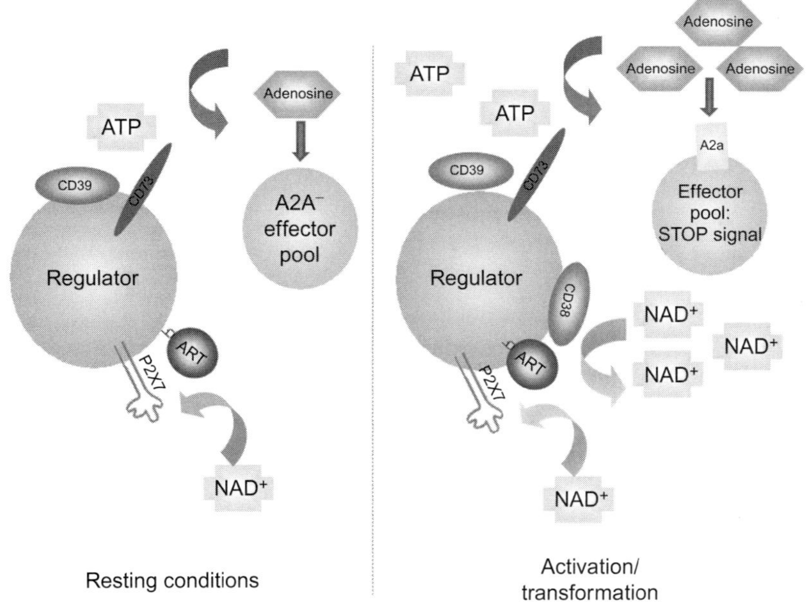

FIGURE 2 *Schematic representation of regulation of T regulatory cell homeostasis by extracellular nucleotides.* Extracellular nucleotide metabolism may be integrated into a homeostatic mechanism for the T cell compartment. Resting conditions are associated with limited ATP or NAD flux, low levels of CD39, CD38, and CD73. Basal mechanisms involve ADP ribosylation of P2X7, NICD, and maintenance of Treg compartment in quiescent state by low-level A_{2A} stimulation (left panel). With acute inflammation and cellular stress, there are increases of pericellular nucleotides that induce cellular activation with resulting abrogation of select Treg responses (secondary to P2X7 activation). Those surviving cells exhibit marked upregulation of CD39, CD38, and CD73. Ultimately, NAD is converted and effectively removed by CD38, while ATP is subjected to hydrolysis by CD39 and CD73 (right panel). As these surviving immune suppressive cells are "protected" by ectonucleotidases and also have potent internal mechanisms to modulate cAMP, such cells, for example, Treg initially expand, and adenosine accumulates to then activate A_{2A} receptors on T effector cells and NK cells (amongst others). This effect has the result of limiting effector cell expansion and blocking immune responses.

Increasing interest in this field will open up several new avenues for investigation. These developments might result in new treatment modalities for thrombotic disorders, vascular inflammation, and disordered immunity.

Acknowledgments

We apologize in advance if, because of space constraints, we have not adequately referenced the work of others and have cited review articles. We thank our colleagues and reviewers of chapter for their invaluable inputs. Support of NIH is recognized.

Conflicts of Interest: The authors have no conflicts of interest to declare.

Abbreviations

ART	ADP ribosyl transferase
cADPR	cyclic ADP ribsose
DC	dendritic cell
EC	endothelial cell
E-NTPDase	ecto-nucleoside triphosphate diphosphohydrolase
IBD	inflammatory bowel disease
IL	interleukin
LC	Langerhans cells
NAD	nicotinamide dinucleotide
NICD	NAD-induced cell death
NPP	nucleotide pyrophosphatase/phosphodiesterase
TF	tissue factor
Treg	T regulatory cells

References

Abbracchio, M. P., & Burnstock, G. (1994). Purinoceptors—Are there families of P2X and P2Y purinoceptors. *Pharmacology & Therapeutics*, *64*(3), 445–475, Review.

Abraham, E. H., Prat, A. G., Gerweck, L., Seneveratne, T., Arceci, R. J., Kramer, R., et al. (1993). The multidrug resistance (mdr1) gene product functions as an ATP channel. *Proceedings of the National Academy of Sciences of the United States of America*, *90*(1), 312–316.

Adriouch, S., Bannas, P., Schwarz, N., Fliegert, R., Guse, A. H., Seman, M., et al. (2008). ADP-ribosylation at R125 gates the P2X7 ion channel by presenting a covalent ligand to its nucleotide binding site. *The FASEB Journal*, *22*(3), 861–869.

Atarashi, K., Nishimura, J., Shima, T., Umesaki, Y., Yamamoto, M., Onoue, M., et al. (2008). ATP drives lamina propria T(H)17 cell differentiation. *Nature*, *455*(7214), 808–812.

Atkinson, B., Dwyer, K., Enjyoji, K., & Robson, S. C. (2006). Ecto-nucleotidases of the CD39/NTPDase family modulate platelet activation and thrombus formation: Potential as therapeutic targets. *Blood Cells, Molecules & Diseases*, *10*, 10.

Aymeric, L., Apetoh, L., Ghiringhelli, F., Tesniere, A., Martins, I., Kroemer, G., et al. (2010). Tumor cell death and ATP release prime dendritic cells and efficient anticancer immunity. *Cancer Research*, *70*(3), 855–858.

Banz, Y., Beldi, G., Wu, Y., Atkinson, B., Usheva, A., & Robson, S. C. (2008). CD39 is incorporated into plasma microparticles where it maintains functional properties and impacts endothelial activation. *British Journal Haematology*, *142*(4), 627–637.

Baurand, A., Eckly, A., Bari, N., Leon, C., Hechler, B., Cazenave, J. P., et al. (2000). Desensitization of the platelet aggregation response to ADP: Differential down-regulation of the P2Y1 and P2cyc receptors. *Thrombosis and Haemostasis*, *84*(3), 484–491.

Beldi, G., Banz, Y., Kroemer, A., Sun, X., Wu, Y., Graubardt, N., et al. (2010). Deletion of CD39 on natural killer cells attenuates hepatic ischemia/reperfusion injury in mice. *Hepatology*, *51*(5), 1702–1711.

Beldi, G., Enjyoji, K., Wu, Y., Miller, L., Banz, Y., Sun, X., et al. (2008). The role of purinergic signaling in the liver and in transplantation: Effects of extracellular nucleotides on hepatic graft vascular injury, rejection and metabolism. *Frontiers in Bioscience*, *13*, 2588–2603.

Beldi, G., Wu, Y., Banz, Y., Nowak, M., Miller, L., Enjyoji, K., et al. (2008). Natural killer T cell dysfunction in CD39-null mice protects against concanavalin A-induced hepatitis. *Hepatology*, *48*(3), 841–852.

Beldi, G., Wu, Y., Sun, X., Imai, M., Enjyoji, K., Csizmadia, E., et al. (2008). Regulated catalysis of extracellular nucleotides by vascular CD39/ENTPD1 is required for liver regeneration. *Gastroenterology*, *135*(5), 1751–1760.

Boden, E. K., & Snapper, S. B. (2008). Regulatory T cells in inflammatory bowel disease. *Current Opinion in Gastroenterology*, *24*(6), 733–741.

Borsellino, G., Kleinewietfeld, M., Di Mitri, D., Sternjak, A., Diamantini, A., Giometto, R., et al. (2007). Expression of ectonucleotidase CD39 by Foxp3+ Treg cells: Hydrolysis of extracellular ATP and immune suppression. *Blood*, *110*(4), 1225–1232.

Bruzzone, S., Franco, L., Guida, L., Zocchi, E., Contini, P., Bisso, A., et al. (2001). A self-restricted CD38-connexin 43 cross-talk affects NAD+ and cyclic ADP-ribose metabolism and regulates intracellular calcium in 3T3 fibroblasts. *The Journal of Biological Chemistry*, *276*(51), 48300–48308.

Buffon, A., Wink, M. R., Ribeiro, B. V., Casali, E. A., Libermann, T. A., Zerbini, L. F., et al. (2007). NTPDase and 5′ ecto-nucleotidase expression profiles and the pattern of extracellular ATP metabolism in the Walker 256 tumor. *Biochimica et Biophysica Acta*, *1770*(8), 1259–1265.

Burnstock, G. (2002). Purinergic signaling and vascular cell proliferation and death. *Arteriosclerosis, Thrombosis, and Vascular Biology*, *22*(3), 364–373.

Burnstock, G., & Knight, G. (2004). Cellular distribution and functions of P2 receptor subtypes in different systems. *International Review of Cytology*, *240*, 31–304.

Chadwick, B. P., & Frischauf, A. M. (1998). The CD39-like gene family—Identification of three new human members (CD39l2, CD39l3, and CD39l4), their murine homologues, and a member of the gene family from Drosophila melanogaster. *Genomics*, *50*(3), 357–367.

Chen, Y., Corriden, R., Inoue, Y., Yip, L., Hashiguchi, N., Zinkernagel, A., et al. (2006). ATP release guides neutrophil chemotaxis via P2Y2 and A3 receptors. *Science*, *314*(5806), 1792–1795.

Chiorazzi, N., Hatzi, K., & Albesiano, E. (2005). B-cell chronic lymphocytic leukemia, a clonal disease of B lymphocytes with receptors that vary in specificity for (auto)antigens. *Annals of the New York Academy of Sciences*, *1062*, 1–12.

Chiorazzi, N., Rai, K. R., & Ferrarini, M. (2005). Chronic lymphocytic leukemia. *The New England Journal of Medicine*, *352*(8), 804–815.

Cho, J., Furie, B. C., Coughlin, S. R., & Furie, B. (2008). A critical role for extracellular protein disulfide isomerase during thrombus formation in mice. *Journal of Clinical Investigation*, *118*(3), 1123–1131.

Chou, J., Mackman, N., Merrill-Skoloff, G., Pedersen, B., Furie, B. C., & Furie, B. (2004). Hematopoietic cell-derived microparticle tissue factor contributes to fibrin formation during thrombus propagation. *Blood*, *104*, 3190–3197.

Cooke, J. P., & Dzau, V. J. (1997). Nitric oxide synthase—Role in the genesis of vascular disease. *Annual Review of Medicine*, *48*(489), 489–509, Review.

Corriden, R., Chen, Y., Inoue, Y., Beldi, G., Robson, S. C., Insel, P. A., et al. (2008). Ecto-nucleoside triphosphate diphosphohydrolase 1 (E-NTPDase1/CD39) regulates neutrophil chemotaxis by hydrolyzing released ATP to adenosine. *The Journal of Biological Chemistry*, *283*(42), 28480–28486.

Deaglio, S., Aydin, S., Grand, M. M., Vaisitti, T., Bergui, L., D'Arena, G., et al. (2010). CD38/CD31 interactions activate genetic pathways leading to proliferation and migration in chronic lymphocytic leukemia cells. *Molecular Medicine*, *16*(3–4), 87–91.

Deaglio, S., Dianzani, U., Horenstein, A. L., Fernandez, J. E., van Kooten, C., Bragardo, M., et al. (1996). Human CD38 ligand. A 120-KDA protein predominantly expressed on endothelial cells. *Journal of Immunology*, *156*(2), 727–734.

Deaglio, S., Dwyer, K. M., Gao, W., Friedman, D., Usheva, A., Erat, A., et al. (2007). Adenosine generation catalyzed by CD39 and CD73 expressed on regulatory T cells mediates immune suppression. *The Journal of Experimental Medicine*, *204*(6), 1257–1265.

Deaglio, S., & Malavasi, F. (2006). The CD38/CD157 mammalian gene family: An evolutionary paradigm for other leukocyte surface enzymes. *Purinergic Signalling*, *2*(2), 431–441.

Deaglio, S., Mehta, K., & Malavasi, F. (2001). Human CD38: A (r)evolutionary story of enzymes and receptors. *Leukemia Research*, *25*(1), 1–12.

Deaglio, S., Morra, M., Mallone, R., Ausiello, C. M., Prager, E., Garbarino, G., et al. (1998). Human CD38 (ADP-ribosyl cyclase) is a counter-receptor of CD31, an Ig superfamily member. *Journal of Immunology*, *160*(1), 395–402.

Deaglio, S., Vaisitti, T., Aydin, S., Ferrero, E., & Malavasi, F. (2006). In-tandem insight from basic science combined with clinical research: CD38 as both marker and key component of the pathogenetic network underlying chronic lymphocytic leukemia. *Blood*, *108*(4), 1135–1144.

Deaglio, S., Vaisitti, T., Zucchetto, A., Gattei, V., & Malavasi, F. (2010). CD38 as a molecular compass guiding topographical decisions of chronic lymphocytic leukemia cells. *Seminars in Cancer Biology*, *20*(6), 416–423.

Deaglio, S., Zubiaur, M., Gregorini, A., Bottarel, F., Ausiello, C. M., Dianzani, U., et al. (2002). Human CD38 and CD16 are functionally dependent and physically associated in natural killer cells. *Blood*, *99*(7), 2490–2498.

Deguchi, H., Takeya, H., Urano, H., Gabazza, E. C., Zhou, H., & Suzuki, K. (1998). Adenosine regulates tissue factor expression on endothelial cells. *Thrombosis Research*, *91*(2), 57–64.

Di Virgilio, F., Boeynaems, J. M., & Robson, S. C. (2009). Extracellular nucleotides as negative modulators of immunity. *Current Opinion in Pharmacology*, *9*(4), 507–513.

Di Virgilio, F., & Robson, S. C. (2009). Identification of novel immunosuppressive pathways paves the way for drug discovery. *Current Opinion in Pharmacology*, *9*(4), 445–446.

Dubyak, G. R., & Elmoatassim, C. (1993). Signal transduction via P2-purinergic receptors for extracellular ATP and other nucleotides. *The American Journal of Physiology*, *265*(3 Part 1), C577–C606, Review.

Dwyer, K., Deaglio, S., Crikis, S., Gao, W., Enjyoji, K., Strom, T. B., et al. (2007). Salutary roles of CD39 in transplantation. *Transplantation Reviews*, *21*, 54–63.

Dwyer, K. M., Hanidziar, D., Putheti, P., Hill, P. A., Pommey, S., McRae, J. L., et al. (2010). Expression of CD39 by human peripheral blood CD4(+) CD25(+) T cells denotes a regulatory memory phenotype. *American Journal of Transplantation*, *10*(11), 2410–2420.

Eltzschig, H. K., Ibla, J. C., Furuta, G. T., Leonard, M. O., Jacobson, K. A., Enjyoji, K., et al. (2003). Coordinated adenine nucleotide phosphohydrolysis and nucleoside signaling in posthypoxic endothelium: Role of ectonucleotidases and adenosine A2B receptors. *The Journal of Experimental Medicine*, *198*(5), 783–796.

Eltzschig, H. K., Kohler, D., Eckle, T., Kong, T., Robson, S. C., & Colgan, S. P. (2009). Central role of Sp1-regulated CD39 in hypoxia/ischemia protection. *Blood*, *113*(1), 224–232.

Eltzschig, H. K., Thompson, L. F., Karhausen, J., Cotta, R. J., Ibla, J. C., Robson, S. C., et al. (2004). Endogenous adenosine produced during hypoxia attenuates neutrophil accumulation: Coordination by extracellular nucleotide metabolism. *Blood*, *104*(13), 3986–3992.

Enjyoji, K., Sevigny, J., Lin, Y., Frenette, P. S., Christie, P. D., Esch, J. S. A., et al. (1999). Targeted disruption of Cd39/ATP diphosphohydrolase results in disordered hemostasis and thromboregulation. *Nature Medicine*, *5*(9), 1010–1017.

Falati, S., Gross, P., Merrill-Skoloff, G., Furie, B. C., & Furie, B. (2002). Real-time in vivo imaging of platelets, tissue factor and fibrin during arterial thrombus formation in the mouse. *Nature Medicine*, *8*(10), 1175–1181.

Falati, S., Liu, Q., Gross, P., Merrill-Skoloff, G., Chou, J., Vandendries, E., et al. (2003). Accumulation of tissue factor into developing thrombi in vivo is dependent upon microparticle P-selectin glycoprotein ligand 1 and platelet P-selectin. *The Journal of Experimental Medicine*, *197*(11), 1585–1598.

Ferrero, E., & Malavasi, F. (1999). The metamorphosis of a molecule: From soluble enzyme to the leukocyte receptor CD38. *Journal of Leukocyte Biology*, *65*(2), 151–161.

Fijnheer, R., Boomgaard, M. N., van den Eertwegh, A. J., Homburg, C. H., Gouwerok, C. W., Veldman, H. A., et al. (1992). Stored platelets release nucleotides as inhibitors of platelet function. *Thrombosis and Haemostasis*, *68*(5), 595–599.

Franceschi, C., Abbracchio, M. P., Barbieri, D., Ceruti, S., Ferrari, D., Iliou, J. P., et al. (1996). Purines and cell death. *Drug Development Research*, *39*(3–4), 442–449.

Friedman, D. J., Kunzli, B. M., YI, A. R., Sevigny, J., Berberat, P. O., Enjyoji, K., et al. (2009). From the Cover: CD39 deletion exacerbates experimental murine colitis and human polymorphisms increase susceptibility to inflammatory bowel disease. *Proceedings of the National Academy of Sciences of the United States of America*, *106*(39), 16788–16793.

Furie, B. (2009). Pathogenesis of thrombosis. *Hematology/the Education Program of the American Society of Hematology*, 255–258.

Furie, B., & Furie, B. C. (2005). Thrombus formation in vivo. *Journal of Clinical Investigation*, *115*(12), 3355–3362.

Furie, B., & Furie, B. C. (2008). Mechanisms of thrombus formation. *The New England Journal of Medicine*, *359*(9), 938–949.

Goding, J. W., & Howard, M. C. (1998). Ecto-enzymes of lymphoid cells. *Immunological Reviews*, *161*, 5–10.

Goepfert, C., Imai, M., Brouard, S., Csizmadia, E., Kaczmarek, E., & Robson, S. C. (2000). CD39 modulates endothelial cell activation and apoptosis. *Molecular Medicine*, *6*(7), 591–603.

Goepfert, C., Sundberg, C., Sevigny, J., Enjyoji, K., Hoshi, T., Csizmadia, E., et al. (2001). Disordered cellular migration and angiogenesis in cd39-null mice. *Circulation*, *104*(25), 3109–3115.

Gordon, E. L., Pearson, J. D., & Slakey, L. L. (1986). The hydrolysis of extracellular adenine nucleotides by cultured endothelial cells from pig aorta. Feed-forward inhibition of adenosine production at the cell surface. *Journal of Biological Chemistry*, *261*(33), 15496–15507.

Grahnert, A., Grahnert, A., Klein, C., Schilling, E., Wehrhahn, J., & Hauschildt, S. (2010). NAD+: A modulator of immune functions. *Innate Immunity*, ePub.

Grierson, J. P., & Meldolesi, J. (1995). Shear stress-induced [Ca2+]i transients and oscillations in mouse fibroblasts are mediated by endogenously released ATP. *The Journal of Biological Chemistry*, *270*(9), 4451–4456.

Haag, F., Adriouch, S., Brass, A., Jung, C., Moller, S., Scheuplein, F., et al. (2007). Extracellular NAD and ATP: Partners in immune cell modulation. *Purinergic Signalling*, *3*(1–2), 71–81.

Harden, T. K., Lazarowski, E. R., & Boucher, R. C. (1997). Release, metabolism and interconversion of adenine and uridine nucleotides: Implications for G protein-coupled P2 receptor agonist selectivity. *Trends in Pharmacological Sciences*, *18*(2), 43–46.

Hechler, B., Lenain, N., Marchese, P., Vial, C., Heim, V., Freund, M., et al. (2003). A role of the fast ATP-gated P2X1 cation channel in thrombosis of small arteries in vivo. *The Journal of Experimental Medicine*, *198*(4), 661–667.

Hilchey, S. P., Kobie, J. J., Cochran, M. R., Secor-Socha, S., Wang, J. C., Hyrien, O., et al. (2009). Human follicular lymphoma CD39+-infiltrating T cells contribute to adenosine-mediated T cell hyporesponsiveness. *Journal of Immunology*, *183*(10), 6157–6166.

Howard, M., Grimaldi, J. C., Bazan, J. F., Lund, F. E., Santos-Argumedo, L., Parkhouse, R. M., et al. (1993). Formation and hydrolysis of cyclic ADP-ribose catalyzed by lymphocyte antigen CD38. *Science*, *262*(5136), 1056–1059.

Hubert, S., Rissiek, B., Klages, K., Huehn, J., Sparwasser, T., Haag, F., et al. (2010). Extracellular NAD+ shapes the Foxp3+ regulatory T cell compartment through the ART2-P2X7 pathway. *The Journal of Experimental Medicine*, *207*(12), 2561–2568.

Ihrcke, N. S., Wrenshall, L. E., Lindman, B. J., & Platt, J. L. (1993). Role of heparan sulfate in immune system-blood vessel interactions. *Immunology Today*, *14*(500), 500–505.

Imai, M., Goepfert, C., Kaczmarek, E., & Robson, S. C. (2000). CD39 modulates IL-1 release from activated endothelial cells. *Biochemical and Biophysical Research Communications*, *270*(1), 272–278.

Jackson, S. W., Hoshi, T., Wu, Y., Sun, X., Enjyoji, K., Cszimadia, E., et al. (2007). Disordered purinergic signaling inhibits pathological angiogenesis in cd39/Entpd1-null mice. *The American Journal of Pathology*, *171*(4), 1395–1404.

Kaczmarek, E., Koziak, K., Sevigny, J., Siegel, J. B., Anrather, J., Beaudoin, A. R., et al. (1996). Identification and characterization of CD39 vascular ATP diphosphohydrolase. *The Journal of Biological Chemistry*, *271*(51), 33116–33122.

Kaneider, N. C., Egger, P., Dunzendorfer, S., Noris, P., Balduini, C. L., Gritti, D., et al. (2002). Reversal of thrombin-induced deactivation of CD39/ATPDase in endothelial cells by HMG-CoA reductase inhibition: Effects on Rho-GTPase and adenosine nucleotide metabolism. *Arteriosclerosis, Thrombosis, and Vascular Biology*, *22*(6), 894–900.

Koshiba, M., Kojima, H., Huang, S., Apasov, S., & Sitkovsky, M. V. (1997). Memory of extracellular adenosine A2A purinergic receptor-mediated signaling in murine T cells. *The Journal of Biological Chemistry*, *272*(41), 25881–25889.

Koziak, K., Kaczmarek, E., Kittel, A., Sevigny, J., Blusztajn, J. K., Esch, J. S. A., et al. (2000). Palmitoylation targets CD39/endothelial ATP diphosphohydrolase to caveolae. *The Journal of Biological Chemistry*, *275*(3), 2057–2062.

Koziak, E., Sevigny, J., Robson, S. C., Siegel, J. B., & Kaczmarek, K. (1999). Analysis of CD39/ATP diphosphohydrolase expression in endothelial cells, platelets and leukocytes. *Thrombosis and Haemostasis*, *82*, 1538–1544.

Krebs, C., Koestner, W., Nissen, M., Welge, V., Parusel, I., Malavasi, F., et al. (2003). Flow cytometric and immunoblot assays for cell surface ADP-ribosylation using a monoclonal antibody specific for ethenoadenosine. *Analytical Biochemistry*, *314*(1), 108–115.

Kunzli, B. M., Berberat, P. O., Dwyer, K., Deaglio, S., Csizmadia, E., Cowan, P., et al. (2010). Variable impact of CD39 in experimental murine colitis. *Digestive Diseases and Sciences*, ePub.

la Sala, A., Ferrari, D., Corinti, S., Cavani, A., Di Virgilio, F., & Girolomoni, G. (2001). Extracellular ATP induces a distorted maturation of dendritic cells and inhibits their capacity to initiate Th1 responses. *Journal of Immunology*, *166*(3), 1611–1617.

la Sala, A., Ferrari, D., Di Virgilio, F., Idzko, M., Norgauer, J., Girolomoni, G., et al. (2003). Alerting and tuning the immune response by extracellular nucleotides. Extracellular ATP induces a distorted maturation of dendritic cells and inhibits their capacity to initiate Th1 responses. *Journal of Leukocyte Biology*, *73*(3), 339–343.

Latz, E. (2010). The inflammasomes: Mechanisms of activation and function. *Current Opinion in Immunology*, *22*(1), 28–33.

Levesque, S. A., Kukulski, F., Enjyoji, K., Robson, S. C., & Sevigny, J. (2010). NTPDase1 governs P2X7-dependent functions in murine macrophages. *European Journal of Immunology*, *40*(5), 1473–1485.

Li, Y., Csizmadia, E., Sevigny, J., Enjyoji, K., & Robson, S. C. (2003). CD39/nucleoside triphosphate diphosphohydrolase-1 (NTPDase-1) modulates allograft rejection and cellular immune responses. *American Journal of Transplantation*, *3*(Suppl. 5), 545.

Liao, H., Hyman, M. C., Baek, A. E., Fukase, K., & Pinsky, D. J. (2010). cAMP/CREB-mediated transcriptional regulation of ectonucleoside triphosphate diphosphohydrolase 1 (CD39) expression. *The Journal of Biological Chemistry*, *285*(19), 14791–14805.

Liu, Q., Kriksunov, I. A., Graeff, R., Munshi, C., Lee, H. C., & Hao, Q. (2005). Crystal structure of human CD38 extracellular domain. *Structure*, *13*(9), 1331–1339.

Luthje, J. (1989). Origin, metabolism and function of extracellular adenine nucleotides in the blood. *Klinische Wochenschrift*, *67*(6), 317–327 published erratum appears in Klin Wochenschr 1989 May 15; 67(10), 558 [Review].

Malavasi, F., Deaglio, S., Funaro, A., Ferrero, E., Horenstein, A. L., Ortolan, E., et al. (2008). Evolution and function of the ADP ribosyl cyclase/CD38 gene family in physiology and pathology. *Physiological Reviews*, *88*(3), 841–886.

Malavasi, F., Funaro, A., Roggero, S., Horenstein, A., Calosso, L., & Mehta, K. (1994). Human CD38: A glycoprotein in search of a function. *Immunology Today*, *15*(3), 95–97.

Maliszewski, C. R., Delespesse, G. J., Schoenborn, M. A., Armitage, R. J., Fanslow, W. C., Nakajima, T., et al. (1994). The CD39 lymphoid cell activation antigen. Molecular cloning and structural characterization. *Journal of Immunology*, *153*(8), 3574–3583.

Mandapathil, M., Hilldorfer, B., Szczepanski, M. J., Czystowska, M., Szajnik, M., Ren, J., et al. (2010). Generation and accumulation of immunosuppressive adenosine by human CD4 +CD25highFOXP3+ regulatory T cells. *The Journal of Biological Chemistry*, *285*(10), 7176–7186.

Marcus, A. J., Broekman, M. J., Drosopoulos, J. H., Islam, N., Pinsky, D. J., Sesti, C., et al. (2003). Metabolic control of excessive extracellular nucleotide accumulation by CD39/ ecto-nucleotidase-1: Implications for ischemic vascular diseases. *The Journal of Pharmacology and Experimental Therapeutics*, *305*(1), 9–16.

Marcus, A. J., Safier, L. B., Hajjar, K. A., Ullman, H. L., Islam, N., Broekman, M. J., et al. (1991). Inhibition of platelet function by an aspirin-insensitive endothelial cell ADPase. Thromboregulation by endothelial cells. *Journal of Clinical Investigation*, *88*(5), 1690–1696.

Mariathasan, S., Weiss, D. S., Newton, K., McBride, J., O'Rourke, K., Roose-Girma, M., et al. (2006). Cryopyrin activates the inflammasome in response to toxins and ATP. *Nature*, *440* (7081), 228–232.

Meghji, P., Pearson, J. D., & Slakey, L. L. (1995). Kinetics of extracellular ATP hydrolysis by microvascular endothelial cells from rat heart. *The Biochemical Journal*, *308*(Pt 3), 725–731.

Mizumoto, N., Kumamoto, T., Robson, S. C., Sevigny, J., Matsue, H., Enjyoji, K., et al. (2002). CD39 is the dominant Langerhans cell associated ecto-NTPDase: Modulatory roles in inflammation and immune responsiveness. *Nature Medicine*, *8*(4), 358–365.

Morabito, F., Damle, R. N., Deaglio, S., Keating, M., Ferrarini, M., & Chiorazzi, N. (2006). The CD38 ectoenzyme family: Advances in basic science and clinical practice. *Molecular Medicine*, *12*(11–12), 342–344.

Moreschi, I., Bruzzone, S., Bodrato, N., Usai, C., Guida, L., Nicholas, R. A., et al. (2008). NAADP+ is an agonist of the human P2Y11 purinergic receptor. *Cell Calcium*, *43*(4), 344–355.

Morote-Garcia, J. C., Rosenberger, P., Nivillac, N. M., Coe, I. R., & Eltzschig, H. K. (2009). Hypoxia-inducible factor-dependent repression of equilibrative nucleoside transporter 2 attenuates mucosal inflammation during intestinal hypoxia. *Gastroenterology*, *136*(2), 607–618.

Mulero, J. J., Yeung, G., Nelken, S. T., & Ford, J. E. (1999). CD39-L4 is a secreted human apyrase, specific for the hydrolysis of nucleoside diphosphates. *The Journal of Biological Chemistry*, *274*(29), 20064–20067.

Ogura, Y., Sutterwala, F. S., & Flavell, R. A. (2006). The inflammasome: First line of the immune response to cell stress. *Cell*, *126*(4), 659–662.

Pacheco, R., Martinez-Navio, J. M., Lejeune, M., Climent, N., Oliva, H., Gatell, J. M., et al. (2005). CD26, adenosine deaminase, and adenosine receptors mediate costimulatory signals in the immunological synapse. *Proceedings of the National Academy of Sciences of the United States of America*, *102*(27), 9583–9588.

Paust, S., & Cantor, H. (2005). Regulatory T cells and autoimmune disease. *Immunological Reviews*, *204*, 195–207.

Plesner, L. (1995). Ecto-ATPases: Identities and functions. *International Review of Cytology*, *158*, 141–214.

Podolsky, D. K. (2002). Inflammatory bowel disease. *The New England Journal of Medicine*, *347*(6), 417–429.

Reinhardt, C., von Bruhl, M. L., Manukyan, D., Grahl, L., Lorenz, M., Altmann, B., et al. (2008). Protein disulfide isomerase acts as an injury response signal that enhances fibrin generation via tissue factor activation. *Journal of Clinical Investigation*, *118*(3), 1110–1122.

Resta, R., Yamashita, Y., & Thompson, L. F. (1998). Ecto-enzyme and signaling functions of lymphocyte CD73. *Immunological Reviews*, *161*, 95–109.

Reutershan, J., Vollmer, I., Stark, S., Wagner, R., Ngamsri, K. C., & Eltzschig, H. K. (2009). Adenosine and inflammation: CD39 and CD73 are critical mediators in LPS-induced PMN trafficking into the lungs. *The FASEB Journal*, *23*(2), 473–482.

Robson, S. C., Daoud, S., Begin, M., Cote, Y. P., Siegel, J. B., Bach, F. H., et al. (1997). Modulation of vascular ATP diphosphohydrolase by fatty acids. *Blood Coagulation & Fibrinolysis*, *8*(1), 21–27.

Robson, S. C., Enjyoji, K., Goepfert, C., Imai, M., Kaczmarek, E., Lin, Y., et al. (2001). Modulation of extracellular nucleotide-mediated signaling by CD39/nucleoside triphosphate diphosphohydrolase-1. *Drug Development Research*, *53*(2–3), 193–207.

Robson, S. C., Kaczmarek, E., Siegel, J. B., Candinas, D., Koziak, K., Millan, M., et al. (1997). Loss of ATP diphosphohydrolase activity with endothelial cell activation. *The Journal of Experimental Medicine*, *185*(1), 153–163.

Robson, S. C., Sevigny, J., Imai, M., & Enjyoji, K. (2000). Thromboregulatory potential of endothelial CD39/nucleoside triphosphate diphosphohydrolase: Modulation of purinergic signalling in platelets. *Emerging Therapeutic Targets*, *4*, 155–171.

Robson, S. C., Sevigny, J., & Zimmermann, H. (2006). The E-NTPDase family of ecto-nucleotidases: Structure function relationships and pathophysiological significance. *Purinergic Signalling*, *2*, 409–430.

Robson, S. C., Wu, Y., Sun, X., Knosalla, C., Dwyer, K., & Enjyoji, K. (2005). Ectonucleotidases of CD39 family modulate vascular inflammation and thrombosis in transplantation. *Seminars in Thrombosis and Hemostasis*, *31*(2), 217–233.

Ross, R. (1995). Cell biology of atherosclerosis. *Annual Review of Physiology*, *57*(791), 791–804.

Saborit-Villarroya, I., Vaisitti, T., Rossi, D., D'Arena, G., Gaidano, G., Malavasi, F., et al. (2011). E2A is a transcriptional regulator of CD38 expression in chronic lymphocytic leukemia. *Leukemia*, *25*, 479–488.

Salmi, M., & Jalkanen, S. (2005). Cell-surface enzymes in control of leukocyte trafficking. *Nature Reviews. Immunology*, *5*(10), 760–771.

Sansom, F. M., Riedmaier, P., Newton, H. J., Dunstone, M. A., Muller, C. E., Stephan, H., et al. (2008). Enzymatic properties of an ecto-nucleoside triphosphate diphosphohydrolase from Legionella pneumophila: Substrate specificity and requirement for virulence. *The Journal of Biological Chemistry*, *283*(19), 12909–12918.

Sansom, F. M., Robson, S. C., & Hartland, E. L. (2008). Possible effects of microbial ecto-nucleoside triphosphate diphosphohydrolases on host-pathogen interactions. *Microbiology and Molecular Biology Reviews*, *72*(4), 765–781.

Sasamura, H., Dzau, V. J., & Pratt, R. E. (1994). Desensitization of angiotensin receptor function. *Kidney International*, *46*(6), 1499–1501.

Schicker, K., Hussl, S., Chandaka, G. K., Kosenburger, K., Yang, J. W., Waldhoer, M., et al. (2009). A membrane network of receptors and enzymes for adenine nucleotides and nucleosides. *Biochimica et Biophysica Acta*, *1793*(2), 325–334.

Sevigny, J., Sundberg, C., Braun, N., Guckelberger, O., Csizmadia, E., Qawi, I., et al. (2002). Differential catalytic properties and vascular topography of murine nucleoside triphosphate diphosphohydrolase 1 (NTPDase1) and NTPDase2 have implications for thromboregulation. *Blood*, *99*(8), 2801–2809.

Sinzinger, H., Virgolini, I., Gazso, A., & Ogrady, J. (1991). Review—Eicosanoids in atherosclerosis. *Experimental Pathology*, *43*(2), 2–19.

Sitkovsky, M., & Lukashev, D. (2005). Regulation of immune cells by local-tissue oxygen tension: HIF1 alpha and adenosine receptors. *Nature Reviews. Immunology*, *5*(9), 712–721.

Sitkovsky, M. V., Lukashev, D., Apasov, S., Kojima, H., Koshiba, M., Caldwell, C., et al. (2004). Physiological control of immune response and inflammatory tissue damage by hypoxia-inducible factors and adenosine A2A receptors. *Annual Review of Immunology*, *22*, 657–682.

Sitkovsky, M., Lukashev, D., Deaglio, S., Dwyer, K., Robson, S. C., & Ohta, A. (2008). Adenosine A2A receptor antagonists: Blockade of adenosinergic effects and T regulatory cells. *British Journal of Pharmacology*, *153*(Suppl. 1), S457–S464.

Stagg, J., Divisekera, U., McLaughlin, N., Sharkey, J., Pommey, S., Denoyer, D., et al. (2010). Anti-CD73 antibody therapy inhibits breast tumor growth and metastasis. *Proceedings of the National Academy of Sciences of the United States of America*, *107*(4), 1547–1552.

Stagg, J., & Smyth, M. J. (2010). Extracellular adenosine triphosphate and adenosine in cancer. *Oncogene*, *29*(39), 5346–5358.

States, D. J., Walseth, T. F., & Lee, H. C. (1992). Similarities in amino acid sequences of Aplysia ADP-ribosyl cyclase and human lymphocyte antigen CD38. *Trends in Biochemical Sciences*, *17*(12), 495.

Sun, X., Wu, Y., Gao, W., Enjyoji, K., Csizmadia, E., Muller, C. E., et al. (2010). CD39/ENTPD1 expression by CD4+Foxp3+ regulatory T cells promotes hepatic metastatic tumor growth in mice. *Gastroenterology*, *139*(3), 1030–1040.

Synnestvedt, K., Furuta, G. T., Comerford, K. M., Louis, N., Karhausen, J., Eltzschig, H. K., et al. (2002). Ecto-5′-nucleotidase (CD73) regulation by hypoxia-inducible factor-1 mediates permeability changes in intestinal epithelia. *The Journal of Clinical Investigation*, *110*(7), 993–1002.

Thammavongsa, V., Kern, J. W., Missiakas, D. M., & Schneewind, O. (2009). Staphylococcus aureus synthesizes adenosine to escape host immune responses. *The Journal of Experimental Medicine*, *206*(11), 2417–2427.

Traut, T. W. (1994). Physiological concentrations of purines and pyrimidines. *Molecular and Cellular Biochemistry*, *140*(1), 1–22, Review.

Tsujimoto, Y. (1997). Apoptosis and necrosis—Intracellular ATP level as a determinant for cell death modes. *Cell Death and Differentiation*, *4*(6), 429–434, Review.

Vaisitti, T., Aydin, S., Rossi, D., Cottino, F., Bergui, L., D'Arena, G., et al. (2010). CD38 increases CXCL12-mediated signals and homing of chronic lymphocytic leukemia cells. *Leukemia*, *24*(5), 958–969.

von Kugelgen, I., & Wetter, A. (2000). Molecular pharmacology of P2Y-receptors. *Naunyn-Schmiedeberg's Archives of Pharmacology*, *362*(4–5), 310–323, Review.

Wang, L., Andersson, M., Karlsson, L., Watson, M. A., Cousens, D. J., Jern, S., et al. (2003). Increased mitogenic and decreased contractile P2 receptors in smooth muscle cells by shear stress in human vessels with intact endothelium. *Arteriosclerosis, Thrombosis, and Vascular Biology*, *23*(8), 1370–1376.

Warny, M., Aboudola, S., Robson, S. C., Sevigny, J., Communi, D., Soltoff, S. P., et al. (2001). P2Y(6) nucleotide receptor mediates monocyte interleukin-8 production in response to UDP or lipopolysaccharide. *The Journal of Biological Chemistry*, *276*(28), 26051–26056.

Weisman, G. A., Erb, L., Garrad, R. C., Theiss, P. M., Santiago-Perez, L. I., Flores, R. V., et al. (1998). P2Y nucleotide receptors in the immune system: Signaling by a P2Y(2) receptor in U937 monocytes. *Drug Development Research*, *45*(3–4), 222–228.

Wihlborg, A. K., Wang, L., Braun, O. O., Eyjolfsson, A., Gustafsson, R., Gudbjartsson, T., et al. (2004). ADP receptor P2Y12 is expressed in vascular smooth muscle cells and stimulates contraction in human blood vessels. *Arteriosclerosis, Thrombosis, and Vascular Biology*, *24*(10), 1810–1815.

Yegutkin, G. G., Henttinen, T., & Jalkanen, S. (2001). Extracellular ATP formation on vascular endothelial cells is mediated by ecto-nucleotide kinase activities via phosphotransfer reactions. *The FASEB Journal*, *15*(1), 251–260.

Yegutkin, G. G., Hytonen, J., Samburski, S. S., Yrjanainen, H., Jalkanen, S., & Viljanen, M. K. (2010). Disordered lymphoid purine metabolism contributes to the pathogenesis of persistent Borrelia garinii infection in mice. *Journal of Immunology*, *184*(9), 5112–5120.

Yegutkin, G. G., Samburski, S. S., & Jalkanen, S. (2003). Soluble purine-converting enzymes circulate in human blood and regulate extracellular ATP level via counteracting pyrophosphatase and phosphotransfer reactions. *The FASEB Journal*, *17*(10), 1328–1330.

Yegutkin, G. G., Samburski, S. S., Jalkanen, S., & Novak, I. (2006). ATP-consuming and ATP-generating enzymes secreted by pancreas. *The Journal of Biological Chemistry*, *281*(40), 29441–29447.

Yoshida, H., & Granger, D. N. (2009). Inflammatory bowel disease: A paradigm for the link between coagulation and inflammation. *Inflammatory Bowel Diseases*, *15*(8), 1245–1255.

Zhang, B. (2010). CD73: A novel target for cancer immunotherapy. *Cancer Research*, *70*(16), 6407–6411.

Zimmermann, H. (1992). 5′-nucleotidase: Molecular structure and functional aspects. *The Biochemical Journal*, *285*, 345–365.

Zimmermann, H. (1999). Two novel families of ectonucleotidases: Molecular structure, catalytic properties and a search for function. *Trends in Pharmacological Sciences*, *20*, 231–236.

G. Burnstock* and C. Kennedy†

*Autonomic Neuroscience Centre, University College Medical School, London, United Kingdom

†Strathclyde Institute of Pharmacy and Biomedical Sciences, University of Strathclyde, Glasgow, United Kingdom

P2X Receptors in Health and Disease

Abstract

Seven P2X receptor subunits have been cloned which form functional homo- and heterotrimers. These are cation-selective channels, equally permeable to Na^+ and K^+ and with significant Ca^{2+} permeability. The three-dimensional structure of the P2X receptor is described. The channel pore is formed by the α-helical transmembrane spanning region 2 of each subunit. When ATP binds to a P2X receptor, the pore opens within milliseconds, allowing the cations to flow. P2X receptors are expressed on both central and peripheral neurons, where they are involved in neuromuscular and synaptic neurotransmission and neuromodulation. They are also expressed in most types of nonneuronal cells and mediate a wide range of actions, such as contraction of smooth muscle, secretion, and immunomodulation. Changes in the expression of P2X receptors have been characterized in many pathological conditions of the cardiovascular, gastrointestinal, respiratory, and urinogenital systems and in the brain and special senses. The therapeutic potential of P2X receptor agonists and antagonists is currently being investigated in a range of disorders, including chronic neuropathic and inflammatory pain, depression, cystic fibrosis, dry eye, irritable bowel syndrome, interstitial cystitis, dysfunctional urinary bladder, and cancer.

1054-3589/11 $35.00
10.1016/B978-0-12-385526-8.00011-4

I. Introduction

P1 (adenosine) and P2 (ATP/ADP) receptor families were proposed in 1978 (Burnstock, 1978). In 1985, on the basis of pharmacology, P2 receptors were divided into two subtypes, P2X and P2Y receptors (Burnstock & Kennedy, 1985). It was not until the early 1990s, however, that P2 receptors for purines and pyrimidines were cloned and characterized and second messenger mechanisms examined (see Ralevic & Burnstock, 1998). $P2Y_1$ (Webb et al., 1993) and $P2Y_2$ (Lustig et al., 1993) G protein-coupled receptors were described first, and a year later, P2X1 and P2X2 ion channel receptors were described (Brake et al., 1994; Valera et al., 1994). Seven P2X receptor subunits have been identified, which interact to form functional homo- and heteromultimers (see Burnstock, 2007a; Ralevic & Burnstock, 1998). P2X receptors have been cloned from many eukaryotic species, including mammals, fish, parasitic trematode worms, slime mold, and green algae, but curiously, there is no evidence of genes encoding P2X receptors in prokaryotes, yeast, *Drosophila melanogaster* or *Caeorhabditis elegans* genomes (see Fountain & Burnstock, 2009). In this chapter, we discuss the current understanding of the molecular physiology, distribution, and physiological and pathophysiological roles of P2X receptors.

II. Molecular Physiology of P2X Receptors

A. Subunit Sequence and Structure

Seven P2X subunits have been cloned, and of these, the P2X1–P2X6 receptors are 379–472 amino acids long, whereas the P2X7 receptor is 595 amino acids long, due to a notably longer COOH terminus. The COOH termini display substantial sequence divergence, but the rest of the sequence has 40–55% pairwise identity, with the P2X4 subunit being the most closely related to the others and the P2X7 subunit being the most dissimilar (see North, 2002). Following the initial cloning of the subunits, hydropathy analysis of their amino acid sequences indicated that each subunit possesses two hydrophobic, transmembrane spanning regions (TMR) that are long enough to span the cell plasma membrane. Further, it predicted that the NH_2 and COOH termini are intracellular and that the bulk of the protein, about 280–300 amino acids, forms an extracellular loop. This structural model has recently been confirmed in recent reports of the three-dimensional crystal structure of a P2X receptor (see Section II.E).

Prior to the determination of the crystal structure, numerous experimental approaches were used to study the structure of P2X receptors. The NH_2 termini of all subunits are quite short (20–30 amino acids), and interestingly, they all contain a consensus site for phosphorylation by protein kinase C

(PKC; Thr-X-Lys/Arg). The role of this site has been studied in some detail in the P2X2 subunit, and a mutation that prevented phosphorylation by PKC reduced the peak current amplitude and sped up desensitization (Ennion & Evans, 2002; Vial et al., 2006). Thus this site may play an important role in controlling the kinetics of currents carried by P2X receptors. All seven subunits also contain 10 conserved cysteine residues in the extracellular loop, which are likely to interact with each other to form disulfide bonds. On the basis of systematic cysteine to alanine mutation analysis, it was proposed that disulfide bonds form in the rat P2X2 receptor between Cys113–Cys164, Cys214–Cys224, Cys258–Cys267, and probably Cys124–Cys130, and Cys147–Cys158 (Clyne et al., 2002). For the human P2X1 receptor, the pairings are Cys117–Cys165, Cys126–Cys149, Cys132–Cys159, Cys217–Cys227, and Cys261–Cys270, though some residues were able to interact promiscuously (Ennion & Evans, 2002). Disrupting these bonds showed that none of the pairings were essential for channel function, but trafficking of the P2X1 subunit was inhibited by disruption of Cys261–Cys270 or Cys117–Cys165 plus another. The extracellular loop of all subunits also contains a variable number of consensus sequences for N-linked glycosylation (P2X1 have five conserved sequences, P2X2— three, P2X3— four, P2X4— six, P2X5— two, P2X6— three, and P2X7— three), and successful trafficking of subunits to the plasma membrane requires some degree of glycosylation (see North, 2002). As noted earlier, the COOH terminal has the most differences in amino acid sequence between subunits, and it also differs greatly in length, from 28 amino acids in the P2X6 subunit to 242 amino acids in the P2X7 subunit. Modulating the length or composition of the COOH terminus modulates the kinetics of currents carried by P2X receptors, as well as their desensitization and permeation properties (Khakh, 2001).

B. Multimerization

As P2X subunits have only two TMR, a single subunit on its own cannot form a functional receptor. A range of techniques, including polyacrylamide gel electrophoresis (Aschrafi et al., 2004; Duckwitz et al., 2006; Nicke, 2008; Nicke et al., 1998), atomic force microscopy (Barrera et al., 2005, 2008), single-particle electron microscopy (Mio et al., 2005), fluorescence resonance energy transfer microscopy (Young et al., 2008), and X-ray crystallography (Gonzales et al., 2009; Kawate et al., 2009), indicate that the functional form is in fact a trimer.

When expressed on their own in expression systems, each of the seven recombinant P2X subunits forms functional receptors, except for the P2X6 subunit, which expresses poorly on its own (Aschrafi et al., 2004; Barrera et al., 2005; Torres et al., 1999). P2X6 homomer expression is greatly improved by N-linked glycosylation, which increases the amount of subunits that are transported from the endoplasmic reticulum to the plasma membrane (Jones et al., 2004). The subunits can also interact with each other, and to date, seven pairs of recombinant subunits have been reported to form functional heteromultimers with pharmacological and biophysical properties that differ from the individual homomultimers: P2X2/3 (Lewis et al., 1995), P2X4/6 (Lê et al., 1998), P2X1/5 (Haines et al., 1999; Lê et al., 1999; Surprenant et al., 2000; Torres et al., 1998), P2X2/6 (King et al., 2000), P2X1/4 (Nicke et al., 2005), and P2X1/2 (Brown et al., 2002; Torres et al., 1999). It is worth noting that on the basis of coimmunoprecipitation studies, the P2X7 subunit was initially thought to be unable to form heteromultimers (North, 2002; Torres et al., 1999), but it has been reported to form a functional complex with P2X4 subunits in heterologous expression systems and mouse macrophages (Guo et al., 2007). A subsequent study could not, however, find such an interaction in *Xenopus laevis* oocytes and rat tissues (Nicke, 2008). The stoichiometry of the subunits within the heteromultimers is not known, except for the P2X1/2 receptor, which comprises one P2X1 and two P2X2 subunits (Aschrafi et al., 2004) and the P2X2/3 receptor, where there is 1:2 ratio of P2X2:P2X3 subunits (Jiang et al., 2003; Wilkinson et al., 2006). Some of the P2X heteromultimers also appear to be expressed in tissues and the best characterized is the native P2X receptor seen in sensory neurones, particularly cells of the nodose ganglia, which has properties that are very similar to those of the recombinant P2X2/3 combination. In addition, some native receptors in the central nervous system (CNS) resemble the P2X4/6 heteromultimer (Lê et al., 1998), while P2X1/5 receptors appear to be expressed in mouse cortical astrocytes (Lalo et al., 2008).

C. Pore Structure

All P2X receptors are cation-selective channels, equally permeable to Na^+ and K^+ and with significant Ca^{2+} permeability (Egan & Khakh, 2004). Substituted cysteine accessibility mutagenesis studies indicated that the α-helical TMR2 lines the pore. Further, the narrowest part of the pore, which likely forms the channel gate, is around halfway through TMR2 (see Khakh, 2001). When ATP binds to a P2X receptor, the gate opens within milliseconds, allowing the cations to flow. Early studies on the P2X7 receptor found that after a short delay, this was followed over hundreds of milliseconds by the development of a second permeability state that allowed larger cations, such as NMDG, dyes such as Lucifer yellow and YO-PRO-1, and ethidium bromide to pass through (Surprenant et al., 1996). Under these conditions, the pore diameter was calculated to increase from about 8 to 40 Å (Zemkova et al., 2008). Subsequently, P2X2 and P2X4 receptors were also shown to undergo a similar change in permeability (Khakh et al., 1999). The transmembrane channel pannexin-1 has been proposed to mediate the influx of large molecules seen after P2X7 receptor stimulation (Pelegrin & Surprenant, 2006), but it plays no role in the response of P2X2 receptors

(Chaumont & Khakh, 2008). Instead, P2X2 pore dilation appears to be intrinsic to the receptor and to be due to a conformational change in its cytosolic domain.

Prolonged exposure of P2X receptors to an agonist leads to desensitization and closure of the receptor pore. On the basis of the time course of desensitization, the P2X receptors are classified as rapidly (P2X1 and P2X3) and slowly (P2X2, P2X4, P2X5, and P2X7) desensitizing receptors (Jarvis & Khakh, 2009; Koshimizu et al., 1999).

D. ATP Binding Site

Analysis of concentration–response curves indicates that three molecules of ATP must bind to the P2X receptor in order for the pore to open (Bean, 1990; Ding & Sachs, 1999; Evans, 2009; Jiang et al., 2003; Vial et al., 2004). P2X receptors, however, lack the common consensus sequences for ATP binding, such as the Walker motif, that are found in other ATP-sensitive proteins, indicating that a novel ATP binding site is present in P2X receptors. This indeed has been shown to be the case through a series of systematic mutagenesis experiments, which identified many of the amino acids involved in ATP binding to the P2X1 subunit (see Evans, 2009). Accordingly, the positively charged lysines present at positions 68, 70, and 309 are proposed to interact with the negatively charged phosphate group of ATP, while its adenine ring is thought to be sandwiched between two regions of aromatic residues at positions 185–186 and 290–292. These amino acids are close to the cytoplasmic ends of TMR1 and TMR2, and recent evidence suggests that the ATP binding site is formed between two adjacent subunits, rather than within a single subunit. Thus, a single ATP molecule binds to two subunits. Analysis of ATP binding in other subtypes further suggests that they contain the same core ATP binding site and that differences in the pharmacological properties of P2X subtypes are due to other sequence differences. According to this structural model, the proposed ATP binding site is not adjacent to the receptor pore, implying that a linker region joins the two and couples agonist binding to pore opening. Mutation analysis indicates the amino acids immediately extracellular to TMR2, which are highly conserved in all P2X subunits, play a crucial, linking role (Evans, 2009; Young, 2010)

E. Three-Dimensional Structure of P2X Receptors

As noted earlier, recent X-ray crystallography studies have finally revealed the three-dimensional structure of a P2X receptor (Gonzales et al., 2009; Kawate et al., 2009) and they support and confirm many of the conclusions discussed based on indirect measurements. These authors created a truncated mutant of the zebrafish p2X4.1 receptor that lacked the most distal 26 and 8 amino acids from the NH_2 and COOH termini, respectively, which could be crystallized and diffracted to a resolution of 3.5 Å. By introducing three further point mutations to decrease nonnative disulfide bond formation and N-linked glycosylation, the resolution was increased to 3.1 Å. Despite these changes in amino acid sequence, these constructs could be expressed in mammalian cells and subsequently activated by ATP, which was more potent than for the wild-type receptors. The peak current amplitudes were, however, much smaller. This retention of functional activity indicates that the structural differences between the mutant and wild-type receptors are relatively minor.

These experiments confirmed that P2X receptors are composed of three subunits, each of which adopt a similar conformation that could be likened to the shape of a leaping dolphin (Fig. 1A). The tail represents the TMR, the upper body, the bulk of the extracellular loop, and the head, the most distal part of the extracellular loop. As a trimer, the subunits wrap around each other to produce a structure that resembles a chalice (Fig. 1B). The TMR are α-helices that extend about 28 Å across the plasma membrane and, as predicted by the hydropathy plots, the bulk of the receptor is extracellular and protrudes from the membrane by about 70 Å. The extracellular domains contain numerous β-strands, which give the structure rigidity and which are the main site of subunit–subunit interactions. Further rigidity is provided by

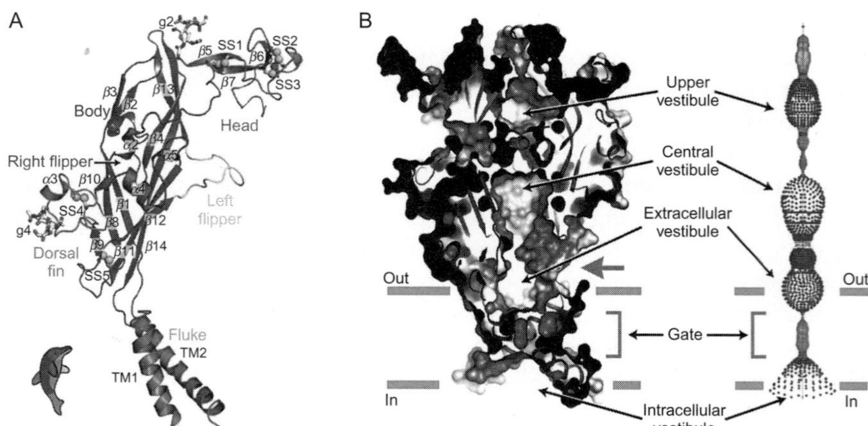

FIGURE 1 *Subunit fold and closed, resting conformation.* (A) The ΔzfP2X4 subunit has a dolphin-like shape. α-helices (TM1–2 and α2–5), β-strands (β1–14), disulfide bonds (SS1–5), and attached glycans (g2 and 4) are indicated. (B) Left-hand panel: a sagittal section reveals a closed conformation of the pore and shows that the gate is located about halfway across the membrane bilayer. Three vestibules (upper, central, and extracellular vestibules) are located on the molecular threefold axis, with the extracellular vestibule connected to the bulk solution through a fenestration (orange arrow). Right-hand panel: pore lining surface calculated by the Hole49 program. Each color represents a different radius range measured from the receptor center (red, <1.15 Å; green, 1.15–2.3 Å; and purple, >2.3 Å). (Reproduced from Kawate et al., 2009, with permission from the Nature Publishing Group.) (For interpretation of the references to color in this figure legend, the reader is referred to the Web version of this chapter.)

the 10 conserved cysteine residues, which were confirmed to interact as described earlier (Fig. 1A). There are fewer subunit interactions in and around the TMR, which is predicted to enable the TMR to move relative to each other when ATP binds and so open the pore.

The TMR cross the membrane at an angle of nearly 45° relative to the plane of the membrane, and within each subunit, they are antiparallel. TMR2 was confirmed as forming the pore and cross each other about half way across the membrane, constricting the pore and giving it the appearance of an hourglass (Fig. 1B). These studies were performed in the absence of ATP and so likely represent the closed state of the channel. Under these conditions, the channel is blocked by a series of mainly hydrophobic residues over two turns of the TMR2 α-helix and about 8 Å in length. Ala344 is at the center of this gate region and is the point at which the TMR2 are closest.

Two pathways are apparent along which extracellular cations may diffuse to reach the transmembrane pore (Fig. 1B). Immediately above the TMR, there is an extracellular vestibule from which three fenestrations, of up to 8 Å in diameter, form a short pathway. The second route is much longer and runs the length of the extracellular domain. This pathway passes through two further vestibules that are lined with acidic (negatively charged) residues, which may act to attract the positively charged Na and Ca ions and so facilitate ion movement. An inner vestibule shaped like an inverted cone, present on the cytoplasmic side of the channel gate, is the likely site through which ions enter and exit the pore.

The structure of the P2X receptor was obtained in the absence of ATP, so its binding site is not directly visible. The eight amino acid residues implicated in ATP binding, as discussed earlier, are, however, located in a deep groove present on the outside of the trimer, which spans the subunits and is 45 Å from the pore (Fig. 2A–C). Three such sites are present in the trimer, consistent with the earlier conclusion that three molecules of ATP must bind to the P2X receptor in order to open the pore. Four of the amino acids proposed to bind ATP are present on one subunit and the other four are found on the neighboring subunit. While six of the residues are orientated toward the grove, two phenylalanines face away from it, suggesting that they may play a role in coupling agonist binding to pore opening. The region immediately extracellular to TMR2 is a β-strand, consistent with a role in coupling the binding of ATP to pore opening (Fig. 2D).

III. Distribution and Physiological Roles of P2X Receptors

Readers are referred to a comprehensive review of the distribution and functions of P2X receptors in all systems of the body (Burnstock & Knight, 2004, and see Table 1, which incorporates further information since 2004; see Burnstock, 2007b; Verkhratsky et al., 2009).

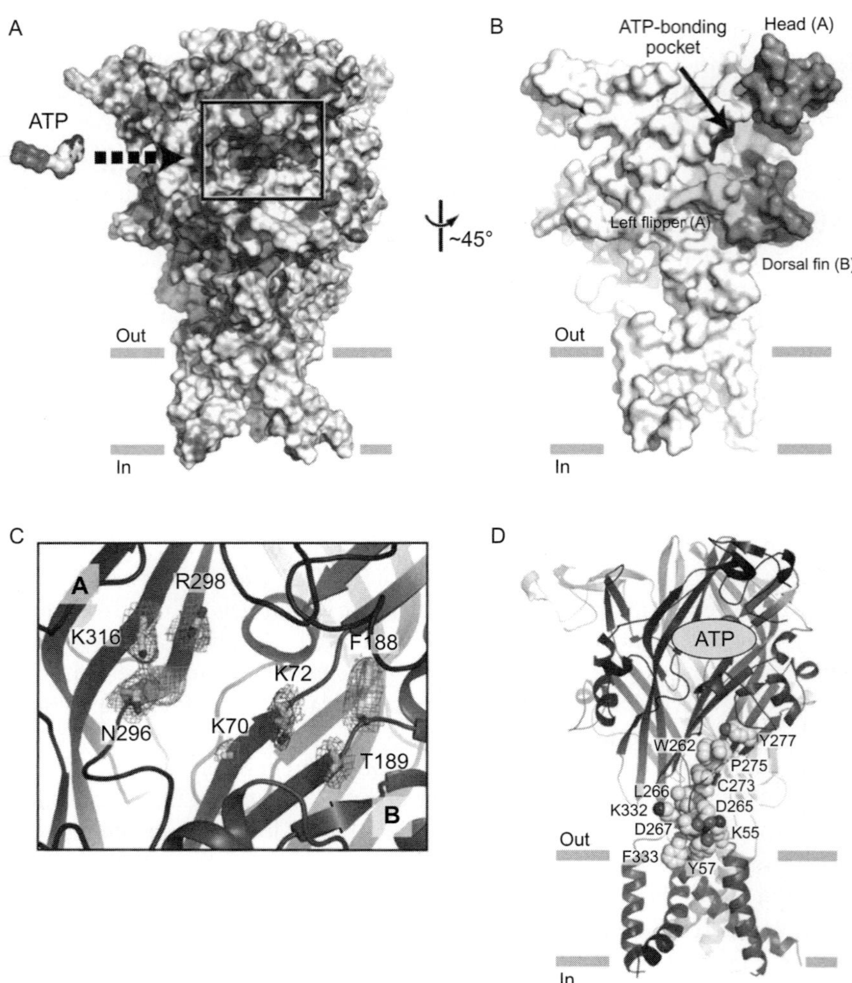

FIGURE 2 *ATP binding site.* (A) A plausible ATP binding pocket located between two neighboring subunits is highlighted in the black rectangle on an electrostatic potential surface representation of the trimeric ΔzfP2X4-B receptor. The surface is colored based on the electrostatic potential contoured from −30 kT (red) to +30 kT (blue). White denotes 0 kT. An ATP molecule, scaled appropriately, is also shown. (B) A surface representation viewed approximately 45° from (a). The head, dorsal fin, and left flipper domains forming the "jaw"-shaped ATP binding pocket are colored as in Fig. 1. The putative ATP binding residues are in blue for subunit A (N296, R298, K316) and in red for subunit B (K70, K72, T189). (C), Close-up view of the highlighted region in an illustrating subunit A (blue) and B (red). Conserved residues implicated in ATP- binding 42–45 are labeled, and side chains are in stick representation. Contors from a $2F_o - F_c$ electron density map drawn around the side chains are in green. The electron density for the side chain of K70 is weak and it has been built as an alanine. (D) Conserved residues, shown in space-filling representation, are located between the ATP binding site and the transmembrane domain—extracellular domain interface. Only residues for a single subunit are shown. (Reproduced from Kawate et al., 2009, with permission from the Nature Publishing Group.) (For interpretation of the references to color in this figure legend, the reader is referred to the Web version of this chapter.)

TABLE 1 Principal P2X Receptors Expressed by Excitable Tissues and Nonneuronal Cells

Neuronal	
Sympathetic neurons	P2X1–7
Parasympathetic neurons	P2X2, P2X3, P2X4, P2X5
Sensory neurons	P2X1–7, predominantly P2X3, and P2X2/3
Enteric neurons	P2X2, P2X3, P2X4, P2X7
Central nervous system	P2X2, P2X4, and P2X6 (including heteromultimers) and P2X7
Retinal neurons	P2X2, P2X3, P2X4, P2X5, P2X7
Muscle cells	
Smooth muscle	P2X1–7, predominantly P2X1
Skeletal muscle	
Developing	P2X2, P2X5, P2X6
Adult	P2X1–7
Cardiac muscle	P2X1, P2X3, P2X4, P2X5, P2X6
Nonneuronal	
Osteoblasts	P2X1, P2X2, P2X5, P2X7
Osteoclasts	P2X1, P2X2, P2X4, P2X7
Cartilage	P2X2
Keratinocytes	P2X2, P2X3, P2X5, P2X7
Fibroblasts	P2X7
Adipocytes	P2X1
Epithelial cells (lung, kidney, trachea, uterus, cornea)	P2X4, P2X5, P2X6, P2X7
Astrocytes	P2X1–7
Oligodendrocytes	P2X1
Microglia	P2X4, P2X7
Müller cells	P2X3, P2X4, P2X5, P2X7
Enteric glial cells	P2X7
Sperm	P2X2, P2X7
Endothelial cells	P2X1, P2X2, P2X3, predominately P2X4
Erythrocytes	P2X2, P2X4, P2X7
Platelets	P2X1
Immune cells (thymocytes, macrophages, neutrophils, eosinophils, lymphocytes, mast cells, dendritic cells)	P2X4 and predominately P2X7, but some P2X1, P2X2, P2X5
Exocrine secretary cells	P2X1, P2X4, P2X7
Endocrine secretory cells (pituitary, pancreas, adrenal, thyroid, testis)	P2X1–7, predominately P2X2/6
Cholangiocytes	P2X2, P2X3, P2X4, P2X6
Interstitial cells of Cajal	P2X2, P2X5
Kupffer cells	P2X1, P2X4, P2X7
Special senses	
Inner ear	P2X1, P2X2, P2X3, P2X7
Eye	P2X2, P2X7
Tongue	P2X2, P2X3
Olfactory organ	P2X2, P2X4
Cochlea hair cells	P2X1, P2X2, P2X7

A. Neuromuscular and Synaptic Transmission

While skeletal neuromuscular transmission in mature animals is mediated by acetylcholine (ACh), in early development, both ACh and ATP, released as cotransmitters from motor nerves, act on postsynaptic nicotinic and P2X receptors (Henning, 1997). In the adult, the ATP released from motor nerves acts postsynaptically to enhance the action of ACh and, after breakdown to adenosine, acts presynaptically via P1 receptors to inhibit ACh release (Ribeiro & Walker, 1975). Rat skeletal muscle transiently expresses first P2X5, then P2X6 receptors during embryological development, before the neuromuscular junction is formed (Ryten et al., 2001).

Sympathetic nerves supplying both visceral and vascular smooth muscle release ATP as a cotransmitter with noradrenaline (NA) to act predominately on postjunctional P2X1 receptors (see Burnstock, 1990). ATP, released as a cotransmitter with ACh in parasympathetic nerves supplying the bladder, also acts via P2X1 receptors (see Burnstock, 2009a). In sympathetically innervated tissues, such as the vas deferens or blood vessels, ATP produces fast responses mediated by P2X1 receptors, followed by a slower component mediated by G protein-coupled α-adrenoceptors. Similarly, in the parasympathetic nerves supplying the urinary bladder, ATP produces a fast transient response via P2X1 receptors, whereas the slower component is mediated by G protein-coupled muscarinic receptors.

Purinergic synaptic transmission in celiac ganglia (Evans et al., 1992; Silinsky et al., 1992) and in the medial habenula (Edwards et al., 1992) was first described in 1992, where postsynaptic membranes expressed P2X receptors. Since then, synaptic cotransmission in several other regions of the CNS has been identified involving postsynaptic P2X receptors (see Burnstock, 2007b).

Presynaptic neuromodulation, mediated by P2X3 receptors, has been described in primary afferent sensory nerve terminals in the dorsal horn of the spinal cord which mediate enhanced release of glutamate (Gu & MacDermott, 1997). Prejunctional P2X receptors have also been described on sympathetic nerve endings in vas deferens mediating enhanced release of NA (see Burnstock & Verkhratsky, 2010).

P2X2, P2X3, P2X2/3, P2X4, and P2X7 receptors are widely distributed in myenteric and submucous plexuses (see Burnstock, 2008a and Fig. 3) mediating synaptic transmission and various reflexes. The intrinsic sensory neurons express P2X3 receptors and are involved in modulation of peristaltic reflexes (Wynn et al., 2003). P2X7 receptors appear to mediate presynaptic neuromodulation.

In 1995, P2X3 receptors were cloned and shown to be expressed predominantly on sensory ganglia and nerves (either P2X3 homomultimer or P2X2/3 heteromultimer receptors; Burnstock, 2009b; Chen et al., 1995, Lewis et al., 1995). All the other P2X receptor subtypes have also been

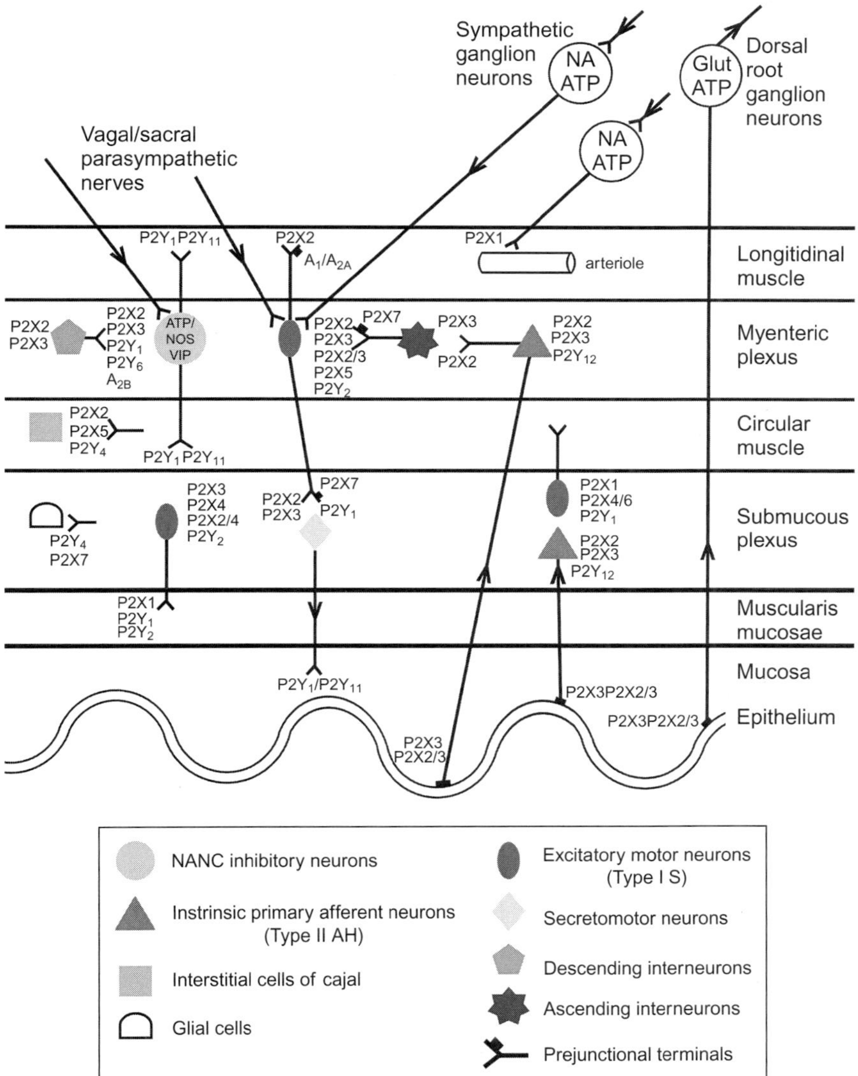

FIGURE 3 *Distribution of P2X receptor subtypes in the gut.* Extrinsic vagal and sacral parasympathetic nerves connect with nonadrenergic, noncholinergic (NANC) inhibitory neurons in the myenteric plexus expressing P2X2 and P2X3 receptors, as well as with cholinergic motor neurons; these neurons are also activated by descending interneurons. Extrinsic sympathetic nerves modulate motility via excitatory motor neurons and constrict blood vessels in the gut via P2X1 receptors. Extrinsic sensory nerves arise from cell bodies in DRG and with subepithelial terminals expressing P2X3 and P2X2/3 receptors and mediate nociception. Intrinsic sensory neurons in both myenteric and submucosal plexuses express P2X2 and P2X3 receptors; they connect with motor pathways involved in peristalsis. Excitatory motor neurons express P2X2, P2X3, P2X2/3, and P2X5 receptors and connect with both interneurons and secretomotor neurons. Interneurons express P2X2 and P2X3 receptors. Enteric glial cells express P2X7 receptors, while interstitial cells of Cajal express P2X2 and P2X5 receptors. P2X7 receptors appear to act as prejunctional modulators of both motor and interneurons (modified from Burnstock, 2008b).

described in nodose ganglia and P2X1–6, but not P2X7, receptors in dorsal root and trigeminal ganglia (see Burnstock & Knight, 2004). Sensory nerves in the brain stem have been shown to express P2X3 receptors. ATP applied to the heart can trigger the vagal reflex via P2X receptors (Pelleg et al., 1993).

There is widespread expression of P2X2, P2X4, and P2X6 receptors in different regions of the CNS, often forming heteromultimer receptors. P2X3 receptors have also been identified in parts of the brain stem. The presence of P2X7 receptors on neurons in the CNS is hotly debated. P2X receptors are involved in neuromodulation as well as neurotransmission and in neuron–glial interactions. The behavioral roles of P2X receptors in learning and memory, locomotion and sleep, and feeding are currently being explored (see Burnstock, 2007b).

B. Nonneuronal Cells

P2X1 and P2X2 receptors on intestinal and uterine smooth muscle mediate contractions. There is dual control of vascular tone: ATP released as a cotransmitter with NA from sympathetic nerves acting mainly on P2X1 receptors on vascular smooth muscle to cause contraction, while ATP released from endothelial cells during shear stress acts on endothelial P2 receptors to release nitric oxide (NO) resulting in vasodilation (see Burnstock, 2009c). While endothelial P2Y receptors appear to be predominant in most blood vessels, there is evidence that P2X receptor subtypes are also present on endothelial cells in some vessels (see Fig. 4; Burnstock, 2010).

mRNA and protein for P2X1/3/4/5/6 receptors have been found to be present in ventricles and P2X1–6 in atria and shown to have excitatory actions (Hansen et al., 1999). Activation of P2X4 receptors leads to increase in cardiac myocyte contractility (Shen et al., 2007).

There is widespread expression of P2X receptor subtypes in different sites in the kidney (see Unwin et al., 2003). Preglomerular arterioles express P2X1 receptors. Glomerular mesangial cells express P2X4, P2X5, and P2X7 receptors, and podocytes P2X1 and P2X7, and most segments of the kidney tubule show immunostaining for P2X receptors: P2X1, P2X4, P2X5, and P2X6 on proximal tubules; P2X4 and P2X6 in distal tubules; and P2X2, P2X2/6, and P2X5 in collecting ducts involved in control of sodium and water transport (see Fig. 5).

P2X1 and P2X7 receptors mediate the uptake of organic cations in canine erythrocytes (Stevenson et al., 2009). The role of the P2X1 receptors expressed by platelets is not clear, but it has been shown that in P2X1 knockout mice there is a decreased level of thrombus formation and increased bleeding times (Nurden, 2007) and P2X1 receptors have also been claimed to have a role in sensing bacteria (Kälvegren et al., 2010).

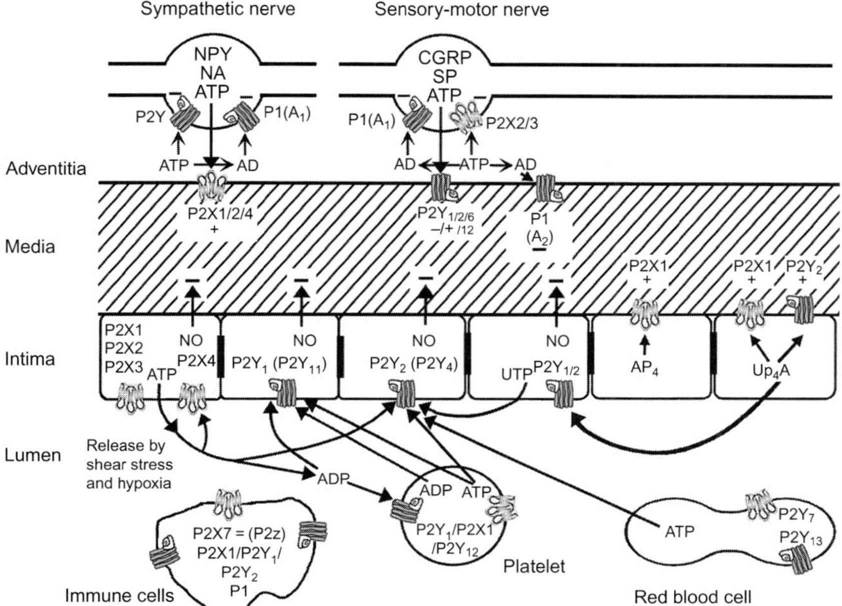

FIGURE 4 *Schematic diagram illustrating the main receptor subtypes for purines and pyrimidines present in blood vessels involved in control of vascular tone.* ATP is released as a cotransmitter with noradrenaline (NA) and neuropeptide Y (NPY) from sympathetic nerves in the adventitia to act at smooth muscle P2X1 receptors and, in some vessels, P2X2, P2X4, and $P2Y_1$, and $P2Y_2$ and $P2Y_6$ receptors, resulting in vasoconstriction (and rarely vasodilation); ATP is released with calcitonin gene-related peptide and substance P from sensory-motor nerves during "axon reflex" activity to act on smooth muscle P2Y receptors, resulting in either vasodilatation or vasoconstriction. P1 (A1) receptors on nerve terminals of sympathetic and sensory nerves mediate adenosine (arising from ecto-enzymatic breakdown of ATP) modulation of transmitter release. P2X2/3 receptors are present on a subpopulation of sensory nerve terminals. P1 (A2) receptors on vascular smooth muscle mediate vasodilatation. Endothelial cells release ATP and UTP during shear stress and hypoxia to act on $P2Y_1$, $P2Y_2$, and sometimes $P2Y_4$, $P2Y_{11}$, P2X1, P2X2, P2X3, and P2X4 receptors, leading to the production of nitric oxide (NO) and subsequent vasodilatation. Adenosine tetraphosphate (AP_4) activates P2X1 receptors to excite smooth muscle. ATP, after its release from aggregating platelets, also acts, together with its breakdown product ADP, on these endothelial receptors. Blood-borne platelets possess $P2Y_1$ and $P2Y_{12}$ ADP-selective receptors as well as P2X1 receptors. Immune cells of various kinds possess P2X7 as well as P1, P2X1, $P2Y_1$, and $P2Y_2$ receptors. ATP released from red blood cells, which express P2X7 and $P2Y_{13}$ receptors, is also involved in some circumstances. The additional involvements of uridine adenosine tetraphosphate (Up_4A) are included (modified from Burnstock, 1996).

The P2X7 receptor is thought to be a major immunomodulator that responds to extracellular ATP at sites of inflammation and tissue damage. P2X1 receptors promote neutrophil chemotaxis and therefore play a significant role in host defense (Lecut et al., 2009). ATP induces P2X7 receptor-independent cytokine and chemokine expression through P2X1

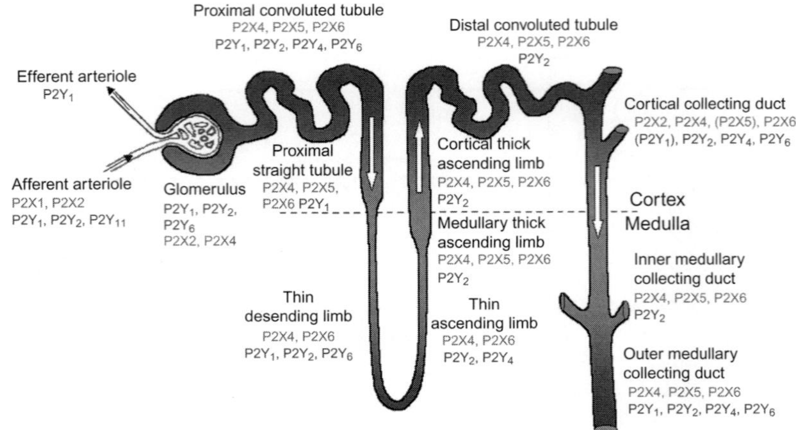

FIGURE 5 Summary of the nephron segments and the distribution of P2 receptor subtypes (updated from Turner et al., 2003, with permission from S. Karger AG, Basel).

and P2X3 receptors in mouse mast cells (Bulanova et al., 2009). Stimulation of P2X7 receptors in human dendritic cells induces the release of tissue factor-bearing microparticles (Baroni et al., 2007).

Purinergic signaling is much involved in keratinocyte turnover in skin epidermis: $P2Y_1$ and $P2Y_2$ receptors in basal and parabasal layers mediate cell proliferation; P2X5 receptors in the granular layer mediate cell differentiation; and P2X7 receptors at the stratum granulosum/stratum corneum border mediate apoptosis (Greig et al., 2003a). The posterior pituitary (PP) expressed protein for P2X2 and P2X6 receptors and P2X2, P2X3, P2X5, and P2X7 receptor channels in anterior pituitary (AP) cells mediating hormone secretion (see Fig. 6 and Stojilkovic et al., 2010). P2X7 receptors are expressed in both osteoblasts, where it enhances differentiation and bone formation and on osteoclasts where activation mediates apoptosis (Grol et al., 2009).

P2X receptors are widely expressed in the special senses mediating a variety of different functions (see Housley et al., 2009). The nasal epithelium expresses P2X2, P2X5, and P2X7 receptors (Gayle & Burnstock, 2005). P2X1, P2X2, P2X3, and P2X2/3 are prominent receptors in the tongue (Bo et al., 1999; Rong et al., 2000) mediating taste sensation and pain.

IV. Pathophysiology of P2X Receptors

The readers are referred to Burnstock (2006a, 2006b, 2006c) for detailed reviews about this topic.

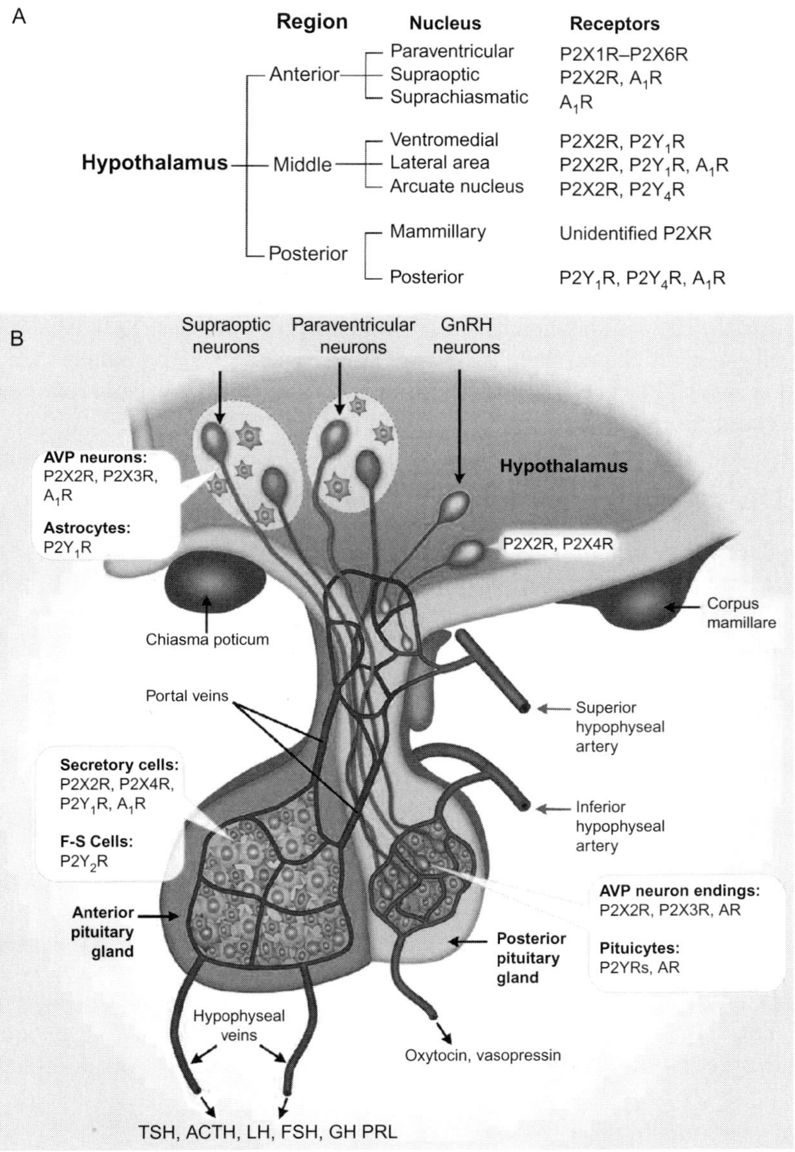

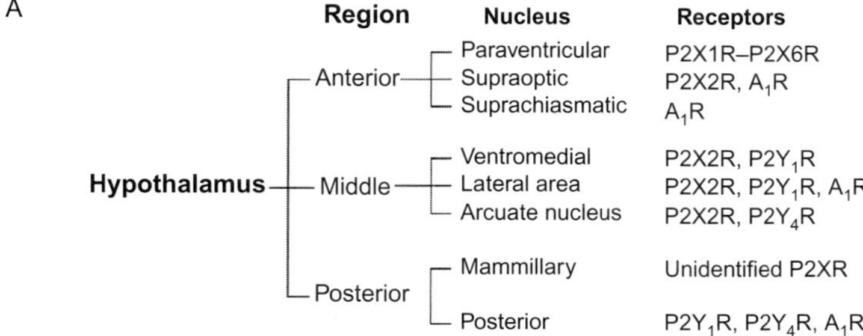

FIGURE 6 *Expression of purinergic receptors in the hypothalamus and pituitary.* (A) Receptors and receptor channels expressed in neurons of nuclei of the hypothalamus. For paraventricular and supraoptic nuclei, receptors expressed in parvocellular areas are listed. (B) Schematic representation of the hypothalamopituitary system. Insets indicate expression of purinergic receptors in secretory and supporting cells in three compartments. Note the pattern of expression of purinergic receptors: P2X2 receptors (R) are expressed in a majority of secretory cells (in anterior and middle hypothalamic neurons, vasopressinergic nerve endings, and anterior pituitary (AP) cells). Supporting cells (astrocytes in the hypothalamus, pituicytes in the posterior pituitary (PP), and folliculostellate (F-S) cells in the anterior lobe) do not express P2XRs.

A. Cardiovascular Diseases

Upregulation of P2X1 receptor mRNA in the hearts of rats with congestive heart failure has been reported and an increase in expression of P2X1 receptors in the atria of patients suffering from dilated cardiomyopathy. P2X4 receptor mRNA was upregulated in ligation-induced heart failure and appears to have a beneficial life-prolonging role (Musa et al., 2009).

ATP, acting via P2X1 receptors, plays a significant cotransmitter role in sympathetic nerves supplying hypertensive blood vessels. The purinergic component is increased in spontaneously hypertensive rats (see Ralevic & Burnstock, 1998). The increase in sympathetic nerve activity in hypertension is well established, and there is an associated hyperplasia and hypertrophy of arterial walls. Placental P2X4 receptors are significantly upregulated in mild preeclampsia (Roberts et al., 2005).

B. Diseases of Special Senses

Purinergic signaling is widespread in the eye, and novel therapeutic strategies are being developed for glaucoma, dry eye, and retinal detachment (Pintor et al., 2003). The formation of P2X7 receptor pores and apoptosis is enhanced in retinal microvessels early in the course of experimental diabetes, suggesting that purinergic vasotoxicity may play a role in microvascular cell death, a feature of diabetic retinopathy (Sugiyama et al., 2004).

P2X receptors have been identified in the vestibular system (Xiang et al., 1999). P2X splice variants are found on the endolymphatic surface of the cochlear endothelium, an area associated with sound transduction. ATP may regulate fluid homeostasis, cochlear blood flow, hearing sensitivity and development, and thus may be useful in the treatment of Ménières disease, tinnitus, and sensorineural deafness (Housley, 2000). Sustained loud noise produces an upregulation of P2X2 receptors in the cochlear, particularly at the site of outer hair cell sound transduction. P2X2 expression is also increased in spiral ganglion neurons (Wang et al., 2003).

Purinergic receptors have been described in the nasal mucosa, including the expression of P2X3 receptors on olfactory neurones. It appears that the

---

Many cells coexpress P2XRs, which facilitate electrical activity, and A_1Rs, which silence electrical activity. P2X7R are also expressed in hypothalamopituitary cells, but the cell types expressing these channels have not been identified. In other brain regions, astroglial cells express P2X7Rs. ATP is cosecreted by neurons making synapses with magnocellular neurons in the hypothalamus and by both vasopressin- and oxytocin-secreting neurons in the PP. ATP is also released by AP cells through still not well-characterized pathways. Green cells, vasopressin (AVP)-secreting neurons; orange cells, oxytocin-secreting neurons; pink cells, GnRH neurons. (Reproduced from Stojilkovic 2009, with permission from Elsevier.) (For interpretation of the references to color in this figure legend, the reader is referred to the Web version of this chapter.)

induction of heat-shock proteins by noxious odor damage can be prevented by the *in vivo* administration of P2 receptor antagonists (Hegg et al., 2003).

C. Immune System and Inflammation

P2X7 receptors located on inflammatory and immune cells play a pivotal role in inflammation and immunomodulation (Di Virgilio, 2007). Purinergic compounds are being explored for the treatment of neurogenic inflammation, rheumatoid arthritis, and periodontitis. ATP-induced apoptosis in macrophages via P2X7 receptors results in killing of the mycobacteria contained within them, in contrast to the macrophage apoptosis produced by other agents (Lammas et al., 1997). The P2X7 receptor plays a fundamental role in lipopolysaccharide signal transduction and activation of macrophages and may therefore represent a therapeutic target for Gram-negative bacterial septicaemia (Sommer et al., 1999). Lesional accumulation of macrophages expressing P2X4 receptors in rat CNS during experimental autoimmune encephalomyelitis (EAE) has been described, and it was suggested that P2X4 receptors might be a valuable marker to dissect the local monocyte heterogeneity in autoimmune disease (Guo & Schluesener, 2005). ATP moderates anti-IgE-induced release of histamine from lung mast cells and may therefore be mechanistically involved in human allergic/asthmatic reactions (Schulman et al., 1999). Alveolar macrophages express P2X7 receptors, which upon stimulation trigger proinflammatory responses, including activation of interleukin (IL)1–6 cytokines and granulomatous reactions (Lemaire & Leduc, 2004).

In addition to the apoptosis mediated by P2X7 receptors, a lower level of activation sometimes results in cell proliferation; it has been suggested that the expression and function of P2X7 receptors on B lymphocytes may correlate with the severity of B-cell chronic lymphocytic leukemia (Adinolfi et al., 2002).

During the acute phase of *Trypanosoma cruzi* infection, the etiologic agent of Chagas disease, thymic atrophy occurs; ATP also induces cell death in $CD4^+/CD8^+$ double-positive thymocytes and may play a central role in thymus atrophy during *T. cruzi* infection (Mantuano-Barradas et al., 2003). Infection by the parasitic blood fluke *Schistosoma mansoni* also leads to thymic atrophy. The cloning and characterization of a P2X receptor (schP2X) from *S. mansoni* provides the first example of a nonvertebrate ATP-gated ion channel and may provide an alternative drug target for the treatment of schistosomiasis (Agboh et al., 2004).

Allopurinal and captopril have a therapeutic effect on granulomatous disorders, such as sarcoidosis, by a direct action on monocyte/macrophage lineage cells partly by downregulation of intracellular adhesion molecular-1 and P2X7 receptors (Mizuno et al., 2004). P2X7 receptors control endocannabinoid production by microglia cells and might constitute promising therapeutics to temper exacerbated microinflammatory responses and allied cell damage (Witting et al., 2004).

D. Diabetes

A feature of diabetic retinopathy is the apoptotic death of microvascular pericytes and endothelial cells; there appears to be an enhancement of P2X7 receptor-induced pore formation and apoptosis in early diabetes on the retinal microvasculature (Sugiyama et al., 2004). In streptozotocin-diabetic animals, P2X7 receptor expression, located in glucagon-containing α cells in pancreatic islets, increases and migrates centrally to take the place of the insulin-containing β cells, although the functional significance of this is not known (Coutinho-Silva et al., 2003).

E. Disorders of the Gut

P2X receptors play major roles in different activities in both healthy and diseased gut (see Burnstock, 2008a, 2008b). During inflammation of the gastrointestinal tract, glial cells proliferate and produce cytokines; thus, P2X7 receptors may play a role in the response of enteric glia to inflammation (Vanderwinden et al., 2003). P2X3 purinergic signaling enhancement in an animal model of colitis has been described (Wynn et al., 2004). P2X3 receptor expression is increased in the enteric plexuses in human irritable bowel syndrome (IBS) suggesting a potential role in dysmotility and pain (Yiangou et al., 2001), and the possibility that P2X receptors are potential targets for the drug treatment of IBS has been raised (Galligan, 2004; Shinoda et al., 2009). Chronic functional visceral hyperalgesia induced in a rat model of IBS is associated with potentiation of ATP-evoked responses and an enhanced expression of P2X3 receptors in colon-specific sensory neurons (Xu et al., 2008). P2X3 immunohistochemistry has been demonstrated in aganglionic bowel in Hirschsprung's disease, suggesting that the sensory nerves may form a significant proportion of its hypertrophic innervation (Facer et al., 2001).

Intrinsic sensory neurones in the submucous plexus of the gut, as well as extrinsic sensory nerves, show positive immunoreactivity for P2X3 receptors (Xiang & Burnstock, 2004). It has been proposed (Burnstock, 2001b) that during moderate distension, low threshold intrinsic enteric sensory fibers may be activated via P2X3 receptors by ATP released from mucosal epithelial cells, leading to reflexes concerned with propulsion of material down the gut. Studies showing that peristalsis is impaired in the small intestine of mice lacking the P2X3 receptor subunit support this view (Bian et al., 2003). In contrast, during substantial (colic) distension associated with pain, higher threshold extrinsic sensory fibers may be activated by ATP released from the mucosal epithelia; these fibers pass messages through the dorsal root ganglia (DRG) to pain centers in the CNS (Wynn et al., 2003, 2004). Peripheral sensitization of P2X3 receptors on vagal and spinal afferents in the stomach may contribute to dyspeptic symptoms and the development of visceral hyperalgesia (Dang et al., 2005).

F. Diseases of the Urinary Tract

1. Kidney and Ureter

There is a substantial presence of purinoceptors in different regions of the nephron, the glomerulus, and renal vascular system in the kidney, including subtypes involved in the regulation of renin secretion, glomerular filtration, and the transport of water, ions, nutrients, and toxins (Unwin et al., 2003). In polycystic kidney disease, tubules are altered, leading to dilated tubules or cysts encapsulated by a single monolayer of renal epithelium. It has been postulated that autocrine purinergic signaling enhances cyst expansion and accelerates disease progression (Schwiebert et al., 2002). An increase in expression of P2X7 receptors has been reported in cystic tissue from the Han:SPRDcy/+rat model of autosomal dominant polycystic kidney disease (Turner et al., 2004a). ATP may inhibit pathological renal cyst growth through P2X7 signaling (Hillman et al., 2004). P2X antagonists and inhibitors of ATP release are being explored as therapeutic agents to treat this disease. There is increased glomerular expression of P2X7 receptors in two rat models of glomerular injury due to diabetes and hypertension (Vonend et al., 2004). A later study of human and experimental glomerulonephritis also showed increase in P2X7 receptor expression in the glomerulus (Turner et al., 2004b).

P2X3 receptors have been found on the suburothelial nerve plexus, and both the human and guinea pig ureter and urothelial cells shown to release ATP in a pressure-dependent fashion when the ureter is distended (Knight et al., 2002). This ATP release is abolished when the urothelium is removed, and sensory nerve-recording studies during ureteral distension demonstrate purinergic involvement, suggesting that specific P2X3 antagonists may have efficacy in alleviating renal colic (Rong & Burnstock, 2004).

2. Lower Urinary Tract

Readers are referred to a recent review on this topic (Burnstock, 2011). In the healthy human bladder, atropine will block at least 95% of parasympathetic nerve-mediated contraction, indicating that its innervation is predominantly cholinergic, although P2X1 receptors are present on the smooth muscle (Burnstock, 2001a). However, there are a number of examples where the purinergic component of cotransmission is increased in pathological conditions (Abbracchio & Burnstock, 1998). For example, purinergic nerve-mediated contraction of the human bladder is increased to 40% in pathophysiological conditions such as interstitial cystitis, outflow obstruction, idiopathic detrusor instability, and probably also neurogenic bladder. ATP release from bladder epithelial cells from patients with interstitial cystitis is significantly greater than from healthy cells and there is change in expression of P2X receptors in urothelial cells (Tempest et al., 2004). P2X1 receptor subtype expression markedly increased in obstructed bladder (Boselli et al., 2001).

Purinergic signaling also plays a role in afferent sensation from the bladder. ATP is released from urothelial cells when the bladder is distended (Vlaskovska et al., 2001). Sensory nerve recording has indicated that P2X3 receptors are involved in mediating the nerve responses to bladder distension, providing mechanosensory feedback involving both the micturition reflex and pain (Cockayne et al., 2000). Purinergic agonists acting on P2X3 receptors in the bladder can sensitize bladder afferent nerves, and these effects mimic the sensitizing effect of cystitis induced by cyclophosphamide (Yu and de Groat, 2004). Thus, P2X3 receptors are a potential target for pharmacological manipulation in the treatment of both pain and detrusor instability. In patients with idiopathic detrusor instability, there is an abnormal purinergic transmission in the bladder (O'Reilly et al., 2002). Voiding dysfunction involves P2X3 receptors in conscious chronic spinal cord injured rats, which raises the possibility that P2X3 antagonists might be useful for the treatment of neurogenic bladder dysfunction (Lu et al., 2002). Drugs that alter ATP release or breakdown might also be therapeutic targets (Chess-Williams, 2004).

G. Diseases of the Reproductive System

In humans, ATP induces a significant increase in sperm fertilizing potential and this provides a rationale for the use of ATP for treatment of spermatozoa during *in vitro* fertilization (Rossato et al., 1999). Knockout mice lacking P2X1 receptors appear normal, but fail to breed, and this is associated with loss of the purinergic component of sympathetic cotransmission in the vas deferens; these findings raise the possibility of developing nonhormonal ways of regulating male fertility (Dunn, 2000). Differential, stage-dependent immunostaining for P2X receptors during spermatogenesis in the adult rat testes has been described (Glass et al., 2001) and opens up another possibility of purinergic targets for both fertility and contraception.

Micromolar concentrations of ATP stimulate biphasic change in transepithelial conductance in the human uterine cervix, phase I mediated by the $P2Y_2$ receptor, phase II by the P2X4 receptor (Gorodeski, 2002). Given the potential role of ATP regulation of cervical paracellular permeability for human fertility, contraception, and health, these findings may lead to the development of drugs that can target specific signaling pathways in the cervix.

H. Diseases of Skin

There is an increase of P2X3 and P2X2/3 nociceptive receptors on sensory nerve endings in inflamed skin (Hamilton et al., 2001), and antagonists are being developed as analgesics. Data have been presented to support a pathogenic role for keratinocyte-derived ATP in irritant dermatitis (Mizumoto et al., 2003). Changes in expression of purinergic receptors in the regenerating epidermis in wound healing have been described (Greig et al., 2003c). P2X receptor antagonists have been shown to accelerate skin barrier repair and prevent epidermal hyperplasia induced by skin barrier disruption (Denda et al., 2002).

I. Diseases of the Airways

P2X4 receptors have been identified on lung epithelial cells and appear to be involved in regulation of ciliary beat, manipulation of which may also be of therapeutic benefit for cystic fibrosis (Zsembery et al., 2003). There is evidence to support the view that vagal afferent purinergic signaling may be involved in the hyperactivity associated with asthma and chronic obstructive pulmonary disease (Adriaensen & Timmermans, 2004).

Erythromycin is widely used for the treatment of upper and lower respiratory tract infections. Erythromycin has been shown to block the P2X receptor-mediated Ca^{2+} influx and may represent one mechanism by which it exerts its antisecretory effects in the treatment of chronic respiratory tract infections (Zhao et al., 2000).

The ventrolateral medulla (VLM) contains a network of respiratory neurons that are responsible for the generation and shaping of respiratory rhythm; it also functions as a chemoreceptive area mediating the ventilating response to hypercapnia. Evidence has been presented that ATP acting on P2X2 receptors expressed in VLM neurones influences these functions (Gourine et al., 2003). A potentially important role for P2 receptor synaptic signaling in respiratory motor control is suggested by the multiple physiological effects of ATP in hypoglossal activity associated with the presence of P2X2, P2X4, and P2X6 receptor mRNA in nucleus ambiguus and the hypoglossal nucleus (Collo et al., 1996), and microinjection of ATP into caudal nucleus of the solitary tract of awake rats produces respiratory responses (Antunes et al., 2005).

Alveolar macrophages play a pivotal role in the development of chronic lung inflammatory reactions such as idiopathic pulmonary fibrosis, silicosis, asbestosis, hypersensitivity pneumonitis, sarcoidosis, and mycobacterium tuberculosis. P2X7 receptors are expressed in alveolar macrophages, which upon stimulation activate the proinflammatory IL-1 to IL-5 cytokine cascade and the formation of multinucleated giant cells, a hallmark of granulomatous reactions (Lemaire & Leduc, 2004).

J. Musculoskeletal Diseases

Purinergic signaling has been implicated in bone development and remodeling (Burnstock and Arnett, 2006; Hoebertz et al., 2003; Orriss et al., 2010). P2X receptors are present on osteoclasts, osteoblasts, and chondrocytes. Deletion of the P2X7 receptor revealed its regulatory roles in bone formation and resorption. It reduces bone resorption by decreasing osteoclast survival, and P2X7 receptors are expressed in a subpopulation of osteoblasts. The multiple purinoceptors on bone and cartilage also represent potential targets for the development of novel therapeutics to inhibit bone resorption in diseases such as rheumatoid arthritis, osteoporosis, tumor-induced osteolysis, and periodontitis (Komarova et al., 2001). Relief of inflammatory pain by the P2X7 receptor antagonist, oxidized ATP, in arthritic rats has been reported (Dell'Antonio et al., 2002).

Lymphoblastoid cells isolated from Duchenne muscular dystrophy patients are highly sensitive to stimulation by extracellular ATP (Ferrari et al., 1994). Evidence for a role of P2X receptor-mediated signaling in muscle regeneration using the mdx mouse model of muscular dystrophy has been presented and raises the possibility of new therapeutic strategies for the treatment of muscle disease (Ryten et al., 2004).

K. Cancer

The anticancer activity of adenine nucleotides was first described by Rapaport (1983). Intraperitoneal injection of ATP into tumor-bearing mice resulted in significant anticancer activity against several fast-growing aggressive carcinomas (Agteresch et al., 2003). In reviews about the use of ATP for the treatment of advanced cancer (Abraham et al., 2003; White & Burnstock, 2006), evidence was presented that extracellular ATP inhibits the growth of a variety of human tumors, including prostate, bladder, breast, colon, liver, ovarian, colorectal, oesophageal, and melanoma cancer cells, partly via P2X7 receptors mediating apoptotic cancer cell death. Analyses have been carried out on the P2 receptor subtypes that contribute to ATP suppression of malignant melanomas (White et al., 2005a, 2005b) in basal and squamous cell tumors (Greig et al., 2003b) and prostate and bladder cancers (Calvert et al., 2004; Shabbir et al., 2004, 2008). P2X5 receptors mediate cell differentiation, which in effect is antiproliferative, and P2X7 receptors mediate apoptotic cell death.

L. Disorders of the CNS

A recent review has focused on purinergic signaling in disorders of the CNS (Burnstock, 2008c).

Accumulation of P2X4 receptor-positive microglia and macrophages following experimental traumatic brain injury and spinal cord injury has been described. Activated microglia also show significant changes in P2X7 receptor expression, which play an important role in controlling microglial proliferation and death. Cerebellar lesions result in upregulation of P2X1 and P2X2 receptors in precerebellar nuclei, and stab wound injury in the nucleus accumbens leads to increased expression of several subtypes of P2X receptors (Franke et al., 2006).

Ischemia can produce and exacerbate many serious insults to the CNS, including stroke and paralysis. Upregulation of P2X2 and P2X4 receptors in cultures of hippocampus, cortex, and striatum is associated with ischemic cell death and was prevented by P2 receptor antagonists (Cavaliere et al., 2003). Following ischemia, P2X7 receptors are upregulated on neurons and glial cells in rat cerebral cortex and become supersensitive in cerebrocortical cell cultures.

M. Neurodegenerative Diseases

P2X receptor involvement in neurodegenerative diseases such as Parkinson's, Alzheimer's, Huntington's, amyotrophic lateral sclerosis (ALS), and multiple sclerosis (MS) has been described.

Release of ATP from disrupted cells may cause cell death in neighboring cells expressing P2X7 receptors, leading to a necrotic volume increase, which has been proposed in the pathogenesis of Parkinson's disease. A facilitatory action by ATP could influence the nigral dopaminergic cell vulnerability in Parkinson's disease.

P2X7 receptors are upregulated in human Alzheimer's diseased brains and in animal models (McLarnon et al., 2006; Parvathenani et al., 2003). Stimulation of P2X7 receptors on human macrophages and microglia enhanced the degenerative lesions observed in Alzheimer's disease.

Changes in P2X receptor-mediated neurotransmission in corticostriatal projections in two different transgenic models of Huntington's disease (Diez-Zaera et al., 2007).

Potentiation of P2X4 receptors by the antiparasite medication ivermectin (22,23-dihydroavermectin B_{1a} + 22,23-dihydroavermectin B_{1b}) extends the life span of the transgenic superoxide dismutase 1 (SOD1) mouse model of ALS (Andries et al., 2007). Increased expression of P2X1 receptors on axotomized facial motoneurons was impaired in SOD1-G93A-mutant mice after injury (Kassa et al., 2007).

P2 receptors on oligodendrocytic progenitor cells regulate migration, proliferation, and differentiation. In MS lesions of autopsied brain tissue, P2X7 receptors were demonstrated on reactive astrocytes, whereas in cultured astrocytes, P2X7 receptor stimulation increased the production of NO synthase activity (Narcisse et al., 2005). Neuronal pathology is an early feature of MS and its animal model, EAE. Lesional accumulation of P2X receptors on macrophages in rat CNS during EAE has been described (Guo & Schluesener, 2005). P2X7 expression is elevated in normal-appearing axon tracts in MS patients, and ATP can kill oligodendrocytes via P2X7 receptor action. Mice deficient in P2X7 receptors are more susceptible to EAE than wild-type mice and show enhanced inflammation in the CNS (Chen & Brosnan, 2006).

Diabetic neuropathy includes central neuropathic complications, including decreased cognitive performance, accompanied by modifications of hippocampal morphology and plasticity (Cox et al., 2005; Trudeau et al., 2004). It has been shown that synaptic ATP signaling is depressed in streptozotocin-induced diabetic rats (Duarte et al., 2007), and the density of P2X3/6/7 receptors was decreased in the diabetic hippocampal nerve terminals.

N. Neuroimmune and Neuroinflammatory Disorders

Microglia are the immune cells in the CNS. Microglial P2X7 receptors are activated by purines to release inflammatory cytokines such as IL-1β, IL-6, and tumor necrosis factor-α (Di Virgilio, 2007). ATP also increases 2-arachidonoylglycerol (2-AG) production via P2X7 receptors on microglial cells and because prolonged increases in 2-AG levels in brain parenchyma are thought to orchestrate neuroinflammation. The P2X7 receptor is involved in the formation of multinucleated giant macrophage-derived cells, a hallmark of chronic inflammatory reactions (Lemaire et al., 2006). Lysophosphatidylcholine, an inflammatory phospholipid, may regulate microglial functions by enhancing the sensitivity of P2X7 receptors. There has been a report that prion infection is associated with hypersensitivity of P2X7 receptors in microglia (Takenouchi et al., 2007). Expression of the P2X4 receptor by lesional activated microglia during formalin-induced inflammatory pain has also been reported (Guo et al., 2005). Activation of microglial cells by proinflammatory bacterial lipopolysaccharide leads to a transient increase in ivermectin-sensitive P2X4 receptor currents, while dominant P2X7 receptor currents remain largely unaffected; both subtypes contribute to neuroinflammatory mechanisms and pathologies (Raouf et al., 2007). P2X7 receptor activation in astrocytes increases chemokine monocyte chemoattractant protein-1 (MCP-1) expression via mitogen-activated protein kinase, and it was suggested that regulation of MCP-1 in astrocytes by ATP may be important in mediating communication with haematopoietic inflammatory cells (Panenka et al., 2001).

O. Epileptic seizures

Microinjection of ATP analogs into the prepiriform cortex induces generalized motor seizures (Knutsen & Murray, 1997). P2X2, P2X4, and P2X6 receptors are expressed in the prepiriform cortex, suggesting that P2X receptor antagonists may have potential as neuroleptic agents. The hippocampus of chronic epileptic rats shows abnormal responses to ATP associated with increased expression of P2X7 receptors, which are substantially upregulated in chronic pilocarpine-induced epilepsy in rats (perhaps in microglia) and may participate in the pathophysiology of temporal lobe epilepsy. In a study of kainate-provoked seizures, enhanced immunoreactivity of the P2X7 receptor was observed in microglia as they are changed from the resting to the activated state (Rappold et al., 2006). There is a decrease of presynaptic P2X receptors in the hippocampus of rats that have suffered a convulsive period, which may be associated with the development of seizures and/or of neurodegeneration during epilepsy (Oses, 2006). Release of glutamate from astrocytes by ATP has been implicated in epileptogenesis (Tian et al., 2005).

P. Neuropsychiatric Disorders

The P2X7 receptor gene has been shown to be involved in both major depressive illness (Lucae et al., 2006) and bipolar affective disorders (Barden et al., 2006).

The involvement of ATP receptors in schizophrenia has been discussed in relation to reports that antipsychotic drugs such as haloperidol, chlorpromazine, and fluspirilene inhibit ATP-evoked responses mediated by P2X receptors (Inoue et al., 1996). It was suggested that ATP may have a facilitating role for dopaminergic neurons and that various antipsychotic drugs express their therapeutic effects by suppression of dopaminergic hyperactivity through inhibition of P2X receptor-mediated effects.

Although ethanol is probably the oldest and most widely used psychoactive drug, the cellular mechanisms by which it affects the nervous system have been poorly understood, although some insights in relation to purinergic P2 receptor signaling have emerged in recent years (Davies et al., 2005). Ethanol inhibits P2X receptor-mediated responses of DRG neurons by an allosteric mechanism. In the case of P2X4 receptors, ethanol inhibition is altered by mutation of histidine 241 in the rat. Further, ethanol differentially affects ATP-gated P2X3 and P2X4 receptor subtypes expressed in *Xenopus* oocytes.

Q. Purinergic Mechanosensory Transduction and Pain

A recent review has been focused on this topic (see Burnstock, 2009d). Visceral pain is one of the most common forms of pain associated with pathological conditions like renal colic, dyspepsia, inflammatory bowel disease, angina, dysmenorrhoea, and interstitial cystitis. P2X3 (homomultimer) and P2X2/3 (heteromultimer) receptors were cloned and shown to be largely located on small nociceptive sensory neurons in the DRG in 1995 (Lewis et al., 1995).

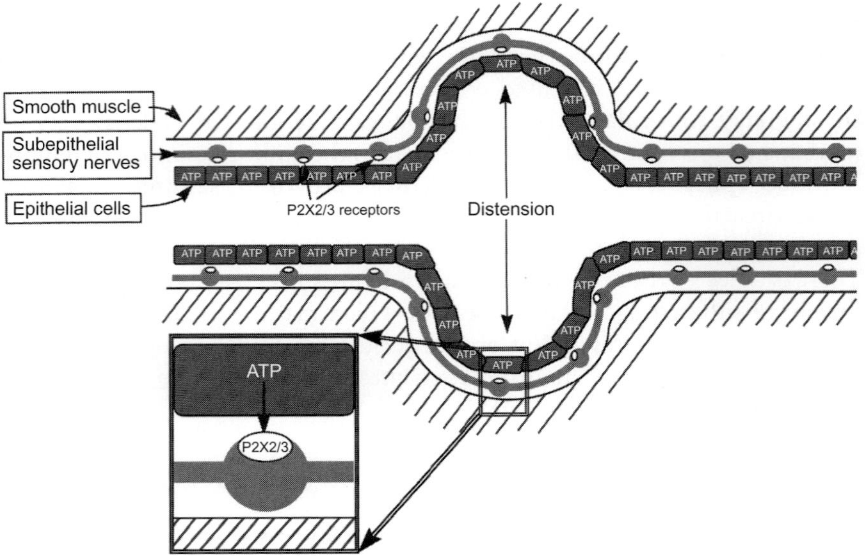

FIGURE 7 Schematic representation of hypothesis for purinergic mechanosensory transduction in tubes (e.g., ureter, vagina, salivary and bile ducts, gut) and sacs (e.g., urinary and gall bladders, lung). It is proposed that distension leads to release of ATP from epithelium lining the tube or sac, which then acts on P2X3 and/or P2X2/3 receptors on subepithelial sensory nerves to convey sensory/nociceptive information to the CNS. (Reproduced from Burnstock 1999.)

A hypothesis was proposed that purinergic mechanosensory transduction occurred in visceral tubes and sacs, including ureter, bladder, and gut, where ATP released from epithelial cells during distension acted on P2X3 homomeric and P2X2/3 heteromeric receptors on subepithelial sensory nerves initiating impulses in sensory pathways to pain centers in the CNS (Burnstock, 1999; Fig. 7). Mice lacking the P2X3 receptor exhibited reduced inflammatory pain and marked urinary bladder hyporeflexia with reduced voiding frequency and increased voiding volume, suggesting that P2X3 receptors are involved in mechanosensory transduction underlying both inflammatory pain and physiological voiding reflexes (Cockayne et al., 2000). In a systematic study of purinergic mechanosensory transduction in the mouse urinary bladder (Vlaskovska et al., 2001), ATP was shown to be released from urothelial cells during distension, and activity initiated in pelvic sensory nerves was mimicked by ATP and α,β-meATP and attenuated by P2X3 antagonists as well as in P2X3 knockout mice.

The uroteric colic that is induced by the passage of a kidney stone causes severe pain. Immunostaining of P2X3 receptors in sensory nerves in the subepithelial region was reported (Lee et al., 2000). Multifiber recordings of ureter afferent nerves were made using a guinea pig preparation perfused *in vitro* (Rong & Burnstock, 2004). Distension of the guinea pig ureter increased spike discharge in sensory neurons, which was mimicked by ATP and reduced by ATP antagonists. Distension of the perfused guinea pig and human ureter caused a pressure-dependent release of ATP from urothelial cells, approximately 10 times the basal release levels (Calvert et al., 2008; Knight et al., 2002).

It was proposed that purinergic mechanosensory transduction in the gut initiated both physiological reflex modulation of peristalsis via intrinsic sensory fibers and nociception via extrinsic sensory fibers (Burnstock, 2001b). Evidence in support of this hypothesis was obtained from a rat pelvic sensory nerve–colorectal preparation (Wynn et al., 2003). Distension of the colorectum led to pressure-dependent increase in release of ATP from mucosalepithelial cells and also evoked pelvic nerve excitation. This excitation was mimicked by application of ATP and α,β-meATP and attenuated by the selective P2X3 and P2X2/3 antagonist 2′,3′-O-trinitrophenyl-ATP and by pyridoxalphosphate-6-azophenyl-2′,4′-disulfonic acid.

The search is on for selective P2X3 and P2X2/3 receptor antagonists that are orally bioavailable and do not degrade *in vivo* for the treatment of pain (see Burnstock, 2006b; Gever et al., 2010). A-317491 (synthesized by Abbott Laboratories) and compound AF-353 (synthesized by Roche Palo Alto) are both effective P2X3 and P2X2/3 antagonists, the latter being orally bioavailable and stable *in vivo*.

R. Neuropathic Pain

P2X3 and P2X2/3 receptors, probably those located on primary afferent nerve terminals in inner lamina 2 of the spinal cord, also play a significant role in neuropathic and inflammatory pain (see Burnstock, 2009e; Wirkner et al., 2007). P2X2, P2X4, and P2X6 receptors have been located on dorsal horn neurons relaying nociceptive information further along the pain pathway (Bardoni et al., 1997). In addition, ATP coreleased with γ-aminobutyric acid in spinal interneurons is probably involved in modulation of nociceptive pathways (Jo & Schlichter, 1999). Importantly, it has been shown that P2X4 and P2X7 receptors on microglia are also involved in neuropathic pain (Hughes et al., 2007; Tsuda et al., 2003), although the underlying mechanisms are still not clear (Inoue, 2007). As neuropathic pain and allodynia are abolished in both P2X4 and P2X7 knockout mice, there is great interest in finding selective antagonists that might be suitable for therapeutic development (see McGaraughty et al., 2007).

S. Migraine

The involvement of ATP in migraine was first suspected in conjunction with the vascular theory of this disorder with ATP released from endothelial cells during reactive hyperaemia associated with pain following cerebral vascular vasospasm (that is not associated with pain; Burnstock, 1989). More recently, P2X3 receptor involvement in neuronal dysfunction in brain areas that mediate nociception such as the trigeminal nucleus and thalamus has been considered (Fabbretti et al., 2006). The interaction of $P2Y_1$ receptors on trigeminal neurons with P2X3 receptors after sensitization of these neurons with algogenic stimuli (e.g., nerve growth factor (NGF), brain-derived neurotrophic factor, or bradykinin) has been proposed and may also represent a new potential target for antimigraine drugs (Fumagalli et al., 2006). In an *in vivo* model of mouse trigeminal pain, anti-NGF treatment suppressed responses evoked by P2X3 receptor activation (D'Arco et al., 2007).

V. Conclusion

P2X receptors comprise a novel family of ligand-gated cation channels, which are widely expressed in nerves and many nonneuronal cells. Following the recent report of their crystal structure, we now have a detailed understanding of how the individual subunits that form the receptor interact with each other. P2X receptors show plasticity in many pathological conditions, suggesting that they may be useful targets for treatment of these diseases. Although in its infancy, the clinical manipulation of purinergic signaling has begun. Several clinically relevant pharmacological interventions are already part of day-to-day practice. However, one of the main reasons why we do not yet have more purinergic therapies in our formularies is the current scarcity of receptor-subtype-specific agonists and antagonists that are stable and effective *in vivo*. In addition to the development of selective agonists and antagonists for the different P2 receptor subtypes, therapeutic strategies are likely to include agents that control the expression of P2 receptors, inhibitors of extracellular breakdown of ATP, and enhancers or inhibitors of ATP transport. Investigating the interactions of purinergic signaling with other established signaling systems will also be very important. Roche have recently developed some small molecules (AF-353 and derivatives) that are orally bioavailable and stable *in vivo* and which are currently in clinical trial (Gever et al., 2006, 2010). There has also been promising development of clinically relevant P2X7 antagonists recently, notably the Abbott compound A438079 (McGaraughty et al., 2007). Antagonists for some of the other P2X subtypes remain, however, to be developed. Nonetheless, the progress made to date suggests that drugs active at P2X receptors are likely to have widespread therapeutic uses.

Conflict of Interest: The authors have no conflict of interest to declare.

Abbreviations

2-AG	2-arachidonoylglycerol
ACh	acetylcholine
ALS	amyotrophic lateral sclerosis
CNS	central nervous system
DRG	dorsal root ganglia
EAE	experimental autoimmune encephalomyelitis
IBS	irritable bowel syndrome
IL	interleukin
MS	multiple sclerosis
NA	noradrenaline
NANC	nonadrenergic, noncholinergic
NGF	nerve growth factor
NO	nitric oxide
PKC	protein kinase C
SOD1	superoxide dismutase 1
TMR	transmembrane spanning regions

References

Abbracchio, M. P., & Burnstock, G. (1998). Purinergic signalling: Pathophysiological roles. *Japanese Journal of Pharmacology*, *78*, 113–145.

Abraham, E. H., Salikhova, A. Y., & Rapaport, E. (2003). ATP in the treatment of advanced cancer. *Current Topics in Membranes*, *54*, 415–452.

Adinolfi, E., Melchiorri, L., Falzoni, S., et al. (2002). P2X7 receptor expression in evolutive and indolent forms of chronic B lymphocytic leukemia. *Blood*, *99*, 706–708.

Adriaensen, D., & Timmermans, J. P. (2004). Purinergic signalling in the lung: Important in asthma and COPD? *Current Opinion in Pharmacology*, *4*, 207–214.

Agboh, K. C., Webb, T. E., Evans, R. J., et al. (2004). Functional characterization of a P2X receptor from *Schistosoma mansoni*. *The Journal of Biological Chemistry*, *279*, 41650–41657.

Agteresch, H. J., van Rooijen, M. H. C., van den Berg, J. W. O., et al. (2003). Growth inhibition of lung cancer cells by adenosine 5-triphosphate. *Drug Development Research*, *60*, 196–203.

Andries, M., Van Damme, P., Robberecht, W., et al. (2007). Ivermectin inhibits AMPA receptor-mediated excitotoxicity in cultured motor neurons and extends the life span of a transgenic mouse model of amyotrophic lateral sclerosis. *Neurobiology of Disease*, *25*, 8–16.

Antunes, V. R., Bonagamba, L. G., & Machado, B. H. (2005). Hemodynamic and respiratory responses to microinjection of ATP into the intermediate and caudal NTS of awake rats. *Brain Research*, *1032*, 85–93.

Aschrafi, A., Sadtler, S., Niculescu, C., et al. (2004). Trimeric architecture of homomeric $P2X_2$ and heteromeric $P2X_{1+2}$ receptor subtypes. *Journal of Molecular Biology*, *342*, 333–343.

Barden, N., Harvey, M., Gagné, B., et al. (2006). Analysis of single nucleotide polymorphisms in genes in the chromosome 12Q24.31 region points to P2RX7 as a susceptibility gene to bipolar affective disorder. *American Journal of Medical Genetics. Part B, Neuropsychiatric Genetics*, *141*, 374–382.

Bardoni, R., Goldstein, P. A., Lee, C. J., et al. (1997). ATP P_{2X} receptors mediate fast synaptic transmission in the dorsal horn of the rat spinal cord. *The Journal of Neuroscience*, *17*, 5297–5304.

Baroni, M., Pizzirani, C., Pinotti, M., et al. (2007). Stimulation of P2 ($P2X_7$) receptors in human dendritic cells induces the release of tissue factor-bearing microparticles. *The FASEB Journal*, *21*, 1926–1933.

Barrera, N. P., Ge, H., Henderson, R. M., et al. (2008). Automated analysis of the architecture of receptors, imaged by atomic force microscopy. *Micron*, *39*, 101–110.

Barrera, N. P., Ormond, S. J., Henderson, R. M., et al. (2005). Atomic force microscopy imaging demonstrates that $P2X_2$ receptors are trimers but that $P2X_6$ receptor subunits do not oligomerize. *The Journal of Biological Chemistry*, *280*, 10759–10765.

Bean, B. P. (1990). ATP-activated channels in rat and bullfrog sensory neurons: Concentration dependence and kinetics. *The Journal of Neuroscience*, *10*, 1–10.

Bian, X., Ren, J., DeVries, M., et al. (2003). Peristalsis is impaired in the small intestine of mice lacking the $P2X_3$ subunit. *The Journal of Physiology*, *551*, 309–322.

Bo, X., Alavi, A., Xiang, Z., et al. (1999). Localization of ATP-gated $P2X_2$ and $P2X_3$ receptor immunoreactive nerves in rat taste buds. *NeuroReport*, *10*, 1107–1111.

Boselli, C., Govoni, S., Condino, A. M., et al. (2001). Bladder instability: A re-appraisal of classical experimental approaches and development of new therapeutic strategies. *Journal of Autonomic Pharmacology*, *21*, 219–229.

Brake, A. J., Wagenbach, M. J., & Julius, D. (1994). New structural motif for ligand-gated ion channels defined by an ionotropic ATP receptor. *Nature*, *371*, 519–523.

Brown, S. G., Townsend-Nicholson, A., Jacobson, K. A., et al. (2002). Heteromultimeric $P2X_{1/2}$ receptors show a novel sensitivity to extracellular pH. *The Journal of Pharmacology and Experimental Therapeutics*, *300*, 673–680.

Bulanova, E., Budagian, V., Orinska, Z., et al. (2009). ATP induces $P2X_7$ receptor-independent cytokine and chemokine expression through $P2X_1$ and $P2X_3$ receptors in murine mast cells. *Journal of Leukocyte Biology*, *85*, 692–702.

Burnstock, G. (1978). A basis for distinguishing two types of purinergic receptor. In R. W. Straub & L. Bolis (Eds.), *Cell membrane receptors for drugs and hormones: A multidisciplinary approach* (pp. 107–118). New York: Raven Press.

Burnstock, G. (1989). The role of adenosine triphosphate in migraine. *Biomedicine and Pharmacotherapy*, *43*, 727–736.

Burnstock, G. (1990). Noradrenaline and ATP as cotransmitters in sympathetic nerves. *Neurochemistry International*, *17*, 357–368.

Burnstock, G. (1996). Development and perspectives of the purinoceptor concept. *Journal of Autonomic Pharmacology*, *16*, 295–302.

Burnstock, G. (1999). Release of vasoactive substances from endothelial cells by shear stress and purinergic mechanosensory transduction. *Journal of Anatomy*, *194*, 335–342.

Burnstock, G. (2001a). Purinergic signalling in lower urinary tract. In M. P. Abbracchio & M. Williams (Eds.), *Handbook of experimental pharmacology, Volume 151/I. Purinergic and pyrimidinergic signalling I—Molecular, nervous and urinogenitary system function* (pp. 423–515). Berlin: Springer-Verlag.

Burnstock, G. (2001b). Purine-mediated signalling in pain and visceral perception. *Trends in Pharmacological Sciences*, *22*, 182–188.

Burnstock, G. (2006a). Pathophysiology and therapeutic potential of purinergic signaling. *Pharmacological Reviews*, *58*, 58–86.

Burnstock, G. (2006b). Purinergic P2 receptors as targets for novel analgesics. *Pharmacology and Therapeutics*, *110*, 433–454.

Burnstock, G. (2007a). Purine and pyrimidine receptors. *Cellular and Molecular Life Sciences*, *64*, 1471–1483.

Burnstock, G. (2007b). Physiology and pathophysiology of purinergic neurotransmission. *Physiological Reviews*, *87*, 659–797.

Burnstock, G. (2008a). The journey to establish purinergic signalling in the gut. *Neurogastroenterology and Motility*, *20*, 8–19.

Burnstock, G. (2008b). Commentary. Purinergic receptors as future targets for treatment of functional GI disorders. *Gut*, *57*, 1193–1194.

Burnstock, G. (2008c). Purinergic signalling and disorders of the central nervous system. *Nature Reviews. Drug Discovery*, *7*, 575–590.

Burnstock, G. (2009a). Purinergic cotransmission. *Experimental Physiology*, *94*, 20–24.

Burnstock, G. (2009b). Purines and sensory nerves. *Handbook of Experimental Pharmacology*, *194*, 333–392.

Burnstock, G. (2009c). Purinergic regulation of vascular tone and remodelling. *Autonomic and Autocoid Pharmacology*, *29*, 63–72.

Burnstock, G. (2009d). Purinergic mechanosensory transduction and visceral pain. *Molecular Pain*, *5*, 69.

Burnstock, G. (2009e). Purinergic receptors and pain. *Current Pharmaceutical Design*, *15*, 1717–1735.

Burnstock, G. (2010). Commentary: Control of vascular tone by purines and pyrimidines. *British Journal of Pharmacology*, *161*, 527–529.

Burnstock, G. (2011). Therapeutic potential of purinergic signalling for diseases of the urinary tract. *British Journal of Urology International*, *107*, 192–204.

Burnstock, G., & Arnett, T. R. (2006). *Edited Monograph: Nucleotides and regulation of bone cell function.* Boca Raton, Florida: Taylor & Francis Group.

Burnstock, G., & Kennedy, C. (1985). Is there a basis for distinguishing two types of P_2-purinoceptor? *General Pharmacology*, *16*, 433–440.

Burnstock, G., & Knight, G. E. (2004). Cellular distribution and functions of P2 receptor subtypes in different systems. *International Review of Cytology*, *240*, 31–304.

Burnstock, G., & Verkhratsky, A. (2010). Vas deferens—A model used to establish sympathetic cotransmission. *Trends in Pharmacological Sciences*, *31*, 131–139.

Calvert, R. C., Shabbir, M., Thompson, C. S., et al. (2004). Immunocytochemical and pharmacological characterisation of P2-purinoceptor-mediated cell growth and death in PC-3 hormone refractory prostate cancer cells. *Anticancer Research*, *24*, 2853–2859.

Calvert, R. C., Thompson, C. S., & Burnstock, G. (2008). ATP release from the human ureter on distension and $P2X_3$ receptor expression on suburothelial sensory nerves. *Purinergic Signalling*, *4*, 377–381.

Cavaliere, F., Florenzano, F., Amadio, S., et al. (2003). Up-regulation of $P2X_2$, $P2X_4$ receptor and ischemic cell death: Prevention by P2 antagonists. *Neuroscience*, *120*, 85–98.

Chaumont, S., & Khakh, B. S. (2008). Patch-clamp coordinated spectroscopy shows P2X2 receptor permeability dynamics require cytosolic domain rearrangements but not Panx-1 channels. *Proceedings of the National Academy of Sciences of the United States of America*, *105*, 12063–12068.

Chen, C. C., Akopian, A. N., Sivilotti, L., et al. (1995). A P2X purinoceptor expressed by a subset of sensory neurons. *Nature*, *377*, 428–431.

Chen, L., & Brosnan, C. F. (2006). Exacerbation of experimental autoimmune encephalomyelitis in $P2X_7R^{-/-}$ mice: Evidence for loss of apoptotic activity in lymphocytes. *Journal of Immunology*, *176*, 3115–3126.

Chess-Williams, R. (2004). Potential therapeutic targets for the treatment of detrusor overactivity. *Expert Opinion on Therapeutic Targets*, *8*, 95–106.

Clyne, J. D., Wang, L. F., & Hume, R. I. (2002). Mutational analysis of the conserved cysteines of the rat $P2X_2$ purinoceptor. *The Journal of Neuroscience*, *22*, 3873–3880.

Cockayne, D. A., Hamilton, S. G., Zhu, Q.-M., et al. (2000). Urinary bladder hyporeflexia and reduced pain-related behaviour in $P2X_3$-deficient mice. *Nature*, *407*, 1011–1015.

Collo, G., North, R. A., Kawashima, E., et al. (1996). Cloning of $P2X_5$ and $P2X_6$ receptors and the distribution and properties of an extended family of ATP-gated ion channels. *The Journal of Neuroscience*, *16*, 2495–2507.

Coutinho-Silva, R., Parsons, M., Robson, T., et al. (2003). P2X and P2Y purinoceptor expression in pancreas from streptozotocin-diabetic rats. *Molecular and Cellular Endocrinology*, *204*, 141–154.

Cox, D. J., Kovatchev, B. P., Gonder-Frederick, L. A., et al. (2005). Relationships between hyperglycemia and cognitive performance among adults with type 1 and type 2 diabetes. *Diabetes Care*, *28*, 71–77.

D'Arco, M., Giniatullin, R., Simonetti, M., et al. (2007). Neutralization of nerve growth factor induces plasticity of ATP-sensitive $P2X_3$ receptors of nociceptive trigeminal ganglion neurons. *The Journal of Neuroscience*, *27*, 8190–8201.

Dang, K., Bielfeldt, K., Lamb, K., et al. (2005). Gastric ulcers evoke hyperexcitability and enhance P2X receptor function in rat gastric sensory neurons. *Journal of Neurophysiology*, *93*, 3112–3119.

Davies, D. L., Kochegarov, A. A., Kuo, S. T., et al. (2005). Ethanol differentially affects ATP-gated $P2X_3$ and $P2X_4$ receptor subtypes expressed in *Xenopus* oocytes. *Neuropharmacology*, *49*, 243–253.

Dell'Antonio, G., Quattrini, A., Cin, E. D., et al. (2002). Relief of inflammatory pain in rats by local use of the selective $P2X_7$ ATP receptor inhibitor, oxidized ATP. *Arthritis and Rheumatism*, *46*, 3378–3385.

Denda, M., Inoue, K., Fuziwara, S., et al. (2002). P2X purinergic receptor antagonist accelerates skin barrier repair and prevents epidermal hyperplasia induced by skin barrier disruption. *Journal of Investigative Dermatology*, *119*, 1034–1040.

Di Virgilio, F. (2007). Liaisons dangereuses: $P2X_7$ and the inflammasome. *Trends in Pharmacological Sciences*, *28*, 465–472.

Diez-Zaera, M., Diaz-Hernandez, M., Alberch, J., et al. (2007). Purinergic system in Huntington's disease: Develop of new therapeutic strategies. *Journal of Neurochemistry*, *101*, 66.

Ding, S., & Sachs, F. (1999). Single channel properties of P2X2 purinoceptors. *The Journal of General Physiology*, *113*, 695–720.

Duarte, J. M., Oses, J. P., Rodrigues, R. J., et al. (2007). Modification of purinergic signaling in the hippocampus of streptozotocin-induced diabetic rats. *Neuroscience*, *149*, 382–391.

Duckwitz, W., Hausmann, R., Aschrafi, A., et al. (2006). $P2X_5$ subunit assembly requires scaffolding by the second transmembrane domain and a conserved aspartate. *The Journal of Biological Chemistry*, *281*, 39561–39572.

Dunn, P. M. (2000). Fertility: Purinergic receptors and the male contraceptive pill. *Current Biology*, *10*, R305–R307.

Edwards, F. A., Gibb, A. J., & Colquhoun, D. (1992). ATP receptor-mediated synaptic currents in the central nervous system. *Nature*, *359*, 144–147.

Egan, T. M., & Khakh, B. S. (2004). Contribution of calcium ions to P2X channel responses. *The Journal of Neuroscience*, *24*, 3413–3420.

Ennion, S. J., & Evans, R. J. (2002). Conserved cysteine residues in the extracellular loop of the human $P2X_1$ receptor form disulfide bonds and are involved in receptor trafficking to the cell surface. *Molecular Pharmacology*, *61*, 303–311.

Evans, R. J. (2009). Orthosteric and allosteric binding sites of P2X receptors. *European Biophysics Journal*, *38*, 319–327.

Evans, R. J., Derkach, V., & Surprenant, A. (1992). ATP mediates fast synaptic transmission in mammalian neurons. *Nature*, *357*, 503–505.

Fabbretti, E., D'Arco, M., Fabbro, A., et al. (2006). Delayed upregulation of ATP $P2X_3$ receptors of trigeminal sensory neurons by calcitonin gene-related peptide. *The Journal of Neuroscience*, *26*, 6163–6171.

Facer, P., Knowles, C. H., Tam, P. K., et al. (2001). Novel capsaicin (VR1) and purinergic (P2X3) receptors in Hirschsprung's intestine. *Journal of Pediatric Research*, *36*, 1679–1684.

Ferrari, D., Munerati, M., Melchiorri, L., et al. (1994). Responses to extracellular ATP of lymphoblastoid cell lines from Duchenne muscular dystrophy patients. *The American Journal of Physiology*, *267*, C886–C892.

Fountain, S. J., & Burnstock, G. (2009). An evolutionary history of P2X receptors. *Purinergic Signalling*, *5*, 269–272.

Franke, H., Krügel, U., & Illes, P. (2006). P2 receptors and neuronal injury. *Pflügers Archiv—European Journal of Physiology*, *452*, 622–644.

Fumagalli, M., Ceruti, S., Verderio, C., et al. (2006). ATP as a neurotransmitter of pain in migraine: A functional role for P2Y receptors in primary cultures from mouse trigeminal sensory ganglia. *Purinergic Signalling*, *2*, 120–121.

Galligan, J. J. (2004). Enteric P2X receptors as potential targets for drug treatment of the irritable bowel syndrome. *British Journal of Pharmacology*, *141*, 1294–1302.

Gayle, S., & Burnstock, G. (2005). Immunolocalisation of P2X and P2Y nucleotide receptors in the rat nasal mucosa. *Cell and Tissue Research*, *319*, 27–36.

Gever, J., Cockayne, D. A., Dillon, M. P., et al. (2006). Pharmacology of P2X channels. *Pflügers Archiv—European Journal of Physiology*, *452*, 513–537.

Gever, J. R., Rothschild, S., Henningsen, R., et al. (2010). AF-353, a novel, potent orally bioavailable P2X3/P2X2/3 receptor antagonist. *British Journal of Pharmacology*, *160*, 1387–1398.

Glass, R., Bardini, M., Robson, T., et al. (2001). Expression of nucleotide P2X receptor subtypes during spermatogenesis in the adult rat testis. *Cells, Tissues, Organs*, *169*, 377–387.

Gonzales, E. B., Kawate, T., & Gouaux, E. (2009). Pore architecture and ion sites in acid-sensing ion channels and P2X receptors. *Nature*, *460*, 599–604.

Gorodeski, G. I. (2002). Regulation of transcervical permeability by two distinct P2 purinergic receptor mechanisms. *American Journal of Physiology. Cell Physiology*, *282*, C75–C83.

Gourine, A. V., Atkinson, L., Deuchars, J., et al. (2003). Purinergic signalling in the medullary mechanisms of respiratory control in the rat: Respiratory neurones express the $P2X_2$ receptor subunit. *The Journal of Physiology*, *552*, 197–211.

Greig, A. V. H., James, S. E., McGrouther, D. A., et al. (2003). Purinergic receptor expression in the regenerating epidermis in a rat model of normal and delayed wound healing. *Experimental Dermatology*, *12*, 860–871.

Greig, A. V. H., Linge, C., Healy, V., et al. (2003). Expression of purinergic receptors in non-melanoma skin cancers and their functional roles in A431 cells. *Journal of Investigative Dermatology*, *121*, 315–327.

Greig, A. V. H., Linge, C., Terenghi, G., et al. (2003). Purinergic receptors are part of a functional signalling system for proliferation and differentiation of human epidermal keratinocytes. *The Journal of Investigative Dermatology*, *120*, 1007–1015.

Grol, M. W., Panupinthu, N., Korcok, J., et al. (2009). Expression, signaling, and function of $P2X_7$ receptors in bone. *Purinergic Signalling*, *5*, 205–221.

Gu, J. G., & MacDermott, A. B. (1997). Activation of ATP P2X receptors elicits glutamate release from sensory neuron synapses. *Nature*, *389*, 749–753.

Guo, C., Masin, M., Qureshi, O. S., et al. (2007). Evidence for functional $P2X_4/P2X_7$ heteromeric receptors. *Molecular Pharmacology*, *72*, 1447–1456.

Guo, L. H., & Schluesener, H. J. (2005). Lesional accumulation of $P2X_4$ receptor$^+$ macrophages in rat CNS during experimental autoimmune encephalomyelitis. *Neuroscience*, *134*, 199–205.

Guo, L. H., Trautmann, K., & Schluesener, H. J. (2005). Expression of $P2X_4$ receptor by lesional activated microglia during formalin-induced inflammatory pain. *Journal of Neuroimmunology*, *163*, 120–127.

Haines, W. R., Torres, G. E., Voigt, M. M., et al. (1999). Properties of the novel ATP-gated ionotropic receptor composed of the $P2X_1$ and $P2X_5$ isoforms. *Molecular Pharmacology*, *56*, 720–727.

Hamilton, S. G., McMahon, S. B., & Lewin, G. R. (2001). Selective activation of nociceptors by P2X receptor agonists in normal and inflamed rat skin. *The Journal of Physiology*, *534*, 437–445.

Hansen, M. A., Dutton, J. L., Balcar, V. J., et al. (1999). P_{2X} (purinergic) receptor distributions in rat blood vessels. *Journal of the Autonomic Nervous System*, *75*, 147–155.

Hegg, C. C., Davis, K., & Lucero, M. T. (2003). Inhibition of heat-shock protein induction in mouse OE by in vivo administration of purinergic receptor antagonists. *Chemical Senses*, *28*, A110.

Henning, R. H. (1997). Purinoceptors in neuromuscular transmission. *Pharmacology and Therapeutics*, *74*, 115–128.

Hillman, K. A., Woolf, A. S., Johnson, T. M., et al. (2004). The $P2X_7$ ATP receptor modulates renal cyst development *in vitro*. *Biochemical and Biophysical Research Communications*, *322*, 434–439.

Hoebertz, A., Arnett, T. R., & Burnstock, G. (2003). Regulation of bone resorption and formation by purines and pyrimidines. *Trends in Pharmacological Sciences*, *24*, 290–297.

Housley, G. D. (2000). Physiological effects of extracellular nucleotides in the inner ear. *Clinical and Experimental Pharmacology and Physiology*, *27*, 575–580.

Housley, G. D., Bringmann, A., & Reichenbach, A. (2009). Purinergic signaling in special senses. *Trends in Neurosciences*, *32*, 128–141.

Hughes, J. P., Hatcher, J. P., & Chessell, I. P. (2007). The role of $P2X_7$ in pain and inflammation. *Purinergic Signalling*, *3*, 163–169.

Inoue, K. (2007). P2 receptors and chronic pain. *Purinergic Signalling*, *3*, 135–144.

Inoue, K., Koizumi, S., & Ueno, S. (1996). Implication of ATP receptors in brain functions. *Progress in Neurobiology*, *50*, 483–492.

Jarvis, M. F., & Khakh, B. S. (2009). ATP-gated P2X cation-channels. *Neuropharmacology*, *56*, 208–215.

Jiang, L. H., Kim, M., Spelta, V., et al. (2003). Subunit arrangement in P2X receptors. *The Journal of Neuroscience*, *23*, 8903–8910.

Jo, Y. H., & Schlichter, R. (1999). Synaptic corelease of ATP and GABA in cultured spinal neurons. *Nature Neuroscience*, *2*, 241–245.

Jones, C. A., Vial, C., Sellers, L. A., et al. (2004). Functional regulation of P2X6 receptors by N-linked glycosylation: Identification of a novel α,β-methylene ATP-sensitive phenotype. *Molecular Pharmacology*, *65*, 979–985.

Kälvegren, H., Skoglund, C., Helldahl, C., et al. (2010). Toll-like receptor 2 stimulation of platelets is mediated by purinergic $P2X_1$-dependent Ca^{2+} mobilisation, cyclooxygenase and purinergic $P2Y_1$ and $P2Y_{12}$ receptor activation. *Thrombosis and Haemostasis*, *103*, 398–407.

Kassa, R. M., Bentivoglio, M., & Mariotti, R. (2007). Changes in the expression of $P2X_1$ and $P2X_2$ purinergic receptors in facial motoneurons after nerve lesions in rodents and correlation with motoneuron degeneration. *Neurobiology of Disease*, *25*, 121–133.

Kawate, T., Michel, J. C., Birdsong, W. T., et al. (2009). Crystal structure of the ATP-gated $P2X_4$ ion channel in the closed state. *Nature*, *460*, 592–598.

Khakh, B. S. (2001). Molecular physiology of P2X receptors and ATP signalling at synapses. *Nature Reviews. Neuroscience*, *2*, 165–174.

Khakh, B. S., Bao, X. R., Labarca, C., et al. (1999). Neuronal P2X transmitter-gated cation channels change their ion selectivity in seconds. *Nature Neuroscience*, *2*, 322–330.

King, B. F., Townsend-Nicholson, A., Wildman, S. S., et al. (2000). Coexpression of rat $P2X_2$ and $P2X_6$ subunits in *Xenopus* oocytes. *The Journal of Neuroscience*, *20*, 4871–4877.

Knight, G. E., Bodin, P., de Groat, W. C., et al. (2002). ATP is released from guinea pig ureter epithelium on distension. *American Journal of Physiology—Renal Physiology*, *282*, F281–F288.

Knutsen, L. J. S., & Murray, T. F. (1997). Adenosine and ATP in epilepsy. In K. A. Jacobson & M. F. Jarvis (Eds.), *Purinergic approaches in experimental therapeutics* (pp. 423–447). New York: Wiley-Liss.

Komarova, S. V., Dixon, S. J., & Sims, S. M. (2001). Osteoclast ion channels: Potential targets for antiresorptive drugs. *Current Pharmaceutical Design*, *7*, 637–654.

Koshimizu, T., Koshimizu, M., & Stojilkovic, S. S. (1999). Contributions of the C-terminal domain to the control of P2X receptor desensitization. *The Journal of Biological Chemistry*, *274*, 37651–37657.

Lalo, U., Pankratov, Y., Wichert, S. P., et al. (2008). P2X1 and P2X5 subunits form the functional P2X receptor in mouse cortical astrocytes. *The Journal of Neuroscience*, *28*, 5473–5480.

Lammas, D. A., Stober, C., Harvey, C. J., et al. (1997). ATP-induced killing of mycobacteria by human macrophages is mediated by purinergic P2Z ($P2X_7$) receptors. *Immunity*, *7*, 433–444.

Lê, K. T., Babinski, K., & Séguéla, P. (1998). Central $P2X_4$ and $P2X_6$ channel subunits coassemble into a novel heteromeric ATP receptor. *The Journal of Neuroscience*, *18*, 7152–7159.

Lê, K. T., Boué-Grabot, E., Archambault, V., et al. (1999). Functional and biochemical evidence for heteromeric ATP-gated channels composed of $P2X_1$ and $P2X_5$ subunits. *The Journal of Biological Chemistry*, *274*, 15415–15419.

Lecut, C., Frederix, K., Johnson, D. M., et al. (2009). $P2X_1$ ion channels promote neutrophil chemotaxis through Rho kinase activation. *Journal of Immunology*, *183*, 2801–2809.

Lee, H. Y., Bardini, M., & Burnstock, G. (2000). Distribution of P2X receptors in the urinary bladder and the ureter of the rat. *The Journal of Urology*, *163*, 2002–2007.

Lemaire, I., Falzoni, S., Leduc, N., et al. (2006). Involvement of the purinergic $P2X_7$ receptor in the formation of multinucleated giant cells. *Journal of Immunology*, *177*, 7257–7265.

Lemaire, I., & Leduc, N. (2004). Purinergic $P2X_7$ receptor function in lung alveolar macrophages: Pharmacologic characterisation and bidirectional regulation by Th1 and Th2 cytokines. *Drug Development Research*, *59*, 118–127.

Lewis, C., Neidhart, S., Holy, C., et al. (1995). Coexpression of $P2X_2$ and $P2X_3$ receptor subunits can account for ATP-gated currents in sensory neurons. *Nature*, *377*, 432–435.

Lu, S. H., Fraser, M. O., Chancellor, M. B., et al. (2002). Voiding dysfunction and purinergic mechanisms in awake chronic spinal cord injured (SCI) rats. *Society for Neuroscience*, Progran No. 68.6.2002. 2002.

Lucae, S., Salyakina, D., Barden, N., et al. (2006). P2RX7, a gene coding for a purinergic ligand-gated ion channel, is associated with major depressive disorder. *Human Molecular Genetics*, *15*, 2438–2445.

Lustig, K. D., Shiau, A. K., Brake, A. J., et al. (1993). Expression cloning of an ATP receptor from mouse neuroblastoma cells. *Proceedings of the National Academy of Sciences of the United States of America*, *90*, 5113–5117.

Mantuano-Barradas, M., Henriques-Pons, A., Araújo-Jorge, T. C., et al. (2003). Extracellular ATP induces cell death in CD4+/CD8+ double-positive thymocytes in mice infected with Trypanosoma cruzi. *Microbes and Infection*, *5*, 1363–1371.

McGaraughty, S., Chu, K. L., Namovic, M. T., et al. (2007). $P2X_7$-related modulation of pathological nociception in rats. *Neuroscience*, *146*, 1817–1828.

McLarnon, J. G., Ryu, J. K., Walker, D. G., et al. (2006). Upregulated expression of purinergic $P2X_7$ receptor in Alzheimer disease and amyloid-β peptide-treated microglia and in peptide-injected rat hippocampus. *Journal of Neuropathology and Experimental Neurology*, *65*, 1090–1097.

Mio, K., Kubo, Y., Ogura, T., et al. (2005). Visualization of the trimeric $P2X_2$ receptor with a crown-capped extracellular domain. *Biochemical and Biophysical Research Communications*, *337*, 998–1005.

Mizumoto, N., Mummert, M. E., Shalhevet, D., et al. (2003). Keratinocyte ATP release assay for testing skin-irritating potentials of structurally diverse chemicals. *Journal of Investigative Dermatology*, *121*, 1066–1072.

Mizuno, K., Okamoto, H., & Horio, T. (2004). Inhibitory influences of xanthine oxidase inhibitor and angiotensin I-converting enzyme inhibitor on multinucleated giant cell formation from monocytes by downregulation of adhesion molecules and purinergic receptors. *The British Journal of Dermatology*, *150*, 205–210.

Musa, H., Tellez, J. O., Chandler, N. J., et al. (2009). P2 purinergic receptor mRNA in rat and human sinoatrial node and other heart regions. *Naunyn-Schmiedeberg's Archives of Pharmacology*, *379*, 541–549.

Narcisse, L., Scemes, E., Zhao, Y., et al. (2005). The cytokine IL-1β transiently enhances $P2X_7$ receptor expression and function in human astrocytes. *Glia*, *49*, 245–258.

Nicke, A. (2008). Homotrimeric complexes are the dominant assembly state of native $P2X_7$ subunits. *Biochemical and Biophysical Research Communications*, *377*, 803–808.

Nicke, A., Baumert, H. G., Rettinger, J., et al. (1998). $P2X_1$ and $P2X_3$ receptors form stable trimers: A novel structural motif of ligand-gated ion channels. *The EMBO Journal*, *17*, 3016–3028.

Nicke, A., Kerschensteiner, D., & Soto, F. (2005). Biochemical and functional evidence for heteromeric assembly of $P2X_1$ and $P2X_4$ subunits. *Journal of Neurochemistry*, *92*, 925–933.

North, R. A. (2002). Molecular physiology of P2X receptors. *Physiological Reviews*, *82*, 1013–1067.

Nurden, A. T. (2007). Does ATP act through $P2X_1$ receptors to regulate platelet activation and thrombus formation? *Journal of Thrombosis and Haemostasis*, *5*, 907–909.

O'Reilly, B. A., Kosaka, A. H., Knight, G. E., et al. (2002). P2X receptors and their role in female idiopathic detrusor instability. *The Journal of Urology*, *167*, 157–164.

Orriss, I. R., Burnstock, G., & Arnett, T. R. (2010). Purinergic signalling and bone remodelling. *Current Opinion in Pharmacology*, *10*, 322–330.

Oses, J. P. (2006). Modification by kainate-induced convulsions of the density of presynaptic P2X receptors in the rat hippocampus. *Purinergic Signalling*, *2*, 252–253.

Panenka, W., Jijon, H., Herx, L. M., et al. (2001). $P2X_7$-like receptor activation in astrocytes increases chemokine monocyte chemoattractant protein-1 expression via mitogen-activated protein kinase. *The Journal of Neuroscience*, *21*, 7135–7142.

Parvathenani, L. K., Tertyshnikova, S., Greco, C. R., et al. (2003). $P2X_7$ mediates superoxide production in primary microglia and is up-regulated in a transgenic mouse model of Alzheimer's disease. *The Journal of Biological Chemistry*, *278*, 13309–13317.

Pelegrin, P., & Surprenant, A. (2006). Pannexin-1 mediates large pore formation and interleukin-1β release by the ATP-gated P2X7 receptor. *The EMBO Journal*, *25*, 5071–5082.

Pelleg, A., Hurt, C. M., Soler-Baillo, J. M., et al. (1993). Electrophysiological-anatomic correlates of ATP-triggered vagal reflex in dogs. *The American Journal of Physiology*, *265*, H681–H690.

Pintor, J., Peral, A., Pelaez, T., et al. (2003). Nucleotides and dinucleotides in ocular physiology: New possibilities of nucleotides as therapeutic agents in the eye. *Drug Development Research*, *59*, 136–145.

Ralevic, V., & Burnstock, G. (1998). Receptors for purines and pyrimidines. *Pharmacological Reviews*, *50*, 413–492.

Raouf, R., Chabot-Dore, A. J., Ase, A. R., et al. (2007). Differential regulation of microglial $P2X_4$ and $P2X_7$ ATP receptors following LPS-induced activation. *Neuropharmacology*, *53*, 496–504.

Rapaport, E. (1983). Treatment of human tumor cells with ADP or ATP yields arrest of growth in the S phase of the cell cycle. *Journal of Cellular Physiology*, *114*, 279–283.

Rappold, P. M., Lynd-Balta, E., & Joseph, S. A. (2006). $P2X_7$ receptor immunoreactive profile confined to resting and activated microglia in the epileptic brain. *Brain Research*, *1089*, 171–178.

Ribeiro, J. A., & Walker, J. (1975). The effects of adenosine triphosphate and adenosine diphosphate on transmission at the rat and frog neuromuscular junctions. *British Journal of Pharmacology*, *54*, 213–218.

Roberts, V. H. J., Brockman, D. E., Pitzer, B. A., et al. (2005). $P2X_4$ purinergic receptor protein expression is upregulated in placenta from preeclamptic pregnancies. *Journal of the Society of Gynecologic Investigation*, *12*, P363A.

Rong, W., & Burnstock, G. (2004). Activation of ureter nociceptors by exogenous and endogenous ATP in guinea pig. *Neuropharmacology*, *47*, 1093–1101.

Rong, W., Burnstock, G., & Spyer, K. M. (2000). P2X purinoceptor-mediated excitation of trigeminal lingual nerve terminals in an *in vitro* intra-arterially perfused rat tongue preparation. *The Journal of Physiology*, *524*, 891–902.

Rossato, M., La Sala, G. B., Balasini, M., et al. (1999). Sperm treatment with extracellular ATP increases fertilization rates in in-vitro fertilization for male factor infertility. *Human Reproduction*, *14*, 694–697.

Ryten, M., Hoebertz, A., & Burnstock, G. (2001). Sequential expression of three receptor subtypes for extracellular ATP in developing rat skeletal muscle. *Developmental Dynamics*, *221*, 331–341.

Ryten, M., Yang, S. Y., Dunn, P. M., et al. (2004). Purinoceptor expression in regenerating skeletal muscle in the mdx mouse model of muscular dystrophy and in satellite cell cultures. *The FASEB Journal*, *18*, 1404–1406.

Schulman, E. S., Glaum, M. C., Post, T., et al. (1999). ATP modulates anti-IgE-induced release of histamine from human lung mast cells. *American Journal of Respiratory Cell and Molecular Biology*, *20*, 530–537.

Schwiebert, E. M., Wallace, D. P., Braunstein, G. M., et al. (2002). Autocrine extracellular purinergic signaling in epithelial cells derived from polycystic kidneys. *American Journal of Physiology—Renal Physiology*, *282*, F763–F775.

Shabbir, M., Thompson, C. S., Jarmulowicz, M., et al. (2008). Effect of extracellular ATP on the growth of hormone refractory prostate cancer *in vivo*. *British Journal of Urology International*, *102*, 108–112.

Shabbir, M., Thompson, C. S., Mikhailidis, D. P., et al. (2004). Adenosine 5′-triphosphate (ATP) increases the cytotoxic effect of mitomycin C in the treatment of high-grade bladder cancer. *The Journal of Urology*, *171*, 260.

Shen, J. B., Cronin, C., Sonin, D., et al. (2007). P2X purinergic receptor-mediated ionic current in cardiac myocytes of calsequestrin model of cardiomyopathy: Implications for the treatment of heart failure. *American Journal of Physiology. Heart and Circulatory Physiology*, *292*, H1077–H1084.

Shinoda, M., Feng, B., & Gebhart, G. F. (2009). Peripheral and central $P2X_3$ receptor contributions to colon mechanosensitivity and hypersensitivity in the mouse. *Gastroenterology*, *137*, 2096–2104.

Silinsky, E. M., Gerzanich, V., & Vanner, S. M. (1992). ATP mediates excitatory synaptic transmission in mammalian neurones. *British Journal of Pharmacology*, *106*, 762–763.

Sommer, J. A., Fisette, P. L., Hu, Y., et al. (1999). Purinergic receptor modulation of LPS-stimulated signaling events and nitric oxide release in RAW 264.7 macrophages. *Journal of Endotoxin Research*, *5*, 70–74.

Stevenson, R. O., Taylor, R. M., Wiley, J. S., et al. (2009). The $P2X_7$ receptor mediates the uptake of organic cations in canine erythrocytes and mononuclear leukocytes: Comparison to equivalent human cell types. *Purinergic Signalling*, *5*, 385–394.

Stojilkovic, S. S. (2009). Purinergic regulation of hypothalamopituitary functions. *Trends in Endocrinology and Metabolism*, *20*, 460–468.

Stojilkovic, S. S., He, M. L., Koshimizu, T. A., et al. (2010). Signaling by purinergic receptors and channels in the pituitary gland. *Molecular and Cellular Endocrinology*, *314*, 184–191.

Sugiyama, T., Kobayashi, M., Kawamura, H., et al. (2004). Enhancement of $P2X_7$-induced pore formation and apoptosis: An early effect of diabetes on the retinal microvasculature. *Investigative Ophthalmology and Visual Science*, *45*, 1026–1032.

Surprenant, A., Rassendren, F., Kawashima, E., et al. (1996). The cytolytic P_{2Z} receptor for extracellular ATP identified as a P_{2X} receptor ($P2X_7$). *Science*, *272*, 735–738.

Surprenant, A., Schneider, D. A., Wilson, H. L., et al. (2000). Functional properties of heteromeric $P2X_{1/5}$ receptors expressed in HEK cells and excitatory junction potentials in guinea-pig submucosal arterioles. *Journal of the Autonomic Nervous System*, *81*, 249–263.

Takenouchi, T., Iwamaru, Y., Imamura, M., et al. (2007). Prion infection correlates with hypersensitivity of P2X7 nucleotide receptor in a mouse microglial cell line. *FEBS Letters*, *581*, 3019–3026.

Tempest, H. V., Dixon, A. K., Turner, W. H., et al. (2004). P2X and P2X receptor expression in human bladder urothelium and changes in interstitial cystitis. *British Journal of Urology International*, *93*, 1344–1348.

Tian, G. F., Azmi, H., Takano, T., et al. (2005). An astrocytic basis of epilepsy. *Natural Medicines*, *11*, 973–981.

Torres, G. E., Egan, T. M., & Voigt, M. M. (1999). Hetero-oligomeric assembly of P2X receptor subunits. Specificities exist with regard to possible partners. *Journal of Biological Chemistry*, *274*, 6653–6659.

Torres, G. E., Haines, W. R., Egan, T. M., et al. (1998). Co-expression of $P2X_1$ and $P2X_5$ receptor subunits reveals a novel ATP-gated ion channel. *Molecular Pharmacology*, *54*, 989–993.

Trudeau, F., Gagnon, S., & Massicotte, G. (2004). Hippocampal synaptic plasticity and glutamate receptor regulation: Influences of diabetes mellitus. *European Journal of Pharmacology*, *490*, 177–186.

Tsuda, M., Shigemoto-Mogami, Y., Koizumi, S., et al. (2003). $P2X_4$ receptors induced in spinal microglia gate tactile allodynia after nerve injury. *Nature*, *424*, 778–783.

Turner, C., Ramesh, B., Srai, S. K. S., et al. (2004). Altered P2 receptor expression in the Han: SPRD cy/$^+$ rat, a model of autosomal dominant polycystic kidney disease. *Cells, Tissues, Organs*, *178*, 168–179.

Turner, C., Tam, F. W. K., Lai, P.-C., et al. (2004b). Glomerular expression of the ATP-sensitive $P2X_7$ receptor is increased in proliferative glomerulonephritis. In: *ASM Meeting, 2004*.

Turner, C., Vonend, O., Chan, C. M., et al. (2003). The pattern of distribution of selected ATP-sensitive P2 receptor subtypes in normal rat kidney: An immunohistological study. *Cells, Tissues, Organs*, *175*, 105–117.

Unwin, R. J., Bailey, M. A., & Burnstock, G. (2003). Purinergic signaling along the renal tubule: The current state of play. *News in Physiological Sciences*, *18*, 237–241.

Valera, S., Hussy, N., Evans, R. J., et al. (1994). A new class of ligand-gated ion channel defined by P_{2X} receptor for extra-cellular ATP. *Nature*, *371*, 516–519.

Vanderwinden, J. M., Timmermans, J. P., & Schiffmann, S. N. (2003). Glial cells, but not interstitial cells, express P2X7, an ionotropic purinergic receptor, in rat gastrointestinal musculature. *Cell and Tissue Research*, *312*, 149–154.

Verkhratsky, A., Krishtal, O. A., & Burnstock, G. (2009). Purinoceptors in neuroglia. *Molecular Neurobiology*, *39*, 190–208.

Vial, C., Rigby, R., & Evans, R. J. (2006). Contribution of $P2X_1$ receptor intracellular basic residues to channel properties. *Biochemical and Biophysical Research Communications*, *350*, 244–248.

Vial, C., Roberts, J. A., & Evans, R. J. (2004). Molecular properties of ATP-gated P2X receptor ion channels. *Trends in Pharmacological Sciences*, *25*, 487–493.

Vlaskovska, M., Kasakov, L., Rong, W., et al. (2001). $P2X_3$ knockout mice reveal a major sensory role for urothelially released ATP. *The Journal of Neuroscience*, *21*, 5670–5677.

Vonend, O., Turner, C., Chan, C. M., et al. (2004). Glomerular expression of the ATP-sensitive $P2X_7$ receptor in diabetic and hypertensive rat models. *Kidney International*, *66*, 157–166.

Wang, J. C., Raybould, N. P., Luo, L., et al. (2003). Noise induces up-regulation of $P2X_2$ receptor subunit of ATP-gated ion channels in the rat cochlea. *NeuroReport*, *14*, 817–823.

Webb, T. E., Simon, J., Krishek, B. J., et al. (1993). Cloning and functional expression of a brain G-protein-coupled ATP receptor. *FEBS Letters*, *324*, 219–225.

White, N., & Burnstock, G. (2006). P2 receptors and cancer. *Trends in Pharmacological Sciences*, *27*, 211–217.

White, N., Butler, P. E. M., & Burnstock, G. (2005a). Human melanomas express functional $P2X_7$ receptors. *Cell and Tissue Research*, *321*, 411–418.

White, N., Ryten, M., Clayton, E., et al. (2005b). P2Y purinergic receptors regulate the growth of human melanomas. *Cancer Letters*, *224*, 81–91.

Wilkinson, W. J., Jiang, L. H., Surprenant, A., et al. (2006). Role of ectodomain lysines in the subunits of the heteromeric $P2X_{2/3}$ receptor. *Molecular Pharmacology*, *70*, 1159–1163.

Wirkner, K., Sperlagh, B., & Illes, P. (2007). $P2X_3$ receptor involvement in pain states. *Molecular Neurobiology*, *36*, 165–183.

Witting, A., Walter, L., Wacker, J., et al. (2004). $P2X_7$ receptors control 2-arachidonoylglycerol production by microglial cells. *Proceedings of the National Academy of Sciences of the United States of America*, *101*, 3214–3219.

Wynn, G., Bei, M., Ruan, H.-Z., et al. (2004). Purinergic component of mechanosensory transduction is increased in a rat model of colitis. *American Journal of Physiology. Gastrointestinal and Liver Physiology*, *287*, G647–G657.

Wynn, G., Rong, W., Xiang, Z., et al. (2003). Purinergic mechanisms contribute to mechanosensory transduction in the rat colorectum. *Gastroenterology*, *125*, 1398–1409.

Xiang, Z., Bo, X., & Burnstock, G. (1999). P2X receptor immunoreactivity in the rat cochlea, vestibular ganglion and cochlear nucleus. *Hearing Research*, *128*, 190–196.

Xiang, Z., & Burnstock, G. (2004). Development of nerves expressing $P2X_3$ receptors in the myenteric plexus of rat stomach. *Histochemistry and Cell Biology*, *122*, 111–119.

Xu, G. Y., Shenoy, M., Winston, J. H., et al. (2008). P2X receptor-mediated visceral hyperalgesia in a rat model of chronic visceral hypersensitivity. *Gut*, *57*, 1230–1237.

Yiangou, Y., Facer, P., Baecker, P. A., et al. (2001). ATP-gated ion channel $P2X_3$ is increased in human inflammatory bowel disease. *Neurogastroenterology and Motility*, *13*, 365–369.

Young, M. T. (2010). P2X receptors: Dawn of the post-structure era. *Trends in Biochemical Sciences*, *35*, 83–90.

Young, M. T., Fisher, J. A., Fountain, S. J., et al. (2008). Molecular shape, architecture, and size of $P2X_4$ receptors determined using fluorescence resonance energy transfer and electron microscopy. *The Journal of Biological Chemistry*, *283*, 26241–26251.

Yu, Y., & de Groat, W. C. (2004). Purinergic agonists sensitize pelvic afferent nerves in the vitro urinary bladder-pelvic nerve preparation of the rat. *The Journal of Urology*, *169*, 47.

Zemkova, H., Balik, A., Jindrichová, M., et al. (2008). Molecular structure of purinergic P2X receptors and their expression in the hypothalamus and pituitary. *Physiological Research*, *57*, S23–S38.

Zhao, D. M., Xue, H. H., Chida, K., et al. (2000). Effect of erythromycin on ATP-induced intracellular calcium response in A549 cells. *American Journal of Physiology. Lung Cellular and Molecular Physiology*, *278*, L726–L736.

Zsembery, A., Boyce, A. T., Liang, L., et al. (2003). Sustained calcium entry through P2X nucleotide receptor channels in human airway epithelial cells. *The Journal of Biological Chemistry*, *278*, 13398–13408.

Ivar von Kügelgen* and T. Kendall Harden†

*Department of Pharmacology and Toxicology, University of Bonn, Bonn, Germany

†Department of Pharmacology, University of North Carolina School of Medicine, Chapel Hill, North Carolina, USA

Molecular Pharmacology, Physiology, and Structure of the P2Y Receptors

Abstract

The P2Y receptors are a widely expressed group of eight nucleotide-activated G protein-coupled receptors (GPCRs). The $P2Y_1$(ADP), $P2Y_2$(ATP/UTP), $P2Y_4$(UTP), $P2Y_6$(UDP), and $P2Y_{11}$(ATP) receptors activate G_q and therefore robustly promote inositol lipid signaling responses. The $P2Y_{12}$(ADP), $P2Y_{13}$(ADP), and $P2Y_{14}$(UDP/UDP-glucose) receptors activate G_i leading to inhibition of adenylyl cyclase and to $G\beta\gamma$-mediated activation of a range of effector proteins including phosphoinositide 3-kinase-γ, inward rectifying K^+ (GIRK) channels, phospholipase C-β2 and -β3, and G protein-receptor kinases 2 and 3. A broad range of physiological responses occur downstream of activation of these receptors ranging from Cl^- secretion by epithelia to aggregation of platelets to neurotransmission. Useful structural models of the P2Y receptors have evolved from extensive genetic analyses coupled with molecular modeling based on three-dimensional structures obtained for rhodopsin and several other GPCRs. Selective ligands have been synthesized for most of the P2Y receptors with the most prominent successes attained with highly selective agonist and antagonist molecules for the ADP-activated $P2Y_1$ and $P2Y_{12}$ receptors. The widely prescribed drug, clopidogrel, which results in irreversible blockade of the platelet $P2Y_{12}$ receptor, is the most important therapeutic agent that targets a P2Y receptor.

Advances in Pharmacology, Volume 61

1054-3589/11 $35.00
10.1016/B978-0-12-385526-8.00012-6

I. Introduction

Drury and Szent-Györgyi (1929) introduced the idea of purines as extracellular signaling molecules, but the concept that adenosine and adenine nucleotides signal through independent P1 and P2 receptors, respectively, was not clearly formulated until the mid-1970s (Burnstock, 1978). Adenosine was eventually shown to act through multiple P1 receptors, and Burnstock and Kennedy proposed in 1985 the existence of two subtypes of nucleotide-activated P2 receptors (Burnstock & Kennedy, 1985). Delineation of these so-called P2X and P2Y receptors initially was based on pharmacological approaches, but molecular cloning and cell-signaling studies subsequently resolved seven P2X receptors that function as fast-responding ligand-gated ion channels (North, 2002) and eight metabotropic P2Y receptors that signal by activating heterotrimeric G proteins (Ralevic & Burnstock, 1998). Appreciation that uridine nucleotides also function as extracellular signaling molecules accrued in the 1980s (Seifert & Schultz, 1989; von Kügelgen et al., 1987), and at least four of the extant P2Y receptors are physiologically regulated by UTP or UDP. This chapter focuses on pharmacological and mechanistic aspects of the P2Y receptors. The reader is referred to other general reviews (Abbracchio et al., 2006; von Kügelgen, 2006) on this topic and to several monographs that provide greater details on medical chemistry (Jacobson et al., 2009; Müller, 2002) and physiology (Burnstock, 2007) related to the P2Y receptors.

II. Signaling Through Heterotrimeric G Proteins

As is the case with a broad range of hormones, neurotransmitters, growth factors, and sensory stimuli, extracellular nucleotides promote physiological effects by activating seven transmembrane-spanning receptors that in turn directly activate heterotrimeric G proteins. The agonist-activated state of a G protein-coupled receptor (GPCR) results in a decrease in affinity of the heterotrimeric G protein for GDP, subsequent binding of GTP, and dissociation of the heterotrimer into GTP-Gα and free G$\beta\gamma$ (Hepler & Gilman, 1992). As initially conceived, GTP-Gα activates downstream effector proteins, whereas G$\beta\gamma$ promotes association of GDP-Gα with the plasma membrane, maintains the G protein in an inactive state, and contributes to receptor/G protein coupling. Approximately 20 different Gα-subunits exist and these can be subdivided into four groups based on structural similarity and effector selectivity (Hepler & Gilman, 1992). Gα-subunits of the Gs family (Gα_s and Gα_{olf}) activate adenylyl cyclase, Gα-subunits of the G$_i$ family (Gα_{i1}, Gα_{i2}, Gα_{i3}, Gα_o, Gα_t, and Gα_{gust}) inhibit adenylyl cyclase and activate cyclic GMP phosphodiesterase, Gα-subunits of the G$_q$ family (Gα_q, Gα_{11}, Gα_{14}, and Gα_{16}) activate phospholipase C-β isozymes, and Gα-subunits of the G12 family ($G\alpha_{12}$ and $G\alpha_{13}$) activate guanine nucleotide exchange factors (GEFs) for Rho family GTPases. It is now clear that Gβγ also directly activates certain effector proteins (e.g., ion channels, phospholipase C-β, phosphatidylinositol 3-kinase-γ, and the Rac guanine nucleotide exchange factor P-Rex1) and, therefore, also fulfills downstream signaling roles upon GPCR activation (Smrcka, 2008). Gβγ-mediated signaling is a particularly important consequence of activation of heterotrimeric G proteins of the G_i family since these proteins are often present at very high concentrations in the plasma membrane.

III. The P2Y Receptors

A. Nomenclature and G Protein Signaling Selectivity

The focus of this chapter is on the eight unambiguously identified P2Y receptors (Fig. 1). These nucleotide-activated GPCR can be divided into two subclasses on the basis of structural similarity (see below) and G protein-signaling selectivity (Abbracchio et al., 2006). The $P2Y_1$ receptor subfamily includes the $P2Y_1$, $P2Y_2$, $P2Y_4$, $P2Y_6$, and $P2Y_{11}$ receptors, which all activate G_q and, therefore, promote phospholipase C-β mediated hydrolysis of PtdIns $(4,5)P_2$ and inositol lipid signaling. Gαq also directly binds and activates a guanine nucleotide exchange factor (p63 RhoGEF) for Rho GTPases (Lutz et al., 2005), and Rho-mediated signaling also is a predictable cellular response to activation of $P2Y_1$-like GPCR. In contrast, the $P2Y_{12}$ receptor subfamily of receptors ($P2Y_{12}$, $P2Y_{13}$, and $P2Y_{14}$ receptors) activate G_i and, therefore, signal through $G\alpha_i$-dependent inhibition of adenylyl cyclase. Activation of heterotrimers of the G_i family also results in release in relatively large amounts of Gβγ, and Gβγ directly binds and activates a diverse group of effectors (Smrcka, 2008), including PtdIns 3-kinase-γ, phospholipase C-β 2 and -β3, inward rectifying K^+ (GIRK) channels, GPCR kinases 2 and 3, a GEF for Rac (P-Rex1) and SNAP-25, which is involved in vesicle membrane fusion. Activation of $P2Y_{12}$, $P2Y_{13}$, or $P2Y_{14}$ receptors has not been unambiguously shown to result in activation of these Gβγ-regulated effectors. However, it is highly likely that the G_i-coupled P2Y receptors activate various of these signaling responses in a cell-specific manner that depends on relative expression of receptor, G protein, and effector isoforms. The awkward nonconsecutive numbering of the confirmed P2Y receptors was the result of premature assignment of a "P2Y" designation to several GPCR that eventually proved to be not activated by nucleotides or were in fact species homologues of previously named P2Y receptors. We first introduce the general pharmacological, signaling, and physiological properties of each receptor and then focus on its structure and mechanisms

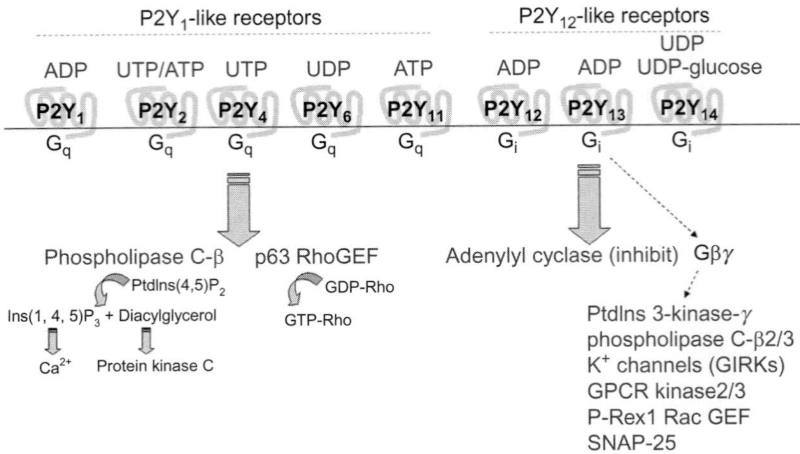

FIGURE 1 Cognate agonists and downstream signaling responses of the P2Y receptors. The $P2Y_1$, $P2Y_2$, $P2Y_4$, $P2Y_6$, and $P2Y_{11}$ receptors belong to the $P2Y_1$-like subgroup of P2Y receptors that selectively activate Gα-subunits ($G\alpha_q$, $G\alpha_{11}$, $G\alpha_{14}$, and $G\alpha_{16}$) of the G_q family of heterotrimeric G proteins. The well-established direct effectors of these Gα-subunits include isozymes (PLC-β1, -β2, -β3, and -β4) of the phospholipase C-β class of lipases as well as p63RhoGEF, which exchanges GTP for GDP (and activates) on RhoA, RhoB, and RhoC. Gαq also directly binds GPCR kinase 2 and kinase 3 (not shown in the figure). The $P2Y_{12}$-like group of GPCR includes the $P2Y_{12}$, $P2Y_{13}$, and $P2Y_{14}$ receptors, which activate Gα-subunits ($G\alpha_{i1}$, $G\alpha_{i2}$, $G\alpha_{i3}$, and $G\alpha_o$) of the G_i family of heterotrimeric G proteins. A major cell-signaling response of the $P2Y_{12}$-like receptors is the inhibition of adenylyl cyclase and, therefore, decreases in cyclic AMP levels. However, G_i heterotrimers are predictably present at high concentrations relative to other heterotrimeric G proteins in the plasma membrane, and their activation results in physiologically important increases in the amount of free Gβγ. Many effector proteins directly bind Gβγ. Therefore, in a cell-dependent fashion, activation of G_i-linked receptors results in Gβγ-dependent activation of PtdIns 3-kinase, certain PLC-β isozymes, inward rectifying K^+ (GIRK) channels, GPCR kinase 2 and 3, a guanine nucleotide exchange factor (GEF) for Rac (P-Rex) and SNAP-25, which is involved in vesicle membrane fusion.

of regulation. The final section considers the structure–activity relationships of synthetic ligands directed at these receptors.

B. $P2Y_1$ Receptor

Webb & her colleagues (1993) cloned the first P2Y receptor from a chick brain library in 1993 and illustrated that expression of this GPCR in *Xenopus* oocytes conferred a Ca^{2+}-activated Cl^- current in response to ATP and other adenine nucleotides. Since the pharmacological selectivity of this response was similar to that first proposed for a P2Y receptor by Burnstock & Kennedy (1985), this receptor was termed the $P2Y_1$ receptor. A phospholipase C-activating P2Y receptor, which was previously studied extensively in turkey erythrocyte membranes (Boyer et al., 1989), was cloned and shown to be essentially identical to the chick $P2Y_1$ receptor

(Filtz et al., 1994). Its stable expression in 1321N1 human astrocytoma cells, which do not exhibit an endogenous second messenger signaling response to extracellular nucleotides, conferred a robust response of phospholipase C to adenine nucleotides. In contrast, activation of this receptor had no effect on cyclic AMP levels. The bovine (Henderson et al., 1995), rat (Tokuyama et al., 1995), and human (Schachter et al., 1996) $P2Y_1$ receptors were subsequently cloned, studied, and shown to activate phospholipase C without an interaction with adenylyl cyclase.

The idea that the $P2Y_1$ receptor signals primarily, if not entirely, through activation of $G\alpha$-subunits of the G_q family was first supported by data illustrating that expression of the receptor conferred effects of nucleotides on inositol lipid signaling but not cyclic AMP accumulation (Filtz et al., 1994; Schachter et al., 1996). Activation of phospholipase C occurs primarily by activation of $G\alpha q$, but some phospholipase C-β isozymes also are activated by release of $G\beta\gamma$-subunits from G_i or G_o (Harden & Sondek, 2006). However, phospholipase C-mediated signaling downstream of recombinant $P2Y_1$ receptors is in general not sensitive to pertussis toxin-mediated inactivation of G_i (Schachter et al., 1996). Moreover, studies with purified human $P2Y_1$ receptor confirmed that whereas robust coupling of this receptor to $G\alpha_q$ or $G\alpha_{11}$ occurs, no interaction with $G\alpha_{i1}$, $G\alpha_{i2}$, $G\alpha_{i3}$, or $G\alpha_o$ could be observed (Waldo & Harden, 2004).

Both ADP and ATP were reported in early studies to be potent agonists at the cloned $P2Y_1$ receptor. However, rigorous studies utilizing highly purified nucleotides and assay conditions in which conversion of ATP to ADP was minimized led to the conclusion that the $P2Y_1$ receptor is an ADP-preferring GPCR (Palmer et al., 1998). Later work applying radioligand-binding assays (Waldo et al., 2002) and purified recombinant human $P2Y_1$ receptor (Waldo & Harden, 2004) confirmed that the affinity of this receptor for ADP is 100-fold greater than for ATP and that ATP is a weak partial agonist and unlikely to be a physiologically important regulator.

ADP plays a major physiological role in platelet aggregation as Born first illustrated in 1962 (Born, 1962). Boyer and his colleagues discovered that adenosine 3′,5′-bisphosphate is a selective antagonist of the $P2Y_1$ receptor (Boyer et al., 1996), and this molecule and a subsequent analog, N^6-methyl adenosine 3′,5′-bisphosphate (MRS2179), were utilized by Kunapuli (Jin et al., 1998) and Gachet (Hechler et al., 1998a,b) and their colleagues to pharmacologically resolve the platelet response to ADP into synergizing signals mediated by two different receptors. Activation of the $P2Y_1$ receptor results in phospholipase C-dependent mobilization of Ca^{2+} and consequential shape change and transient aggregation (Gachet, 2006). Complete aggregation in response to ADP requires activation of the G_i/adenylyl cyclase-coupled $P2Y_{12}$ receptor as is discussed in more detail below.

The $P2Y_1$ receptor is widely expressed with the highest expression found in brain, prostate, and placenta (Abbracchio et al., 2006).

Nucleotide-promoted responses occurring through the "P2Y receptor" as initially defined by Burnstock & Kennedy (1985) probably most often occur through activation of the $P2Y_1$ receptor. Thus, in addition to its clearly delineated role in platelets, pharmacological studies strongly implicate the $P2Y_1$ receptor in a variety of functional responses in, for example, smooth muscle, endothelial cells, and neural tissue.

The most prominent phenotype of $P2Y_1(-/-)$ mice occurs as a loss of platelet shape change and aggregation in response to low concentrations of ADP (Fabre et al., 1999; Léon et al., 1999). There is an increase in bleeding time and resistance to localized arterial thrombosis as well as to systemic thromboembolism. Other changes occurring in the $P2Y_1$ receptor knockout mouse include elevated weight, higher blood glucose levels (Laplante et al., 2010), and a reduction in inflammatory nociceptive hyperalgesia (Malin & Molliver, 2009). Very selective high-affinity antagonists have been developed for the $P2Y_1$ receptor (see section below), and pharmacological studies with ADP analogs as well as these antagonists implicated the $P2Y_1$ receptor in a number of physiological effects regulated by adenine nucleotides. For example, vasodilation, bone resorption, and astrocyte-dependent regulation of neuronal signaling all appear to occur downstream of activation of $P2Y_1$ receptors by ADP.

C. $P2Y_2$ Receptor

Lustig and his colleagues used expression cloning in *Xenopus* oocytes to identify the coding sequence of the mouse $P2Y_2$ receptor in an epithelial cell library (Lustig et al., 1993). Nucleotides promoted a Ca^{2+}-activated Cl^- current in oocytes expressing this GPCR. The pharmacological selectivity of this receptor was similar to that previously observed for a receptor natively expressed in epithelial cells and referred to as the P2U-receptor due to its sensitivity to both uridine and adenosine triphosphate. Stable expression of the human $P2Y_2$ receptor in 1321N1 human astrocytoma cells conferred a response of phospholipase C, but not adenylyl cyclase, to UTP and ATP (Lazarowski et al., 1995; Parr et al., 1994). The similar high potencies of these nucleotides together with observations that both ATP and UTP are released from cells physiologically (Lazarowski et al., 1997) indicate that both nucleotides are cognate agonists of this receptor. The existence of the $P2Y_4$ receptor as a receptor that is selectively activated by UTP in human tissues (but equipotently by UTP and ATP in rodent tissues) and the $P2Y_6$ receptor, which is selectively activated by UDP, complicates assignment of uridine nucleotide-promoted responses to a given uridine nucleotide-activated receptor. $P2Y_2$, $P2Y_4$, and $P2Y_6$ receptors are all commonly expressed in many different types of epithelial cells (Abbracchio et al., 2006; Leipziger, 2003), and lack of antagonists that selectively block individual uridine nucleotide-activated receptors makes their pharmacological resolution difficult. For example, suramin has been shown to be an antagonist of the $P2Y_2$ receptor (see below), but the low affinity of this antagonism together with weak activity at the $P2Y_4$ and $P2Y_6$ receptors makes it unsuitable for clear pharmacological distinction among the three receptors.

The $P2Y_2$ receptor primarily signals through G_q and phospholipase C-β isozymes. Modest pertussis toxin-sensitivity of signaling through this receptor was observed in several studies suggesting that the receptor may also signal through G_i in some cell types (Abbracchio et al., 2006).

Message for the $P2Y_2$ receptor is widely expressed with high levels found in lung, heart, skeletal muscle, spleen, kidney, bone marrow, lymphocytes, and macrophages (Abbracchio et al., 2006). A broad range of pharmacological studies (Burnstock, 2007; Leipziger, 2003) are consistent with a major role of this receptor in regulation of ion transport in most epithelial tissues, i.e., lung, eye, intestine, kidney, although the aforementioned difficulties in pharmacologically resolving this receptor from other uridine nucleotide-activated P2Y receptors should be acknowledged. Disruption of the $P2Y_2$ receptor gene unambiguously illustrated the importance of this receptor in nucleotide-promoted Cl^- secretion in mouse airway epithelia (Cressman et al., 1999). Clinical studies also strongly support the idea that the $P2Y_2$ receptor will be important as a therapeutic target in treatment of cystic fibrosis using agonists of this receptor (Yerxa, 2001).

Activation of the $P2Y_2$ receptor promotes proliferation of many cell types (Burnstock, 2007). The receptor also is important in smooth muscle and endothelial cell function (Burnstock, 2007; Ralevic & Burnstock, 1998), chemotaxis of neutrophils and other cell types (Chen et al., 2006), mediation of "find-me" signals to nucleotides released from apoptotic cells (Elliott et al., 2009), regulation of bone formation (Hoebertz et al., 2002), and regulation of inflammatory responses (Burnstock, 2007).

D. $P2Y_4$ Receptor

Communi et al. (1995) and Nguyen et al. (1995) cloned a GPCR from human genomic DNA that is 51% homologous to the $P2Y_2$ receptor and 35% homologous to the $P2Y_1$ receptor. This receptor was designated the $P2Y_4$ receptor, and its stable expression in 1321N1 astrocytoma cells conferred uridine nucleotide-promoted inositol lipid hydrolysis with no effect on cyclic AMP accumulation observed. Although the human receptor was initially proposed to be activated by both UTP and UDP, examination of the action of uridine nucleotides under conditions that maximized the purity and stability of agonists during the assay revealed that whereas UTP is a very potent agonist, UDP is essentially inactive (Nicholas et al., 1996). Interestingly, the rat $P2Y_4$ receptor subsequently was shown to be equipotently activated by both UTP and ATP (Bogdanov et al., 1998), whereas ATP was illustrated to be a potent competitive antagonist at the human receptor (Kennedy et al., 2000).

The $P2Y_4$ receptor primarily signals through G_q/phospholipase C-β (Communi et al., 1995). In addition to mobilizing intracellular Ca^{2+}, this receptor also has been shown to couple to N-type Ca^{2+} channels and M-type K^+ channels (Filippov et al., 2003).

The highest levels of $P2Y_4$ receptor expression are found in the intestine, brain, and pituitary with lower levels found in bone marrow, monocytes, lymphocytes, and liver (Abbracchio et al., 2006). Disruption of the $P2Y_4$ receptor gene resulted in loss of nucleotide-promoted Cl^- responses in jejunal epithelium (Robaye et al., 2003), and although both $P2Y_4$ and $P2Y_2$ receptors apparently regulate this response in the upper intestinal tract, regulation of Cl^- secretion in the colon is through the $P2Y_4$ receptor alone (Ghanem et al., 2005).

E. $P2Y_6$ Receptor

Chang et al. (1995) initially cloned the $P2Y_6$ receptor from a rat aortic smooth muscle library. This receptor is 42%, 45%, and 35% homologous to the $P2Y_1$, $P2Y_2$, and $P2Y_4$ receptors, respectively. Stable expression of the $P2Y_6$ receptor in 1321N1 human astrocytoma cells or CHO-K1 cells conferred nucleotide-promoted activation of phospholipase C and no effect on adenylyl cyclase activity was observed. UTP initially was reported to be the most potent agonist, but subsequent work established that UDP is the most potent native agonist (Nicholas et al., 1996). A GPCR initially cloned from chick and designated the "$P2Y_3$" receptor (Webb et al., 1996) was later shown to be a species homologue of the mammalian $P2Y_6$ receptor (Li et al., 1998). This receptor selectively couples to G_q and promotes inositol lipid signaling through phospholipase C-β isozymes.

The $P2Y_6$ receptor is widely expressed in placenta, thymus, spleen, kidney, vascular smooth muscle, lung, intestine, bone, fat cells, and some brain regions (Abbracchio et al., 2006). However, physiological and pharmacological studies have revealed fewer roles for this receptor than for several of the other G_q-coupled P2Y receptors. This may be in part due to the difficulty of discerning between action of UTP and UDP at the $P2Y_6$ versus $P2Y_2$ and $P2Y_4$ receptors. Specific antagonist probes for the $P2Y_6$ receptor also are not available.

Contractile responses of vascular smooth muscle to $P2Y_6$ receptor activation are well established (Malmsjö et al., 2003). Epithelial cells, including basolateral responses of rat colonic cells, mouse gallbladder, and human nasal epithelium, exhibit $P2Y_6$ receptor-promoted Cl^- secretion (Lazarowski et al., 2001; Leipziger, 2003). Monocytes produce chemokines in response to UDP (Warny et al., 2001).

F. P2Y$_{11}$ Receptor

A GPCR, the P2Y$_{11}$ receptor, was cloned from a human placental library that exhibited 33% identity with the P2Y$_1$ receptor and less than 30% identity with other P2Y receptors (Communi et al., 1997). Stable expression of this receptor conferred ATP-promoted activation of inositol lipid hydrolysis in 1321N1 human astrocytoma cells. However, in contrast to results with previously cloned P2Y receptors, activation of this receptor also resulted in activation of adenylyl cyclase albeit with lower efficiency than observed with activation of phospholipase C, i.e., higher concentrations of agonist were required (Communi et al., 1999; Qi et al., 2001a). The physiological significance of this bifurcation of signal has not been firmly established. This receptor is the only P2Y receptor identified to date that shows high selectivity for ATP—ADP is a weak agonist, and UTP and UDP are inactive. Interestingly, this GPCR is not present in the mouse or rat genome, and the canine P2Y$_{11}$ receptor, which is 70% homologous to the human receptor, is more potently activated by ADP than ATP.

The P2Y$_{11}$ receptor is expressed in human spleen, liver, intestine, brain, and pituitary (Abbracchio et al., 2006). It is also present in B lymphocytes and dendritic cells. The absence of this receptor in the mouse genome together with the facts that it is activated by ATP and no selective P2Y$_{11}$ receptor probes are available has made resolution of the functions of this receptor very difficult. Potential roles in granulocyte differentiation and dendritic cell maturation and migration remain the best established functions for the P2Y$_{11}$ receptor (Wilkin et al., 2001).

G. P2Y$_{12}$ Receptor

Cooper and Rodbell discovered in 1979 that ADP promotes inhibition of adenylyl cyclase in platelet membranes (Cooper & Rodbell, 1979), and pharmacological studies and data from the P2Y$_1$(−/−) mouse in the late 1990s confirmed that the G$_q$/phospholipase C-activating P2Y$_1$ receptor and a second receptor that inhibits adenylyl cyclase mediate the actions of ADP in platelet aggregation (Gachet, 2006). A clever expression cloning strategy was utilized by Hollopeter et al. (2001) to identify the ADP-activated P2Y$_{12}$ receptor, which had previously eluded identification due to its very low (<25%) homology to the P2Y$_1$ receptor and other previously identified P2Y receptors. The expressed receptor, as well as purified recombinant human P2Y$_{12}$ receptor (Bodor et al., 2003), is specifically activated by ADP. ATP has little, if any, agonist activity, and ATP and a number of ATP analogs were shown to be antagonists of the receptor in studies with platelets (Gachet, 2006; Kauffenstein et al., 2004). However, it should be pointed out that ATP and several ATP analogs may be potent agonists at the P2Y$_{12}$ receptor natively expressed in several other cell types (Simon et al., 2001).

Studies with recombinant $P2Y_{12}$ receptor illustrate that this receptor couples to members of the G_i family of hetertrimeric G proteins. Thus, inhibition of cyclic AMP accumulation is a prominent effect of $P2Y_{12}$ receptor activation, but other downstream signaling events, that is, activation of phosphoinositide 3-kinase or K^+ channels, almost certainly follow from release of $G\beta\gamma$ from heterotrimeric G_i (Gachet, 2006). The platelet receptor selectively couples to $G\alpha_{i2}$, has been shown by biochemical and genetic studies (Jantzen et al., 2001) as well as with purified human $P2Y_{12}$ receptor reconstituted with purified G proteins in model phospholipid vesicles (Bodor et al., 2003).

The $P2Y_{12}$ receptor is expressed in megakaryocytes, platelets, and neural tissue (Abbracchio et al., 2006). Its physiological role in the thrombotic response to ADP has been established by over four decades of research (Gachet, 2006). Moreover, it is targeted therapeutically by clopidogrel (and newer clinical agents), which through an active metabolite covalently modifies and inactivates the $P2Y_{12}$ receptor (Savi & Herbert, 2005; von Kügelgen, 2006). Studies with the $P2Y_{12}(-/-)$ mouse confirmed the function of the $P2Y_{12}$ receptor in platelets and extended knowledge of the interplay that occurs between this receptor and the $P2Y_1$ receptor during ADP-promoted platelet aggregation (Andre et al., 2003; Foster et al., 2001). Although functional studies have suggested a role for the $P2Y_{12}$ receptor in neuronal function, the neural expression of this receptor occurs mostly in glial cells. Nucleotides released at sites of injury act as extracellular stimuli to signal recruitment of microglia to the site of damage, and microglia of $P2Y_{12}(-/-)$ mice are defective in capacity to polarize, migrate, or generate process extensions toward nucleotides both *in vitro* and *in vivo* (Haynes et al., 2006).

H. $P2Y_{13}$ Receptor

Identification of the $P2Y_{12}$ receptor also led to discovery of a GPCR, the $P2Y_{13}$ receptor, which exhibits approximately 50% identity to the $P2Y_{12}$ receptor (Communi et al., 2001). Expression of this receptor in 1321N1 cells and other cell lines revealed that ADP also is a potent agonist of this receptor. It also is of interest that the naturally occurring nucleoside polyphosphate, Ap_3A, exhibited high agonist potency at the $P2Y_{13}$ receptor. As with the $P2Y_{12}$ receptor, the $P2Y_{13}$ receptor also signals through G_i to inhibit adenylyl cyclase and promote other G_i-dependent cell-signaling responses.

The $P2Y_{13}$ receptor is expressed in spleen, bone marrow, peripheral leukocytes, brain, liver, pancreas, and heart (Communi et al., 2001). The physiological functions of this receptor remain to be established.

I. $P2Y_{14}$ Receptor

Chambers and his collaborators (Chambers et al., 2000; Freeman et al., 2001) expressed a number of orphan GPCR in yeast strains engineered to exhibit signals from a pheromone receptor-induced reporter gene with the goal of identifying receptor-activating agonists. An orphan GPCR, KIAA0001 (subsequently referred to as the $P2Y_{14}$ receptor and approximately 47% identical to the $P2Y_{12}$ and $P2Y_{13}$ receptors), was identified that is activated by UDP-glucose. UDP-galactose, UDP-glucuronic acid, and UDP-*N*-acetylglucosamine also were activators of KIAA0001 in the yeast test system whereas a series of nucleoside diphosphates and triphosphates and ADP-, CDP-, and GDP-sugars were not. Similar results were obtained after transient or stable expression of the receptor with the promiscuous Gα-subunit $G\alpha_{16}$ or with the chimeric G protein $G\alpha_{q/i}$ in which the last five amino acids at the carboxyl terminus of $G\alpha_q$ are substituted with those of $G\alpha_i$. Subsequent work has confirmed that this receptor primarily couples to $G_{i/o}$, and the human $P2Y_{14}$ receptor stably expressed in a range of cell types robustly inhibits adenylyl cyclase.

Although most studies only have focused on activation of the $P2Y_{14}$ receptor by UDP-sugars, Fricks and her colleagues illustrated that UDP itself is a cognate ligand for this receptor (Fricks et al., 2008, 2009). Transient coexpression of the human $P2Y_{14}$ receptor with $G\alpha_{q/i}$ in COS-7 cells first suggested that UDP is a partial agonist/competitive antagonist at this receptor. However, UDP was a potent full agonist at the rat receptor in this engineered test system. Moreover, recent studies measuring $P2Y_{14}$ receptor-promoted inhibition of adenylyl cyclase in HEK293, CHO, or C6 glioma cells have shown that UDP is full agonist with a potency three to five-fold greater than that of UDP-glucose (Carter et al., 2009). Thus, both UDP and UDP-sugars appear to be physiological agonists of the $P2Y_{14}$ receptor.

The $P2Y_{14}$ receptor is relatively widely distributed with expression found in placenta, spleen, bone marrow, thymus, stomach, intestine, adipose tissue, and brain (Abbracchio et al., 2006). It is prominently found in glial cells as well as in lymphocytes, neutrophils, and megakaryocytes. A broad range of studies indicate a role for this signaling protein in immune and inflammatory responses including neuroimmune responses in the brain. The $P2Y_{14}$ receptor is expressed in many epithelial tissues, and inflammatory responses of both respiratory and uterine epithelium apparently involve $P2Y_{14}$ receptor-promoted release of proinflammatory cytokines (Harden et al., 2010). The capacity of UDP-sugars to decrease baseline muscle tension is lost in isolated forestomach prepared from the $P2Y_{14}(-/-)$ mouse (Bassil et al., 2009).

IV. P2Y Receptor Structure

A. General Features

All P2Y receptors belong to the largest subfamily of GPCRs, the class A or rhodopsin receptor family (consisting of about 670 different receptor proteins; Fredriksson et al., 2003; Lagerström & Schiöth, 2008). Crystal structures of P2Y receptors are not available, but modeling predicts the typical features of GPCRs including seven hydrophobic transmembrane regions (TMs) connected by three extracellular loops (ELs) and three intracellular loops (ILs; see Fig. 2 for the predicted two-dimensional structure of the human $P2Y_{12}$-receptor; for a three-dimensional model of the $P2Y_{12}$ receptor see Jacobson et al., 2005). The recently published crystal structures of β-adrenoceptors (Cherezov et al., 2007; Rasmussen et al., 2007; Warne et al., 2008) and the adenosine A_{2A} receptor (Jaakola et al., 2008) confirm the general view of GPCR architecture of GPCRs although differences exist in regions involved in receptor activation compared with the inactive and active structure of rhodopsin/opsin (Park et al., 2008).

Sequence comparison shows that P2Y receptors and adenosine receptors belong to different receptor subgroups within class A GPCRs (see Fredriksson et al., 2003). The same is true when P2Y receptors are compared with the recently identified GPCRs for the nucleobase adenine (Bender et al., 2002; Gorzalka et al., 2005; von Kügelgen et al., 2008). The P2Y receptor family evolved in vertebrates. The cartilaginous fish *Raja erinacea* (little skate) possesses a primitive P2Y receptor responding to a variety of extracellular nucleotides including ATP, ADP, UTP, and UDP (Dranoff et al., 2000). More specialized P2Y-receptor subtypes evolved in mammals

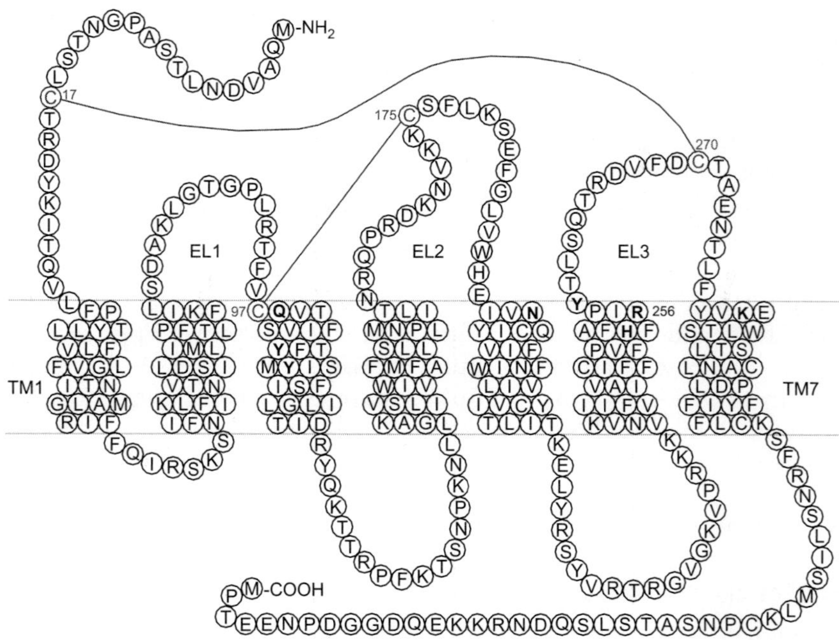

FIGURE 2 Predicted secondary structure of the human $P2Y_{12}$-receptor. TM, transmembrane region; EL, extracellular loop. Modified from von Kügelgen (2006).

enabling detection of specific extracellular nucleotides to promote cell- and tissue-specific signaling. Some of the human subtypes discussed above were lost during evolution in other mammalian species. Rodents, for example, do not possess $P2Y_{11}$ receptors, but the expression of a $P2Y_{11}$-like receptor in Xenopus indicates that this receptor developed early in evolution (Devader et al., 2007). $P2Y_4$ receptors are present in man and rodents but not in the dog (Haitina et al., 2009). The physiologically very important $P2Y_{12}$ receptor appears to be maintained in all species. A sequence comparison of 74 $P2Y_{12}$ receptor orthologs from evolutionary old bony and cartilaginous fishes to the more modern mammals shows that the $P2Y_{12}$ receptor was more rigidly conserved during evolution than other P2Y-like receptors (Schöneberg et al., 2007).

B. Receptor Proteins

P2Y receptors expressed at the level of the cell membrane are modified by N-linked glycosylation (Erb et al., 1993; Zhong et al., 2004), which has been shown to be essential for signal transduction by the $P2Y_{12}$ receptor, but not for its ligand binding or cell surface expression (Zhong et al., 2004). All P2Y receptors possess at their extracellular domains four cysteine residues, which, as has been shown for the $P2Y_1$, $P2Y_2$, and $P2Y_{12}$ receptors, are likely to form two disulfide bridges: the first one between the N-terminal domain and EL3 and the second bridge between EL1 and EL2 (Algaier et al., 2008; Ding et al., 2003; Hillmann et al., 2009; Hoffmann et al., 1999; see Fig. 2). Receptor domains involved in ligand binding and signaling transduction are discussed below.

The receptors of the P2Y-receptor family show a relatively high diversity in the amino acid composition. The human $P2Y_1$ and $P2Y_{12}$ receptors, for example, share only 19% identical amino acid residues, despite some similarities in their pharmacological profiles (Table I). Species homologues of one subtype have a much higher homology: 95% of the amino acids of the human and mouse $P2Y_1$ receptors are identical. Sequence comparison also shows that the predicted TMs of these receptors are more conserved than EL or IL regions (von Kügelgen & Wetter, 2000). Some highly conserved motifs encode for parts of the predicted TM 3, 6, and 7 that are in close apposition with the ILs. Interestingly, these parts of the receptor proteins are likely to be involved in ligand binding as discussed below.

C. Ligand-Binding Sites

Mutational analysis of the P2Y receptors combined with molecular modeling has helped identify the putative ligand-binding sites of these receptors (Costanzi et al., 2004; Erb et al., 1995; Jiang et al., 1997; Moro et al., 1998; van Rhee et al., 1995). Protein structure models were developed

TABLE I Agonists Acting at Functionally Defined Mammalian P2Y Receptor Subtypes

Type	*Principle agonists*	*Selected references*
$P2Y_1$	MRS2365 > 2-MeSADP > ADP = ADPβS	Henderson et al. (1995), Tokuyama et al. (1995), Ayyanthan et al. (1996), Janssens et al. (1996), Léon et al. (1996, 1997), Palmer et al. (1998), Chhatriwala et al. (2004), Waldo & Harden (2004)
$P2Y_2$	UTP MRS2698 ≥ ATP > INS37217 > Ap4A > ATPγS	Lustig et al. (1993), Parr et al. (1994), Lazarowski et al. (1995), Rice et al. (1995), Chen et al. (1996), Nicholas et al. (1996), Zambon et al. (2000), Yerxa et al. (2002), Wildman et al. (2003), Shen et al. (2004), Ivanov et al. (2007a, 2007b)
$P2Y_4$	UTP > UTPγS (human) UTP = ATP (rat, mouse)	Communi et al. (1995, 1996a), Nguyen et al. (1995), Nicholas et al. (1996), Bogdanov et al. (1998), Webb et al. (1998), Kennedy et al. (2000), Suarez-Huerta et al. (2001); Wildman et al. (2003); Herold et al. (2004)
$P2Y_6$	MRS2693 > PSB0474 > UDP = 5Br-UDP > UTP >> ATP	Chang et al. (1995), Communi et al. (1996b), Nicholas et al. (1996), Southey et al. (1996), Maier et al. (1997), Lazarowski et al. (2001), Besada et al. (2006), El-Tayeb et al. (2006)
$P2Y_{11}$	ATPγS = BzATP > NF546 > ARC67085 > ATP (human) ADPβS = 2-MeSADP > ATP (canine)	Communi et al. (1997, 1999), Qi et al. (2001b), Zambon et al. (2001), White et al. (2003), Meis et al. (2010).
$P2Y_{12}$	2-Methylthio-ADP > ADP > ATP	Foster et al. (2001), Hollopeter et al. (2001), Takasaki et al. (2001), Zhang et al. (2001), Chhatriwala et al. (2004), Ennion et al. (2004), Pausch et al. (2004)
$P2Y_{13}$	ADP = 2-methylthio-ADP > ADPβS	Communi et al. (2001), Zhang et al. (2002), Marteau et al. (2003), Fumagalli et al. (2004)

TABLE I *(continued)*

Type	*Principle agonists*	*Selected references*
P2Y$_{14}$	α,β-Methylene-2-thio-UDP > MRS2690 > UDP = UDP-glucose	Charlton et al. (1997), Chambers et al. (2000), Freeman et al. (2001), Ko et al. (2007), Fricks et al. (2009); Das et al. (2010), Gao et al. (2010).

Only mammalian receptors are listed (see von Kügelgen, 2006). ARC67085, 2-propylthio-β,γ-difluoromethylene-D-ATP; Ap4A, diadenosine-tetraphosphate; ATPγS, adenosine-(O-3-thiotriphosphate); 5-Br-UDP, 5-bromo-UDP; BzATP, benzoyl-benzoyl-ATP; INS37217, P^1-(uridine 5′)-P^4-(2′-deoxycytidine-5′)tetraphosphate; 2-MeSADP, 2-methylthio-ADP; MRS2365, (N)-methanocarba-2-methylthio-ADP; 2-MeSATP, 2-methylthio-ATP; MRS2690, diphosphoric acid 1-a-D-glucopyranosyl ester 2-[(4′-methylthio)uridin-5′-yl] ester; MRS2693, 5-iodo-UdP; MRS2698, 2′-amino-2′-deoxy-2-thio-UTP; NF546, 4,4′-(carbonylbis(imino-3,1-phenylene-carbonylimino-3,1-(4-methyl-phenylene)carbonylimino))bis(1,3-xylene-α,α′-diphosphonic acid); PSB0474, 3-phenacyl-UDP; UTPγS, uridine-(O-3-thiotriphosphate).

for the P2Y$_1$ receptor (Costanzi et al., 2004; Major & Fischer, 2004), the P2Y$_2$ receptor (Hillmann et al., 2009; Ivanov et al., 2007a; Jacobson et al., 2006), the P2Y$_4$ receptor (Jacobson et al., 2006), the P2Y$_6$ receptor (Costanzi et al., 2005), the P2Y$_{11}$ receptor (Zylberg et al., 2007), the P2Y$_{12}$ receptor (Jacobson et al., 2005), and the P2Y$_{14}$ receptor (Ivanov et al., 2007b). As typical for rhodopsin-like GPCRs, these models indicate involvement of several amino acid residues in the extracellular portions of TMs in ligand binding. Site-directed mutagenesis confirmed crucial roles of several polar residues within TMs 3, 5, 6, and 7 in ligand recognition (Ecke et al., 2008a; Erb et al., 1995; Guo et al., 2002; Hillmann et al., 2009; Hoffmann et al., 1999, 2008a; Jiang et al., 1997; Mao et al., 2010; Qi et al., 2001b). Histidine in TM6 at position 6.52 of the P2Y receptor protein (nomenclature according to Ballesteros & Weinstein, 1995) is highly conserved within species and across P2Y receptor subtypes and contributes to ligand recognition as shown for the P2Y$_1$, P2Y$_2$, and P2Y$_{12}$ receptor (Erb et al., 1995; Hoffmann et al., 2008a; Jiang et al., 1997; Mao et al., 2010). Basic residues (Arg or Lys) in TMs 6 and 7 are also highly conserved within species and P2Y receptor subtypes (see Schöneberg et al., 2007; von Kügelgen, 2006) and are likely involved in interaction with the negatively charged phosphate groups of nucleotide ligands as indicated by the results of mutagenesis studies. For example, replacement of Arg-256 of the human P2Y$_{12}$ receptor (TM6; position 6.55) with the basic amino acid lysine was well tolerated. In contrast, replacement with alanine caused an 80-fold decrease in agonist potency and replacement with the acidic amino acid aspartate essentially abolished receptor function (Hoffmann et al., 2008a). A dominant role of the basic residue at position 6.55 in TM6 was also demonstrated for the $P2Y_1$ receptor (Lys-280; Guo et al., 2002, Jiang et al., 1997), for the $P2Y_2$ receptor (Arg-265; Erb et al., 1995), and for the $P2Y_{11}$ receptor (Arg-265; Ecke et al., 2008a, Qi et al., 2001b). Orphan receptors (including P2YR5 and GPR23) with sequences related to those of P2Y receptors exist, but these proteins lack a basic residue at position 6.55. Several of these receptors are activated by lysophosphatidic acid or leukotriene B4, but not by extracellular nucleotides (Li et al., 1997; Noguchi et al., 2003; Pasternack et al., 2008; Yanagida et al., 2009; Yokomizo et al., 1997) underlining the functional importance of a basic residue at position 6.55 in TM6 for P2Y receptor action. As shown for the $P2Y_1$ and the $P2Y_{12}$ receptors, this basic residue is part of a common site of interaction of agonists and competitive antagonists. Its replacement also affected the potencies of chemically unrelated receptor antagonists including PPADS, NF023, and reactive blue 2 (Guo et al., 2002; Hoffmann et al., 2008a). A more detailed analysis using a series of novel derivatives of reactive blue 2 showed that the sulfonic acid at ring D of these antagonists interacts with Arg-256 of the human $P2Y_{12}$ receptor (position 6.55 in TM6; Hoffmann et al., 2009). Among these compounds, a highly potent (pA_2-value 9.8) and selective $P2Y_{12}$ receptor antagonist was identified (Hoffmann et al., 2009; see below).

In addition to TMs, the ELs of the P2Y receptors have been shown to contribute to the ligand recognition process. Glu-209 within the second and Arg-287 within EL3 of the $P2Y_1$ receptor were proposed to form additional, low-affinity binding sites (meta-binding sites; Hoffmann et al., 1999; Moro et al., 1999) that may also represent sites of interaction with allosteric modulators. Asp-204 in EL2 of the $P2Y_1$ receptor was suggested to coordinate a Mg^{2+} ion involved in nucleotide binding (Major et al., 2004). A study analyzing a chimeric receptor combining regions of the $P2Y_1$ receptor with regions of the $P2Y_6$ receptor suggested contribution of Tyr-110 in EL1 of the human $P2Y_1$ receptor in binding of the nucleobase (Hoffmann et al., 2004). Arg-95, Gly-96, and Asp-97 in EL1 of the human $P2Y_2$ receptor couple the receptor to integrins and are involved in apical targeting (Bagchi et al., 2005). Leu-108 in EL1 of the $P2Y_2$ receptor also contributes to apical targeting (Qi et al., 2005). Moreover, Arg-272 in EL3 of the human $P2Y_2$ receptor was proposed to play a gatekeeper role, possibly responsible for the recognition and orientation of nucleotide ligands (Hillmann et al., 2009). Chimeric and mutational analysis revealed that agonism of ATP at the rat $P2Y_4$ receptor depends on three residues (Asn-177, Ile-183, and Leu-190) in EL2 (Herold et al., 2004). Glu-186 in EL2 of the $P2Y_{11}$ receptor contributes to the ligand recognition process of the receptor (Ecke et al., 2008a), and a similar role was proposed for Lys-174 within EL2 and Arg-265 within EL3 of the human $P2Y_{12}$ receptor (Cattaneo et al., 2003; Daly et al., 2009; Mao et al., 2010).

D. Intracellular Receptor Domains

Intracellular domains of P2Y receptors control dimerization, G protein activation, and receptor internalization and sorting by mechanisms similar to those described for other GPCRs. Arg-333 and Arg-334 in the C-terminal portion of the human $P2Y_1$ receptor were demonstrated to be crucial for coupling to G_q (Ding et al., 2005). A chimeric human $P2Y_{12}$ receptor construct with its C-terminus replaced by the corresponding C-terminus of the human $P2Y_1$ receptor displayed a high level of constitutive activity (Ding et al., 2006). Moreover, Leu-115 (position 3.43) and the DRY motif at the intracellular portion of TM3 of the $P2Y_{12}$ receptor were reported to be involved in modifying basal activity of the receptor protein in activating G proteins (Schöneberg et al., 2007). Residues of the C-terminus are involved in interaction with beta-arrestin, regulation of receptor internalization, and receptor dimerization of the $P2Y_1$, $P2Y_2$, and $P2Y_4$ receptors (Brinson & Harden, 2001; Choi et al., 2008; Hoffmann et al., 2008b; Reiner et al., 2009; for mechanisms involved in cellular $P2Y_2$ receptor distribution, see Morris et al., 2010; Norambuena et al., 2010). Seven of the eight mammalian P2Y-receptor subtypes are expressed in a polarized manner in epithelial cells (Wolff et al., 2005), and charged residues of the C-terminus of the $P2Y_1$ receptor contribute to the basolateral-sorting of the receptor in polarized cells such as Madin–Darby canine kidney cells (Wolff et al., 2010).

E. Receptor Polymorphism

Polymorphisms and mutations in human P2Y receptors have been described and associated with diseases. This work mainly has focused on the $P2Y_1$, $P2Y_2$, and $P2Y_{12}$ receptors and their roles in cardiovascular function.

Conflicting reports exist concerning the influence of polymorphisms in the $P2Y_1$ receptor on platelet reactivity to ADP (Fontana et al., 2005; Hetherington et al., 2005; Storey et al., 2009). A haplotype ($P2Y_{12}$ H2) with differences in the nontranslated 5′-region of the $P2Y_{12}$ receptor gene also has been described, and conflicting reports discuss the potential influence of this haplotype on an increased risk of ischemic incidents and a reduced clinical efficacy of the $P2Y_{12}$ receptor antagonist clopidogrel (Amisten et al., 2008, Angiolillo et al., 2005; Bonello et al., 2010; Cavallari et al., 2007; Fontana et al., 2003; Rudez et al., 2009; Schettert et al., 2006; Staritz et al., 2009; Zee et al., 2008). A recently published study described the occurrence of a novel $P2Y_{12}$ H2/$P2Y_{13}$ Thr-158 haplotype without any association with acute myocardial infarction and classical cardiovascular risk factors (Amisten et al., 2008). Variations in the $P2Y_{12}$ receptor gene affect the *in vitro* potency of the competitive $P2Y_{12}$ receptor antagonist cangrelor in blocking ADP-induced platelet aggregation (Bouman et al., 2010). In contrast, the $P2Y_{12}$ receptor antagonist ticagrelor (which has a distinct chemical structure and displays a noncompetitive mode of action; see below) blocked ADP-induced platelet aggregation without any influence of variations in the $P2Y_{12}$ receptor gene (Storey et al., 2009). Patients with congenital bleeding disorder have been shown to carry polymorphisms of the $P2Y_{12}$ receptor. Variations include Arg-256-Gln and Arg-265-Trp (Cattaneo et al., 2003), Pro-258-Thr (Remijn et al., 2007), and Lys-174-Glu (Daly et al., 2009). All these variations reduced receptor function when studied *in vitro*. Moreover, in several cases, congenital bleeding disorders were shown to be caused by either base pair deletions within the $P2Y_{12}$ receptor gene preventing the expression of functional receptor proteins (Hollopeter et al., 2001; Watson et al., 2010) or by absence of the translation initiation codon (Shiraga et al., 2005).

A polymorphism (Arg-334-Cys) of the human $P2Y_2$ receptor with an Arg allele frequency of 0.8 (66% Arg/Arg, 29% heterozygous, and 5% Cys/Cys) has been described (Janssens et al., 1999). The response of the Cys variant to stimulation by an agonist had a slower time course, but no differences in agonist potency were observed (Janssens et al., 1999). Additional $P2Y_2$ receptor polymorphisms include Leu-46-Pro and Arg-312-Ser, of which Arg-312-Ser was significantly more common in patients with cystic fibrosis (Büscher et al., 2006). ATP-induced increases in Ca^{2+}-concentration were more pronounced in cells expressing the Ser-312 variant when compared with cells expressing the Arg-312 variant (Büscher et al., 2006). Moreover, variations in the $P2Y_2$ receptor gene were reported to be associated with an increased risk of hypertension (Wang et al., 2010), myocardial infarction (Wang et al., 2009a), and cerebral infarction (Wang et al., 2009b). Finally, the common (present in about 20% of the population) Ala-87-Thr polymorphism of the $P2Y_{11}$ receptor is associated with increased risk of acute myocardial infarction and increased levels of C-reactive protein (Amisten et al., 2007).

F. Receptor Dimers

P2Y receptors form homo- and heterodimers as has been described for other GPCRs (Albizu et al., 2010; Milligan, 2009). The first example reported in the literature was a dimer composed of the $P2Y_1$ receptor and the adenosine A_1-receptor (Yoshioka et al., 2001, 2002). These dimers showed marked differences in their pharmacological properties when compared with the original receptors. For example, the agonist activity of ADPβS was blocked by DPCPX (cyclopentyl-dipropyl-xanthine; Yoshioka et al., 2001). Interestingly, blockade of effects of nucleotides by adenosine A_1-receptor antagonists had been previously observed in many studies analyzing effects of metabolically stable adenine nucleotides on transmitter release (e.g., von Kügelgen et al., 1992; alternative explanations for these observations include the breakdown of nucleotides to adenosine).

Homodimerization of the $P2Y_1$ receptor has been proposed to be induced by agonist activation and to be involved in receptor internalization (Choi et al., 2008). $P2Y_{11}$ receptors by themselves do not undergo internalization, but when present in a $P2Y_1$/$P2Y_{11}$ heterodimer also internalize in an agonist-dependent manner (Ecke et al., 2008b). Heterodimerization of the $P2Y_2$ receptor with the adenosine A_1-receptor also occurred, and showed an impaired signaling via $G_{i/o}$ and an enhanced signaling via G_q (Suzuki et al., 2006). Evidence for dimerization of $P2Y_4$ and $P2Y_6$ receptors also has been reported (D'Ambrosi et al., 2007).

V. Selective P2Y Receptor Ligands as Pharmacological Tools

A. $P2Y_1$ Receptor

As outlined above, ADP is the physiological agonist of the $P2Y_1$ receptor. Its analog 2-methylthio-ADP has a higher potency at the human $P2Y_1$ receptor than ADP (Waldo & Harden, 2004; Table I). The Northern (N) conformation of the ribose is preferred, and the (*N*)-methanocarba analog of 2-methylthio-ADP (MRS2365) is much more potent ($EC_{50}=0.4$ nM) than 2-methylthio-ADP and displays strict selectivity for the $P2Y_1$ receptor over the ADP-activated $P2Y_{12}$ and $P2Y_{13}$ receptors (Chhatriwala et al., 2004). 2-Methylthio-ATP and ATPγS act as agonists at the $P2Y_1$ receptor with potencies similar to that of ADP. ATP itself is a relatively weak partial agonist (Waldo & Harden, 2004).

The human $P2Y_1$ receptor is blocked by suramin, PPADS, reactive blue-2 (Table II) and, in addition, by NF023 (8′-[carbonylbis(imino-3,1-phenylene-carbonylimino)]bis-1,3,5-naphthalene-trisulphonic acid; an analog of suramin) and MRS2210 (6-[2′-chloro-azophenyl]-pyridoxal-α5-phosphate; an analog of PPADS; see Guo et al., 2002). Adenosine analogs with phosphate substitutions in the 3′- and 5′-positions were the first selective competitive antagonists of the $P2Y_1$ receptor identified (Boyer et al., 1996), and bisphosphate analogs with higher affinity and selectivity for the $P2Y_1$ receptor including 2′-deoxy-N^6-methyladenosine-3′,5′-bisphosphate (MRS2179; Camaioni et al., 1998; Table II) have been synthesized. MRS2179 acts as a competitive antagonist at the turkey $P2Y_1$ receptor with a pA_2-value of about 7 (Boyer et al., 1998), and a similar affinity constant was observed at the human $P2Y_1$ receptor (Table II and Waldo et al., 2002). Replacement of the ribose with a cyclopentane fused in the N-conformation with a propane bridge resulted in nonnucleotide bisphosphate analogs that retained high affinity for the $P2Y_1$ receptor. For example, the chloro- and iodo-analogs MRS2279 (2-chloro-N^6-methyl-(*N*)-methanocarba-2′-deoxyadenosine 3′,5′-bisphosphate) and MRS2500 (2-iodo-N^6-methyl-(*N*)-methanocarba-2′-deoxyadenosine 3′,5′-bisphosphate) exhibit affinity constants of about 4 and 1 nM, respectively.

TABLE II Affinities (K_B in micromolar) or Half Maximal Concentrations of Selected Antagonists at Recombinant Human P2Y Receptors

Compound	*P2Y₁*	*P2Y₂*	*P2Y₄*	*P2Y₆*	*P2Y₁₁*	*P2Y₁₂*	*P2Y₁₃*	*P2Y₁₄*	*Selected references*
Suramin	3	50[#]	–(300 μM)	↓27% (100 μM)	0.8	3	↓80% (10 μM)		Charlton et al. (1996a, 1996b), O'Grady et al. (1996), Communi et al. (1999), Zhang et al. (2001), Marteau et al. (2003)
PPADS	4–12	–(30 μM)	↓30% (100 μM)	↓69% (100 μM)	–(100 μM)	–(100 μM)	↓50% (10 μM)		Charlton et al. (1996a, 1996b), Communi et al. (1996a, 1999), O'Grady et al. (1996), Robaye et al. (1997), Qi et al. (2001b), Takasaki et al. (2001), Marteau et al. (2003)
RB-2	0.8	2[a]	↓33% (100 μM)	1	↓80% (100 μM)	0.025	↓80% (10 μM)		Boyer et al. (1994), Communi et al. (1996a, 1999), Robaye et al. (1997), Takasaki et al. (2001), Guo et al. (2002), Marteau et al. (2003), Hoffmann et al. (2008a), Hillmann et al. (2009)
MRS2179	0.15	–(30 μM)	–(30 μM)	–(30 μM)		–(10 μM)	–(100 μM)		Boyer et al. (1998), Camaioni et al. (1998), Moro et al. (1998), Savi et al. (2001), Zhang et al. (2001), Marteau et al. (2003)

MRS2279	0.004	–(30 μM)	–(30 μM)	–(30 μM)	–(30 μM)			Nandanan et al. (2000), Boyer et al. (2002)
MRS2500	0.002					–(100 μM)		Kim et al. (2003), Cattaneo et al. (2004)
MRS2578	–(10 μM)	–(10 μM)	–(10 μM)	0.04	–(10 μM)			Mamedova et al. (2004)
NF340	–(10 μM)	–(10 μM)	–(10 μM)	–(10 μM)	0.01	–(10 μM)		Meis et al. (2010)
Cangrelor						0.0008	↓80% (0.01 μM)	Takasaki et al. (2001), Marteau et al. (2003), Vasiljev et al. (2003), Hoffmann et al. (2008a)
Clopidogrel m.						0.1[b]	–(2 μM)	Savi et al. (2001), Marteau et al. (2003)
MRS2211	>10					–(10 μM)	0.5	Kim et al. (2005)

The table summarizes studies analyzing the potencies (affinity constant in μM) of P2-receptor antagonists at recombinant human P2Y-receptors or inhibitory effects mediated by these antagonists on responses to receptor stimulation. PPADS, pyridoxal-5′-phosphate-6-azophenyl-2′,4′-disulfonate; RB-2, reactive blue 2; MRS2179, 2′-deoxy-N^6-methyladenosine-3′,5′-bisphosphate; MRS2279, 2-chloro-N^6-methyl-(*N*)-methanocarba-2′-deoxyadenosine 3′,5′-bisphosphate; MRS2500, 2-iodo-N^6-methyl-(*N*)-methanocarba-2′-deoxyadenosine 3′,5′-bisphosphate; MRS2578, *N*,*N*′-1,4-Butanediylbis[*N*′-(3-isothiocyanatophenyl)thio urea; NF340, 4,4′-(carbonylbis(imino-3,1-(4-methyl-phenylene)carbonylimino))bis(naphthalene-2,6-disulfonic acid); cangrelor = ARC69931MX, N^6-(2-methylthioethyl)-2-(3,3,3-trifluoropropylthio)-β,γ-dichloromethylene-ATP; Clopidogrel m., active metabolites of clopidogrel; MRS2211, 2-[(2-chloro-5-nitrophenyl)azo]-5-hydroxy-6-methyl-3-[(phosphonooxy)methyl]-4-pyridinecarboxaldehyde. #, slope different from unity. –(30 μM), no antagonistic effect at concentrations up to 30 μM. ↓30%(100 μM), decrease by 30% at the concentration of 100 μM.

[a] Half maximal concentration.

[b] Estimated from published data.

These bisphosphate analogs show no interaction with other P2Y receptors (Table II), and therefore, MRS2500 blocks platelet aggregation by a selective interaction with the $P2Y_1$ receptor (Cattaneo et al., 2004; Hechler et al., 2006). [^{32}P]-labeled (Houston et al., 2006) and [^{125}I]-labeled (Ohlmann et al., 2010) MRS2500 have been developed as selective and high-affinity radioligands for recombinant and natively expressed $P2Y_1$ receptors. An interesting and novel approach for development of targeted pharmacological tools is the synthesis of multivalent congeners by attaching, for example, multiple bisphosphate $P2Y_1$ receptor antagonists to a polyamidoamine dendrimer carrier (de Castro et al., 2010; Jacobson, 2010).

The $P2Y_1$ receptor antagonists described thus far are polar structures that are expected to exhibit poor absorption when given orally. Several nonnucleotide antagonists developed recently circumvent this limitation. A diaryl urea analog was demonstrated to act as a selective and orally bioavailable antagonist inhibiting $P2Y_1$ receptor-promoted platelet aggregation with an EC_{50} of 4 μM (Pfefferkorn et al., 2008). A tetrahydro-4-quinolinamine analog showed even higher potency inhibiting $P2Y_1$ receptor-promoted platelet aggregation with an EC_{50} of 0.5 μM (Morales-Ramos et al., 2008). In addition to antagonists, 2,2′-pyridylisatogen tosylate has been proposed to act as an allosteric modulator of the $P2Y_1$ receptor (Gao et al., 2004).

B. $P2Y_2$ Receptor

Triphosphate nucleotides including UTP, ATP, UTPγS, and ATPγS act as full agonists at this receptor (Table I). 2-Thio-UTP and its 2′-amino-2′-deoxy-analog (MRS2698) are selective $P2Y_2$-receptor agonists (El-Tayeb et al., 2006; Ivanov et al., 2007a; Jacobson et al., 2006). In addition to triphosphate nucleotides, the receptor also responds to dinucleoside polyphosphates including diadenosine-tetraphosphate (Ap4A; Lazarowski et al., 1995; Patel et al., 2001) as well as to Up4U (diquafosol, INS365; Pendergast et al., 2001). INS365 is undergoing phase III clinical trials for the treatment of dry eye disease (Tauber et al., 2004). The dinucleoside analog P^1-(uridine 5′)-P^4-(2′-deoxycytidine-5′) tetraphosphate (INS37217) is a potent agonist at the $P2Y_2$ receptor with some effects at the $P2Y_4$ receptor (Table I).

Suramin blocks the $P2Y_2$ receptor with an affinity about 20 times lower than observed at the $P2Y_1$ receptor (Table II). Reactive blue-2 acted as an antagonist at the recombinant human $P2Y_2$ receptor with an IC_{50} of 2 μM (Hillmann et al., 2009). Flavonoid derivatives and reactive blue-2 derivatives have been studied for their antagonistic action at the $P2Y_2$ receptor (Brunschweiger & Müller, 2006; Weyler et al., 2008). Among these compounds, the reactive blue-2 analog PSB416 displayed an IC_{50} of 22 μM (Hillmann et al., 2009). Some acyclic nucleotide analogs of UTP showed antagonistic properties at the $P2Y_2$-receptor; however, their affinities are low (Sauer et al., 2009).

C. $P2Y_4$ Receptor

As discussed above, the human $P2Y_4$ receptor is activated by UTP, but not by ATP. UDP and ADP also are inactive (Nicholas et al., 1996). Selective agonists for the $P2Y_4$ receptor are not available, although the agonist 2′-azido-2′-deoxy-UTP shows some preference for this receptor (Jacobson et al., 2009).

Suramin does not block the $P2Y_4$ receptor even at high concentrations (Table II). PPADS reduced maximal agonist responses of the human $P2Y_4$ receptor but was without effect at the rat $P2Y_4$ receptor (Wildman et al., 2003). Reactive blue-2 caused a modest reduction of agonist-induced responses of the human $P2Y_4$ receptor and abolished responses at the rat $P2Y_4$ receptor (Bogdanov et al., 1998; Wildman et al., 2003).

D. $P2Y_6$ Receptor

The $P2Y_6$ receptor is a uridine diphosphate-preferring receptor with UDP much more potent than UTP; adenine nucleotides are essentially inactive (Nicholas et al., 1996; Table I). UDPβS and Up_3U also act as agonists at the $P2Y_6$ receptor (Hou et al., 2002; Shaver et al., 2005). A number of potent and selective agonists have been developed for this receptor including 3-phenacyl-UDP (PSB0474; El-Tayeb et al., 2006), 5-iodo-UDP (MRS2693; Besada et al., 2006), α,β-methylene-UDP (MRS2782; Ko et al., 2008), INS48823 (Brunschweiger & Müller, 2006; Korcok et al., 2005), and 5-O-methyl-UDP (Ginsburg-Shmuel et al., 2010). In contrast to the $P2Y_1$ receptor, the $P2Y_6$ receptor prefers the Southern (S) conformation of the ribose (Maruoka et al., 2010).

The $P2Y_6$ receptor is blocked by reactive blue-2, PPADS, and suramin (Table II). 4,4′-Diisothiocyanatostilbene-2,2′-disulfonate (DIDS; see von Kügelgen, 2006) and its analog MRS2578 (*N*,*N*′-1,4-butanediylbis [*N*′-(3-isothiocyanatophenyl)thiourea]; Mamedova et al., 2004 act as irreversible or slowly reversible antagonists at human and rat $P2Y_6$ receptors. MRS2578 shows no interaction with $P2Y_1$, $P2Y_2$, $P2Y_4$, and $P2Y_{11}$ receptors (Mamedova et al., 2004).

E. $P2Y_{11}$ Receptor

ATP is the cognate agonist of the human $P2Y_{11}$ receptor (Table I). This receptor is also activated by NAD^+ and $NAADP^+$ (Moreschi et al., 2006, 2008). ATPγS and the $P2Y_{12}$ receptor antagonist, 2-propylthio-β,γ-dichlormethylene-D-ATP (ARC67085), act as potent synthetic agonists of the $P2Y_{11}$ receptor (Table I). Adenosine-5′-O-(α-boranotriphosphate) diastereoisomers activate $P2Y_{11}$ receptors, but not $P2Y_2$ receptors and therefore, can be used to discriminate between these receptors (Ecke et al., 2006).

Interestingly, the suramin analog NF546 recently was shown to act as a potent and selective nonnucleotide receptor agonist at the $P2Y_{11}$ receptor (Meis et al., 2010) and is likely to facilitate research on the yet-to-be-defined roles of this receptor in physiology and pathophysiology. Iantherans isolated from the marine sponge *Ianthella quadrangulata* also activate the human $P2Y_{11}$ receptor (Greve et al., 2007). In contrast to the human receptor, ADP analogs act as agonists at the canine P2Y $receptor_{11}$ (Torres et al., 2002).

Suramin acts as an antagonist at the human $P2Y_{11}$ receptor with a moderate affinity (pA_2-value of 6.1; Table II). Its analog NF157 has a slightly higher potency at the $P2Y_{11}$ receptor, but this molecule also blocks other P2-receptor subtypes (Moreschi et al., 2006; Ullmann et al., 2005). In contrast, the novel analog NF340 displays selectivity for the $P2Y_{11}$ receptor and exhibits a pA_2-value of 8 for this receptor (Meis et al., 2010). The bisphosphate derivative adenosine-3′-phosphate-5′-phosphosulfate was reported to act as a partial agonist at the $P2Y_{11}$ receptor (Communi et al., 1999).

F. $P2Y_{12}$ Receptor

ADP is the primary physiological agonist of the $P2Y_{12}$ receptor, which also is activated by adenine diphosphate derivatives, including the very potent agonist 2-methylthio-ADP (Table I). The receptor is blocked by suramin and, with a relatively high potency, by reactive blue-2 (Boyer et al., 1994; Hoffmann et al., 2008a). An analog of reactive blue-2, PSP0739 (see Baqi & Müller, 2010; Baqi et al., 2009), was shown recently to act as a very potent, competitive, and selective $P2Y_{12}$ receptor antagonist (pA_2-value at the recombinant human $P2Y_{12}$ receptor of 9.8; Hoffmann et al., 2009).

The thienopyridine compounds ticlopidine (Maffrand et al., 1988), clopidogrel (Herbert et al., 1993), and prasugrel (CS-747; Sugidachi et al., 2000) are used in pharmacotherapy to inhibit platelet aggregation. Indeed, clopidogrel is one highest prescribed drug in the world. These compounds act only *in vivo* where they are metabolized to active metabolites that interact in a covalent manner of the cysteine residues Cys-97 and Cys-175 within the $P2Y_{12}$ receptor (Algaier et al., 2008; Ding et al., 2009; Savi et al., 2006). The active metabolite of prasugrel has been shown to modify the $P2Y_{12}$ receptor without an effect on the $P2Y_1$ receptor (Hashimoto et al., 2007; Niitsu et al., 2005).

2-Methylthio-AMP and ATP are low-affinity antagonists (Hollopeter et al., 2001; Kauffenstein et al., 2004) of the $P2Y_{12}$ receptor. The acyclic analog of adenosine bisphosphate, MRS2395, inhibited the ADP-induced aggregation of human platelets without effect on the $P2Y_1$ receptor-promoted phospholipase C activity (Xu et al., 2002). Triphosphate analogs, including cangrelor (AR-C69931MX, N^6-(2-methylthioethyl)-2-

(3,3,3-trifluoropropylthio)-β,γ-dichloromethylene-ATP; Table II), and AR-C67085 (2-propylthio-β,γ-dichloromethylene-D-ATP), act as very potent competitive $P2Y_{12}$-antagonists (Ingall et al., 1999). The pA_2-value of cangrelor (AR-C69931MX) at the recombinant human $P2Y_{12}$ receptor was 9.2 (Hoffmann et al., 2008a), and AR-C67085 exhibited a pA_2-value of 8.2 at the native human $P2Y_{12}$ receptor (Vasiljev et al., 2003). However, these compounds are not subtype selective since AR-C67085 is an agonist at the human $P2Y_{11}$ receptor (see above), and both AR-C67085 and cangrelor also block human and rat $P2Y_{13}$ receptors (see below). [^{3}H]PSB0413 is an analog of AR-C67085 with a K_D value of 5 nM that has been developed as a radioligand for the $P2Y_{12}$ receptor (El-Tayeb et al., 2005). Novel nucleotide antagonists of this receptor also include dinucleoside polyphosphates and the AMP analog INS50589 (Douglass et al., 2008; Johnson et al., 2007).

Several nucleoside analogs block the $P2Y_{12}$ receptor. Ticagrelor (AZD6140) is an orally active, reversible $P2Y_{12}$ receptor antagonist (James et al., 2009; Springthorpe et al., 2007) that was recently approved for therapy by the European Medicines Agency. Interestingly, ticagrelor binds to the human $P2Y_{12}$ receptor by a mechanism distinct from that of ADP and antagonizes ADP-induced receptor activation and platelet aggregation in a noncompetitive manner (van Giezen et al., 2009). Another nucleoside derivative with a carba structure also acts as a potent $P2Y_{12}$ receptor antagonist (Ye et al., 2008). Moreover, even an adenine derivative (BF0801) has been reported to block the $P2Y_{12}$ receptor (Zhang et al., 2010). Ethoxycarbonyl-piperazin-oxopentanoic acid derivatives (Wang et al., 2007), 6-amino-2-mercapto-3H-pyrimidin-4-one derivatives (Crepaldi et al., 2009), and piperazinyl-glutamate-pyridines derivatives (Parlow et al., 2009, 2010) represent novel chemical structures acting as $P2Y_{12}$ receptor antagonists.

G. $P2Y_{13}$ Receptor

The $P2Y_{13}$ receptor is an ADP-activated GPCR that responds to adenine diphosphate analogs similarly to the $P2Y_{12}$ receptor. ATP and 2-methylthio-ATP appear to be partial agonists with a weak potency at the $P2Y_{13}$ receptor (Marteau et al., 2003).

The human $P2Y_{13}$ receptor is blocked by suramin, reactive blue 2, and high concentrations of PPADS (Table II; cf. Marteau et al., 2003). The 2-chloro-5-nitro analog of PPADS (MRS2211) has been shown to act as a competitive antagonist at the human $P2Y_{13}$ receptor with a pA_2-value of 6.3 (Table I; Kim et al., 2005). Moreover, cangrelor also blocks the human $P2Y_{13}$ receptor with a noncompetitive mode of interaction (Marteau et al., 2003). The same appears to be true for the interaction of cangrelor with the rat $P2Y_{13}$ receptor (Fumagalli et al., 2004; see also Wirkner et al., 2004).

H. P2Y$_{14}$ Receptor

UDP and UDP-glucose are physiological agonists of the P2Y$_{14}$ receptor (see above and Harden et al., 2010). The 2-thio analog of UDP-glucose, MRS2690, is much more potent at the P2Y$_{14}$ receptor and inactive at the P2Y$_2$ receptor (Table I and Ko et al., 2007). The UDP analog α,β-difluoromethylene-UDP (MRS2802) also potently ($EC_{50}=50$ nM) activates the P2Y$_{14}$ receptor and shows no activity at the P2Y$_6$ receptor (Carter et al., 2009). The novel derivative α,β-methylene-2-thio-UDP acts with a very high potency ($EC_{50}=0.92$ nM) and high selectivity (2160-fold selective vs. P2Y$_6$; Das et al., 2010) for the P2Y$_{14}$ receptor. Depending on the experimental conditions, UDP may also act as a partial agonist or an antagonist (Carter et al., 2009; Fricks et al., 2008, 2009).

VI. Conclusions

The widely expressed P2Y receptors play important roles in physiology and pathophysiology. Mammalian P2Y receptor subtypes are highly diverse in both their amino acid sequences and their pharmacological profiles. Although advances have been made recently in the synthesis of selective ligands, additional progress is needed in generation of highly selective agonists for each of the P2Y receptors, and selective antagonists are currently available for only two of the eight P2Y receptor subtypes. The molecular models currently available for these receptors are useful, but these would be greatly strengthened by knowledge that would be provided by three-dimensional crystal structures of these signaling proteins. The success of clopidogrel as a therapeutic agent illustrates the health potential of targeting these receptors. Several candidate drugs are in the final stages of clinical testing, and increased emphasis on development of therapeutic agents that target these receptors is expected.

Acknowledgments

Research on P2Y receptors by T. K. H. is supported by NIGMS grant GM38213.
Conflict of Interest statement: The authors have no conflicts of interest to declare.

Abbreviations

GPCR	G protein-coupled receptor
EL	extracellular loop
IL	intracellular loop
TM	transmembrane

References

Abbracchio, M. P., Burnstock, G., Boeynaems, J. M., Barnard, E. A., Boyer, J. L., Kennedy, C., et al. (2006). International Union of Pharmacology LVIII: Update on the P2Y G protein-coupled nucleotide receptors: From molecular mechanisms and pathophysiology to therapy. *Pharmacological Reviews*, *58*, 281–341.

Albizu, L., Cottet, M., Kralikova, M., Stoev, S., Seyer, R., Brabet, I., et al. (2010). Time-resolved FRET between GPCR ligands reveals oligomers in native tissues. *Nature Chemical Biology*, *6*, 587–594.

Algaier, I., Jakubowski, J. A., Asai, F., & von Kügelgen, I. (2008). Interaction of the active metabolite of prasugrel, R-138727, with cysteine 97 and cysteine 175 of the human $P2Y_{12}$ receptor. *Journal of Thrombosis and Haemostasis*, *6*, 1908–1914.

Amisten, S., Braun, O. O., Johansson, L., Ridderstråle, M., Melander, O., & Erlinge, D. (2008). The $P2Y_{13}$ Met-158-Thr polymorphism, which is in linkage disequilibrium with the $P2Y_{12}$ locus, is not associated with acute myocardial infarction. *PLoS ONE*, *3*, e1462. 10.1371/journal.pone.0001462.

Amisten, S., Melander, O., Wihlborg, A. K., Berglund, G., & Erlinge, D. (2007). Increased risk of acute myocardial infarction and elevated levels of C-reactive protein in carriers of the Thr-87 variant of the ATP receptor $P2Y_{11}$. *European Heart Journal*, *28*, 13–18.

Andre, P., Delaney, S. M., LaRocca, T., Vincent, D., DeGuzman, F., Jurek, M., et al. (2003). $P2Y_{12}$ regulates platelet adhesion/activation, thrombus growth, and thrombus stability in injured arteries. *Journal of Clinical Investigation*, *112*, 398–406.

Angiolillo, D. J., Fernandez-Ortiz, A., Bernardo, E., Ramírez, C., Cavallari, U., Trabetti, E., et al. (2005). Lack of association between the $P2Y_{12}$ receptor gene polymorphism and platelet response to clopidogrel in patients with coronary artery disease. *Thrombosis Research*, *116*, 491–497.

Ayyanathan, K., Webb, T. E., Sandhu, A. K., Athwal, R. S., Barnard, E. A., & Kunapuli, S. P. (1996). Cloning and chromosomal localization of the human $P2Y_1$ purinoceptor. *Biochemical and Biophysical Research Communications*, *218*, 783–788.

Bagchi, S., Liao, Z., Gonzalez, F. A., Chorna, N. E., Seye, C. I., Weisman, G. A., et al. (2005). The $P2Y_2$ nucleotide receptor interacts with alpha integrins to activate G_o and induce cell migration. *Journal of Biological Chemistry*, *280*, 39050–39057.

Ballesteros, J. A., & Weinstein, H. (1995). Integrated methods for the construction of three-dimensional models and computational probing of structure-function relations in G protein-coupled receptors. In P. M. Conn & S. C. Sealfon (Eds.), *Methods in Neurosciences*. Vol. 25 (pp. 366–428). USA: Academic Press.

Baqi, Y., Atzler, K., Köse, M., Glänzel, M., & Müller, C. E. (2009). High-affinity, non-nucleotide-derived competitive antagonists of platelet $P2Y_{12}$ receptors. *Journal of Medicinal Chemistry*, *52*, 3784–3793.

Baqi, Y., & Müller, C. E. (2010). Synthesis of alkyl- and aryl-amino-substituted anthraquinone derivatives by microwave-assisted copper(0)-catalyzed Ullmann coupling reactions. *Nature Protocols*, *5*, 945–953.

Bassil, A. K., Bourdu, S., Townson, K. A., Wheeldon, A., Jarvie, E. M., Zebda, N., et al. (2009). UDP-glucose modulates gastric function through $P2Y_{14}$ receptor-dependent and -independent mechanisms. *American Journal of Physiology. Gastrointestinal and Liver Physiology*, *296*, G923–G930.

Bender, E., Buist, A., Jurzak, M., Langlois, X., Baggerman, G., Verhasselt, P., et al. (2002). Characterization of an orphan G protein-coupled receptor localized in the dorsal root ganglia reveals adenine as a signaling molecule. *Proceedings of the National Academy of Sciences of the United States of America*, *99*, 8573–8578.

Besada, P., Shin, D. H., Costanzi, S., Ko, H., Mathé, C., Gagneron, J., et al. (2006). Structure-activity relationships of uridine 5′-diphosphate analogues at the human $P2Y_6$ receptor. *Journal of Medicinal Chemistry*, *49*, 5532–5543.

Bodor, E. T., Waldo, G. L., Hooks, S. B., Corbitt, J., Boyer, J. L., & Harden, T. K. (2003). Purification and functional reconstitution of the human $P2Y_{12}$ receptor. *Molecular Pharmacology*, *64*, 1210–1216.

Bogdanov, Y. D., Wildman, S. S., Clements, M. P., King, B. F., & Burnstock, G. (1998). Molecular cloning and characterization of rat $P2Y_4$ nucleotide receptor. *British Journal of Pharmacology*, *124*, 428–430.

Bonello, L., Bonello-Palot, N., Armero, S., Bonello, C., Mokhtar, O. A., Arques, S., et al. (2010). Impact of $P2Y_{12}$-ADP receptor polymorphism on the efficacy of clopidogrel dose-adjustment according to platelet reactivity monitoring in coronary artery disease patients. *Thrombosis Research*, *125*, e167–e170.

Born, G. V. (1962). Aggregation of blood platelets by adenosine diphosphate and its reversal. *Nature*, *194*, 927–929.

Bouman, H. J., van Werkum, J. W., Rudez, G., Leebeek, F. W., Kruit, A., Hackeng, C. M., et al. (2010). The influence of variation in the $P2Y_{12}$ receptor gene on in vitro platelet inhibition with the direct $P2Y_{12}$ antagonist cangrelor. *Thrombosis and Haemostasis*, *103*, 379–386.

Boyer, J. L., Adams, M., Ravi, R. G., Jacobson, K. A., & Harden, T. K. (2002). 2-Chloro N^6-methyl-N-methanocarba-2′-deoxyadenosine-3′, 5′-bisphosphate is a selective high affinity $P2Y_1$ receptor antagonist. *British Journal of Pharmacology*, *135*, 2004–2010.

Boyer, J. L., Downes, C. P., & Harden, T. K. (1989). Kinetics of activation of phospholipase C by P_2-purinergic receptor agonists and guanine nucleotides. *Journal of Biological Chemistry*, *264*, 884–890.

Boyer, J. L., Mohanram, A., Camaioni, E., Jacobson, K. A., & Harden, T. K. (1998). Competitive and selective antagonism of $P2Y_1$ receptors by N^6-methyl 2′-deoxyadenosine 3′, 5′-bisphosphate. *British Journal of Pharmacology*, *124*, 1–3.

Boyer, J. L., Romero-Avila, T., Schachter, J. B., & Harden, T. K. (1996). Identification of competitive antagonists of the $P2Y_1$ receptor. *Molecular Pharmacology*, *50*, 1323–1329.

Boyer, J. L., Zohn, I. E., Jacobson, K. A., & Harden, T. K. (1994). Differential effects of P2-purinoceptor antagonists on phospholipase C- and adenylyl cyclase-coupled P2Y-purinoceptors. *British Journal of Pharmacology*, *113*, 614–620.

Brinson, A. E., & Harden, T. K. (2001). Differential regulation of the uridine nucleotide-activated $P2Y_4$ and $P2Y_6$ receptors. SER-333 and SER-334 in the carboxyl terminus are involved in agonist-dependent phosphorylation desensitization and internalization of the $P2Y_4$ receptor. *Journal of Biological Chemistry*, *276*, 11939–11948.

Brunschweiger, A., & Müller, C. E. (2006). P2 receptors activated by uracil nucleotides–An update. *Current Medicinal Chemistry*, *13*, 289–312.

Burnstock, G. (1978). A.basis for distinguishing two types of purinergic receptor. In R. W. Straub & L. Bolis (Eds.), *Cell membrane receptors for drugs and hormones: A multidisciplinary approach* (pp. 107–118). USA: Raven Press.

Burnstock, G. (2007). Physiology and pathophysiology of purinergic neurotransmission. *Physiological Reviews*, *87*, 659–797.

Burnstock, G., & Kennedy, C. (1985). Is there a basis for distinguishing two types of P2-purinoceptor? *General Pharmacology*, *16*, 433–440.

Büscher, R., Hoerning, A., Patel, H. H., Zhang, S., Arthur, D. B., Grasemann, H., et al. (2006). $P2Y_2$ receptor polymorphisms and haplotypes in cystic fibrosis and their impact on Ca^{2+} influx. *Pharmacogenetics and Genomics*, *16*, 199–205.

Camaioni, E., Boyer, J. L., Mohanram, A., Harden, T. K., & Jacobson, K. A. (1998). Deoxyadenosine-bisphosphate derivatives as potent antagonists at $P2Y_1$ receptors. *Journal of Medicinal Chemistry*, *41*, 183–190.

Carter, R. L., Fricks, I. P., Barrett, M. O., Burianek, L. E., Zhou, Y., Ko, H., et al. (2009). Quantification of G_i-mediated inhibition of adenylyl cyclase activity reveals that UDP is a potent agonist of the human $P2Y_{14}$ receptor. *Molecular Pharmacology*, *76*, 1341–1348.

Cattaneo, M., Lecchi, A., Ohno, M., Joshi, B. V., Besada, P., Tchilibon, S., et al. (2004). Antiaggregatory activity in human platelets of potent antagonists of the $P2Y_1$ receptor. *Biochemical Pharmacology*, *68*, 1995–2002.

Cattaneo, M., Zighetti, M. L., Lombardi, R., Martinez, C., Lecchi, A., Conley, P. B., et al. (2003). Molecular bases of defective signal transduction in the platelet $P2Y_{12}$ receptor of a patient with congenital bleeding. *Proceedings of the National Academy of Sciences of the United States of America*, *100*, 1978–1983.

Cavallari, U., Trabetti, E., Malerba, G., Biscuola, M., Girelli, D., Olivieri, O., et al. (2007). Gene sequence variations of the platelet $P2Y_{12}$ receptor are associated with coronary artery disease. *BMC Medical Genetics*, *8*, 59. 10.1186/1471-2350-8-59.

Chambers, J. K., Macdonald, L. E., Sarau, H. M., Ames, R. S., Freeman, K., Foley, J. J., et al. (2000). A G protein-coupled receptor for UDP-glucose. *Journal of Biological Chemistry*, *275*, 10767–10771.

Chang, K., Hanaoka, K., Kumada, M., & Takuwa, Y. (1995). Molecular cloning and functional analysis of a novel P2 nucleotide receptor. *Journal of Biological Chemistry*, *270*, 26152–26158.

Charlton, S. J., Brown, C. A., Weisman, G. A., Turner, J. T., Erb, L., & Boarder, M. R. (1996a). PPADS and suramin as antagonists at cloned P2Y- and P2U-purinoceptors. *British Journal of Pharmacology*, *118*, 704–710.

Charlton, S. J., Brown, C. A., Weisman, G. A., Turner, J. T., Erb, L., & Boarder, M. R. (1996b). Cloned and transfected $P2Y_4$ receptors: Characterization of a suramin and PPADS-insensitive response to UTP. *British Journal of Pharmacology*, *119*, 1301–1303.

Charlton, M. E., Williams, A. S., Fogliano, M., Sweetnam, P. M., & Duman, R. S. (1997). The isolation and characterization of a novel G protein-coupled receptor regulated by immunologic challenge. *Brain Research*, *764*, 141–148.

Chen, Y., Corriden, R., Inoue, Y., Yip, L., Hashiguchi, N., Zinkernagel, A., et al. (2006). ATP release guides neutrophil chemotaxis via $P2Y_2$ and A_3 receptors. *Science*, *314*, 1792–1795.

Chen, Z. P., Krull, N., Xu, S., Levy, A., & Lightman, S. L. (1996). Molecular cloning and functional characterization of a rat pituitary G protein-coupled adenosine triphosphate (ATP) receptor. *Endocrinology*, *137*, 1833–1840.

Cherezov, V., Rosenbaum, D. M., Hanson, M. A., Rasmussen, S. G., Thian, F. S., Kobilka, T. S., et al. (2007). High-resolution crystal structure of an engineered human $beta_2$-adrenergic G protein-coupled receptor. *Science*, *318*, 1258–1265.

Chhatriwala, M., Ravi, R. G., Patel, R. I., Boyer, J. L., Jacobson, K. A., & Harden, T. K. (2004). Induction of novel agonist selectivity for the ADP-activated $P2Y_1$ receptor versus the ADP-activated $P2Y_{12}$ and $P2Y_{13}$ receptors by conformational constraint of an ADP analog. *Journal of Pharmacology and Experimental Therapeutics*, *311*, 1038–1043.

Choi, R. C., Simon, J., Tsim, K. W., & Barnard, E. A. (2008). Constitutive and agonist-induced dimerizations of the $P2Y_1$ receptor: Relationship to internalization and scaffolding. *Journal of Biological Chemistry*, *283*, 11050–11063.

Communi, D., Gonzalez, N. S., Detheux, M., Brezillon, S., Lannoy, V., Parmentier, M., et al. (2001). Identification of a novel human ADP receptor coupled to Gi. *Journal of Biological Chemistry*, *276*, 41479–41485.

Communi, D., Govaerts, C., Parmentier, M., & Boeynaems, J. M. (1997). Cloning of a human purinergic P2Y receptor coupled to phospholipase C and adenylyl cyclase. *Journal of Biological Chemistry*, *272*, 31969–31973.

Communi, D., Motte, S., Boeynaems, J. M., & Pirotton, S. (1996). Pharmacological characterization of the human $P2Y_4$ receptor. *European Journal of Pharmacology*, *317*, 383–389.

Communi, D., Parmentier, M., & Boeynaems, J. M. (1996). Cloning, functional expression and tissue distribution of the human P2Y_6 receptor. *Biochemical and Biophysical Research Communications*, *222*, 303–308.

Communi, D., Pirotton, S., Parmentier, M., & Boeynaems, J. M. (1995). Cloning and functional expression of a human uridine nucleotide receptor. *Journal of Biological Chemistry*, *270*, 30849–30852.

Communi, D., Robaye, B., & Boeynaems, J. M. (1999). Pharmacological characterization of the human P2Y_{11} receptor. *British Journal of Pharmacology*, *128*, 1199–1206.

Cooper, D. M., & Rodbell, M. (1979). ADP is a potent inhibitor of human platelet plasma membrane adenylate cyclase. *Nature*, *282*, 517–518.

Costanzi, S., Joshi, B. V., Maddileti, S., Mamedova, L., Gonzalez-Moa, M. J., Marquez, V. E., et al. (2005). Human P2Y_6 receptor: Molecular modeling leads to the rational design of a novel agonist based on a unique conformational preference. *Journal of Medicinal Chemistry*, *48*, 8108–8111.

Costanzi, S., Mamedova, L., Gao, Z. G., & Jacobson, K. A. (2004). Architecture of P2Y nucleotide receptors: Structural comparison based on sequence analysis, mutagenesis, and homology modeling. *Journal of Medicinal Chemistry*, *47*, 5393–5404.

Crepaldi, P., Cacciari, B., Bonache, M. C., Spalluto, G., Varani, K., Borea, P. A., et al. (2009). 6-Amino-2-mercapto-3H-pyrimidin-4-one derivatives as new candidates for the antagonism at the P2Y_{12} receptors. *Bioorganic & Medicinal Chemistry*, *17*, 4612–4621.

Cressman, V. L., Lazarowski, E., Homolya, L., Boucher, R. C., Koller, B. H., & Grubb, B. R. (1999). Effect of loss of P2Y_2 receptor gene expression on nucleotide regulation of murine epithelial Cl^- transport. *Journal of Biological Chemistry*, *274*, 26461–26468.

Daly, M. E., Dawood, B. B., Lester, W. A., Peake, I. R., Rodeghiero, F., Goodeve, A. C., et al. (2009). Identification and characterization of a novel P2Y_{12} variant in a patient diagnosed with type 1 von Willebrand disease in the European MCMDM-1VWD study. *Blood*, *113*, 4110–4113.

D'Ambrosi, N., Iafrate, M., Saba, E., Rosa, P., & Volonté, C. (2007). Comparative analysis of P2Y_4 and P2Y_6 receptor architecture in native and transfected neuronal systems. *Biochimica et Biophysica Acta*, *1768*, 1592–1599.

Das, A., Ko, H., Burianek, L. E., Barrett, M. O., Harden, T. K., & Jacobson, K. A. (2010). Human P2Y_{14} receptor agonists: Truncation of the hexose moiety of uridine-5′-diphosphoglucose and its replacement with alkyl and aryl groups. *Journal of Medicinal Chemistry*, *53*, 471–480.

de Castro, S., Maruoka, H., Hong, K., Kilbey, S. M., Costanzi, S., Hechler, B., et al. (2010). Functionalized congeners of P2Y_1 receptor antagonists: 2-alkynyl (N)-methanocarba 2′-deoxyadenosine 3′, 5′-bisphosphate analogues and conjugation to a polyamidoamine (PAMAM) dendrimer carrier. *Bioconjugate Chemistry*, *21*, 1190–1205.

Devader, C., Drew, C. M., Geach, T. J., Tabler, J., Townsend-Nicholson, A., & Dale, L. (2007). A novel nucleotide receptor in Xenopus activates the cAMP second messenger pathway. *FEBS Letters*, *581*, 5332–5336.

Ding, Z., Bynagari, Y. S., Mada, S. R., Jakubowski, J. A., & Kunapuli, S. P. (2009). Studies on the role of the extracellular cysteines and oligomeric structures of the P2Y_{12} receptor when interacting with antagonists. *Journal of Thrombosis and Haemostasis*, *7*, 232–234.

Ding, Z., Kim, S., Dorsam, R. T., Jin, J., & Kunapuli, S. P. (2003). Inactivation of the human P2Y_{12} receptor by thiol reagents requires interaction with both extracellular cysteine residues, Cys17 and Cys270. *Blood*, *101*, 3908–3914.

Ding, Z., Kim, S., & Kunapuli, S. P. (2006). Identification of a potent inverse agonist at a constitutively active mutant of human P2Y_{12} receptor. *Molecular Pharmacology*, *69*, 338–345.

Ding, Z., Tuluc, F., Bandivadekar, K. R., Zhang, L., Jin, J., & Kunapuli, S. P. (2005). Arg333 and Arg334 in the COOH terminus of the human P2Y_1 receptor are crucial for G_q coupling. *American Journal of Physiology. Cell Physiology*, *288*, C559–C567.

Douglass, J. G., Patel, R. I., Yerxa, B. R., Shaver, S. R., Watson, P. S., Bednarski, K., et al. (2008). Lipophilic modifications to dinucleoside polyphosphates and nucleotides that confer antagonist properties at the platelet $P2Y_{12}$ receptor. *Journal of Medicinal Chemistry*, *51*, 1007–1025.

Dranoff, J. A., O'Neill, A. F., Franco, A. M., Cai, S. Y., Connolly, G. C., Ballatori, N., et al. (2000). A primitive ATP receptor from the little skate *Raja erinacea*. *Journal of Biological Chemistry*, *275*, 30701–30706.

Drury, A. N., & Szent-Györgyi, A. (1929). The physiological activity of adenine compounds with especial reference to their action upon the mammalian heart. *Journal de Physiologie*, *68*, 213–237.

Ecke, D., Fischer, B., & Reiser, G. (2008). Diastereoselectivity of the $P2Y_{11}$ nucleotide receptor: Mutational analysis. *British Journal of Pharmacology*, *155*, 1250–1255.

Ecke, D., Hanck, T., Tulapurkar, M. E., Schäfer, R., Kassack, M., Stricker, R., et al. (2008). Hetero-oligomerization of the $P2Y_{11}$ receptor with the $P2Y_1$ receptor controls the internalization and ligand selectivity of the $P2Y_{11}$ receptor. *Biochemical Journal*, *409*, 107–116.

Ecke, D., Tulapurkar, M. E., Nahum, V., Fischer, B., & Reiser, G. (2006). Opposite diastereoselective activation of $P2Y_1$ and $P2Y_{11}$ nucleotide receptors by adenosine 5′-O-(alpha-boranotriphosphate) analogues. *British Journal of Pharmacology*, *149*, 416–423.

Elliott, M. R., Chekeni, F. B., Trampont, P. C., Lazarowski, E. R., Kadl, A., Walk, S. F., et al. (2009). Nucleotides released by apoptotic cells act as a find-me signal to promote phagocytic clearance. *Nature*, *461*, 282–286.

El-Tayeb, A., Griessmeier, K. J., & Müller, C. E. (2005). Synthesis and preliminary evaluation of [^{3}H]PSB-0413, a selective antagonist radioligand for platelet $P2Y_{12}$ receptors. *Bioorganic & Medicinal Chemistry Letters*, *15*, 5450–5452.

El-Tayeb, A., Qi, A., & Müller, C. E. (2006). Synthesis and structure-activity relationships of uracil nucleotide derivatives and analogues as agonists at human $P2Y_2$, $P2Y_4$, and $P2Y_6$ receptors. *Journal of Medicinal Chemistry*, *49*, 7076–7087.

Ennion, S. J., Powell, A. D., & Seward, E. P. (2004). Identification of the $P2Y_{12}$ receptor in nucleotide inhibition of exocytosis from bovine chromaffin cells. *Molecular Pharmacology*, *66*, 601–611.

Erb, L., Garrad, R., Wang, Y., Quinn, T., Turner, J. T., & Weisman, G. A. (1995). Site-directed mutagenesis of P2U purinoceptors. Positively charged amino acids in transmembrane helices 6 and 7 affect agonist potency and specificity. *Journal of Biological Chemistry*, *270*, 4185–4188.

Erb, L., Lustig, K. D., Sullivan, D. M., Turner, J. T., & Weisman, G. A. (1993). Functional expression and photoaffinity labeling of a cloned P2U purinergic receptor. *Proceedings of the National Academy of Sciences of the United States of America*, *90*, 10449–10453.

Fabre, J. E., Nguyen, M., Latour, A., Keifer, J. A., Audoly, L. P., Coffman, T. M., et al. (1999). Decreased platelet aggregation, increased bleeding time and resistance to thromboembolism in $P2Y_1$-deficient mice. *Nature Medicine*, *5*, 1199–1202.

Filippov, A. K., Simon, J., Barnard, E. A., & Brown, D. A. (2003). Coupling of the nucleotide $P2Y_4$ receptor to neuronal ion channels. *British Journal of Pharmacology*, *138*, 400–406.

Filtz, T. M., Li, Q., Boyer, J. L., Nicholas, R. A., & Harden, T. K. (1994). Expression of a cloned P2Y purinergic receptor that couples to phospholipase C. *Molecular Pharmacology*, *46*, 8–14.

Fontana, P., Dupont, A., Gandrille, S., Bachelot-Loza, C., Reny, J. L., Aiach, M., et al. (2003). Adenosine diphosphate-induced platelet aggregation is associated with $P2Y_{12}$ gene sequence variations in healthy subjects. *Circulation*, *108*, 989–995.

Fontana, P., Remones, V., Reny, J. L., Aiach, M., & Gaussem, P. (2005). $P2Y_1$ gene polymorphism and ADP-induced platelet response. *Journal of Thrombosis and Haemostasis*, *3*, 2349–2350.

Foster, C. J., Prosser, D. M., Agans, J. M., Zhai, Y., Smith, M. D., Lachowicz, J. E., et al. (2001). Molecular identification and characterization of the platelet ADP receptor targeted by thienopyridine antithrombotic drugs. *Journal of Clinical Investigation*, *107*, 1591–1598.

Fredriksson, R., Lagerstrom, M. C., Lundin, L. G., & Schioth, H. B. (2003). The G-protein-coupled receptors in the human genome form five main families. Phylogenetic analysis, paralogon groups, and fingerprints. *Molecular Pharmacology*, *63*, 1256–1272.

Freeman, K., Tsui, P., Moore, D., Emson, P. C., Vawter, L., Naheed, S., et al. (2001). Cloning, pharmacology, and tissue distribution of G-protein-coupled receptor GPR105 (KIAA0001) rodent orthologs. *Genomics*, *78*, 124–128.

Fricks, I. P., Carter, R. L., Lazarowski, E. R., & Harden, T. K. (2009). Gi-dependent cell signaling responses of the human $P2Y_{14}$ receptor in model cell systems. *Journal of Pharmacology and Experimental Therapeutics*, *330*, 162–168.

Fricks, I. P., Maddileti, S., Carter, R. L., Lazarowski, E. R., Nicholas, R. A., Jacobson, K. A., et al. (2008). UDP is a competitive antagonist at the human $P2Y_{14}$ receptor. *Journal of Pharmacology and Experimental Therapeutics*, *325*, 588–594.

Fumagalli, M., Trincavelli, L., Lecca, D., Martini, C., Ciana, P., & Abbracchio, M. P. (2004). Cloning, pharmacological characterisation and distribution of the rat G-protein-coupled $P2Y_{13}$ receptor. *Biochemical Pharmacology*, *68*, 113–124.

Gachet, C. (2006). Regulation of platelet functions by P2 receptors. *Annual Review of Pharmacology and Toxicology*, *46*, 277–300.

Gao, Z. G., Ding, Y., & Jacobson, K. A. (2010). UDP-glucose acting at $P2Y_{14}$ receptors is a mediator of mast cell degranulation. *Biochemical Pharmacology*, *79*, 873–879.

Gao, Z. G., Mamedova, L., Tchilibon, S., Gross, A. S., & Jacobson, K. A. (2004). 2, 2′-Pyridylisatogen tosylate antagonizes $P2Y_1$ receptor signaling without affecting nucleotide binding. *Biochemical Pharmacology*, *68*, 231–237.

Ghanem, E., Robaye, B., Leal, T., Leipziger, J., Van Driessche, W., Beauwens, R., et al. (2005). The role of epithelial $P2Y_2$ and $P2Y_4$ receptors in the regulation of intestinal chloride secretion. *British Journal of Pharmacology*, *146*, 364–369.

Ginsburg-Shmuel, T., Haas, M., Schumann, M., Reiser, G., Kalid, O., Stern, N., et al. (2010). 5-OMe-UDP is a potent and selective $P2Y_6$-receptor agonist. *Journal of Medicinal Chemistry*, *53*, 1673–1685.

Gorzalka, S., Vittori, S., Volpini, R., Cristalli, G., von Kügelgen, I., & Müller, C. E. (2005). Evidence for the functional expression and pharmacological characterization of adenine receptors in native cells and tissues. *Molecular Pharmacology*, *67*, 955–964.

Greve, H., Meis, S., Kassack, M. U., Kehraus, S., Krick, A., Wright, A. D., et al. (2007). New iantherans from the marine sponge Ianthella quadrangulata: Novel agonists of the $P2Y_{11}$ receptor. *Journal of Medicinal Chemistry*, *50*, 5600–5607.

Guo, D., von Kügelgen, I., Moro, S., Kim, Y. C., & Jacobson, K. A. (2002). Evidence for the recognition of non-nucleotide antagonists within the transmembrane domains of the human $P2Y_1$ receptor. *Drug Develepmont Research*, *57*, 173–181.

Haitina, T., Fredriksson, R., Foord, S. M., Schiöth, H. B., & Gloriam, D. E. (2009). The G protein-coupled receptor subset of the dog genome is more similar to that in humans than rodents. *BMC Genomics*, *10*, 24. 10.1186/1471-2164-10-24.

Harden, T. K., Sesma, J. I., Fricks, I. P., & Lazarowski, E. R. (2010). Signaling and pharmacological properties of the $P2Y_{14}$ receptor. *Acta Physiologica (Oxford)*, *199*, 149–160.

Harden, T. K., & Sondek, J. (2006). Regulation of phospholipase C isozymes by Ras superfamily GTPases. *Annual Review of Pharmacology and Toxicology*, *46*, 355–379.

Hashimoto, M., Sugidachi, A., Isobe, T., Niitsu, Y., Ogawa, T., Jakubowski, J. A., et al. (2007). The influence of $P2Y_{12}$ receptor deficiency on the platelet inhibitory activities of prasugrel in a mouse model: Evidence for specific inhibition of $P2Y_{12}$ receptors by prasugrel. *Biochemical Pharmacology*, *74*, 1003–1009.

Haynes, S. E., Hollopeter, G., Yang, G., Kurpius, D., Dailey, M. E., Gan, W. B., et al. (2006). The $P2Y_{12}$ receptor regulates microglial activation by extracellular nucleotides. *Nature Neuroscience*, *9*, 1512–1519.

Hechler, B., Eckly, A., Ohlmann, P., Cazenave, J. P., & Gachet, C. (1998). The $P2Y_1$ receptor, necessary but not sufficient to support full ADP-induced platelet aggregation, is not the target of the drug clopidogrel. *British Journal Haematology*, *103*, 858–866.

Hechler, B., Nonne, C., Roh, E. J., Cattaneo, M., Cazenave, J. P., Lanza, F., et al. (2006). MRS2500 [2-iodo-N6-methyl-(N)-methanocarba-2′-deoxyadenosine-3′, 5′-bisphosphate], a potent, selective, and stable antagonist of the platelet $P2Y_1$ receptor with strong antithrombotic activity in mice. *Journal of Pharmacology and Experimental Therapeutics*, *316*, 556–563.

Hechler, B., Vigne, P., Léon, C., Breittmayer, J. P., Gachet, C., & Frelin, C. (1998). ATP derivatives are antagonists of the $P2Y_1$ receptor: Similarities to the platelet ADP receptor. *Molecular Pharmacology*, *53*, 727–733.

Henderson, D. J., Elliot, D. G., Smith, G. M., Webb, T. E., & Dainty, I. A. (1995). Cloning and characterisation of a bovine P2Y receptor. *Biochemical and Biophysical Research Communications*, *212*, 648–656.

Hepler, J. R., & Gilman, A. G. (1992). G proteins. *Trends in Biochemical Science*, *17*, 383–387.

Herbert, J. M., Tissinier, A., Defreyn, G., & Maffrand, J. P. (1993). Inhibitory effect of clopidogrel on platelet adhesion and intimal proliferation after arterial injury in rabbits. *Arteriosclerosis and Thrombosis*, *13*, 1171–1179.

Herold, C. L., Qi, A. D., Harden, T. K., & Nicholas, R. A. (2004). Agonist versus antagonist action of ATP at the $P2Y_4$ receptor is determined by the second extracellular loop. *Journal of Biological Chemistry*, *279*, 11456–11464.

Hetherington, S. L., Singh, R. K., Lodwick, D., Thompson, J. R., Goodall, A. H., & Samani, N. J. (2005). Dimorphism in the $P2Y_1$ ADP receptor gene is associated with increased platelet activation response to ADP. *Arteriosclerosis, Thrombosis, and Vascular Biology*, *25*, 252–257.

Hillmann, P., Ko, G. Y., Spinrath, A., Raulf, A., von Kügelgen, I., Wolff, S. C., et al. (2009). Key determinants of nucleotide-activated G protein-coupled $P2Y_2$ receptor function revealed by chemical and pharmacological experiments, mutagenesis and homology modeling. *Journal of Medicinal Chemistry*, *52*, 2762–2775.

Hoebertz, A., Mahendran, S., Burnstock, G., & Arnett, T. R. (2002). ATP and UTP at low concentrations strongly inhibit bone formation by osteoblasts: A novel role for the $P2Y_2$ receptor in bone remodeling. *Journal of Cellular Biochemistry*, *86*, 413–419.

Hoffmann, K., Baqi, Y., Morena, M. S., Glänzel, M., Müller, C. E., & von Kügelgen, I. (2009). Interaction of new, very potent non-nucleotide antagonists with Arg256 of the human platelet $P2Y_{12}$ receptor. *Journal of Pharmacology and Experimental Therapeutics*, *331*, 648–655.

Hoffmann, C., Moro, S., Nicholas, R. A., Harden, T. K., & Jacobson, K. A. (1999). The role of amino acids in extracellular loops of the human $P2Y_1$ receptor in surface expression and activation processes. *Journal of Biological Chemistry*, *274*, 14639–14647.

Hoffmann, K., Sixel, U., Di Pasquale, F., & von Kügelgen, I. (2008). Involvement of basic amino acid residues in transmembrane regions 6 and 7 in agonist and antagonist recognition of the human platelet $P2Y_{12}$-receptor. *Biochemical Pharmacology*, *76*, 1201–1213.

Hoffmann, C., Soltysiak, K., West, P. L., & Jacobson, K. A. (2004). Shift in purine/pyrimidine base recognition upon exchanging extracellular domains in P2$Y_{1/6}$ chimeric receptors. *Biochemical Pharmacology*, *68*, 2075–2086.

Hoffmann, C., Ziegler, N., Reiner, S., Krasel, C., & Lohse, M. J. (2008). Agonist-selective, receptor-specific interaction of human P2Y receptors with beta-arrestin-1 and -2. *Journal of Biological Chemistry*, *283*, 30933–30941.

Hollopeter, G., Jantzen, H. M., Vincent, D., Li, G., England, L., Ramakrishnan, V., et al. (2001). Identification of the platelet ADP receptor targeted by antithrombotic drugs. *Nature*, *409*, 202–207.

Hou, M., Harden, T. K., Kuhn, C. M., Baldetorp, B., Lazarowski, E., Pendergast, W., et al. (2002). UDP acts as a growth factor for vascular smooth muscle cells by activation of P2Y_6 receptors. *American Journal of Physiology. Heart and Circulatory Physiology*, *282*, H784–H792.

Houston, D., Ohno, M., Nicholas, R. A., Jacobson, K. A., & Harden, T. K. (2006). [32P]2-iodo-N^6-methyl-(N)-methanocarba-2′-deoxyadenosine-3′, 5′-bisphosphate ([^{32}P]MRS2500), a novel radioligand for quantification of native P2Y_1 receptors. *British Journal of Pharmacology*, *147*, 459–467.

Ingall, A. H., Dixon, J., Bailey, A., Coombs, M. E., Cox, D., McInally, J. I., et al. (1999). Antagonists of the platelet P2T receptor: A novel approach to antithrombotic therapy. *Journal of Medicinal Chemistry*, *42*, 213–220.

Ivanov, A. A., Fricks, I., Harden, T. K., & Jacobson, K. A. (2007). Molecular dynamics simulation of the P2Y_{14} receptor. Ligand docking and identification of a putative binding site of the distal hexose moiety. *Bioorganic & Medicinal Chemistry Letters*, *17*, 761–766.

Ivanov, A. A., Ko, H., Cosyn, L., Maddileti, S., Besada, P., Fricks, I., et al. (2007). Molecular modeling of the human P2Y_2 receptor and design of a selective agonist, 2′-amino-2′-deoxy-2-thiouridine 5′-triphosphate. *Journal of Medicinal Chemistry*, *50*, 1166–1176.

Jaakola, V. P., Griffith, M. T., Hanson, M. A., Cherezov, V., Chien, E. Y., Lane, J. R., et al. (2008). The 2.6 angstrom crystal structure of a human A_{2A} adenosine receptor bound to an antagonist. *Science*, *322*, 1211–1217.

Jacobson, K. A. (2010). GPCR ligand-dendrimer (GLiDe) conjugates: Future smart drugs? *Trends in Pharmacological Sciences*, *31*, 575–579.

Jacobson, K. A., Costanzi, S., Ivanov, A. A., Tchilibon, S., Besada, P., Gao, Z. G., et al. (2006). Structure activity and molecular modeling analyses of ribose- and base-modified uridine 5′-triphosphate analogues at the human P2Y_2 and P2Y_4 receptors. *Biochemical Pharmacology*, *71*, 540–549.

Jacobson, K. A., Ivanov, A. A., de Castro, S., Harden, T. K., & Ko, H. (2009). Development of selective agonists and antagonists of P2Y receptors. *Purinergic Signalling*, *5*, 75–89.

Jacobson, K. A., Mamedova, L., Joshi, B. V., Besada, P., & Costanzi, S. (2005). Molecular recognition at adenine nucleotide (P2) receptors in platelets. *Seminars in Thrombosis and Hemostasis*, *31*, 205–216.

James, S., Akerblom, A., Cannon, C. P., Emanuelsson, H., Husted, S., Katus, H., et al. (2009). Comparison of ticagrelor, the first reversible oral P2Y_{12} receptor antagonist, with clopidogrel in patients with acute coronary syndromes: Rationale, design, and baseline characteristics of the PLATelet inhibition and patient Outcomes (PLATO) trial. *American Heart Journal*, *157*, 599–605.

Janssens, R., Communi, D., Pirotton, S., Samson, M., Parmentier, M., & Boeynaems, J. M. (1996). Cloning and tissue distribution of the human P2Y_1 receptor. *Biochemical and Biophysical Research Communications*, *221*, 588–593.

Janssens, R., Paindavoine, P., Parmentier, M., & Boeynaems, J. M. (1999). Human P2Y_2 receptor polymorphism: Identification and pharmacological characterization of two allelic variants. *British Journal of Pharmacology*, *127*, 709–716.

Jantzen, H. M., Milstone, D. S., Gousset, L., Conley, P. B., & Mortensen, R. M. (2001). Impaired activation of murine platelets lacking G alpha$_{i2}$. *Journal of Clinical Investigation*, *108*, 477–483.

Jiang, Q., Guo, D., Lee, B. X., Van Rhee, A. M., Kim, Y. C., Nicholas, R. A., et al. (1997). A mutational analysis of residues essential for ligand recognition at the human P2Y$_1$ receptor. *Molecular Pharmacology*, *52*, 499–507.

Jin, J., Daniel, J. L., & Kunapuli, S. P. (1998). Molecular basis for ADP-induced platelet activation. II. The P2Y$_1$ receptor mediates ADP-induced intracellular calcium mobilization and shape change in platelets. *Journal of Biological Chemistry*, *273*, 2030–2034.

Johnson, F. L., Boyer, J. L., Leese, P. T., Crean, C., Krishnamoorthy, R., Durham, T., et al. (2007). Rapid and reversible modulation of platelet function in man by a novel P2Y$_{12}$ ADP-receptor antagonist, INS50589. *Platelets*, *18*, 346–356.

Kauffenstein, G., Hechler, B., Cazenave, J. P., & Gachet, C. (2004). Adenine triphosphate nucleotides are antagonists at the P2Y receptor. *Journal of Thrombosis and Haemostasis*, *2*, 1980–1988.

Kennedy, C., Qi, A. D., Herold, C. L., Harden, T. K., & Nicholas, R. A. (2000). ATP, an agonist at the rat P2Y$_4$ receptor, is an antagonist at the human P2Y$_4$ receptor. *Molecular Pharmacology*, *57*, 926–931.

Kim, Y. C., Lee, J. S., Sak, K., Marteau, F., Mamedova, L., Boeynaems, J. M., et al. (2005). Synthesis of pyridoxal phosphate derivatives with antagonist activity at the P2Y$_{13}$ receptor. *Biochemical Pharmacology*, *70*, 266–274.

Kim, H. S., Ohno, M., Xu, B., Kim, H. O., Choi, Y., Ji, X. D., et al. (2003). 2-Substitution of adenine nucleotide analogues containing a bicyclo[3.1.0]hexane ring system locked in a northern conformation: Enhanced potency as P2Y$_1$ receptor antagonists. *Journal of Medicinal Chemistry*, *46*, 4974–4987.

Ko, H., Carter, R. L., Cosyn, L., Petrelli, R., de Castro, S., Besada, P., et al. (2008). Synthesis and potency of novel uracil nucleotides and derivatives as P2Y$_2$ and P2Y$_6$ receptor agonists. *Bioorganic & Medicinal Chemistry*, *16*, 6319–6332.

Ko, H., Fricks, I., Ivanov, A. A., Harden, T. K., & Jacobson, K. A. (2007). Structure-activity relationship of uridine 5′-diphosphoglucose analogues as agonists of the human P2Y$_{14}$ receptor. *Journal of Medicinal Chemistry*, *50*, 2030–2039.

Korcok, J., Raimundo, L. N., Du, X., Sims, S. M., & Dixon, S. J. (2005). P2Y$_6$ nucleotide receptors activate NF-kappaB and increase survival of osteoclasts. *Journal of Biological Chemistry*, *280*, 16909–16915.

Lagerström, M. C., & Schiöth, H. B. (2008). Structural diversity of G protein-coupled receptors and significance for drug discovery. *Nature Reviews. Drug Discovery*, *7*, 339–357.

Laplante, M.-A., Monassier, L., Freund, M., Bousquet, P., & Gachet, C. (2010). The purinergic P2Y$_1$ receptor supports leptin secretion in adipose tissue. *Endocrinology*, *151*, 2060–2070.

Lazarowski, E. R., Homolya, L., Boucher, R. C., & Harden, T. K. (1997). Direct demonstration of mechanically induced release of cellular UTP and its implication for uridine nucleotide receptor activation. *Journal of Biological Chemistry*, *272*, 24348–24354.

Lazarowski, E. R., Rochelle, L. G., O'Neal, W. K., Ribeiro, C. M., Grubb, B. R., Zhang, V., et al. (2001). Cloning and functional characterization of two murine uridine nucleotide receptors reveal a potential target for correcting ion transport deficiency in cystic fibrosis gallbladder. *Journal of Pharmacology and Experimental Therapeutics*, *297*, 43–49.

Lazarowski, E. R., Watt, W. C., Stutts, M. J., Boucher, R. C., & Harden, T. K. (1995). Pharmacological selectivity of the cloned human P2U-purinoceptor: Potent activation by diadenosine tetraphosphate. *British Journal of Pharmacology*, *116*, 1619–1627.

Leipziger, J. (2003). Control of epithelial transport via luminal P2 receptors. *American Journal of Physiology - Renal Physiology*, *284*, F419–F432.

Léon, C., Hechler, B., Freund, M., Eckly, A., Vial, C., Ohlmann, P., et al. (1999). Defective platelet aggregation and increased resistance to thrombosis in purinergic $P2Y_1$ receptor-null mice. *Journal of Clinical Investigation*, *104*, 1731–1737.

Léon, C., Hechler, B., Vial, C., Leray, C., Cazenave, J. P., & Gachet, C. (1997). The $P2Y_1$ receptor is an ADP receptor antagonized by ATP and expressed in platelets and megakaryoblastic cells. *FEBS Letters*, *403*, 26–30.

Léon, C., Vial, C., Cazenave, J. P., & Gachet, C. (1996). Cloning and sequencing of a human cDNA encoding endothelial $P2Y_1$ purinoceptor. *Gene*, *171*, 295–297.

Li, Q., Olesky, M., Palmer, R. K., Harden, T. K., & Nicholas, R. A. (1998). Evidence that the p2y3 receptor is the avian homologue of the mammalian $P2Y_6$ receptor. *Molecular Pharmacology*, *54*, 541–546.

Li, Q., Schachter, J. B., Harden, T. K., & Nicholas, R. A. (1997). The 6H1 orphan receptor, claimed to be the p2y5 receptor, does not mediate nucleotide-promoted second messenger responses. *Biochemical and Biophysical Research Communications*, *236*, 455–460.

Lustig, K. D., Shiau, A. K., Brake, A. J., & Julius, D. (1993). Expression cloning of an ATP receptor from mouse neuroblastoma cells. *Proceedings of the National Academy of Sciences of the United States of America*, *90*, 5113–5117.

Lutz, S., Freichel-Blomquiest, A., Yang, Y., Rumenapp, U., Jakobs, K. H., Schmidt, M., & Wieland, T. (2005). The guanine nucleotide exchange factor p63RhoGEF, a specific link between Gq/11-coupled receptor signaling and Rho. *Journal of Biological Chemistry*, *280*, 1134–1139.

Maffrand, J. P., Bernat, A., Delebassee, D., Defreyn, G., Cazenave, J. P., & Gordon, J. L. (1988). ADP plays a key role in thrombogenesis in rats. *Thrombosis and Haemostasis*, *59*, 225–230.

Maier, R., Glatz, A., Mosbacher, J., & Bilbe, G. (1997). Cloning of $P2Y_6$ cDNAs and identification of a pseudogene: Comparison of P2Y receptor subtype expression in bone and brain tissues. *Biochemical and Biophysical Research Communications*, *240*, 298–302.

Major, D. T., & Fischer, B. (2004). Molecular recognition in purinergic receptors. 1. A comprehensive computational study of the $hP2Y_1$-receptor. *Journal of Medicinal Chemistry*, *47*, 4391–4404.

Major, D. T., Nahum, V., Wang, Y., Reiser, G., & Fischer, B. (2004). Molecular recognition in purinergic receptors. 2. Diastereoselectivity of the $hP2Y_1$-receptor. *Journal of Medicinal Chemistry*, *47*, 4405–4416.

Malin, S. A., & Molliver, D. C. (2009). Gi- and Gq-coupled ADP (P2Y) receptors act in opposition to modulate nociceptive signaling and inflammatory pain behavior. *Molecular Pain*, *6*, 21.

Malmsjö, M., Hou, M., Pendergast, W., Erlinge, D., & Edvinsson, L. (2003). Potent $P2Y_6$ receptor mediated contractions in human cerebral arteries. *BMC Pharmacology*, *3*, 4. 10.1186/1471-2210-3-4.

Mamedova, L. K., Joshi, B. V., Gao, Z. G., von Kügelgen, I., & Jacobson, K. A. (2004). Diisothiocyanate derivatives as potent, insurmountable antagonists of $P2Y_6$ nucleotide receptors. *Biochemical Pharmacology*, *67*, 1763–1770.

Mao, Y., Zhang, L., Jin, J., Ashby, B., & Kunapuli, S. P. (2010). Mutational analysis of residues important for ligand interaction with the human $P2Y_{12}$ receptor. *European Journal of Pharmacology*, *644*, 10–16.

Marteau, F., Le Poul, E., Communi, D., Communi, D., Labouret, C., Savi, P., et al. (2003). Pharmacological characterization of the human $P2Y_{13}$ receptor. *Molecular Pharmacology*, *64*, 104–112.

Maruoka, H., Barrett, M. O., Ko, H., Tosh, D. K., Melman, A., Burianek, L. E., et al. (2010). Pyrimidine ribonucleotides with enhanced selectivity as $P2Y_6$ receptor agonists: Novel 4-alkyloxyimino, (S)-methanocarba, and 5′-triphosphate gamma-ester modifications. *Journal of Medicinal Chemistry*, *53*, 4488–4501.

Meis, S., Hamacher, A., Hongwiset, D., Marzian, C., Wiese, M., Eckstein, N., et al. (2010). NF546 [4, 4′-(carbonylbis(imino-3, 1-phenylene-carbonylimino-3, 1-(4-methyl-phenylene)-carbonylimino))-bis(1, 3-xylene-alpha, alpha′-diphosphonic acid) tetrasodium salt] is a non-nucleotide P2Y11 agonist and stimulates release of interleukin-8 from human monocyte-derived dendritic cells. *Journal of Pharmacology and Experimental Therapeutics*, *332*, 238–247.

Milligan, G. (2009). G protein-coupled receptor hetero-dimerization: Contribution to pharmacology and function. *British Journal of Pharmacology*, *158*, 5–14.

Morales-Ramos, A. I., Mecom, J. S., Kiesow, T. J., Graybill, T. L., Brown, G. D., Aiyar, N. V., et al. (2008). Tetrahydro-4-quinolinamines identified as novel $P2Y_1$ receptor antagonists. *Bioorganic & Medicinal Chemistry Letters*, *18*, 6222–6226.

Moreschi, I., Bruzzone, S., Bodrato, N., Usai, C., Guida, L., Nicholas, R. A., et al. (2008). $NAADP^+$ is an agonist of the human $P2Y_{11}$ purinergic receptor. *Cell Calcium*, *43*, 344–355.

Moreschi, I., Bruzzone, S., Nicholas, R. A., Fruscione, F., Sturla, L., Benvenuto, F., et al. (2006). Extracellular NAD^+ is an agonist of the human $P2Y_{11}$ purinergic receptor in human granulocytes. *Journal of Biological Chemistry*, *281*, 31419–31429.

Moro, S., Guo, D., Camaioni, E., Boyer, J. L., Harden, T. K., & Jacobson, K. A. (1998). Human $P2Y_1$ receptor: Molecular modeling and site-directed mutagenesis as tools to identify agonist and antagonist recognition sites. *Journal of Medicinal Chemistry*, *41*, 1456–1466.

Moro, S., Hoffmann, C., & Jacobson, K. A. (1999). Role of the extracellular loops of G protein-coupled receptors in ligand recognition: A molecular modeling study of the human $P2Y_1$ receptor. *Biochemistry*, *38*, 3498–3507.

Morris, G. E., Nelson, C. P., Everitt, D., Brighton, P. J., Standen, N. B., Challiss, R. A., et al. (2010). G protein-coupled receptor kinase 2 and arrestin2 regulate arterial smooth muscle P2Y-purinoceptor signalling. *Cardiovascular Research*, in press.

Müller, C. E. (2002). P2-pyrimidinergic receptors and their ligands. *Current Pharmaceutical Design*, *8*, 2353–2369.

Nandanan, E., Jang, S. Y., Moro, S., Kim, H. O., Siddiqui, M. A., Russ, P., et al. (2000). Synthesis, biological activity, and molecular modeling of ribose-modified deoxyadenosine bisphosphate analogues as $P2Y_1$ receptor ligands. *Journal of Medicinal Chemistry*, *43*, 829–842.

Nguyen, T., Erb, L., Weisman, G. A., Marchese, A., Heng, H. H., Garrad, R. C., et al. (1995). Cloning, expression, and chromosomal localization of the human uridine nucleotide receptor gene. *Journal of Biological Chemistry*, *270*, 30845–30848.

Nicholas, R. A., Watt, W. C., Lazarowski, E. R., Li, Q., & Harden, K. (1996). Uridine nucleotide selectivity of three phospholipase C-activating P2 receptors: Identification of a UDP-selective, a UTP-selective, and an ATP- and UTP-specific receptor. *Molecular Pharmacology*, *50*, 224–229.

Niitsu, Y., Jakubowski, J. A., Sugidachi, A., & Asai, F. (2005). Pharmacology of CS-747 (prasugrel, LY640315), a novel, potent antiplatelet agent with in vivo $P2Y_{12}$ receptor antagonist activity. *Seminars in Thrombosis and Hemostasis*, *31*, 184–194.

Noguchi, K., Ishii, S., & Shimizu, T. (2003). Identification of p2y9/GPR23 as a novel G protein-coupled receptor for lysophosphatidic acid, structurally distant from the Edg family. *Journal of Biological Chemistry*, *278*, 25600–25606.

Norambuena, A., Palma, F., Poblete, M. I., Donoso, M. V., Pardo, E., González, A., et al. (2010). UTP controls cell surface distribution and vasomotor activity of the human $P2Y_2$ receptor through an epidermal growth factor receptor-transregulated mechanism. *Journal of Biological Chemistry*, *285*, 2940–2950.

North, R. A. (2002). Molecular physiology of P2X receptors. *Physiological Reviews*, *82*, 1013–1067.

O'Grady, S. M., Elmquist, E., Filtz, T. M., Nicholas, R. A., & Harden, T. K. (1996). A guanine nucleotide-independent inwardly rectifying cation permeability is associated with P2Y$_1$ receptor expression in Xenopus oocytes. *Journal of Biological Chemistry*, *271*, 29080–29087.

Ohlmann, P., de Castro, S., Brown, G. G., Gachet, C., Jacobson, K. A., & Harden, T. K. (2010). Quantification of recombinant and platelet P2Y$_1$ receptors utilizing a [^{125}I]-labeled high-affinity antagonist 2-iodo-N^6-methyl-(N)-methanocarba-2′-deoxyadenosine-3′, 5′-bisphosphate ([^{125}I]MRS2500). *Pharmacological Research*, *62*, 344–351.

Palmer, R. K., Boyer, J. L., Schachter, J. B., Nicholas, R. A., & Harden, T. K. (1998). Agonist action of adenosine triphosphates at the human P2Y$_1$ receptor. *Molecular Pharmacology*, *54*, 1118–1123.

Park, J. H., Scheerer, P., Hofmann, K. P., Choe, H. W., & Ernst, O. P. (2008). Crystal structure of the ligand-free G-protein-coupled receptor opsin. *Nature*, *454*, 183–187.

Parlow, J. J., Burney, M. W., Case, B. L., Girard, T. J., Hall, K. A., Harris, P. K., et al. (2010). Piperazinyl glutamate pyridines as potent orally bioavailable P2Y$_{12}$ antagonists for inhibition of platelet aggregation. *Journal of Medicinal Chemistry*, *53*, 2010–2037.

Parlow, J. J., Burney, M. W., Case, B. L., Girard, T. J., Hall, K. A., Hiebsch, R. R., et al. (2009). Piperazinyl-glutamate-pyridines as potent orally bioavailable P2Y$_{12}$ antagonists for inhibition of platelet aggregation. *Bioorganic & Medicinal Chemistry Letters*, *19*, 4657–4663.

Parr, C. E., Sullivan, D. M., Paradiso, A. M., Lazarowski, E. R., Burch, L. H., Olsen, J. C., et al. (1994). Cloning and expression of a human P2U nucleotide receptor, a target for cystic fibrosis pharmacotherapy. *Proceedings of the National Academy of Sciences of the United States of America*, *91*, 3275–3279.

Pasternack, S. M., von Kügelgen, I., Aboud, K. A., Lee, Y. A., Rüschendorf, F., Voss, K., et al. (2008). G protein-coupled receptor P2Y5 and its ligand LPA are involved in maintenance of human hair growth. *Nature Genetics*, *40*, 329–334.

Patel, K., Barnes, A., Camacho, J., Paterson, C., Boughtflower, R., Cousens, D., et al. (2001). Activity of diadenosine polyphosphates at P2Y receptors stably expressed in 1321N1 cells. *European Journal of Pharmacology*, *430*, 203–210.

Pausch, M. H., Lai, M., Tseng, E., Paulsen, J., Bates, B., & Kwak, S. (2004). Functional expression of human and mouse P2Y$_{12}$ receptors in Saccharomyces cerevisiae. *Biochemical and Biophysical Research Communications*, *324*, 171–177.

Pendergast, W., Yerxa, B. R., Douglass, J. G., Shaver, S. R., Dougherty, R. W., Redick, C. C., et al. (2001). Synthesis and P2Y receptor activity of a series of uridine dinucleoside 5′-polyphosphates. *Bioorganic & Medicinal Chemistry Letters*, *11*, 157–160.

Pfefferkorn, J. A., Choi, C., Winters, T., Kennedy, R., Chi, L., Perrin, L. A., et al. (2008). P2Y$_1$ receptor antagonists as novel antithrombotic agents. *Bioorganic & Medicinal Chemistry Letters*, *18*, 3338–3343.

Qi, A. D., Kennedy, C., Harden, T. K., & Nicholas, R. A. (2001). Differential coupling of the human P2Y$_{11}$ receptor to phospholipase C and adenylyl cyclase. *British Journal of Pharmacology*, *132*, 318–326.

Qi, A. D., Wolff, S. C., & Nicholas, R. A. (2005). The apical targeting signal of the P2Y$_2$ receptor is located in its first extracellular loop. *Journal of Biological Chemistry*, *280*, 29169–29175.

Qi, A. D., Zambon, A. C., Insel, P. A., & Nicholas, R. A. (2001). An arginine/glutamine difference at the juxtaposition of transmembrane domain 6 and the third extracellular loop contributes to the markedly different nucleotide selectivities of human and canine P2Y$_{11}$ receptors. *Molecular Pharmacology*, *60*, 1375–1382.

Ralevic, V., & Burnstock, G. (1998). Receptors for purines and pyrimidines. *Pharmacological Reviews*, *50*, 413–492.

Rasmussen, S. G., Choi, H. J., Rosenbaum, D. M., Kobilka, T. S., Thian, F. S., Edwards, P. C., et al. (2007). Crystal structure of the human beta2 adrenergic G-protein-coupled receptor. *Nature*, *450*, 383–387.

Reiner, S., Ziegler, N., Léon, C., Lorenz, K., von Hayn, K., Gachet, C., et al. (2009). beta-Arrestin-2 interaction and internalization of the human $P2Y_1$ receptor are dependent on C-terminal phosphorylation sites. *Molecular Pharmacology*, *76*, 1162–1171.

Remijn, J. A., Ijsseldijk, M. J., Strunk, A. L., Abbes, A. P., Engel, H., Dikkeschei, B., et al. (2007). Novel molecular defect in the platelet ADP receptor $P2Y_{12}$ of a patient with haemorrhagic diathesis. *Clinical Chemistry and Laboratory Medicine*, *45*, 187–189.

Rice, W. R., Burton, F. M., & Fiedeldey, D. T. (1995). Cloning and expression of the alveolar type II cell P2u-purinergic receptor. *American Journal of Respiratory Cell and Molecular Biology*, *12*, 27–32.

Robaye, B., Boeynaems, J. M., & Communi, D. (1997). Slow desensitization of the human $P2Y_6$ receptor. *European Journal of Pharmacology*, *329*, 231–236.

Robaye, B., Ghanem, E., Wilkin, F., Fokan, D., Van Driessche, W., Schurmans, S., et al. (2003). Loss of nucleotide regulation of epithelial chloride transport in the jejunum of $P2Y_4$-null mice. *Molecular Pharmacology*, *63*, 777–783.

Rudez, G., Bouman, H. J., van Werkum, J. W., Leebeek, F. W., Kruit, A., Ruven, H. J., et al. (2009). Common variation in the platelet receptor $P2Y_{12}$ gene is associated with residual on-clopidogrel platelet reactivity in patients undergoing elective percutaneous coronary interventions. *Circulation. Cardiovascular Genetics*, *2*, 515–521.

Sauer, R., El-Tayeb, A., Kaulich, M., & Müller, C. E. (2009). Synthesis of uracil nucleotide analogs with a modified, acyclic ribose moiety as $P2Y_2$ receptor antagonists. *Bioorganic & Medicinal Chemistry*, *17*, 5071–5079.

Savi, P., & Herbert, J. M. (2005). Clopidogrel and ticlopidine: $P2Y_{12}$ adenosine diphosphate-receptor antagonists for the prevention of atherothrombosis. *Seminars in Thrombosis and Hemostasis*, *31*, 174–183.

Savi, P., Labouret, C., Delesque, N., Guette, F., Lupker, J., & Herbert, J. M. (2001). $P2Y_{12}$, a new platelet ADP receptor, target of clopidogrel. *Biochemical and Biophysical Research Communications*, *283*, 379–383.

Savi, P., Zachayus, J. L., Delesque-Touchard, N., Labouret, C., Hervé, C., Uzabiaga, M. F., et al. (2006). The active metabolite of clopidogrel disrupts $P2Y_{12}$ receptor oligomers and partitions them out of lipid rafts. *Proceedings of the National Academy of Sciences of the United States of America*, *103*, 11069–11074.

Schachter, J. B., Li, Q., Boyer, J. L., Nicholas, R. A., & Harden, T. K. (1996). Second messenger cascade specificity and pharmacological selectivity of the human $P2Y_1$-purinoceptor. *British Journal of Pharmacology*, *118*, 167–173.

Schettert, I. T., Pereira, A. C., Lopes, N. H., Hueb, W. A., & Krieger, J. E. (2006). Association between platelet $P2Y_{12}$ haplotype and risk of cardiovascular events in chronic coronary disease. *Thrombosis Research*, *118*, 679–683.

Schöneberg, T., Hermsdorf, T., Engemaier, E., Engel, K., Liebscher, I., Thor, D., et al. (2007). Structural and functional evolution of the $P2Y_{12}$-like receptor group. *Purinergic Signalling*, *3*, 255–268.

Seifert, R., & Schultz, G. (1989). Involvement of pyrimidinoceptors in the regulation of cell functions by uridine and by uracil nucleotides. *Trends in Pharmacological Sciences*, *10*, 365–369.

Shaver, S. R., Rideout, J. L., Pendergast, W., Douglass, J. G., Brown, E. G., Boyer, J. L., et al. (2005). Structure-activity relationships of dinucleotides: Potent and selective agonists of P2Y receptors. *Purinergic Signalling*, *1*, 183–191.

Shen, J., Seye, C. I., Wang, M., Weisman, G. A., Wilden, P. A., & Sturek, M. (2004). Cloning, up-regulation, and mitogenic role of porcine $P2Y_2$ receptor in coronary artery smooth muscle cells. *Molecular Pharmacology*, *66*, 1265–1274.

Shiraga, M., Miyata, S., Kato, H., Kashiwagi, H., Honda, S., Kurata, Y., et al. (2005). Impaired platelet function in a patient with $P2Y_{12}$ deficiency caused by a mutation in the translation initiation codon. *Journal of Thrombosis and Haemostasis*, *3*, 2315–2323.

Simon, J., Vigne, P., Eklund, K. M., Michel, A. D., Carruthers, A. M., Humphrey, P. P., et al. (2001). Activity of adenosine diphosphates and triphosphates on a $P2Y_T$ -type receptor in brain capillary endothelial cells. *British Journal of Pharmacology*, *132*, 173–182.

Smrcka, A. V. (2008). G protein betagamma subunits: Central mediators of G protein-coupled receptor signaling. *Cellular and Molecular Life Sciences*, *65*, 2191–2214.

Southey, M. C., Hammet, F., Hutchins, A. M., Paidhungat, M., Somers, G. R., & Venter, D. J. (1996). Molecular cloning and sequencing of a novel human P2 nucleotide receptor. *Biochimica et Biophysica Acta*, *1309*, 77–80.

Springthorpe, B., Bailey, A., Barton, P., Birkinshaw, T. N., Bonnert, R. V., Brown, R. C., et al. (2007). From ATP to AZD6140: The discovery of an orally active reversible $P2Y_{12}$ receptor antagonist for the prevention of thrombosis. *Bioorganic & Medicinal Chemistry Letters*, *17*, 6013–6018.

Staritz, P., Kurz, K., Stoll, M., Giannitsis, E., Katus, H. A., & Ivandic, B. T. (2009). Platelet reactivity and clopidogrel resistance are associated with the H2 haplotype of the $P2Y_{12}$-ADP receptor gene. *International Journal of Cardiology*, *133*, 341–345.

Storey, R. F., Thornton, M. S., Lawrance, R., Husted, S., Wickens, M., Emanuelsson, H., et al. (2009). Ticagrelor yields consistent dose-dependent inhibition of ADP-induced platelet aggregation in patients with atherosclerotic disease regardless of genotypic variations in $P2RY_{12}$, $P2RY_1$, and ITGB3. *Platelets*, *20*, 341–348.

Suarez-Huerta, N., Pouillon, V., Boeynaems, J., & Robaye, B. (2001). Molecular cloning and characterization of the mouse $P2Y_4$ nucleotide receptor. *European Journal of Pharmacology*, *416*, 197–202.

Sugidachi, A., Asai, F., Ogawa, T., Inoue, T., & Koike, H. (2000). The in vivo pharmacological profile of CS-747, a novel antiplatelet agent with platelet ADP receptor antagonist properties. *British Journal of Pharmacology*, *129*, 1439–1446.

Suzuki, T., Namba, K., Tsuga, H., & Nakata, H. (2006). Regulation of pharmacology by hetero-oligomerization between A_1 adenosine receptor and $P2Y_2$ receptor. *Biochemical and Biophysical Research Communications*, *351*, 559–565.

Takasaki, J., Kamohara, M., Saito, T., Matsumoto, M., Matsumoto, S., Ohishi, T., et al. (2001). Molecular cloning of the platelet $P2T_{AC}$ ADP receptor: Pharmacological comparison with another ADP receptor, the $P2Y_1$ receptor. *Molecular Pharmacology*, *60*, 432–439.

Tauber, J., Davitt, W. F., Bokosky, J. E., Nichols, K. K., Yerxa, B. R., Schaberg, A. E., et al. (2004). Double-masked, placebo-controlled safety and efficacy trial of diquafosol tetrasodium (INS365) ophthalmic solution for the treatment of dry eye. *Cornea*, *23*, 784–792.

Tokuyama, Y., Hara, M., Jones, E. M., Fan, Z., & Bell, G. I. (1995). Cloning of rat and mouse P2Y purinoceptors. *Biochemical and Biophysical Research Communications*, *211*, 211–218.

Torres, B., Zambon, A. C., & Insel, P. A. (2002). $P2Y_{11}$ receptors activate adenylyl cyclase and contribute to nucleotide-promoted cAMP formation in MDCK-D1 cells. A mechanism for nucleotide-mediated autocrine-paracrine regulation. *Journal of Biological Chemistry*, *277*, 7761–7765.

Ullmann, H., Meis, S., Hongwiset, D., Marzian, C., Wiese, M., Nickel, P., et al. (2005). Synthesis and structure-activity relationships of suramin-derived $P2Y_{11}$ receptor antagonists with nanomolar potency. *Journal of Medicinal Chemistry*, *48*, 7040–7048.

van Giezen, J. J., Nilsson, L., Berntsson, P., Wissing, B. M., Giordanetto, F., Tomlinson, W., et al. (2009). Ticagrelor binds to human $P2Y_{12}$ independently from ADP but antagonizes ADP-induced receptor signaling and platelet aggregation. *Journal of Thrombosis and Haemostasis*, *7*, 1556–1565.

van Rhee, A. M., Fischer, B., van Galen, P. J., & Jacobson, K. A. (1995). Modelling the P2Y purinoceptor using rhodopsin as template. *Drug Design & Discovery*, *13*, 133–154.

Vasiljev, K. S., Uri, A., & Laitinen, J. T. (2003). 2-Alkylthio-substituted platelet $P2Y_{12}$ receptor antagonists reveal pharmacological identity between the rat brain G_i-linked ADP receptors and $P2Y_{12}$. *Neuropharmacology*, *45*, 145–154.

von Kügelgen, I. (2006). Pharmacological profiles of cloned mammalian P2Y-receptor subtypes. *Pharmacology & Therapeutics*, *110*, 415–432.

von Kügelgen, I., Häussinger, D., & Starke, K. (1987). Evidence for a vasoconstriction-mediating receptor for UTP, distinct from the P2 purinoceptor, in rabbit ear artery. *Naunyn-Schmiedeberg's Archives of Pharmacology*, *336*, 556–560.

von Kügelgen, I., Schiedel, A. C., Hoffmann, K., Alsdorf, B. B., Abdelrahman, A., & Müller, C. E. (2008). Cloning and functional expression of a novel G_i protein-coupled receptor for adenine from mouse brain. *Molecular Pharmacology*, *73*, 469–477.

von Kügelgen, I., Späth, L., & Starke, K. (1992). Stable adenine nucleotides inhibit [^{3}H]-noradrenaline release in rabbit brain cortex slices by direct action at presynaptic adenosine A_1-receptors. *Naunyn-Schmiedeberg's Archives of Pharmacology*, *346*, 187–196.

von Kügelgen, I., & Wetter, A. (2000). Molecular pharmacology of P2Y-receptors. *Naunyn-Schmiedeberg's Archives of Pharmacology*, *362*, 310–323.

Waldo, G. L., Corbitt, J., Boyer, J. L., Ravi, G., Kim, H. S., Ji, X. D., et al. (2002). Quantitation of the $P2Y_1$ receptor with a high affinity radiolabeled antagonist. *Molecular Pharmacology*, *62*, 1249–1257.

Waldo, G. L., & Harden, T. K. (2004). Agonist binding and G_q-stimulating activities of the purified human $P2Y_1$ receptor. *Molecular Pharmacology*, *65*, 426–436.

Wang, Z. X., Nakayama, T., Sato, N., Izumi, Y., Kasamaki, Y., Ohta, M., et al. (2009). Association of the purinergic receptor P2Y, G-protein coupled, 2 ($P2RY_2$) gene with myocardial infarction in Japanese men. *Circulation Journal*, *73*, 2322–2329.

Wang, Z., Nakayama, T., Sato, N., Izumi, Y., Kasamaki, Y., Ohta, M., et al. (2010). The purinergic receptor P2Y, G-protein coupled, 2 ($P2RY_2$) gene associated with essential hypertension in Japanese men. *Journal of Human Hypertension*, *24*, 327–335.

Wang, Z., Nakayama, T., Sato, N., Yamaguchi, M., Izumi, Y., Kasamaki, Y., et al. (2009). Purinergic receptor P2Y, G-protein coupled, 2 ($P2RY_2$) gene is associated with cerebral infarction in Japanese subjects. *Hypertension Research*, *32*, 989–996.

Wang, Y. X., Vincelette, J., da Cunha, V., Martin-McNulty, B., Mallari, C., Fitch, R. M., et al. (2007). A novel $P2Y_{12}$ adenosine diphosphate receptor antagonist that inhibits platelet aggregation and thrombus formation in rat and dog models. *Thrombosis and Haemostasis*, *97*, 847–855.

Warne, T., Serrano-Vega, M. J., Baker, J. G., Moukhametzianov, R., Edwards, P. C., Henderson, R., et al. (2008). Structure of a beta1-adrenergic G-protein-coupled receptor. *Nature*, *454*, 486–491.

Warny, M., Aboudola, S., Robson, S. C., Sévigny, J., Communi, D., Soltoff, S. P., et al. (2001). $P2Y_6$ nucleotide receptor mediates monocyte interleukin-8 production in response to UDP or lipopolysaccharide. *Journal of Biological Chemistry*, *276*, 26051–26056.

Watson, S., Daly, M., Dawood, B., Gissen, P., Makris, M., Mundell, S., et al. (2010). Phenotypic approaches to gene mapping in platelet function disorders—Identification of new variant of $P2Y_{12}$, TxA_2 and GPVI receptors. *Hämostaseologie*, *30*, 29–38.

Webb, T. E., Henderson, D., King, B. F., Wang, S., Simon, J., Bateson, A. N., et al. (1996). A novel G protein-coupled P2 purinoceptor ($P2Y_3$) activated preferentially by nucleoside diphosphates. *Molecular Pharmacology*, *50*, 258–265.

Webb, T. E., Henderson, D. J., Roberts, J. A., & Barnard, E. A. (1998). Molecular cloning and characterization of the rat P2Y$_4$ receptor. *Journal of Neurochemistry*, *71*, 1348–1357.

Webb, T. E., Simon, J., Krishek, B. J., Bateson, A. N., Smart, T. G., King, B. F., et al. (1993). Cloning and functional expression of a brain G-protein-coupled ATP receptor. *FEBS Letters*, *324*, 219–225.

Weyler, S., Baqi, Y., Hillmann, P., Kaulich, M., Hunder, A. M., Müller, I. A., et al. (2008). Combinatorial synthesis of anilinoanthraquinone derivatives and evaluation as non-nucleotide-derived P2Y$_2$ receptor antagonists. *Bioorganic & Medicinal Chemistry Letters*, *18*, 223–227.

White, P. J., Webb, T. E., & Boarder, M. R. (2003). Characterization of a Ca^{2+} response to both UTP and ATP at human P2Y$_{11}$ receptors: Evidence for agonist-specific signaling. *Molecular Pharmacology*, *63*, 1356–1363.

Wildman, S. S., Unwin, R. J., & King, B. F. (2003). Extended pharmacological profiles of rat P2Y$_2$ and rat P2Y$_4$ receptors and their sensitivity to extracellular H^+ and Zn^{2+} ions. *British Journal of Pharmacology*, *140*, 1177–1186.

Wilkin, F., Duhant, X., Bruyns, C., Suarez-Huerta, N., Boeynaems, J. M., & Robaye, B. (2001). The P2Y$_{11}$ receptor mediates the ATP-induced maturation of human monocyte-derived dendritic cells. *Journal of Immunology*, *166*, 7172–7177.

Wirkner, K., Schweigel, J., Gerevich, Z., Franke, H., Allgaier, C., Barsoumian, E. L., et al. (2004). Adenine nucleotides inhibit recombinant N-type calcium channels via G protein-coupled mechanisms in HEK 293 cells; involvement of the P2Y$_{13}$ receptor-type. *British Journal of Pharmacology*, *141*, 141–151.

Wolff, S. C., Qi, A. D., Harden, T. K., & Nicholas, R. A. (2005). Polarized expression of human P2Y receptors in epithelial cells from kidney, lung, and colon. *American Journal of Physiology. Cell Physiology*, *288*, C624–C632.

Wolff, S. C., Qi, A. D., Harden, T. K., & Nicholas, R. A. (2010). Charged residues in the C-terminus of the P2Y$_1$ receptor constitute a basolateral-sorting signal. *Journal of Cell Science*, *123*, 2512–2520.

Xu, B., Stephens, A., Kirschenheuter, G., Greslin, A. F., Cheng, X., Sennelo, J., et al. (2002). Acyclic analogues of adenosine bisphosphates as P2Y receptor antagonists: Phosphate substitution leads to multiple pathways of inhibition of platelet aggregation. *Journal of Medicinal Chemistry*, *45*, 5694–5709.

Yanagida, K., Masago, K., Nakanishi, H., Kihara, Y., Hamano, F., Tajima, Y., et al. (2009). Identification and characterization of a novel lysophosphatidic acid receptor, p2y5/LPA6. *Journal of Biological Chemistry*, *284*, 17731–17741.

Ye, H., Chen, C., Zhang, H. C., Haertlein, B., Parry, T. J., Damiano, B. P., et al. (2008). Carbanucleosides as potent antagonists of the adenosine 5′-diphosphate (ADP) purinergic receptor P2Y$_{12}$ on human platelets. *ChemMedChem*, *3*, 732–736.

Yerxa, B. R. (2001). Therapeutic use of nucleotides in respiratory and ophthalmic diseases. *Drug Development Research*, *52*, 196–201.

Yerxa, B. R., Sabater, J. R., Davis, C. W., Stutts, M. J., Lang-Furr, M., Picher, M., et al. (2002). Pharmacology of INS37217 [P1-(uridine 5′)-P4-(2′-deoxycytidine 5′)tetraphosphate, tetrasodium salt], a next-generation P2Y$_2$ receptor agonist for the treatment of cystic fibrosis. *Journal of Pharmacology and Experimental Therapeutics*, *302*, 871–880.

Yokomizo, T., Izumi, T., Chang, K., Takuwa, Y., & Shimizu, T. (1997). A G-protein-coupled receptor for leukotriene B4 that mediates chemotaxis. *Nature*, *387*, 620–624.

Yoshioka, K., Saitoh, O., & Nakata, H. (2001). Heteromeric association creates a P2Y-like adenosine receptor. *Proceedings of the National Academy of Sciences of the United States of America*, *98*, 7617–7622.

Yoshioka, K., Saitoh, O., & Nakata, H. (2002). Agonist-promoted heteromeric oligomerization between adenosine A$_1$ and P2Y$_1$ receptors in living cells. *FEBS Letters*, *523*, 147–151.

Zambon, A. C., Brunton, L. L., Barrett, K. E., Hughes, R. J., Torres, B., & Insel, P. A. (2001). Cloning, expression, signaling mechanisms, and membrane targeting of $P2Y_{11}$ receptors in Madin Darby canine kidney cells. *Molecular Pharmacology*, *60*, 26–35.

Zambon, A. C., Hughes, R. J., Meszaros, J. G., Wu, J. J., Torres, B., Brunton, L. L., et al. (2000). $P2Y_2$ receptor of MDCK cells: Cloning, expression, and cell-specific signaling. *American Journal of Physiology. Renal Physiology*, *279*, F1045–F1052.

Zee, R. Y., Michaud, S. E., Diehl, K. A., Chasman, D. I., Emmerich, J., Gaussem, P., et al. (2008). Purinergic receptor P2Y, G-protein coupled, 12 gene variants and risk of incident ischemic stroke, myocardial infarction, and venous thromboembolism. *Atherosclerosis*, *197*, 694–699.

Zhang, S., Hu, L., Du, H., Guo, Y., Zhang, Y., Niu, H., et al. (2010). BF0801, a novel adenine derivative, inhibits platelet activation via phosphodiesterase inhibition and $P2Y_{12}$ antagonism. *Thrombosis and Haemostasis*, *104*, 845–857.

Zhang, F. L., Luo, L., Gustafson, E., Lachowicz, J., Smith, M., Qiao, X., et al. (2001). ADP is the cognate ligand for the orphan G protein-coupled receptor SP1999. *Journal of Biological Chemistry*, *276*, 8608–8615.

Zhang, F. L., Luo, L., Gustafson, E., Palmer, K., Qiao, X., Fan, X., et al. (2002). $P2Y_{13}$: Identification and characterization of a novel Galpha$_i$-coupled ADP receptor from human and mouse. *Journal of Pharmacology and Experimental Therapeutics*, *301*, 705–713.

Zhong, X., Kriz, R., Seehra, J., & Kumar, R. (2004). N-linked glycosylation of platelet $P2Y_{12}$ ADP receptor is essential for signal transduction but not for ligand binding or cell surface expression. *FEBS Letters*, *562*, 111–117.

Zylberg, J., Ecke, D., Fischer, B., & Reiser, G. (2007). Structure and ligand-binding site characteristics of the human $P2Y_{11}$ nucleotide receptor deduced from computational modelling and mutational analysis. *Biochemical Journal*, *405*, 277–286.

David Erlinge

Department of Cardiology, Lund University, Skane University Hospital, Lund, Sweden

P2Y Receptors in Health and Disease

Abstract

The purine- and pyrimidine-sensitive P2Y receptors belong to the large group of G-protein-coupled receptors that are the target of approximately one-third of the pharmaceutical drugs used in the clinic today. It is therefore not unexpected that the P2Y receptors could be useful targets for drug development. This chapter will discuss P2Y receptor-based therapies currently used, in development and possible future developments.

The platelet inhibitors blocking the ADP-receptor $P2Y_{12}$ reduce myocardial infarction, stroke, and mortality in patients with cardiovascular disease. Clopidogrel (Plavix) was for many years the second most selling drug in the world. The improved $P2Y_{12}$ inhibitors prasugrel, ticagrelor, and elinogrel are now entering the clinic with even more pronounced protective effects.

The UTP-activated $P2Y_2$ receptor stimulates ciliary movement and secretion from epithelial cells. Cystic fibrosis is a monogenetic disease where reduced chloride ion secretion results in a severe lung disease and early death. No specific treatment has been available, but the $P2Y_2$ agonist Denufosol has been shown to improve lung function and is expected to be introduced as treatment for cystic fibrosis soon.

In preclinical studies, there are indications that P2Y receptors can be important for diabetes, osteoporosis, cardiovascular, and atherosclerotic disease. In conclusion, P2Y receptors are important for the health of humans for many diseases, and we can expect even more beneficial drugs targeting P2Y receptors in the future.

Advances in Pharmacology, Volume 61

1054-3589/11 $35.00
10.1016/B978-0-12-385526-8.00013-8

I. Introduction

Nucleotides were probably some of the first transmitters used for cell communication in early evolution. Their receptors are present on all cell types in the human body. It is therefore not surprising to find that they are important for disease development. The purine- and pyrimidine-sensitive P2Y receptors belong to the large group of G-protein-coupled receptors that are the target of approximately one-third of the pharmaceutical drugs used in the clinic today. P2Y receptors could therefore be useful targets for drug development. This chapter will discuss P2Y receptor-based therapies currently used, in development and possible future developments.

So far, eight P2Y receptors have been cloned: the ADP-sensitive $P2Y_1$, $P2Y_{12}$, and $P2Y_{13}$; UTP- and ATP-sensitive $P2Y_2$ and $P2Y_4$; UDP-sensitive $P2Y_6$; ATP-sensitive $P2Y_{11}$; and the UDP and UDP-glucose sensitive $P2Y_{14}$. To terminate signaling, ectonucleotidases are present in the circulation and on cell surfaces, rapidly degrading extracellular ATP into ADP, AMP, and adenosine (Gordon, 1986; Zimmermann, 2006). This multitude of receptors and degrading enzymes gives opportunities for organ-specific drug actions and reduced side effects.

This review first concentrates on the already widespread use of $P2Y_{12}$-receptor blockers for the treatment of cardiovascular disease. Then discuss the promising use of $P2Y_2$ agonists as stimulators of chloride, water, and mucus secretion for the treatment of cystic fibrosis and dry eye disease. After this, some diseases will be discussed in which P2Y receptor involvement is probable from preclinical or genetic data. The role of P2Y receptors in inflammatory and central nervous system diseases is covered in other chapters and will not be discussed here.

II. P2Y Receptors in Platelet Activation

Thrombus formation in the arteries is dependent on platelets and their ability to attach to the injured wall despite the rapid arterial blood flow. Once attached, the platelets initiate thrombus formation by initiating both platelet aggregation and coagulation.

The platelet undergoes a series of activation steps before degranulation. The primary step is activation by collagen, von Willebrand factor, and other integrins. Two important positive feedback loops are then activated: cyclo-oxygenase-1 (COX-1) generates thromboxane, and adenosine diphosphate (ADP) is released from vesicles. Thromboxane activates the thromboxane receptor, and ADP activates the $P2Y_{12}$ and $P2Y_1$ receptors on the platelet extracellular surface resulting in platelet shape change, with annexin expression generating thrombin, in turn activating the coagulation system. Further, the activated platelet excretes alpha and dense granulae resulting in release of growth factors, vasocontractile agents, prothrombotic, and inflammatory proteins. The platelet forms platelet–monocyte complexes leading to activation of inflammatory cells. The final step is the activation of GpIIb/IIIa receptors which bind fibrinogen resulting in platelet aggregation (Michelson, 2003).

ADP released from erythrocytes was found to cause platelet aggregation already in 1961, before the concept of P2 receptors was conceived (Gaarder et al., 1961). Later, the effects were attributed to a single receptor designated P2T but it was not until 1998 that the three receptors involved were characterized (Daniel et al., 1998; Jin & Kunapuli, 1998; Jin et al., 1998). Two ADP receptors ($P2Y_{12}$ and $P2Y_1$) and one ion channel-coupled ATP receptor (P2X1) are expressed on platelets. The $P2Y_{12}$ receptor is coupled to inhibition of cAMP stimulating and is the most important P2 receptor on platelets.

III. Clopidogrel for the Treatment of ACS and STEMI

The first indication for using ADP-receptor blockers came from the early stent era. Stenting of the blood vessel after balloon angioplasty decreased the problem with dissection and acute occlusion of the coronary artery. However, stents created a new problem—stent thrombosis. The platelets recognize the metal as a foreign surface to the body, attach, and aggregate. Two factors helped to ameliorate this problem. First, the use of intravascular ultrasound enabled PCI operators to size the stents appropriately so that they were well positioned into the vessel wall. Second, the ADP-receptor blocker ticlodipine was used in combination with aspirin to inhibit platelet aggregation (Berger et al., 1999).

Ticlopidine belongs to the class of thienopyridines and was later followed by clopidogrel and prasugrel. The discovery of ticlopidine did not follow today's high-throughput screening strategies. In fact, if modern methods for drug development would have been used, it would not have been discovered at all since it is not active at the receptor before conversion in the liver. Instead, drug candidates were given per oral to rats, and the effects were measured on platelets in plasma. Thereby, it was converted into its active metabolite, and the ADP blocking effects could be discovered. Ticlopidine has the disadvantage that it causes neutropenia in a small percentage of patients. It was therefore further developed into clopidogrel (Plavix) which lacks this side effect. Clopidogrel actually has a slightly weaker platelet inhibitory effect compared to ticlopidine, but the better safety profile without the need to monitor neutrophil counts resulted in widespread use of clopidogrel instead of ticlopidine.

When comparing clopidogrel to aspirin, clopidogrel was only marginally better in the CAPRIE study ("A randomized, blinded, trial of clopidogrel vs. aspirin in patients at risk of ischaemic events (CAPRIE; CAPRIE Steering

Committee, 1996")). However, when used as an adjunct to aspirin in patients with acute coronary syndromes (ACS) in the CURE trial, clopidogrel reduced the incidence of especially myocardial infarction by 20% (Yusuf et al., 2001). The CURE trial included patients treated both medically and invasively, and the effect was similar for both groups. The CREDO study examined an elective PCI population and demonstrated superior long-term effect of pretreatment and prolonged therapy compared to only 1 month clopidogrel poststenting. If pretreatment was initiated at least 6 h before PCI, outcome at 28 days was favorable. A subsequent meta-analysis has shown that pretreatment has a major clinical effect in an ACS population.[8]

Clopidogrel treatment for ST-elevation myocardial infarction (STEMI) patients has never been tested with the now recommended primary PCI treatment strategy. However, it has been studied in the large (40,000 patients) COMMIT study where patients were treated medically with or without thrombolysis.[9] Here, clopidogrel as an adjunct to aspirin reduced the incidence of myocardial infarction as well as mortality. The benefit for STEMI patients was also seen in the CLARITY trial, in which patients were transferred after thrombolysis to PCI, some after a few days.[10] Extrapolations of these studies have led to a recommendation to give clopidogrel as soon as possible after diagnosing a STEMI, often in the ambulance.

The duration of clopidogrel treatment after ACS has been discussed and based on the CURE study, treatment for 12 months is currently recommended. In the CHARISMA trial, patients with manifest atherosclerotic disease as well as patients with only risk factors but no current ACS were randomized to clopidogrel or placebo.[11] The CHARISMA trial failed to show improvement in the entire population. Subgroup analysis of the atherosclerotic arm did, however, show a benefit, but in the risk factor group, clopidogrel harmed the patients. Long-term clopidogrel should not generally be used, but since there was a benefit in patients with an established vascular disease, long-term treatment could be considered in secondary prevention for selected cases and patients with drug eluting stents.

IV. Aspirin and Clopidogrel Resistance

"Resistance" is a commonly used word to describe insufficient effect of aspirin or clopidogrel. For some clinicians, it means that a patient suffers from a new myocardial infarction despite aspirin treatment. For a basic researcher, true aspirin resistance is lack of inhibition of arachidonic acid-induced platelet aggregation even when aspirin is added *ex vivo*. This latter resistance is highly unusual (Hillarp et al., 2003). However, it is common to see weak or absent effect of aspirin for many other platelet agonists in patients on aspirin treatment. This may depend on low compliance, reduced uptake, or that platelet activators use other positive feedback mechanisms such as ADP release, instead of COX-1 and generation of thromboxane.

Aspirin resistance can also be defined as high urinary levels of thromboxane degradation products. But thromboxane can be generated by other cells than the platelets, for example, inflammatory cells. A patient with a high level of atherosclerotic inflammation will therefore continue to have high metabolite levels in urine even if the platelets are inhibited. It was therefore not surprising that these patients did not benefit from addition of clopidogrel.[13] Overall, we do not have any causal treatment for aspirin resistance. Increasing the dose beyond 100 mg does not give additional effect in larger meta-analyses or in the recent CURRENT/OASIS-7 trial; instead, it may increase bleeding complications (Mehta et al., 2010).

Clopidogrel has a very variable response, and a large group of patients can be referred to as clopidogrel resistant. The pathogenesis for clopidogrel resistance is more precisely defined biochemically than that for aspirin resistance. Several studies have now shown that the inhibitory effect of clopidogrel on ADP-induced platelet aggregation is strongly correlated with the plasma level of its active metabolite (Wallentin et al., 2008). Clopidogrel is a prodrug that needs to be converted by cytochrome p450-enzymes (CYP) in the liver into its active metabolite, which then binds irreversibly to the ADP-receptor $P2Y_{12}$. Several esterases inactivate up to 85% of the prodrug. Common (20–25% of the population) genetic polymorphisms in the *CYP2C19* gene and the *ABCB1* gene coding for an intestinal transporter result in lower plasma levels of active metabolite, weaker ADP inhibition (stronger platelet aggregation), and higher incidence of stent thrombosis and myocardial infarction (Mega et al., 2009; Wallentin et al., 2008). Clopidogrel resistance is thus a "pharmacokinetic disease."

Poor responders to clopidogrel defined by pharmacodynamic platelet aggregation tests have been shown to have an increased risk of stent thrombosis and myocardial infarction in a large number of clinical studies (Gurbel et al., 2007). When using the discrimination levels of these pharmacodynamic tests to define clinically poor responders, lower levels of active metabolite in the poor responder groups were found.[19] Interestingly, another important clopidogrel resistant group, the diabetic patients, also displayed lower levels of active metabolite. The difference did not seem to rely on changes on the receptor level, since both poor responder and diabetic platelets responded normally to ADP before clopidogrel treatment. Further, they responded fully to active metabolite added *ex vivo* (Erlinge et al., 2008).

Some register studies have indicated that proton pump inhibitors can cause clopidogrel resistance by interaction on liver enzymes. However, in larger randomized trials, this could not be confirmed.

Increased doses of clopidogrel can improve its effect. However, the recently presented CURRENT/OASIS-7 trial could not demonstrate any significant effect of 600 mg loading dose combined with 150 mg maintenance dose for 7 days of clopidogrel compared to 300/75 mg despite a population of more than 25,000 patients (Mehta et al., 2010). Subgroup analysis (performed even if the primary endpoint was negative) did, however, find a positive effect of higher doses in the PCI population. But since the protocol used upstream clopidogrel loading and treatment, the PCI population cannot be identified beforehand. The currently ongoing GRAVITAS trial will shed further light on this issue by randomizing clopidogrel nonresponders to either higher dose or lower dose chronic clopidogrel treatment.

V. Novel ADP-Receptor Blockers for the Treatment of ACS

The high prevalence of clopidogrel resistance has made it important to develop more efficient ADP-receptor blockers. Unfortunately, Cangrelor, an interesting short acting ADP antagonist for intravenous use, recently failed in clinical trials. INS50589 has been tested as adjunct to coronary artery bypass grafting (CABG), with negative results. However, the two oral drugs prasugrel and ticagrelor have succeeded in large phase III clinical trials (Table I).

TABLE I Investigational Antiplatelet $P2Y_{12}$ Receptor Antagonist Agents

Drug	*Mechanism of action*	*Route of administration*	*Frequency*	*Stage of development*
Ticlopidine	Prodrug, irreversible $P2Y_{12}$ antagonist	Oral	Twice daily	Clinical use
Clopidogrel	Prodrug, irreversible $P2Y_{12}$ antagonist	Oral	Once daily	Clinical use
Prasugrel	Prodrug, irreversible $P2Y_{12}$ antagonist	Oral	Once daily	Clinical use
Ticagrelor	Direct-acting reversible $P2Y_{12}$ antagonist	Oral	Twice daily	Completed phase III—undergoing regulatory review
Cangrelor	Direct-acting reversible $P2Y_{12}$ antagonist	Intravenous	Infusion	Phase III (negative)
Elinogrel	Direct-acting reversible $P2Y_{12}$ antagonist	Oral or intravenous	Once daily	Phase II
INS50589	Direct-acting reversible $P2Y_{12}$ antagonist	Intravenous	Infusion	Phase I CABG (negative)

A number of antiplatelet agents are in clinical development aiming at the $P2Y_{12}$ receptor has been developed.

Prasugrel (Efient) is a thienopyridine like clopidogrel, but instead of 85% of the prodrug being converted to an inactive metabolite, esterases convert the prodrug into an intermediate metabolite, which is then converted to active metabolite by one single CYP step. This results in a much higher and consistent level of active metabolite in the circulation, with a stronger and more rapid inhibition of ADP-induced platelet aggregation (Wallentin et al., 2008). In the TRITON study, including both STEMI and non-ST-elevation-ACS (NSTE-ACS) patients, prasugrel markedly reduced myocardial infarction and stent thrombosis compared to clopidogrel (Wiviott et al., 2007). One of the most important findings was the potent effect in patients with diabetes, a group known to be resistant to both aspirin and clopidogrel. Thus, for the first time, we have an oral platelet inhibitor with a clear clinical effect on diabetic patients (Wiviott et al., 2008).

Not surprisingly, prasugrel was markedly better than clopidogrel for patients with genetic polymorphisms for low enzymatic activity in CYP2C19 (Mega et al., 2009). The STEMI patients also did benefit markedly with prasugrel, and in this group, a reduction in mortality was found (Montalescot et al., 2009). Interestingly, although the prasugrel-treated population as a whole suffered from increased bleeding complications, neither the subgroups consisting of diabetic nor STEMI patient suffered from this. Prasugrel does not seem to have any poor responders and seems to solve the problem with clopidogrel resistance. Further, prasugrel reduces platelet–monocyte complex formation and annexin exposure on the platelet surface (a prerequisite for coagulation activation; Braun et al., 2008).

Bleeding complications were otherwise the major limitation for prasugrel in TRITON. They were clearly increased in the whole population and patients with a previous TIA/stroke suffered from significantly increased intracranial and fatal bleedings. Older patients (>75 years) and those with low weight had increased bleeding complications, resulting in a net neutral effect on the combined endpoint of cardiovascular events and bleeding complications.

Ticagrelor (Brilinta, Brilique) is not a prodrug like clopidogrel and prasugrel, thereby being independent of conversion in the liver for activation. Further, it is a classical competitive antagonist without irreversible binding, a property that can be important for patients in need of urgent surgery or with bleeding complications since the effects of the drug are lost much more rapidly compared to the 5–7 days needed to restore platelet function for other thienopyridines.

In the phase III trial PLATO, ticagrelor significantly reduced cardiovascular events compared to clopidogrel, and surprisingly, without any major increase in bleeding complications (Wallentin et al., 2009). This is probably one of the reasons why ticagrelor also resulted in a significant reduction in total mortality, a very rare finding in antiplatelet studies. However, other pleiotrophic effects could also have contributed. Ticagrelor also inhibits the adenosine reuptake protein ENT1 in red blood cells (Bjorkman et al., 2007). Increased adenosine levels could improve microcirculation via its vasodilatory actions, and ticagrelor has been shown to improve reactive hyperaemia after coronary ischemia (Bjorkman et al., 2007). Adenosine is anti-inflammatory and cardioprotective, actions that could be important for patients with cardiovascular disease (Hasko & Cronstein, 2004; Hasko et al., 2008). However, adenosine may contribute to some unexpected side effects of ticagrelor. Ticagrelor causes dyspnea and bradycardia, both well-known side effects of adenosine infusion. Interestingly, dyspnea is also induced by two other competitive $P2Y_{12}$ antagonists, cangrelor and elinogrel.

Vascular smooth muscle cells express $P2Y_{12}$ receptors, which mediate contractile function after stimulation with ADP (Wihlborg et al., 2004). At mRNA level, the $P2Y_{12}$ receptor is the highest expressed ADP receptor and the second highest expressed P2 receptor in human smooth muscle cells (Wihlborg et al., 2004). The contractions are not inhibited in patients medicated with clopidogrel. However, drugs with antagonistic effects on $P2Y_{12}$ receptors that reach the peripheral circulation (AZD6140 and INS50589), affecting both platelets and VSMC, could be of double therapeutic benefit in their prevention of both thrombosis and vasospasm (Wihlborg et al., 2004).

Elinogrel is also a reversible competitive antagonist of $P2Y_{12}$ receptors. The phase II trial INNOVATE-PCI was recently presented (European Society of Cardiology, 2010). The safety profile was acceptable, but the study was not empowered to see effects on efficacy. A major phase III trial is planned.

In conclusion, blocking the $P2Y_{12}$ receptor is the most important antiplatelet strategy since aspirin was discovered and has protected millions of patients from new myocardial infarctions and prolonged life. Improved $P2Y_{12}$ blockers (prasugrel and ticagrelor) are now introduced and will be even better in protecting patients from cardiovascular disease.

A. Other Platelet P2 Receptors

The platelets also express $P2Y_1$ receptors that have been shown in knockout mice to be of similar importance as $P2Y_{12}$ receptors (Fabre et al., 1999; Leon et al., 1999).The P2X1 receptor is involved in platelet shape change and activation by collagen under shear conditions.[124]

The physiological importance of the P2X1 receptor is not fully elucidated. However, recent studies have demonstrated that P2X1 receptors can generate significant functional platelet responses alone and in synergy with other receptor pathways. Further, experiments in transgenic animals indicate an important role for P2X1 receptors in platelet activation, particularly under conditions of shear stress and thus during arterial thrombosis (Mahaut-Smith et al., 2004).

VI. Pulmonary Disease

Activation of $P2Y_2$ receptors by UTP and ATP on bronchial epithelia stimulates transepithelial chloride, water, and mucus secretion. Further, it stimulates ciliary movement and beating frequency, which is essential for removal of mucus and bacteria from the airways. ATP and UTP can be released when airways are stretched, and it has been suggested that one of the important reasons for coughing is to release ATP and UTP, which in turn stimulate secretion and ciliary movement.

Cystic fibrosis is one of the most common monogenic heritable diseases we know. DNA polymorphisms create a nonfunctional chloride channel which causes a dysfunctional, dry mucus which leads to chronic infections and to early death in respiratory insufficiency. Inspire Pharmaceuticals has developed more stable UTP analogs for inhalation which are simulators of the $P2Y_2$ receptor (Kellerman et al., 2008). The drug Denufosol has received orphan drug designation, and clinical trials for cystic fibrosis have succeeded up to phase III. The TIGER-1 trial met the primary endpoint of improved lung function (FEV1) and showed continued increasing lung function in the open label safety extension. Interestingly, it also reduced the occurrence of sinusitis and headache, suggesting that also mucociliary clearance from the sinuses was improved. The FDA has demanded a confirmatory phase III trial (TIGER-2), and recruitment has been completed. Results are expected early 2011, and if it is successful, Denufosol will be available for the treatment of CF. This will be the first causal treatment for cystic fibrosis and a major breakthrough for this suffering patient group.

VII. Dry Eye Disease

Reduced secretion of mucus and water in other organs is indeed a major problem for millions of people in the world, suffering from, for example, dry eyes or dryness of the mouth. Based on similar principles as for cystic fibrosis treatment, UTP analogs have been developed. Prolacria (diquafosol tetrasodium) has been shown to be effective in the treatment of dry eye disease reducing corneal bruises but failed to meet the primary endpoint in a phase III trial (Peral et al., 2008). It is already approved for the Japanese market under the name Diquas Ophthalmic Solution. It is a selective $P2Y_2$ receptor agonist, and activation of $P2Y_2$ receptors on the ocular surface stimulates release of the three natural tear components involved in tear secretion—mucin, lipids, and fluid.

VIII. Hypertension

P2 receptor-mediated blood pressure regulation is the net result of balancing contractile and dilatory effects. ATP and UTP released on the luminal side of endothelial cells, and erythrocytes stimulate vasodilatation in contrast to release from nerves on the adventitial side which results in vasoconstriction. The most important dilatory P2Y receptors on the endothelium are the $P2Y_1$ and $P2Y_2$ receptors, although a small dilatory effect of $P2Y_6$ has also been shown (Erlinge & Burnstock, 2008).

ATP plays a significantly greater role as a sympathetic cotransmitter in spontaneously hypertensive rats (SHR; Brock & Van Helden, 1995; Vidal et al., 1986), and there is increased responsiveness of the renal vasculature of isolated perfused rat kidneys to α,β-meATP in SHR (Fernandez et al., 2000), while mesenteric vascular contraction in response to ATP via P2X1 and $P2Y_2$ receptors is not altered in DOCA-salt hypertension (Galligan et al., 2001). The endothelium-dependent relaxation to ATP is impaired in SHR because of the simultaneous generation of an endothelium-derived contracting factor (EDCF). ATP-induced vasoconstriction is significantly potentiated in SHR aorta (Yang et al., 2004).

Diadenosine polyphosphates such as AP_4A, AP_5A, and AP_6A are combinations of two adenosine molecules connected with four to six phosphate groups. They are more stable in the circulation compared to ATP. They have been identified as vasocontractile agents (Schluter et al., 1994), mediating their effects through actions on P2X1 and $P2Y_2$ receptors. AP_5A and AP_6A are stored at higher levels in platelets from patients with hypertension and could contribute to their increased peripheral vascular resistance (Hollah et al., 2001). The combined purine and pyrimidine UP_4A is a novel endothelium-derived vasoconstrictive factor more potent than endothelin in renal vasoconstriction (Jankowski et al., 2005). It is released when the endothelium is stimulated by thrombin, acetylcholine, or mechanical stress and can be cleaved into either ATP or UTP to stimulate both P2X1 and $P2Y_2$ receptors on smooth muscle cells causing increased blood pressure (Jankowski et al., 2005).

A. Pulmonary Hypertension

Red blood cell release of ATP is important for the regulation of pulmonary resistance (Sprague et al., 2003), and patients with pulmonary hypertension have impaired release of ATP from red blood cells (Sprague et al., 2001). Endothelium-dependent relaxation to ATP has been shown in human pulmonary arteries (Greenberg et al., 1987). ATP is a mitogen for pulmonary artery cells, which may be relevant for the media hypertrophy seen in lung arteries in patients with pulmonary hypertension (Zhang et al., 2004). Infusions of ATP–$MgCl_2$ have been claimed to be clinically useful in the treatment of children with pulmonary hypertension, but it is possible that the effect is adenosine mediated (Brook et al., 1994).

IX. Migraine and Vascular Pain

Migraine headache is caused by distinct vascular responses, first a vasoconstriction, which is then followed by a vasodilatation when the patient experiences pain. It has been suggested that purinergic signaling is involved in these changes. ATP released from perivascular sympathetic nerves may participate in producing the initial vasospasm mediated by P2 receptors in vascular smooth muscle. Cerebral arteries are strongly contracted by UTP and especially UDP via $P2Y_6$ receptors. A UDP analog might therefore have similar effects as the $5\text{-}HT_{1D}$ agonists that currently are used in the treatment of migraine (Malmsjo et al., 2003).

X. Atherosclerosis

Atherosclerosis is the most important cause of ischemic stroke, myocardial infarction, and peripheral artery disease and is now considered to be an inflammatory disease (Hansson, 2005). The formation of an atherosclerotic plaque begins with the accumulation of cholesterol followed by invasion of macrophages phagocytosing cholesterol and becoming foam cells. Smooth muscle cells can form a fibrous cap which ideally stabilizes the plaque and covers the lipid-rich region. The dangerous path is if activation of inflammation by oxidized LDL stimulates macrophages and dendritic cells into antigen presenting cells which activates T-lymphocytes releasing cytokines and metalloproteinases. These processes degrade the fibrous cap. In the end, a vulnerable plaque is formed, and when it ruptures, the thrombogenic content recruits and stimulates platelets. The platelets aggregate and cause the formation of a local thrombus occluding the artery or embolizing, resulting in ischemic stroke or myocardial infarction. As described above, antagonists to the ADP-receptor $P2Y_{12}$ have been very successful in the treatment of cardiovascular disease by preventing platelet aggregation. Evidence both from basic research and from clinical studies indicates important involvement of P2Y receptors in many other parts in the atherosclerotic process.

Fish oil components are preventive against cardiovascular disease, and they have been shown to increase the release of ATP from the caudal artery in aged rats (Hashimoto et al., 1998a). High cholesterol diet which is a major risk factor decreases ATP release from arteries (Hashimoto et al., 1998b; Karoon & Burnstock, 1998). It is thus possible that P2Y receptors are mediating some parts of their mechanisms.

The most important inflammatory cells in development of atherosclerosis are monocytes that in the plaque differentiate into macrophages or dendritic cells and T-lymphocytes that are coordinators of the inflammatory reaction in the plaque (Hansson, 2005). T-lymphocytes and macrophages express a large number of P2 receptors and have been suggested to be important regulators of atherosclerosis (Di Virgilio & Solini, 2002; Wang et al., 2004). The P2 receptors most clearly implicated in atherosclerotic inflammation are P2X7, $P2Y_2$, and $P2Y_{11}$. The P2X7 receptor is antiapoptotic and mitogenic for T-lymphocytes (Baricordi et al., 1996; Kawamura et al., 2005). P2X7 is important for release of IL-1(Ferrari et al., 1997), tumor necrosis factor (Suzuki et al., 2004), and L-selectin (Gu et al., 1998). ATP and UTP are chemotactic for dendritic cells via the $P2Y_2$ receptor and can attract inflammatory cells to the atherosclerotic plaque (Idzko et al., 2002). The $P2Y_2$ receptors release free radicals in human macrophages (Schmid-Antomarchi et al., 1997). $P2Y_1$ receptor knockout mice exhibit reduced plaque area occupied by macrophages and the decreased total amount of atherosclerotic lesions in ApoE knockout mice (Hechler et al., 2008). The effect is not mediated by platelet inhibition. Selective knockout of bone marrow or the rest of the most have clearly shown that the protective effect is not dependent on bone marrow derived cells (white blood cells or platelets). Instead, the effect is dependent on $P2Y_1$ receptors on the endothelium which triggers expression of adhesive leukocyte receptors, an important early event in atherosclerosis which accounts for recruitment of inflammatory cells to the plaque.

ATP inhibits T cell activation (CD4+) by an elevation of cAMP after stimulation of $P2Y_{11}$ receptors (Duhant et al., 2002). ATP acting on $P2Y_{11}$ receptors stimulates maturation of human monocyte-derived dendritic cells and causes immune suppression by attenuating T-helper 1 cytokines and stimulating T-helper 2 cytokines (Kaufmann et al., 2005; Marteau et al., 2005).

A polymorphism in the $P2Y_{11}$ receptor has been shown to have clinical importance by increasing the risk of myocardial infarction (Amisten et al., 2007). The G-459-A polymorphism which is carried by 20% of the population causes an Ala-87-Thr change in the $P2Y_{11}$ receptor and increases the risk of acute myocardial infarction by 21%. The odds ratio increased stepwise depending on how many Thr-87 alleles were present. Further, in subgroups with a family history and in patients suffering from myocardial infarction early in life, the risk is further increased up to 100%. The reason is that the genetic component of the risk is higher in these groups. The mechanism by which the polymorphism causes myocardial infarction seems to be coupled to increased inflammation because patients with the Thr-87 variant also had increased levels of the inflammatory marker C-reactive protein. Even small increases of C-reactive protein are a strong marker of inflammation and an independent prognostic risk factor for the development of myocardial infarction (Hansson, 2005). In conclusion, the $P2Y_{11}$ receptor is important in the development of atherosclerosis through an anti-inflammatory effect mediated via T-lymphocytes, macrophage, or dendritic cells.

A dysfunctional endothelium can be proinflammatory, and recruit monocytes to the atherosclerotic plaque and cause the extravasation and loss of peripheral resistance that is so detrimental in sepsis. ATP causes neutrophil adherence to endothelial cells (Dawicki et al., 1995) In human microvascular endothelial cells, ATP stimulates release of IL-6, IL-8, monocyte chemoattractant protein-1 and triggers the expression of intercellular adhesion molecule-1 (Seiffert et al., 2006). In human coronary endothelial cells, UTP and ATP stimulate the expression of proinflammatory vascular cell adhesion molecule-1 through activation of the $P2Y_2$ receptor and increase the adherence of monocytic cells (Seye et al., 2003). These effects have been confirmed in an *in vivo* neointima model, in which perivascular infusion of UTP enhanced infiltration by macrophages (Seye et al., 2002).

In conclusion, ATP and UTP stimulate several inflammatory responses known to be important for atherosclerosis development. $P2Y_1$ and $P2Y_2$ antagonist and $P2Y_{11}$ agonists are possible future P2Y-based drugs that could be developed for the treatment of atherosclerosis.

XI. Osteoporosis

Osteoporosis is a major problem in the elderly population, leading to fractures and pain. Several P2Y receptors are present and have effects on both osteoclasts and osteoblasts. Physical activity stimulates bone formation and reduces osteoporosis. Evidence suggests that compression and tension of bone cells induced by exercise release nucleotides, in turn activating P2Y receptors, leading to bone formation (Turner & Robling, 2004).

Bisphosphonates are the most important drugs used in the treatment of osteoporosis as they inhibit osteoclast resorption and stimulate proliferation of osteoblasts. It has been shown that treatment with bisphosphonates promotes nonlytic ATP release leading to activation of ERKs through the involvement of P2Y receptors. This leads to increased proliferation of osteoblast-like cells. A major role in this signal transduction pathway seems to be the involvement of $P2Y_1$ and $P2Y_2$ receptors (Romanello et al., 2006). Recently, the $P2Y_{13}$ knockout mice exhibited a bone phenotype with increased bone density in females. A major EU-grant-stimulated collaboration effort tries to sort out the involved receptors and suggest future treatment strategies (Blom et al., 2010).

XII. Diabetes

A large number of studies indicate that extracellular ATP and ADP have a key role in regulating insulin secretion by purinoreceptors (Bertrand et al., 1991; Fernandez-Alvarez et al., 2001; Loubatieres-Mariani et al., 1979;

Petit et al., 1989; Poulsen et al., 1999). Insulin-secretory granules contain 3.5 mM ATP and 5 mM ADP (Hutton et al., 1983), and with a novel biosensor, it has been demonstrated that glucose stimulation releases ATP from a single pancreatic beta cell to a local extracellular ATP concentration exceeding 25 μM (Hazama et al., 1998).

ADP has been shown to increase insulin release via P2Y receptors *in vitro*, *in vivo*, and in human beta cells (Bertrand et al., 1991; Fernandez-Alvarez et al., 2001; Loubatieres-Mariani et al., 1979). Beta cells are activated by ADP and can also propagate this signal via intermittent release of ATP (Hellman et al., 2004). However, ADP has also been shown to decrease insulin release (Petit et al., 1989; Poulsen et al., 1999). The reason for these discrepancies is probably in part due to involvement of several different purinoreceptors (Blachier & Malaisse, 1988; Farret et al., 2006; Grapengiesser et al., 2004, 2005; Hellman et al., 2004; Hillaire-Buys et al., 1994; Poulsen et al., 1999; Salehi et al., 2005; Squires et al., 1994). Therefore, we wanted to examine if endogenous release of purines may regulate insulin secretion and to determine the P2Y receptors involved.

Interestingly, P2Y receptors seem to exert a yin and yang effect on insulin. ADP acting on $P2Y_1$ receptors stimulates insulin secretion, while ADP acting on $P2Y_{13}$ receptors is an inhibitor of insulin secretion. ADP acting on $P2Y_{13}$ receptors act as an autocrine inhibitor of insulin release both *in vitro* and *in vivo* (Amisten et al., 2010). Pancreatic beta cell loss represents a key factor in the pathogenesis of diabetes. ADP induces apoptosis in beta cells where $P2Y_{13}$ has been shown to be the proapoptotic receptor and the $P2Y_{13}$ receptor antagonist MRS2211 protects the cells from ADP-induced apoptosis (Tan et al., 2010). $P2Y_{13}$ antagonists could therefore be suitable for long-term diabetes treatment, increasing insulin secretion and protecting the beta cells.

UDP is a stimulator of insulin secretion by activation of $P2Y_6$ receptors (Parandeh et al., 2008). A variety of neurotransmitters, gastrointestinal hormones, and metabolic signals are known to potentiate insulin secretion through GPCRs. It has been demonstrated that beta cell-specific inactivation of the genes encoding the G-protein alpha subunits Galphaq and Galpha11 resulted in impaired glucose tolerance and insulin secretion in mice, and the autocrine response to glucose *per se* was also diminished. Among the autocrine mediators involved in this process, extracellular nucleotides such as uridine diphosphate acting through the Gq/G11-coupled $P2Y_6$ receptor were involved. Thus, the Gq/G11-mediated signaling pathway potentiates insulin secretion in response to glucose by integrating systemic as well as autocrine/paracrine mediators (Sassmann et al., 2010).

High glucose levels stimulate the release of ATP into the extracellular space from several tissues and cell types such as endothelial cells, blood vessels, mesangial cells, macrophages, and beta cells (Chen et al., 2006;

Hazama et al., 1998; Nilsson et al., 2006; Parodi et al., 2002; Solini et al., 2005). An increase in glucose from 5 to 15 mM results in a marked increase in the proatherogenic NFAT signaling pathway in vascular smooth muscle cells (Nilsson et al., 2006). The effect is mediated by via glucose-induced release of ATP and UTP that subsequently activate $P2Y_2$ but also $P2Y_6$ receptors (after degradation to UDP; Nilsson et al., 2006). Thus, nucleotide release is a potential metabolic sensor for the arterial smooth muscle response to high glucose. Diabetic patients experience microvascular disease characterized by increased wall–lumen ratio mainly because of increased amounts of vascular smooth muscle cells and have higher rates of restenosis after coronary angioplasty. High glucose-induced release of extracellular nucleotides acting on P2Y receptors to stimulate smooth muscle cell growth via NFAT activation may provide a link between diabetes and diabetic vascular disease (Nilsson et al., 2006).

In summary, several P2Y receptors could be targets for diabetes treatment for improving insulin secretion, preventing beta cell death and protect the microcirculation.

XIII. Cardioprotection

ATP is released in the heart as a sympathetic cotransmitter with catecholamines but it may also be released from other sources in the heart such as endothelium, platelets, red blood cells, and ischemic myocardium (Clemens & Forrester, 1981; Gordon, 1986; Vassort, 2001). ATP in the interstitial space has been measured with microdialysis and found to have basic levels of 40 nM. However, the levels can increase markedly during electrical stimulation, ischemia, challenge with ischemia, or increased work load (Vassort, 2001). ATP is released from cardiomyocytes upon hypoxia (Vassort, 2001) as well as UTP (Erlinge et al., 2005). Both UTP and ATP are released from the heart in coronary ischemia (Erlinge et al., 2005), and patients with acute STEMI have higher plasma levels of both ATP and UTP (Borna et al., 2005; Wihlborg et al., 2006).

Opposing effects of ATP and UTP have been uncovered for their effects on cardiomyocyte growth. UTP but not ATP causes hypertrophic growth in neonatal cardiomyocytes (Pham et al., 2003). On the contrary, ATP inhibits hypertrophy and can even induce apoptotic cell death and necrosis (Zheng et al., 1994, 1996). Similarly, UTP but not ATP protects cultured cardiomyocytes against hypoxic stress (Yitzhaki et al., 2005). Since UTP is released during preconditioning (Erlinge et al., 2005), a role for UTP in the protective effects of preconditioning is plausible. UTP has prominent cardioprotective effects salvaging heart muscle during myocardial infarction and improving heart function *in vivo* (Yitzhaki et al., 2006).

XIV. Oncology

ATP infusions have clearly been shown to be beneficial for end-stage lung cancer by reducing weight loss and improving muscle strength and quality of life, but it is unclear if this is a P2Y-mediated mechanism (Beijer et al., 2009).

XV. Conclusion

The P2 receptor family provides many opportunities for drug development. The large number of receptor subtypes and the increasing knowledge of their tissue distributions open up for selective specific drug development with limited side effects. So far, several $P2Y_{12}$ receptor antagonists have been established for treating patients with ACS with extraordinary success. Clopidogrel is one of the most prescribed drugs in the world, protecting millions of patients from myocardial infarction. The followers, prasugrel and ticagrelor, are even more potent $P2Y_{12}$ antagonists with markedly better effect and are currently introduced into clinical practice. However, improved platelet inhibition can also be achieved with antagonists against the other two P2 receptors on platelets: $P2Y_1$ and P2X1. Knockout mice and experimental thrombosis models using selective $P2Y_1$ and P2X1 antagonists have shown that, depending on the conditions, these receptors could also be potential targets for new antithrombotic drugs.

Denufosol is a promising $P2Y_2$ agonist that improves ciliary movement and the production of better mucus in the lungs. It has proven to be of benefit for patients with cystic fibrosis, and hopefully, it can be used in the clinic within a few years.

When the exact actions of P2Y receptors on osteoblasts and osteoclasts have been sorted out, there are good hopes that a drug targeting a P2Y receptor could be useful for the treatment of osteoporosis.

$P2Y_{13}$ antagonists or $P2Y_6$ agonists could be used in the treatment of diabetes by increasing insulin release or even protect the beta cell from apoptosis. ATP/UTP/UDP on $P2Y_2$ and $P2Y_6$ are potent stimulators of inflammation, smooth muscle cell proliferation, and migration. Blockers of these receptors could protect against atherosclerosis, chronic transplant rejection, and diabetic microvascular disease.

In conclusion, the extracellular nucleotides and their P2Y receptors are important for disease development and drugs targeting these receptors are already among the most widely used pharmaceuticals in the world. New treatment areas are expected to be developed within the coming years.

Conflict of Interest: I have received speakers and advisory committee fee's from Lilly and AstraZeneca.

Abbreviations

ACS	acute coronary syndromes
ADP	adenosine 5′-diphosphate
2-MeSADP	2′-methylene adenosine 5′-diphosphate
ASA	ascorbic acid
ATP	adenosine 5′-triphosphate
CABG	coronary artery bypass grafting
cAMP	adenosine 3′,5′-monophosphate
COX-1	cyclooxygenase-1
GPIIb/IIIa	glycoprotein IIb/IIIa
GPCR	G-protein-coupled receptor
NFAT	nuclear factor of activated T cells
PCI	percutaneous coronary intervention
STEMI	ST-elevation myocardial infarction
UDP	uridine triphosphate
UTP	uridine diphosphate
VSMC	vascular smooth muscle cells

References

Amisten, S., Meidute-Abaraviciene, S., Tan, C., Olde, B., Lundquist, I., Salehi, A., et al. (2010). ADP mediates inhibition of insulin secretion by activation of P2Y13 receptors in mice. *Diabetologia*, *53*(9), 1927–1934.

Amisten, S., Melander, O., Wihlborg, A. K., Berglund, G., & Erlinge, D. (2007). Increased risk of acute myocardial infarction and elevated levels of C-reactive protein in carriers of the Thr-87 variant of the ATP receptor P2Y11. *European Heart Journal*, *28*(1), 13–18.

Baricordi, O. R., Ferrari, D., Melchiorri, L., Chiozzi, P., Hanau, S., Chiari, E., et al. (1996). An ATP-activated channel is involved in mitogenic stimulation of human T lymphocytes. *Blood*, *87*(2), 682–690.

Beijer, S., Hupperets, P. S., van den Borne, B. E., Eussen, S. R., van Henten, A. M., van den Beuken-van Everdingen, M., et al. (2009). Effect of adenosine 5′-triphosphate infusions on the nutritional status and survival of preterminal cancer patients. *Anti-Cancer Drugs*, *20*(7), 625–633.

Berger, P. B., Bell, M. R., Rihal, C. S., Ting, H., Barsness, G., Garratt, K., et al. (1999). Clopidogrel versus ticlopidine after intracoronary stent placement. *Journal of the American College of Cardiology*, *34*(7), 1891–1894.

Bertrand, G., Chapal, J., Puech, R., & Loubatieres-Mariani, M. M. (1991). Adenosine-5′-O-(2-thiodiphosphate) is a potent agonist at P2 purinoceptors mediating insulin secretion from perfused rat pancreas. *British Journal of Pharmacology*, *102*(3), 627–630.

Bjorkman, J.-A., Kirk, I., & van Giezen, J. J. (2007). AZD6140 inhibits adenosine uptake into erythrocytes and enhances coronary blood flow after local ischemia or intracoronary adenosine infusion. *Circulation*, *116*:II_28.

Blachier, F., & Malaisse, W. J. (1988). Effect of exogenous ATP upon inositol phosphate production, cationic fluxes and insulin release in pancreatic islet cells. *Biochimica et Biophysica Acta*, *970*(2), 222–229.

Blom, D., Yamin, T. T., Champy, M. F., Selloum, M., Bedu, E., Carballo-Jane, E., et al. (2010). Altered lipoprotein metabolism in P2Y(13) knockout mice. *Biochimica et Biophysica Acta*, *1801*(12), 1349–1360.

Borna, C., Lazarowski, E., van Heusden, C., Ohlin, H., & Erlinge, D. (2005). Resistance to aspirin is increased by ST-elevation myocardial infarction and correlates with adenosine diphosphate levels. *Thrombosis Journal*, *3*, 10.

Braun, O. O., Johnell, M., Varenhorst, C., James, S., Brandt, J. T., Jakubowski, J. A., et al. (2008). Greater reduction of platelet activation markers and platelet-monocyte aggregates by prasugrel compared to clopidogrel in stable coronary artery disease. *Thrombosis and Haemostasis*, *100*(4), 626–633.

Brock, J. A., & Van Helden, D. F. (1995). Enhanced excitatory junction potentials in mesenteric arteries from spontaneously hypertensive rats. *Pflugers Archiv*, *430*(6), 901–908.

Brook, M. M., Fineman, J. R., Bolinger, A. M., Wong, A. F., Heymann, M. A., & Soifer, S. J. (1994). Use of ATP-MgCl2 in the evaluation and treatment of children with pulmonary hypertension secondary to congenital heart defects. *Circulation*, *90*(3), 1287–1293.

CAPRIE Steering Committee. (1996). A randomised, blinded, trial of clopidogrel versus aspirin in patients at risk of ischaemic events (CAPRIE). *Lancet*, *348*(9038), 1329–1339.

Chen, Y. J., Hsu, K. W., & Chen, Y. L. (2006). Acute glucose overload potentiates nitric oxide production in lipopolysaccharide-stimulated macrophages: The role of purinergic receptor activation. *Cell Biology International*, *30*(10), 817–822.

Clemens, M. G., & Forrester, T. (1981). Appearance of adenosine triphosphate in the coronary sinus effluent from isolated working rat heart in response to hypoxia. *The Journal of Physiology*, *312*, 143–158.

Daniel, J. L., Dangelmaier, C., Jin, J., Ashby, B., Smith, J. B., & Kunapuli, S. P. (1998). Molecular basis for ADP-induced platelet activation I. Evidence for three distinct ADP receptors on human platelets. *The Journal of Biological Chemistry*, *273*(4), 2024–2029.

Dawicki, D. D., McGowan-Jordan, J., Bullard, S., Pond, S., & Rounds, S. (1995). Extracellular nucleotides stimulate leukocyte adherence to cultured pulmonary artery endothelial cells. *The American Journal of Physiology*, *268*(4 Pt 1), L666–L673.

Di Virgilio, F., & Solini, A. (2002). P2 receptors: New potential players in atherosclerosis. *British Journal of Pharmacology*, *135*(4), 831–842.

Duhant, X., Schandene, L., Bruyns, C., Gonzalez, N. S., Goldman, M., Boeynaems, J. M., et al. (2002). Extracellular adenine nucleotides inhibit the activation of human CD4+ T lymphocytes. *Journal of Immunology*, *169*(1), 15–21.

Erlinge, D., & Burnstock, G. (2008). P2 receptors in cardiovascular regulation and disease. *Purinergic Signalling*, *4*(1), 85–87.

Erlinge, D., Harnek, J., van Heusden, C., Olivecrona, G., Jern, S., & Lazarowski, E. (2005). Uridine triphosphate (UTP) is released during cardiac ischemia. *International Journal of Cardiology*, *100*(3), 427–433.

Erlinge, D., Varenhorst, C., Braun, O. O., James, S., Winters, K. J., Jakubowski, J. A., et al. (2008). Patients with poor responsiveness to thienopyridine treatment or with diabetes have lower levels of circulating active metabolite, but their platelets respond normally to active metabolite added ex vivo. *Journal of the American College of Cardiology*, *52*(24), 1968–1977.

Fabre, J. E., Nguyen, M., Latour, A., Keifer, J. A., Audoly, L. P., Coffman, T. M., et al. (1999). Decreased platelet aggregation, increased bleeding time and resistance to thromboembolism in P2Y1-deficient mice. *Natural Medicines*, *5*(10), 1199–1202.

Farret, A., Filhol, R., Linck, N., Manteghetti, M., Vignon, J., Gross, R., et al. (2006). P2Y receptor mediated modulation of insulin release by a novel generation of 2-substituted-5′-O-(1-boranotriphosphate)-adenosine analogues. *Pharmaceutical Research*, *23*(11), 2665–2671.

Fernandez, O., Wangensteen, R., Osuna, A., & Vargas, F. (2000). Renal vascular reactivity to P(2)-purinoceptor activation in spontaneously hypertensive rats. *Pharmacology*, *60*(1), 47–50.

Fernandez-Alvarez, J., Hillaire-Buys, D., Loubatieres-Mariani, M. M., Gomis, R., & Petit, P. (2001). P2 receptor agonists stimulate insulin release from human pancreatic islets. *Pancreas*, *22*(1), 69–71.

Ferrari, D., Chiozzi, P., Falzoni, S., Dal Susino, M., Melchiorri, L., Baricordi, O. R., et al. (1997). Extracellular ATP triggers IL-1 beta release by activating the purinergic P2Z receptor of human macrophages. *Journal of Immunology*, *159*(3), 1451–1458.

Gaarder, A., Jonsen, J., Laland, S., Hellem, A., & Owren, P. A. (1961). Adenosine diphosphate in red cells as a factor in the adhesiveness of human blood platelets. *Nature*, *192*, 531–532.

Galligan, J. J., Hess, M. C., Miller, S. B., & Fink, G. D. (2001). Differential localization of P2 receptor subtypes in mesenteric arteries and veins of normotensive and hypertensive rats. *The Journal of Pharmacology and Experimental Therapeutics*, *296*(2), 478–485.

Gordon, J. L. (1986). Extracellular ATP: Effects, sources and fate. *The Biochemical Journal*, *233*(2), 309–319.

Grapengiesser, E., Dansk, H., & Hellman, B. (2004). Pulses of external ATP aid to the synchronization of pancreatic beta-cells by generating premature Ca(2+) oscillations. *Biochemical Pharmacology*, *68*(4), 667–674.

Grapengiesser, E., Dansk, H., & Hellman, B. (2005). External ATP triggers Ca2+ signals suited for synchronization of pancreatic beta-cells. *The Journal of Endocrinology*, *185*(1), 69–79.

Greenberg, B., Rhoden, K., & Barnes, P. J. (1987). Endothelium-dependent relaxation of human pulmonary arteries. *The American Journal of Physiology*, *252*(2 Pt 2), H434–H438.

Gu, B., Bendall, L. J., & Wiley, J. S. (1998). Adenosine triphosphate-induced shedding of CD23 and L-selectin (CD62L) from lymphocytes is mediated by the same receptor but different metalloproteases. *Blood*, *92*(3), 946–951.

Gurbel, P. A., Becker, R. C., Mann, K. G., Steinhubl, S. R., & Michelson, A. D. (2007). Platelet function monitoring in patients with coronary artery disease. *Journal of the American College of Cardiology*, *50*(19), 1822–1834.

Hansson, G. K. (2005). Inflammation, atherosclerosis, and coronary artery disease. *The New England Journal of Medicine*, *352*(16), 1685–1695.

Hashimoto, M., Shinozuka, K., Shahdat, H. M., Kwon, Y. M., Tanabe, Y., Kunitomo, M., et al. (1998). Antihypertensive effect of all-cis-5,8,11,14,17-icosapentaenoate of aged rats is associated with an increase in the release of ATP from the caudal artery. *Journal of Vascular Research*, *35*(1), 55–62.

Hashimoto, M., Shinozuka, K., Tanabe, Y., Shahdat, H. M., Gamoh, S., Kwon, Y. M., et al. (1998). Long-term supplementation with a high cholesterol diet decreases the release of ATP from the caudal artery in aged rats. *Life Sciences*, *63*(21), 1879–1885.

Hasko, G., & Cronstein, B. N. (2004). Adenosine: An endogenous regulator of innate immunity. *Trends in Immunology*, *25*(1), 33–39.

Hasko, G., Linden, J., Cronstein, B., & Pacher, P. (2008). Adenosine receptors: Therapeutic aspects for inflammatory and immune diseases. *Nature Reviews. Drug Discovery*, *7*(9), 759–770.

Hazama, A., Hayashi, S., & Okada, Y. (1998). Cell surface measurements of ATP release from single pancreatic beta cells using a novel biosensor technique. *Pflugers Archiv*, *437*(1), 31–35.

Hechler, B., Freund, M., Ravanat, C., Magnenat, S., Cazenave, J. P., & Gachet, C. (2008). Reduced atherosclerotic lesions in P2Y1/apolipoprotein E double-knockout mice: The contribution of non-hematopoietic-derived P2Y1 receptors. *Circulation*, *118*(7), 754–763.

Hellman, B., Dansk, H., & Grapengiesser, E. (2004). Pancreatic beta-cells communicate via intermittent release of ATP. *American Journal of Physiology. Endocrinology and Metabolism*, *286*(5), E759–E765.

Hillaire-Buys, D., Chapal, J., Bertrand, G., Petit, P., & Loubatieres-Mariani, M. M. (1994). Purinergic receptors on insulin-secreting cells. *Fundamental & Clinical Pharmacology*, *8*(2), 117–127.

Hillarp, A., Lethagen, S., & Mattiasson, I. (2003). Aspirin resistance is not a common biochemical phenotype explained by unblocked cyclooxygenase-1 activity. *Journal of Thrombosis and Haemostasis*, *1*(1), 196–197.

Hollah, P., Hausberg, M., Kosch, M., Barenbrock, M., Letzel, M., Schlatter, E., et al. (2001). A novel assay for determination of diadenosine polyphosphates in human platelets: Studies in normotensive subjects and in patients with essential hypertension. *Journal of Hypertension*, *19*(2), 237–245.

Hutton, J. C., Penn, E. J., & Peshavaria, M. (1983). Low-molecular-weight constituents of isolated insulin-secretory granules. Bivalent cations, adenine nucleotides and inorganic phosphate. *The Biochemical Journal*, *210*(2), 297–305.

Idzko, M., Dichmann, S., Ferrari, D., Di Virgilio, F., la Sala, A., Girolomoni, G., et al. (2002). Nucleotides induce chemotaxis and actin polymerization in immature but not mature human dendritic cells via activation of pertussis toxin-sensitive P2y receptors. *Blood*, *100*(3), 925–932.

Jankowski, V., Tolle, M., Vanholder, R., Schonfelder, G., van der Giet, M., Henning, L., et al. (2005). Uridine adenosine tetraphosphate: A novel endothelium-derived vasoconstrictive factor. *Natural Medicines*, *11*(2), 223–227.

Jin, J., Daniel, J. L., & Kunapuli, S. P. (1998). Molecular basis for ADP-induced platelet activation. II. The P2Y1 receptor mediates ADP-induced intracellular calcium mobilization and shape change in platelets. *The Journal of Biological Chemistry*, *273*(4), 2030–2034.

Jin, J., & Kunapuli, S. P. (1998). Coactivation of two different G protein-coupled receptors is essential for ADP-induced platelet aggregation. *Proceedings of the National Academy of Sciences of the United States of America*, *95*(14), 8070–8074.

Karoon, P., & Burnstock, G. (1998). Reduced sympathetic noradrenergic neurotransmission in the tail artery of Donryu rats fed with high cholesterol-supplemented diet. *British Journal of Pharmacology*, *123*, 1016–1021.

Kaufmann, A., Musset, B., Limberg, S. H., Renigunta, V., Sus, R., Dalpke, A. H., et al. (2005). "Host tissue damage" signal ATP promotes non-directional migration and negatively regulates toll-like receptor signaling in human monocytes. *The Journal of Biological Chemistry*, *280*(37), 32459–32467.

Kawamura, H., Aswad, F., Minagawa, M., Malone, K., Kaslow, H., Koch-Nolte, F., et al. (2005). P2X7 receptor-dependent and -independent T cell death is induced by nicotinamide adenine dinucleotide. *Journal of Immunology*, *174*(4), 1971–1979.

Kellerman, D., Rossi Mospan, A., Engels, J., Schaberg, A., Gorden, J., & Smiley, L. (2008). Denufosol: A review of studies with inhaled P2Y(2) agonists that led to Phase 3. *Pulmonary Pharmacology & Therapeutics*, *21*(4), 600–607.

Leon, C., Hechler, B., Freund, M., Eckly, A., Vial, C., Ohlmann, P., et al. (1999). Defective platelet aggregation and increased resistance to thrombosis in purinergic P2Y(1) receptor-null mice. *Journal of Clinical Investigation*, *104*(12), 1731–1737.

Loubatieres-Mariani, M. M., Chapal, J., Lignon, F., & Valette, G. (1979). Structural specificity of nucleotides for insulin secretory action from the isolated perfused rat pancreas. *European Journal of Pharmacology*, *59*(3–4), 277–286.

Mahaut-Smith, M. P., Tolhurst, G., & Evans, R. J. (2004). Emerging roles for P2X1 receptors in platelet activation. *Platelets*, *15*(3), 131–144.

Malmsjo, M., Hou, M., Pendergast, W., Erlinge, D., & Edvinsson, L. (2003). Potent P2Y6 receptor mediated contractions in human cerebral arteries. *BMC Pharmacology*, *3*, 4.

Marteau, F., Gonzalez, N. S., Communi, D., Goldman, M., & Boeynaems, J. M. (2005). Thrombospondin-1 and indoleamine 2,3-dioxygenase are major targets of extracellular ATP in human dendritic cells. *Blood*, *106*(12), 3860–3866.

Mega, J. L., Close, S. L., Wiviott, S. D., Shen, L., Hockett, R. D., Brandt, J. T., et al. (2009). Cytochrome p-450 polymorphisms and response to clopidogrel. *The New England Journal of Medicine*, *360*(4), 354–362.

Mehta, S. R., Bassand, J. P., Chrolavicius, S., Diaz, R., Eikelboom, J. W., Fox, K. A., et al. (2010). Dose comparisons of clopidogrel and aspirin in acute coronary syndromes. *The New England Journal of Medicine*, *363*(10), 930–942.

Michelson, A. D. (2003). How platelets work: Platelet function and dysfunction. *Journal of Thrombosis and Thrombolysis*, *16*(1–2), 7–12.

Montalescot, G., Wiviott, S. D., Braunwald, E., Murphy, S. A., Gibson, C. M., McCabe, C. H., et al. (2009). Prasugrel compared with clopidogrel in patients undergoing percutaneous coronary intervention for ST-elevation myocardial infarction (TRITON-TIMI 38): Double-blind, randomised controlled trial. *Lancet*, *373*(9665), 723–731.

Nilsson, J., Nilsson, L. M., Chen, Y. W., Molkentin, J. D., Erlinge, D., & Gomez, M. F. (2006). High glucose activates nuclear factor of activated T cells in native vascular smooth muscle. *Arteriosclerosis, Thrombosis, and Vascular Biology*, *26*(4), 794–800.

Parandeh, F., Abaraviciene, S. M., Amisten, S., Erlinge, D., & Salehi, A. (2008). Uridine diphosphate (UDP) stimulates insulin secretion by activation of P2Y6 receptors. *Biochemical and Biophysical Research Communications*, *370*(3), 499–503.

Parodi, J., Flores, C., Aguayo, C., Rudolph, M. I., Casanello, P., & Sobrevia, L. (2002). Inhibition of nitrobenzylthioinosine-sensitive adenosine transport by elevated D-glucose involves activation of P2Y2 purinoceptors in human umbilical vein endothelial cells. *Circulation Research*, *90*(5), 570–577.

Peral, A., Dominguez-Godinez, C. O., Carracedo, G., & Pintor, J. (2008). Therapeutic targets in dry eye syndrome. *Drug News & Perspectives*, *21*(3), 166–176.

Petit, P., Bertrand, G., Schmeer, W., & Henquin, J. C. (1989). Effects of extracellular adenine nucleotides on the electrical, ionic and secretory events in mouse pancreatic beta-cells. *British Journal of Pharmacology*, *98*(3), 875–882.

Pham, T. M., Morris, J. B., Arthur, J. F., Post, G. R., Brown, J. H., & Woodcock, E. A. (2003). UTP but not ATP causes hypertrophic growth in neonatal rat cardiomyocytes. *Journal of Molecular and Cellular Cardiology*, *35*(3), 287–292.

Poulsen, C. R., Bokvist, K., Olsen, H. L., Hoy, M., Capito, K., Gilon, P., et al. (1999). Multiple sites of purinergic control of insulin secretion in mouse pancreatic beta-cells. *Diabetes*, *48*(11), 2171–2181.

Romanello, M., Bivi, N., Pines, A., Deganuto, M., Quadrifoglio, F., Moro, L., et al. (2006). Bisphosphonates activate nucleotide receptors signaling and induce the expression of Hsp90 in osteoblast-like cell lines. *Bone*, *39*(4), 739–753.

Salehi, A., Qader, S. S., Grapengiesser, E., & Hellman, B. (2005). Inhibition of purinoceptors amplifies glucose-stimulated insulin release with removal of its pulsatility. *Diabetes*, *54*(7), 2126–2131.

Sassmann, A., Gier, B., Grone, H. J., Drews, G., Offermanns, S., & Wettschureck, N. (2010). The Gq/G11-mediated signaling pathway is critical for autocrine potentiation of insulin secretion in mice. *Journal of Clinical Investigation*, *120*(6), 2184–2193.

Schluter, H., Offers, E., Bruggemann, G., van der Giet, M., Tepel, M., Nordhoff, E., et al. (1994). Diadenosine phosphates and the physiological control of blood pressure. *Nature*, *367*(6459), 186–188.

Schmid-Antomarchi, H., Schmid-Alliana, A., Romey, G., Ventura, M. A., Breittmayer, V., Millet, M. A., et al. (1997). Extracellular ATP and UTP control the generation of reactive oxygen intermediates in human macrophages through the opening of a charybdotoxin-sensitive Ca2+-dependent K+ channel. *Journal of Immunology*, *159*(12), 6209–6215.

Seiffert, K., Ding, W., Wagner, J. A., & Granstein, R. D. (2006). ATPgammaS enhances the production of inflammatory mediators by a human dermal endothelial cell line via purinergic receptor signaling. *Journal of Investigative Dermatology*, *126*(5), 1017–1027.

Seye, C. I., Kong, Q., Erb, L., Garrad, R. C., Krugh, B., Wang, M., et al. (2002). Functional P2Y2 nucleotide receptors mediate uridine 5′-triphosphate-induced intimal hyperplasia in collared rabbit carotid arteries. *Circulation*, *106*(21), 2720–2726.

Seye, C. I., Yu, N., Jain, R., Kong, Q., Minor, T., Newton, J., et al. (2003). The P2Y2 nucleotide receptor mediates UTP-induced vascular cell adhesion molecule-1 expression in coronary artery endothelial cells. *The Journal of Biological Chemistry*, *278*(27), 24960–24965.

Solini, A., Iacobini, C., Ricci, C., Chiozzi, P., Amadio, L., Pricci, F., et al. (2005). Purinergic modulation of mesangial extracellular matrix production: Role in diabetic and other glomerular diseases. *Kidney International*, *67*(3), 875–885.

Sprague, R. S., Olearczyk, J. J., Spence, D. M., Stephenson, A. H., Sprung, R. W., & Lonigro, A. J. (2003). Extracellular ATP signaling in the rabbit lung: Erythrocytes as determinants of vascular resistance. *American Journal of Physiology. Heart and Circulatory Physiology*, *285*(2), H693–H700.

Sprague, R. S., Stephenson, A. H., Ellsworth, M. L., Keller, C., & Lonigro, A. J. (2001). Impaired release of ATP from red blood cells of humans with primary pulmonary hypertension. *Experimental Biology and Medicine (Maywood)*, *226*(5), 434–439.

Squires, P. E., James, R. F., London, N. J., & Dunne, M. J. (1994). ATP-induced intracellular Ca2+ signals in isolated human insulin-secreting cells. *Pflugers Archiv*, *427*(1–2), 181–183.

Suzuki, T., Hide, I., Ido, K., Kohsaka, S., Inoue, K., & Nakata, Y. (2004). Production and release of neuroprotective tumor necrosis factor by P2X7 receptor-activated microglia. *The Journal of Neuroscience*, *24*(1), 1–7.

Tan, C., Salehi, A., Svensson, S., Olde, B., & Erlinge, D. (2010). ADP receptor P2Y(13) induce apoptosis in pancreatic beta-cells. *Cellular and Molecular Life Sciences*, *67*(3), 445–453.

Turner, C. H., & Robling, A. G. (2004). Exercise as an anabolic stimulus for bone. *Current Pharmaceutical Design*, *10*(21), 2629–2641.

Vassort, G. (2001). Adenosine 5′-triphosphate: A P2-purinergic agonist in the myocardium. *Physiological Reviews*, *81*(2), 767–806.

Vidal, M., Hicks, P. E., & Langer, S. Z. (1986). Differential effects of alpha-beta-methylene ATP on responses to nerve stimulation in SHR and WKY tail arteries. *Naunyn Schmiedeberg's Archives of Pharmacology*, *332*(4), 384–390.

Wallentin, L., Becker, R. C., Budaj, A., Cannon, C. P., Emanuelsson, H., Held, C., et al. (2009). Ticagrelor versus clopidogrel in patients with acute coronary syndromes. *The New England Journal of Medicine*, *361*(11), 1045–1057.

Wallentin, L., Varenhorst, C., James, S., Erlinge, D., Braun, O. O., Jakubowski, J. A., et al. (2008). Prasugrel achieves greater and faster P2Y12receptor-mediated platelet inhibition than clopidogrel due to more efficient generation of its active metabolite in aspirin-treated patients with coronary artery disease. *European Heart Journal*, *29*, 21–30.

Wang, L., Jacobsen, S. E., Bengtsson, A., & Erlinge, D. (2004). P2 receptor mRNA expression profiles in human lymphocytes, monocytes and CD34+ stem and progenitor cells. *BMC Immunology*, *5*, 16.

Wihlborg, A. K., Balogh, J., Wang, L., Borna, C., Dou, Y., Joshi, B. V., et al. (2006). Positive inotropic effects by uridine triphosphate (UTP) and uridine diphosphate (UDP) via P2Y2 and P2Y6 receptors on cardiomyocytes and release of UTP in man during myocardial infarction. *Circulation Research*, *98*(7), 970–976.

Wihlborg, A. K., Wang, L., Braun, O. O., Eyjolfsson, A., Gustafsson, R., Gudbjartsson, T., et al. (2004). ADP receptor P2Y12 is expressed in vascular smooth muscle cells and stimulates contraction in human blood vessels. *Arteriosclerosis, Thrombosis, and Vascular Biology*, *24*(10), 1810–1815.

Wiviott, S. D., Braunwald, E., Angiolillo, D. J., Meisel, S., Dalby, A. J., Verheugt, F. W., et al. (2008). Greater clinical benefit of more intensive oral antiplatelet therapy with prasugrel in patients with diabetes mellitus in the trial to assess improvement in therapeutic outcomes by optimizing platelet inhibition with prasugrel-thrombolysis in Myocardial Infarction 38. *Circulation*, *118*(16), 1626–1636.

Wiviott, S. D., Braunwald, E., McCabe, C. H., Montalescot, G., Ruzyllo, W., Gottlieb, S., et al. (2007). Prasugrel versus clopidogrel in patients with acute coronary syndromes. *The New England Journal of Medicine*, *357*(20), 2001–2015.

Yang, D., Gluais, P., Zhang, J. N., Vanhoutte, P. M., & Feletou, M. (2004). Endothelium-dependent contractions to acetylcholine, ATP and the calcium ionophore A 23187 in aortas from spontaneously hypertensive and normotensive rats. *Fundamental & Clinical Pharmacology*, *18*(3), 321–326.

Yitzhaki, S., Shainberg, A., Cheporko, Y., Vidne, B. A., Sagie, A., Jacobson, K. A., et al. (2006). Uridine-5′-triphosphate (UTP) reduces infarct size and improves rat heart function after myocardial infarct. *Biochemical Pharmacology*, *72*(8), 949–955.

Yitzhaki, S., Shneyvays, V., Jacobson, K. A., & Shainberg, A. (2005). Involvement of uracil nucleotides in protection of cardiomyocytes from hypoxic stress. *Biochemical Pharmacology*, *69*(8), 1215–1223.

Yusuf, S., Zhao, F., Mehta, S. R., Chrolavicius, S., Tognoni, G., & Fox, K. K. (2001). Effects of clopidogrel in addition to aspirin in patients with acute coronary syndromes without ST-segment elevation. *The New England Journal of Medicine*, *345*(7), 494–502.

Zhang, S., Remillard, C. V., Fantozzi, I., & Yuan, J. X. (2004). ATP-induced mitogenesis is mediated by cyclic AMP response element-binding protein-enhanced TRPC4 expression and activity in human pulmonary artery smooth muscle cells. *American Journal of Physiology. Cell Physiology*, *287*(5), C1192–C1201.

Zheng, J. S., Boluyt, M. O., Long, X., O'Neill, L., Lakatta, E. G., & Crow, M. T. (1996). Extracellular ATP inhibits adrenergic agonist-induced hypertrophy of neonatal cardiac myocytes. *Circulation Research*, *78*(4), 525–535.

Zheng, J. S., Boluyt, M. O., O'Neill, L., Crow, M. T., & Lakatta, E. G. (1994). Extracellular ATP induces immediate-early gene expression but not cellular hypertrophy in neonatal cardiac myocytes. *Circulation Research*, *74*(6), 1034–1041.

Zimmermann, H. (2006). Ectonucleotidases in the nervous system. *Novartis Foundation Symposium*, *276*, 113–128, discussion 128–130, 233–117, 275–181.

Laszlo Köles*,†, Anna Leichsenring*, Patrizia Rubini*, and Peter Illes*

*Rudolph-Boehm-Institute of Pharmacology and Toxicology, University of Leipzig, Leipzig, Germany

†Department of Pharmacology and Pharmacotherapy, Semmelweis University, Budapest, Hungary

P2 Receptor Signaling in Neurons and Glial Cells of the Central Nervous System

Abstract

Purine and pyrimidine nucleotides are extracellular signaling molecules in the central nervous system (CNS) leaving the intracellular space of various CNS cell types via nonexocytotic mechanisms. In addition, ATP is a neuro- and gliotransmitter released by exocytosis from neurons and neuroglia. These nucleotides activate P2 receptors of the P2X (ligand-gated cationic channels) and P2Y (G protein-coupled receptors) types. In mammalians, seven P2X and eight P2Y receptor subunits occur; three P2X subtypes form homomeric or heteromeric P2X receptors. P2Y subtypes may also hetero-oligomerize with each other as well as with other G protein-coupled receptors. P2X receptors are able to physically associate with various types of ligand-gated ion channels and thereby to interact with them. The P2 receptor homomers or heteromers exhibit specific sensitivities against pharmacological ligands and have preferential functional roles. They may be situated at both presynaptic (nerve terminals) and postsynaptic (somatodendritic) sites of neurons, where they modulate either transmitter release or the postsynaptic sensitivity to neurotransmitters. P2 receptors exist at neuroglia (e.g., astrocytes, oligodendrocytes) and microglia in the CNS. The neuroglial P2 receptors subserve the neuron–glia cross talk especially via their end-feets projecting to neighboring synapses. In addition, glial networks are able to communicate through coordinated oscillations of their intracellular Ca^{2+} over considerable distances. P2 receptors are involved in the physiological regulation of CNS functions as well

Advances in Pharmacology, Volume 61

1054-3589/11 $35.00
10.1016/B978-0-12-385526-8.00014-X

as in its pathophysiological dysregulation. Normal (motivation, reward, embryonic and postnatal development, neuroregeneration) and abnormal regulatory mechanisms (pain, neuroinflammation, neurodegeneration, epilepsy) are important examples for the significance of P2 receptor-mediated/ modulated processes.

I. Introduction

The first report on the extracellular effects of purines appeared in the late 1920s of the past century by Drury and Szent-Györgyi. The following four decades in purine research were characterized by sporadic publications showing further extracellular actions of nucleotides or nucleosides. The purinergic research field got new impulses 40 years ago when Burnstock (1972) proposed the purinergic neurotransmission hypothesis. Burnstock's revolutionary idea was followed by a major boom in this research area. It became evident that adenosine 5′-triphosphate (ATP) is an important signaling molecule of cell-to-cell communication in the central and peripheral nervous system, acting as a neuro- or gliotransmitter in the dialog of neurons with each other or with glial cells (Fields & Stevens, 2000; Franke & Illes, 2006). In a number of pathophysiological conditions, purines turned out to be attractive therapeutic targets for both peripheral and central nervous system (CNS) diseases (Burnstock, 2007; Köles et al., 2005). This review focuses on the role of the purinergic transmission in the CNS including the localization of P2 receptors and the actions of purines in neurons as well as in glial cells, their role in neuron–glia communication, and the modulation/ integration of neurotransmission.

II. The Source and Fate of Extracellular ATP

To fulfill its extracellular functions, ATP must reach the extracellular space. Intracellular ATP concentrations are in the millimolar range, whereas extracellular ATP concentrations range from nanomolar to micromolar (e.g., Frenguelli et al., 2007; Lazarowski et al., 2000). Driven by this chemical concentration gradient, virtually all cell types can be a source of ATP by a number of mechanisms: vesicular release at both synaptic and extrasynaptic sites; membrane transport involving ATP binding cassette proteins; permeation through connexin or pannexin hemichannels, plasmalemmal voltage-dependent anion channels and even P2X7 receptors operating as ATP-permeable channels or osmotic transporters linked to anion channels (Abbracchio et al., 2009; Bodin & Burnstock, 2001; Stout et al., 2002). Besides this activity-regulated ATP release, dramatic leakage of purines occurs from injured or dying cells via the damaged cell membrane, and so cells are exposed to high concentrations of purines after cell death in the neighboring areas (Cook & McCleskey, 2002). The high ATP level may activate protective mechanisms mediating survival and regeneration. However, it can initiate and aggregate harmful mechanisms as well, leading to further destruction and damage (Burnstock, 2007; Franke & Illes, 2006; Köles et al., 2005; Volonte et al., 2003).

Extracellular nucleotides exert different effects by interacting with plasma membrane receptors named P2 receptors. A complex family of ectoenzymes (ecto-ATPases, ectoapyrases, and ecto-5′-nucleotidases) rapidly hydrolyzes or interconverts extracellular nucleotides, thereby either terminating their signaling action or producing an active metabolite of altered receptor selectivity (Zimmermann, 2006). While ATP and adenosine 5′-diphosphate (ADP) directly stimulate subtypes of the P2 receptors, adenosine stimulates the P1 receptor class (the so-called adenosine receptors). Since P1 and P2 receptors are often functionally antagonistic, the breakdown of ATP not simply limits its extracellular actions by enhancing its removal but brings new players, with different actions into the game as well. Accordingly, we have essentially in all tissues a complex extracellular regulatory system with the involvement of the P2 receptors, the nucleotide hydrolyzing and interconverting enzymes, and the P1 receptors (Abbracchio et al., 2009).

III. Recombinant P2 Receptors

The P2 receptors have been divided into two types: G protein-coupled receptors (P2Y) and ligand-gated cation channels (P2X) (Burnstock & Kennedy, 1985). Seven different P2X subunits (P2X1–7) and eight distinct P2Y subunits ($P2Y_1$, $P2Y_2$, $P2Y_4$, $P2Y_6$, $P2Y_{11–14}$) have been cloned to date from mammalian cells (Abbracchio et al., 2006; North, 2002). The P2 receptors are among the most abundant receptors in mammalian tissues, they are expressed in all types of cells including neurons and glial cells. The expression profile of P2 receptor subtypes varies depending on the cell type (Burnstock, 2007; Burnstock & Knight, 2004).

When the cloned receptors are expressed in heterologous expression systems (e.g., *Xenopus laevis* oocytes), their individual phenotype can be well characterized. However, in most cases, the phenotypes observed in native tissues do not closely resemble those reported for the cloned subunits (North, 2002; Ralevic & Burnstock, 1998). They may have different pharmacological profiles (e.g., selectivity of the ligands for the P2 receptor) or functional properties (e.g., channel kinetics, coupling to signaling pathways). One of the possible explanations for this divergence is that, for example, various P2X subunits can coassemble to form a channel with distinct functional properties. Indeed, both ionotropic P2X channels and the G protein-coupled P2Y receptors have been reported to exist in oligomeric (homomeric or heteromeric) assembly of more than one subunit (Burnstock, 2007; Ralevic & Burnstock, 1998).

A. P2X Receptors

P2X receptor channels comprise the third family of ligand-gated ion channels, in addition to Cys-loop and glutamate receptor families (North, 1996). The P2X1 and P2X2 subunit proteins were cloned in 1994, and the P2X3 subtype 1 year later (Brake et al., 1994; Chen et al., 1995; Valera et al., 1994). The P2X subunit proteins are 379 (P2X6) to 595 (P2X7) amino acids long, possessing intracellular cytoplasmic N- and C-termini and two transmembrane α-helices connected by a large extracellular loop (Newbolt et al., 1998). Three P2X receptor subunits assemble into an ATP-activated ion channel, by forming a central pore (Nörenberg & Illes, 2000; North, 2002; Ralevic & Burnstock, 1998).

Among subunits, the C-terminal domain is the least conserved part in amino acid composition, indicating that it may confer subunit-specific properties. C-termini play a major role in direct physical interactions with Cys-loop ion channels (see Section IV.A). Both the N- and C-termini are targets for posttranscriptional modifications, RNA splicing, phosphorylation, and protein–protein interactions with other regulatory molecules (Koshimizu et al., 2006; North, 2002). These modifications and interactions may have critical influences on the channel function, for example, regulating the rate of desensitization (Boué-Grabot et al., 2000; Smith et al., 1999) or receptor trafficking (Royle et al., 2002).

The ectodomain of 280 amino acids contains the ATP binding pocket and binding sites for antagonists and modulators. It is glycosylated and contains 10 conserved cysteine residues forming a series of disulfide bridges and hydrophobic regions close to the pore vestibule, for possible receptor/channel modulation by cations (magnesium, calcium, zinc, copper ions, and protons; Ennion & Evans, 2002; North, 2002; Rettinger et al., 2000). Putative phosphorylation of the ectodomain by ecto-protein kinase C (PKC) has also been reported to result in facilitation of P2X3 receptor function (Wirkner et al., 2005). However, this direct phosphorylation is a matter of dispute (Brown & Yule, 2007). Eventually, the transmembrane domains are involved in the heteromerization (Jiang et al., 2003).

1. Homomeric P2X Receptors

Biochemical evidence indicates that both homomeric and heteromeric receptors occur as stable trimers (Aschrafi et al., 2004; Illes & Ribeiro, 2004; Nicke et al., 1998). The three subunit composition of P2X receptors was supported in addition by a wealth of further data: (1) single channel analysis of P2X receptor currents indicated two to three sequential binding steps (Ding & Sachs, 1999; Riedel et al., 2007); (2) the kinetic behavior of

P2X receptor-currents was simulated with an allosteric model describing channel opening in the di- or triliganded state (Karoly et al., 2008; Sokolova et al., 2004); (3) atomic force microscopy provided evidence for three receptor-subunits, which moved away from the central pore as the channel opened (Nakazawa et al., 2005; Shinozaki et al., 2009); (4) fluorescence resonance energy transfer and electron microscopy supplied a rough structure of three interacting subunits (Young, 2010); and (5) the crystal structure of a zebrafish P2X4 receptor mutant supported the existence of corresponding intersubunit pockets as the binding site for ATP (Kawate et al., 2009; Young, 2010).

Six stable homomeric P2X trimers are formed from the individual subunits; the P2X6 subunits obviously do not oligomerize with each other. P2X1 and P2X3 homomers are rapidly desensitizing, while other homomeric channels show moderate (P2X4) or slow (P2X2, P2X5, P2X6, P2X7) desensitization. All P2X receptors are permeable to small cations such as Na^+, K^+, and Ca^{2+}; the Ca^{2+} permeability is variable among the receptors (Abbracchio et al., 2009; Köles et al., 2007).

The homomeric P2X1 receptor has a relatively high permeability to Ca^{2+}. It is activated by both ATP and α,β-methylene ATP (α,β-meATP), shows rapid desensitization, and undergoes agonist-dependent internalization and recycling. It is likely to be basally phosphorylated at a number of sites (Dutton et al., 2000; Evans et al., 1996; Vial et al., 2004).

The homomeric P2X2 receptor is activated by ATP but not by α,β-meATP. It is somewhat less permeable to Ca^{2+} than P2X1 and is inhibited by extracellular Ca^{2+} (Evans et al., 1996). It is potentiated by acidification and inhibited by alkalization. Zinc and copper also potentiate P2X2 currents, which undergo little or no desensitization during a constant exposure to agonists. The P2X2 pore is dilated during prolonged ATP application, and the change is prominent when the density of P2X2 receptors in the plasma membrane is high (Fujiwara & Kubo, 2004; Virginio et al., 1999). In contrast to P2X1 receptors, P2X2 receptors appear to be more stable at the cell surface (Khakh et al., 2001).

P2X3 channels—like their P2X1 counterparts—desensitize rapidly and respond to α,β-meATP (Chen et al., 1995). However, unlike P2X1 receptors, P2X3 receptors are potentiated by zinc. Their Ca^{2+} permeability is lower than that of the P2X1 receptors (North, 2002). P2X3 receptors may be positively modulated by inflammatory mediators such as bradykinin and substance P (Paukert et al., 2001). Phosphorylation at PKC sites at the ectodomain may be involved in the regulation of the P2X3 receptor current (Stanchev et al., 2006; Wirkner et al., 2005). Histidin 206 residue of the extracellular loop is involved in the dual effect of acidic pH on P2X3 receptor currents. Acidification decreases the current amplitude at low agonist concentrations (slower activation) and increases it at high concentrations (slower desensitization; Gerevich et al., 2007a). Conserved lysin and arginine residues in the extracellular loop are involved in agonist binding to the P2X3 receptor (Fischer et al., 2007).

Homomeric P2X4 receptors are activated by ATP but not by α,β-meATP. A specific feature of P2X4 receptors is their potentiation by ivermectin, an allosteric modulator, probably by interfering with receptor internalization (Toulme et al., 2006). These receptors are also characterized by their unique refractoriness to the nonspecific antagonists suramin and pyridoxal-phosphate-6-azophenyl-2′,4′-disulfonate acid (PPADS). Zinc facilitates, copper inhibits the P2X4 receptor currents, and the Ca^{2+} permeability of the channel is relatively high (Buell et al., 1996; Köles et al., 2007). The homomeric P2X4 receptor desensitizes, but with a moderate speed and a current decline typically within 5–10 s at maximal ATP concentrations; it shows pore dilation in case of prolonged exposure to agonists (Khakh et al., 1999a). In neurons, it undergoes rapid internalization and subsequent reinsertion into the plasma membrane; the surface expression and function of synaptic P2X4 receptors is probably controlled by interactions with the endocytic machinery (Royle et al., 2002). In microglia, it is localized predominantly in lysosomes, but it can also quickly traffic out of lysosomes to upregulate its exposure at the cell surface (Qureshi et al., 2007; see below).

The homomeric P2X5 receptor desensitizes slowly, it is not activated by α,β-meATP, but is potentiated by zinc, and inhibited by high Ca^{2+} concentration. Although P2X channels were described as essentially selective to cations, P2X5 receptors are also chloride permeable. Their permeability to large cations does not require slow pore dilation. A single-nucleotide polymorphism determines whether an individual organism forms a functional or a nonfunctional P2X5 receptor (Bo et al., 2003a; Roberts et al., 2006).

Unlike other P2X subtypes, the P2X6 receptor is not able to form functional homomers, possibly due to problems with glycosylation (Jones et al., 2004).

The P2X7 receptor requires much higher ATP concentrations ($>100\ \mu M$) for activation than other P2X channels, it has an extremely long intracellular C-terminal tail (240 amino acids) and the composition of a large pore accompanies prolonged agonist activity. The mechanism of pore formation by the P2X7 receptor and its importance in *in vivo* circumstances are not completely clear, but it may have a (patho)physiological significance. Since the receptor is nondesensitizing, the pore stays open as long as it binds ATP, and P2X7 receptor activation causes a massive disturbance of cytoplasmic ion homeostasis (Ferrari et al., 2006; Sperlágh et al., 2006). P2X7 receptor function can be modulated by cations; external calcium, magnesium, zinc, and copper inhibit the receptor (Virginio et al., 1997). More than 260 single-nucleotide polymorphisms have been described in the human P2X7 receptor gene (Ferrari et al., 2006).

2. *Heteromeric P2X Receptors*

Hitherto six heteromeric P2X channels have been identified and one more is claimed to exist. The P2X2/3 receptor was the first recognized heteromeric P2X channel (Lewis et al., 1995; Radford et al., 1997). Subsequently, the existence of P2X4/6 and P2X1/5 heteromers has been reported (Le et al., 1998; Torres et al., 1998). Then, functional P2X2/6 (King et al., 2000), P2X1/2 (Brown et al., 2002), and P2X1/4 (Nicke et al., 2005) heteromers were also described. The P2X7 subunit was postulated not to be involved in receptor heteromerization (Torres et al., 1999). However, most recently, the existence of the P2X4/7 channels was also reported, at least for the recombinant receptor in heterologic expression systems (Guo et al., 2007).

In most cases, the heteromeric P2X receptors showed a pharmacological and functional profile distinct from those of the cloned homomeric P2X receptors often explaining the discrepancies and filling the gap between the properties of the cloned and the native P2X receptors. For instance, the P2X2/3 heteromeric channel shares with the homomeric P2X3 receptors mostly its pharmacological profile (e.g., sensitivity to α,β-meATP); however, it resembles the homomeric P2X2 receptor in its slow desensitization kinetics, pore formation, potentiation by zinc, and blockade by calcium (Chizh & Illes, 2001; Köles et al., 2007).

The P2X6 subunit, which does not form homomers, readily forms heteromers with P2X2 and P2X4 subunits (Nörenberg & Illes, 2000; Rubio & Soto, 2001). Since the P2X6 subunit can increase the Ca^{2+} permeability of the P2X2 receptor and change the pharmacology of both P2X2 and P2X4, it may operate as a modulatory subunit of the heteromer, which potentiates the function of its counterparts (Egan & Khakh, 2004).

The P2X1/5 heteromer has unique properties. It is more sensitive to ATP than the other P2X channels such as the homomeric P2X1 or P2X5 receptors (currents are activated at nanomolar concentrations), and the kinetics of the response are diverse at different ATP concentrations. At low concentrations, the current persists with unaltered amplitude over several seconds, but when the ATP concentration exceeds 300 nM, after an initial peak it declines followed by a sustained component. α,β-meATP also elicits a sustained current by activating this channel, which is not seen for either of the homomers when expressed separately (North, 2002; Verkhratsky et al., 2009). The contribution of heteromeric P2X1/5 receptor channels to the excitability of astrocytes (see below) is positively modulated by phosphoinositides through the P2X1 lipid binding domain (Ase et al., 2010).

B. P2Y Receptors

P2Y receptors, in common with other G protein-coupled receptors, have seven transmembrane domains, an extracellular N-terminus containing several potential glycosylation sites, and an intracellular C-terminus containing several consensus binding/phosphorylation sites for protein kinases. Some positively charged residues in transmembrane domains 3, 6, and 7 seem to be crucial for receptor activation by nucleotides (Abbracchio et al., 2006; Erb et al., 2006).

1. Homomeric P2Y Receptors

The sorting of the P2Y receptor subtypes can be based on their pharmacological profile or on their G protein subtype coupling preference. The simplified pharmacological classification differentiates between purine- and pyrimidine-selective P2Y receptors. The purine-selective $P2Y_1$, $P2Y_{11}$, $P2Y_{12}$, and $P2Y_{13}$ receptors are activated by ATP or ADP. However, the pyrimidine-selective $P2Y_4$, $P2Y_6$, and $P2Y_{14}$ receptors are activated by uridine 5′-triphosphate (UTP), uridine 5′-diphosphate (UDP) or in the case of $P2Y_{14}$, UDP-glucose. The $P2Y_2$ receptor is activated approximately equally by both ATP and UTP, and is often placed in the pyrimidine-sensitive subgroup (Abbracchio et al., 2006; Ralevic & Burnstock, 1998). Recently, an additional uracil nucleotide recognizing P2Y-like deorphanized receptor has also been reported (Ciana et al., 2006).

The classification of the P2Y receptors based on the G protein subtype preference is the following: $P2Y_1$, $P2Y_2$, $P2Y_4$, $P2Y_6$, and $P2Y_{11}$ belong to the G_q protein-coupled subfamily, while the G_i-coupled subfamily comprises the $P2Y_{12–14}$ receptors (Abbracchio et al., 2006; Köles et al., 2005). However, coupling of the same P2Y receptor to different G proteins is also possible (see Section IV.B).

The conventional idea until the 1990s was that a G protein-coupled receptor is a monomeric transmembrane protein. Such a single receptor upon ligand binding interacts allosterically with the heterotrimeric single G protein and initiates the events characteristic for receptor activation. However, this idea was revised as a growing body of biochemical and biophysical evidence indicated that G protein-coupled receptors exist as homo- or hetero-oligomeric complexes (Bouvier, 2001).

Indeed, it was suggested that $P2Y_1$ subtypes need to be oligomerized to be active in smooth muscle cells and endothelial cells (Wang et al., 2002). Further, fluorescence resonance energy transfer analysis demonstrated the existence of homo-oligomeric complexes of $P2Y_2$ subunits in heterologous expression systems (Kotevic et al., 2005). The $P2Y_4$ receptor has been reported to form homo-oligomeric complexes and appear as stable dimers in several CNS and peripheral neuronal cells (D'Ambrosi et al., 2006). A functional study revealed that $P2Y_{12}$ receptors exist also predominantly as homo-oligomers situated in lipid rafts, and this state is essential for their functionality. An active metabolite of clopidogrel (an irreversible $P2Y_{12}$ antagonist) dissociated the oligomers into dimeric receptors that are partitioned out of lipid rafts. Thus, after the application of clopidogrel, $P2Y_{12}$ was no longer able to bind ATP/ADP (Savi et al., 2006).

2. Hetero-Oligomeric Assembly of P2Y Receptors

P2Y receptor subtypes may form heteromers with each other. Hetero-oligomerization between the $P2Y_1$ and $P2Y_{11}$ receptors which alters the ligand selectivity and is necessary for the subsequent internalization has recently been reported. It has also an important impact on $P2Y_{11}$ receptor desensitization (Ecke et al., 2008). The dynamic architecture of $P2Y_4$ and $P2Y_6$ proteins involves the formation of complex hetero-oligomers as well. Such complexes comprise $P2Y_4$ (dimeric) and $P2Y_6$ (monomeric) receptors in native neuronal phenotypes. The monomeric/dimeric protomers are differently distributed in specialized membrane microdomains, and the homo- and hetero-oligomeric complexes are differently modulated by ligand activation (D'Ambrosi et al., 2007).

Hetero-oligomeric assembly of P2Y receptors with other G protein-coupled receptors has also been reported (for review, see Fischer & Krügel, 2007). For instance, A_1 adenosine receptors interact with the $P2Y_1$ or $P2Y_2$ receptor, respectively. The A_1–$P2Y_1$ hetero-oligomer has $P2Y_1$-like agonist selectivity but a preferential signaling pathway characteristic for the A_1 receptors (Yoshioka et al., 2001). A functional cross talk between the $P2Y_1$ and the A_1 receptors involving their hetero-oligomerization was also confirmed in CNS synapses. $P2Y_1$ receptor stimulation impaired the potency of A_1 receptor-coupling to G protein, whereas the stimulation of A_1 receptors increased the functional responsiveness of $P2Y_1$ receptors (Tonazzini et al., 2007; Yoshioka et al., 2002).

The association of the A_1 receptor with the $P2Y_2$ receptor did not seem to affect the ligand selectivity of these receptors, and the stimulation of the $P2Y_2$ receptor with UTP in the A_1–$P2Y_2$ receptor complex caused uncoupling of the A_1 receptor from G_i proteins. On the contrary, the simultaneous activation of the two receptors enhanced signaling via $G_{q/11}$ protein, characteristic for $P2Y_2$ activation (Suzuki et al., 2006). The homo- and hetero-oligomerization of the P2Y receptor subtypes with each other or with the adenosine receptors diversify the agonist and antagonist selectivity, signaling, and functional properties of the P2Y receptors and may explain some unexpected data obtained from experiments in native preparations. P2 proteins are considered not as separate entities but as dynamic and continuously changing and interacting cell constituents instead, and a combinatory calculation may allow the prediction of their complex dynamic architecture and sophisticated nature (Volonte et al., 2006).

A close colocalization of $P2Y_2$ and β_2 adrenergic receptors was also suggested in mouse pineal gland tumor cells indicating that a direct physical interaction/receptor heteromerization may exist not only among the members of the P2 receptor family but also between the purinergic and other G protein-coupled receptors (Suh et al., 2001). Further, $P2Y_4$ receptors have been reported to be colocalized at the membrane level with NMDAR1 receptors. P2Y subunits are able to modulate the functions of various voltage- and/or ligand-gated ion channels, in a manner probably requiring localization of the receptor in close physical proximity to the channel (Cavaliere et al., 2004; Köles et al., 2008). Therefore, it cannot be excluded that metabotropic and ionotropic purinergic receptors may colocalize in higher-order complexes thereby further complicating the purinergic (patho) physiology and pharmacology.

IV. Signaling via P2 Receptors

A. P2X Receptors

The P2X channels in response to agonist challenge allow rapid, nonselective passage of cations across the cell membrane. All P2X receptors are permeable to Na^+, K^+, and Ca^{2+}. Especially, certain central neurons (medial habenula, somatosensory cortex) show an exceptionally high Ca^{2+} permeability in response to ATP (Edwards et al., 1997; Pankratov et al., 2002). P2X5 receptors allow Cl^- to pass as well.

The time course of the effect of ATP at P2X receptors is strongly influenced by receptor desensitization. Recombinant P2X receptors display varying degrees of desensitization (see above). This process does not depend on the production and diffusion of second messengers within the cytosol or the membrane, and therefore, the cellular response time is generally very rapid (Burnstock, 2007; Ralevic & Burnstock, 1998).

However, the overall consequence of P2X activation may be much more complex than a simple transient current flow through the membrane. If the calcium permeability is high, it may result in an increase of intracellular calcium concentration and depolarization of the cell membrane, subsequently activating voltage-gated calcium channels. Thus, calcium ions may accumulate in the cytoplasm leading to activation of several intracellular kinases, for example, PKC, mitogen-activated protein kinases (MAPKs), or the Ca^{2+}/calmodulin-dependent protein kinases II (CaMKII; Erb et al., 2006).

Further, some P2X receptors (especially P2X7, but under certain conditions, recombinant P2X2, P2X2/3, and P2X4 receptors as well as some native neuronal P2X receptors) seem to be unique in their ability to produce a large conductance pathway (pore), that is, they may become permeable for organic cations (e.g., *N*-methyl-D-glucamine) and fluorescent dyes following long-lasting exposure to agonists. However, the pore formation and change of permeability seem to be variable in their occurrence between different cell types, and some details appear to be different between various receptors (Khakh et al., 1999b; Robertson et al., 2001; Virginio et al., 1999).

Among the extracellular ATP-gated ion channels, especially, the P2X7 receptor-mediated effects seem to be very divergent. Activation of the P2X7 receptor not only opens a cation-permeable ion channel (and later a larger pore), but it also results in the activation of several fundamentally different downstream signaling pathways (e.g., by leading to the secretion of interleukin-1β (IL-1β) and other cytokines, the activation of phospholipase D and nuclear factor κ-light-chain-enhancer of activated B cells (NF-κB), the stimulation of MAPKs, the generation of reactive oxygen species, and apoptosis). P2X7 receptor stimulation also results in shedding of membrane proteins and rearrangement of the cytoskeleton such as membrane blebbing (Ferrari et al., 2006; Sperlágh et al., 2006). Several studies revealed the interactions of P2X7 receptor with macromolecules influencing signaling and trafficking (the P2X7 signaling complex). For more details, see recent reviews Köles et al. (2008) and Sperlágh et al. (2006).

Physical cross talk of P2X receptors with other ion channels such as nicotinic acetylcholine (ACh), $GABA_A$, and 5-HT_3 receptor channels may further complicate their signaling (e.g., via current occlusion). This issue as well as the interactions of P2X receptors with cell adhesion molecules and gap junction proteins has been detailed in a recent review (Köles et al., 2008). Further, a cross-inhibition between TRPV1 and P2X3 receptors was also reported (Stanchev et al., 2009).

B. P2Y Receptors

The coupling of P2Y receptors to G proteins was briefly described above. The heterotrimeric G proteins are composed of α and the tightly associated βγ subunits. P2Y receptor agonists by binding to the receptor cause the dissociation of the α subunit from the βγ dimer, and according to the traditional view, the Gα subunit initiates the further downstream events. However, the βγ subunit is also acknowledged as an active participant in P2Y receptor signaling; for instance, it seems to be important in the regulation of channel activities.

Based on the identity of the α subunit in the trimeric complex, four main heterotrimeric G protein subfamilies have been characterized (G_s, $G_{i/o}$, $G_{q/11}$, and $G_{12/13}$), and individual P2Y receptor subtypes may be linked to one or more of them (Abbracchio et al., 2006; Erb et al., 2006).

As it was mentioned earlier, $G_{q/11}$ and $G_{i/o}$ proteins appear to be important in P2Y receptor signaling. Coupling to $G_{q/11}$ proteins stimulates membrane-bound phospholipase C (PLC) which then cleaves phosphatidylinositol-bisphosphate (PIP_2) in the membrane into two second messengers, inositol trisphosphate (IP_3) and diacylglycerol (DAG). Membrane-bound PIP_2 itself may fulfill important biological functions such as regulation of ion channel activity. IP_3 mobilizes intracellular calcium, while DAG—in the presence of calcium—activates PKC leading to phosphorylation of intracellular macromolecules. Ca^{2+} is able to build complexes with the calcium binding protein calmodulin thereby activating the CaMK and can lead to various other intracellular events. This was observed in case of the $P2Y_1$, $P2Y_2$, $P2Y_4$, $P2Y_6$, and $P2Y_{11}$ receptors (Abbracchio et al., 2006; Erb et al., 2006; Volonte et al., 2006).

$P2Y_{12}$, $P2Y_{13}$, and $P2Y_{14}$ receptors bind preferentially G proteins containing the $G_{i/o}$ subunit. Further, coupling of $P2Y_2$, $P2Y_4$, and $P2Y_{11}$ to G_o was also reported. G_i activation is classically associated with the inhibition of adenylate cyclase and decreased cyclic AMP (cAMP), but the activation of the $G_{i/o}$ subunit may also have other consequences. For instance, $P2Y_2$, $P2Y_4$, $P2Y_{13}$, and $P2Y_{14}$ receptors via $G_{i/o}$ can cause the activation of the PLC–IP_3–DAG–Ca^{2+}-release and PKC and/or CaMK activation mechanism (Abbracchio et al., 2006; Erb et al., 2006). Agonist-specific signaling was also reported for P2Y receptors. In response to ATP, the $P2Y_{11}$ receptor couples to G_s protein activating adenylate cyclase, while its stimulation by UTP results in coupling to G_q protein and PLC activation (Communi et al., 1997; White et al., 2003). In other cases, G_s proteins do not seem to play an important role in P2Y signaling.

Certain P2Y receptors may interact with monomeric G proteins (Erb et al., 2001, 2006). For instance, $P2Y_2$ or $P2Y_{12}$ receptors may couple to $G\alpha_{12/13}$ subunits known to be involved in activation of small homomeric GTPases. Further, P2 receptors can couple to the MAPK/extracellular signal-regulated kinase (ERK) pathway or interact with macromolecules at the MAPK signaling level. Integrins can be involved in P2Y signaling, P2Y receptors can exhibit a cross talk with tyrosine kinases and with receptor tyrosine kinases. P2Y receptors may interact with PDZ proteins. These P2Y receptor-mediated downstream events as well as their direct or indirect interactions with ion channels were described in a recent review (Köles et al., 2008).

V. Pharmacology of the P2 Receptors

The conventional receptor ligands are not suitable for human therapeutic interventions due to their pharmacokinetic disadvantages or poor selectivity. Hence, we need synthesis of more potent and selective chemicals with drug-like properties. The progress in the past few years promises a breakthrough in the purine pharmacology.

A. P2X Receptors

ATP is a general, natural agonist at each P2X receptor (although at P2X7 at high concentrations only), α,β-meATP activates P2X1, P2X3, and P2X2/3, as well as P2X1/4, P2X1/5, and P2X4/6 receptors and the rest of the ionotropic receptor family is relatively insensitive to this agonist. 3′-O-(4-benzoyl)benzoyl ATP (BzATP) is a potent agonist at the P2X7 and P2X1 receptors, with a minor activity at P2X2–4 receptors (Jarvis & Khakh, 2009).

The conventional antagonists of P2X receptors, such as suramin and PPADS, are nonselective (although P2X4 and P2X7 receptors are relatively insensitive to these antagonists); they block several P2Y receptor subtypes, and inhibit ecto-ATPases. Suramin even inhibits *N*-methyl-D-aspartate (NMDA) glutamate receptors. Brilliant blue G is a preferential P2X7 receptor antagonist, but at higher concentrations also blocks P2X2, P2X4, and P2X5 (Jarvis & Khakh, 2009). Oxidized ATP at high concentrations and long incubation periods is an irreversible antagonist of the P2X7 receptor (Köles et al., 2007).

Over the past decade, P2X receptor pharmacology has speeded up. 2′,3′-O-(2,4,6-trinitrophenyl) ATP (TNP-ATP) was identified as an antagonist for the P2X receptor subtypes P2X1–4. Diinosine pentaphosphate is a noncompetitive antagonist at the rapidly desensitizing P2X1 and P2X3 subtypes and influences to a much lesser extent the slowly desensitizing P2X2 and P2X2/3 (Gever et al., 2006). More recently, potent antagonists of the P2X1, P2X3, and P2X7 receptors have been introduced. Suramin derivatives such as NF023, NF279, and NF449 have been identified as potent P2X1-selective blockers. However, the poor pharmacokinetics of these compounds may limit their *in vivo* usefulness (Gever et al., 2006).

A-317491 is a highly potent, small molecule P2X3 and P2X2/3 antagonist, but it has also a limited oral bioavailability and CNS penetration (Jarvis, 2010). However, some heterocyclic amides and diaminopyrimidines proved to be drug-like P2X3 or dual P2X3/P2X2/3 antagonists, with improved pharmacokinetics, such as RO-3, RO-4, RO-51, or RO-85 (Carter et al., 2009; Jahangir et al., 2009; Jarvis, 2010).

The new era of the selective P2X7 receptor antagonists was introduced by the discovery of KN-62, known to inhibit the CaMKII as well. Subsequently, several chemical series of selective P2X7 antagonists with improved drug-like properties have also been discovered, such as several 4,5-diarylimidazolines, cyclic imides (e.g., AZ116453743), arylhydrazides (e.g., A-847227), aryltetrazoles/aryltriazoles (e.g., A-438079), cycloguanidines (e.g., A-740003), and adamantanes (e.g., GSK314181A) (Carroll et al., 2009). For more details and overview of the enormous development on the P2X receptor pharmacology, and patents on novel P2X7 and other P2X receptor-selective ligands, see recent reviews (Friedle et al., 2010; Gunosewoyo & Kassiou, 2010; Jarvis & Khakh, 2009).

B. P2Y Receptors

The basic agonist profile of the P2Y receptors was mentioned earlier. The most potent agonists at the $P2Y_1$ receptors are ADP and its analogs (such as

ADP-β-S). The $P2Y_2$ receptor is activated approximately equally by both ATP and UTP. The $P2Y_4$ receptors show UTP preference, while UDP is the most potent agonist at $P2Y_6$ receptors. $P2Y_{11}$ receptors are activated by ATP, $P2Y_{12}$ and $P2Y_{13}$ receptors prefer ADP and its analogs, and the $P2Y_{14}$ receptor is preferentially activated by UDP-glucose (Abbracchio et al., 2009; Köles et al., 2008).

The classical antagonists show the following profile: suramin antagonizes most P2Y receptors but not $P2Y_4$; PPADS antagonizes most potently the $P2Y_1$ receptors, but it blocks other P2Y receptors only weakly, or not at all; reactive blue-2 is not effective as an antagonist at $P2Y_2$ receptors but more ($P2Y_6$) or less effectively (other P2Y receptors) antagonizes the residual ones.

Regarding the antagonists, MRS2179 is a specific antagonist of $P2Y_1$ receptors, AR-C126313 is selective for $P2Y_2$ receptors, MRS2578 is specific for $P2Y_6$, NF157 antagonizes $P2Y_{11}$, CT50547, clopidogrel, or the related antithrombotic compounds are selective for $P2Y_{12}$, while MRS2211 for $P2Y_{13}$. AR-C69931MX is a specific antagonist for both $P2Y_{12}$ and $P2Y_{13}$ receptors. For more detailed pharmacology, selective agonists, and antagonists of the P2Y receptors, see recent reviews (Fischer & Krügel, 2007; Jacobson et al., 2009; von Kügelgen, 2006).

VI. Distribution of P2 Receptors in the CNS

A. P2X Receptors

All P2X subunits are expressed in the CNS, with the preferential expression of P2X2, and P2X4. The expression of P2X subtypes varies in different regions and cell types of the CNS (Nörenberg & Illes, 2000). Although all P2X receptor subtypes except P2X6 were identified in rat astrocyte cultures (Fumagalli et al., 2003), the evidence for the presence of functional P2X receptors in glial cells is limited, mostly restricted to P2X1/5 and P2X7 receptors in astroglia. P2X4 and P2X7 receptors are the dominant P2X receptors in microglia (see Section VII.C.5).

The expression and functional role of P2X1 receptors in central neurons seems to be limited. P2X1 subunits have been described to occur in cerebellar astrocytes (Loesch & Burnstock, 1998). P2X1 and P2X5 subtypes are colocalized in astrocytes and form a functional P2X1/5 heteromer. Further, hitherto the cortical astrocyte is the only "real" cell type (i.e., not a heterologous expression system), where this heteromeric receptor type was found (Lalo et al., 2008; Verkhratsky et al., 2009).

An extensive expression pattern for P2X2 receptors was observed in the CNS; they are distributed in neuronal structures including cerebral and cerebellar cortex, hippocampus, habenula, nigrostriatal system, various brain nuclei, ventrolateral medulla, area postrema, nucleus of solitary tract, and spinal cord (Kanjhan et al., 1999; Nörenberg & Illes, 2000). They readily heteromerize and often form P2X2/6 or P2X2/3 heteromers. They are involved in several neuronal functions, for example, in the presynaptic regulation of transmitter release in the CNS and in sensory functions (in this latter ones mostly as P2X2/3 heteromers; Nörenberg & Illes, 2000; Roberts et al., 2006).

cDNAs encoding P2X3 subunits were first isolated from dorsal root ganglia (DRGs; Chen et al., 1995; Lewis et al., 1995). The central projections of P2X3 receptor-labeled nerves in DRG neurons are located in inner lamina II of the dorsal horn of the spinal cord (Vulchanova et al., 1997). Presence of the presynaptic P2X3 subunits has been reported on rat brain synaptic terminals as well (Diaz-Hernandez et al., 2001).

Among the P2X receptors, the P2X4 subtype is the most widely distributed in the brain, including various CNS sites such as hippocampus, cerebellum, and brain stem, in part displaying overlapping distribution with P2X2 and/or P2X6 subunits (Bo et al., 2003b; Nörenberg & Illes, 2000). P2X4 may play a role in fast synaptic transmission or the modulation of neurotransmitter release (Nörenberg & Illes, 2000; Rubio & Soto, 2001). Important localization of the P2X4 receptors is the microglia, where the P2X4 receptor protein expression is increased following nerve injury. Heteromer formation between P2X4 and P2X7 receptors has recently been reported. Both the homomeric and heteromeric forms may be involved in certain pain situations (Guo et al., 2007; Jarvis, 2010; Raouf et al., 2007; Tsuda et al., 2003).

P2X5 receptors show restricted localization in the CNS (Nörenberg & Illes, 2000); the P2X1/5 heteromer in astrocytes was mentioned above. P2X6 subunits are present in the brain, but only in heteromeric form together with the P2X2 or P2X4 subtypes (see above).

Although histochemical data argued for the presence of the P2X7 receptor in neuronal structures, later, the value of antibodies in identifying this target was seriously questioned (Sim et al., 2004). However, functional results strongly support a role for P2X7 or P2X7-like receptors in neuronal and astroglial functions and argue for their localization on these cell types, as it was recently clearly demonstrated (Carrasquero et al., 2009; Nörenberg et al., 2010; Oliveira et al., 2011; Sperlágh et al., 2006). Neuronal and astroglial P2X7 receptors can be involved, for example, in the regulation of neuro- and gliotransmitter release (see below). Nevertheless, P2X7 receptors are predominantly present at immunocompetent cells in the body, including abundant expression in microglia in the CNS. They may form an important link between microglial and neuronal/astroglial responses in physiological and pathophysiological states (Ferrari et al., 2006; Fields & Stevens, 2000; Sperlágh et al., 2006).

B. P2Y Receptors

P2Y receptors are expressed in both neurons and glial cell types in the CNS (Abbracchio et al., 2009; Fischer & Krügel, 2007). $P2Y_1$ receptors are widely distributed in the brain, including cerebral and cerebellar cortex, hippocampus, caudate nucleus, nucleus accumbens, the basal ganglia, subthalamic nucleus, and midbrain (Moran-Jimenez & Matute, 2000; Simon et al., 1997). $P2Y_1$ receptors seem to be the dominant P2Y receptors in neurons but they are also critically involved in astrocyte functions (Illes & Ribeiro, 2004; Verkhratsky et al., 2009). $P2Y_2$ receptors are expressed at lower levels in all regions of the human brain, but they are the other prominent P2Y receptor type involved in ATP-induced signaling in astrocytes. Similarly, $P2Y_4$, $P2Y_6$, $P2Y_{11}$, $P2Y_{12}$, $P2Y_{13}$, and $P2Y_{14}$ receptors were detected in both neuronal and glial cell types of the CNS (Moore et al., 2001; Verkhratsky et al., 2009). Microglia also express multiple P2Y receptor subtypes ($P2Y_1$, $P2Y_2$, $P2Y_{2/4}$, $P2Y_6$, and $P2Y_{12}$); especially, the $P2Y_6$ and $P2Y_{12}$ receptors seem to be functionally important (Boucsein et al., 2003; Inoue, 2008; Sasaki et al., 2003).

VII. Role of P2 Receptors in Neuronal and Glial Functions in the CNS

A. P2X Receptor-Mediated Synaptic Currents in the CNS

The widespread CNS distribution of P2X receptors on neurons supports the role for extracellular ATP acting as a fast neurotransmitter. Postsynaptic P2X receptor-mediated fast synaptic currents have been first described in the medial habenula (Edwards et al., 1992). Then, ATP-induced currents have been registered in various CNS regions including spinal cord (Bardoni et al., 1997), locus coeruleus (Nieber et al., 1997), hypothalamic arcuate nucleus (Wakamori & Sorimachi, 2004), hippocampal CA1 (Pankratov et al., 1998), CA3 regions (Mori et al., 2001), and somatosensory cortex (Pankratov et al., 2002). A typical situation is that ATP-mediated synaptic currents account for a 5–15% of total excitatory postsynaptic currents (EPSCs) largely mediated by glutamate (Abbracchio et al., 2009; Pankratov et al., 2007). The exception from this rule is the medial habenula, where EPSCs solely mediated by ATP have been observed (Abbracchio et al., 2009; Robertson et al., 2001).

ATP-induced currents in neurons of medial habenula and somatosensory cortex have a very high Ca^{2+} permeability, and P2X receptors may be an important route for Ca^{2+} influx at resting membrane potentials, when NMDA receptors are blocked by Mg^{2+} (Abbracchio et al., 2009; Pankratov et al., 2002). Therefore, the ATP component can bring heterogeneity to the function of excitatory synapses, given that P2X receptors provide significant calcium entry which does not require postsynaptic depolarization such as in case of the NMDA receptors. The calcium entry itself may strengthen the synaptic connection over a longer period of time (long-term potentiation, LTP or long-term depression, LTD), particularly, if the changes in intracellular calcium can lead to the insertion of either P2X or AMPA receptors in the membrane (Pankratov et al., 2007). Indeed, activation of P2X receptors has been implicated in regulation of synaptic plasticity including the control of LTP and LTD in various brain regions (Pankratov et al., 2009; see Section VIII.A).

The postsynaptic P2X receptors may interact with several ionotropic receptors thereby modulating the ion flux through the separate ligand-gated channels. The current occlusion between ionotropic nucleotide-gated and other channels was first reported in the early 1990s, that is, ATP- and nicotine-activated currents were nonadditive in several neuronal cell types (Nakazawa, 1994). Subsequently, it was elaborated that P2X receptors may interact not only with the nicotinic ACh channels but also with other members of the Cys-loop family, that is, with the $GABA_A$ and $5\text{-}HT_3$ receptor channels. Further, most experiments devoted to this issue indicated that the physical interaction occurs at the intracellular domains, between the large cytoplasmic loop of the Cys-loop channels and the C-terminus of the P2X receptors. Nevertheless, the mechanism of this interaction probably cannot be completely unified and simplified because besides the physical association of the receptors, activity-dependent changes may also play an important role in current occlusion (Boué-Grabot et al., 2003; Khakh et al., 2005; Sokolova et al., 2001). These interactions might contribute to the diversity of both the postsynaptic and the presynaptic P2X-mediated actions.

B. Neuromodulation via P2 Receptors

Besides the discovery of the transmitter role of ATP in the CNS, it was revealed that P2X and P2Y receptor subtypes are found at presynaptic sites at the synapses as well, and the activation of these presynaptic P2X or P2Y receptors can modulate the release of the major neurotransmitters (ACh, norepinephrine, dopamine, serotonin, glutamate, GABA) (Cunha & Ribeiro, 2000; Sperlágh & Vizi, 1991; Sperlágh et al., 2007; von Kügelgen et al., 1989; Westfall et al., 1990). Further, activation of postsynaptic P2 receptors can also result in modulation of the effects of neurotransmitters (Fischer & Krügel, 2007; Hussl & Boehm, 2006; Illes & Ribeiro, 2004).

1. Neuromodulation via Presynaptic P2 Receptors

Since P2X receptors have relatively high Ca^{2+} permeability (see above), they can facilitate Ca^{2+}-dependent neurotransmitter release. Indeed, ATP or its analogs acting via P2X receptors have been reported to increase ACh release, although these data refer to peripheral neurons or to the neuromuscular junction (e.g., Deuchars et al., 2001; Sperlágh & Vizi, 1991). It is unclear whether such facilitatory P2X receptors also exist on the terminals of central cholinergic neurons (Sperlágh et al., 2007).

The presynaptic facilitatory action of ATP on noradrenergic transmission was described not only in peripheral neurons (e.g., Boehm, 1999; Sperlágh & Vizi, 1991; Sperlágh et al., 2000) but also in several brain areas such as locus coeruleus (Fröhlich et al., 1996) and hippocampus (Papp et al., 2004).

Numerous studies demonstrated the facilitatory effect of P2X receptor activation on glutamate release in central synapses, including spinal cord (Gu & MacDermott, 1997; Li & Perl, 1995; Nakatsuka & Gu, 2001), brain stem nuclei such as nucleus tractus solitarii, nucleus ambiguous (Jameson et al., 2008; Jin et al., 2004; Khakh & Henderson, 1998; Shigetomi & Kato, 2004; Watano et al., 2004), hippocampus (Rodrigues et al., 2005; Sperlágh et al., 2002), and cortical synaptosomes (Patti et al., 2006).

Similarly, P2X receptor activation in various CNS regions has been reported to facilitate the release of the inhibitory transmitter GABA, for instance, in the spinal cord (Hugel & Schlichter, 2000), brain stem (Watano et al., 2004), midbrain synaptosomes (Gomez-Villafuertes et al., 2001), and hippocampus (Aihara et al., 2002; Sperlágh et al., 2002). Glycin release was also augmented via presynaptic P2X receptor activation in the spinal cord (Rhee et al., 2000) and the trigeminal nucleus (Wang et al., 2001).

The presynaptic metabotropic receptors are generally considered as inhibitory modulators; however, in some cases, the P2Y receptor subtypes exert stimulatory presynaptic influence on other transmitter systems. For instance, ATP has been reported to facilitate dopamine release via P2Y receptors in the striatum (Zhang et al., 1995), or the nucleus accumbens (Krügel et al., 1999, 2001a). Interestingly, the activation of P2Y receptors is also implicated to elicit and potentiate glutamate release in the medial habenula nucleus (Price et al., 2003). In addition to P2X receptors, activation of $P2Y_1$ receptors has also been reported to increase the inhibitory postsynaptic current (IPSC) frequency (a sign of GABA release) in hippocampal slices (Kawamura et al., 2004).

Nevertheless, other reports demonstrated inhibitory effects via presynaptic P2Y receptors. Early publications showed that ATP inhibits ACh release in the peripheral nervous system, although this effect was probably due to the enzymatically generated adenosine (Sperlágh et al., 2007; Vizi & Knoll, 1976). Recently, the involvement of $P2Y_1$ receptors in the inhibition of ACh release was shown in neuromuscular junctions (De Lorenzo et al., 2006). Inhibitory P2 receptors involved in the modulation of ACh release have been demonstrated in the CNS (rat cerebral cortex) as well (Cunha et al., 1994).

Similarly, several studies indicated the presynaptic inhibitory role of ATP via P2Y or P2Y-like receptors on noradrenergic transmission in the peripheral nervous system (Sperlágh et al., 2007). Regarding the CNS, similar inhibitory P2Y receptors have also been reported in the rat brain cortex (von Kügelgen et al., 1994) and hippocampus (Koch et al., 1997). In the CNS, ATP was also shown to inhibit the release of serotonin (von Kügelgen et al., 1997) and dopamine (Trendelenburg & Bültmann, 2000) via activation of P2Y receptors.

ATP has been reported to inhibit glutamate release by acting at P2Y receptors in pyramidal neurons of hippocampal slices (Mendoza-Fernández et al., 2000); ATP and its metabolically stable analog ATP-γ-S inhibited depolarization-evoked glutamate release from rat brain cortical slices (Bennett & Boarder, 2000). Rodrigues et al. (2005) demonstrated that single hippocampal pyramidal neurons express $P2Y_1$, $P2Y_2$, and $P2Y_4$ receptors, and the release of glutamate is inhibited by these receptors. A glia-driven synaptic depression was observed in hippocampal cell cultures, and it was partly mediated by ATP itself acting on P2Y receptors and partly by adenosine acting on A_1 adenosine receptors (Koizumi et al., 2003). A similar mechanism has also been demonstrated in intact hippocampal slices, where ATP released from neurons and astrocytes acted on $P2Y_1$ receptors to excite interneurons, resulting in increased synaptic inhibition within intact hippocampal circuits (Bowser & Khakh, 2004).

The activation of P2Y receptors causes blockade of the N-type calcium channels in DRG cells (Borvendeg et al., 2003), and this effect may decrease the release of glutamate from DRG terminals in the spinal cord and thereby partly counterbalance the algogenic effect of ATP (Gerevich & Illes, 2004; Gerevich et al., 2004).

According to our present knowledge, there is no information regarding whether the release of GABA and other inhibitory amino acids is subject to modulation by inhibitory P2 receptors (Sperlágh et al., 2007).

It is widely accepted that glia are active partners at the synapse, dynamically regulating synaptic transmission (Newman, 2003). Glial cells may regulate synaptic transmission and modulate neuronal activity by releasing neuroactive substances including ATP. Conversely, ATP may act at glial P2 receptors influencing glial functions. The role of ATP as gliotransmitter involved in the control of neuronal or neuroglial circuits will be discussed later.

2. *Neuromodulation at the Postsynaptic Level*

Several data indicate that the postsynaptic P2 receptors may also be involved in the neuromodulatory role of purines. For instance, $P2Y_1$ receptor activation in cerebellar Purkinje cells postsynaptically increased the colocalized $GABA_A$ receptor-sensitivity through G protein-coupled intracellular Ca^{2+} ($[Ca^{2+}]_i$) elevation (Saitow et al., 2005). In cultured striatal neurons of the rat, ATP was reported to release Ca^{2+} from an intracellular pool by $P2Y_1$ receptor stimulation (Rubini et al., 2006). The mixed D_1/D_2 receptor agonist dopamine increased the ATP-induced $[Ca^{2+}]_i$ transients in a subpopulation of neurons (Rubini et al., 2008). At the same time, dopamine did not affect the response to K^+ in these cells, excluding a modulation of voltage-sensitive ion channels. Selective D_1 (SKF 83566) and D_2 (sulpiride) receptor antagonists failed to modify the effect of ATP but unmasked in the previously unresponsive neurons an inhibitory and facilitatory effect of dopamine, respectively. Thus, P2Y receptors facilitate via the increase of $[Ca^{2+}]_i$ the conductance of $GABA_A$ receptors, whereas D_1/D_2 receptors, in contrast, alter the $[Ca^{2+}]_i$ transients induced by P2Y receptor stimulation.

A complex interaction between multiple P2Y receptor subtypes and ionotropic glutamate NMDA receptors in the prefrontal cortex (PFC) involving both direct and indirect mechanisms was also revealed. On the one hand, $P2Y_1$ receptor activation via a membrane-delimited, direct postsynaptic cross talk inhibited NMDA receptor function (Luthardt et al., 2003), while on the other hand, ATP via $P2Y_4$ receptors indirectly facilitated the conductance of NMDA receptors at a subpopulation of PFC pyramidal neurons. This latter effect involved ATP-induced glutamate release from astrocytes, and this glutamate via metabotropic glutamate receptors facilitated the postsynaptic NMDA-mediated currents (Wirkner et al., 2002, 2007a).

Potentiating interactions between ATP via G_q-coupled P2Y receptors and TRPV channels were observed in the nociceptive system (Lakshmi & Joshi, 2005; Moriyama et al., 2003; Tominaga et al., 2001). Further, P2 receptors have been reported to interact with each other in this system. A G_q protein-dependent inhibitory interaction was reported between $P2Y_1$ and P2X3 receptors in DRGs (Gerevich et al., 2005). More recently, it has been revealed that the underlying mechanism of this interaction is a G protein-dependent facilitation of the desensitization of P2X3 receptors and the suppression of recovery from the desensitized state (Gerevich et al., 2007b).

P2Y receptor activation might result in the inhibition of several types of voltage-activated Ca^{2+} channels (for review, see Köles et al., 2008). In neuronal structures such as DRG cells, inhibition of N-type Ca^{2+} channels was reported (e.g., Borvendeg et al., 2003; Gerevich et al., 2004). Regarding the role of N-type Ca^{2+} channels in transmitter release, these receptors might be involved in the presynaptic inhibitory function of ATP/ADP described earlier.

Agonists of various G protein-coupled P2Y receptors have been shown to inhibit the M-type potassium current in various neuronal cell types including hippocampal neurons. Closure of this channel facilitates membrane depolarization and fast excitatory transmission. Recent data indicated that PLC-dependent PIP_2 depletion of the cell membrane might be an essential step in the P2Y receptor-mediated inhibition of M-type currents (Filippov et al., 2006; Nakazawa et al., 1994).

Several reports demonstrated that P2Y receptor agonists activate outwardly rectifying potassium channels in neurons of various brain regions (for instance, striatal or hippocampal neurons). This interaction seems to involve a pertussis toxin (PTX)-resistant G protein. Either the $\beta\gamma$-subunit of this G protein interacts with the channel by a membrane-delimited mechanism, or its α-subunit initiates a signaling cascade involving cytoplasmic second messengers (Ikeuchi & Nishizaki, 1996; Ikeuchi et al., 1996).

Cross talk of P2X receptors with different ion channels such as Cys-loop ionotropic receptors were described earlier. These are mostly inhibitory interactions such as current occlusion (Köles et al., 2008). Such phenomena might be involved in both presynaptic and postsynaptic effects of purines.

C. ATP as a Gliotransmitter

Glial cells have traditionally been viewed as passive elements in the CNS, providing structural and metabolic support to neurons but playing little role in information processing. However, it has been demonstrated that neurons and glia are not independent cellular entities of the nervous system but strongly interconnected components.

1. Role of ATP in Glia–Glia Communication

Information processing is not an exclusive property of neurons but it is shared by astrocytes, the most abundant glial cells in the CNS. Astrocytes, besides their other functions (biochemical support of endothelial cells that form the blood–brain barrier, supply of nutrients to the nervous tissue, maintenance of extracellular ion balance, etc.), may directly modulate neuronal activity by releasing neuroactive substances, and they should be considered as active partners at the synapse (Newman, 2003). They participate in physiological events such as neuronal development and synaptic activity. Further, astroglial cells respond to brain injuries with reactive gliosis, characterized by astrocytic proliferation and hypertrophy, responses that can ameliorate brain damage from injury but paradoxically may contribute to neuronal cell death (Ciccarelli et al., 2001; James & Butt, 2002).

ATP is one of the most important glial signaling molecules. It was reported to be released from astrocytes via exocytosis or diffusion through hemichannels, pannexins, volume-regulated channels, or P2X7 receptors (Bowser & Khakh, 2007; Coco et al., 2003; Guthrie et al., 1999); a stimulus-dependent outflow of ATP from astrocytic lysosomes is a further possibility (Zhang et al., 2007). More recently, the exocytotic release of ATP from astrocytes has been suggested to be a property of cultured glial cells only, with probably no relevance for their mature counterparts in the brain (Fiacco et al., 2009; Hamilton & Attwell, 2010). However, regulated

ATP release from astrocytes is certainly important in glial–glial signaling (propagation of Ca^{2+} waves within the astroglial syncytium), and in glial–neuronal reciprocal communications (modulation of transmitter release, regulation of synaptic plasticity) (Abbracchio et al., 2009; Verkhratsky et al., 2009).

Astrocyte function and activation regulated by extracellular nucleotides is dependent on the concentration of extracellular nucleotides and on the composition of P2 receptor subtypes expressed by the cell. During physiological signaling, small and transient increases in extracellular levels of ATP principally activate the respective glial receptors involved in physiological interglial signaling, coupling and coordination of glial and neuronal functions. However, in pathological states (after injury or under inflammatory conditions), massive release of ATP causes larger and more prolonged increases in intracellular Ca^{2+} which is sufficient to trigger further ATP release, to initiate and amplify the inflammatory responses—with the involvement of activated microglia—contributing to neural damage (James & Butt, 2002; Verkhratsky et al., 2009).

As it was briefly described earlier, astroglia express multiple P2 receptors. There is ample evidence for functional astrocytic $P2Y_1$ and $P2Y_2$ receptors (Idestrup & Salter, 1998; King et al., 1996), and astrocytes express functional $P2Y_4$, $P2Y_6$ and $P2Y_{12}$, and $P2Y_{13}$ and $P2Y_{14}$ receptors as well (Fischer et al., 2009; Fumagalli et al., 2003; Jimenez et al., 2000; Moore et al., 2003). Further, all P2X receptor subtypes except P2X6 were identified in rat astrocyte cultures (Fumagalli et al., 2003).

The consequence of the P2Y receptor activation in astrocytes can be PLC-dependent Ca^{2+} mobilization, that is, generation of $[Ca^{2+}]_i$ transients (Pearce et al., 1989). These Ca^{2+} elevations can propagate to neighboring astrocytes (Dani et al., 1992). Release of ATP from astrocytes may be an important mechanism contributing to the propagation of glial Ca^{2+} waves (Guthrie et al., 1999). $P2Y_1$ receptors located in astrocytes are involved in ATP-mediated Ca^{2+} wave propagation in various CNS regions (Fam et al., 2000; John et al., 2001; Salter & Hicks, 1994; Verkhratsky et al., 2009). Both $P2Y_1$ and $P2Y_2$ receptors have been reported to be involved in propagation of Ca^{2+} waves in response to released ATP in spinal cord astrocytes. However, Ca^{2+} waves propagated via $P2Y_2$ receptors travel faster and further than those propagated via $P2Y_1$ receptors. Further, they show differential frequency dependence. Therefore, alteration in the expression or function of these receptor subtypes may control the rate and extent of astrocytic Ca^{2+} waves (Fam et al., 2003; Gallagher & Salter, 2003).

Ca^{2+} waves in astroglial syncytium can be generated both by IP_3 diffusion through gap junctions and by a regenerative wave of ATP release. The contribution of these mechanisms differs between various brain regions. Further, remodeling of P2Y receptor profile and Ca^{2+} wave propagation was reported in certain conditions such as in response to IL-1β treatment or inhibition of connexin 43 synthesis (John et al., 1999; Verkhratsky et al., 2009).

In spite of the numerous reports about the expression of P2X subtypes in astrocytes, their functional role is not fully characterized (Verkhratsky et al., 2009). One of the remarkable P2X receptor types localized in cortical astrocytes is the P2X1/5 heteromer. It has a very high sensitivity to ATP (nanomolar concentrations), so it can detect extremely low amounts of extracellular ATP. It is a slowly desensitizing receptor type, and the cortical astrocyte seems to be the main localization of this receptor subtype (Lalo et al., 2008; Torres et al., 1998). Another peculiar P2X receptor type of the cortical astrocytes is the P2X7 receptor (Sperlágh et al., 2006). Recent reports clearly demonstrated that cortical and cerebellar astroglia of rats and mice possess functional P2X7 receptors (Carrasquero et al., 2009; Nörenberg et al., 2010; Oliveira et al., 2011). They can be mostly involved in the control of gliotransmitter release and the remodeling of astroglial, microglial, and neuronal functions in response to pathological stimuli. These functions will be detailed later.

2. Role of ATP in Bidirectional Glia–Neuron Communication

Besides the glia–glia communication, propagation of astroglial Ca^{2+} waves couples glial and neuronal functions as well (James & Butt, 2002). A cross talk between neurons and glial cells plays an important role in CNS functions. Glial cells seem to be involved in certain feedback mechanisms. They reply to neuronal activity by increasing their intracellular Ca^{2+} which subsequently stimulates the release of agents from the glia themselves to modulate the neuronal activity (Araque et al., 1999). Glutamate and ATP are suggested to be important mediators in the cross talk between neurons and glial cells (Gebicke-Haerter et al., 1988; Koizumi et al., 2003).

ATP acting via P2Y or P2X7 receptors is one of the major regulators or triggers of gliotransmitter release from astrocytes. It results in the modulation of synaptic transmission of the brain, and it fits to the widely acknowledged modulatory role of ATP. In many cases, the modulatory influence of ATP on synaptic strength occurs not by neuronal presynaptic action, but involves regulation of transmitter release of glial origin.

Indeed, ATP via P2Y receptors facilitates glutamate release from astrocytes in various brain regions or in the spinal cord (Domercq et al., 2006; Wirkner et al., 2007a; Zeng et al., 2009). Similarly, several reports showed that opening of P2X7 channels triggers release of gliotransmitters such as glutamate, GABA, or ATP itself from astrocytes in various brain regions (Duan et al., 2003; Fellin et al., 2006a; Khakpay et al., 2010; Sperlágh et al., 2006; Suadicani et al., 2006). Further, astrocytes in the brainstem released ATP in response to a decrease in pH (Gourine et al., 2010). This ATP led to propagated astrocytic $[Ca^{2+}]_i$ excitation and eventually to a $P2Y_1$ receptor-mediated activation of respiratory neurons; a low pH of the blood is normally a consequence of an elevated partial pressure of pCO_2 known to be the physiological stimulus of the respiratory neuronal network. It is noteworthy that astrocytic ATP released by any of the above mechanisms can be interconverted to adenosine. Considering that glutamate and ATP can be coreleased, transmitters of astrocytic origin may provide a coordinated regulation of the synaptic transmission where glutamate plays the excitatory role, and adenosine is the inhibitory substance (Fellin et al., 2006b; Verkhratsky et al., 2009).

3. Role of ATP in Astrocytes under Pathophysiological Conditions

As a response to CNS damage, astrocytes proliferate, release inflammatory mediators, and are involved in the inflammatory response as well (James & Butt, 2002). ATP stimulates the proliferation of primary astrocytes (Abbracchio et al., 1994). Tissue injury or administration of ATP analogs induces astrogliosis (Abbracchio et al., 1999; Franke et al., 2001a; Verkhratsky et al., 2009). Purine-induced gliosis involves the activation of a P2Y receptor linked to MAPK/ERK and cyclooxygenase-2 (Brambilla et al., 2002; Neary et al., 1999). IL-1β enhances ATP-evoked release of arachidonic acid, mediated through $P2Y_2$ receptors in mouse astrocytes (Stella et al., 1997). $P2Y_1$ (and P2X7) receptors mediate the effects of ATP on IL-1β-induced transcription factors, NF-κB and activator protein-1 activation (John et al., 2001). Activation of the $P2Y_{12}$ receptor has been reported to inhibit cAMP-dependent differentiation and thereby to switch from differentiation into enhanced proliferation in rat C6 glioma cells (van Kolen & Slegers, 2004).

P2X7 receptors also contribute to pathological events; probably, they are even upregulated in the diseased tissue (Franke et al., 2001b, 2004; Verkhratsky et al., 2009). Prolonged activation of P2X7 receptors leads to a sustained glutamate release, contributing to the exacerbation of pathological events (Fellin et al., 2006a). Activation of P2X7 receptors in astroglial cells also decreased glutamate uptake and reduced the expression and activity of glutamine synthetase (Lo et al., 2008). High concentrations of ATP, acting through P2X7 receptors in astroglia, increased the production of endocannabinoid 2-arachidonoylglycerol, modulated the release of tumor necrosis factor α (TNFα), stimulated nitric oxide production, induced phosphorylation of Akt and p38MAPK/ERK1/ERK2, stimulated transmembrane transport of nicotinamide adenine dinucleotide (NADH), and regulated NF-κB signaling (Gendron et al., 2003; Jacques-Silva et al., 2004; Kucher & Neary, 2005; Lu et al., 2007; Murakami et al., 2003; Verkhratsky et al., 2009; Walter et al., 2004). Further, stimulation of P2X7 receptors increased the production of leukotrienes (Ballerini et al., 2005). P2X7 receptors are also involved in the control of expression of other receptors and channels; in particular, P2X7 stimulation upregulates expression of $P2Y_2$ receptors (D'Alimonte et al., 2007;

Verkhratsky et al., 2009). Altogether, the activation of P2Y and P2X7 receptors is associated with astroglial responses to brain lesions, and they are a part of the functional remodeling accompanying astrogliosis and neuroinflammation (Verkhratsky et al., 2009).

At the same time, some P2Y receptors may be involved in glio- and neuroprotective events. $P2Y_6$ receptors expressed in a human astrocytoma cell line are suggested to interact with the TNFα-related intracellular signals to prevent apoptotic cell death (Kim et al., 2003). Expression of leukemia inhibitory factor, a cytokine involved in the survival and differentiation of the neuronal cells, is intensely induced by ATP in astrocytes. $P2Y_2$ and $P2Y_4$ receptors are involved in this effect (Yamakuni et al., 2002). $P2Y_2$ receptors were also reported to upregulate antiapoptotic proteins and activate several cell survival mechanisms in astrocytes (Chorna et al., 2004). $P2Y_1$ receptor activation was reported to enhance neuroprotection by astrocytes against oxidative stress via IL-6 release in hippocampal cultures (Fujita et al., 2009). P2Y receptor activation enhanced the neuroprotection in old mice by increasing astrocyte mitochondrial metabolism (Wu et al., 2007).

Taken together, astrocyte function and activation regulated by extracellular nucleotides is probably dependent on their concentration and on the composition of P2 receptor subtypes expressed by the cell. During physiological signaling, small and transient increases in extracellular levels of ATP principally activate glial P2Y and rapidly desensitizing P2X receptors. Activation of $P2Y_1$ and $P2Y_2$ receptors evoke propagated Ca^{2+} waves that serve as interglial signaling, and couple/coordinate glial and neuronal functions. After injury or under inflammatory conditions, larger and more prolonged increases in extracellular ATP cause larger and more prolonged increases in intracellular Ca^{2+} which is sufficient to trigger further ATP release, to initiate and amplify the inflammatory response including proliferation of astrocytes, and activation of microglia (see Section VII.C.5) resulting in excessive inflammatory events and cell death (James & Butt, 2002). Nevertheless, extracellular nucleotides can stimulate autocrine or paracrine signaling pathways, permitting modulation of the inflammatory response as well. The P2 receptor system probably constitutes a mechanism whereby activation of the proinflammatory signaling cascade can be coordinated with information from the extracellular environment (John et al., 2001; Liu et al., 2000).

4. *Role of ATP in Oligodendroglia*

Oligodendroglia, another type of neuroglial cells involved in the insulation of axons in the CNS, also express multiple P2 receptors. Although most P2X receptor subtypes are expressed in oligodendroglial precursor cells, in mature oligodendrocytes, mostly, the P2X7 receptors may be functionally important (Agresti et al., 2005; Verkhratsky et al., 2009). Activation of this receptor type was reported to trigger sustained inward currents and oligodendroglial death. It might be relevant in demyelinating diseases. P2X7 receptor expression is increased in multiple sclerosis, and P2X7 receptor antagonists prevented ATP-induced excitotoxicity in oligodendrocytes and inhibited demyelination in an animal model of multiple sclerosis (Matute et al., 2007).

Oligodendroglial precursor cells express $P2Y_1$, $P2Y_2$, and $P2Y_4$ receptors, but mostly the $P2Y_1$ seems to be functionally important. P2Y receptor activation may be involved in the control of migration and maturation of these cells (Agresti et al., 2005; Verkhratsky et al., 2009).

5. Role of ATP in Microglia

Although microglia, immune cells in the CNS, may play an important role against infection, overstimulation of this cell type is involved in pathological conditions as well. Ischemic brain damage, trauma, or neurodegenerative diseases are characterized by microglial activation, migration to the site of injury, release of proinflammatory substances (e.g., nitric oxide, superoxide radicals, and several kinds of cytokines), and phagocytosis of damaged cells. Microglial proliferation has also been observed after spinal cord damage (Inoue, 2008; Parvathenani et al., 2003; Tsuda et al., 2003).

Extracellular nucleotides have a central role in the regulation of microglial functions, and ATP is also secreted by microglia (Illes et al., 1996). Microglial cells are known to bear both P2X and P2Y nucleotide receptors. Under resting conditions, the functional expression of P2X7 as well as $P2Y_1$, $P2Y_2$, and $P2Y_4$ was dominant, but $P2Y_6$, $P2Y_{12}$, $P2Y_{13}$, and $P2Y_{14}$ receptors also contributed to the nucleotide responses. Lipopolysaccharide (LPS)-induced microglial activation has been reported to change the functional expression of P2 receptors; especially, the density of $P2Y_6$, $P2Y_{12}$, and P2X4 receptors was increased (Bianco et al., 2005; Kobayashi et al., 2008; Raouf et al., 2007).

Microglial responses are promoted by extracellular nucleotides (Ferrari et al., 1997). First of all, extracellular ATP functions as a chemoattractant, and $P2Y_{12}$ microglial receptors play a dominant role in the microglial chemotaxis (Haynes et al., 2006; Honda et al., 2001). Gene expression of $P2Y_{12}$ in spinal microglia has been reported to increase dramatically in a neuropathic pain model, and the increased $P2Y_{12}$ worked as a gateway of the following events in microglia after nerve injury (Kobayashi et al., 2008).

Neuronal injury results in the release or leakage of ATP, and it can be the danger signal from damaged neurons to microglia. In addition to microglial migration by ATP, another nucleotide, UTP, interconverted to UDP, an endogenous agonist of the $P2Y_6$ receptor, greatly activates the motility of microglia and orders microglia to engulf damaged neurons. Further, activation of $P2Y_6$ receptors that are upregulated in microglia (see previous paragraph) in response to neuronal injury results in phagocytosis. Thus, UDP could be a diffusible molecule that signals the crisis of damaged neurons to microglia and triggers phagocytosis (Inoue, 2008; Inoue et al., 2009; Koizumi et al., 2007).

The P2X7 receptor is involved in excessive inflammatory events and is the main candidate among P2 receptors to induce cell death in the immune cells. P2X7 receptors expressed in microglia (see above) may regulate cytokine production and early inflammatory gene expression. For instance, in microglial cells, stimulation of P2X7 receptors potently activates NF-κB of activated T cells, a central transcription factor involved in cytokine gene expression (Ferrari et al., 1999). ATP-activated ATP secretion and binding to the P2X7 receptor mediate IL-1β secretion from microglia (Ferrari et al., 1997). ATP stimulates *de novo* synthesis of TNFα via P2X7 receptors in a primary culture of rat brain microglia (Hide et al., 2000). P2X7 receptor activation increased the secretion of IL-1α, IL-1β, and IL-18 and reduced levels of IL-6 in fetal brain-derived preactivated microglia (Rampe et al., 2004). P2X7 receptors are also involved in the regulation of 2-arachydonoylglycerol—the most abundant endocannabinoid—production in microglial cells. Prolonged increases in 2-arachydonoylglycerol levels in brain are thought to play an important role in neuroinflammation (Witting et al., 2004). P2X7 receptors were shown to be upregulated in microglia in neuroinflammatory/neurodegenerative disorders (McLarnon et al., 2006; Parvathenani et al., 2003).

In contrast, deletion of P2X7 receptors or administration of P2X7 antagonists was not cytoprotective in mice *in vivo* (Le Feuvre et al., 2003). P2X7 receptor-activated microglia was reported to release TNFs protecting neurons against glutamate toxicity (Suzuki et al., 2004). Nevertheless, P2X7 receptors, especially those located on microglia, are hot candidates as mediators of pathological neuroinflammatory events and cell death.

Recently, a role for purinergic microglial receptors in tactile allodynia after nerve injury has been suggested. Following nerve injury, expression of P2X4 was reported to increase strikingly in hyperactive microglia (but not in neurons or in astrocytes) in the spinal cord (Tsuda et al., 2003). Pharmacological blockade of spinal P2X4 receptors reversed tactile allodynia caused by peripheral nerve injury without affecting acute pain behaviors in rats. Conversely, intraspinal administration of microglia, in which P2X4 had been induced and stimulated, produced tactile allodynia in these animals.

Taken together, these findings suggest that microglial P2X4 and P2X7 receptors may represent novel interesting therapeutic targets for pain and inflammatory neurological diseases characterized by abnormal microglial response (Jarvis, 2010; Tsuda et al., 2003). Functional interactions are likely to exist between the P2X4 and P2X7 receptors in microglia, and the possible heteromerization of P2X4 and P2X7 receptors, reported recently in mouse macrophages, also might have pathophysiological importance (Guo et al., 2007).

VIII. Involvement of the P2 Receptors in Physiological and Pathophysiological CNS Functions—Possible Therapeutic Consequences

A. Synaptic Plasticity and Cognitive Functions

LTP and LTD are forms of synaptic plasticity, and potential mechanisms for memory formation and learning. Activation of P2X and P2Y receptors has been implicated in regulation of synaptic plasticity including the control of LTP and LTD in various brain regions (Fujii et al., 1999; Guzman et al., 2010; Pankratov et al., 2009; Saitow et al., 2005). The data are controversial, including reports about the facilitation, or inhibition of LTP, or no effect at all in the same brain region (hippocampus) by P2X receptor activation (Pankratov et al., 2009; Sim et al., 2006; Wang et al., 2004). These discrepancies might reflect the complex nature of LTP induction and/or maintenance (Abbracchio et al., 2009).

B. Motivation, Behavior and Reward, Psychiatric Disorders

The mesolimbic–mesocortical dopaminergic pathway is considered to play a number of roles in behavioral and psychiatric functions such as locomotion, motivation, reward, and the pathogenesis of schizophrenia. P2 receptors in the nucleus accumbens play an essential role in the mediation of locomotion and reward processes. The neuronal activity of the nucleus accumbens and thereby the mesolimbic dopaminergic system is positively modulated by P2 purinoceptors (Krügel et al., 2003).

Both P2 receptor agonists and antagonists increased the locomotor response in an open field situation in rats, but the characteristics of these changes were different. In the case of P2 receptor activation, the locomotion induced by a new environment was longer lasting and more consistent, while the blockade of P2 receptors led to an increased running speed accompanied by more stops and more changes of the movement direction (Kittner et al., 2004a).

Food intake was reduced after microinjection of PPADS into the nucleus accumbens. Stimulation of hypothalamic $P2Y_1$ receptors enhanced food intake in rats in the mammalian brain (Kittner et al., 2004b, 2006). Recently, P2X2 receptor immunoreactivity was observed in hypothalamic neurons associated with the regulation of food intake (Colldén et al., 2010).

It has been shown that endogenous ATP is involved in mediating amphetamine-induced sensitization and reward-motivated behavior (Kittner et al., 2001; Krügel et al., 2001a, 2001b). Repeated amphetamine-treatment has been reported to enhance $P2Y_1$ receptor expression (Franke et al., 2003).

Anxiolytic-like effects were also reported in rats after stimulation of $P2Y_1$ receptors (Kittner et al., 2003), and the P2X7 receptor gene was implicated in anxiety disorders (Erhardt et al., 2007). Genetic analysis of a French population highlighted a Gln640Arg single-nucleotide polymorphism of the P2X7 receptor gene located at the C-terminal domain of this receptor as a potential susceptibility gene for bipolar affective disorder (Barden et al., 2006) and major depression (Lucae et al., 2006). In support of this assumption, when the behavioral profile of P2X7 gene knockout mice was examined in models of depression and anxiety, the authors found an antidepressant-like phenotype together with a higher responsiveness to a subefficacious dose of the antidepressant imipramine (Basso et al., 2009).

Proper functioning of the PFC is necessary for higher order cognitive functions, such as attention, memory, and learning. ATP-activating $P2Y_4$ receptors located on astrocytes released glutamate. This glutamate acting on metabotropic glutamate (mGlu) receptors positively interacted with NMDA receptors, and thereby facilitated the monosynaptically evoked EPSC amplitudes in the PFC (Wirkner et al., 2007a). However, activation of the $P2Y_1$ receptor subtype inhibited NMDA receptor channels (Luthardt et al., 2003), suggesting that different P2Y receptors in the PFC may be involved in fine tuning of cognitive functions.

These data provide strong evidence that purinergic transmission plays an essential role in the modulation of the mesolimbic dopaminergic system and thereby in locomotion, motivation, feeding behavior, certain memory functions, and reward. It might provide new therapeutic strategies for the pathological changes of the mesolimbic dopaminergic functions in the future.

C. Pain

ATP released from different cell types is involved in the initiation of pain by acting on P2 receptors on sensory nerve terminals (Abbracchio et al., 2009; Burnstock, 1996). Although the P2Y receptor subtypes are also involved in pain processes (e.g., Andó et al., 2010; Borvendeg et al., 2003; Gerevich et al., 2004, 2007b), the major purinoceptors involved in nociception are the P2X, especially the P2X3 and the P2X2/3 subtypes (Abbracchio et al., 2009; Chizh & Illes, 2001). These receptors detect extracellular ATP and initiate pain at the periphery, and they are involved in the transmission and modulation of sensory inputs in the spinal cord (Chizh & Illes, 2001). These receptors are potential therapeutic targets for the management of pathological pain conditions. It is suggested that either an accumulation of ATP in the periphery or a functional upregulation of P2X3 and P2X2/3 receptors is underlying mechanisms of chronic pain and hyperalgesia (Barclay et al., 2002; Jarvis, 2010; Souslova et al., 2000; Wirkner et al., 2007b). P2X3 and P2X2/3 receptors appear to play overlapping but slightly different roles in nociception. Homomeric P2X3 receptors mainly mediate transient, while heteromeric P2X2/3 mainly mediate sustained nociceptive responses (Cockayne et al., 2005).

While P2X3 and P2X2/3 receptors on neuronal cells are the main targets for exploring P2X receptor functions in pain, other P2X receptor subtypes (P2X4 and P2X7), located on nonneuronal cells, might also be involved in pain mechanisms (Chessell et al., 2005; Tsuda et al., 2005). Injury of primary sensory neurons produces long-lasting abnormal hypersensitivity to normally innocuous stimuli, a phenomenon known as tactile allodynia. Spinal administration of P2X4 receptor stimulated microglia caused allodynia, while blockade of the P2X4 receptors reversed tactile allodynia after spinal nerve injury (Tsuda et al., 2003). Inflammatory and neuropathic hypersensitivity to both mechanical and thermal stimuli are completely absent in P2X7 knockout mice while normal nociceptive processing is preserved (Chessell et al., 2005). Interestingly, spinal microglial P2X7 receptors were involved in the induction but not in the maintenance of morphine tolerance, and the blockade of P2X7 receptors prevented the development of morphine tolerance (Zhou et al., 2010). The new P2X2/3 or P2X3 as well as P2X7 antagonists might provide a new therapeutic strategy in certain pain modalities (see Section VII.C.5; Jarvis, 2010).

Recently, the microglial $P2Y_{12}$ receptor was also implicated in neuropathic pain. Pharmacological blockade of $P2Y_{12}$ receptors prevented the development or produced a striking alleviation of existing tactile allodynia (Kobayashi et al., 2008; Tozaki-Saitoh et al., 2008). In contrast, robust expression of G_i-coupled P2Y receptors ($P2Y_{12}$, $P2Y_{13}$, and $P2Y_{14}$) was reported in sensory neurons and their activation inhibited nociceptive signaling in isolated neurons; agonists for these receptors reduced behavioral hyperalgesia *in vivo* (Malin & Molliver, 2010).

D. CNS Injury, Ischemia, Neuroinflammation, and Neurodegenerative Disorders

During physiological conditions, the concentration of extracellular ATP is in the micromolar range. However, under pathological conditions such as ischemia and injury, it can reach millimolar concentrations. At lower levels ($\leq 100\ \mu M$), ATP acts via P2Y and P2X1–6 receptors to promote physiological functions. However, due to a higher ATP concentration (>1 mM), especially activation of P2X7 receptors in microglia can mediate inflammatory responses initiated at sites of damage (Fields & Stevens, 2000; Verkhratsky et al., 2009; see Section VII.C.5).

ATP may play a key role in survival, repair, and remodeling in the nervous system. Purines are involved in neurite outgrowth and regeneration. ATP stimulates the synthesis and release of protein trophic factors and can act in combination with growth factors to stimulate astrocyte proliferation and contributes to the process of reactive gliosis, a hypertrophic/hyperplastic reaction, which enables the injured brain to restore its damaged functions (D'Ambrosi et al., 2001; Franke & Illes, 2006).

However, ATP can mediate not only regeneration and survival but also cell death, apoptotic and necrotic features of degeneration. P2 receptors can mediate and aggravate toxic signaling in many CNS neurons (Amadio et al., 2002; Cavaliere et al., 2002; Volonte et al., 2003). Besides ionotropic glutamate receptors, P2X receptors can also contribute to neuronal injury (Franke & Illes, 2006). Interplay between the purinergic and glutamatergic system was also reported; P2X (e.g., P2X7) receptor activation facilitated glutamate release (Duan et al., 2003; Fellin et al., 2006a; Sperlágh et al., 2002).

Especially, P2X7 receptors appear to be important elements in the mechanisms of cellular damage induced by hypoxia/ischemia. Their role in apoptosis or other pathological events was repeatedly pointed out, but other receptors such as P2X4 were also implicated (Cavaliere et al., 2003; Franke & Illes, 2006; Franke et al., 2004; Sperlágh et al., 2006; see Sections VII.C.5. and VIII.C).

Since P2X7 receptor activation-dependent mechanisms may contribute to inflammatory responses observed in neurodegenerative and autoimmune/neuroinflammatory diseases, P2X7 receptor antagonists could have therapeutic utility in the treatment of CNS injury, hypoxia, or neurodegenerative/neuroinflammatory disorders such as Alzheimer's disease, Parkinson's disease, or multiple sclerosis (Abbracchio et al., 2009; Friedle et al., 2010; Skaper et al., 2010). In contrast, a recent report provided evidence for a trophic role of P2X7 receptors in microglia and raised the question whether inhibition of the P2X7 receptor can be a real therapeutic avenue for the inhibition of neuroinflammation (Monif et al., 2009).

Recently, GPR17, a P2Y-like receptor responding to both uracil nucleotides and leukotrienes, was shown to act as a "sensor" that is activated upon brain injury playing a crucial role in the early phases of tissue damage and orchestrating the local remodeling/repair response. It also seems to play a role in oligodendrocyte differentiation, and it was suggested as a potential therapeutic target after spinal cord injury or in case of demyelinating neurodegenerative conditions such as multiple sclerosis (Ceruti et al., 2009; Lecca et al., 2008).

Several P2 receptors (e.g., P2X1, P2X2, P2X4, P2X7, and $P2Y_1$ receptors) are upregulated in neurons or astroglial cells in various brain regions after ischemia, or tissue damage (Cavaliere et al., 2003; Florenzano et al., 2002; Franke & Illes, 2006; Franke et al., 2001b, 2004; James & Butt, 2001). Therefore, not only the extracellular ATP concentration but also changes in the P2 receptor subunit expression may be important factors of purinergic signaling after CNS injury. Extracellular ATP seems to fulfill a regulatory role in response to CNS damage. The extent of reactive gliosis, the modulatory influences on the production of cytokines and trophic factors in ischemia or trauma, may be dependent on the activation of different P2 receptor subtypes by different concentrations of extracellular ATP. It can explain how ATP is involved in mechanisms of both neuronal degeneration and repair.

E. Epilepsy

The intraventricular injection of high doses of ATP in rats evoked severe clonic–tonic convulsions. Unilateral microinjection of nucleotides into the rat prepiriform cortex caused a convulsant response antagonized by suramin (Knutsen & Murray, 1997). Alterations of expressions of P2X receptor subtypes (P2X2, P2X4, and P2X7) were repeatedly reported in various brain regions in animal models of epilepsy (Doná et al., 2009; Kang et al., 2003; Kim et al., 2009; Vianna et al., 2002). Prolonged inflammatory responses in the CNS may significantly contribute to the pathology seen in epilepsy (reactive gliosis, excitotoxicity, and cell death). Purine (P2X7)-mediated microglial responses may contribute to neurodegenerative consequences during status epilepticus (Rappold et al., 2006). Further, status epilepticus induced a particular microglial activation state, characterized by enhanced purinergic signaling (Avignone et al., 2008). Blockade of P2X7 receptors prevented astroglial death in the dentate gyrus following pilocarpine-induced status epilepticus (Kim et al., 2009).

F. Embryonal and Early Postnatal Development

Abundant and dynamic expression of P2X and P2Y receptors during embryonic and postnatal development has been repeatedly demonstrated. Purinergic control may be one of the earliest to develop in the embryos of higher species. Different purinoceptors may participate in neurite outgrowth (involving P2X3), postnatal neurogenesis (related to P2X4 and P2X5), and cell death (possibly involving P2X7 receptors) (Cheung et al., 2003, 2005; Heine et al., 2007).

Especially, the P2X3 receptor subtype shows an abundant distribution in embryonic and neonatal brain, in contrast to its limited presence in adult brain. ATP may function as a signaling molecule to inhibit motor axon outgrowth in the embryonic neural tube, most likely via P2X3 receptors (Cheung et al., 2005). The transient functional expression of P2X3 receptors on neurons may be related to synapse formation (Dunn et al., 2005).

Based on the role of purinergic receptors in embryonal and early postnatal development, it cannot be excluded that alterations of the purinergic regulation of embryonal growth might be involved in the onset of morphological malformations. This possibility must be very carefully considered in the design and development of potential P2 receptor ligands and in their proper indications and contraindications for potential drugs.

IX. Conclusion

P2 purinoceptors exhibit a wide distribution in the CNS both in neuronal and nonneuronal cells. Although they appear to be involved in the modulation of diverse functions primarily mediated by the classic monoamine and amino acid transmitters, their functional significance is still enormous. Due to the polarity of most P2 receptor-ligands, they fail to pass the blood–brain barrier and have to be applied therefore by intracerebroventricular or intrathecal routes under experimental conditions. This is a serious limitation to their therapeutic applicability. However, new pharmaceutical formulations such as nanoparticle-based application systems as well as molecules with unchanged pharmacological profile but better permeability may circumvent this complicating factor. The inhibition of ecto-nucleotidases by small molecules and the ensuing manipulation of the endogenous levels of purine or pyrimidine nucleotide concentrations in the CNS are additional possibilities. Eventually, brain-permeable molecules may help to selectively up- or down-regulate certain P2X or P2Y receptors and thereby to exert pharmacological effects. Last but not least, not only drugs that have an immediate neuronal target but also interventions targeting neurons indirectly via neuro- or microglial modulatory functions have to be considered.

The number of neurological and psychiatric illnesses, where P2 receptors are involved in the etiology, is impressive and includes just as an example neuropathic pain, migraine, ischemia/stroke, neurodegenerative disorders (Alzheimer's and Parkinson's disease, amyotrophic lateralsclerosis), epilepsy, drug dependence, mood disorders such as depression, and schizophrenia. It is interesting to note that the same receptors (e.g., P2X7) may have a dual function; in the early phase of an ischemic damage, it is apoptotic/necrotic, whereas at its later phase, it may promote neuroregeneration. However, the restitution of neuronal connections by promoting axonal outgrowth may on the one hand help to establish new beneficial synaptic connections but on the other hand it may support the development of epileptic foci giving rise to reverberating neuronal circuits. In addition, glial proliferation may be beneficial by substituting the damaged brain tissue, but glial scars may hinder the reinnervation of neuronal targets by axonal outgrowth. Thus, both the choice of the pharmacological intervention and the time window in which it can be used have to be considered in detail.

Acknowledgments

The authors' laboratories were sponsored by OTKA, MÖB-DAAD, DFG, VW-Foundation, and CARIPLO-Foundation.

Conflict of Interest: None of the authors have a conflict of interest.

Abbreviations

$[Ca^{2+}]_i$	intracellular calcium concentration
5-HT	5-hydroxytryptamine
ACh	acetylcholine
ADP	adenosine 5′-diphosphate
ADP-β-S	adenosine 5′-*O*-(2-thiodiphosphate)
AMP	adenosine 5′-monophosphate
AMPA	α-amino-3-hydroxyl-5-methyl-4-isoxazole-propionic acid
ATP	adenosine 5′-triphosphate
ATP-γ-S	adenosine 5′-*O*-(3-thiotriphosphate)
BzATP	3′-*O*-(4-benzoyl)benzoyl-ATP
CaMK	calmodulin-dependent protein kinase
cAMP	cyclic AMP
CNS	central nervous system
DAG	diacylglycerol
DRG	dorsal root ganglion
EPSC	excitatory postsynaptic current
ERK	extracellular signal-regulated kinase
GABA	γ-aminobutyric acid
GTP	guanosine 5′-triphosphate
IL	interleukin
IP_3	inositol trisphosphate
IPSC	inhibitory postsynaptic current
LPS	lipopolysaccharide
LTD	long-term depression
LTP	long-term potentiation
MAPK	mitogen-activated protein kinase
NADH	nicotinamide adenine dinucleotide
NF-κB	nuclear factor κ-light-chain-enhancer of activated B cells
NMDA	*N*-methyl-D-aspartate
PFC	prefrontal cortex
PIP_2	phosphatidylinositol-bisphosphate
PKC	protein kinase C
PLC	phospholipase C
PPADS	pyridoxal-phosphate-6-azophenyl-2′,4′-disulfonate
PTX	pertussis toxin
TNFα	tumor necrosis factor α
TNP-ATP	2′,3′-*O*-(2,4,6-trinitrophenyl) ATP
TRPV	transient receptor potential vanilloid
UDP	uridine 5′-diphosphate
UTP	uridine 5′-triphosphate
α,β-meATP	α,β-methylene ATP

References

Abbracchio, M. P., Brambilla, R., Ceruti, S., & Cattabeni, F. (1999). Signalling mechanisms involved in P2Y receptor-mediated reactive astrogliosis. *Progress in Brain Research*, *120*, 333–342.

Abbracchio, M. P., Burnstock, G., Boeynaems, J. M., Barnard, E. A., Boyer, J. L., Kennedy, C., et al. (2006). International Union of Pharmacology LVIII: Update on the P2Y G protein-coupled nucleotide receptors: From molecular mechanisms and pathophysiology to therapy. *Pharmacological Reviews*, *58*, 281–341.

Abbracchio, M. P., Burnstock, G., Verkhratsky, A., & Zimmermann, H. (2009). Purinergic signalling in the nervous system: An overview. *Trends in Neurosciences*, *32*, 19–29.

Abbracchio, M. P., Saffrey, M. J., Höpker, V., & Burnstock, G. (1994). Modulation of astroglial cell proliferation by analogues of adenosine and ATP in primary cultures of rat striatum. *Neuroscience*, *59*, 67–76.

Agresti, C., Meomartini, M. E., Amadio, S., Ambrosini, E., Volonté, C., Aloisi, F., et al. (2005). ATP regulates oligodendrocyte progenitor migration, proliferation, and differentiation: Involvement of metabotropic P2 receptors. *Brain Research. Brain Research Reviews*, *48*, 157–165.

Aihara, H., Fujiwara, S., Mizuta, I., Tada, H., Kanno, T., Tozaki, H., et al. (2002). Adenosine triphosphate accelerates recovery from hypoxic/hypoglycemic perturbation of guinea pig hippocampal neurotransmission via a P2 receptor. *Brain Research*, *952*, 31–37.

Amadio, S., D'Ambrosi, N., Cavaliere, F., Murra, B., Sancesario, G., Bernardi, G., et al. (2002). P2 receptor modulation and cytotoxic function in cultured CNS neurons. *Neuropharmacology*, *42*, 489–501.

Andó, R. D., Méhész, B., Gyires, K., Illes, P., & Sperlágh, B. (2010). A comparative analysis of the activity of ligands acting at P2X and P2Y receptor subtypes in models of neuropathic, acute and inflammatory pain. *British Journal of Pharmacology*, *159*, 1106–1117.

Araque, A., Parpura, V., Sanzgiri, R. P., & Haydon, P. G. (1999). Tripartite synapses: Glia, the unacknowledged partner. *Trends in Neurosciences*, *22*, 208–215.

Aschrafi, A., Sadtler, S., Niculescu, C., Rettinger, J., & Schmalzing, G. (2004). Trimeric architecture of homomeric $P2X_2$ and heteromeric $P2X_{1+2}$ receptor subtypes. *Journal of Molecular Biology*, *342*, 333–343.

Ase, A. R., Bernier, L. P., Blais, D., Pankratov, Y., & Séguéla, P. (2010). Modulation of heteromeric $P2X_{1/5}$ receptors by phosphoinositides in astrocytes depends on the $P2X_1$ subunit. *Journal of Neurochemistry*, *113*, 1676–1684.

Avignone, E., Ulmann, L., Levavasseur, F., Rassendren, F., & Audinat, E. (2008). Status epilepticus induces a particular microglial activation state characterized by enhanced purinergic signaling. *The Journal of Neuroscience*, *28*, 9133–9144.

Ballerini, P., Ciccarelli, R., Caciagli, F., Rathbone, M. P., Werstiuk, E. S., Traversa, U., et al. (2005). $P2X_7$ receptor activation in rat brain cultured astrocytes increases the biosynthetic release of cysteinyl leukotrienes. *International Journal of Immunopathology and Pharmacology*, *18*, 417–430.

Barclay, J., Patel, S., Dorn, G., Wotherspoon, G., Moffatt, S., Eunson, L., et al. (2002). Functional downregulation of $P2X_3$ receptor subunit in rat sensory neurons reveals a significant role in chronic neuropathic and inflammatory pain. *The Journal of Neuroscience*, *22*, 8139–8147.

Barden, N., Harvey, M., Gagné, B., Shink, E., Tremblay, M., Raymond, C., et al. (2006). Analysis of single nucleotide polymorphisms in genes in the chromosome 12Q24.31 region points to $P2RX_7$ as a susceptibility gene to bipolar affective disorder. *American Journal of Medical Genetics Part B: Neuropsychiatric Genetics*, *141B*, 374–382.

Bardoni, R., Goldstein, P. A., Lee, C. J., Gu, J. G., & MacDermott, A. B. (1997). ATP P2X receptors mediate fast synaptic transmission in the dorsal horn of the rat spinal cord. *The Journal of Neuroscience*, *17*, 5297–5304.

Basso, A. M., Bratcher, N. A., Harris, R. R., Jarvis, M. F., Decker, M. W., & Rueter, L. E. (2009). Behavioral profile of $P2X_7$ receptor knockout mice in animal models of depression and anxiety: Relevance for neuropsychiatric disorders. *Behavioural Brain Research*, *198*, 83–90.

Bennett, G. C., & Boarder, M. R. (2000). The effect of nucleotides and adenosine on stimulus-evoked glutamate release from rat brain cortical slices. *British Journal of Pharmacology*, *131*, 617–623.

Bianco, F., Fumagalli, M., Pravettoni, E., D'Ambrosi, N., Volonte, C., Matteoli, M., et al. (2005). Pathophysiological roles of extracellular nucleotides in glial cells: Differential expression of purinergic receptors in resting and activated microglia. *Brain Research. Brain Research Reviews*, *48*, 144–156.

Bo, X., Jiang, L. H., Wilson, H. L., Kim, M., Burnstock, G., Surprenant, A., et al. (2003). Pharmacological and biophysical properties of the human $P2X_5$ receptor. *Molecular Pharmacology*, *63*, 1407–1416.

Bo, X., Kim, M., Nori, S. L., Schoepfer, R., Burnstock, G., & North, R. A. (2003). Tissue distribution of $P2X_4$ receptors studied with an ectodomain antibody. *Cell and Tissue Research*, *313*, 159–165.

Bodin, P., & Burnstock, G. (2001). Purinergic signalling: ATP release. *Neurochemical Research*, *26*, 959–969.

Boehm, S. (1999). ATP stimulates sympathetic transmitter release via presynaptic P2X purinoceptors. *The Journal of Neuroscience*, *19*, 737–746.

Borvendeg, S. J., Gerevich, Z., Gillen, C., & Illes, P. (2003). P2Y receptor-mediated inhibition of voltage-dependent Ca^{2+} channels in rat dorsal root ganglion neurons. *Synapse*, *47*, 159–161.

Boucsein, C., Zacharias, R., Färber, K., Pavlovic, S., Hanisch, U. K., & Kettenmann, H. (2003). Purinergic receptors on microglial cells: Functional expression in acute brain slices and modulation of microglial activation in vitro. *The European Journal of Neuroscience*, *17*, 2267–2276.

Boué-Grabot, E., Archambault, V., & Seguela, P. (2000). A protein kinase C site highly conserved in P2X subunits controls the desensitization kinetics of $P2X_2$ ATP-gated channels. *The Journal of Biological Chemistry*, *275*, 10190–10195.

Boué-Grabot, E., Barajas-López, C., Chakfe, Y., Blais, D., Bélanger, D., Emerit, M. B., et al. (2003). Intracellular cross talk and physical interaction between two classes of neurotransmitter-gated channels. *Journal of Neuroscience*, *23*, 1246–1253.

Bouvier, M. (2001). Oligomerization of G-protein-coupled transmitter receptors. *Nature Reviews. Neuroscience*, *2*, 274–286.

Bowser, D. N., & Khakh, B. S. (2004). ATP excites interneurons and astrocytes to increase synaptic inhibition in neuronal networks. *The Journal of Neuroscience*, *24*, 8606–8620.

Bowser, D. N., & Khakh, B. S. (2007). Vesicular ATP is the predominant cause of intercellular calcium waves in astrocytes. *The Journal of General Physiology*, *129*, 485–491.

Brake, A. J., Wagenbach, M. J., & Julius, D. (1994). New structural motif for ligand-gated ion channels defined by an ionotropic ATP receptor. *Nature*, *371*, 519–523.

Brambilla, R., Neary, J. T., Cattabeni, F., Cottini, L., D'Ippolito, G., Schiller, P. C., et al. (2002). Induction of COX-2 and reactive gliosis by P2Y receptors in rat cortical astrocytes is dependent on ERK1/2 but independent of calcium signalling. *Journal of Neurochemistry*, *83*, 1285–1296.

Brown, S. G., Townsend-Nicholson, A., Jacobson, K. A., Burnstock, G., & King, B. F. (2002). Heteromultimeric $P2X_{1/2}$ receptors show a novel sensitivity to extracellular pH. *The Journal of Pharmacology and Experimental Therapeutics*, *300*, 673–680.

Brown, D. A., & Yule, D. I. (2007). Protein kinase C regulation of $P2X_3$ receptors is unlikely to involve direct receptor phosphorylation. *Biochimica et Biophysica Acta*, *1773*, 166–175.

Buell, G., Lewis, C., Collo, G., North, R. A., & Surprenant, A. (1996). An antagonist-insensitive P2X receptor expressed in epithelia and brain. *The EMBO Journal*, *15*, 55–62.

Burnstock, G. (1972). Purinergic nerves. *Pharmacological Reviews*, *24*, 509–581.

Burnstock, G. (1996). A unifying purinergic hypothesis for the initiation of pain. *Lancet*, *347*, 1604–1605.

Burnstock, G. (2007). Physiology and pathophysiology of purinergic neurotransmission. *Physiological Reviews*, *87*, 659–797.

Burnstock, G., & Kennedy, C. (1985). Is there a basis for distinguishing two types of P2-purinoceptor? *General Pharmacology*, *16*, 433–440.

Burnstock, G., & Knight, G. E. (2004). Cellular distribution and functions of P2 receptor subtypes in different systems. *International Review of Cytology*, *240*, 31–304.

Carrasquero, L. M., Delicado, E. G., Bustillo, D., Gutiérrez-Martín, Y., Artalejo, A. R., & Miras-Portugal, M. T. (2009). $P2X_7$ and $P2Y_{13}$ purinergic receptors mediate intracellular calcium responses to BzATP in rat cerebellar astrocytes. *Journal of Neurochemistry*, *110*, 879–889.

Carroll, W. A., Donnelly-Roberts, D., & Jarvis, M. F. (2009). Selective $P2X_7$ receptor antagonists for chronic inflammation and pain. *Purinergic Signalling*, *5*, 63–73.

Carter, D. S., Alam, M., Cai, H., Dillon, M. P., Ford, A. P., Gever, J. R., et al. (2009). Identification and SAR of novel diaminopyrimidines. Part 1: The discovery of RO-4, a dual $P2X_3/P2X_{2/3}$ antagonist for the treatment of pain. *Bioorganic & Medicinal Chemistry Letters*, *19*, 1628–1631.

Cavaliere, F., Amadio, S., Angelini, D. F., Sancesario, G., Bernardi, G., & Volonté, C. (2004). Role of the metabotropic $P2Y_4$ receptor during hypoglycemia: Cross talk with the ionotropic NMDAR1 receptor. *Experimental Cell Research*, *300*, 149–158.

Cavaliere, F., Florenzano, F., Amadio, S., Fusco, F. R., Viscomi, M. T., D'Ambrosi, N., et al. (2003). Up-regulation of $P2X_2$, $P2X_4$ receptor and ischemic cell death prevention by P2 antagonists. *Neuroscience*, *120*, 85–98.

Cavaliere, F., Sancesario, G., Bernardi, G., & Volonté, C. (2002). Extracellular ATP and nerve growth factor intensify hypoglycemia-induced cell death in primary neurons: Role of P2 and NGFRp75 receptors. *Journal of Neurochemistry*, *83*, 1129–1138.

Ceruti, S., Villa, G., Genovese, T., Mazzon, E., Longhi, R., Rosa, P., et al. (2009). The P2Y-like receptor GPR17 as a sensor of damage and a new potential target in spinal cord injury. *Brain*, *132*, 2206–2218.

Chen, C. C., Akopian, A. N., Sivilotti, L., Colquhoun, D., Burnstock, G., & Wood, J. N. (1995). A P2X purinoceptor expressed by a subset of sensory neurons. *Nature*, *77*, 428–431.

Chessell, I. P., Hatcher, J. P., Bountra, C., Michel, A. D., Hughes, J. P., Green, P., et al. (2005). Disruption of the $P2X_7$ purinoceptor gene abolishes chronic inflammatory and neuropathic pain. *Pain*, *114*, 386–396.

Cheung, K. K., Chan, W. Y., & Burnstock, G. (2005). Expression of P2X purinoceptors during rat brain development and their inhibitory role on motor axon outgrowth in neural tube explant cultures. *Neuroscience*, *133*, 937–945.

Cheung, K. K., Ryten, M., & Burnstock, G. (2003). Abundant and dynamic expression of G protein-coupled P2Y receptors in mammalian development. *Developmental Dynamics*, *228*, 254–266.

Chizh, B. A., & Illes, P. (2001). P2X receptors and nociception. *Pharmacological Reviews*, *53*, 553–568.

Chorna, N. E., Santiago-Pérez, L. I., Erb, L., Seye, C. I., Neary, J. T., Sun, G. Y., et al. (2004). P2Y receptors activate neuroprotective mechanisms in astrocytic cells. *Journal of Neurochemistry*, *91*, 119–132.

Ciana, P., Fumagalli, M., Trincavelli, M. L., Verderio, C., Rosa, P., Lecca, D., et al. (2006). The orphan receptor GPR17 identified as a new dual uracil nucleotides/cysteinyl-leukotrienes receptor. *The EMBO Journal*, *25*, 4615–4627.

Ciccarelli, R., Ballerini, P., Sabatino, G., Rathbone, M. P., D'Onofrio, M., Caciagli, F., et al. (2001). Involvement of astrocytes in purine-mediated reparative processes in the brain. *International Journal of Developmental Neuroscience*, *19*, 395–414.

Cockayne, D. A., Dunn, P. M., Zhong, Y., Rong, W., Hamilton, S. G., Knight, G. E., et al. (2005). $P2X_2$ knockout mice and $P2X_2/P2X_3$ double knockout mice reveal a role for the $P2X_2$ receptor subunit in mediating multiple sensory effects of ATP. *Journal de Physiologie*, *567*, 621–639.

Coco, S., Calegari, F., Pravettoni, E., Pozzi, D., Taverna, E., Rosa, P., et al. (2003). Storage and release of ATP from astrocytes in culture. *The Journal of Biological Chemistry*, *278*, 1354–1362.

Colldén, G., Mangano, C., & Meister, B. (2010). $P2X_2$ purinoreceptor protein in hypothalamic neurons associated with the regulation of food intake. *Neuroscience*, *171*, 62–78.

Communi, D., Govaerts, C., Parmentier, M., & Boeynaems, J. M. (1997). Cloning of a human purinergic P2Y receptor coupled to phospholipase C and adenylyl cyclase. *The Journal of Biological Chemistry*, *272*, 31969–31973.

Cook, S. P., & McCleskey, E. W. (2002). Cell damage excites nociceptors through release of cytosolic ATP. *Pain*, *95*, 41–47.

Cunha, R. A., & Ribeiro, J. A. (2000). ATP as a presynaptic modulator. *Life Sciences*, *68*, 119–137.

Cunha, R. A., Ribeiro, J. A., & Sebastião, A. M. (1994). Purinergic modulation of the evoked release of [^{3}H]acetylcholine from the hippocampus and cerebral cortex of the rat: Role of the ectonucleotidases. *The European Journal of Neuroscience*, *6*, 33–42.

D'Alimonte, I., Ciccarelli, R., Di Iorio, P., Nargi, E., Buccella, S., Giuliani, P., et al. (2007). Activation of $P2X_7$ receptors stimulates the expression of $P2Y_2$ receptor mRNA in astrocytes cultured from rat brain. *International Journal of Immunopathology and Pharmacology*, *20*, 301–316.

D'Ambrosi, N., Iafrate, M., Saba, E., Rosa, P., & Volonte, C. (2007). Comparative analysis of $P2Y_4$ and $P2Y_6$ receptor architecture in naive and transfected neuronal systems. *Biochimica et Biophysica Acta*, *1768*, 1592–1599.

D'Ambrosi, N., Iafrate, M., Vacca, F., Amadio, S., Tozzi, A., Mercuri, N. B., et al. (2006). The $P2Y_4$ receptor forms homo-oligomeric complexes in several CNS and PNS neuronal cells. *Purinergic Signalling*, *2*, 575–582.

D'Ambrosi, N., Murra, B., Cavaliere, F., Amadio, S., Bernardi, G., Burnstock, G., et al. (2001). Interaction between ATP and nerve growth factor signaling in the survival and neuritic outgrowth from PC12 cells. *Neuroscience*, *108*, 527–534.

Dani, J. W., Chernjavsky, A., & Smith, S. J. (1992). Neuronal activity triggers calcium waves in hippocampal astrocyte networks. *Neuron*, *8*, 429–440.

De Lorenzo, S., Veggetti, M., Muchnik, S., & Losavio, A. (2006). Presynaptic inhibition of spontaneous acetylcholine release mediated by P2Y receptors at the mouse neuromuscular junction. *Neuroscience*, *142*, 71–85.

Deuchars, S. A., Atkinson, L., Brooke, R. E., Musa, H., Milligan, C. J., Batten, T. F., et al. (2001). Neuronal $P2X_7$ receptors are targeted to presynaptic terminals in the central and peripheral nervous systems. *The Journal of Neuroscience*, *21*, 7143–7152.

Diaz-Hernandez, M., Gomez-Villafuertes, R., Hernando, F., Pintor, J., & Miras-Portugal, M. T. (2001). Presence of different ATP receptors on rat midbrain single synaptic terminals. Involvement of the $P2X_3$ subunits. *Neuroscience Letters*, *301*, 159–162.

Ding, S., & Sachs, F. (1999). Single channel properties of $P2X_2$ purinoceptors. *The Journal of General Physiology*, *113*, 695–720.

Domercq, M., Brambilla, L., Pilati, E., Marchaland, J., Volterra, A., & Bezzi, P. (2006). P2Y$_1$ receptor-evoked glutamate exocytosis from astrocytes: Control by tumor necrosis factor-α and prostaglandins. *The Journal of Biological Chemistry*, *281*, 30684–30696.

Doná, F., Ulrich, H., Persike, D. S., Conceição, I. M., Blini, J. P., Cavalheiro, E. A., et al. (2009). Alteration of purinergic P2X$_4$ and P2X$_7$ receptor expression in rats with temporal-lobe epilepsy induced by pilocarpine. *Epilepsy Research*, *83*, 157–167.

Duan, S., Anderson, C. M., Keung, E. C., Chen, Y., Chen, Y., & Swanson, R. A. (2003). P2X$_7$ receptor-mediated release of excitatory amino acids from astrocytes. *The Journal of Neuroscience*, *23*, 1320–1328.

Dunn, P. M., Gever, J., Ruan, H. Z., & Burnstock, G. (2005). Developmental changes in heteromeric P2X$_{2/3}$ receptor expression in rat sympathetic ganglion neurons. *Developmental Dynamics*, *234*, 505–511.

Dutton, J. L., Poronnik, P., Li, G. H., Holding, C. A., Worthington, R. A., Vandenberg, R. J., et al. (2000). P2X$_1$ receptor membrane redistribution and down-regulation visualized by using receptor-coupled green fluorescent protein chimeras. *Neuropharmacology*, *39*, 2054–2066.

Ecke, D., Hanck, T., Tulapurkar, M. E., Schäfer, R., Kassack, M., Stricker, R., et al. (2008). Hetero-oligomerization of the P2Y$_{11}$ receptor with the P2Y$_1$ receptor controls the internalization and ligand selectivity of the P2Y$_{11}$ receptor. *The Biochemical Journal*, *409*, 107–116.

Edwards, F. A., Gibb, A. J., & Colquhoun, D. (1992). ATP receptor-mediated synaptic currents in the central nervous system. *Nature*, *359*, 144–147.

Edwards, F. A., Robertson, S. J., & Gibb, A. J. (1997). Properties of ATP receptor-mediated synaptic transmission in the rat medial habenula. *Neuropharmacology*, *36*, 1253–1268.

Egan, T. M., & Khakh, B. S. (2004). Contribution of calcium ions to P2X channel responses. *The Journal of Neuroscience*, *24*, 3413–3420.

Ennion, S. J., & Evans, R. J. (2002). Conserved cysteine residues in the extracellular loop of the human P2X$_1$ receptor form disulfide bonds and are involved in receptor trafficking to the cell surface. *Molecular Pharmacology*, *61*, 303–311.

Erb, L., Liao, Z., Seye, C. I., & Weisman, G. A. (2006). P2 receptors: Intracellular signaling. *Pflügers Archiv: European Journal of Physiology*, *452*, 552–562.

Erb, L., Liu, J., Ockerhausen, J., Kong, Q., Garrad, R. C., Griffin, K., et al. (2001). An RGD sequence in the P2Y$_2$ receptor interacts with $\alpha_V\beta_3$ integrins and is required for G$_o$-mediated signal transduction. *The Journal of Cell Biology*, *153*, 491–501.

Erhardt, A., Lucae, S., Unschuld, P. G., Ising, M., Kern, N., Salyakina, D., et al. (2007). Association of polymorphisms in P2RX7 and CaMKKb with anxiety disorders. *Journal of Affective Disorders*, *101*, 159–168.

Evans, R. J., Lewis, C., Virginio, C., Lundstrom, K., Buell, G., Surprenant, A., et al. (1996). Ionic permeability of, and divalent cation effects on, two ATP-gated cation channels (P2X receptors) expressed in mammalian cells. *The Journal of Physiology*, *497*, 413–422.

Fam, S. R., Gallagher, C. J., Kalia, L. W., & Salter, M. W. (2003). Differential frequency dependence of P2Y$_1$- and P2Y$_2$- mediated Ca^{2+} signaling in astrocytes. *The Journal of Neuroscience*, *23*, 4437–4444.

Fam, S. R., Gallagher, C. J., & Salter, M. W. (2000). P2Y$_1$ purinoceptor-mediated Ca^{2+} signaling and Ca^{2+} wave propagation in dorsal spinal cord astrocytes. *The Journal of Neuroscience*, *20*, 2800–2808.

Fellin, T., Pascual, O., & Haydon, P. G. (2006). Astrocytes coordinate synaptic networks: Balanced excitation and inhibition. *Physiology (Bethesda)*, *21*, 208–215.

Fellin, T., Pozzan, T., & Carmignoto, G. (2006). Purinergic receptors mediate two distinct glutamate release pathways in hippocampal astrocytes. *The Journal of Biological Chemistry*, *281*, 4274–4284.

Ferrari, D., Chiozzi, P., Falzoni, S., Dal Susino, M., Collo, G., Buell, G., et al. (1997). ATP-mediated cytotoxicity in microglial cells. *Neuropharmacology*, *36*, 1295–1301.

Ferrari, D., Pizzirani, C., Adinolfi, E., Lemoli, R. M., Curti, A., Idzko, M., et al. (2006). The $P2X_7$ receptor: A key player in IL-1 processing and release. *Journal of Immunology*, *176*, 3877–3883.

Ferrari, D., Stroh, C., & Schulze-Osthoff, K. (1999). $P2X_7/P2_Z$ purinoreceptor-mediated activation of transcription factor NFAT in microglial cells. *The Journal of Biological Chemistry*, *274*, 13205–13210.

Fiacco, T. A., Agulhon, C., & McCarthy, K. D. (2009). Sorting out astrocyte physiology from pharmacology. *Annual Review of Pharmacology and Toxicology*, *49*, 151–174.

Fields, R. D., & Stevens, B. (2000). ATP: An extracellular signaling molecule between neurons and glia. *Trends in Neurosciences*, *23*, 625–633.

Filippov, A. K., Choi, R. C., Simon, J., Barnard, E. A., & Brown, D. A. (2006). Activation of $P2Y_1$ nucleotide receptors induces inhibition of the M-type K^+ current in rat hippocampal pyramidal neurons. *The Journal of Neuroscience*, *26*, 9340–9348.

Fischer, W., Appelt, K., Grohmann, M., Franke, H., Nörenberg, W., & Illes, P. (2009). Increase of intracellular Ca^{2+} by P2X and P2Y receptor-subtypes in cultured cortical astroglia of the rat. *Neuroscience*, *160*, 767–783.

Fischer, W., & Krügel, U. (2007). P2Y receptors: Focus on structural, pharmacological and functional aspects in the brain. *Current Medicinal Chemistry*, *14*, 2429–2455.

Fischer, W., Zadori, Z., Kullnick, Y., Gröger-Arndt, H., Franke, H., Wirkner, K., et al. (2007). Conserved lysin and arginin residues in the extracellular loop of $P2X_3$ receptors are involved in agonist binding. *European Journal of Pharmacology*, *576*, 7–17.

Florenzano, F., Viscomi, M. T., Cavaliere, F., Volonté, C., & Molinari, M. (2002). Cerebellar lesion up-regulates $P2X_1$ and $P2X_2$ purinergic receptors in precerebellar nuclei. *Neuroscience*, *115*, 425–434.

Franke, H., Grosche, J., Schadlich, H., Krügel, U., Allgaier, C., & Illes, P. (2001). P2X receptor expression on astrocytes in the nucleus accumbens of rats. *Neuroscience*, *108*, 421–429.

Franke, H., Günther, A., Grosche, J., Schmidt, R., Rossner, S., Reinhardt, R., et al. (2004). $P2X_7$ receptor expression after ischemia in the cerebral cortex of rats. *Journal of Neuropathology and Experimental Neurology*, *63*, 686–699.

Franke, H., & Illes, P. (2006). Involvement of P2 receptors in the growth and survival of neurons in the CNS. *Pharmacology & Therapeutics*, *109*, 297–324.

Franke, H., Kittner, H., Grosche, J., & Illes, P. (2003). Enhanced $P2Y_1$ receptor expression in the brain after sensitisation with *d*-amphetamine. *Psychopharmacology*, *167*, 187–194.

Franke, H., Krügel, U., Schmidt, R., Grosche, J., Reichenbach, A., & Illes, P. (2001). P2 receptor-types involved in astrogliosis in vivo. *British Journal of Pharmacology*, *134*, 1180–1189.

Frenguelli, B. G., Wigmore, G., Llaudet, E., & Dale, N. (2007). Temporal and mechanistic dissociation of ATP and adenosine release during ischaemia in the mammalian hippocampus. *Journal of Neurochemistry*, *101*, 1400–1413.

Friedle, S. A., Curet, M. A., & Watters, J. J. (2010). Recent patents on novel $P2X_7$ receptor antagonists and their potential for reducing central nervous system inflammation. *Recent Patents on CNS Drug Discovery*, *5*, 35–45.

Fröhlich, R., Boehm, S., & Illes, P. (1996). Pharmacological characterization of P2 purinoceptor types in rat locus coeruleus neurons. *European Journal of Pharmacology*, *315*, 255–261.

Fujii, S., Kato, H., & Kuroda, Y. (1999). Extracellular adenosine 5′-triphosphate plus activation of glutamatergic receptors induces long-term potentiation in CA1 neurons of guinea pig hippocampal slices. *Neuroscience Letters*, *276*, 21–24.

Fujita, T., Tozaki-Saitoh, H., & Inoue, K. (2009). $P2Y_1$ receptor signaling enhances neuroprotection by astrocytes against oxidative stress via IL-6 release in hippocampal cultures. *Glia*, *57*, 244–257.

Fujiwara, Y., & Kubo, Y. (2004). Density-dependent changes of the pore properties of the $P2X_2$ receptor channel. *The Journal of Physiology*, *558*, 31–43.

Fumagalli, M., Brambilla, R., D'Ambrosi, N., Volonté, C., Matteoli, M., Verderio, C., et al. (2003). Nucleotide-mediated calcium signaling in rat cortical astrocytes: Role of P2X and P2Y receptors. *Glia*, *43*, 218–230.

Gallagher, C. J., & Salter, M. W. (2003). Differential properties of astrocyte calcium waves mediated by $P2Y_1$ and $P2Y_2$ receptors. *The Journal of Neuroscience*, *23*, 6728–6739.

Gebicke-Haerter, P. J., Wurster, S., Schobert, A., & Hertting, G. (1988). P2-purinoceptor induced prostaglandin synthesis in primary rat astrocyte cultures. *Naunyn-Schmiedeberg's Archives of Pharmacology*, *338*, 704–707.

Gendron, F. P., Neary, J. T., Theiss, P. M., Sun, G. Y., Gonzalez, F. A., & Weisman, G. A. (2003). Mechanisms of $P2X_7$ receptor-mediated ERK1/2 phosphorylation in human astrocytoma cells. *American Journal of Physiology. Cell Physiology*, *284*, 571–581.

Gerevich, Z., Borvendeg, S. J., Schröder, W., Franke, H., Wirkner, K., Nörenberg, W., et al. (2004). Inhibition of N-type voltage-activated calcium channels in rat dorsal root ganglion neurons by P2Y receptors is a possible mechanism of ADP-induced analgesia. *The Journal of Neuroscience*, *24*, 797–807.

Gerevich, Z., & Illes, P. (2004). P2Y receptors and pain transmission. *Purinergic Signalling*, *1*, 3–10.

Gerevich, Z., Müller, C., & Illes, P. (2005). Metabotropic $P2Y_1$ receptors inhibit $P2X_3$ receptor-channels in rat dorsal root ganglion neurons. *European Journal of Pharmacology*, *521*, 34–38.

Gerevich, Z., Zadori, Z. S., Köles, L., Kopp, L., Milius, D., Wirkner, K., et al. (2007). Dual effect of acid pH on purinergic $P2X_3$ receptors depends on the histidine 206 residue. *The Journal of Biological Chemistry*, *282*, 33949–33957.

Gerevich, Z., Zadori, Z. S., Müller, C., Wirkner, K., Schröder, W., Rubini, P., et al. (2007). Metabotropic P2Y receptors inhibit $P2X_3$ receptor-channels via G protein-dependent facilitation of their desensitization. *British Journal of Pharmacology*, *151*, 226–236.

Gever, J. R., Cockayne, D. A., Dillon, M. P., Burnstock, G., & Ford, A. P. (2006). Pharmacology of P2X channels. *Pflügers Archiv: European Journal of Physiology*, *452*, 513–537.

Gomez-Villafuertes, R., Gualix, J., & Miras-Portugal, M. T. (2001). Single GABAergic synaptic terminals from rat midbrain exhibit functional P2X and dinucleotide receptors, able to induce GABA secretion. *Journal of Neurochemistry*, *77*, 84–93.

Gourine, A. V., Kasymov, V., Marina, N., Tang, F., Figueiredo, M. F., Lane, S., et al. (2010). Astrocytes control breathing through pH-dependent release of ATP. *Science*, *329*, 571–675.

Gu, J. G., & MacDermott, A. B. (1997). Activation of ATP P2X receptors elicits glutamate release from sensory neuron synapses. *Nature*, *389*, 749–753.

Gunosewoyo, H., & Kassiou, M. (2010). P2X purinergic receptor ligands: Recently patented compounds. *Expert Opinion on Therapeutic Patents*, *20*, 625–646.

Guo, C., Masin, M., Qureshi, O. S., & Murrell-Lagnado, R. D. (2007). Evidence for functional $P2X_4/P2X_7$ heteromeric receptors. *Molecular Pharmacology*, *72*, 1447–1456.

Guthrie, P. B., Knappenberger, J., Segal, M., Bennett, M. V., Charles, A. C., & Kater, S. B. (1999). ATP released from astrocytes mediates glial calcium waves. *The Journal of Neuroscience*, *19*, 520–528.

Guzman, S. J., Schmidt, H., Franke, H., Krügel, U., Eilers, J., Illes, P., et al. (2010). $P2Y_1$ receptors inhibit long-term depression in the prefrontal cortex. *Neuropharmacology*, *59*, 406–415.

Hamilton, N. B., & Attwell, D. (2010). Do astrocytes really exocytose neurotransmitters? *Nature Reviews. Neuroscience*, *11*, 227–238.

Haynes, S. E., Hollopeter, G., Yang, G., Kurpius, D., Dailey, M. E., Gan, W. B., et al. (2006). The $P2Y_{12}$ receptor regulates microglial activation by extracellular nucleotides. *Nature Neuroscience*, *9*, 1512–1519.

Heine, C., Wegner, A., Grosche, J., Allgaier, C., Illes, P., & Franke, H. (2007). P2 receptor expression in the dopaminergic system of the rat brain during development. *Neuroscience*, *149*, 165–181.

Hide, I., Tanaka, M., Inoue, A., Nakajima, K., Kohsaka, S., Inoue, K., et al. (2000). Extracellular ATP triggers tumor necrosis factor α release from rat microglia. *Journal of Neurochemistry*, *75*, 965–972.

Honda, S., Sasaki, Y., Ohsawa, K., Imai, Y., Nakamura, Y., Inoue, K., et al. (2001). Extracellular ATP or ADP induce chemotaxis of cultured microglia through $G_{i/o}$-coupled P2Y receptors. *The Journal of Neuroscience*, *21*, 1975–1982.

Hugel, S., & Schlichter, R. (2000). Presynaptic P2X receptors facilitate inhibitory GABAergic transmission between cultured rat spinal cord dorsal horn neurons. *The Journal of Neuroscience*, *20*, 2121–2130.

Hussl, S., & Boehm, S. (2006). Functions of neuronal P2Y receptors. *Pflügers Archiv: European Journal of Physiology*, *452*, 538–551.

Idestrup, C. P., & Salter, M. W. (1998). P2Y and P2U receptors differentially release intracellular Ca^{2+} via the phospholipase C/inositol 1,4,5-triphosphate pathway in astrocytes from the dorsal spinal cord. *Neuroscience*, *86*, 913–923.

Ikeuchi, Y., & Nishizaki, T. (1996). ATP-regulated K^+ channel and cytosolic Ca^{2+} mobilization in cultured rat spinal neurons. *European Journal of Pharmacology*, *302*, 163–169.

Ikeuchi, Y., Nishizaki, T., Mori, M., & Okada, Y. (1996). Adenosine activates the K^+ channel and enhances cytosolic Ca^{2+} release via a P2Y purinoceptor in hippocampal neurons. *European Journal of Pharmacology*, *304*, 191–199.

Illes, P., Nörenberg, W., & Gebicke-Haerter, P. J. (1996). Molecular mechanisms of microglial activation. B. Voltage- and purinoceptor-operated channels in microglia. *Neurochemistry International*, *29*, 13–24.

Illes, P., & Ribeiro, J. A. (2004). Molecular physiology of P2 receptors in the central nervous system. *European Journal of Pharmacology*, *483*, 5–17.

Inoue, K. (2008). Purinergic systems in microglia. *Cellular and Molecular Life Sciences*, *65*, 3074–3080.

Inoue, K., Koizumi, S., Kataoka, A., Tozaki-Saitoh, H., & Tsuda, M. (2009). $P2Y_6$-evoked microglial phagocytosis. *International Review of Neurobiology*, *85*, 159–163.

Jacobson, K. A., Ivanov, A. A., de Castro, S., Harden, T. K., & Ko, H. (2009). Development of selective agonists and antagonists of P2Y receptors. *Purinergic Signalling*, *5*, 75–89.

Jacques-Silva, M. C., Rodnight, R., Lenz, G., Liao, Z., Kong, Q., Tran, M., et al. (2004). $P2X_7$ receptors stimulate AKT phosphorylation in astrocytes. *British Journal of Pharmacology*, *141*, 1106–1117.

Jahangir, A., Alam, M., Carter, D. S., Dillon, M. P., Bois, D. J., Ford, A. P., et al. (2009). Identification and SAR of novel diaminopyrimidines. Part 2: The discovery of RO-51, a potent and selective, dual $P2X_3/P2X_{2/3}$ antagonist for the treatment of pain. *Bioorganic & Medicinal Chemistry Letters*, *19*, 1632–1635.

James, G., & Butt, A. M. (2001). Changes in P2Y and P2X purinoceptors in reactive glia following axonal degeneration in the rat optic nerve. *Neuroscience Letters*, *312*, 33–36.

James, G., & Butt, A. M. (2002). P2Y and P2X purinoceptor mediated Ca^{2+} signalling in glial cell pathology in the central nervous system. *European Journal of Pharmacology*, *447*, 247–260.

Jameson, H. S., Pinol, R. A., Kamendi, H., & Mendelowitz, D. (2008). ATP facilitates glutamatergic neurotransmission to cardiac vagal neurons in the nucleus ambiguus. *Brain Research*, *1201*, 88–92.

Jarvis, M. F. (2010). The neural–glial purinergic receptor ensemble in chronic pain states. *Trends in Neurosciences*, *33*, 48–57.

Jarvis, M. F., & Khakh, B. S. (2009). ATP-gated P2X cation-channels. *Neuropharmacology*, *56*, 208–215.

Jiang, L. H., Kim, M., Spelta, V., Bo, X., Surprenant, A., & North, R. A. (2003). Subunit arrangement in P2X receptors. *The Journal of Neuroscience*, *23*, 8903–8910.

Jimenez, A. I., Castro, E., Communi, D., Boeynaems, J. M., Delicado, E. G., & Miras-Portugal, M. T. (2000). Coexpression of several types of metabotropic nucleotide receptors in single cerebellar astrocytes. *Journal of Neurochemistry*, *75*, 2071–2079.

Jin, Y. H., Bailey, T. W., Li, B. Y., Schild, J. H., & Andresen, M. C. (2004). Purinergic and vanilloid receptor activation releases glutamate from separate cranial afferent terminals in nucleus tractus solitarius. *The Journal of Neuroscience*, *24*, 4709–4717.

John, G. R., Scemes, E., Suadicani, S. O., Liu, J. S., Charles, P. C., Lee, S. C., et al. (1999). IL-1β differentially regulates calcium wave propagation between primary human fetal astrocytes via pathways involving P2 receptors and gap junction channels. *Proceedings of the National Academy of Sciences of the United States of America*, *96*, 11613–11618.

John, G. R., Simpson, J. E., Woodroofe, M. N., Lee, S. C., & Brosnan, C. F. (2001). Extracellular nucleotides differentially regulate interleukin-1β signaling in primary human astrocytes: Implications for inflammatory gene expression. *The Journal of Neuroscience*, *21*, 4134–4142.

Jones, C. A., Vial, C., Sellers, L. A., Humphrey, P. P., Evans, R. J., & Chessell, I. P. (2004). Functional regulation of $P2X_6$ receptors by N-linked glycosylation: Identification of a novel $\alpha\beta$-methylene ATP-sensitive phenotype. *Molecular Pharmacology*, *65*, 979–985.

Kang, T. C., An, S. J., Park, S. K., Hwang, I. K., & Won, M. H. (2003). $P2X_2$ and $P2X_4$ receptor expression is regulated by a $GABA_A$ receptor-mediated mechanism in the gerbil hippocampus. *Brain Research. Molecular Brain Research*, *116*, 168–175.

Kanjhan, R., Housley, G. D., Burton, L. D., Christie, D. L., Kippenberger, A., Thorne, P. R., et al. (1999). Distribution of the $P2X_2$ receptor subunit of the ATP-gated ion channels in the rat central nervous system. *The Journal of Comparative Neurology*, *407*, 11–32.

Karoly, R., Mike, A., Illes, P., & Gerevich, Z. (2008). The unusual state-dependent affinity of $P2X_3$ receptors can be explained by an allosteric two-open-state model. *Molecular Pharmacology*, *73*, 224–234.

Kawamura, M., Gachet, C., Inoue, K., & Kato, F. (2004). Direct excitation of inhibitory interneurons by extracellular ATP mediated by $P2Y_1$ receptors in the hippocampal slice. *The Journal of Neuroscience*, *24*, 10835–10845.

Kawate, T., Michel, J. C., Birdsong, W. T., & Gouaux, E. (2009). Crystal structure of the ATP-gated $P2X_4$ ion channel in the closed state. *Nature*, *460*, 592–598.

Khakh, B. S., Bao, X. R., Labarca, C., & Lester, H. A. (1999). Neuronal P2X transmitter-gated cation channels change their ion selectivity in seconds. *Nature Neuroscience*, *2*, 322–330.

Khakh, B. S., Fisher, J. A., Nashmi, R., Bowser, D. N., & Lester, H. A. (2005). An angstrom scale interaction between plasma membrane ATP-gated $P2X_2$ and $\alpha_4\beta_2$ nicotinic channels measured with fluorescence resonance energy transfer and total internal reflection fluorescence microscopy. *The Journal of Neuroscience*, *25*, 6911–6920.

Khakh, B. S., & Henderson, G. (1998). ATP receptor-mediated enhancement of fast excitatory neurotransmitter release in the brain. *Molecular Pharmacology*, *54*, 372–378.

Khakh, B. S., Proctor, W. R., Dunwiddie, T. V., Labarca, C., & Lester, H. A. (1999). Allosteric control of gating and kinetics at $P2X_4$ receptor channels. *The Journal of Neuroscience*, *19*, 7289–7299.

Khakh, B. S., Smith, W. B., Chiu, C. S., Ju, D., Davidson, N., & Lester, H. A. (2001). Activation-dependent changes in receptor distribution and dendritic morphology in hippocampal neurons expressing $P2X_2$-green fluorescent protein receptors. *Proceedings of the National Academy of Sciences of the United States of America*, *98*, 5288–5293.

Khakpay, R., Polster, D., Köles, L., Skorinkin, A., Szabo, B., Wirkner, K., et al. (2010). Potentiation of the glutamatergic synaptic input to rat locus coeruleus neurons by $P2X_7$ receptors. *Purinergic Signalling*, *6*, 349–359.

Kim, J. E., Kwak, S. E., Jo, S. M., & Kang, T. C. (2009). Blockade of P2X receptor prevents astroglial death in the dentate gyrus following pilocarpine-induced status epilepticus. *Neurological Research*, *31*, 982–988.

Kim, S. G., Soltysiak, K. A., Gao, Z. G., Chang, T. S., Chung, E., & Jacobson, K. A. (2003). Tumor necrosis factor α-induced apoptosis in astrocytes is prevented by the activation of $P2Y_6$, but not $P2Y_4$ nucleotide receptors. *Biochemical Pharmacology*, *65*, 923–931.

King, B. F., Neary, J. T., Zhu, Q., Wang, S., Norenberg, M. D., & Burnstock, G. (1996). P2 purinoceptors in rat cortical astrocytes: Expression, calcium-imaging and signalling studies. *Neuroscience*, *74*, 1187–1196.

King, B. F., Townsend-Nicholson, A., Wildman, S. S., Thomas, T., Spyer, K. M., & Burnstock, G. (2000). Coexpression of rat $P2X_2$ and $P2X_6$ subunits in *Xenopus* oocytes. *The Journal of Neuroscience*, *20*, 4871–4877.

Kittner, H., Franke, H., Fischer, W., Schultheis, N., Krügel, U., & Illes, P. (2003). Stimulation of $P2Y_1$ receptors causes anxiolytic-like effects in the rat elevated plus-maze: Implications for the involvement of $P2Y_1$ receptor-mediated nitric oxide production. *Neuropsychopharmacology*, *28*, 435–444.

Kittner, H., Franke, H., Harsch, J. I., El-Ashmawy, I. M., Seidel, B., Krügel, U., et al. (2006). Enhanced food intake after stimulation of hypothalamic $P2Y_1$ receptors in rats: Modulation of feeding behaviour by extracellular nucleotides. *The European Journal of Neuroscience*, *24*, 2049–2056.

Kittner, H., Hoffmann, E., Krügel, U., & Illes, P. (2004). P2 receptor-mediated effects on the open field behaviour of rats in comparison with behavioural responses induced by the stimulation of dopamine D_2-like and by the blockade of ionotrophic glutamate receptors. *Behavioural Brain Research*, *149*, 197–208.

Kittner, H., Krügel, U., Hoffmann, E., & Illes, P. (2004). Modulation of feeding behaviour by blocking purinergic receptors in the rat nucleus accumbens: A combined microdialysis, electroencephalographic and behavioural study. *The European Journal of Neuroscience*, *19*, 396–404.

Kittner, H., Krügel, U., & Illes, P. (2001). The purinergic P2 receptor antagonist pyridoxalphosphate-6-azophenyl-2′,4′-disulphonic acid prevents both the acute locomotor effects of amphetamine and the behavioural sensitization caused by repeated amphetamine injections in rats. *Neuroscience*, *102*, 241–243.

Knutsen, L. J. S., & Murray, T. F. (1997). Adenosine and ATP in epilepsy. In K. A. Jacobson & M. F. Jarvis (Eds.), *Purinergic Approaches in Experimental Therapeutics* (pp. 423–447). New York: Wiley-Liss.

Kobayashi, K., Yamanaka, H., Fukuoka, T., Dai, Y., Obata, K., & Noguchi, K. (2008). $P2Y_{12}$ receptor upregulation in activated microglia is a gateway of p38 signaling and neuropathic pain. *The Journal of Neuroscience*, *28*, 2892–2902.

Koch, H., von Kügelgen, I., & Starke, K. (1997). P2-receptor-mediated inhibition of noradrenaline release in the rat hippocampus. *Naunyn-Schmiedeberg's Archives of Pharmacology*, *355*, 707–715.

Koizumi, S., Fujishita, K., Tsuda, M., Shigemoto-Mogami, Y., & Inoue, K. (2003). Dynamic inhibition of excitatory synaptic transmission by astrocyte-derived ATP in hippocampal cultures. *Proceedings of the National Academy of Sciences of the United States of America*, *100*, 11023–11028.

Koizumi, S., Shigemoto-Mogami, Y., Nasu-Tada, K., Shinozaki, Y., Ohsawa, K., Tsuda, M., et al. (2007). UDP acting at $P2Y_6$ receptors is a mediator of microglial phagocytosis. *Nature*, *446*, 1091–1095.

Köles, L., Fürst, S., & Illes, P. (2005). P2X and P2Y receptors as possible targets of therapeutic manipulations in CNS illnesses. *Drug News & Perspectives*, *18*, 85–101.

Köles, L., Fürst, S., & Illes, P. (2007). Purine ionotropic (P2X) receptors. *Current Pharmaceutical Design*, *13*, 2368–2384.

Köles, L., Gerevich, Z., Oliveira, J. F., Zadori, Z. S., Wirkner, K., & Illes, P. (2008). Interaction of P2 purinergic receptors with cellular macromolecules. *Naunyn-Schmiedeberg's Archives of Pharmacology*, *377*, 1–33.

Koshimizu, T. A., Kretschmannova, K., He, M. L., Ueno, S., Tanoue, A., Yanagihara, N., et al. (2006). Carboxyl-terminal splicing enhances physical interactions between the cytoplasmic tails of purinergic P2X receptors. *Molecular Pharmacology*, *69*, 1588–1598.

Kotevic, I., Kirschner, K. M., Porzig, H., & Baltensperger, K. (2005). Constitutive interaction of the $P2Y_2$ receptor with the hematopoietic cell-specific G protein $G_{\alpha 16}$ and evidence for receptor oligomers. *Cellular Signalling*, *17*, 869–880.

Krügel, U., Kittner, H., Franke, H., & Illes, P. (2001). Stimulation of P2 receptors in the ventral tegmental area enhances dopaminergic mechanisms in vivo. *Neuropharmacology*, *40*, 1084–1093.

Krügel, U., Kittner, H., Franke, H., & Illes, P. (2003). Purinergic modulation of neuronal activity in the mesolimbic dopaminergic system in vivo. *Synapse*, *47*, 134–142.

Krügel, U., Kittner, H., & Illes, P. (1999). Adenosine 5′-triphosphate-induced dopamine release in the rat nucleus accumbens in vivo. *Neuroscience Letters*, *265*, 49–52.

Krügel, U., Kittner, H., & Illes, P. (2001). Mechanisms of adenosine 5′-triphosphate-induced dopamine release in the rat nucleus accumbens in vivo. *Synapse*, *39*, 222–232.

Kucher, B. M., & Neary, J. T. (2005). Bi-functional effects of ATP/P2 receptor activation on tumor necrosis factor-α release in lipopolysaccharide-stimulated astrocytes. *Journal of Neurochemistry*, *92*, 525–535.

Lakshmi, S., & Joshi, P. G. (2005). Co-activation of $P2Y_2$ receptor and TRPV channel by ATP: Implications for ATP induced pain. *Cellular and Molecular Neurobiology*, *25*, 819–832.

Lalo, U., Pankratov, Y., Wichert, S. P., Rossner, M. J., North, R. A., Kirchhoff, F., et al. (2008). $P2X_1$ and $P2X_5$ subunits form the functional P2X receptor in mouse cortical astrocytes. *The Journal of Neuroscience*, *28*, 5473–5480.

Lazarowski, E. R., Boucher, R. C., & Harden, T. K. (2000). Constitutive release of ATP and evidence for major contribution of ecto-nucleotide pyrophosphatase and nucleoside diphosphokinase to extracellular nucleotide concentrations. *The Journal of Biological Chemistry*, *275*, 31061–31068.

Le Feuvre, R. A., Brough, D., Touzani, O., & Rothwell, N. J. (2003). Role of $P2X_7$ receptors in ischemic and excitotoxic brain injury in vivo. *Journal of Cerebral Blood Flow and Metabolism*, *23*, 381–384.

Le, K. T., Babinski, K., & Seguela, P. (1998). Central $P2X_4$ and $P2X_6$ channel subunits coassemble into a novel heteromeric ATP receptor. *The Journal of Neuroscience*, *18*, 7152–7159.

Lecca, D., Trincavelli, M. L., Gelosa, P., Sironi, L., Ciana, P., Fumagalli, M., et al. (2008). The recently identified P2Y-like receptor GPR17 is a sensor of brain damage and a new target for brain repair. *PLoS One*, *3*, e3579.

Lewis, C., Neidhart, S., Holy, C., North, R. A., Buell, G., & Surprenant, A. (1995). Coexpression of $P2X_2$ and $P2X_3$ receptor subunits can account for ATP-gated currents in sensory neurons. *Nature*, *377*, 432–435.

Li, J., & Perl, E. R. (1995). ATP modulation of synaptic transmission in the spinal substantia gelatinosa. *The Journal of Neuroscience*, *15*, 3357–3365.

Liu, J. S., John, G. R., Sikora, A., Lee, S. C., & Brosnan, C. F. (2000). Modulation of interleukin-1β and tumor necrosis factor α signaling by P2 purinergic receptors in human fetal astrocytes. *The Journal of Neuroscience*, *20*, 5292–5299.

Lo, J. C., Huang, W. C., Chou, Y. C., Tseng, C. H., Lee, W. L., & Sun, S. H. (2008). Activation of $P2X_7$ receptors decreases glutamate uptake and glutamine synthetase activity in RBA-2 astrocytes via distinct mechanisms. *Journal of Neurochemistry*, *105*, 151–164.

Loesch, A., & Burnstock, G. (1998). Electron-immunocytochemical localization of $P2X_1$ receptors in rat cerebellum. *Cell and Tissue Research*, *294*, 253–260.

Lu, H., Burns, D., Garnier, P., Wei, G., Zhu, K., & Ying, W. (2007). $P2X_7$ receptors mediate NADH transport across the plasma membranes of astrocytes. *Biochemical and Biophysical Research Communications*, *362*, 946–950.

Lucae, S., Salyakina, D., Barden, N., Harvey, M., Gagné, B., Labbé, M., et al. (2006). P2RX7, a gene coding for a purinergic ligand-gated ion channel, is associated with major depressive disorder. *Human Molecular Genetics*, *15*, 2438–2445.

Luthardt, J., Borvendeg, S. J., Sperlagh, B., Poelchen, W., Wirkner, K., & Illes, P. (2003). $P2Y_1$ receptor activation inhibits NMDA receptor-channels in layer V pyramidal neurons of the rat prefrontal and parietal cortex. *Neurochemistry International*, *42*, 161–172.

Malin, S. A., & Molliver, D. C. (2010). G_i- and G_q-coupled ADP (P2Y) receptors act in opposition to modulate nociceptive signaling and inflammatory pain behavior. *Molecular Pain*, *6*, 21.

Matute, C., Torre, I., Pérez-Cerdá, F., Pérez-Samartín, A., Alberdi, E., Etxebarria, E., et al. (2007). $P2X_7$ receptor blockade prevents ATP excitotoxicity in oligodendrocytes and ameliorates experimental autoimmune encephalomyelitis. *The Journal of Neuroscience*, *27*, 9525–9533.

McLarnon, J. G., Ryu, J. K., Walker, D. G., & Choi, H. B. (2006). Upregulated expression of purinergic $P2X_7$ receptor in Alzheimer disease and amyloid-β peptide-treated microglia and in peptide-injected rat hippocampus. *Journal of Neuropathology and Experimental Neurology*, *65*, 1090–1097.

Mendoza-Fernández, V., Andrew, R. D., & Barajas-López, C. (2000). ATP inhibits glutamate synaptic release by acting at P2Y receptors in pyramidal neurons of hippocampal slices. *The Journal of Pharmacology and Experimental Therapeutics*, *293*, 172–179.

Monif, M., Reid, C. A., Powell, K. L., Smart, M. L., & Williams, D. A. (2009). The $P2X_7$ receptor drives microglial activation and proliferation: A trophic role for $P2X_7$R pore. *The Journal of Neuroscience*, *29*, 3781–3791.

Moore, D. J., Chambers, J. K., Wahlin, J. P., Tan, K. B., Moore, G. B., Jenkins, O., et al. (2001). Expression pattern of human P2Y receptor subtypes: A quantitative reverse transcription-polymerase chain reaction study. *Biochimica et Biophysica Acta*, *1521*, 107–119.

Moore, D. J., Murdock, P. R., Watson, J. M., Faull, R. L., Waldvogel, H. J., Szekeres, P. G., et al. (2003). GPR105, a novel $G_{i/o}$-coupled UDP-glucose receptor expressed on brain glia and peripheral immune cells, is regulated by immunologic challenge: Possible role in neuroimmune function. *Brain Research. Molecular Brain Research*, *118*, 10–23.

Moran-Jimenez, M. J., & Matute, C. (2000). Immunohistochemical localization of the $P2Y_1$ purinergic receptor in neurons and glial cells of the central nervous system. *Brain Research. Molecular Brain Research*, *78*, 50–58.

Mori, M., Heuss, C., Gahwiler, B. H., & Gerber, U. (2001). Fast synaptic transmission mediated by P2X receptors in CA3 pyramidal cells of rat hippocampal slice cultures. *The Journal of Physiology*, *535*, 115–123.

Moriyama, T., Iida, T., Kobayashi, K., Higashi, T., Fukuoka, T., Tsumura, H., et al. (2003). Possible involvement of $P2Y_2$ metabotropic receptors in ATP-induced transient receptor potential vanilloid receptor 1-mediated thermal hypersensitivity. *The Journal of Neuroscience*, *23*, 6058–6062.

Murakami, K., Nakamura, Y., & Yoneda, Y. (2003). Potentiation by ATP of lipopolysaccharide-stimulated nitric oxide production in cultured astrocytes. *Neuroscience*, *117*, 37–42.

Nakatsuka, T., & Gu, J. G. (2001). ATP P2X receptor-mediated enhancement of glutamate release and evoked EPSCs in dorsal horn neurons of the rat spinal cord. *The Journal of Neuroscience*, *21*, 6522–6531.

Nakazawa, K. (1994). ATP-activated current and its interaction with acetylcholine-activated current in rat sympathetic neurons. *The Journal of Neuroscience*, *14*, 740–750.

Nakazawa, K., Inoue, K., & Inoue, K. (1994). ATP reduces voltage-activated K^+ current in cultured rat hippocampal neurons. *Pflügers Archiv: European Journal of Physiology*, *429*, 143–145.

Nakazawa, K., Yamakoshi, Y., Tsuchiya, T., & Ohno, Y. (2005). Purification and aqueous phase atomic force microscopic observation of recombinant $P2X_2$ receptor. *European Journal of Pharmacology*, *518*, 107–110.

Neary, J. T., Kang, Y., Bu, Y., Yu, E., Akong, K., & Peters, C. M. (1999). Mitogenic signaling by ATP/P2Y purinergic receptors in astrocytes: Involvement of a calcium-independent protein kinase C, extracellular signal-regulated protein kinase pathway distinct from the phosphatidylinositol-specific phospholipase C/calcium pathway. *The Journal of Neuroscience*, *19*, 4211–4220.

Newbolt, A., Stoop, R., Virginio, C., Surprenant, A., North, R. A., Buell, G., et al. (1998). Membrane topology of an ATP-gated ion channel (P2X receptor). *The Journal of Biological Chemistry*, *273*, 15177–15182.

Newman, E. A. (2003). New roles for astrocytes: Regulation of synaptic transmission. *Trends in Neurosciences*, *26*, 536–542.

Nicke, A., Bäumert, H. G., Rettinger, J., Eichele, A., Lambrecht, G., Mutschler, E., et al. (1998). $P2X_1$ and $P2X_3$ receptors form stable trimers: A novel structural motif of ligand-gated ion channels. *The EMBO Journal*, *17*, 3016–3028.

Nicke, A., Kerschensteiner, D., & Soto, F. (2005). Biochemical and functional evidence for heteromeric assembly of $P2X_1$ and $P2X_4$ subunits. *Journal of Neurochemistry*, *92*, 925–933.

Nieber, K., Poelchen, W., & Illes, P. (1997). Role of ATP in fast excitatory synaptic potentials in locus coeruleus neurones of the rat. *British Journal of Pharmacology*, *122*, 423–430.

Nörenberg, W., & Illes, P. (2000). Neuronal P2X receptors: Localisation and functional properties. *Naunyn-Schmiedeberg's Archives of Pharmacology*, *362*, 324–339.

Nörenberg, W., Schunk, J., Fischer, W., Sobottka, H., Riedel, T., Oliveira, J. F., et al. (2010). Electrophysiological classification of $P2X_7$ receptors in rat cultured neocortical astroglia. *British Journal of Pharmacology*, *160*, 1941–1952.

North, R. A. (1996). P2X receptors: A third major class of ligand-gated ion channels. *Ciba Foundation Symposium*, *198*, 91–105.

North, R. A. (2002). Molecular physiology of P2X receptors. *Physiological Reviews*, *82*, 1013–1067.

Oliveira, J. F., Riedel, T., Leichsenring, A., Heine, C., Franke, H., Krügel, U., et al. (2011). Rodent cortical astroglia express in situ functional $P2X_7$ receptors sensing pathologically high ATP concentrations. *Cereb Cortex*, *21*, 806–820.

Pankratov, Y., Castro, E., Miras-Portugal, M. T., & Krishtal, O. (1998). A purinergic component of the excitatory postsynaptic current mediated by P2X receptors in the CA1 neurons of the rat hippocampus. *The European Journal of Neuroscience*, *10*, 3898–3902.

Pankratov, Y., Lalo, U., Krishtal, O., & Verkhratsky, A. (2002). Ionotropic P2X purinoreceptors mediate synaptic transmission in rat pyramidal neurones of layer II/III of somato-sensory cortex. *The Journal of Physiology*, *542*, 529–536.

Pankratov, Y., Lalo, U., Krishtal, O. A., & Verkhratsky, A. (2009). P2X receptors and synaptic plasticity. *Neuroscience*, *158*, 137–148.

Pankratov, Y., Lalo, U., Verkhratsky, A., & North, R. A. (2007). Quantal release of ATP in mouse cortex. *The Journal of General Physiology*, *129*, 257–265.

Papp, L., Balázsa, T., Köfalvi, A., Erdélyi, F., Szabó, G., Vizi, E. S., et al. (2004). P2X receptor activation elicits transporter-mediated noradrenaline release from rat hippocampal slices. *The Journal of Pharmacology and Experimental Therapeutics*, *310*, 973–980.

Parvathenani, L. K., Tertyshnikova, S., Greco, C. R., Roberts, S. B., Robertson, B., & Posmantur, R. (2003). $P2X_7$ mediates superoxide production in primary microglia and is up-regulated in a transgenic mouse model of Alzheimer's disease. *The Journal of Biological Chemistry*, *278*, 13309–13317.

Patti, L., Raiteri, L., Grilli, M., Parodi, M., Raiteri, M., & Marchi, M. (2006). $P2X_7$ receptors exert a permissive role on the activation of release-enhancing presynaptic α_7 nicotinic receptors co-existing on rat neocortex glutamatergic terminals. *Neuropharmacology*, *50*, 705–713.

Paukert, M., Osteroth, R., Geisler, H. S., Brandle, U., Glowatzki, E., Ruppersberg, J. P., et al. (2001). Inflammatory mediators potentiate ATP-gated channels through the $P2X_3$ subunit. *The Journal of Biological Chemistry*, *276*, 21077–21082.

Pearce, B., Murphy, S., Jeremy, J., Morrow, C., & Dandona, P. (1989). ATP-evoked Ca^{2+} mobilisation and prostanoid release from astrocytes: P2-purinergic receptors linked to phosphoinositide hydrolysis. *Journal of Neurochemistry*, *52*, 971–977.

Price, G. D., Robertson, S. J., & Edwards, F. A. (2003). Long-term potentiation of glutamatergic synaptic transmission induced by activation of presynaptic P2Y receptors in the rat medial habenula nucleus. *The European Journal of Neuroscience*, *17*, 844–850.

Qureshi, O. S., Paramasivam, A., Yu, J. C., & Murrell-Lagnado, R. D. (2007). Regulation of $P2X_4$ receptors by lysosomal targeting, glycan protection and exocytosis. *Journal of Cell Science*, *120*, 3838–3849.

Radford, K. M., Virginio, C., Surprenant, A., North, R. A., & Kawashima, E. (1997). Baculovirus expression provides direct evidence for heteromeric assembly of $P2X_2$ and $P2X_3$ receptors. *The Journal of Neuroscience*, *17*, 6529–6533.

Ralevic, V., & Burnstock, G. (1998). Receptors for purines and pyrimidines. *Pharmacological Reviews*, *50*, 413–492.

Rampe, D., Wang, L., & Ringheim, G. E. (2004). $P2X_7$ receptor modulation of β-amyloid- and LPS-induced cytokine secretion from human macrophages and microglia. *Journal of Neuroimmunology*, *147*, 56–61.

Raouf, R., Chabot-Doré, A. J., Ase, A. R., Blais, D., & Séguéla, P. (2007). Differential regulation of microglial $P2X_4$ and $P2X_7$ ATP receptors following LPS-induced activation. *Neuropharmacology*, *53*, 496–504.

Rappold, P. M., Lynd-Balta, E., & Joseph, S. A. (2006). $P2X_7$ receptor immunoreactive profile confined to resting and activated microglia in the epileptic brain. *Brain Research*, *1089*, 171–178.

Rettinger, J., Aschrafi, A., & Schmalzing, G. (2000). Roles of individual N-glycans for ATP potency and expression of the rat $P2X_1$ receptor. *The Journal of Biological Chemistry*, *275*, 33542–33547.

Rhee, J. S., Wang, Z. M., Nabekura, J., Inoue, K., & Akaike, N. (2000). ATP facilitates spontaneous glycinergic IPSC frequency at dissociated rat dorsal horn interneuron synapses. *The Journal of Physiology*, *524*, 471–483.

Riedel, T., Lozinsky, I., Schmalzing, G., & Markwardt, F. (2007). Kinetics of $P2X_7$ receptor-operated single channels currents. *Biophysical Journal*, *92*, 2377–2391.

Roberts, J. A., Vial, C., Digby, H. R., Agboh, K. C., Wen, H., Atterbury-Thomas, A., et al. (2006). Molecular properties of P2X receptors. *Pflügers Archiv: European Journal of Physiology*, *452*, 486–500.

Robertson, S. J., Ennion, S. J., Evans, R. J., & Edwards, F. A. (2001). Synaptic P2X receptors. *Current Opinion in Neurobiology*, *11*, 378–386.

Rodrigues, R. J., Almeida, T., Richardson, P. J., Oliveira, C. R., & Cunha, R. A. (2005). Dual presynaptic control by ATP of glutamate release via facilitatory $P2X_1$, $P2X_{2/3}$, and $P2X_3$ and inhibitory $P2Y_1$, $P2Y_2$, and/or $P2Y_4$ receptors in the rat hippocampus. *The Journal of Neuroscience*, *25*, 6286–6295.

Royle, S. J., Bobanovic, L. K., & Murrell-Lagnado, R. D. (2002). Identification of a non-canonical tyrosine-based endocytic motif in an ionotropic receptor. *The Journal of Biological Chemistry*, *277*, 35378–35385.

Rubini, P., Engelhardt, J., Wirkner, K., & Illes, P. (2008). Modulation by D_1 and D_2 dopamine receptors of ATP-induced release of intracellular Ca^{2+} in cultured rat striatal neurons. *Neurochemistry International*, *52*, 113–118.

Rubini, P., Pinkwart, C., Franke, H., Gerevich, Z., Nörenberg, W., & Illes, P. (2006). Regulation of intracellular Ca^{2+} by $P2Y_1$ receptors may depend on the developmental stage of cultured rat striatal neurons. *Journal of Cellular Physiology*, *209*, 81–93.

Rubio, M. E., & Soto, F. (2001). Distinct localization of P2X receptors at excitatory postsynaptic specializations. *The Journal of Neuroscience*, *21*, 641–653.

Saitow, F., Murakoshi, T., Suzuki, H., & Konishi, S. (2005). Metabotropic P2Y purinoceptor-mediated presynaptic and postsynaptic enhancement of cerebellar GABAergic transmission. *The Journal of Neuroscience*, *25*, 2108–2116.

Salter, M. W., & Hicks, J. L. (1994). ATP-evoked increases in intracellular calcium in neurons and glia from the dorsal spinal cord. *The Journal of Neuroscience*, *14*, 1563–1575.

Sasaki, Y., Hoshi, M., Akazawa, C., Nakamura, Y., Tsuzuki, H., Inoue, K., et al. (2003). Selective expression of $G_{i/o}$-coupled ATP receptor $P2Y_{12}$ in microglia in rat brain. *Glia*, *44*, 242–250.

Savi, P., Zachayus, J. L., Delesque-Touchard, N., Labouret, C., Hervé, C., Uzabiaga, M. F., et al. (2006). The active metabolite of Clopidogrel disrupts $P2Y_{12}$ receptor oligomers and partitions them out of lipid rafts. *Proceedings of the National Academy of Sciences of the United States of America*, *103*, 11069–11074.

Shigetomi, E., & Kato, F. (2004). Action potential-independent release of glutamate by Ca^{2+} entry through presynaptic P2X receptors elicits postsynaptic firing in the brainstem autonomic network. *The Journal of Neuroscience*, *24*, 3125–3135.

Shinozaki, Y., Sumitomo, K., Tsuda, M., Koizumi, S., Inoue, K., & Torimitsu, K. (2009). Direct observation of ATP-induced conformational changes in single $P2X_4$ receptors. *PLoS Biology*, *7*, e103.

Sim, J. A., Chaumont, S., Jo, J., Ulmann, L., Young, M. T., Cho, K., et al. (2006). Altered hippocampal synaptic potentiation in $P2X_4$ knock-out mice. *The Journal of Neuroscience*, *26*, 9006–9009.

Sim, J. A., Young, M. T., Sung, H. Y., North, R. A., & Surprenant, A. (2004). Reanalysis of $P2X_7$ receptor expression in rodent brain. *The Journal of Neuroscience*, *24*, 6307–6314.

Simon, J., Webb, T. E., & Barnard, E. A. (1997). Distribution of $[^{35}S]dATP\alpha S$ binding sites in the adult rat neuraxis. *Neuropharmacology*, *36*, 1243–1251.

Skaper, S. D., Debetto, P., & Giusti, P. (2010). The $P2X_7$ purinergic receptor: From physiology to neurological disorders. *The FASEB Journal*, *24*, 337–345.

Smith, F. M., Humphrey, P. P., & Murrell-Lagnado, R. D. (1999). Identification of amino acids within the $P2X_2$ receptor C-terminus that regulate desensitization. *The Journal of Physiology*, *520*, 91–99.

Sokolova, E., Nistri, A., & Giniatullin, R. (2001). Negative cross talk between anionic GABAA and cationic P2X ionotropic receptors of rat dorsal root ganglion neurons. *The Journal of Neuroscience*, *21*, 4958–4968.

Sokolova, E., Skorinkin, A., Fabbretti, E., Masten, L., Nistri, A., & Giniatullin, R. (2004). Agonist-dependence of recovery from desensitization of $P2X_3$ receptors provides a novel and sensitive approach for their rapid up or downregulation. *British Journal of Pharmacology*, *141*, 1048–1058.

Souslova, V., Cesare, P., Ding, Y., Akopian, A. N., Stanfa, L., Suzuki, R., et al. (2000). Warm-coding deficits and aberrant inflammatory pain in mice lacking $P2X_3$ receptors. *Nature*, *407*, 1015–1017.

Sperlágh, B., Erdélyi, F., Szabó, G., & Vizi, E. S. (2000). Local regulation of $[^{(3)}H]$-noradrenaline release from the isolated guinea-pig right atrium by P2X-receptors located on axon terminals. *British Journal of Pharmacology*, *131*, 1775–1783.

Sperlágh, B., Heinrich, A., & Csölle, C. (2007). P2 receptor-mediated modulation of neurotransmitter release—An update. *Purinergic Signalling*, *3*, 269–284.

Sperlágh, B., Köfalvi, A., Deuchars, J., Atkinson, L., Milligan, C. J., Buckley, N. J., et al. (2002). Involvement of $P2X_7$ receptors in the regulation of neurotransmitter release in the rat hippocampus. *Journal of Neurochemistry*, *81*, 1196–1211.

Sperlágh, B., & Vizi, E. S. (1991). Effect of presynaptic P2 receptor stimulation on transmitter release. *Journal of Neurochemistry*, *56*, 1466–1470.

Sperlágh, B., Vizi, E. S., Wirkner, K., & Illes, P. (2006). $P2X_7$ receptors in the nervous system. *Progress in Neurobiology*, *78*, 327–346.

Stanchev, D., Blosa, M., Milius, D., Gerevich, Z., Rubini, P., Schmalzing, G., et al. (2009). Cross-inhibition between native and recombinant TRPV1 and $P2X_3$ receptors. *Pain*, *143*, 26–36.

Stanchev, D., Flehmig, G., Gerevich, Z., Nörenberg, W., Dihazi, H., Fürst, S., et al. (2006). Decrease of current responses at human recombinant $P2X_3$ receptors after substitution by Asp of Ser/Thr residues in protein kinase C phosphorylation sites of their ecto-domains. *Neuroscience Letters*, *393*, 78–83.

Stella, N., Estelles, A., Siciliano, J., Tencé, M., Desagher, S., Piomelli, D., et al. (1997). Interleukin-1 enhances the ATP-evoked release of arachidonic acid from mouse astrocytes. *The Journal of Neuroscience*, *17*, 2939–2946.

Stout, C. E., Costantin, J. L., Naus, C. C., & Charles, A. C. (2002). Intercellular calcium signaling in astrocytes via ATP release through connexin hemichannels. *The Journal of Biological Chemistry*, *277*, 10482–10488.

Suadicani, S. O., Brosnan, C. F., & Scemes, E. (2006). $P2X_7$ receptors mediate ATP release and amplification of astrocytic intercellular Ca^{2+} signaling. *The Journal of Neuroscience*, *26*, 1378–1385.

Suh, B. C., Kim, J. S., Namgung, U., Han, S., & Kim, K. T. (2001). Selective inhibition of β_2-adrenergic receptor-mediated cAMP generation by activation of the $P2Y_2$ receptor in mouse pineal gland tumor cells. *Journal of Neurochemistry*, *77*, 1475–1485.

Suzuki, T., Hide, I., Ido, K., Kohsaka, S., Inoue, K., & Nakata, Y. (2004). Production and release of neuroprotective tumor necrosis factor by $P2X_7$ receptor-activated microglia. *The Journal of Neuroscience*, *24*, 1–7.

Suzuki, T., Namba, K., Tsuga, H., & Nakat, H. (2006). Regulation of pharmacology by hetero-oligomerization between A_1 adenosine receptor and $P2Y_2$ receptor. *Biochemical and Biophysical Research Communications*, *351*, 559–565.

Tominaga, M., Wada, M., & Masu, M. (2001). Potentiation of capsaicin receptor activity by metabotropic ATP receptors as a possible mechanism for ATP-evoked pain and hyperalgesia. *Proceedings of the National Academy of Sciences of the United States of America*, *98*, 6951–6956.

Tonazzini, I., Trincavelli, M. L., Storm-Mathisen, J., Martini, C., & Bergersen, L. H. (2007). Co-localization and functional cross-talk between A_1 and $P2Y_1$ purine receptors in rat hippocampus. *The European Journal of Neuroscience*, *26*, 890–902.

Torres, G. E., Egan, T. M., & Voigt, M. M. (1999). Hetero-oligomeric assembly of P2X receptor subunits. Specificities exist with regard to possible partners. *The Journal of Biological Chemistry*, *274*, 6653–6659.

Torres, G. E., Haines, W. R., Egan, T. M., & Voigt, M. M. (1998). Co-expression of $P2X_1$ and $P2X_5$ receptor subunits reveals a novel ATP-gated ion channel. *Molecular Pharmacology*, *54*, 989–993.

Toulme, E., Soto, F., Garret, M., & Boué-Grabot, E. (2006). Functional properties of internalization-deficient $P2X_4$ receptors reveal a novel mechanism of ligand-gated channel facilitation by ivermectin. *Molecular Pharmacology*, *69*, 576–587.

Tozaki-Saitoh, H., Tsuda, M., Miyata, H., Ueda, K., Kohsaka, S., & Inoue, K. (2008). $P2Y_{12}$ receptors in spinal microglia are required for neuropathic pain after peripheral nerve injury. *The Journal of Neuroscience*, *28*, 4949–4956.

Trendelenburg, A. U., & Bültmann, R. (2000). P2 receptor-mediated inhibition of dopamine release in rat neostriatum. *Neuroscience*, *96*, 249–252.

Tsuda, M., Inoue, K., & Salter, M. W. (2005). Neuropathic pain and spinal microglia: A big problem from molecules in "small" glia. *Trends in Neurosciences*, *28*, 101–107.

Tsuda, M., Shigemoto-Mogami, Y., Koizumi, S., Mizokoshi, A., Kohsaka, S., Salter, M. W., et al. (2003). $P2X_4$ receptors induced in spinal microglia gate tactile allodynia after nerve injury. *Nature*, *424*, 778–783.

Valera, S., Hussy, N., Evans, R. J., Adami, N., North, R. A., Surprenant, A., et al. (1994). A new class of ligand-gated ion channel defined by P2X receptor for extracellular ATP. *Nature*, *371*, 516–519.

van Kolen, K., & Slegers, H. (2004). $P2Y_{12}$ receptor stimulation inhibits β-adrenergic receptor-induced differentiation by reversing the cyclic AMP-dependent inhibition of protein kinase B. *Journal of Neurochemistry*, *89*, 442–453.

Verkhratsky, A., Krishtal, O. A., & Burnstock, G. (2009). Purinoceptors on neuroglia. *Molecular Neurobiology*, *39*, 190–208.

Vial, C., Tobin, A. B., & Evans, R. J. (2004). G-protein-coupled receptor regulation of $P2X_1$ receptors does not involve direct channel phosphorylation. *The Biochemical Journal*, *382*, 101–110.

Vianna, E. P., Ferreira, A. T., Naffah-Mazzacoratti, M. G., Sanabria, E. R., Funke, M., Cavalheiro, E. A., et al. (2002). Evidence that ATP participates in the pathophysiology of pilocarpine-induced temporal lobe epilepsy: Fluorimetric, immunohistochemical, and Western blot studies. *Epilepsia*, *5*, 227–229.

Virginio, C., Church, D., North, R. A., & Surprenant, A. (1997). Effects of divalent cations, protons and calmidazolium at the rat $P2X_7$ receptor. *Neuropharmacology*, *36*, 1285–1294.

Virginio, C., MacKenzie, A., Rassendren, F. A., North, R. A., & Surprenant, A. (1999). Pore dilation of neuronal P2X receptor channels. *Nature Neuroscience*, *2*, 315–321.

Vizi, E. S., & Knoll, J. (1976). The inhibitory effect of adenosine and related nucleotides on the release of acetylcholine. *Neuroscience*, *1*, 391–398.

Volonte, C., Amadio, S., Cavaliere, F., D'Ambrosi, N., Vacca, F., & Bernardi, G. (2003). Extracellular ATP and neurodegeneration. *Current Drug Targets. CNS and Neurological Disorders*, *2*, 403–412.

Volonte, C., Amadio, S., D'Ambrosi, N., Colpi, M., & Burnstock, G. (2006). P2 receptor web: Complexity and fine-tuning. *Pharmacology & Therapeutics*, *112*, 264–280.

von Kügelgen, I. (2006). Pharmacological profiles of cloned mammalian P2Y-receptor subtypes. *Pharmacology & Therapeutics*, *110*, 415–432.

von Kügelgen, I., Koch, H., & Starke, K. (1997). P2-receptor-mediated inhibition of serotonin release in the rat brain cortex. *Neuropharmacology*, *36*, 1221–1227.

von Kügelgen, I., Schöffel, E., & Starke, K. (1989). Inhibition by nucleotides acting at presynaptic P2-receptors of sympathetic neuro-effector transmission in the mouse isolated vas deferens. *Naunyn-Schmiedeberg's Archives of Pharmacology*, *340*, 522–532.

von Kügelgen, I., Späth, L., & Starke, K. (1994). Evidence for P2-purinoceptor-mediated inhibition of noradrenaline release in rat brain cortex. *British Journal of Pharmacology, 113*, 815–822.

Vulchanova, L., Riedl, M. S., Shuster, S. J., Buell, G., Surprenant, A., North, R. A., et al. (1997). Immunohistochemical study of the $P2X_2$ and $P2X_3$ receptor subunits in rat and monkey sensory neurons and their central terminals. *Neuropharmacology, 36*, 1229–1242.

Wakamori, M., & Sorimachi, M. (2004). Properties of nativenaive P2X receptors in large multipolar neurons dissociated from rat hypothalamic arcuate nucleus. *Brain Research, 1005*, 51–59.

Walter, L., Dinh, T., & Stella, N. (2004). ATP induces a rapid and pronounced increase in 2-arachidonoylglycerol production by astrocytes, a response limited by monoacylglycerol lipase. *The Journal of Neuroscience, 24*, 8068–8074.

Wang, Y., Haughey, N. J., Mattson, M. P., & Furukawa, K. (2004). Dual effects of ATP on rat hippocampal synaptic plasticity. *NeuroReport, 15*, 633–636.

Wang, L., Karlsson, L., Moses, S., Hultgårdh-Nilsson, A., Andersson, M., Borna, C., et al. (2002). P2 receptor expression profiles in human vascular smooth muscle and endothelial cells. *Journal of Cardiovascular Pharmacology, 40*, 841–853.

Wang, Z. M., Katsurabayashi, S., Rhee, J. S., Brodwick, M., & Akaike, N. (2001). Substance P abolishes the facilitatory effect of ATP on spontaneous glycine release in neurons of the trigeminal nucleus pars caudalis. *The Journal of Neuroscience, 21*, 2983–2991.

Watano, T., Calvert, J. A., Vial, C., Forsythe, I. D., & Evans, R. J. (2004). P2X receptor subtype-specific modulation of excitatory and inhibitory synaptic inputs in the rat brainstem. *The Journal of Physiology, 558*, 745–757.

Westfall, D. P., Shinozuka, K., Forsyth, K. M., & Bjur, R. A. (1990). Presynaptic purine receptors. *Annals of the New York Academy of Sciences, 604*, 130–135.

White, P. J., Webb, T. E., & Boarder, M. R. (2003). Characterization of a Ca^{2+} response to both UTP and ATP at human $P2Y_{11}$ receptors: Evidence for agonist-specific signaling. *Molecular Pharmacology, 63*, 1356–1363.

Wirkner, K., Günther, A., Weber, M., Guzman, S. J., Krause, T., Fuchs, J., et al. (2007). Modulation of NMDA receptor current in layer V pyramidal neurons of the rat prefrontal cortex by P2Y receptor activation. *Cerebral Cortex, 17*, 621–631.

Wirkner, K., Köles, L., Thümmler, S., Luthardt, J., Poelchen, W., Franke, H., et al. (2002). Interaction between P2Y and NMDA receptors in layer V pyramidal neurons of the rat prefrontal cortex. *Neuropharmacology, 42*, 476–488.

Wirkner, K., Sperlagh, B., & Illes, P. (2007). $P2X_3$ receptor involvement in pain states. *Molecular Neurobiology, 36*, 165–183.

Wirkner, K., Stanchev, D., Köles, L., Klebingat, M., Dihazi, H., Flehmig, G., et al. (2005). Regulation of human recombinant $P2X_3$ receptors by ecto-protein kinase C. *The Journal of Neuroscience, 25*, 7734–7742.

Witting, A., Walter, L., Wacker, J., Möller, T., & Stella, N. (2004). $P2X_7$ receptors control 2-arachidonoylglycerol production by microglial cells. *Proceedings of the National Academy of Sciences of the United States of America, 101*, 3214–3219.

Wu, J., Holstein, J. D., Upadhyay, G., Lin, D. T., Conway, S., Muller, E., et al. (2007). Purinergic receptor-stimulated IP_3-mediated Ca^{2+} release enhances neuroprotection by increasing astrocyte mitochondrial metabolism during aging. *The Journal of Neuroscience, 27*, 6510–6520.

Yamakuni, H., Kawaguchi, N., Ohtani, Y., Nakamura, J., Katayama, T., Nakagawa, T., et al. (2002). ATP induces leukemia inhibitory factor mRNA in cultured rat astrocytes. *Journal of Neuroimmunology, 129*, 43–50.

Yoshioka, K., Hosoda, R., Kuroda, Y., & Nakata, H. (2002). Hetero-oligomerization of adenosine A_1 receptors with $P2Y_1$ receptors in rat brains. *FEBS Letters, 531*, 299–303.

Yoshioka, K., Saitoh, O., & Nakata, H. (2001). Heteromeric association creates a P2Y-like adenosine receptor. *Proceedings of the National Academy of Sciences of the United States of America*, *98*, 7617–7622.

Young, M. T. (2010). P2X receptors: Dawn of the post-structure era. *Trends in Biochemical Sciences*, *35*, 83–90.

Zeng, J. W., Liu, X. H., Zhao, Y. D., Xiao, Z., He, W. J., Hu, Z. A., et al. (2009). Role of $P2Y_1$ receptor in astroglia-to-neuron signaling at dorsal spinal cord. *Journal of Neuroscience Research*, *87*, 2667–2676.

Zhang, Z., Chen, G., Zhou, W., Song, A., Xu, T., Luo, Q., et al. (2007). Regulated ATP release from astrocytes through lysosome exocytosis. *Nature Cell Biology*, *9*, 945–953.

Zhang, Y. X., Yamashita, H., Ohshita, T., Sawamoto, N., & Nakamura, S. (1995). ATP increases extracellular dopamine level through stimulation of P2Y purinoceptors in the rat striatum. *Brain Research*, *691*, 205–212.

Zhou, D., Chen, M. L., Zhang, Y. Q., & Zhao, Z. Q. (2010). Involvement of spinal microglial $P2X_7$ receptor in generation of tolerance to morphine analgesia in rats. *The Journal of Neuroscience*, *30*, 8042–8047.

Zimmermann, H. (2006). Ectonucleotidases in the nervous system. *Novartis Foundation Symposium*, *276*, 113–128.

Hidetoshi Tozaki-Saitoh, Makoto Tsuda, and Kazuhide Inoue

Department of Molecular and System Pharmacology, Graduate School of Pharmaceutical Sciences, Kyushu University, Higashi, Fukuoka, Japan

Role of Purinergic Receptors in CNS Function and Neuroprotection

Abstract

The purinergic receptor family contains some of the most abundant receptors in living organisms. A growing body of evidence indicates that extracellular nucleotides play important roles in the regulation of neuronal and glial functions in the nervous system through purinergic receptors. Nucleotides are released from or leaked through nonexcitable cells and neurons during normal physiological and pathophysiological conditions. Ionotropic P2X and metabotropic P2Y purinergic receptors are expressed in the central nervous system (CNS), participate in the synaptic processes, and mediate intercellular communications between neuron and gila and between glia and other glia. Glial cells in the CNS are classified into astrocytes, oligodendrocytes, and microglia. Astrocytes express many types of purinergic receptors, which are integral to their activation. Astrocytes release adenosine triphosphate (ATP) as a "gliotransmitter" that allows communication with neurons, the vascular walls of capillaries, oligodendrocytes, and microglia. Oligodendrocytes are myelin-forming cells that construct insulating layers of myelin sheets around axons, and using purinergic receptor signaling for their development and for myelination. Microglia also express many types of purinergic receptors and are known to function as immunocompetent cells in the CNS. ATP and other nucleotides work as "warning molecules" especially by activating microglia

Advances in Pharmacology, Volume 61

1054-3589/11 $35.00
10.1016/B978-0-12-385526-8.00015-1

in pathophysiological conditions. Studies on purinergic signaling could facilitate the development of novel therapeutic strategies for disorder of the CNS.

I. Introduction

In 1972, Burnstock first proposed a role for nucleotides as neurotransmitters (Burnstock, 1972). In 1993, the first receptors for nucleotides, called P2 purinoceptors, were cloned (Lustig et al., 1993; Webb et al., 1993). During the past two decades, evidences have accumulated for the participation of extracellular nucleotides and nucleosides as neurotransmitters in neuronal signaling (Burnstock, 2007a). Numerous subtypes of these receptors were cloned, and subsequently, the field of purinergic nervous system has been widely accepted by scientists. Purine-sensitive receptors were first classified as P1 G-coupled receptors, which are activated by adenosine and purinergic receptors, and respond to adenosine triphosphate (ATP) stimulation (Burnstock & Kennedy, 1985). Based on receptor cloning and receptor-induced signal transduction, purinergic receptors were subdivided into ATP-gated ionotropic receptors (P2X) and G protein-coupled metabotropic receptors (P2Y; Abbracchio et al., 2006; Khakh et al., 2001). Accumulating evidences indicate that nucleotides are released from and leaked through nonexcitable cells and neurons and are involved in cell-to-cell communication during normal physiological and pathophysiological conditions (Abbracchio et al., 2009; Fields & Burnstock, 2006). We now have vast amount of information indicating the ubiquitous presence of purinergic signaling throughout the body.

ATP and uridine triphosphate (UTP) and their metabolites are the main purinergic agonists that activate P2X or P2Y receptors. It is known that activation of purinergic receptors often results in increased intracellular free calcium concentrations. Changes in calcium are involved in the regulation of many physiological processes in the central nervous system (CNS; Burnstock, 2007a). In the extracellular space, ecto-enzymes rapidly degrade these nucleotides to ADP or UDP, subsequently activating distinct P2Y receptors. Alternatively, these nucleotides can be finally degraded to adenosine, which activates P1 receptors (Burnstock, 2007b).

Glial cells make up over 70% of the total cell population in the CNS. Astrocytes express many types of P2X/Y receptors and release ATP spontaneously or in response to various stimuli (Verkhratsky et al., 2009). Further, astrocytes communicate with neurons at synapses, with oligodendrocytes, microglial cells, and the vascular walls of capillaries (Abbracchio et al., 2009; Iadecola & Nedergaard, 2007; Inoue et al., 2007). Microglia are known as the resident macrophages of the CNS and express many types of purinergic receptors (Farber & Kettenmann, 2006; Inoue, 2008). Oligodendrocytes are myelin-forming cells that construct insulating layers of myelin sheets around axons in the CNS and utilize both P1 and purinergic receptors for their development (Fields & Stevens-Graham, 2002). Although it was traditionally believed that glial cells are the only source of physical and metabolic supports in the nervous system for neurons, a growing body of evidence has dramatically changed this classical view. An emerging conceptualization of glial cells indicates that neuron–glia interactions are a key component of the CNS functions, and purinergic signaling is one of its mediators (Abbracchio et al., 2009). In addition to crucial roles in normal physiological conditions, purinergic signaling plays an important role in pathophysiological conditions of the CNS (Burnstock, 2008) including physical trauma, cerebral ischemia, neurodegenerative diseases, neuroinflammatory disorders, neuropsychiatric disorders, and neuropathic pain (Table I). In this chapter, an overview of studies on the role of the purinergic receptors in physiological and pathophysiological processes of the cell types forming CNS will be presented. Large amounts of hints for developing new drugs for neuroprotection may be buried in these fundamental studies.

II. The Role of Purinergic Receptors in Neuronal Cell Function and Survival

A. Synaptic Transmission

Synaptic currents mediated by activation of P2X receptors are generally found in the CNS. The first conceptualization of ATP-mediated transmission in the peripheral nervous system was introduced by Burnstock (Burnstock, 1972; Burnstock et al., 1970). Thereafter, it was demonstrated that ATP is released from brain synaptosomal preparations by stimulation with high extracellular concentrations of potassium (White, 1978) and is released from Schaffer collateral–commissural afferents of hippocampal slices by electrical stimulation (Wieraszko & Seyfried, 1989). Later, studies have shown that exogenously applied ATP depolarizes dorsal horn neurons and elicits currents in sensory neurons (Jahr & Jessell, 1983; Krishtal et al., 1983). ATP induces fast synaptic currents in cultured neurons from the hippocampus (Inoue et al., 1992; Mori et al., 2001), spinal cord (Bardoni et al., 1997), locus coeruleus (Nieber et al., 1997), and somatosensory cortex (Pankratov et al., 2002), as well as in slices from the medial habenula (Edwards et al., 1992). Activation of P2X receptors has been implicated in the regulation of synaptic plasticity (Pankratov et al., 2009). However, a consensus opinion regarding the role of P2X receptors in long-term potentiation and long-term depression has not yet been gained. Obtaining a better understanding of the role of P2X receptors in synaptic plasticity, it would require consideration of complex factors around the

TABLE I Involvement of Purinergic Receptors in *In Vivo* Models of CNS Disorders

CNS disorders	*Receptor*	*Primary observation*	*References*
Spinal cord injury	A_{2A}	Agonists reduce tissue damage	Genovese et al. (2010)
	P2	ATP-$MgCl_2$ protected the spinal cord from secondary injury after the trauma	Cakir et al. (2003)
	P2X7	*o*-ATP and PPADS improved functional recovery and diminished cell death in the peritraumatic zone	Wang et al. (2004)
	P2X7	Systemic administration of BBG reduced damage and improved motor recovery	Peng et al. (2009)
	$P2Y_2$	Increased expression of the receptor	Rodriguez-Zayas et al. (2010)
Alzheimer's disease	A_1 or A_{2A}	Caffeine administration resulted in lower hippocampal Aβ levels	Arendash et al. (2006)
	A_{2A}	SCH58261and caffeine prevented the Aβ-induced cognitive impairment	Dall'Igna et al. (2007)
	A_{2A}	SCH58261 and A_{2A} receptor knockout prevented Aβ-induced synaptotoxicity and memory dysfunction	Canas et al. (2009)
	P2X7	Increased expression of the receptor	McLarnon et al. (2006) and Parvathenani et al. (2003)
	P2X7	BBG reduced purinergic receptor expression, attenuated gliosis, diminished leakiness of blood–brain barrier in Aβ-injected brain	Ryu and McLarnon (2008)
	P2X7	KO mice did not show Aβ-induced accumulation of IL-1β in hippocampus	Sanz et al. (2009)
	$P2Y_1$	Colocalized with neurofibrillary tangles, neuritic plaques, and neuropil threads	Moore et al. (2000)
	$P2Y_2$	Immunoreactivity was reduced in the AD parietal cortex	Lai et al. (2008)
Parkinson's disease	A_1	Antagonist reversed the neuroprotective and antineuroinflammatory effects of paeoniflorin on MPTP-treated mice	Liu et al. (2006)
	A_{2A}	Antagonist reduces postsynaptic effects of dopamine depletion	Schwarzschild et al. (2006) *review*
	P2X7	A-438079 partially prevented the 6-OHDA-induced depletion of striatal DA stores	Marcellino et al. (2010)
Huntington's disease	A_{2A}	Agonist and antagonist have beneficial data	Popoli et al. (2007, 2008) *review*

TABLE I *(continued)*

CNS disorders	*Receptor*	*Primary observation*	*References*
	A_{2A}	A genetic variation in the receptor gene influence age at onset in HD	Dhaenens et al. (2009)
Inflammatory pain	A_{2A}	KO mice showed a reduction in inflammatory hyperalgesia	Ledent et al. (1997), Li et al. (2010)
	A_3	Carrageenan-induced paw edema and heat hyperalgesia was reduced in KO mice	Wu et al. (2002)
	P2X2/3	Double KO mice had reduced pain-related behaviors after intraplantar injection of formalin	Cockayne et al. (2005)
	P2X4	KO mice do not show pain hypersensitivity and inflammatory PGE2	Ulmann et al. (2010)
	P2X7	Severity of mAb-induced arthritis is attenuated in KO mice	Labasi et al. (2002)
	P2X7	Inflammatory and neuropathic hypersensitivity are completely absent in KO mice, while normal nociceptive processing is preserved	Chessell et al. (2005)
	$P2Y_1$	Sensitization of nociceptors by inflammatory injury is modestly impaired in the KO mice	Malin and Molliver (2010)
	$P2Y_2$	KO mice showed decreased thermal hyperalgesia, but mechanical hypersensitivity was not altered	Malin et al. (2008)
Neuropathic pain	A_1	Intrathecal administration of adenosine reduces allodynia	Taylor (2009) *review*
	A_{2A}	A single intrathecal injection of the agonists produced a reversal of mechanical allodynia	Loram et al. (2009)
	A_{2A}	KO mice showed decreased mechanical allodynia and thermal hyperalgesia	Bura et al. (2008)
	P2X2/3	cPLA2 activation in dorsal root ganglia	Hasegawa et al. (2009)
	P2X2/3	A-317491 attenuated mechanical allodynia	McGaraughty et al. (2003)
	P2X4	P2X4 upregulation in spinal microglia in developing neuropathic pain	Tsuda et al. (2003), Ulmann et al. (2008), and Coull et al. (2005)
	P2X4	KO mice showed no defects in acute physiological pain or tissue but did in neuropathic pain	Tsuda et al. (2009a)
	P2X7	Antagonist attenuated mechanical allodynia	Donnelly-Roberts and Jarvis (2007) and Honore et al. (2006) *review*

(continued)

TABLE I *(continued)*

CNS disorders	*Receptor*	*Primary observation*	*References*
	$P2Y_{12}$	Antagonist prevents the development of allodynia	Kobayashi et al. (2008) Tozaki-Saitoh et al. (2008)
	UTP/UDP sensitive	UTP and UDP have antiallodynic effects	Okada et al. (2002)
Trauma	A_1	AIT-082 enhances nerve fiber regeneration	Rathbone et al. (1999)
	A_1	KO mice showed enhanced microglial response	Haselkorn et al. (2010)
	P2	Apyrase had a significant reduction in the transplanted glioma size	Morrone et al. (2006)
	P2X1 and P2X2	Increased expression of the receptor	Florenzano et al. (2002)
	P2X1-4, P2X7	Increased expression of the receptor	Franke et al. (2001)
	P2X	Neuronal cell death	Ryu et al. (2002)
	$P2Y_1$	Regulated expression of the receptor activation of PI3K/Akt pathway	Franke et al. (2006, 2009)
	$P2Y_6$	Microglia phagocytosis	Koizumi et al. (2007)
	$P2Y_{12}$	Microglia chemotaxis	Haynes et al. (2006)
Amyotrophic lateral sclerosis	A_{2A}	Protect motor neurons from ALS model	Mojsilovic-Petrovic et al. (2006)
	P2X1	P2X1 induction in axotomized facial motoneurons was impaired in familial ALS model mice	Kassa et al. (2007)
Multiple sclerosis (EAE)	P1	Caffeine decreases the incidence of EAE	Chen et al. (2010)
	P2X4	Local accumulation of the receptor	Guo and Schluesener (2005)
	P2X7	Antagonist (BBG, o-ATP) reduces demyelination and ameliorates the associated neurological symptoms	Matute et al. (2007)
	P2X7	KO mice developed more severe clinical and pathological expression of EAE	Chen and Brosnan (2006)
Ischemia	P2X	The cotreatments of vigabatrin and P2X receptor antagonists effectively prevent ischemia-induced neurodegeneration	Kim et al. (2006)
	P2X	ATP protecting against hypoxic/hypoglycemic perturbation of hippocampal neurotransmission	Aihara et al. (2002)
	P2Y	Extracellular ATP levels was increased in cortical spreading depression	Schock et al. (2007)
Epilepsy	P1	Adenosine in ketogenic diet therapy for epilepsy	Masino et al. (2009)

TABLE 1 *(continued)*

CNS disorders	*Receptor*	*Primary observation*	*References*
	A_1	Adenosine hyperpolarizes neuronal membrane potential	Kawamura et al. (2010)
	A_1	R-PIA has neuroprotective and anticonvulsant roles	Vianna et al. (2005)
	A_1	The receptor gene variants associated with posttraumatic seizures	Wagner et al. (2010)

synapse (i.e., purinergic regulation of release of both excitatory and inhibitory neurotransmitters), extracellular purinergic metabolism, receptor phosphorylation by modulation of ecto-kinase activity (Fujii et al., 2002), and purinergic effects on perisynaptic glia (also known as the tripartite synapse) (Araque et al., 1999).

B. Hypoxic Insult

In hippocampal slices, administration of ATP or its metabolically stable ATP-analog ATPγS exerts an inhibitory action on fast excitatory postsynaptic currents and population spikes (Coppi et al., 2007; Pedata et al., 2007) under normoxic conditions. Further, after removal, ATPγS induces a significant potentiation of evoked responses. Although the application of DPCPX, an A_1 adenosine receptor inhibitor, reduces this inhibitory activity, ecto-NTPDase blockade still elicits a decrease in evoked synaptic responses, which is even more pronounced than that induced by ATP alone. This result suggests that in the CA1 area of the rat hippocampus a P2 component, in addition to P1 receptor activation, is responsible for ATP-mediated decrease in synaptic transmission. After transient oxygen and glucose deprivation, an ischemia-like insult, synaptic responses in the hippocampal CA1 region were irreversibly lost. Purinergic receptor antagonist—such as suramin, pyridoxal-phosphate-6-azophenyl-2′,4′-disulfonate (PPADS), MRS2179, and brilliant blue G (BBG)—was reported to have protective effects against the insult. Significant recovery of the amplitude of field excitatory postsynaptic potential was observed within 15 min of recovery after deprivation. Under similar ischemic-like conditions, efflux of radiolabeled ATP and its metabolites from hippocampal slices was observed (Juranyi et al., 1999), and ATP outflow also increased *in vivo* during focal cerebral ischemia (Melani et al., 2005). Thus, at the ischemic site, levels of extracellular nucleotides may remain elevated after injuries. Accordingly, purinergic signaling may be an important mechanism in brain ischemia.

III. The Role of Purinergic Receptors in Astrocyte Function and Survival

Glia are classified into astrocytes, oligodendrocytes, and microglia. Astrocytes are important intermediaries in the CNS, play a pivotal role in brain homeostasis, and have been recently shown to be active modulators of neural activity (Koizumi et al., 2005). Astrocytic processes closely ensheath synapses and express a wide range of receptors for neurotransmitters (Belanger & Magistretti, 2009). In addition, astrocytic endfeet come in close contact to cerebral capillary vessels (Iadecola & Nedergaard, 2007; Kacem et al., 1998).

A. Intercellular Signal Transmission

Multiple P2X and P2Y receptors are also expressed by astrocytes (Verkhratsky et al., 2009). These receptors are differentially expressed depending on the specific conditions present, indicating differential roles in both physiological CNS function and in disease. Unlike neurons, astrocytes lack the ability to propagate regenerative electrical signals; however, they can produce regenerative calcium waves that spread within astrocyte networks (Guthrie et al., 1999; Koizumi et al., 2002; Newman, 2001). $P2Y_1$ and $P2Y_2$ receptors may play major roles in generating and maintaining propagating calcium waves in cultured astrocytes (Bowser & Khakh, 2007; Fam et al., 2003; Gallagher & Salter, 2003). ATP has been identified as a mediator of calcium waves between astrocytes and in reciprocal signaling between neurons and glia (Cotrina et al., 1998a, 1998b; Koizumi et al., 2003). Astrocytic ATP has been shown to decrease the excitability of neurons in the retina (Newman, 2003) and mediate presynaptic inhibition in cultured hippocampal neurons (Koizumi et al., 2003; Zhang et al., 2003). In hippocampal cultures, endogenous ATP released from astrocytes dynamically suppresses the frequency of the spontaneous neuronal calcium oscillations (Koizumi et al., 2003) and excitatory postsynaptic currents (Zhang et al., 2003), by inhibiting presynaptic functions of glutamatergic neurons.

It has been reported that astrocytic calcium signaling causes cerebrovascular constrictions (Mulligan & MacVicar, 2004; Takano et al., 2006). This calcium signal can be evoked by activity-dependent glutamate release from neurons (Zonta et al., 2003). Further study suggested that local extracellular potassium concentrations appeared to work as vasoactive switches for dilation and constriction (Girouard et al., 2010). As mentioned previously, astrocytic endfeet cover the intracerebral blood vessels, and in other sites, astrocytes are positioned to sense neuronal activity at the synapse. Collectively, this indicates that astrocytes mediate neurovascular coupling, and thus control the appropriate metabolic supply for neurons, and purinergic signaling may be involved in this system (Xu et al., 2008).

B. Astrogliosis

Astrocytes are known to undergo reactive astrogliosis, which is characterized by cell proliferation, hypertrophy, enhanced synthesis of neurotrophins and inflammatory mediators, and enhanced integrity of glial fibrillary acidic protein (GFAP; Abbracchio & Ceruti, 2006). These phenomena are widely observed in various kinds of traumatic or ischemic events as well as chronic neurodegenerative and demyelinating disorders, such as Alzheimer's, Parkinson's, and Huntington's diseases as well as multiple sclerosis. A_{2A} adenosine receptor inhibition suppresses basic fibroblast growth factor (bFGF)-induced astrogliosis in rat striatal primary astrocytes (Brambilla et al., 2003). Interactions between purinergic receptors and growth factor receptors can be synergistic (Neary et al., 1996). The release of cytokines and growth factors can be affected by purinergic receptors and can influence cell proliferation. Activation of A_1 adenosine receptors enhances the release of nerve growth factor (NGF) and S100-β protein from cultured astrocytes (Ciccarelli et al., 1999). The purinergic receptor is known to induce astrogliosis via activation of the extracellular-regulated kinases (ERKs), which involves induction of inflammatory and anti-inflammatory genes, cyclins, adhesion molecules, and antiapoptotic factors (Neary et al., 2006). Injection of the P2Y agonist 2-methylthio ATP into the rat nucleus accumbens triggers increases in proliferating GFAP-positive cells and hypertrophic morphological changes (Franke et al., 1999, 2004). These effects are inhibited by several purinergic receptor antagonists, which suggest significant involvement of $P2Y_1$ receptors in activation of astrocytes. Further detailed *in vivo* analysis revealed that $P2Y_1$ receptor is the main contributor to activation of astrocytic ERK/Akt signaling and stimulation of the generation of astrogliosis (Franke et al., 2009). In primary mouse astrocytes, stimulation of $P2Y_1$ and $P2Y_6$ receptors leads to an intracellular calcium-dependent activation of the nuclear factor of activated T cells (NFAT), a transcription factor believed to activate astrogliosis (Perez-Ortiz et al., 2008). It was also reported that janus kinase 2 (JAK2) and signal transducers and activators of transcription 3 (STAT3; Washburn & Neary, 2006) and phosphoinositide 3-kinase (PI3K), Akt and cAMP response element binding protein (CREB) were activated by $P2Y_1$ receptor activation, which, respectively, play a role in the production of GFAP and glial cell line-derived neurotrophic factor (GDNF) in astrocytes under transient ischemic conditions (Sun et al., 2008).

C. Cytoprotection

Metabotropic P2Y receptors are linked to several cytoprotective pathways. $P2Y_1$ receptor activation rescues hydrogen peroxide-induced (an oxidative stressor) cell death of astrocytes. This receptor activation appears to induce thioredoxin reductase and protein tyrosine phosphatase, which acts as a reductant and a suppressor of ERK1/2 hyperactivation (Shinozaki et al., 2005, 2006). Another study suggested that IP_3-mediated intracellular calcium release protects astrocytes from oxidative stress (Wu et al., 2007a). Activation of $P2Y_2$ receptors has also been reported to inhibit cell death in an astrocyte cell line (Burgos et al., 2007). $P2Y_6$ but not $P2Y_4$ receptors interact rapidly with tumor necrosis factor α (TNFα)-related intracellular signals to prevent apoptotic cell death in an astrocyte cell line (Kim et al., 2003). This astrocytic modulation by purinergic signaling also affects the viability of neuronal cells (Wu et al., 2007a). In hippocampal–astroglial cultures, stimulation of $P2Y_1$ receptors protects the cells against hydrogen peroxide-induced oxidative damage (Fujita et al., 2009). The mechanism implicated in this phenomenon includes interleukin-6 (IL-6) release from ATP-stimulated astrocytes. Microglial cells are generally considered as immunocompetent cells in the brain. However, astrocytes also have important roles in immune responses. Activated astrocytes can release various immune mediators, such as cytokines, chemokines, and growth factors, that may exert either neuroprotective or neurotoxic effects. Whether activation of astrocytes is beneficial or detrimental to neighboring neurons is still unresolved, and the final effect may result from a complex interplay between pro- and anti-inflammatory factors.

IV. The Role of Purinergic Receptors in Microglial Function

When brain dysfunction or damage occurs, it is generally thought to be caused by a disturbance of neural networks through the loss or malfunction of neurons. However, at the same time, there should also be loss of cohesion in glial networks and emergent activation of glia. In neural diseases, microglia have attracted attention as sensors and effectors of brain-damaging events. It has been suggested that, in response to injury, microglia migrate to the site of the injury, release proinflammatory substances, and phagocytose damaged cells and their remnants. Purinergic participations have been reported in these distinct processes (Inoue, 2008). Microglia also perform various functions by releasing trophic factors, such as bFGF, NGF, and brain-derived neurotrophic factor (BDNF), as well as cytokines, chemokines, and plasminogen (Heese et al., 1997; Hide et al., 2000; Inoue et al., 1998; Kataoka et al., 2009; Liu et al., 1998; Nakajima et al., 2001; Shigemoto-Mogami et al., 2001; Shimojo et al., 1991). These releasable factors may contribute to both neuronal injuries and neuroprotective effects. It has been reported that plasminogen is a neuroprotective substance (Nagata et al., 1993). It is also known that IL-6 regulates neuronal survival (de Araujo et al., 2009; Hama et al., 1991). These properties of microglia can be modulated by cytokines and neurotransmitters, including ATP via their specific receptors in the CNS (Kreutzberg, 1996).

A. Releasable Factors Regulated by Purinergic Receptors

1. Adenosine Receptors

Microglial cells can release a broad spectrum of substances, including neurotrophic factors, cytokines, and chemokines. Genetic deletion of A_1 adenosine receptors results in enhanced activation of microglia and subsequently severe demyelination and the progressive-relapsing form of experimental allergic encephalomyelitis, which is generally considered to be a model for multiple sclerosis (Tsutsui et al., 2004). Pharmacological assessments have revealed that stimulation of A_{2A} receptors by P1 receptor agonists induces microglial NGF mRNA synthesis and NGF protein release through cyclic AMP-mediated mechanism (Heese et al., 1997). It is widely accepted that neurotrophins regulate neuronal functions, support the survival, and enhance the growth of various types of neurons (Lindsay et al., 1994; Nakajima et al., 2007). Thus, the NGF released from microglia may participate in the survival and regeneration of various types of neurons. It has been shown that microglial NGF is necessary for apoptosis to occur during the development of the retina (Frade & Barde, 1998). Similarly, in the cerebellum, microglia induce developing Purkinje cells to die through respiratory bursts (Bessis et al., 2007). A_{2A} receptors and the downstream cAMP pathway also regulate cyclooxygenase-2 (COX-2) mRNA, synthesis of prostaglandin E2 (PGE2; Fiebich et al., 1996), and increase inducible nitric oxide synthase (iNOS) and nitric oxide (NO) levels (Saura et al., 2005). Accordingly, a possible mechanism of A_{2A}-induced brain damage may involve the generation of inflammatory mediators. Blockade of A_{2A} receptor signaling may be neuroprotective in models of cerebral ischemia (Marcoli et al., 2004; Phillis, 1995). Increased expression of ecto-5′-nucleotidase after ischemia (Braun et al., 1997) could increase A_{2A} receptor activation in microglia by enhanced production of adenosine. The interactions between receptors of the adenosine receptor family seem to be complicated. Notably, van der Putten et al. (2009) reported that toll-like receptor (TLR)-activated primary microglia from the rhesus monkey showed an increased expression of A_{2A} receptors and decreased expression of A_3 receptors. These alterations enhanced A_{2A}-mediated suppression of lipopolysaccharide (LPS)-induced cytokine responses following adenosine exposure. This same study also showed that A_{2A} receptor-mediated inhibitory signaling in unstimulated microglia was effectively counteracted by A_3 receptor-mediated signaling. These results demonstrate the dynamic nature of adenosine receptor signaling and its effects on the proinflammatory responses of activated microglia.

2. P2X4 Receptors

Activation of P2X4 receptors by ATP evokes a biphasic release of BDNF from microglia (Trang et al., 2009). P2X4 receptor-mediated extracellular calcium influxes induce BDNF release by soluble *N*-ethylmaleimide-sensitive factor attachment protein receptor (SNARE)-dependent exocytosis. Simultaneously, p38 mitogen-activated protein kinase (MAPK) is phosphorylated in an extracellular calcium-dependent manner, which is critical for BDNF release and synthesis. P2X4-mediated BDNF has been shown to be a key molecule mediating this microglia–neuron signaling (Beggs & Salter, 2010; Keller et al., 2007; Tsuda et al., 2005). Nerve injury-induced allodynia in the rat was effectively suppressed by 2′,3′-O-(2,4,6-trinitrophenyl)adenosine-5′-triphosphate (TNP-ATP) but not by PPADS (Tsuda et al., 2003). Further, both rats treated intrathecally with a P2X4 antisense oligonucleotide and mice lacking P2X4 receptors show attenuated tactile allodynia after nerve injuries (Tsuda et al., 2003, 2009a; Ulmann et al., 2008). These findings indicate that tactile allodynia depends on P2X4 receptors in the spinal cord. Surprisingly, intrathecal administration of ATP-stimulated microglia is sufficient to cause allodynia in normal rats, whereas microglia pretreated with siRNA for BDNF failed to induce allodynia. BDNF released from microglia acts on neuronal tropomyosin receptor kinase B (TrkB) in spinal lamina I to cause a rise in intracellular chloride ions, thereby suppressing gamma-aminobutyric acid (GABA)- and glycine-mediated inhibition in these cells (Coull et al., 2005). This disinhibition unmasks innocuous inputs to lamina I neurons and facilitates their responses to noxious inputs (Keller et al., 2007). Expression of P2X4 receptors in the spinal cord, after nerve injuries, is markedly and specifically increased in microglia. Based on the above evidence, upregulation of P2X4 receptor expression in microglia may be a key process in neuropathic pain. Activating both TLRs and the nucleotide binding oligomerization domain 2 receptor (NOD2; another pattern recognition receptor) in cultured microglia increases the expression of P2X4 receptor mRNA (Guo et al., 2006). The extracellular matrix protein fibronectin increases P2X4 receptor mRNA and protein levels in primary cultured microglial cells (Nasu-Tada et al., 2006; Tsuda et al., 2009b). The level of fibronectin protein was elevated in the dorsal horn 3–7 days after nerve injury (Nasu-Tada et al., 2006), and this matched the time when P2X4 receptor protein levels started to increase (Tsuda et al., 2003). It has also been found that upregulation of P2X4 receptors in the spinal cord and allodynia after spinal nerve injuries was significantly suppressed by intrathecal administration of echistatin, an Arg-Gly-Asp (RGD)-containing snake venom peptide that interacts with integrin proteins and antagonizes $\alpha_{IIb}\beta_3$, $\alpha_v\beta_3$, and $\alpha_5\beta_1$. Further, intrathecal delivery of fibronectin increased P2X4 receptor expression and produced allodynia, a behavior that was not produced in P2X4 receptor-deficient mice (Tsuda et al., 2008a). These results suggest that the fibronectin–integrin signaling system participates in the upregulation of P2X4 receptor expression after nerve injuries and subsequent neuropathic pain. Signaling under fibronectin–integrin interactions may involve the lyn src family kinase. Lyn-knockout mice exhibit a striking reduction in upregulation of P2X4 and tactile allodynia after nerve injuries (Tsuda et al., 2008b). Lyn-deficient microglial cells do not show increases in P2X4 receptor expression when challenged with fibronectin. Lyn tyrosine kinase distinctly activates the PI3K-Akt, MAPK, MAPK/ERK kinase (MEK), and ERK signaling cascades (Tsuda et al., 2009b). Signaling through the PI3K-Akt pathway induces degradation of p53 via the corresponding ubiquitin ligase mouse double minute 2 (MDM2) in a proteasome-dependent manner (Vassilev et al., 2004). Since a p53 inhibitor, pifithrin-α, and an MDM2 antagonist nutlin-3 suppress fibronectin-stimulated P2X4 receptor upregulation, the consequences of an attenuated repressive effect of p53 may be enhanced P2X4 gene expression. However, activated MEK-ERK signaling in microglia exposed to fibronectin enhances eukaryotic translation initiation factor 4E (eIF4E) phosphorylation by activating MAPK-interacting protein kinase-1 (MNK1), which may play a role in regulating P2X4 receptor expression at translational levels (Tsuda et al., 2009b).

There is evidence that P2X4 receptors in microglial cells and peripheral macrophages are located predominantly within lysosomal compartments, and in these compartments, they are targeted by their N- and C-terminal motifs (Qureshi et al., 2007). Notably, P2X4 receptors remain stable within lysosomes because of their N-linked glycosylation. Lysosomal exocytosis induced by calcium ionophores results in the trafficking of P2X4 receptors to the plasma membrane (Qureshi et al., 2007). Conventional microglial activation by LPS enhances functional surface expression of P2X4 receptors (Boumechache et al., 2009; Toulme et al., 2010). These results highlight the importance of the mechanisms underlying P2X4 recycling between the lysosome and the plasma membrane.

3. P2X7 Receptors

ATP stimulates the release of plasminogen in a concentration-dependent manner, from 10 to 100 μM, with a peak response approximately 5–10 min after stimulation (Inoue et al., 1998). This ATP-induced plasminogen release depends on an increase in the intracellular calcium concentration. Because pharmacological profiles indicate that long-lasting intracellular calcium increases appear to be dependent on P2X7 receptors, it has been suggested that ATP causes an increase in intracellular calcium concentrations via the ionotropic P2X7 receptor, which stimulates the release of plasminogen from the microglia. It has been demonstrated that microglia-derived plasminogen interactions with neurons promote the development of mesencephalic dopaminergic neurons and enhance neurite outgrowth from explants of neocortical tissue (Nagata et al., 1993; Nakajima et al., 1993).

It has been reported that ATP potently stimulates TNFα release and TNFα mRNA expression in primary cultures of rat brain microglia via P2X7 receptors (Hide et al., 2000). TNFα release was dependent on calcium influx and mediated by MAPKs. In a detailed mechanistic description, it was suggested that P2X7 receptor activation upregulates JNK-induced TNFα transcription, and p38-mediated mRNA nuclear export leads to *de novo* TNFα synthesis and release (Hide et al., 2000). ERK was also involved in ATP-induced TNFα release, but it seems to be regulated by other unidentified purinergic receptors (Hide, 2003; Suzuki et al., 2004). This P2X7 receptor-induced TNFα release from microglia provides cocultured neurons with effective protection against glutamate-induced cell death. However, this protective effect of microglia-derived TNFα was not seen following LPS-induced massive TNFα release (Suzuki et al., 2004). It is still not entirely clear whether TNF is beneficial or toxic (Fontaine et al., 2002). Indeed, ischemic damage causes microglia to secrete excessive amounts of TNFα and induce inflammatory destruction in the brain (Barone et al., 1997; Meistrell et al., 1997). Most neurodegenerative diseases, such as Alzheimer's disease and Parkinson's disease, are also associated with neuronal inflammation and over activation of microglia (Akiyama et al., 2000; Gao et al., 2003). Nevertheless, several lines of evidence indicate that TNFα can provide protection to neurons because it is able to facilitate the expression of anti-apoptotic and antioxidative proteins. Moreover, it has also been reported that TNFα plays a role in both the long-term behavioral recovery and the histological repair of damaged tissue in TNFα-deficient mice, although TNFα has a deleterious effect during the acute brain response to trauma (Scherbel et al., 1999).

It has been shown that extracellular ATP causes a large release of interleukin-1β (IL-1β) from LPS-primed microglial cell lines and primary microglia by activating the P2X7 receptor (Ferrari et al., 1996). Since LPS-stimulated cells release ATP, the presence of an autocrine loop has been suggested (Ferrari et al., 1997). ATP itself has no effect on the accumulation of intracellular pro-IL-1β in the absence of LPS, whereas in LPS-treated cells, ATP slightly increases the synthesis of pro-IL-1β. In ATP-stimulated microglia, the p20 proteolytic fragment derived from the activation of the IL-1β-converting enzyme can be detected by immunoblotting. Since IL-1β release can be inhibited by increasing the extracellular K^+ concentration and also inhibited by caspase inhibitors, it is suggested that ATP triggers accelerated maturation of pro-IL-1β and release of intracellularly accumulated IL-1β by activating the IL-1β-converting enzyme/caspase 1 in mouse microglia. Extracellular ATP is the only endogenous compound known to cause a significant reduction in intracellular K^+ and consequent release of IL-1β (Perregaux & Gabel, 1994; Sanz & Di Virgilio, 2000). However, it should be noted that not only ATP but also metabolites like ADP and AMP can stimulate microglial IL-1 release, when cells are primed by ATP before ADP and AMP application (Chakfe et al., 2002; Ferrari et al., 1997). These results suggest the existence of more a complex interaction between purinergic receptor subtypes in LPS-induced IL-1β release. IL-1β is one of the principal proinflammatory cytokines produced in the CNS following systemic or local insults (White & Jones, 2008). Microglia and meningeal macrophages produce IL-1β in the ischemic area after strokes and brain damage (Minami et al., 1992), and this is thought to be the major source of IL-1β secreted in response to neuronal damage. It has been suggested that IL-1β plays a direct role in the pathophysiology of stroke because an IL-1β receptor antagonist or IL-1β antibodies significantly reduce cerebral ischemia and neuronal loss in rats (Relton & Rothwell, 1992; Touzani et al., 1999; Yamasaki et al., 1992). IL-1β is also implicated in several diseases, specifically multiple sclerosis (Clerico et al., 2007; Martin & Near, 1995), amyotrophic lateral sclerosis (Meissner et al., 2010; Pasinelli et al., 1998), and epilepsy (Vezzani et al., 2010).

Several chemokines are also known to be released from microglia, as release of CCL3 and CXCL2 is induced by extracellular ATP via the P2X7 receptor (Kataoka et al., 2009; Shiratori et al., 2010). The P2X7 receptor can function as a calcium permeable channel and thereby elicit long-lasting calcium influx into intracellular space (North, 2002). This calcium signal activates a calcineurin-NFAT signal cascade that leads to upregulation of CCL3 and CXCL2 transcription and subsequent release (Kataoka et al., 2009; Shiratori et al., 2010). The P2X7 receptor also activates several MAPKs via a protein kinase C (PKC)-dependent pathway that induces CXCL2 transcription (Shiratori et al., 2010). CCL3 is a CC chemokine characterized as a chemoattractant for monocytes and T cells and may be necessary for recruitment of these cells to inflammatory sites. *In vitro* experiments show that CCL3 increases the transmigration of bone marrow-derived dendritic cells across brain microvessel endothelial monolayers (Zozulya et al., 2007). The functional involvement of CCL3 has been suggested in several CNS disorders, including multiple sclerosis (Balashov et al., 1999; Zang et al., 2001), Wallerian degeneration (Perrin et al., 2005), and Sandhoff disease (Wu & Proia, 2004). In addition, CCL3 enhances the response of transient receptor potential vanilloid 1 (TRPV1) expressed on DRG neurons and thermal sensitivity, which implies involvement of CCL3 in thermal hyperalgesia during inflammation (Zhang et al., 2005). CXCL2, a CXC chemokine family member, interacts with a chemokine receptor, CXCR2. CXCR2 is involved in several CNS disorders. Based on the collective body of works, the regulation of microglia activation warrants study, as it may benefit the treatment of neurodegenerative diseases.

4. P2Y Receptors

Extracellular ATP evokes the release of IL-6 in the microglial cell line MG-5 (Ohsawa et al., 1997; Shigemoto-Mogami et al., 2001). Although it was reported that TNFα and IL-1β stimulate IL-6 production in other glial cells (Norris et al., 1994), ATP may directly evoke the release of IL-6, since TNFα alone did not induce the release of IL-6 (Sawada et al., 1992), and release of IL-1β is not evoked by ATP alone in MG-5 cells. At the transcription level, ATP induced a sevenfold increase in the amount of IL-6 mRNA, which was inhibited by suramin and the p38 inhibitor SB203580 but not by the MEK1 inhibitor PD98059. Although ATP activated two distinct MAPK, ERK1/2 and p38, in MG-5 cells, ERK1/2 did not appear to be involved in IL-6 production (Shigemoto-Mogami et al., 2001). ATP-evoked p38 activation is dependent on extracellular calcium. P2X7 receptor is known to induce sustained calcium entry and, in fact, 2′(3′)-O-(4-benzoylbenzoyl)adenosine-5′-triphosphate (BzATP) evokes sustained calcium entry (North, 2002), leading to the phosphorylation of p38 in MG-5 cells. However, BzATP induced only a very small amount of IL-6 production in MG-5 cells, and BBG, a specific P2X7 antagonist, does not inhibit ATP-induced release of IL-6 in MG-5 cells (Shigemoto-Mogami et al., 2001). A calcium-dependent PKC may be an additional signal because ATP-evoked IL-6 production is attenuated by Gö6976, an inhibitor of calcium-dependent PKC (Shigemoto-Mogami et al., 2001). The P2Y receptor responsible for these responses is insensitive to pertussis toxin (PTX) and is linked to phospholipase C (PLC). Accordingly, these data suggest that the P2Y receptor rather than the P2X7 receptor seems to have a major role in IL-6 production in MG-5 cells. IL-6 is an important mediator of inflammatory and immune responses in the periphery. In addition, several studies indicate that IL-6 is also produced in the CNS and may play an important role in a variety of CNS functions, such as cell-to-cell signaling, coordination of neuroimmune responses, the protection of neurons from insults, as well as neuronal differentiation, growth, and survival (Gruol & Nelson, 1997; Marz et al., 1999). IL-6 may also contribute to the etiology of neuropathological disorders. Elevated levels of IL-6 in the CNS are found in several neurological disorders, including AIDS, dementia complex, Alzheimer's disease, multiple sclerosis, systemic lupus erythematosus, CNS traumas, and both viral and bacterial meningitis.

B. Purinergic Regulation of Chemotaxis

Microglia have a highly branched morphology and monitor the brain parenchyma under physiological conditions. The initial microglial responses that occur after brain injury and in various neurological diseases are characterized by microglial accumulation in the affected sites of the brain, which results from the migration and proliferation of these cells. In the presence of an injury, microglia extend their processes toward the injury site, retract their branching processes, migrate, and help clear cellular debris by phagocytosis (Inoue, 2008). Cell injury results in a leakage of nucleotides, which affects the motility of adjacent cells, including microglia (Davalos et al., 2005; Nimmerjahn et al., 2005). In addition, cells release or leak uridine nucleotides (Lazarowski et al., 1997) and nucleotide sugars (Lazarowski et al., 2003) in response to various stimuli or ischemic injuries (Erlinge et al., 2005). The first observation of this phenomenon was made by Honda et al. (2001). They examined the possibility that ATP released from injured neurons and nerve terminals affects cell motility in rat primary cultured microglia and found that extracellular ATP and ADP-induced membrane ruffling and markedly enhanced chemotaxis in a Boyden chamber assay. Further analyses using the Dunn chemotaxis chamber assay, which permits direct observation of cell movement, showed that both ATP and ADP induced chemotaxis of microglia. This activity appeared to be mediated by G(i/o)-coupled P2Y receptors ($P2Y_{12}$ receptor), since AR-C69931MX, a $P2Y_{12}$ receptor blocker, or PTX treatments clearly inhibited this ruffling. As an intracellular signaling molecule underlying these phenomena, the small G-protein Rac was activated by ATP and ADP stimulation, its activation inhibited by pretreatment with PTX, and its migration required activation of PI3K and PLC signaling pathways (Irino et al., 2008; Ohsawa et al., 2007). Integrin β1 translocation to membrane ruffles was also required for microglial chemotaxis toward ATP, and this translocation was regulated by protein kinase A (PKA) activity (Nasu-Tada et al., 2005). Another study indicated that ATP-induced chemotaxis of microglial processes requires P2Y receptor-activated initiation of outward potassium (Wu et al., 2007b). Two recent *in vivo* studies obtained data from microglia in the undisturbed brain. By imaging GFP-labeled microglia through the thinned skull, they observed rapid movements of cellular processes without apparent translocation of the cell bodies (Nimmerjahn et al., 2005). This movement is controlled by extracellular ATP and the $P2Y_{12}$ receptor. Local injection of ATP induced a rapid chemotactic response of the microglial processes (Davalos et al., 2005), and microglia in $P2Y_{12}$ receptor knockout mice significantly delayed directional process extension toward sites of damage *in vivo* (Haynes et al., 2006). Process extension could also be observed in primary microglia cultured on collagen gels (Ohsawa et al., 2010). Microglial activation is characterized by a transformation of the resting ramified microglia to an amoeboid form, and it is a well-established concept that microglial cells can rapidly accumulate near injured neurons where they transform into phagocytes (Sierra et al., 2010). Haynes et al. (2006) reported that microglial $P2Y_{12}$ receptors were downregulated upon microglial activation by tissue damage or LPS stimulation, and that loss of $P2Y_{12}$ receptor expression accompanied microglial transformation from the highly ramified to the amoeboid state. In LPS-activated microglia, ATP induced process retraction and repulsive migration. These repulsive effects of ATP are mediated by the G_s-coupled adenosine A_{2A} receptor and depend on the breakdown of ATP to adenosine (Orr et al., 2009). Thus, ATP-induced repulsion by activated microglia involves upregulation of the adenosine A_{2A} receptor and coincident downregulation of the $P2Y_{12}$ receptor.

C. Purinergic Regulation of Phagocytosis

UDP, an agonist for $P2Y_6$ receptors, facilitates phagocytosis of primary cultured microglial cells (Koizumi et al., 2007). Neuronal injury induced by administration of kainic acid (KA) causes upregulation of $P2Y_6$ receptors in microglia in the hippocampus. KA-evoked neuronal injury results in an increase in extracellular UTP, which is immediately metabolized into UDP. In the injured hippocampus, engulfment of exogenous fluorescent microspheres could be the observed cause, and this effect is significantly suppressed by disturbing $P2Y_6$ receptor signaling. Dying cells in the CNS need to recruit phagocytes that are either in close proximity or distant from the lesion. Nucleotides such as ATP and UTP that are released by damaged neurons may be utilized as mediators of microglial movement and facilitate the recruitment of microglia. Excess production of neurons and loss by apoptosis are evident in most regions of the developing CNS. It has been suggested that microglial phagocytosis contributes to the arrangement of brain architecture in the developing mouse brain by clearing apoptotic neuronal cells (Sierra et al., 2010).

V. The Role of Purinergic Receptors in Oligodendrocyte Function and Survival

Oligodendrocytes are the myelin-forming cells in the CNS, and they construct layers of myelin sheets around axons to insulate them in support of impulse conduction. Both P1 and P2 receptors contribute to the modulation of oligodendrocyte development (Agresti et al., 2005b). In culture, oligodendrocyte progenitor cells obtained from the neonatal rat brain express several P2X (P2X1,2,3,4,7) and P2Y ($P2Y_{1,2,4}$) receptors. Further functional analysis determined that P2X7 and $P2Y_1$ receptors were the main ionotropic and metabotropic purinergic receptors active in oligodendrocyte progenitor cells (Agresti et al., 2005a). ATP control of oligodendrocyte development is exerted mainly by activation of metabotropic $P2Y_1$ receptors (Agresti et al., 2005a). ATP and ADP acting on $P2Y_1$ receptors show antiproliferative effects and induce expression of differentiated oligodendrocyte markers. However, adenosine also has a similar effect to ATP. Thus, P1 receptors may be also involved in oligodendrocyte differentiation. All four P1 adenosine receptors at the mRNA level have been identified in cultured oligodendrocytes and their precursors cells (Stevens et al., 2002). In cocultures of dorsal root ganglion neurons and oligodendrocyte precursor cells, adenosine and ATP are released during action potentials, and this release triggers calcium signals in immature oligodendrocytes (Stevens et al., 2002). Activation of adenosine receptors inhibits oligodendrocyte progenitor cell proliferation, promotes their differentiation, and initiates myelination, representing a purinergic signal interaction between axon activity and oligodendrocytes (Stevens et al., 2002). In already differentiated oligodendrocytes, ATP stimulates myelination through a mechanism involving astrocytes. Leukemia inhibitory factor (LIF) is released by astrocytes in response to ATP released from axons firing action potentials, and LIF promotes myelination by mature oligodendrocytes (Ishibashi et al., 2006). These results demonstrate mechanisms that involve interactions between astrocytes and myelinating oligodendrocytes by means of purinergic and cytokine signaling through which neuronal activity affects myelination. P2X7 receptors expressed in oligodendrocyte may cause cell death and can be implicated in the pathogenesis of ischemia and multiple sclerosis (Domercq et al., 2010; Matute et al., 2007). Because of their high permeability to calcium and prolonged activation, P2X7 receptors in oligodendrocytes are lethal to differentiated oligodendrocytes in culture and to mature oligodendrocytes in isolated optic nerves both *in vitro* and *in vivo* (Matute et al., 2007).

VI. Conclusion

Since Burnstock first published a hypothesis regarding purinergic neurotransmission in 1972, the field of purinergic signaling has matured immensely. Today, there is interest in purinergic signaling not just in the nervous system but also in many other physiological events. Most importantly, purines and pyrimidines mediate interactions between different types of brain components, such as neurons, astrocytes, microglia, oligodendrocytes, and the vasculature. All of these neural and nonneural cells can release nucleotides and have appropriate sets of membrane receptors for extracellular ATP, UTP, and their metabolites. We must not ignore the existence of extracellular enzymes that regulate ATP hydrolysis and even synthesis. The activities of ecto-enzymes could be packaged with P2 and P1 receptors at membrane microdomain levels. These complex interactions between various types of receptors and ecto-enzymes that are coexpressed in cells may appear in the overlapping specificities of many agonists and antagonists.

There are an increasing number of investigations into the therapeutic potential of purinergic signaling in various CNS disorders. Due to the lack of selective agonists and antagonists for purinergic receptors, cooperative bilateral developments between medicinal chemists and pathological researchers are utilized to facilitate drug discovery. A variety of purinergic receptors are universally expressed in glial cells. Purinergic signaling cascades mediated by these receptors control glial physiological and pathological responses. Detailed analysis of these signaling cascades may extend our knowledge regarding mechanisms mediating neuroglial networks and provide new strategies for treating neural disorders. More selective pharmacological tools may allow breakthroughs that uncover the physiological and pathological functions of purinergic signaling in the CNS. In turn, this will contribute to the development of more selective drugs for the treatment of nervous system disorders.

Conflict of Interest. The authors have no conflicts of interest to declare.

Abbreviations

6-OHDA	6-hydroxydopamine
AD	Alzheimer's disease
ALS	amyotrophic lateral sclerosis
Aβ	amyloid β
BBG	brilliant blue G
BDNF	brain-derived neurotrophic factor
bFGF	basic fibroblast growth factor
BzATP	2′(3′)-O-(4-benzoylbenzoyl)adenosine-5′-triphosphate
CNS	central nervous system
COX	cyclooxygenase
cPLA2	cytosolic phospholipase A2
CREB	cAMP response element binding protein
DA	dopamine
DPCPX	8-cyclopentyl-1,3-dipropylxanthine
EAE	experimental autoimmune encephalomyelitis
ecto-NTPDase	ecto-nucleoside triphosphate diphosphohydrolase
eIF4E	eukaryotic translation initiation factor 4E
ERK	extracellular-regulated kinases
GABA	gamma-aminobutyric acid
GDNF	glial cell line-derived neurotrophic factor
GFAP	glial fibrillary acidic protein
IL-1β	interleukin-1β
IL-6	interleukin-6
iNOS	inducible nitric oxide synthase
JAK	Janus kinase
KA	kainic acid
KO	knockout
LIF	leukemia inhibitory factor
LPS	lipopolysaccharide
MAPK	mitogen-activated protein kinase
MDM2	mouse double minute 2
MEK	MAPK/ERK kinase
MNK1	MAPK-interacting protein kinase-1
MPTP	1-methyl-4-phenyl-1,2,4,5-tetrahydropyridine
NFAT	nuclear factor of activated T cells
NGF	nerve growth factor
NO	nitric oxide
o-ATP	oxidized ATP
PGE2	prostaglandin E2
PI3K	phosphoinositide 3-kinase
PKA	protein kinase A

PKC	protein kinase C
PLC	phospholipase C
PPADS	pyridoxal-phosphate-6-azophenyl-2′,4′-disulfonate
PTX	pertussis toxin
R-PIA	(−)-*N*(6)-(2-phenylisopropyl)adenosine
SNARE	soluble *N*-ethylmaleimide-sensitive factor attachment protein receptor
STAT	signal transducers and activators of transcription
TLR	toll-like receptor
TNFα	tumor necrosis factor α
TNP-ATP	2′,3′-*O*-(2,4,6-trinitrophenyl)adenosine-5′-triphosphate
TrkB	tropomyosin receptor kinase B
TRPV1	transient receptor potential vanilloid 1

References

Abbracchio, M. P., Burnstock, G., Boeynaems, J. M., Barnard, E. A., Boyer, J. L., Kennedy, C., et al. (2006). International Union of Pharmacology LVIII: Update on the P2Y G protein-coupled nucleotide receptors: from molecular mechanisms and pathophysiology to therapy. *Pharmacological Reviews*, *58*(3), 281–341.

Abbracchio, M. P., Burnstock, G., Verkhratsky, A., & Zimmermann, H. (2009). Purinergic signalling in the nervous system: An overview. *Trends in Neurosciences*, *32*(1), 19–29.

Abbracchio, M. P., & Ceruti, S. (2006). Roles of P2 receptors in glial cells: Focus on astrocytes. *Purinergic Signalling*, *2*(4), 595–604.

Agresti, C., Meomartini, M. E., Amadio, S., Ambrosini, E., Serafini, B., Franchini, L., et al. (2005). Metabotropic P2 receptor activation regulates oligodendrocyte progenitor migration and development. *Glia*, *50*(2), 132–144.

Agresti, C., Meomartini, M. E., Amadio, S., Ambrosini, E., Volonte, C., Aloisi, F., et al. (2005). ATP regulates oligodendrocyte progenitor migration, proliferation, and differentiation: Involvement of metabotropic P2 receptors. *Brain Research. Brain Research Reviews*, *48* (2), 157–165.

Aihara, H., Fujiwara, S., Mizuta, I., Tada, H., Kanno, T., Tozaki, H., et al. (2002). Adenosine triphosphate accelerates recovery from hypoxic/hypoglycemic perturbation of guinea pig hippocampal neurotransmission via a P(2) receptor. *Brain Research*, *952*(1), 31–37.

Akiyama, H., Barger, S., Barnum, S., Bradt, B., Bauer, J., Cole, G. M., et al. (2000). Inflammation and Alzheimer's disease. *Neurobiology of Aging*, *21*(3), 383–421.

Araque, A., Parpura, V., Sanzgiri, R. P., & Haydon, P. G. (1999). Tripartite synapses: Glia, the unacknowledged partner. *Trends in Neurosciences*, *22*(5), 208–215.

Arendash, G. W., Schleif, W., Rezai-Zadeh, K., Jackson, E. K., Zacharia, L. C., Cracchiolo, J. R., et al. (2006). Caffeine protects Alzheimer's mice against cognitive impairment and reduces brain beta-amyloid production. *Neuroscience*, *142*(4), 941–952.

Balashov, K. E., Rottman, J. B., Weiner, H. L., & Hancock, W. W. (1999). CCR5(+) and CXCR3(+) T cells are increased in multiple sclerosis and their ligands MIP-1alpha and IP-10 are expressed in demyelinating brain lesions. *Proceedings of the National Academy of Sciences of the United States of America*, *96*(12), 6873–6878.

Bardoni, R., Goldstein, P. A., Lee, C. J., Gu, J. G., & MacDermott, A. B. (1997). ATP P2X receptors mediate fast synaptic transmission in the dorsal horn of the rat spinal cord. *The Journal of Neuroscience*, *17*(14), 5297–5304.

Barone, F. C., Arvin, B., White, R. F., Miller, A., Webb, C. L., Willette, R. N., et al. (1997). Tumor necrosis factor-alpha. A mediator of focal ischemic brain injury. *Stroke*, *28*(6), 1233–1244.

Beggs, S., & Salter, M. W. (2010). Microglia–neuronal signalling in neuropathic pain hypersensitivity 2.0. *Current Opinion in Neurobiology*, *20*(4), 474–480.

Belanger, M., & Magistretti, P. J. (2009). The role of astroglia in neuroprotection. *Dialogues in Clinical Neuroscience*, *11*(3), 281–295.

Bessis, A., Bechade, C., Bernard, D., & Roumier, A. (2007). Microglial control of neuronal death and synaptic properties. *Glia*, *55*(3), 233–238.

Boumechache, M., Masin, M., Edwardson, J. M., Gorecki, D. C., & Murrell-Lagnado, R. (2009). Analysis of assembly and trafficking of native P2X4 and P2X7 receptor complexes in rodent immune cells. *The Journal of Biological Chemistry*, *284*(20), 13446–13454.

Bowser, D. N., & Khakh, B. S. (2007). Vesicular ATP is the predominant cause of intercellular calcium waves in astrocytes. *The Journal of General Physiology*, *129*(6), 485–491.

Brambilla, R., Cottini, L., Fumagalli, M., Ceruti, S., & Abbracchio, M. P. (2003). Blockade of A2A adenosine receptors prevents basic fibroblast growth factor-induced reactive astrogliosis in rat striatal primary astrocytes. *Glia*, *43*(2), 190–194.

Braun, N., Lenz, C., Gillardon, F., Zimmermann, M., & Zimmermann, H. (1997). Focal cerebral ischemia enhances glial expression of ecto-5′-nucleotidase. *Brain Research*, *766* (1–2), 213–226.

Bura, S. A., Nadal, X., Ledent, C., Maldonado, R., & Valverde, O. (2008). A 2A adenosine receptor regulates glia proliferation and pain after peripheral nerve injury. *Pain*, *140*(1), 95–103.

Burgos, M., Neary, J. T., & Gonzalez, F. A. (2007). P2Y2 nucleotide receptors inhibit trauma-induced death of astrocytic cells. *Journal of Neurochemistry*, *103*(5), 1785–1800.

Burnstock, G. (1972). Purinergic nerves. *Pharmacological Reviews*, *24*(3), 509–581.

Burnstock, G. (2007a). Physiology and pathophysiology of purinergic neurotransmission. *Physiological Reviews*, *87*(2), 659–797.

Burnstock, G. (2007b). Purine and pyrimidine receptors. *Cellular and Molecular Life Sciences*, *64*(12), 1471–1483.

Burnstock, G. (2008). Purinergic signalling and disorders of the central nervous system. *Nature Reviews. Drug Discovery*, *7*(7), 575–590.

Burnstock, G., Campbell, G., Satchell, D., & Smythe, A. (1970). Evidence that adenosine triphosphate or a related nucleotide is the transmitter substance released by non-adrenergic inhibitory nerves in the gut. *British Journal of Pharmacology*, *40*(4), 668–688.

Burnstock, G., & Kennedy, C. (1985). Is there a basis for distinguishing two types of P2-purinoceptor? *General Pharmacology*, *16*(5), 433–440.

Cakir, E., Baykal, S., Karahan, S. C., Kuzeyli, K., & Uydu, H. (2003). Acute phase effects of ATP-$MgCl_2$ on experimental spinal cord injury. *Neurosurgical Review*, *26*(1), 67–70.

Canas, P. M., Porciuncula, L. O., Cunha, G. M., Silva, C. G., Machado, N. J., Oliveira, J. M., et al. (2009). Adenosine A2A receptor blockade prevents synaptotoxicity and memory dysfunction caused by beta-amyloid peptides via p38 mitogen-activated protein kinase pathway. *The Journal of Neuroscience*, *29*(47), 14741–14751.

Chakfe, Y., Seguin, R., Antel, J. P., Morissette, C., Malo, D., Henderson, D., et al. (2002). ADP and AMP induce interleukin-1beta release from microglial cells through activation of ATP-primed P2X7 receptor channels. *The Journal of Neuroscience*, *22*(8), 3061–3069.

Chen, L., & Brosnan, C. F. (2006). Exacerbation of experimental autoimmune encephalomyelitis in P2X7R−/− mice: Evidence for loss of apoptotic activity in lymphocytes. *Journal of Immunology*, *176*(5), 3115–3126.

Chen, G. Q., Chen, Y. Y., Wang, X. S., Wu, S. Z., Yang, H. M., Xu, H. Q., et al. (2010). Chronic caffeine treatment attenuates experimental autoimmune encephalomyelitis induced by guinea pig spinal cord homogenates in Wistar rats. *Brain Research*, *1309*, 116–125.

Chessell, I. P., Hatcher, J. P., Bountra, C., Michel, A. D., Hughes, J. P., Green, P., et al. (2005). Disruption of the P2X7 purinoceptor gene abolishes chronic inflammatory and neuropathic pain. *Pain*, *114*(3), 386–396.

Ciccarelli, R., Di Iorio, P., Bruno, V., Battaglia, G., D'Alimonte, I., D'Onofrio, M., et al. (1999). Activation of A(1) adenosine or mGlu3 metabotropic glutamate receptors enhances the release of nerve growth factor and S-100beta protein from cultured astrocytes. *Glia*, *27*(3), 275–281.

Clerico, M., Contessa, G., & Durelli, L. (2007). Interferon-beta1a for the treatment of multiple sclerosis. *Expert Opinion on Biological Therapy*, *7*(4), 535–542.

Cockayne, D. A., Dunn, P. M., Zhong, Y., Rong, W., Hamilton, S. G., Knight, G. E., et al. (2005). P2X2 knockout mice and P2X2/P2X3 double knockout mice reveal a role for the P2X2 receptor subunit in mediating multiple sensory effects of ATP. *The Journal of Physiology*, *567*(Pt 2), 621–639.

Coppi, E., Pugliese, A. M., Stephan, H., Muller, C. E., & Pedata, F. (2007). Role of P2 purinergic receptors in synaptic transmission under normoxic and ischaemic conditions in the CA1 region of rat hippocampal slices. *Purinergic Signalling*, *3*(3), 203–219.

Cotrina, M. L., Lin, J. H., Alves-Rodrigues, A., Liu, S., Li, J., Azmi-Ghadimi, H., et al. (1998). Connexins regulate calcium signaling by controlling ATP release. *Proceedings of the National Academy of Sciences of the United States of America*, *95*(26), 15735–15740.

Cotrina, M. L., Lin, J. H., & Nedergaard, M. (1998). Cytoskeletal assembly and ATP release regulate astrocytic calcium signaling. *The Journal of Neuroscience*, *18*(21), 8794–8804.

Coull, J. A., Beggs, S., Boudreau, D., Boivin, D., Tsuda, M., Inoue, K., et al. (2005). BDNF from microglia causes the shift in neuronal anion gradient underlying neuropathic pain. *Nature*, *438*(7070), 1017–1021.

Dall'Igna, O. P., Fett, P., Gomes, M. W., Souza, D. O., Cunha, R. A., & Lara, D. R. (2007). Caffeine and adenosine A(2a) receptor antagonists prevent beta-amyloid (25-35)-induced cognitive deficits in mice. *Experimental Neurology*, *203*(1), 241–245.

Davalos, D., Grutzendler, J., Yang, G., Kim, J. V., Zuo, Y., Jung, S., et al. (2005). ATP mediates rapid microglial response to local brain injury *in vivo*. *Nature Neuroscience*, *8*(6), 752–758.

de Araujo, E. G., da Silva, G. M., & Dos Santos, A. A. (2009). Neuronal cell survival: The role of interleukins. *Annals of the New York Academy of Sciences*, *1153*, 57–64.

Dhaenens, C. M., Burnouf, S., Simonin, C., Van Brussel, E., Duhamel, A., Defebvre, L., et al. (2009). A genetic variation in the ADORA2A gene modifies age at onset in Huntington's disease. *Neurobiology of Disease*, *35*(3), 474–476.

Domercq, M., Perez-Samartin, A., Aparicio, D., Alberdi, E., Pampliega, O., & Matute, C. (2010). P2X7 receptors mediate ischemic damage to oligodendrocytes. *Glia*, *58*(6), 730–740.

Donnelly-Roberts, D. L., & Jarvis, M. F. (2007). Discovery of P2X7 receptor-selective antagonists offers new insights into P2X7 receptor function and indicates a role in chronic pain states. *British Journal of Pharmacology*, *151*(5), 571–579.

Edwards, F. A., Gibb, A. J., & Colquhoun, D. (1992). ATP receptor-mediated synaptic currents in the central nervous system. *Nature*, *359*(6391), 144–147.

Erlinge, D., Harnek, J., van Heusden, C., Olivecrona, G., Jern, S., & Lazarowski, E. (2005). Uridine triphosphate (UTP) is released during cardiac ischemia. *International Journal of Cardiology*, *100*(3), 427–433.

Fam, S. R., Gallagher, C. J., Kalia, L. V., & Salter, M. W. (2003). Differential frequency dependence of P2Y1- and P2Y2-mediated Ca^{2+} signaling in astrocytes. *The Journal of Neuroscience*, *23*(11), 4437–4444.

Farber, K., & Kettenmann, H. (2006). Purinergic signaling and microglia. *Pflügers Archiv: European Journal of Physiology*, *452*(5), 615–621.

Ferrari, D., Chiozzi, P., Falzoni, S., Hanau, S., & Di Virgilio, F. (1997). Purinergic modulation of interleukin-1 beta release from microglial cells stimulated with bacterial endotoxin. *The Journal of Experimental Medicine*, *185*(3), 579–582.

Ferrari, D., Villalba, M., Chiozzi, P., Falzoni, S., Ricciardi-Castagnoli, P., & Di Virgilio, F. (1996). Mouse microglial cells express a plasma membrane pore gated by extracellular ATP. *Journal of Immunology*, *156*(4), 1531–1539.

Fiebich, B. L., Biber, K., Lieb, K., van Calker, D., Berger, M., Bauer, J., et al. (1996). Cyclooxygenase-2 expression in rat microglia is induced by adenosine A2a-receptors. *Glia*, *18*(2), 152–160.

Fields, R. D., & Burnstock, G. (2006). Purinergic signalling in neuron–glia interactions. *Nature Reviews. Neuroscience*, *7*(6), 423–436.

Fields, R. D., & Stevens-Graham, B. (2002). New insights into neuron–glia communication. *Science*, *298*(5593), 556–562.

Florenzano, F., Viscomi, M. T., Cavaliere, F., Volonte, C., & Molinari, M. (2002). Cerebellar lesion up-regulates P2X1 and P2X2 purinergic receptors in precerebellar nuclei. *Neuroscience*, *115*(2), 425–434.

Fontaine, V., Mohand-Said, S., Hanoteau, N., Fuchs, C., Pfizenmaier, K., & Eisel, U. (2002). Neurodegenerative and neuroprotective effects of tumor necrosis factor (TNF) in retinal ischemia: Opposite roles of TNF receptor 1 and TNF receptor 2. *Journal of Neuroscience*, *22*(7) RC216.

Frade, J. M., & Barde, Y. A. (1998). Microglia-derived nerve growth factor causes cell death in the developing retina. *Neuron*, *20*(1), 35–41.

Franke, H., Grosche, J., Schadlich, H., Krugel, U., Allgaier, C., & Illes, P. (2001). P2X receptor expression on astrocytes in the nucleus accumbens of rats. *Neuroscience*, *108*(3), 421–429.

Franke, H., Grummich, B., Hartig, W., Grosche, J., Regenthal, R., Edwards, R. H., et al. (2006). Changes in purinergic signaling after cerebral injury—Involvement of glutamatergic mechanisms? *International Journal of Developmental Neuroscience*, *24*(2–3), 123–132.

Franke, H., Krugel, U., Grosche, J., Heine, C., Hartig, W., Allgaier, C., et al. (2004). P2Y receptor expression on astrocytes in the nucleus accumbens of rats. *Neuroscience*, *127*(2), 431–441.

Franke, H., Krugel, U., & Illes, P. (1999). P2 receptor-mediated proliferative effects on astrocytes *in vivo*. *Glia*, *28*(3), 190–200.

Franke, H., Sauer, C., Rudolph, C., Krugel, U., Hengstler, J. G., & Illes, P. (2009). P2 receptor-mediated stimulation of the PI3-K/Akt-pathway *in vivo*. *Glia*, *57*(10), 1031–1045.

Fujii, S., Kato, H., & Kuroda, Y. (2002). Cooperativity between extracellular adenosine 5′-triphosphate and activation of *N*-methyl-D-aspartate receptors in long-term potentiation induction in hippocampal CA1 neurons. *Neuroscience*, *113*(3), 617–628.

Fujita, T., Tozaki-Saitoh, H., & Inoue, K. (2009). P2Y1 receptor signaling enhances neuroprotection by astrocytes against oxidative stress via IL-6 release in hippocampal cultures. *Glia*, *57*(3), 244–257.

Gallagher, C. J., & Salter, M. W. (2003). Differential properties of astrocyte calcium waves mediated by P2Y1 and P2Y2 receptors. *The Journal of Neuroscience*, *23*(17), 6728–6739.

Gao, H. M., Liu, B., Zhang, W., & Hong, J. S. (2003). Novel anti-inflammatory therapy for Parkinson's disease. *Trends in Pharmacological Sciences*, *24*(8), 395–401.

Genovese, T., Melani, A., Esposito, E., Paterniti, I., Mazzon, E., Di Paola, R., et al. (2010). Selective adenosine A(2a) receptor agonists reduce the apoptosis in an experimental model of spinal cord trauma. *Journal of Biological Regulators and Homeostatic Agents*, *24*(1), 73–86.

Girouard, H., Bonev, A. D., Hannah, R. M., Meredith, A., Aldrich, R. W., & Nelson, M. T. (2010). Astrocytic endfoot Ca^{2+} and BK channels determine both arteriolar dilation and constriction. *Proceedings of the National Academy of Sciences of the United States of America*, *107*(8), 3811–3816.

Gruol, D. L., & Nelson, T. E. (1997). Physiological and pathological roles of interleukin-6 in the central nervous system. *Molecular Neurobiology*, *15*(3), 307–339.

Guo, L. H., Guo, K. T., Wendel, H. P., & Schluesener, H. J. (2006). Combinations of TLR and NOD2 ligands stimulate rat microglial P2X4R expression. *Biochemical and Biophysical Research Communications*, *349*(3), 1156–1162.

Guo, L. H., & Schluesener, H. J. (2005). Lesional accumulation of P2X(4) receptor(+) macrophages in rat CNS during experimental autoimmune encephalomyelitis. *Neuroscience*, *134*(1), 199–205.

Guthrie, P. B., Knappenberger, J., Segal, M., Bennett, M. V., Charles, A. C., & Kater, S. B. (1999). ATP released from astrocytes mediates glial calcium waves. *The Journal of Neuroscience*, *19*(2), 520–528.

Hama, T., Kushima, Y., Miyamoto, M., Kubota, M., Takei, N., & Hatanaka, H. (1991). Interleukin-6 improves the survival of mesencephalic catecholaminergic and septal cholinergic neurons from postnatal, two-week-old rats in cultures. *Neuroscience*, *40*(2), 445–452.

Hasegawa, S., Kohro, Y., Tsuda, M., & Inoue, K. (2009). Activation of cytosolic phospholipase A2 in dorsal root ganglion neurons by Ca^{2+}/calmodulin-dependent protein kinase II after peripheral nerve injury. *Molecular Pain*, *5*, 22.

Haselkorn, M. L., Shellington, D. K., Jackson, E. K., Vagni, V. A., Janesko-Feldman, K., Dubey, R. K., et al. (2010). Adenosine A1 receptor activation as a brake on the microglial response after experimental traumatic brain injury in mice. *Journal of Neurotrauma*, *27* (5), 901–910.

Haynes, S. E., Hollopeter, G., Yang, G., Kurpius, D., Dailey, M. E., Gan, W. B., et al. (2006). The P2Y12 receptor regulates microglial activation by extracellular nucleotides. *Nature Neuroscience*, *9*(12), 1512–1519.

Heese, K., Fiebich, B. L., Bauer, J., & Otten, U. (1997). Nerve growth factor (NGF) expression in rat microglia is induced by adenosine A2a-receptors. *Neuroscience Letters*, *231*(2), 83–86.

Hide, I. (2003). Mechanism of production and release of tumor necrosis factor implicated in inflammatory diseases. *Nippon Yakurigaku Zasshi*, *121*(3), 163–173.

Hide, I., Tanaka, M., Inoue, A., Nakajima, K., Kohsaka, S., Inoue, K., et al. (2000). Extracellular ATP triggers tumor necrosis factor-alpha release from rat microglia. *Journal of Neurochemistry*, *75*(3), 965–972.

Honda, S., Sasaki, Y., Ohsawa, K., Imai, Y., Nakamura, Y., Inoue, K., et al. (2001). Extracellular ATP or ADP induce chemotaxis of cultured microglia through Gi/o-coupled P2Y receptors. *The Journal of Neuroscience*, *21*(6), 1975–1982.

Honore, P., Donnelly-Roberts, D., Namovic, M. T., Hsieh, G., Zhu, C. Z., Mikusa, J. P., et al. (2006). A-740003 [N-(1-{[(cyanoimino)(5-quinolinylamino) methyl]amino}-2,2-dimethylpropyl)-2-(3,4-dimethoxyphenyl)acetamide], a novel and selective P2X7 receptor antagonist, dose-dependently reduces neuropathic pain in the rat. *The Journal of Pharmacology and Experimental Therapeutics*, *319*(3), 1376–1385.

Iadecola, C., & Nedergaard, M. (2007). Glial regulation of the cerebral microvasculature. *Nature Neuroscience*, *10*(11), 1369–1376.

Inoue, K. (2008). Purinergic systems in microglia. *Cellular and Molecular Life Sciences*, *65*(19), 3074–3080.

Inoue, K., Koizumi, S., & Tsuda, M. (2007). The role of nucleotides in the neuron–glia communication responsible for the brain functions. *Journal of Neurochemistry*, *102*(5), 1447–1458.

Inoue, K., Nakajima, K., Morimoto, T., Kikuchi, Y., Koizumi, S., Illes, P., et al. (1998). ATP stimulation of Ca^{2+}-dependent plasminogen release from cultured microglia. *British Journal of Pharmacology*, *123*(7), 1304–1310.

Inoue, K., Nakazawa, K., Fujimori, K., Watano, T., & Takanaka, A. (1992). Extracellular adenosine 5′-triphosphate-evoked glutamate release in cultured hippocampal neurons. *Neuroscience Letters*, *134*(2), 215–218.

Irino, Y., Nakamura, Y., Inoue, K., Kohsaka, S., & Ohsawa, K. (2008). Akt activation is involved in P2Y12 receptor-mediated chemotaxis of microglia. *Journal of Neuroscience Research*, *86*(7), 1511–1519.

Ishibashi, T., Dakin, K. A., Stevens, B., Lee, P. R., Kozlov, S. V., Stewart, C. L., et al. (2006). Astrocytes promote myelination in response to electrical impulses. *Neuron*, *49*(6), 823–832.

Jahr, C. E., & Jessell, T. M. (1983). ATP excites a subpopulation of rat dorsal horn neurones. *Nature*, *304*(5928), 730–733.

Juranyi, Z., Sperlagh, B., & Vizi, E. S. (1999). Involvement of P2 purinoceptors and the nitric oxide pathway in [3H]purine outflow evoked by short-term hypoxia and hypoglycemia in rat hippocampal slices. *Brain Research*, *823*(1–2), 183–190.

Kacem, K., Lacombe, P., Seylaz, J., & Bonvento, G. (1998). Structural organization of the perivascular astrocyte endfeet and their relationship with the endothelial glucose transporter: A confocal microscopy study. *Glia*, *23*(1), 1–10.

Kassa, R. M., Bentivoglio, M., & Mariotti, R. (2007). Changes in the expression of P2X1 and P2X2 purinergic receptors in facial motoneurons after nerve lesions in rodents and correlation with motoneuron degeneration. *Neurobiology of Disease*, *25*(1), 121–133.

Kataoka, A., Tozaki-Saitoh, H., Koga, Y., Tsuda, M., & Inoue, K. (2009). Activation of P2X7 receptors induces CCL3 production in microglial cells through transcription factor NFAT. *Journal of Neurochemistry*, *108*(1), 115–125.

Kawamura, M., Jr., Ruskin, D. N., & Masino, S. A. (2010). Metabolic autocrine regulation of neurons involves cooperation among pannexin hemichannels, adenosine receptors, and KATP channels. *The Journal of Neuroscience*, *30*(11), 3886–3895.

Keller, A. F., Beggs, S., Salter, M. W., & De Koninck, Y. (2007). Transformation of the output of spinal lamina I neurons after nerve injury and microglia stimulation underlying neuropathic pain. *Molecular Pain*, *3*, 27.

Khakh, B. S., Burnstock, G., Kennedy, C., King, B. F., North, R. A., Seguela, P., et al. (2001). International union of pharmacology. XXIV. Current status of the nomenclature and properties of P2X receptors and their subunits. *Pharmacological Reviews*, *53*(1), 107–118.

Kim, D. S., Kwak, S. E., Kim, J. E., Won, M. H., & Kang, T. C. (2006). The co-treatments of vigabatrin and P2X receptor antagonists protect ischemic neuronal cell death in the gerbil hippocampus. *Brain Research*, *1120*(1), 151–160.

Kim, S. G., Soltysiak, K. A., Gao, Z. G., Chang, T. S., Chung, E., & Jacobson, K. A. (2003). Tumor necrosis factor alpha-induced apoptosis in astrocytes is prevented by the activation of P2Y6, but not P2Y4 nucleotide receptors. *Biochemical Pharmacology*, *65*(6), 923–931.

Kobayashi, K., Yamanaka, H., Fukuoka, T., Dai, Y., Obata, K., & Noguchi, K. (2008). P2Y12 receptor upregulation in activated microglia is a gateway of p38 signaling and neuropathic pain. *The Journal of Neuroscience*, *28*(11), 2892–2902.

Koizumi, S., Fujishita, K., & Inoue, K. (2005). Regulation of cell-to-cell communication mediated by astrocytic ATP in the CNS. *Purinergic Signalling*, *1*(3), 211–217.

Koizumi, S., Fujishita, K., Tsuda, M., Shigemoto-Mogami, Y., & Inoue, K. (2003). Dynamic inhibition of excitatory synaptic transmission by astrocyte-derived ATP in hippocampal cultures. *Proceedings of the National Academy of Sciences of the United States of America*, *100*(19), 11023–11028.

Koizumi, S., Saito, Y., Nakazawa, K., Nakajima, K., Sawada, J. I., Kohsaka, S., et al. (2002). Spatial and temporal aspects of Ca^{2+} signaling mediated by P2Y receptors in cultured rat hippocampal astrocytes. *Life Sciences*, *72*(4–5), 431–442.

Koizumi, S., Shigemoto-Mogami, Y., Nasu-Tada, K., Shinozaki, Y., Ohsawa, K., Tsuda, M., et al. (2007). UDP acting at P2Y6 receptors is a mediator of microglial phagocytosis. *Nature*, *446*(7139), 1091–1095.

Kreutzberg, G. W. (1996). Microglia: a sensor for pathological events in the CNS. *Trends in Neurosciences*, *19*(8), 312–318.

Krishtal, O. A., Marchenko, S. M., & Pidoplichko, V. I. (1983). Receptor for ATP in the membrane of mammalian sensory neurones. *Neuroscience Letters*, *35*(1), 41–45.

Labasi, J. M., Petrushova, N., Donovan, C., McCurdy, S., Lira, P., Payette, M. M., et al. (2002). Absence of the P2X7 receptor alters leukocyte function and attenuates an inflammatory response. *Journal of Immunology*, *168*(12), 6436–6445.

Lai, M. K., Tan, M. G., Kirvell, S., Hobbs, C., Lee, J., Esiri, M. M., et al. (2008). Selective loss of P2Y2 nucleotide receptor immunoreactivity is associated with Alzheimer's disease neuropathology. *Journal of Neural Transmission*, *115*(8), 1165–1172.

Lazarowski, E. R., Homolya, L., Boucher, R. C., & Harden, T. K. (1997). Direct demonstration of mechanically induced release of cellular UTP and its implication for uridine nucleotide receptor activation. *The Journal of Biological Chemistry*, *272*(39), 24348–24354.

Lazarowski, E. R., Shea, D. A., Boucher, R. C., & Harden, T. K. (2003). Release of cellular UDP-glucose as a potential extracellular signaling molecule. *Molecular Pharmacology*, *63* (5), 1190–1197.

Ledent, C., Vaugeois, J. M., Schiffmann, S. N., Pedrazzini, T., El Yacoubi, M., Vanderhaeghen, J. J., et al. (1997). Aggressiveness, hypoalgesia and high blood pressure in mice lacking the adenosine A2a receptor. *Nature*, *388*(6643), 674–678.

Li, L., Hao, J. X., Fredholm, B. B., Schulte, G., Wiesenfeld-Hallin, Z., & Xu, X. J. (2010). Peripheral adenosine A2A receptors are involved in carrageenan-induced mechanical hyperalgesia in mice. *Neuroscience*, *170*(3), 923–928.

Lindsay, R. M., Wiegand, S. J., Altar, C. A., & DiStefano, P. S. (1994). Neurotrophic factors: From molecule to man. *Trends in Neurosciences*, *17*(5), 182–190.

Liu, X., Mashour, G. A., Webster, H. F., & Kurtz, A. (1998). Basic FGF and FGF receptor 1 are expressed in microglia during experimental autoimmune encephalomyelitis: Temporally distinct expression of midkine and pleiotrophin. *Glia*, *24*(4), 390–397.

Liu, H. Q., Zhang, W. Y., Luo, X. T., Ye, Y., & Zhu, X. Z. (2006). Paeoniflorin attenuates neuroinflammation and dopaminergic neurodegeneration in the MPTP model of Parkinson's disease by activation of adenosine A1 receptor. *British Journal of Pharmacology*, *148*(3), 314–325.

Loram, L. C., Harrison, J. A., Sloane, E. M., Hutchinson, M. R., Sholar, P., Taylor, F. R., et al. (2009). Enduring reversal of neuropathic pain by a single intrathecal injection of adenosine 2A receptor agonists: A novel therapy for neuropathic pain. *The Journal of Neuroscience*, *29*(44), 14015–14025.

Lustig, K. D., Shiau, A. K., Brake, A. J., & Julius, D. (1993). Expression cloning of an ATP receptor from mouse neuroblastoma cells. *Proceedings of the National Academy of Sciences of the United States of America*, *90*(11), 5113–5117.

Malin, S. A., Davis, B. M., Koerber, H. R., Reynolds, I. J., Albers, K. M., & Molliver, D. C. (2008). Thermal nociception and TRPV1 function are attenuated in mice lacking the nucleotide receptor P2Y2. *Pain*, *138*(3), 484–496.

Malin, S. A., & Molliver, D. C. (2010). Gi- and Gq-coupled ADP (P2Y) receptors act in opposition to modulate nociceptive signaling and inflammatory pain behavior. *Molecular Pain*, *6*, 21.

Marcellino, D., Suarez-Boomgaard, D., Sanchez-Reina, M. D., Aguirre, J. A., Yoshitake, T., Yoshitake, S., et al. (2010). On the role of P2X(7) receptors in dopamine nerve cell degeneration in a rat model of Parkinson's disease: Studies with the P2X(7) receptor antagonist A-438079. *Journal of Neural Transmission*, *117*(6), 681–687.

Marcoli, M., Bonfanti, A., Roccatagliata, P., Chiaramonte, G., Ongini, E., Raiteri, M., et al. (2004). Glutamate efflux from human cerebrocortical slices during ischemia: Vesicular-like mode of glutamate release and sensitivity to A(2A) adenosine receptor blockade. *Neuropharmacology*, *47*(6), 884–891.

Martin, D., & Near, S. L. (1995). Protective effect of the interleukin-1 receptor antagonist (IL-1ra) on experimental allergic encephalomyelitis in rats. *Journal of Neuroimmunology*, *61*(2), 241–245.

Marz, P., Heese, K., Dimitriades-Schmutz, B., Rose-John, S., & Otten, U. (1999). Role of interleukin-6 and soluble IL-6 receptor in region-specific induction of astrocytic differentiation and neurotrophin expression. *Glia*, *26*(3), 191–200.

Masino, S. A., Kawamura, M., Wasser, C. A., Pomeroy, L. T., & Ruskin, D. N. (2009). Adenosine, ketogenic diet and epilepsy: The emerging therapeutic relationship between metabolism and brain activity. *Current Neuropharmacology*, *7*(3), 257–268.

Matute, C., Torre, I., Perez-Cerda, F., Perez-Samartin, A., Alberdi, E., Etxebarria, E., et al. (2007). P2X(7) receptor blockade prevents ATP excitotoxicity in oligodendrocytes and ameliorates experimental autoimmune encephalomyelitis. *The Journal of Neuroscience*, *27* (35), 9525–9533.

McGaraughty, S., Wismer, C. T., Zhu, C. Z., Mikusa, J., Honore, P., Chu, K. L., et al. (2003). Effects of A-317491, a novel and selective P2X3/P2X2/3 receptor antagonist, on neuropathic, inflammatory and chemogenic nociception following intrathecal and intraplantar administration. *British Journal of Pharmacology*, *140*(8), 1381–1388.

McLarnon, J. G., Ryu, J. K., Walker, D. G., & Choi, H. B. (2006). Upregulated expression of purinergic P2X(7) receptor in Alzheimer disease and amyloid-beta peptide-treated microglia and in peptide-injected rat hippocampus. *Journal of Neuropathology and Experimental Neurology*, *65*(11), 1090–1097.

Meissner, F., Molawi, K., & Zychlinsky, A. (2010). Mutant superoxide dismutase 1-induced IL-1beta accelerates ALS pathogenesis. *Proceedings of the National Academy of Sciences of the United States of America*, *107*(29), 13046–13050.

Meistrell, M. E., 3rd, Botchkina, G. I., Wang, H., Di Santo, E., Cockroft, K. M., Bloom, O., et al. (1997). Tumor necrosis factor is a brain damaging cytokine in cerebral ischemia. *Shock*, *8*(5), 341–348.

Melani, A., Turchi, D., Vannucchi, M. G., Cipriani, S., Gianfriddo, M., & Pedata, F. (2005). ATP extracellular concentrations are increased in the rat striatum during *in vivo* ischemia. *Neurochemistry International*, *47*(6), 442–448.

Minami, M., Kuraishi, Y., Yabuuchi, K., Yamazaki, A., & Satoh, M. (1992). Induction of interleukin-1 beta mRNA in rat brain after transient forebrain ischemia. *Journal of Neurochemistry*, *58*(1), 390–392.

Mojsilovic-Petrovic, J., Jeong, G. B., Crocker, A., Arneja, A., David, S., Russell, D. S., et al. (2006). Protecting motor neurons from toxic insult by antagonism of adenosine A2a and Trk receptors. *The Journal of Neuroscience*, *26*(36), 9250–9263.

Moore, D., Iritani, S., Chambers, J., & Emson, P. (2000). Immunohistochemical localization of the P2Y1 purinergic receptor in Alzheimer's disease. *NeuroReport*, *11*(17), 3799–3803.

Mori, M., Heuss, C., Gahwiler, B. H., & Gerber, U. (2001). Fast synaptic transmission mediated by P2X receptors in CA3 pyramidal cells of rat hippocampal slice cultures. *The Journal of Physiology*, *535*(Pt 1), 115–123.

Morrone, F. B., Oliveira, D. L., Gamermann, P., Stella, J., Wofchuk, S., Wink, M. R., et al. (2006). *In vivo* glioblastoma growth is reduced by apyrase activity in a rat glioma model. *BMC Cancer*, *6*, 226.

Mulligan, S. J., & MacVicar, B. A. (2004). Calcium transients in astrocyte endfeet cause cerebrovascular constrictions. *Nature*, *431*(7005), 195–199.

Nagata, K., Nakajima, K., Takemoto, N., Saito, H., & Kohsaka, S. (1993). Microglia-derived plasminogen enhances neurite outgrowth from explant cultures of rat brain. *International Journal of Developmental Neuroscience*, *11*(2), 227–237.

Nakajima, K., Honda, S., Tohyama, Y., Imai, Y., Kohsaka, S., & Kurihara, T. (2001). Neurotrophin secretion from cultured microglia. *Journal of Neuroscience Research*, *65*(4), 322–331.

Nakajima, K., Nagata, K., Hamanoue, M., Takemoto, N., & Kohsaka, S. (1993). Microglia-derived elastase produces a low-molecular-weight plasminogen that enhances neurite outgrowth in rat neocortical explant cultures. *Journal of Neurochemistry*, *61*(6), 2155–2163.

Nakajima, K., Tohyama, Y., Maeda, S., Kohsaka, S., & Kurihara, T. (2007). Neuronal regulation by which microglia enhance the production of neurotrophic factors for GABAergic, catecholaminergic, and cholinergic neurons. *Neurochemistry International*, *50*(6), 807–820.

Nasu-Tada, K., Koizumi, S., & Inoue, K. (2005). Involvement of beta1 integrin in microglial chemotaxis and proliferation on fibronectin: Different regulations by ADP through PKA. *Glia*, *52*(2), 98–107.

Nasu-Tada, K., Koizumi, S., Tsuda, M., Kunifusa, E., & Inoue, K. (2006). Possible involvement of increase in spinal fibronectin following peripheral nerve injury in upregulation of microglial P2X4, a key molecule for mechanical allodynia. *Glia*, *53*(7), 769–775.

Neary, J. T., Kang, Y., Shi, Y. F., Tran, M. D., & Wanner, I. B. (2006). P2 receptor signalling, proliferation of astrocytes, and expression of molecules involved in cell–cell interactions. *Novartis Foundation Symposium*, *276*, 131–143, discussion 143–137, 233–137, 275–181.

Neary, J. T., Rathbone, M. P., Cattabeni, F., Abbracchio, M. P., & Burnstock, G. (1996). Trophic actions of extracellular nucleotides and nucleosides on glial and neuronal cells. *Trends in Neurosciences*, *19*(1), 13–18.

Newman, E. A. (2001). Propagation of intercellular calcium waves in retinal astrocytes and Muller cells. *The Journal of Neuroscience*, *21*(7), 2215–2223.

Newman, E. A. (2003). Glial cell inhibition of neurons by release of ATP. *The Journal of Neuroscience*, *23*(5), 1659–1666.

Nieber, K., Poelchen, W., & Illes, P. (1997). Role of ATP in fast excitatory synaptic potentials in locus coeruleus neurones of the rat. *British Journal of Pharmacology*, *122*(3), 423–430.

Nimmerjahn, A., Kirchhoff, F., & Helmchen, F. (2005). Resting microglial cells are highly dynamic surveillants of brain parenchyma *in vivo*. *Science*, *308*(5726), 1314–1318.

Norris, J. G., Tang, L. P., Sparacio, S. M., & Benveniste, E. N. (1994). Signal transduction pathways mediating astrocyte IL-6 induction by IL-1 beta and tumor necrosis factor-alpha. *Journal of Immunology*, *152*(2), 841–850.

North, R. A. (2002). Molecular physiology of P2X receptors. *Physiological Reviews*, *82*(4), 1013–1067.

Ohsawa, K., Imai, Y., Nakajima, K., & Kohsaka, S. (1997). Generation and characterization of a microglial cell line, MG5, derived from a p53-deficient mouse. *Glia*, *21*(3), 285–298.

Ohsawa, K., Irino, Y., Nakamura, Y., Akazawa, C., Inoue, K., & Kohsaka, S. (2007). Involvement of P2X4 and P2Y12 receptors in ATP-induced microglial chemotaxis. *Glia*, *55*(6), 604–616.

Ohsawa, K., Irino, Y., Sanagi, T., Nakamura, Y., Suzuki, E., Inoue, K., et al. (2010). P2Y12 receptor-mediated integrin-beta1 activation regulates microglial process extension induced by ATP. *Glia*, *58*(7), 790–801.

Okada, M., Nakagawa, T., Minami, M., & Satoh, M. (2002). Analgesic effects of intrathecal administration of P2Y nucleotide receptor agonists UTP and UDP in normal and neuropathic pain model rats. *The Journal of Pharmacology and Experimental Therapeutics*, *303*(1), 66–73.

Orr, A. G., Orr, A. L., Li, X. J., Gross, R. E., & Traynelis, S. F. (2009). Adenosine A(2A) receptor mediates microglial process retraction. *Nature Neuroscience*, *12*(7), 872–878.

Pankratov, Y., Lalo, U., Krishtal, O., & Verkhratsky, A. (2002). Ionotropic P2X purinoreceptors mediate synaptic transmission in rat pyramidal neurones of layer II/III of somato-sensory cortex. *The Journal of Physiology*, *542*(Pt 2), 529–536.

Pankratov, Y., Lalo, U., Krishtal, O. A., & Verkhratsky, A. (2009). P2X receptors and synaptic plasticity. *Neuroscience*, *158*(1), 137–148.

Parvathenani, L. K., Tertyshnikova, S., Greco, C. R., Roberts, S. B., Robertson, B., & Posmantur, R. (2003). P2X7 mediates superoxide production in primary microglia and is up-regulated in a transgenic mouse model of Alzheimer's disease. *The Journal of Biological Chemistry*, *278*(15), 13309–13317.

Pasinelli, P., Borchelt, D. R., Houseweart, M. K., Cleveland, D. W., & Brown, R. H. Jr. (1998). Caspase-1 is activated in neural cells and tissue with amyotrophic lateral sclerosis-associated mutations in copper–zinc superoxide dismutase. *Proceedings of the National Academy of Sciences of the United States of America*, *95*(26), 15763–15768.

Pedata, F., Melani, A., Pugliese, A. M., Coppi, E., Cipriani, S., & Traini, C. (2007). The role of ATP and adenosine in the brain under normoxic and ischemic conditions. *Purinergic Signalling*, *3*(4), 299–310.

Peng, W., Cotrina, M. L., Han, X., Yu, H., Bekar, L., Blum, L., et al. (2009). Systemic administration of an antagonist of the ATP-sensitive receptor P2X7 improves recovery after spinal cord injury. *Proceedings of the National Academy of Sciences of the United States of America*, *106*(30), 12489–12493.

Perez-Ortiz, J. M., Serrano-Perez, M. C., Pastor, M. D., Martin, E. D., Calvo, S., Rincon, M., et al. (2008). Mechanical lesion activates newly identified NFATc1 in primary astrocytes: Implication of ATP and purinergic receptors. *The European Journal of Neuroscience*, *27*(9), 2453–2465.

Perregaux, D., & Gabel, C. A. (1994). Interleukin-1 beta maturation and release in response to ATP and nigericin. Evidence that potassium depletion mediated by these agents is a necessary and common feature of their activity. *The Journal of Biological Chemistry*, *269* (21), 15195–15203.

Perrin, F. E., Lacroix, S., Aviles-Trigueros, M., & David, S. (2005). Involvement of monocyte chemoattractant protein-1, macrophage inflammatory protein-1alpha and interleukin-1beta in Wallerian degeneration. *Brain*, *128*(Pt 4), 854–866.

Phillis, J. W. (1995). The effects of selective A1 and A2a adenosine receptor antagonists on cerebral ischemic injury in the gerbil. *Brain Research*, *705*(1–2), 79–84.

Popoli, P., Blum, D., Domenici, M. R., Burnouf, S., & Chern, Y. (2008). A critical evaluation of adenosine A2A receptors as potentially "druggable" targets in Huntington's disease. *Current Pharmaceutical Design*, *14*(15), 1500–1511.

Popoli, P., Blum, D., Martire, A., Ledent, C., Ceruti, S., & Abbracchio, M. P. (2007). Functions, dysfunctions and possible therapeutic relevance of adenosine A2A receptors in Huntington's disease. *Progress in Neurobiology*, *81*(5–6), 331–348.

Qureshi, O. S., Paramasivam, A., Yu, J. C., & Murrell-Lagnado, R. D. (2007). Regulation of P2X4 receptors by lysosomal targeting, glycan protection and exocytosis. *Journal of Cell Science*, *120*(Pt 21), 3838–3849.

Rathbone, M. P., Middlemiss, P. J., Crocker, C. E., Glasky, M. S., Juurlink, B. H., Ramirez, J. J., et al. (1999). AIT-082 as a potential neuroprotective and regenerative agent in stroke and central nervous system injury. *Expert Opinion on Investigational Drugs*, *8* (8), 1255–1262.

Relton, J. K., & Rothwell, N. J. (1992). Interleukin-1 receptor antagonist inhibits ischaemic and excitotoxic neuronal damage in the rat. *Brain Research Bulletin*, *29*(2), 243–246.

Rodriguez-Zayas, A. E., Torrado, A. I., & Miranda, J. D. (2010). P2Y2 receptor expression is altered in rats after spinal cord injury. *International Journal of Developmental Neuroscience*, *28*(6), 413–421.

Ryu, J. K., Kim, J., Choi, S. H., Oh, Y. J., Lee, Y. B., Kim, S. U., et al. (2002). ATP-induced *in vivo* neurotoxicity in the rat striatum via P2 receptors. *NeuroReport*, *13*(13), 1611–1615.

Ryu, J. K., & McLarnon, J. G. (2008). Block of purinergic P2X(7) receptor is neuroprotective in an animal model of Alzheimer's disease. *NeuroReport*, *19*(17), 1715–1719.

Sanz, J. M., Chiozzi, P., Ferrari, D., Colaianna, M., Idzko, M., Falzoni, S., et al. (2009). Activation of microglia by amyloid beta requires P2X7 receptor expression. *Journal of Immunology*, *182*(7), 4378–4385.

Sanz, J. M., & Di Virgilio, F. (2000). Kinetics and mechanism of ATP-dependent IL-1 beta release from microglial cells. *Journal of Immunology*, *164*(9), 4893–4898.

Saura, J., Angulo, E., Ejarque, A., Casado, V., Tusell, J. M., Moratalla, R., et al. (2005). Adenosine A2A receptor stimulation potentiates nitric oxide release by activated microglia. *Journal of Neurochemistry*, *95*(4), 919–929.

Sawada, M., Suzumura, A., & Marunouchi, T. (1992). TNF alpha induces IL-6 production by astrocytes but not by microglia. *Brain Research*, *583*(1–2), 296–299.

Scherbel, U., Raghupathi, R., Nakamura, M., Saatman, K. E., Trojanowski, J. Q., Neugebauer, E., et al. (1999). Differential acute and chronic responses of tumor necrosis factor-deficient mice to experimental brain injury. *Proceedings of the National Academy of Sciences of the United States of America*, *96*(15), 8721–8726.

Schock, S. C., Munyao, N., Yakubchyk, Y., Sabourin, L. A., Hakim, A. M., Ventureyra, E. C., et al. (2007). Cortical spreading depression releases ATP into the extracellular space and purinergic receptor activation contributes to the induction of ischemic tolerance. *Brain Research*, *1168*, 129–138.

Schwarzschild, M. A., Agnati, L., Fuxe, K., Chen, J. F., & Morelli, M. (2006). Targeting adenosine A2A receptors in Parkinson's disease. *Trends in Neurosciences*, *29*(11), 647–654.

Shigemoto-Mogami, Y., Koizumi, S., Tsuda, M., Ohsawa, K., Kohsaka, S., & Inoue, K. (2001). Mechanisms underlying extracellular ATP-evoked interleukin-6 release in mouse microglial cell line, MG-5. *Journal of Neurochemistry*, *78*(6), 1339–1349.

Shimojo, M., Nakajima, K., Takei, N., Hamanoue, M., & Kohsaka, S. (1991). Production of basic fibroblast growth factor in cultured rat brain microglia. *Neuroscience Letters*, *123* (2), 229–231.

Shinozaki, Y., Koizumi, S., Ishida, S., Sawada, J., Ohno, Y., & Inoue, K. (2005). Cytoprotection against oxidative stress-induced damage of astrocytes by extracellular ATP via P2Y1 receptors. *Glia*, *49*(2), 288–300.

Shinozaki, Y., Koizumi, S., Ohno, Y., Nagao, T., & Inoue, K. (2006). Extracellular ATP counteracts the ERK1/2-mediated death-promoting signaling cascades in astrocytes. *Glia*, *54*(6), 606–618.

Shiratori, M., Tozaki-Saitoh, H., Yoshitake, M., Tsuda, M., & Inoue, K. (2010). P2X7 receptor activation induces CXCL2 production in microglia through NFAT and PKC/MAPK pathways. *Journal of Neurochemistry*, *114*(3), 810–819.

Sierra, A., Encinas, J. M., Deudero, J. J., Chancey, J. H., Enikolopov, G., Overstreet-Wadiche, L. S., et al. (2010). Microglia shape adult hippocampal neurogenesis through apoptosis-coupled phagocytosis. *Cell Stem Cell*, *7*(4), 483–495.

Stevens, B., Porta, S., Haak, L. L., Gallo, V., & Fields, R. D. (2002). Adenosine: A neuron–glial transmitter promoting myelination in the CNS in response to action potentials. *Neuron*, *36*(5), 855–868.

Sun, J. J., Liu, Y., & Ye, Z. R. (2008). Effects of P2Y1 receptor on glial fibrillary acidic protein and glial cell line-derived neurotrophic factor production of astrocytes under ischemic condition and the related signaling pathways. *Neuroscience Bulletin*, *24*(4), 231–243.

Suzuki, T., Hide, I., Ido, K., Kohsaka, S., Inoue, K., & Nakata, Y. (2004). Production and release of neuroprotective tumor necrosis factor by P2X7 receptor-activated microglia. *The Journal of Neuroscience*, *24*(1), 1–7.

Takano, T., Tian, G. F., Peng, W., Lou, N., Libionka, W., Han, X., et al. (2006). Astrocyte-mediated control of cerebral blood flow. *Nature Neuroscience*, *9*(2), 260–267.

Taylor, B. K. (2009). Spinal inhibitory neurotransmission in neuropathic pain. *Current Pain and Headache Reports*, *13*(3), 208–214.

Toulme, E., Garcia, A., Samways, D., Egan, T. M., Carson, M. J., & Khakh, B. S. (2010). P2X4 receptors in activated C8-B4 cells of cerebellar microglial origin. *The Journal of General Physiology*, *135*(4), 333–353.

Touzani, O., Boutin, H., Chuquet, J., & Rothwell, N. (1999). Potential mechanisms of interleukin-1 involvement in cerebral ischaemia. *Journal of Neuroimmunology*, *100*(1–2), 203–215.

Tozaki-Saitoh, H., Tsuda, M., Miyata, H., Ueda, K., Kohsaka, S., & Inoue, K. (2008). P2Y12 receptors in spinal microglia are required for neuropathic pain after peripheral nerve injury. *The Journal of Neuroscience*, *28*(19), 4949–4956.

Trang, T., Beggs, S., Wan, X., & Salter, M. W. (2009). P2X4-receptor-mediated synthesis and release of brain-derived neurotrophic factor in microglia is dependent on calcium and p38-mitogen-activated protein kinase activation. *The Journal of Neuroscience*, *29*(11), 3518–3528.

Tsuda, M., Inoue, K., & Salter, M. W. (2005). Neuropathic pain and spinal microglia: A big problem from molecules in "small" glia. *Trends in Neurosciences*, *28*(2), 101–107.

Tsuda, M., Kuboyama, K., Inoue, T., Nagata, K., Tozaki-Saitoh, H., & Inoue, K. (2009). Behavioral phenotypes of mice lacking purinergic P2X4 receptors in acute and chronic pain assays. *Molecular Pain*, *5*, 28.

Tsuda, M., Shigemoto-Mogami, Y., Koizumi, S., Mizokoshi, A., Kohsaka, S., Salter, M. W., et al. (2003). P2X4 receptors induced in spinal microglia gate tactile allodynia after nerve injury. *Nature*, *424*(6950), 778–783.

Tsuda, M., Toyomitsu, E., Komatsu, T., Masuda, T., Kunifusa, E., Nasu-Tada, K., et al. (2008). Fibronectin/integrin system is involved in P2X(4) receptor upregulation in the spinal cord and neuropathic pain after nerve injury. *Glia*, *56*(5), 579–585.

Tsuda, M., Toyomitsu, E., Kometani, M., Tozaki-Saitoh, H., & Inoue, K. (2009). Mechanisms underlying fibronectin-induced up-regulation of P2X4R expression in microglia: Distinct roles of PI3K-Akt and MEK-ERK signalling pathways. *Journal of Cellular and Molecular Medicine*, *13*(9B), 3251–3259.

Tsuda, M., Tozaki-Saitoh, H., Masuda, T., Toyomitsu, E., Tezuka, T., Yamamoto, T., et al. (2008). Lyn tyrosine kinase is required for P2X(4) receptor upregulation and neuropathic pain after peripheral nerve injury. *Glia*, *56*(1), 50–58.

Tsutsui, S., Schnermann, J., Noorbakhsh, F., Henry, S., Yong, V. W., Winston, B. W., et al. (2004). A1 adenosine receptor upregulation and activation attenuates neuroinflammation and demyelination in a model of multiple sclerosis. *The Journal of Neuroscience*, *24*(6), 1521–1529.

Ulmann, L., Hatcher, J. P., Hughes, J. P., Chaumont, S., Green, P. J., Conquet, F., et al. (2008). Up-regulation of P2X4 receptors in spinal microglia after peripheral nerve injury mediates BDNF release and neuropathic pain. *The Journal of Neuroscience*, *28*(44), 11263–11268.

Ulmann, L., Hirbec, H., & Rassendren, F. (2010). P2X4 receptors mediate PGE2 release by tissue-resident macrophages and initiate inflammatory pain. *The EMBO Journal*, *29*(14), 2290–2300.

van der Putten, C., Zuiderwijk-Sick, E. A., van Straalen, L., de Geus, E. D., Boven, L. A., Kondova, I., et al. (2009). Differential expression of adenosine A3 receptors controls adenosine A2A receptor-mediated inhibition of TLR responses in microglia. *Journal of Immunology*, *182*(12), 7603–7612.

Vassilev, L. T., Vu, B. T., Graves, B., Carvajal, D., Podlaski, F., Filipovic, Z., et al. (2004). *In vivo* activation of the p53 pathway by small-molecule antagonists of MDM2. *Science*, *303*(5659), 844–848.

Verkhratsky, A., Krishtal, O. A., & Burnstock, G. (2009). Purinoceptors on neuroglia. *Molecular Neurobiology*, *39*(3), 190–208.

Vezzani, A., Balosso, S., Maroso, M., Zardoni, D., Noe, F., & Ravizza, T. (2010). ICE/caspase 1 inhibitors and IL-1beta receptor antagonists as potential therapeutics in epilepsy. *Current Opinion in Investigational Drugs*, *11*(1), 43–50.

Vianna, E. P., Ferreira, A. T., Dona, F., Cavalheiro, E. A., & da Silva Fernandes, M. J. (2005). Modulation of seizures and synaptic plasticity by adenosinergic receptors in an experimental model of temporal lobe epilepsy induced by pilocarpine in rats. *Epilepsia*, *46*(Suppl. 5), 166–173.

Wagner, A. K., Miller, M. A., Scanlon, J., Ren, D., Kochanek, P. M., & Conley, Y. P. (2010). Adenosine A1 receptor gene variants associated with post-traumatic seizures after severe TBI. *Epilepsy Research*, *90*(3), 259–272.

Wang, X., Arcuino, G., Takano, T., Lin, J., Peng, W. G., Wan, P., et al. (2004). P2X7 receptor inhibition improves recovery after spinal cord injury. *Natural Medicines*, *10*(8), 821–827.

Washburn, K. B., & Neary, J. T. (2006). P2 purinergic receptors signal to STAT3 in astrocytes: Difference in STAT3 responses to P2Y and P2X receptor activation. *Neuroscience*, *142*(2), 411–423.

Webb, T. E., Simon, J., Krishek, B. J., Bateson, A. N., Smart, T. G., King, B. F., et al. (1993). Cloning and functional expression of a brain G-protein-coupled ATP receptor. *FEBS Letters*, *324*(2), 219–225.

White, T. D. (1978). Release of ATP from a synaptosomal preparation by elevated extracellular K^+ and by veratridine. *Journal of Neurochemistry*, *30*(2), 329–336.

White, F. A., & Jones, K. J. (2008). IL-1beta signaling initiates inflammatory hypernociception. *Brain, Behavior, and Immunity*, *22*(7), 1014–1015.

Wieraszko, A., & Seyfried, T. N. (1989). ATP-induced synaptic potentiation in hippocampal slices. *Brain Research*, *491*(2), 356–359.

Wu, W. P., Hao, J. X., Halldner-Henriksson, L., Xu, X. J., Jacobson, M. A., Wiesenfeld-Hallin, Z., et al. (2002). Decreased inflammatory pain due to reduced carrageenan-induced inflammation in mice lacking adenosine A3 receptors. *Neuroscience*, *114*(3), 523–527.

Wu, J., Holstein, J. D., Upadhyay, G., Lin, D. T., Conway, S., Muller, E., et al. (2007). Purinergic receptor-stimulated IP3-mediated Ca^{2+} release enhances neuroprotection by increasing astrocyte mitochondrial metabolism during aging. *The Journal of Neuroscience*, *27*(24), 6510–6520.

Wu, Y. P., & Proia, R. L. (2004). Deletion of macrophage-inflammatory protein 1 alpha retards neurodegeneration in Sandhoff disease mice. *Proceedings of the National Academy of Sciences of the United States of America*, *101*(22), 8425–8430.

Wu, L. J., Vadakkan, K. I., & Zhuo, M. (2007). ATP-induced chemotaxis of microglial processes requires P2Y receptor-activated initiation of outward potassium currents. *Glia*, *55*(8), 810–821.

Xu, H. L., Mao, L., Ye, S., Paisansathan, C., Vetri, F., & Pelligrino, D. A. (2008). Astrocytes are a key conduit for upstream signaling of vasodilation during cerebral cortical neuronal activation *in vivo*. *American Journal of Physiology. Heart and Circulatory Physiology*, *294*(2), H622–H632.

Yamasaki, Y., Suzuki, T., Yamaya, H., Matsuura, N., Onodera, H., & Kogure, K. (1992). Possible involvement of interleukin-1 in ischemic brain edema formation. *Neuroscience Letters*, *142*(1), 45–47.

Zang, Y. C., Halder, J. B., Samanta, A. K., Hong, J., Rivera, V. M., & Zhang, J. Z. (2001). Regulation of chemokine receptor CCR5 and production of RANTES and MIP-1alpha by interferon-beta. *Journal of Neuroimmunology*, *112*(1–2), 174–180.

Zhang, N., Inan, S., Cowan, A., Sun, R., Wang, J. M., Rogers, T. J., et al. (2005). A proinflammatory chemokine, CCL3, sensitizes the heat- and capsaicin-gated ion channel TRPV1. *Proceedings of the National Academy of Sciences of the United States of America*, *102*(12), 4536–4541.

Zhang, J. M., Wang, H. K., Ye, C. Q., Ge, W., Chen, Y., Jiang, Z. L., et al. (2003). ATP released by astrocytes mediates glutamatergic activity-dependent heterosynaptic suppression. *Neuron*, *40*(5), 971–982.

Zonta, M., Angulo, M. C., Gobbo, S., Rosengarten, B., Hossmann, K. A., Pozzan, T., et al. (2003). Neuron-to-astrocyte signaling is central to the dynamic control of brain microcirculation. *Nature Neuroscience*, *6*(1), 43–50.

Zozulya, A. L., Reinke, E., Baiu, D. C., Karman, J., Sandor, M., & Fabry, Z. (2007). Dendritic cell transmigration through brain microvessel endothelium is regulated by MIP-1alpha chemokine and matrix metalloproteinases. *Journal of Immunology*, *178*(1), 520–529.

Index

Recent Volumes in the Serial

Volume 58

$GABA_B$ Receptor Pharmacology: A Tribute to Norman Bowery
Edited by Thomas P. Blackburn

Volume 59

Cardiovascular Pharmacology - Heart and Circulation
Edited by Paul M. Vanhoutte

Volume 60

Cardiovascular Pharmacology - Endothelial Control
Edited by Paul M. Vanhoutte